Geophysical Monograph Series

Including

IUGG Volumes
Maurice Ewing Volumes
Mineral Physics Volumes

78 **Climate Change in Continental Isotopic Records** *P. K. Swart, K. C. Lohmann, J. McKenzie, and S. Savin (Eds.)*

79 **The Tornado: Its Structure, Dynamics, Prediction, and Hazards** *C. Church, D. Burgess, C. Doswell, R. Davies-Jones (Eds.)*

80 **Auroral Plasma Dynamics** *R. L. Lysak (Ed.)*

81 **Solar Wind Sources of Magnetospheric Ultra-Low Frequency Waves** *M. J. Engebretson, K. Takahashi, and M. Scholer (Eds.)*

82 **Gravimetry and Space Techniques Applied to Geodynamics and Ocean Dynamics (IUGG Volume 17)** *Bob E. Schutz, Allen Anderson, Claude Froidevaux, and Michael Parke (Eds.)*

83 **Nonlinear Dynamics and Predictability of Geophysical Phenomena (IUGG Volume 18)** *William I. Newman, Andrei Gabrielov, and Donald L. Turcotte (Eds.)*

84 **Solar System Plasmas in Space and Time** *J. Burch, J. H. Waite, Jr. (Eds.)*

85 **The Polar Oceans and Their Role in Shaping the Global Environment** *O. M. Johannessen, R. D. Muench, and J. E. Overland (Eds.)*

86 **Space Plasmas: Coupling Between Small and Medium Scale Processes** *Maha Ashour-Abdalla, Tom Chang, and Paul Dusenbery (Eds.)*

87 **The Upper Mesosphere and Lower Thermosphere: A Review of Experiment and Theory** *R. M. Johnson and T. L. Killeen (Eds.)*

88 **Active Margins and Marginal Basins of the Western Pacific** *Brian Taylor and James Natland (Eds.)*

89 **Natural and Anthropogenic Influences in Fluvial Geomorphology** *John E. Costa, Andrew J. Miller, Kenneth W. Potter, and Peter R. Wilcock (Eds.)*

90 **Physics of the Magnetopause** *Paul Song, B.U.Ö. Sonnerup, and M.F. Thomsen (Eds.)*

91 **Seafloor Hydrothermal Systems: Physical, Chemical, Biological, and Geological Interactions** *Susan E. Humphris, Robert A. Zierenberg, Lauren S. Mullineaux, and Richard E. Thomson (Eds.)*

92 **Mauna Loa Revealed: Structure, Composition, History, and Hazards** *J. M. Rhodes and John P. Lockwood (Eds.)*

93 **Cross-Scale Coupling in Space Plasmas** *James L. Horwitz, Nagendra Singh, and James L. Burch (Eds.)*

94 **Double-Diffusive Convection** *Alan Brandt and H. J. S. Fernando (Eds.)*

95 **Earth Processes: Reading the Isotopic Code** *Asish Basu and Stan Hart (Eds.)*

96 **Subduction Top to Bottom** *Gray E. Bebout, David Scholl, Stephen Kirby, and John Platt (Eds.)*

97 **Radiation Belts: Models and Standards** *J. F. Lemaire, D. Heynderickx, and D. N. Baker (Eds.)*

98 **Magnetic Storms Bruce** *T. Tsurutani, Walter D. Gonzalez, Yohsuke Kamide, and John K. Arballo (Eds.)*

99 **Coronal Mass Ejections** *Nancy Crooker, Jo Ann Joselyn, and Joan Feynman (Eds.)*

100 **Large Igneous Provinces** *John J. Mahoney and Millard F. Coffin (Eds.)*

101 **Properties of Earth and Planetary Materials at High Pressure and Temperature** *Murli Manghnani and Takehiki Yagi (Eds.)*

102 **Measurement Techniques in Space Plasmas: Particles** *Robert F. Pfaff, Joseph E. Borovsky, and David T. Young (Eds.)*

103 **Measurement Techniques in Space Plasmas: Fields** *Robert F. Pfaff, Joseph E. Borovsky, and David T. Young (Eds.)*

104 **Geospace Mass and Energy Flow: Results From the International Solar-Terrestrial Physics Program** *James L. Horwitz, Dennis L. Gallagher, and William K. Peterson (Eds.)*

105 **New Perspectives on the Earth's Magnetotail** *A. Nishida, D. N. Baker, and S. W. H. Cowley (Eds.)*

106 **Faulting and Magmatism at Mid-Ocean Ridges** *W. Roger Buck, Paul T. Delaney, Jeffrey A. Karson, and Yves Lagabrielle (Eds.)*

107 **Rivers Over Rock: Fluvial Processes in Bedrock Channels** *Keith J. Tinkler and Ellen E. Wohl (Eds.)*

108 **Assessment of Non-Point Source Pollution in the Vadose Zone** *Dennis L. Corwin, Keith Loague, and Timothy R. Ellsworth (Eds.)*

109 **Sun-Earth Plasma Interactions** *J. L. Burch, R. L. Carovillano, and S. K. Antiochos (Eds.)*

110 **The Controlled Flood in Grand Canyon** *Robert H. Webb, John C. Schmidt, G. Richard Marzolf, and Richard A. Valdez (Eds.)*

111 **Magnetic Helicity in Space and Laboratory Plasmas** *Michael R. Brown, Richard C. Canfield, and Alexei A. Pevtsov (Eds.)*

112 **Mechanisms of Global Climate Change at Millennial Time Scales** *Peter U. Clark, Robert S. Webb, and Lloyd D. Keigwin (Eds.)*

113 **Faults and Subsurface Fluid Flow in the Shallow Crust** *William C. Haneberg, Peter S. Mozley, J. Casey Moore, and Laurel B. Goodwin (Eds.)*

114 **Inverse Methods in Global Biogeochemical Cycles** *Prasad Kasibhatla, Martin Heimann, Peter Rayner, Natalie Mahowald, Ronald G. Prinn, Dana E. Hartley (Eds.)*

Geophysical Monograph 115

Atlantic Rifts and Continental Margins

Webster Mohriak
Manik Talwani
Editors

American Geophysical Union
Washington, DC

ISBN 0-87590-098-4
ISSN 0065-8448

Printed in the United States of America.

CONTENTS

CONTENTS

PREFACE

In compiling this volume, we have aimed to develop and enhance our current understanding of the structural evolution and sedimentation processes along divergent continental margins. To counteract the unfortunate situation of a lack of modern seismic and potential fields data on circum-Atlantic passive margins in the literature, we have linked new data from oil companies with that of research institutions. To update the data offered in most volumes used as reference works for the study of continental margins, now upwards of 20 years old, and to remedy the dispersal of important, more recent contributions in specialized journals, we present a current synthesis of materials in one volume focused on the deeper geology of the sedimentary basins along continental margins.

In the early 1990's, as oil companies and other institutions developed tools to probe deeper into the architecture of passive margin sedimentary basins, a great amount of data based on regional deep seismic profiles evolved rapidly from its specialized niche as geophysical interpretation of the Earth's interior to widespread use by those same companies and institutions. At the same time, these findings demonstrated that some breakthroughs in data acquisition, processing and interpretation initially achieved by research institutions could almost instantaneously be globalized throughout different research groups, thereby influencing the thinking of geoscientists worldwide.

As such, we designed this book to meet the need for a reference work on the geology and geophysics of continental margins. The range of topics is broad enough to have continuing value, but not so broad as to be diffuse and disconnected, while the level of detail and the data content ensure lasting value.

In organizing this book, we were inspired by the special session "Geology and Geophysics of Continental Margins," presented at the Fifth International Congress of the Brazilian Geophysical Society, São Paulo, Brazil, in 1997. Papers developed from that session appear in the present volume, whose scope is enlarged by incorporating other relevant papers, selected to ensure a full spectrum of themes, ranging from rift to salt tectonics, to drift-phase structures along continental margins.

A brief introductory historical review summarizes the rapidly expanding knowledge of the subject to help illuminate the regional background of the areas discussed in more detail by the subsequent chapters. Their subjects range from the enigmatic influence of Precambrian compressional events to extensional structures whose roots may be linked deep in the continental crust, and from basement-involved orogenic belts to deep water foldbelts that can also be interpreted in sandbox models. Geographically, the papers cover a wide spectrum of continental margins at both sides of the Atlantic, north and south of the Equator. The geodynamic mechanisms discussed vary from the origin of syn-rift structures, modifying elements related to transpression and transtension, to magmatic episodes during the breakup and, later, during the thermal phase of subsidence. A detailed discussion of several aspects of deep water fold belts both in West African and in the Gulf of Mexico has important bearings on the exploration of frontier regions in the 21st century. Of special interest both for academic purposes and for hydrocarbon exploration are comparisons between conjugate margins across the Atlantic, particularly between the Eastern Brazilian margin and the Western African margin, notably on several aspects of salt tectonics, rift geometry, and exploration strategies.

We hope that this volume will be of use to established scientists working in this field and will provide a valuable introduction for non-specialists in the geology and geophysics of Atlantic rifts and continental margins.

We are indebted to all the authors, who toiled through several stages of producing the first drafts and the final papers. The volume has greatly benefited from countless hours of extraordinary efforts by both the authors and the reviewers, whose invaluable assistance and insightful suggestions were instrumental in improving the presentations. To all these people the editors express their gratitude for their whole-hearted dedication. (Reviewers: Albert W. Bally, Richard Buffler, Carlos Cramez, Robert Fryklund, Joseph Gallart, Steven Henry, Barry Katz, Webster Mohriak, Benjamin B. B. Neves, Dave Roberts, Bruce Rosendahl, Mark G. Rowan, Peter Szatmari, Manik Talwani, Anthony Tankard, and Gabor Tari.)

Special thanks are due to the Executive Committee of the Fifth International Geophysical Congress, particularly to Icaro Vitorello, head of the Technical Program Committee, whose help was absolutely essential to the success of the conference and to the session "Geology and Geophysics of Continental Margins," on which this book is based. We also thank members of the organizing committee of the SBGf and also the Petrobras staff at CENSUD who helped arrange several short course and brainstorming sessions held in São Paulo and in Rio de Janeiro, particularly F. Taioli, Y. R. Marangoni, M. S. M. Mantovani, M. A. L. Latgé, F. B. Silva, M. C. Lana, and F. Feijó. Special thanks to D. Matos and R. L. Silveira for the incentive to prepare the volume, and to S. R. Michelucci, L. R. Guardado and C. F. Lucchesi for their management support for the conference short courses and brainstorming sessions sponsored by Petrobras.

We are grateful to Marcelo Bassetto, who helped organize the special session, and also helped in the selection of the papers to be included in this volume. We also thank Marcelo for acting as the co-convener contact and for replacing the main editor during absences. His companionship and friendliness are greatly appreciated and, needless to say, also his remarkable good taste in sharing the same strong passion for Sharon Stone movies. We extend thanks also to Ines S. Vieira, who keenly and actively participated in the basin analysis projects that were conducted in the Northeastern Brazilian margin, part of which are included in this volume. Special thanks are also due to A.W. Bally, whose assistance and renowned expertise both as co-author on several contributions and as a referee were invaluable to the final production of this book.

Webster Mohriak

Manik Talwani

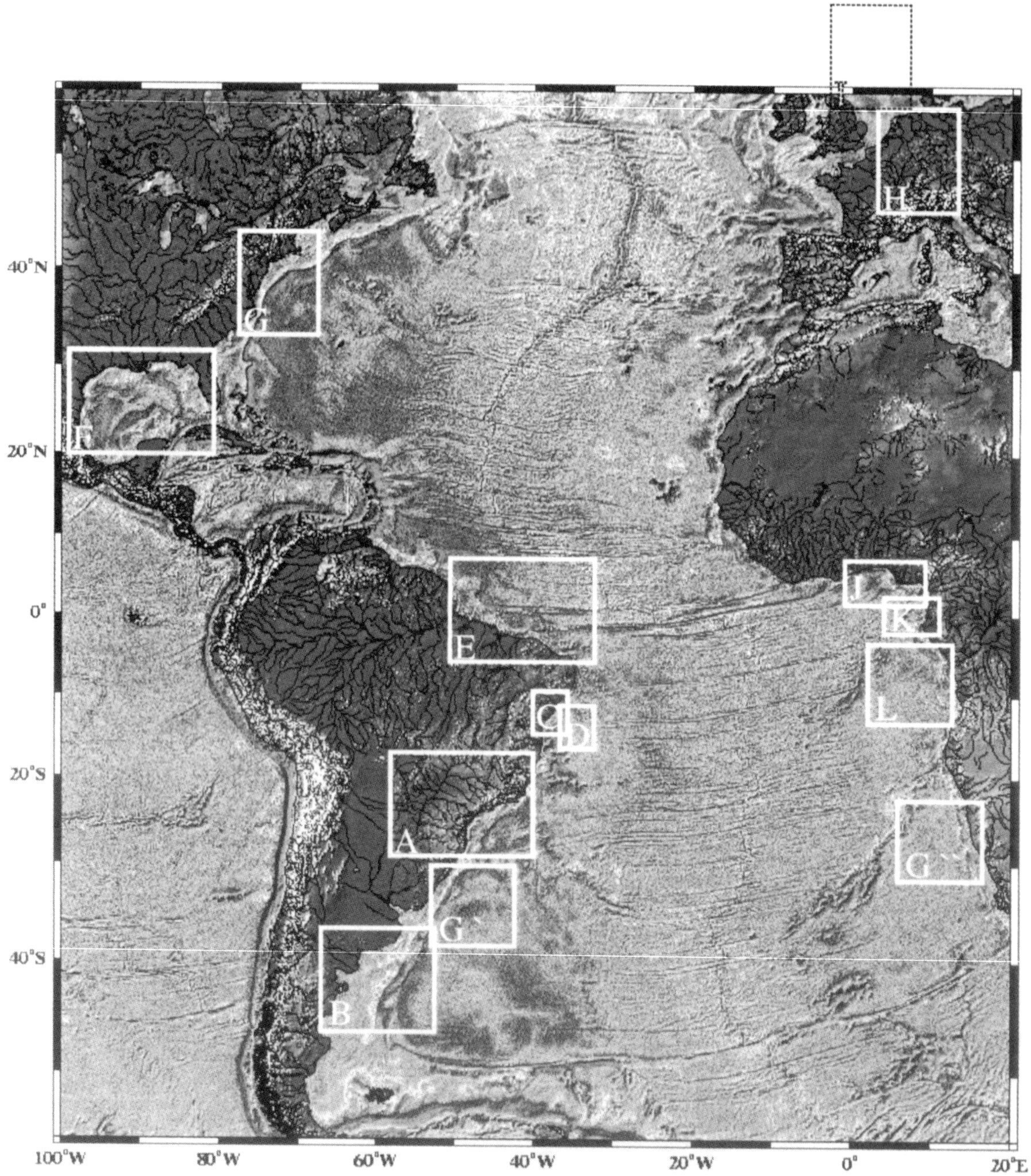

Figure 1. Geosat map showing the areas around the Atlantic Basin covered in this volume. The following authors have the areas below as their main focus of discussion: M. Heilbron et al., area A; R. Meissner, area H; R. England, area I; B. Rosendahl e t al., area K; P. Dailly, area K; Marton et al., area L; S. Wu and W. Bally, areas F and J/K; M. Rowan et al., area F; J. Soares et al., area B; M. Talwani and Abreu, areas G, G' and G"; M. Bassetto et al., area A.; P. Gomes et al., area D; W. Mohriak et al., area C; H. Koyi, areas A and D; and R. Darros, area D.

INTRODUCTION

In this introduction, we present a summary of topics discussed in subsequent chapters; a visualization of the regions, discussed in a plate-tectonics framework using Geosat data; and some puzzling questions still open to debate on the geodynamic evolution of rifts and continental margins. By presenting a very large dataset and plausible explanations for some of the questions detailed below, we hope that the volume provides an incentive for all researchers to devise new or alternative models in their effort to understand Atlantic rifts and continental margins.

The image on the cover of this book and Figure 1 both present a visualization of the regions discussed in this volume. Our aim in choosing this map was to provide focus on megaregional tectonic elements that may be significant for the interpretation of more particular geological and geophysical datasets. The Geosat image (courtesy D. Sandwell) depicts topography on land and bathymetry offshore. Also included in this image are recent earthquake epicenters (large dots) and presently active volcanoes (triangles).

The 15 papers that form the heart of this volume explore a wide range of themes, from the enigmatic influence of Precambrian compressional events (associated with lithospheric convergence) to Paleozoic and Mesozoic extensional structures (whose roots may be linked deep in the continental crust), and from basement-involved orogenic belts to deep water foldbelts that can also be interpreted in sandbox models. The majority of the papers address topics relevant to basin analysis and exploratory interpretation of the Atlantic rifts and continental margins. Geographically, the papers cover a wide spectrum of continental margins at both sides of the Atlantic, north and south of the Equator. The geodynamic mechanisms discussed in these papers vary from the origin of the syn-rift structures, modifying elements related to transpression and transtension, as well as to magmatic episodes during the breakup and also during the thermal phase of subsidence.

A detailed discussion of several aspects of deep water foldbelts both in West Africa and in the Gulf of Mexico has important bearing on the exploration of frontier regions in the 21st century. Of special interest both for academic purposes and for hydrocarbon exploration are comparisons between the geology and geophysical framework of conjugate margins across the Atlantic, particularly between the Eastern Brazilian margin and the Western African margin, and notably on several aspects of salt tectonics, rift geometry, and exploration strategies.

Our pursuit of a balanced representation of academic and exploration works led us to select papers written in a larger than usual format, and which would include detailed discussions of regional structures based on seismic profiles and potential field data, as well as analyses of conceptual and alternative models of geodynamic evolution. By this approach, we wished to provide insights for interpreting key factors related to these regions and to avoid a larger number of short papers that could not address puzzling questions, and whose elucidation remains elusive for many geoscientists worldwide.

Atlantic Rifts and Continental Margins
Geophysical Monograph 115

For example, conventional models most often explain the subsidence of passive margins as a combination of lithospheric stretching and cooling, assuming a homogeneous continental crust. Are these models too simplistic? Why do we have the Atlantic rifts and margins located near the regions affected by earlier lithospheric convergence episodes ? And how are early subsidence phases in the intracratonic basins related to on-going convergence processes in nearby areas ? Heilbron et al. address these topics by analyzing the onshore and offshore geology of the southeastern Brazilian region (Figure 1, area A), from the Precambrian foldbelts to Paleozoic and Mesozoic basins near the Atlantic seaboard. The ultradeep water province of this region is characterized by sedimentary sequences resting on oceanic crust that is not as tectonically frozen as thought in the recent past. Bassetto et al. go on to discuss the geomorphology and main structural elements of ultradeep water along the southeastern Brazilian margin by integrating potential field data and regional seismic profiles on the oceanic crust.

The rift phase in the sedimentary basins along the Atlantic continental margins is usually separated from the drift-phase sediments by a breakup unconformity associated with thermal cooling. At the same time, tectonic uplift associated with transtensional and transpressional movements might also be involved in the process. In this regard, Soares et al. discuss spectacular seismic and outcrop images of these unconformities in the southernmost region of the South American continent (area B).

The Atlantic margins are also characterized by several aborted rifts along the continental border. Because these rifts show a lack of thermal subsidence phase and extremely negative gravity anomalies in the depocenters, there is evidence to indicate that isostatic compensation might be related to mantle uplifts along the adjacent basins, which then evolved to form the oceanic crust. Mohriak et al. discuss some aspects of the onshore and offshore geology of the northeastern Brazilian region (area C), while Gomes et al. discuss the offshore geology of the northeastern Brazilian region (area D) by integrating regional seismic profiles in the ultradeep water region extending to the oceanic crust.

The transform segments of Atlantic rifts and continental margins are less understood than the pure orthogonal segments. Some questions still under debate here refer to the shearing zones between plates affected by transform movements. Do oceanic transform faults simply segment passive margins, or do they continue for a long distance into the continental crust? Do the transform faults on passive margins step onshore to form major lithospheric or continental strike-slip fault systems? Are transfer faults that segment halfgraben systems predecessors of oceanic transform faults? Darros addresses some of these questions in relation to the equatorial Atlantic margins, incorporating palinspastic reconstructions both along the Brazilian (area D) and the African margins (area J).

Salt tectonics is an important factor controlling sedimentation processes along Atlantic continental margins. Several questions remain to be elucidated in this field, which is still in flux due to important conceptual breakthroughs made during the past decade. What is the role of basement tectonics in salt basins, and is basement tectonics an additional driving force for salt tectonics? Which processes are involved in the formation of deep water foldbelts found in the Gulf of Mexico and other sedimentary basins?

The origin and kinematics of deep-water folded belts (Niger Delta toes) differ from conventional "in sequence" foreland folded belts. In deep-water folded belts the deformation appears to move away from the "oceanic" foreland, and lower units appear to deform and to be stabilized prior to the deformation of higher, often more deformed units. What can we learn from a direct and detailed comparison of salt and shale tectonics, and how do they differ in their seismic expression?

Rowan et al. tackle some of the questions involving the salt tectonics in the Gulf of Mexico (area F). For their part, Wu and Bally discusses both salt and shale tectonics along continental margins by comparing the Gulf of Mexico with the Niger Delta (area J/K). Koyi complements these papers by reviewing the physical models developed to simulate some of the structural styles associated with extensional and compressional tectonic processes along continental margins.

The breakup of continental plates is directly related to rifts and the formation of divergent margins. The details of the processes related to oceanic crust inception, however, are still a subject of debate among geoscientists. Are stretched margins (Galicia type) and volcanic margins (Voring Plateau, Baltimore Canyon, Pelotas, Namibia) distinct types or just end-members of a spectrum that includes numerous intermediate types? What is the temporal relationship of rifting, the formation of oceanward-diverging volcanic wedges and the formation of typical oceanic crust? Are flood basalts and associated radial dyke systems simply precursory events of the rift phase? Are they coeval with oceanward diverging wedges, or do they merely coincide with the termination or else perhaps with the inception of the seaward-dipping volcanic wedges? What is the composition of the material lying between the base of seaward-dipping reflectors and the Moho? Is it thinned continental crust or newly initiated oceanic crust? The term "underplating" is widely used by some experts. What does it mean, and what does this layer underplate? If it underplates newly formed ocean crust, is the term "underplating" still valid? What is the role of hot-spots and plumes in the initiation of continental splitting? Talwani and Abreu address some of these topics in their study of the North American Eastern Atlantic margin (area G), which they compare with counterpart conjugate margins of Brazil (area G') and Namibia (area G").

Crossing the Atlantic, we turn to Meissner, who discusses the European mainland (area H). Here main topics include the geological and geophysical signatures of the deep structures associated with the formation of "tectonically induced" accommodation space in the development of rifts and continental margins, particularly in terranes related to several episodes of lithospheric convergence. England then focuses his discussion on the relationship between basement structures, rift architecture and sedimentation processes by integrating potential field and seismic data from the North Atlantic European margin (area I).

Moving to the south, the African margin is characterized by the Niger Delta (discussed by Wu and Bally), which is illustrated on the map as the interplay of areas J and K. The Rio Muni Basin, as discussed by Dailly, is a new frontier for hydrocarbon exploration. Rosendahl et al. add the Gabon Basin to this section of the volume with a number of hydrocarbon discoveries already established there,

The deep water region of this segment of the Western African margin (area K) is an important scenario in which to discuss the deep structure of the rifts and the economic importance of passive margins as a function of the richness or absence of hydrocarbon source beds. What are the oceanographic and biological conditions that enhance the deposition of these source beds in the distal portions of the rifts? What are the respective roles of upwelling currents, volcanic activity, size and restriction of basins, deepwater versus shallow water currents, and the water depths in which source beds are deposited? Predicting with confidence the occurrence of source beds and their maturation profiles for underexplored passive margins will be a most important key to the discovery of significant new reserves.

Tilted rift blocks are common features in the northwestern segment of the African margin, and the occurrence of major unconformities in the deep water region is a puzzling enigma. How common is "post-rifting" deepwater onlap of and burial by sediments on erosional scarps and associated horsts? What criteria do we use to distinguish them from synrift onlaps ? What is the geomorphological setting associated with the widespread erosion associated with "breakup" unconformities? What processes are responsible for the formation of ubiquitous continental slope unconformities? What are the respective roles of slope instability, marine currents and the indirect impact of sea-level changes? The dataset presented and discussed by Dailly and Rosendahl et al. provides opportunities for analyzing these questions.

The West African margin may also provide a laboratory to analyze the relationship of salt deposition to rifting and to volcanic processes during the initial stages of oceanic crust formation. Widespread salt is often deposited in post-rift sag basins, but are there other events responsible for salt deposits within the rift sequence, and are there examples of salt deposits on volcanic oceanic crust? Is salt entirely the product of sea-water evaporation, or are volcanic hydrothermal processes significant contributors? The association of diapirs (presumably salt) with the marginal magnetic anomalies in some basins, such as the Sergip/Alagoas in Eastern Brazil and in the Carolina Trough offshore the U.S. East Coast is particularly puzzling, indicating that salt deposition advanced towards the crustal limit, in contrast to some models that advocate its deposition on proximal lagoons and along restricted gulfs inboard of the continental margin. As such, Marton et al. discuss extensional processes and salt tectonics in the ultradeep water regions of the Atlantic continental margins based on the interpretation of the Kwanza Basin (area L).

We have designed this book as a state-of-the-art reference on the geology and geophysics of Atlantic rifts and continental margins. We hope that the papers mentioned in this introduction, and those not mentioned but presented within, will be a solid and inspiring contribution. We also hope that the many insights derived from the analysis of the papers individually and as a whole will contribute to the never-ending quest for elucidating some of the questions presented above and, in doing so, expand the horizons of the field.

Webster Mohriak

From Collision to Extension: The Roots of the Southeastern Continental Margin of Brazil

Monica Heilbron[1], Webster U. Mohriak[2,1] , Cláudio M. Valeriano[1], Edison J. Milani[2,1] , Julio Almeida[1], and Miguel Tupinambá[1].

The South Atlantic Meso-Cenozoic continental margins are located in regions characterized by a long-lived history of Proterozoic extension, structural inversion and compressional remobilization of basement and supracrustal rocks. The roots of the present-day southeastern Brazilian continental margin (e.g. Santos and Campos basins) are associated with terranes directly affected by the Brasiliano orogenic collage. This event was responsible for the Ribeira fold belt, which is characterized by compressional, metamorphic and magmatic episodes from Late Precambrian to the earliest Paleozoic. The initial phases of subsidence of the intracratonic Paraná basin, located west of the Ribeira fold belt, correspond to early Paleozoic siliciclastic rocks deposited in depocenters that were probably controlled by Brasiliano fabrics. The basin-forming stress fields may be related to the lithospheric convergence between Panthalassa oceanic crust and cratonic blocks of western Gondwana. The last phase of subsidence in the Paraná basin is marked by Late Jurassic / Early Cretaceous tholeiitic continental flood basalts. These basalts heralded the breakup of Gondwana. They were also deposited on the Precambrian basement offshore, and are believed to be part of the rift succession. The breakup of western Gondwana and the onset of a new phase of plate divergence in the South Atlantic were marked by thick wedges of seaward-dipping reflectors near the incipient oceanic-ridge spreading center. Subsequently, a few episodes of intraplate tectonic and magmatic activity are also possibly related to compressional stresses resulting from subduction in the Andean margin and ridge push in the mid-Atlantic spreading ridge.

1. INTRODUCTION

The origin of passive margin sedimentary basins is related to extensional processes in the lithosphere that form rifts and culminate with ocean opening. An intriguing aspect of the South Atlantic continental margins is that they developed along the grain of a major Neoproterozoic orogenic belt. This thermo-tectonic episode (called Brasiliano orogeny in Brazil, or the Pan-African orogeny in Africa) affected infracrustal and supracrustal rocks, which are in turn related to older episodes of rifting, inversion and compressional tectonics. The roots of the present-day South Atlantic continental margin, which do not allow direct study because of the thick post-breakup sedimentary cover, may only be analysed within the deformed belts located adjacent to these Mesozoic basins. Older Precambrian belts profoundly influenced the Brasiliano/Pan Afri-

[1] - UERJ - Rio de Janeiro State University, Faculty of Geology, Rio de Janeiro - RJ

[2] - PETROBRAS - Petróleo Brasileiro S.A. – E&P, Rio de Janeiro – RJ

Atlantic Rifts and Continental Margins
Geophysical Monograph 115

can events. These Neoproterozoic belts also have tectono-sedimentary signatures of previous episodes of rift formation and basin development, with a complete sequence of rift-passive continental margin evolution. This tectonic framework is generally attributed to a Wilson cycle of basin formation, structural inversion, and compressional tectonics within orogenic belts (e.g., Wilson, 1966; Dewey and Bird, 1973).

The objectives of this paper are:

a) to interpret the preserved remnants of the geological units of Ribeira fold belt in terms of basin formation, basin destruction and compressional tectonics;

b) to discuss a tectonic model that would explain the evolution of the Neoproterozoic-earliest Paleozoic Ribeira belt, at eastern Brazil;

c) to review the stratigraphic evolution of the Paraná basin within a global tectonic framework;

d) to relate extensional tectonics and subsidence in the early stages of the Paraná basin with collision or compressional tectonics along its borders;

e) to relate the basaltic volcanism in the Paraná basin to the early stages of Mesozoic extension in the Gondwana;

f) to relate continental breakup and inception of oceanic crust with the migration of the volcanism towards the continental margin, where the crustal limit is characterized by thick wedges of seaward-dipping reflectors.

2. TECTONIC UNITS OF SOUTHEASTERN BRAZIL

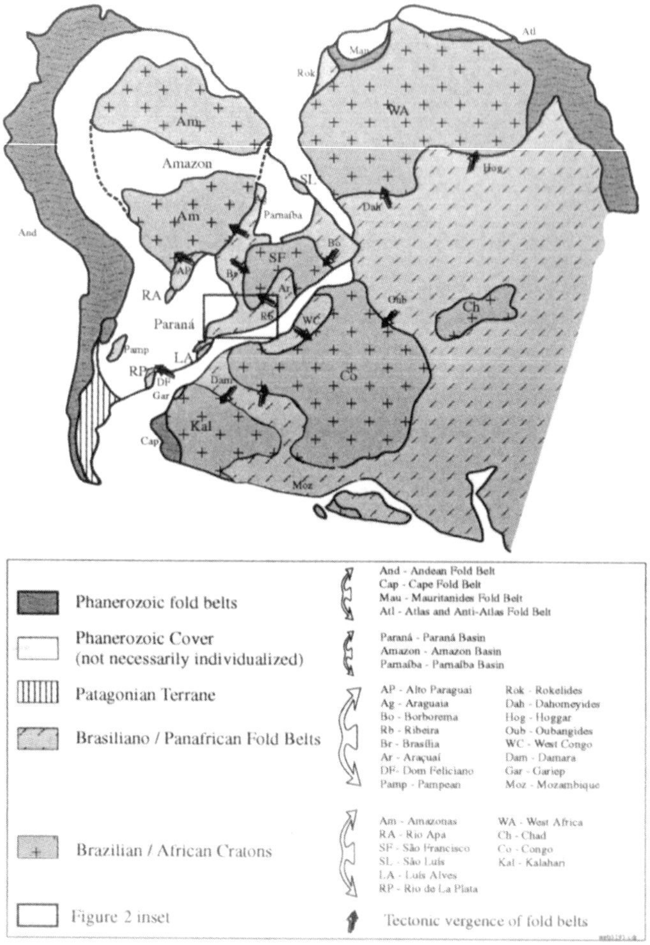

Figure 1. Schematic tectonic map of part of west Gondwana, simplified from Trompette (1994), Almeida and Hasui (1984). The studied segment of the Ribeira belt (Fig. 2) is comprised within the rectangle.

The South American platform (Almeida, 1967; Cordani et al., 1967) comprises the following geological units (Fig. 1): (1) Precambrian shields composed by cratons and Panafrican/Brasiliano orogenic belts; 2) the Paraná, Amazonas and Parnaíba intracratonic basins that constitute the Paleozoic to Mesozoic platform cover; 3) the Jurassic-Cretaceous interior rifts and Atlantic passive margin basins associated with the breakup of Gondwana and formation of the South Atlantic; and 4) post-breakup rift basins and related alkaline plutons developed during Neo-Cretaceous/Tertiary reactivation (Fig. 2).

In the Atlantic shield of southeastern Brazil, the Precambrian basement was affected by several accretional/collisional episodes indicative of lithospheric convergence and extensional episodes that formed sedimentary basins. The Brasiliano orogenic collage, of late Precambrian/Cambrian age, was the final stage in the evolution of the basement platform (Almeida and Hasui, 1984), which overprinted major cratonic blocks and intervening orogenic belts chiefly composed of metasedimentary rocks (Fig. 1). The final stages of formation and cratonization of the Gondwana supercontinent basement took place during Cambrian to Ordovician times. The main Precambrian tectonic units of southeastern Brazil, that will be discussed in this work, are the São Francisco craton and the Ribeira fold belt, located at the southeastern cratonic border (Figs. 1 and 2).

The widespread Brasiliano (Pan African) collage is related to the convergence of previously dispersed fragments of the Rodinia supercontinent (Brito Neves and Cordani, 1991; Hoffman 1992; Brito Neves, 1993) or to closing and structural inversion of isolated but tectonically linked rift segments (Tankard et al., 1995). The record of the different orogenic periods at southeastern Brazil is distributed between the 700-450 Ma interval, and indicates that convergence was largely diachronic in the different fold belts.

Figure 2 shows a simplified geological map of southeastern Brazil with the principal tectonic elements and

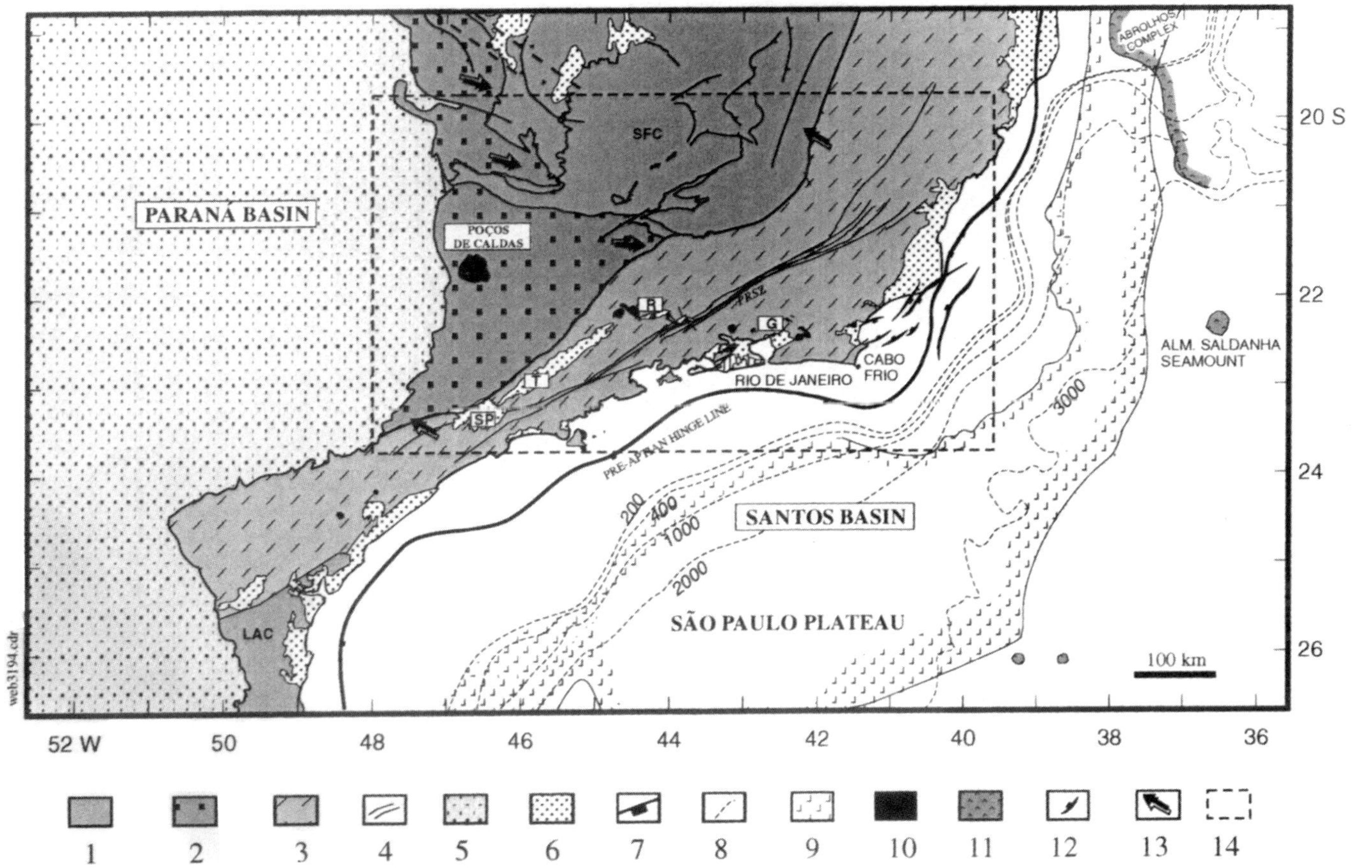

Figure 2. Simplified tectonic map of southeastern Brazil, modified from Heilbron et al. (1995), Campos Neto and Figueiredo (1995). The tectonic features of the marginal basins were extracted from Mohriak et al. (1995). Legend: 1- São Francisco craton (SFC) and foreland zone of the Brasiliano fold belts; 2- Brasília belt; 3- Ribeira belt; 4- Paraíba do Sul shear zone (PRSZ); 5- Paraná basin; 6- Cenozoic cover and São Paulo (SP), Taubaté (T), Resende (R) and Guanabara (G) interior rifts; 7- Pre-Aptian hinge line; 8- Bathymetric contour lines (depth in meters); 9- Aptian Evaporites; 10- Meso-Cenozoic Alkaline rocks; 11- Meso-Cenozoic volcanic rocks; 12- Major normal faults; 13- Tectonic vergence of fold belts; 14 - Location of Figure 3.

sedimentary basins. The Paraná basin represents the intracratonic Paleozoic-Mesozoic stage, the Santos and Campos basins reflect the extensional processes related to break-up and dispersion of Gondwana fragments; the São Paulo-Taubaté-Resende-Guanabara-Barra de São João rift system corresponds to the Neo-Cretaceous/Tertiary activation of the platform (Almeida, 1976).

3. THE RIBEIRA BELT AND GONDWANA AMALGAMATION

The Ribeira belt extends for 1,400 km along the Atlantic margin (Fig. 2). It is a complex orogenic belt developed along the southern and southeastern borders of the São Francisco craton (SFC). Some authors attribute it to the convergence between the São Francisco/Congo plate and another plate or microplate located at the coastal region of Brazil (Campos Neto and Figueiredo, 1995; Heilbron et al., 1998b). The limits between the Ribeira and the Araçuaí belts along the eastern border of the SFC are poorly constrained. The distinction of the two belts is probably more of a geographic character: both belts display the same continuous structural trend, the sedimentary successions are related to post-1.8 Ga extensional processes, and they share a common Neoproterozoic collisional history. Relics of ocean-floor petrotectonic assemblage have been recently documented within the Araçuaí belt (Pedrosa-Soares et al., 1998), and testify a Neoproterozoic continental dispersal around 800 Ma, which may be common to the Ribeira belt.

The Ribeira belt resulted from the youngest lithospheric convergence of the Pan African-Brasiliano collage, with the principal development during Neoproterozoic-

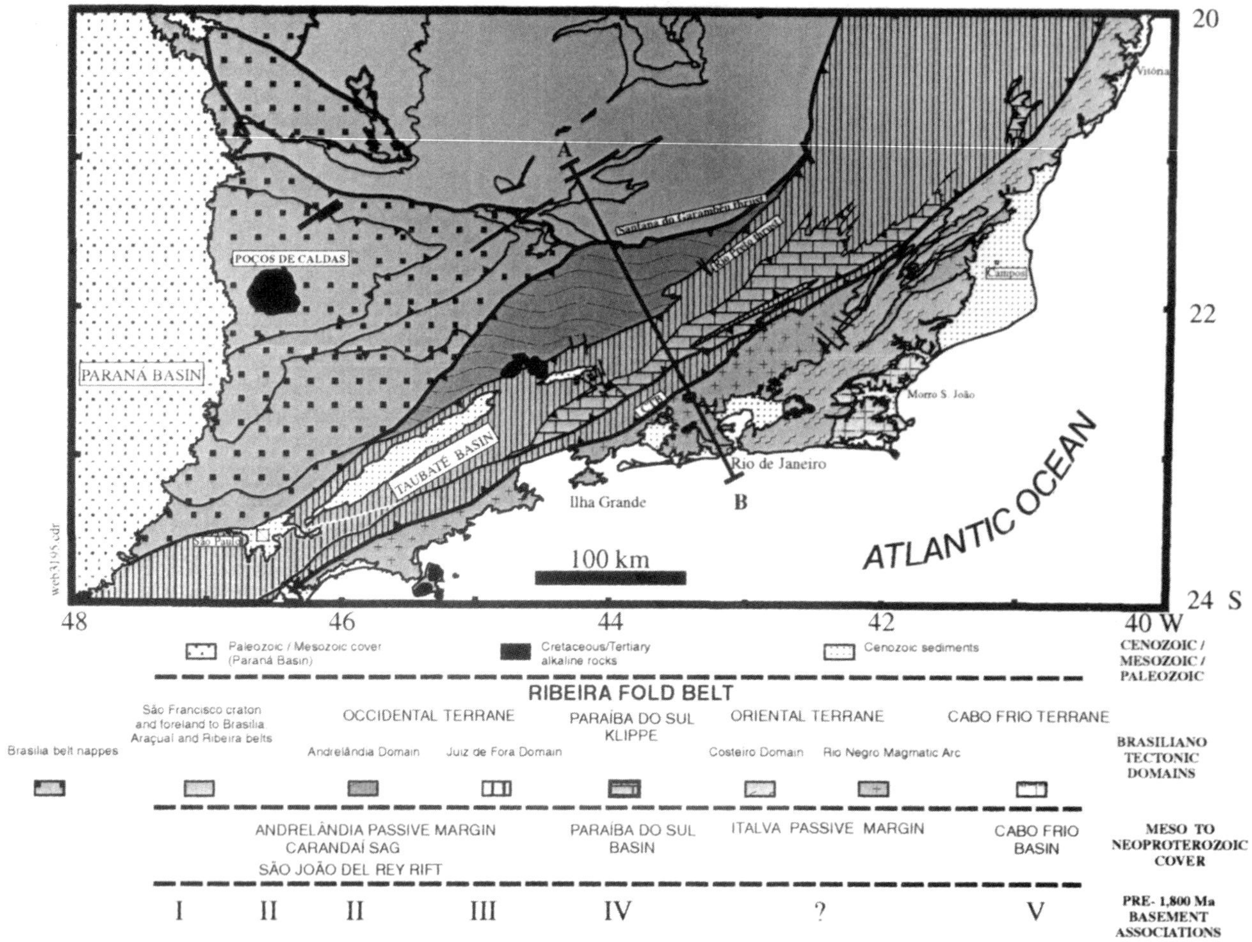

Figure 3. Detailed tectonic map of the southeastern region of Brazil with the major tectonic domains of the Ribeira belt. See text for lithology of basement association (roman numbers); CTB - Central Tectonic Boundary; Line A-B corresponds to the section of Figure 4.

Cambrian times, with the latest granites extending into the Ordovician.

3.1 Tectonic Organization and Main Discontinuities

The present crustal structure of the central segment of the Ribeira belt (Heilbron et al., 1995, 1998b, 1999) is defined by four different tectonostratigraphic terranes, in the sense used by Howell (1989). From NW to SE, these are (Fig. 3): a) the reworked margin of the SFC, defined as the Occidental terrane; b) The Paraíba do Sul Klippe; (c) the Oriental (Costeiro or Serra do Mar) terrane, which probably includes another cratonic block or microplate (Campos Neto and Figueiredo, 1995); and d) the Cabo Frio terrane (Fonseca et al., 1984; Fonseca, 1993)

The Occidental terrane displays an internal tectonic organization with two crustal scale thrust sheets (Andrelândia and Juiz de Fora) which override the foreland of the SFC (Heilbron et al., 1995, 1998b). Important shear zones and associated mylonites and tectonic "mixing" envelope both thrust sheets. The Juiz de Fora domain is regarded as a crustal scale duplex which resulted from the docking of the Oriental terrane (Fig. 2).

The Paraíba do Sul Klippe is the uppermost thrust slice of the central segment of the belt. This klippe occupied the hinge zone of the Paraíba do Sul megasynform, which is a major structure that overprints thrust-stacking and is associated with steep mylonite zones of dextral strike-slip motion (Fig. 4). Some lithological units of the cover are similar to those described in the Oriental terrane, but it has been difficult to restore the paleogeography of this terrane.

The Oriental terrane, based on available structural and geochronological data, is the locus of the Ribeira magmatic arc (Tupinambá et al., 1998; Almeida et al., 1998).

Table 1: Litho-stratigraphic units of the Central Segment of the Ribeira belt

Lithological associations / Tectonic units	Pre-1.8 Ga basement	Post-1.8 Ga cover	Brasiliano granitoid rocks
Occidental terrane *Autochthonous domain*	Barbacena and Mantiqueira complexes (Associations I and II)	-Andrelândia basin -Carandaí basin -Lenheiro and Tiradentes Depositional cycles (S. João del Rei basin)	-----------------------
Andrelândia domain	Mantiqueira complex (Association II)	Andrelândia basin	syn- to late collisional granitoids
Juiz de Fora domain	Juiz de Fora complex (Association III)	Andrelândia basin	syn- to post-collisional granitoids
Paraíba do Sul Klippe	Paraíba do Sul basin (?) Quirino complex (Association IV)	Paraíba do Sul basin (?)	syn- to post-collisional granitoids
Oriental/Costeiro terrane	?	Italva basin	yn- to post-collisional and post-tectonic granitoids
Cabo Frio terrane	Região dos Lagos complex (Association V)	Búzios basin	leucogranites

The contact between the Occidental and Oriental terranes is given by a conspicuous NW-dipping shear zone that can be traced continuously for at least 200 km from the coast of São Paulo state to the Serra dos Orgãos region in the Rio de Janeiro state (Figs. 3 and 4). This structure was developed during the late stages of the Brasiliano collage and was named Almeida et al. (1998) as the Central Tectonic Boundary (CTB).

The Cabo Frio terrane occupies a small coastal area in the Rio de Janeiro state. Ongoing structural and geochronological investigation, including new U/Pb data (Schmitt et al., 1999), indicates a relatively late docking event of this terrane.

3.2- Precambrian Lithotectonic Units

Within each terrane, a three-fold classification of lithological associations has been mapped (Fig. 4, Tab. 1): a) pre-1.8 Ga basement rocks; b) post-1.8 Ga metasedimentary cover including metabasic rocks of continental to Morb geochemical affinity; and c) Brasiliano/Pan African granitoids. These units are tentatively correlated between the terranes, considering lithologic and tectono-stratigraphic similarities and their identical structural-metamorphic evolution during the Brasiliano collage. Geochronological and geochemical data from the central segment of Ribeira belt are also considered for correlation. The lithotectonic units of this segment of the belt are summarized in Table 1.

3.2.1- Basement (pre-1.8 Ga) Units

The basement of the supracrustal sequences of the Ribeira belts is similar to that of the cratonic area, but with different degrees of superimposed Brasiliano reworking. In a review of the São Francisco craton basement, Teixeira and Figueiredo (1991) emphasized the importance of the Transamazonian Collage (ca. 2.35-1.85 Ga, Brito Neves and Sato, 1998) in the generation and widespread reworking of the crust, which left only minor relics of Archean nuclei. Cooling, uplift and erosion during the final stages of the Transamazonian cycle generated a continent-scale unconformity separating the post-1.8 Ga sedimentary basins from their basement. Five litho-tectonic associations are represented in Table 1 and in Figs. 4, 5 and 6:

Association I: Archean Barbacena greenstone belt and associated Paleoproterozoic orthogneisses, which crop out only in the northern part of the autochthonous domain of the Occidental terrane (Cordani and Brito Neves, 1982;

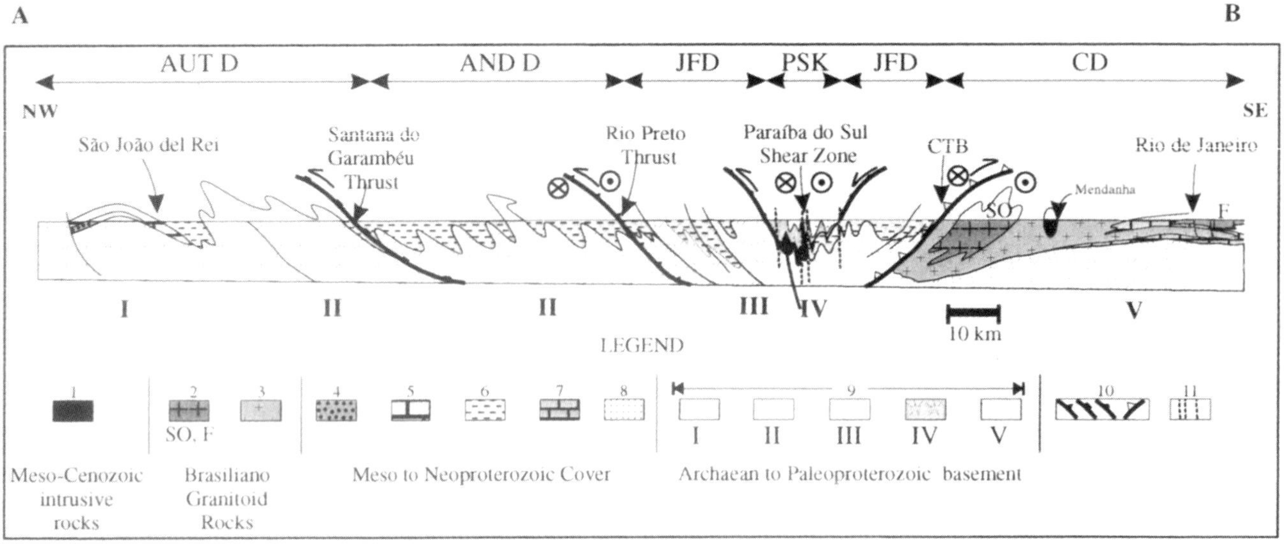

Figure 4. Geological cross section in the Central segment of the Ribeira belt between São João del Rei and Rio de Janeiro. AUTD- Autochthonous domain; ANDD- Andrelândia domain; JFD- Juiz de Fora domain; PSK- Paraíba do Sul Klippe; CD- Costeiro domain.

Ribeiro et al., 1990; Machado et al., 1992; Teixeira and Figueiredo, 1991; Machado and Carneiro, 1992).

Association II: Tonalite-granodiorite migmatitic orthogneisses with abundant mafic enclaves, minor granulitic and alkaline intrusive rocks (Mantiqueira Complex). This association crops out at the southern part of the Autochthonous and the Andrelândia domains. Rb/Sr ages of 2.8 and 2.07-1.98 Ga were reported by Teixeira and Figueiredo (1991) and by Cordani et al. (1973a).

Association III: Archean to Paleoproterozoic granulitic orthogneisses predominate at the Juiz de Fora Complex (Oliveira, 1982; Grossi Sad and Barbosa, 1985; Duarte et al., 1997; Heilbron et al., 1997, 1998a). Migmatitic structures predate the granulitization (Sad and Dutra, 1985; Heilbron et al., 1998). U-Pb and Rb-Sr data (Delhal et al., 1969; Oliveira 1980; Teixeira and Figueiredo, 1991; Machado et al, 1996) indicate Paleoproterozoic ages for both protoliths and granulitic metamorphism. Few Archean ages indicate the existence of older protoliths (Oliveira, 1980).

Association IV: Paleoproterozoic granitic to granodioritic orthogneisses with basic and calc-silicatic (tremolite-rich) enclaves integrate the Quirino complex within the Paraíba do Sul Klippe (Machado, 1984; Almeida et al., 1993; Heilbron et al., 1995; Valladares et al., 1997). The orthogneisses are structurally overlain by the supracrustal rocks of the Paraíba do Sul Group (Fig. 4). Ages around 2.2 Ga were obtained on zircons by the U-Pb method (Machado et al., 1996, Valladares et al., 1997).

Association V: Paleoproterozoic (ca. 2.0 Ga) orthogneisses of granitic to granodioritic composition and basic lenses (Região dos Lagos) constitute the basement lithologic association of the Cabo Frio terrane (Heilbron et al., 1982; Zimbres et al., 1990; Fonseca, 1994; Schmitt et al., 1999). Sm/Nd model ages (T_{DM}) between 2.6 and 2.3 Ga (Fonseca, 1994) and $\varepsilon Nd_{2.0}$ around -6.0 suggest reworking of an older crustal component.

3.2.2 Meso to Neoproterozoic Cover

The cover of the Occidental terrane

The Meso/Neoproterozoic cover of the external tectonic domains of the Occidental terrane (Autochthonous and Andrelândia domains) records the fragmentation and dispersion of the Rodinia supercontinent, from continental rift basins to well-developed passive margins around the southern São Francisco craton.

Post-Transamazonian sedimentation began with the Tiradentes and Lenheiro depositional cycles (Andreis et al., 1989; Ribeiro et al., 1990) in the São João del Rei basin (Figs. 4, 5 and 7). The Tiradentes unit is a transgressive sequence characterized by orthoquartzitic metarenites, with lenses of metaconglomerates and metapelites. The Lenheiro unit consists of channelised and cross-bedded metarenites, sedimentary breccias, metaconglomerates and metapelites, suggestive of deltaic character. These facies display abundant primary structures, and were interpreted by Andreis et al. (1989) and Ribeiro et al. (1990, 1995) as shallow, tide-influenced shelf sedimentation, followed by progressive shoaling and by braided fluvial deposition. The São João del Rei basin (Tiradentes and Lenheiro cycles) may correspond to an interior rift developed in the interval between 1.7 - 1.3 Ga (Fig. 5).

An extensional episode characterized by a mafic dike swarm, block tilting and erosion, generated an unconformity that separates the Tiradentes-Lenheiro sequence from the overlying Carandaí sequence (Ebert, 1958). Geochemical and isotopic data from the mafic dikes (Ribeiro, 1992) indicate Sm-Nd model ages (T_{DM}) between 1.4 and 1.8 Ga and asthenospheric source. The Carandaí depositional cycle is characterized by laminated black pelites, followed by marls containing carbonatic lenses. Rare diamictites were interpreted as debris-flows from raised basement areas (Andreis et al., 1989). According to Ribeiro et al. (1995) and Ribeiro (1997), this cycle may correspond to an interior sag basin developed between 1.3 – 1.0 Ga (Fig. 5).

The Andrelândia depositional cycle (Ebert, 1957; Andreis et al., 1989; Paciullo, 1997) is the uppermost sedimentary sequence in Autochthonous and Andrelândia domains of the Occidental terrane. It consists of a basal psammitic association with meta-arkosic gneisses with orthoquartzitic intercalations; plagioclase schists and paragneisses; and a pelitic association at the top, with schists and gneisses (locally graphitic), with minor calc-silicatic and Mn-rich rocks. Deformed and mostly disrupted mafic dikes of predominantly continental tholeiite character (Gonçalves and Figueiredo, 1992; Paciullo, 1992) are common within the Andrelândia rocks. Sm-Nd model ages (T_{DM}) of 1184 and 1053 Ma, and positive values of ε_{Nd} (+ 4.8 and 3.11 for 1.0 Ga) support their derivation from a depleted mantle reservoir and a short crustal residence time span (Heilbron et al., 1990). Ultramafic metamorphic rocks have also been described as deformed lenses within the Andrelândia sequence (Trouw et al., 1986; Ribeiro et al., 1990). The mafic/ultramafic association is interpreted as the record of lithospheric extension period(s) during the evolution of the Andrelândia basin. This cycle may correspond to a passive margin developed during the Neoproterozoic (Fig. 5).

The geological units of the Andrelândia depositional cycle constitute the post-1.8 Ga cover of the Juiz de Fora domain. They have similar lithological characteristics (Figs. 4, 5 and 6): a basal psamo-pelitic succession and a top pelitic unit. Even with the observed complex deformation caused by intense tectonic mixing with basement rocks, careful mapping and geochronology allow distinction of the metasedimentary sequence from its basement. In the Serra do Mar region, the paragneisses with minor

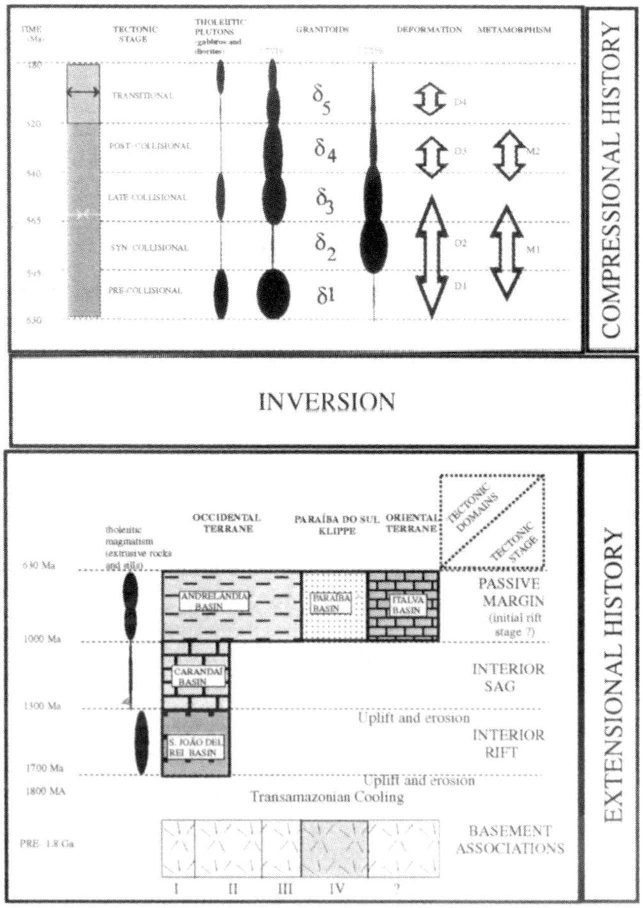

Figure 5. Schematic evolution of the Ribeira belt from the post-1.8 Ga extensional period to the Neoproterozoic III-Cambrian orogenic period. The pre-1.8 Ga basement associations are indicated by the same roman numbers as in figure 4. As recorded by several sedimentary basins, the extensional history shows the transition from the initial rifting (ca. 1.7 Ga) to passive margin construction related to the dispersion of Rodinia after 1.0 Ga. The tectonic stages of the orogeny and related deformational (D), metamorphic (M) and magmatic (γ) episodes are correlated. The syn-collisional stage (ca. 580 Ma) involved docking of the Oriental terrane.

quartzites and calc-silicate rocks were considered by Heilbron et al. (1991) as equivalent to the Andrelândia depositional Cycle, although several local denominations are also used. The occurrence of this unit at the Serra do Mar region is a consequence of fold-repetition at the southern limb of the Paraíba do Sul megasynform (Figs. 3 and 4).

Supracrustals of the Paraíba do Sul Klippe

The metasediments of the uppermost allochthonous sheet in the central segment of the Ribeira belt, named the

Paraíba do Sul Group (Ebert, 1955; Rosier, 1965), crop out in the core of the megasynformal structure along the Paraíba do Sul River (Figs. 3 to 6). It is composed of metapelitic schists and gneisses, arkosic gneisses with abundant dolomitic marbles, and calcsilicatic intercalations. Rare feldspathic quartzites also occur. At the southern segment of the Rio de Janeiro state, a subdivision of the Paraíba do Sul Group was proposed by Almeida et al. (1993), with three units: basal psammites with calcsilicatic, marble and pelitic intercalations, intermediate pelites, and a stratified top unit with psammitic and pelitic layers with abundant marble, calc-silicatic and Mn-rich rocks. Grossi Sad and Dutra (1998) also proposed a regional subdivision for the Paraíba do Sul Group into different formations, and envisaged a back-arc tectonic setting during sedimentation and volcanism. Safe inferences on their relative ages are hampered by the absence of geochronological data, by the tectonic character of contacts with the Andrelândia Depositional Cycle, and by the lack of defined stratigraphic relationships between the Paraíba do Sul supracrustals and the Paleoproterozoic Quirino unit.

The cover of the Oriental or Costeiro terrane

High-grade gneisses with calcsilicate, quartzite and calcitic marble lenses, named Italva Group by Machado Filho et al. (1983), comprise the supracrustal units of the Oriental terrane. Several lithological associations have been recognized in the Italva Group (Rosier, 1965; Heilbron et al 1982, 1993; Grossi Sad and Dutra, 1988; Tupinambá, 1993a):

a) meta-psammitic banded biotite gneiss with meta-mafic bands, decametric calcitic marble lenses and quartzite layers;
b) meta-pelitic gneisses with dolomitic marble lens, metric calcsilicatic layers and para-amphibolite (marls) lenses;
c) thick layers of calcitic marble and amphibolitic rocks.

These associations are suggestive of shallow carbonate platform developed at the margin of the Oriental terrane (Fig. 5).

The cover of the Cabo Frio terrane

Pelitic gneisses with thick layers of calcsilicatic rocks, amphibolitic layers and garnet quartzite lenses represent the supracrustal association of this terrane (Búzios group). Migmatitic structures and kyanite/sillimanite metamorphic parageneses are characteristic secondary features.

Rb/Sr ages dispersed between 580 and 520 Ma and by more recent U/Pb data around 520 Ma (Zimbres et al., 1990; Fonseca et al., 1994; Fonseca, 1994; Schmitt et al., 1999) indicate a Paleoproterozoic basement with Brasiliano overprinting. Sm/Nd model ages of 1.3 - 1.0 Ga are reported from the metasediments (Fonseca, 1994).

3.3- Brasiliano Collage and the Inversion of the Meso to Neoproterozoic Basins

3.3.1-Deformation

A similar metamorphic and structural evolution is recorded in all tectonic domains of the Occidental terrane. Different structural styles and metamorphic facies are related to deformation at specific crustal levels before thrust stacking (Figs. 4 and 6, Table 2). Three main deformational pulses were defined (Fig. 5) based on time relationships with metamorphism and magmatism:

a) the main deformation (D1+D2), coeval with the M1 metamorphic stage, is consistent with an oblique collision model (NW/W with vergence toward the SFC) with development of the most penetrative structures and fabrics (e.g. ductile thrusts, folds, main foliation, stretching and mineral lineations) as a consequence of thrust stacking of the different tectonic domains (Heilbron, 1995; Heilbron et al., 1998);
b) the late compressive deformation (D3) generated subvertical folds and NE trending transpressive shear zones, coeval with M2 metamorphic stage. Outstanding D3 map-scale folds are the Paraíba do Sul synform, the Rio de Janeiro antiform (Figs. 4 and 6) and the Paraíba do Sul transpressive shear zone (Campanha. 1981, Chrispim and Tupinambá, 1989; Ebert et al., 1991; Correia Neto et al., 1993, Trouw, 1995), expressed by a mylonite zone running along the hinge zone of the Paraíba do Sul synform for tens of kilometers (Fig. 2);
c) the transtensive deformation (D4), probably associated to cooling and collapse processes of the orogen (Heilbron, 1993; Machado, 1984).

Both sets of D3 and D4 shear zones acted as magmatic conduits for the ascension of late-collisional to post-tectonic granites (Valladares et al., 1995; Ebert et al.,1995).

Throughout the different sectors of Oriental and Cabo Frio terranes, the geometrical and kinematic analyses are not homogeneous in detail. Detailed structural data is available in sectors such as the coastal region (Heilbron et al., 1982; Machado and Demange, 1990; Silva et al., 1991; Schmitt and Trouw, 1997) and at the Serra do Órgãos (Almeida, et al., 1998).

The roof thrust limiting the Oriental terrane is a planar discrete mylonitic zone with NE/SW trend, gently dipping to NNW, associated with a high-rake stretching lineation. The internal deformational fabric of the hanging wall rocks

Table 2- Structures related to the Main Deformation within he different tectonic domains of the Ribeira belt.

Tectonic units	Structural Style
Occidental terrane	
Autochthonous domain	Open to tight folds, clear polarity and vergence to SFC S1 slaty cleavage and S2 crenulation cleavage, subhorizontal attitude, L2 mineral lineation Reactivation of syn-sedimentary faults
Andrelândia domain Figure 6d	Nappe structure with mylonites and basement lenses at the sole thrust, internal structure: isoclinal to tight D2 folds S2 crenulation schistosity, L2 mineral lineation, D1 folds, S1 schistosity NW vergence and low angle structures at the base of the domain Dextral transpressional component at the top of the domain
Juiz de Fora domain Figure 6c	Crustal duplex or C-type thrust Internal structure: tectonic intercalation of basement and cover, numerous basement duplexes. Tectonic mélange at the basal and upper contacts Mylonitc S2 foliation, stretching lineation, intrafolial D2 folds, intense transposition. Previous NW thrusting followed by dextral NE/SW transpression
Paraíba do Sul Klippe Figure 6b	The uppermost tectonic unit of the belt with mylonitic rocks and tectonic intercalation at the base of the thrust sheet Internal structure: Coarse S2 foliation, tight D2 folds, rare S2 crenulation schistosity in pelitic gneisses, NE/SW subhorizontal L2 (mineral) lineation Late NE/SW extension parallel to the trend of the belt
Oriental/Costeiro terrane Figure 6a	Underthrusts the Occidental terrane. Limited by a discrete mylonitic zone with a constant NW/N down dip mineral lineation Internal structure: coarse gentle dipping schistosity, D2 recumbent folding. Few kinematic indicators indicate top movement to NW. Late ENE vergence were reported at the Serra dos Orgãos Region.
Cabo Frio terrane	Overthrusts the Oriental terrane. Internal Structure: coarse gentle dipping schistosity, D2 and recumbent folding.

(Occidental terrane) of this shear zone display kinematic indicators with clear indications of tectonic transport of top to the north (between NE and NW) with subordinate dextral strike-slip component.

The internal fabric of the Oriental terrane is quite different. The foliation is not mylonitic, except in a few shear zones, best preserved within the orthogneisses. In these outcrops, opposite shear sense indications, parallel to the lineation, were described with movement both of top to NW and top to SE. Large scale recumbent isoclinal folding of the metamorphic foliation is the most conspicuous internal structure of the Oriental terrane. The regional pattern of this recumbent folding is symmetric, with highly deformed hinge zones, without evident vergence (Fig. 6a). Two sets of orthogonal normal folds and vertical shear zones also overprint the structures of the main deformation of the Oriental terrane.

The Cabo Frio terrane overthrusts the Oriental terrane (Fonseca 1998). The main structural characteristics are a gently dipping coarse schistosity impressed on both the basement and cover rocks and the recumbent folding of the basement-cover contact along the coast of Rio de Janeiro state, with NNW-trending axes. At cape Búzios, asymmetric recumbent folds indicate a local ENE vergence. Orthogonal subvertical folding related to shear zones completes the structural evolution of this terrane.

3.3.2 Metamorphism

The compressional history of the belt is characterized by two metamorphic events (Fig. 5). These events were discriminated and dated by use of microtectonic criteria and by the application of U-Pb techniques on metamorphic minerals (Machado et al., 1996). The M1 stage is Neopro-

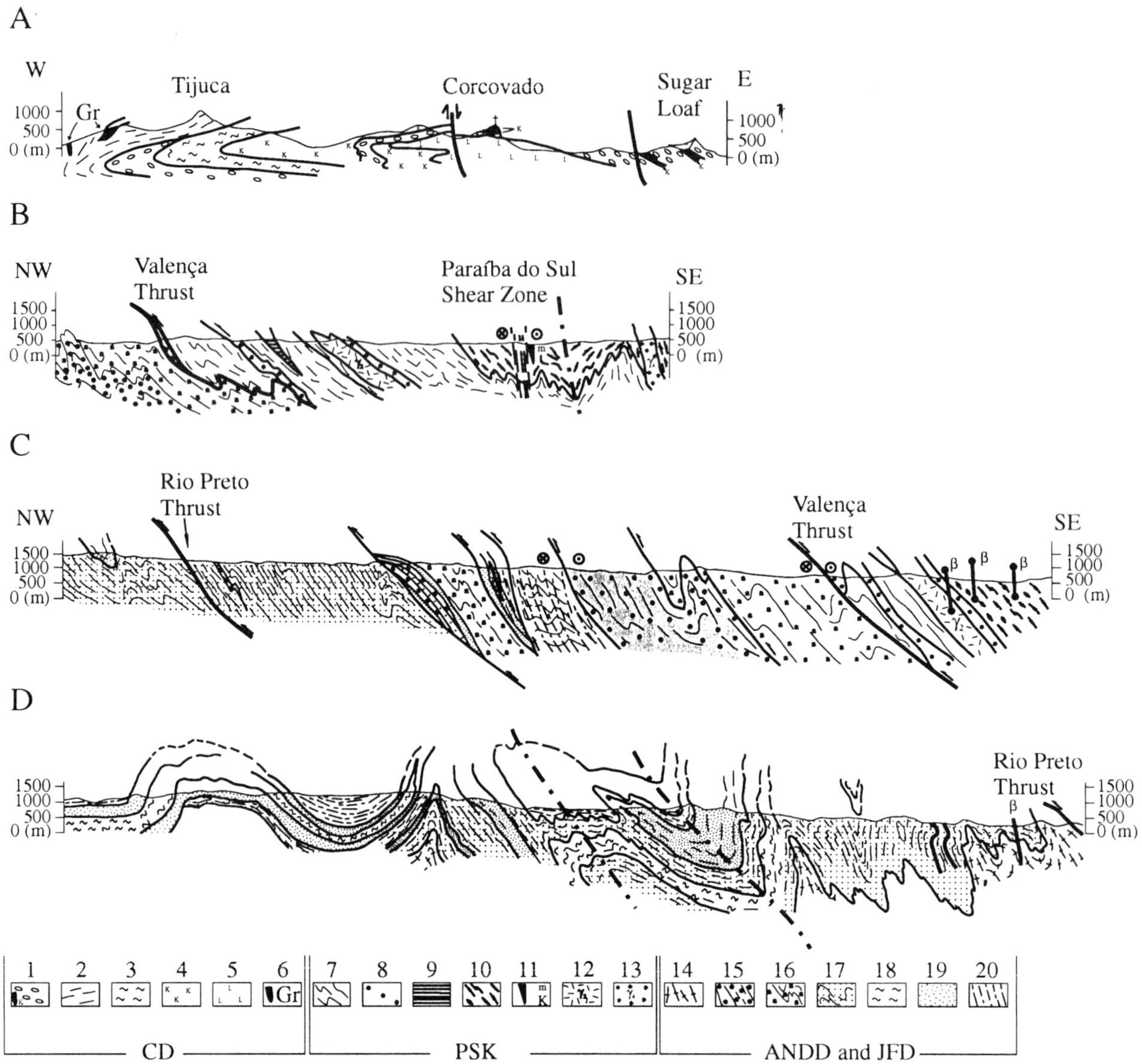

Figure 6. Detailed cross sections showing different structural styles of the tectonic domains. A- the Costeiro domain in Rio de Janeiro city with large-scale recumbent folds without clear vergence. Normal faults are related to Meso-Cenozoic tectonism; B- the Paraíba do Sul domain, with D1/D2 tight folds and thrusts refolded by the Paraíba do Sul synform and associated shear zone (D3), C- the Juiz de Fora domain with thrust-imbrication toward the SFC and scarce folding; D- the Andrelândia domain, with NW-vergent D1/D2 recumbent folds refolded by the vertical to inclined D3 folds. Legend: 1- Augen- gneiss with kinzigite lenses (K); 2- Rio Negro magmatic arc-rocks; 3 to 5 metasedimentary units of the Italva Sequence, Biotite paragneiss with quartzite layers (3); Kinzigite (4); Leucogneiss (5); 6- post-tectonic granites; 7- Hornblende-biotite orthogneisses of the Quirino complex; 8 to 11 metasedimentary units of the Paraíba do Sul group, garnet-sillimanite-biotite gneiss (8), biotite gneiss with quartzite layers (9), Biotite gneiss (10), Marble; (11); 12- Foliated S-type syn-collisional granite; 13- Late-collisional I-type leucogranite; 14- Hornblende gneisses and migmatites of the Mantiqueira complex; 15- Orthogneisses of the Juiz de Fora; 16 Orthogranulites of the Juiz de Fora complex, 17 to 20 metasedimentary units of the Andrelândia Sequence, Biotite gneiss with calcsilicate and quartzite layers (17), Kinzigite (18), Quartzitic unit (19), sillimanite-garnet- biotite gneisses and schists (20).

terozoic (595-565 Ma), and the M2 stage is early Paleozoic (540-520 Ma) in age (Fig. 5).

The M1 stage produced intermediate to high pressure mineral parageneses. Microstructural observations indicate that M1 parageneses form the main foliation. The peak metamorphic temperature increases eastwards, from the Occidental to Oriental terranes, displaying successive metamorphic zones: biotite, garnet, staurolite-kyanite, kyanite-sillimanite, K-feldspar, cordierite (only at the Oriental terrane), and locally orthopyroxene-sillimanite zones (Heilbron, 1985, 1993, 1995; Trouw et al., 1986; Tupinambá, 1993b). The spatial distribution of the M1 metamorphic zones, with high-grade metamorphic zones over low-grade zones, suggests an inverted gradient. Geothermobarometric data from the northern sector of the Andrelândia domain indicate maximum temperatures of 700-900°C and pressures of 8-10 Kb for early M1 stage (Trouw, 1992). Late M1 stage metamorphic conditions of the Juiz de Fora domain, around 700 – 750°C and 6-7 Kb, were reported by Duarte (1998). The re-equilibration at low temperatures is suggestive of an isothermal decompression P – T path.

The M2 stage shows syn-to-late-D3 relationship with deformation, and generated high temperature/low-pressure parageneses, generally of retrogressive character. Estimated conditions of 500-600 °C and 5-6 Kb were reported in the northern sector of Andrelândia domain (Trouw, 1992). In the Oriental terrane, the M2 stage reached higher temperatures, resulting in migmatization and generation of several I and S-type granitoid intrusions, which are preferentially located along antiformal structures and steep shear zones.

The cover of the Cabo Frio terrane reached higher metamorphic pressures when compared with the metamorphic evolution of the Oriental terrane, and peak temperatures around the transition between amphibolite and granulite facies. The high pressure parageneses (with kyanite, garnet and sillimanite) were impressed at the D1 foliation, subsequently refolded by at least three phases of deformation (Heilbron et al., 1982). New U/Pb data reported by Schmitt et al. (1999) indicate younger ages (ca. 520 Ma) for the metamorphic pulse of the Cabo Frio terrane

3.3.3- Magmatism

The Brasiliano magmatism was separated into five tectonic stages of igneous emplacement, based on available geochronological data and the age relationships with the deformation episodes (Heilbron, 1993; Machado and Demange, 1994; Heilbron et al., 1995, Machado et al., 1996; Tupinambá et al., 1998). Figure 5 and Table 1 show the distribution of the magmatic episodes in southeastern Brazil. The general characteristics, structures and tectonic setting of emplacement are given in Table 3.

The pre-collisional arc-related magmatism occurs only in the Oriental terrane (Figueiredo and Campos Neto, 1993; Tupinambá et al., 1998). The U/Pb geochronological and geochemical data suggest that a magmatic arc was active at the Oriental terrane of the Ribeira Belt during the pre-collisional phase of the Brasiliano-Pan African collage. The proposed magmatic arc occupies 2/3 in area of the Oriental terrane (Fig. 3), running for almost 600 km along the Atlantic Coast, from northern São Paulo State to southern Espírito Santo State. It is noteworthy that neither at the Occidental terrane nor at the Paraíba do Sul klippe the pre-collisional magmatic rocks have been previously described. In contrast, syn- to late-collisional granitoids are abundant within these two tectonic units.

The Brasiliano syn- to late and post-collisional granitoids occur in all terranes, showing a spatial and temporal polarity within the belt. They are more abundant in the Juiz de Fora domain and toward the coast, indicating thickening of the crust as the result of collision (Figs. 4, 6 and 7).

The late-collisional period is characterized by metaluminous to slightly peraluminous leucogranites related to the Late Deformation subvertical shear zones. Elongate syn-D3 batholiths and stocks of leucogranite are widespread at the Paraíba do Sul klippe and at the Oriental terrane

Small plutons and stocks of high-K calc-alkaline to alkali-calcic trends intrude all previously described lithological units of the Costeiro terrane. They have not been reported yet within any of the other terranes of the central Ribeira belt. Textural patterns and structural relationships with country rocks point to post-tectonic emplacement.

3.4- From Post-1.8 Ga to Gondwana Assembly: An Evolutionary Tectonic Model

An evolutionary tectonic model for the central segment of the belt is presented below. Although other models have been previously proposed (Ebert et al., 1991; Machado & Endo 1993; Tankard et al., 1995; Ebert e Hasui, 1998), mostly based on structural data, the present model (Figs. 5 and 7) is believed to be compatible with all the observed relationships between structural evolution and metamorphic/magmatic features, besides associated geochronological and geochemical data. The geochronological investigation of the Cabo Frio terrane is still in course (Schmitt et al., 1999) and therefore the evolution of this tectonic unit is not discussed here.

At end of the Transamazonian collage (ca. 1.9 Ga) the amalgamation of terranes such as microplates and magmatic arcs generated a major continent comprising the pre-

Table 3- General characteristics of the Brasiliano granitoids of the Occidental and Oriental terranes, classified according to time relationships with the deformation phases (Heilbron et al., 1995)

Time relationship with deformation	Age (Ma)	Geochemistry	Structures	Tectonic Domains	Envisaged tectonic setting within the work	References
Post- tectonic Syn- to post-D4	< 520	• Calc-alkaline to alkali-calcic plutons • Tholeiitic gabbros and diorites	Non-foliated and mostly zoned stocks and sheets, primary flux foliation and layering, dikes, proposition of mixing and mingling process	Only detected at the Oriental terrane	Extensional Collapse of the orogenic belt	Pires et al., 1982; Wiedeman, 1989; 1993; Junho 1993; Figueiredo and Campos Neto, 1993;
Post-collisional Syn-D3	540-520	• Alkali-calcic stocks of slightly peraluminous leucogranites	Subvertical dikes and plutons related to D3 shear zones, weakly foliated at the borders	Recorded at both tectonic terranes	Crustal reworking of basement and cover rocks	Machado and Demange, 1994, Heilbron, 1993, 1995, Heilbron et al., 1995, Machado et al., 1996, Valladares 1996.
Late-collisional Late-D2	565-540	• Abundant metaluminous I-type granites with basic enclaves • Subordinated two-mica peraluminous S- leucogranites	Weakly foliated plutons and sheets	Recorded at both tectonic terranes	Crustal reworking of basement and cover rocks	Grossi Sad and Barbosa, 1985; Tupinambá, 1993a; Machado and Demange, 1994, Heilbron, 1995, Heilbron et al., 1995; Junho et al., 1999.
Syn-collisional Syn-D1+D2	595-565	• Abundant peraluminous S granites • Metaluminous I-type granites	Foliated and mylonitic plutons	Recorded at both tectonic terranes	Crustal reworking of basement and cover rocks	Tupinambá, 1993a; Machado and Demange, 1994, Heilbron, 1995, Heilbron et al., 1995, Machado et al., 1996.
Pre-collisional Pre-D1	630-595	• Tonalites to granodiorites and tholeiitic gabbros	Foliated plutons, mylonitic structures nearby tectonic boundaries	Only detected at the Oriental terrane	Cordilleran Magmatic Arc	Figueiredo and Campos Neto, 1993; Tupinambá et al., 1998

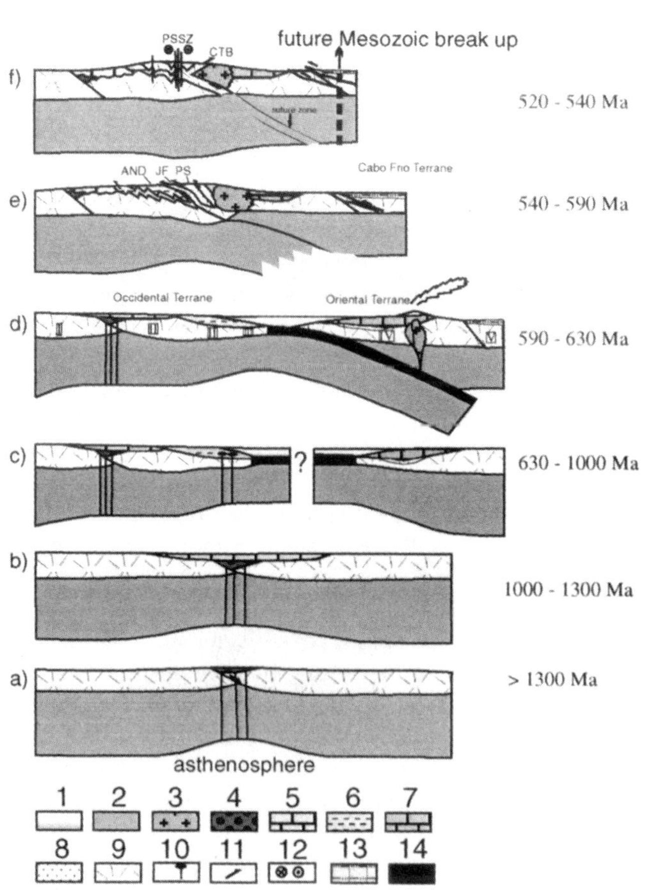

Figure 7. Schematic tectonic evolution of the central segment of the Ribeira belt, from the post-1.8 Ga interior rifting to the Pan African/Brasiliano orogeny during the amalgamation of the Gondwana. a) block-faulting, associated basic dike swarm and subsequent uplift and erosion of the São João del Rei rift; b) subsidence and unconformable deposition of the Carandaí interior sag basin; c) the continental fragmentation and dispersion associated to Rodinia break-up generated the Andrelândia and Italva passive margins of the Occidental and Oriental terranes, respectively. Two pulses of tholeiitic magmatism mark extensional episodes within the Andrelândia basin; d) tectonic inversion and construction of the Oriental terrane active margin initiating the pre-collisional stage of the Brasiliano orogeny, with generation of the Rio Negro magmatic arc intruding the Italva passive margin and its pre-1.8 Ga basement; e) the collisional stage results in intense shortening of the Occidental margin with the development of NW-vergent crustal thrusts-sheets (Andrelândia, Juiz de Fora and Paraíba do Sul domains); f) continued convergence during the post-collisional stage generated a transpressional regime with dextral shear zones and associated steep folding, which involved back-thrusting along the Central Tectonic Boundary (CTB), now the surface expression of a cryptic suture. Intense magmatism characterize the post-collisional stage and subsequent transition to cratonization. Legend: 1- sea-water; 2- lithospheric mantle; 3- Rio Negro magmatic Arc; 4- São João del Rei rift basin; 5- Carandaí interior sag basin; 6- Andrelândia passive margin; 7- Italva passive margin; 8- Paraíba do Sul fragment; 9- pre-1.8 Ga continental crust; 10- basic magmatism; 11- major thrusts (CTB- central tectonic boundary); 12- Subvertical dextral shear zones (PSSZ- Paraíba do Sul shear zone); 13- Cabo Frio terrane; 14- Neoproterozoic oceanic crust. Roman numbers refer to basement associations as in Figure 3.

sent southeastern Brazil. A variety of geological and geophysical evidence indicates the existence of large continental masses during the Orosirian period of the Paleoproterozoic (Van Schmus et al., 1993; Ledru et al., 1994; Brito Neves et al., 1995; Rogers, 1996). Pieces of the Transamazonian amalgamated paleocontinent can partly be restored from the pre-1.8 Ga basement units in the Ribeira belt.

The Staterian period is characterized by widespread taphrogenesis in the São Francisco-Congo paleocontinent. Important rift systems developed at this time, e.g. the Espinhaço and São João del Rei rifts (Brito Neves et al., 1995). Felsic volcanism around 1.8/1.75 Ga is related to the initial rift stage.

According to Ribeiro et al. (1995) and to Trouw et al (1997), another extensional event, poorly constrained but tentatively related to the Ectasian period (ca. 1.3 Ga) of the Mesoproterozoic, led to the deformation and tilting of the São João del Rei rift basin sedimentary successions, predating the Carandaí basin (Ribeiro et al., 1995; Ribeiro, 1997). Basic dikes, related to this extensional episode, intruded the rift successions of the São João del Rei basin (Fig. 7). According to Ribeiro et al. (1995), after an initial rift stage, the Carandaí basin developed as an interior sag with deposition of carbonatic (limestones; carbonatic pelites, and marls) and pelitic facies. The repetitive sedimentary successions and the absence of structures and magmatism suggest conditions of tectonic stability during the development of the Carandaí basin.

At the end of the Mesoproterozoic, the Grenville collage led to the formation of the Rodinia supercontinent (Hoffman, 1991). Clear evidence of the imprint of the Greenville orogeny has only been reported in the central, northern and northeastern regions of Brazil (Brito Neves et al., 1995). Probably, the southeastern region of Brazil was located relatively far from the Greenvillian front.

The Mesoproterozoic/Neoproterozoic transition is characterized by the break up and dispersion of the Rodinia supercontinent and of other minor continental masses. These processes were probably diachronic, and generated important passive margin basins. Both the Brasília and Ribeira passive margins, located westward and southeastward of the São Francisco-Congo continent, respectively, developed at this time. Although poorly constrained, this tectonic phase is probably related to the 1.0-0.9 Ga interval, based on Sm/Nd model ages of juvenile basic dikes (Heilbron et al., 1990; Trouw and Pankhurst, 1993). The evolution of the Brasília margin at the Alto Rio Grande interference zone was described by Campos Neto (1992), Paciullo (1997) and Ribeiro et al. (1995). The Ribeira passive margin may be tentatively restored within the different tectonic domains of the Occidental terrane (Autochthonous, Andrelândia and Juiz de Fora tectonic domains). A basal psammitic unit associated with tholeiitic mafic lenses was interpreted by Paciullo (1997) as the initial rift stage of the Andrelândia basin. A sandy braided river system concomitant with tholeiitic magmatism is the suggested sedimentation environment for this stage. The passive margin stage of the Andrelândia basin is represented by the following lithofacies associations: a) pelites and orthoquartzitic arenites related to a shallow marine transgression; b) thin-bedded turbidites with scarce dropstones and debris flow deposits that record a low-stand regime associated to a glacial period; and c) pelites with calc-silicatic rocks and manganese rich meta-chert (gondite) that represent the post-glacial high-stand marine facies (Paciullo 1997, Paciullo et al., 1998).

Another Neoproterozoic passive margin is inferred, probably at the same time, at the border of the Oriental terrane (Serra do Mar plate or microplate). High-grade metamorphism (granulite facies or the transition between amphibolite and granulite facies) and intense tectonic transposition restrict paleo-environmental reconstructions. Nevertheless, the predominance of (meta) pelite-carbonate lithological association of the Italva group (northern Rio de Janeiro state) may be regarded as indicative of a former shelf.

The Brasiliano collage at the central segment of the Ribeira belt was the result of continued convergence of the São Francisco continent (Occidental terrane) and the Oriental terrane (Serra do Mar or Costeiro microplate). Based on the tectono-metamorphic and magmatic evolution and on U/Pb data, the Brasiliano orogeny was subdivided into five tectonic stages (Figs. 5 and 7).

a) As SE-vergent subduction of the São Francisco plate started, the pre-collisional stage (ca. 630-595 Ma) is recorded by the D1 deformation of the cover and the basement of the Andrelândia passive margin; a cordilleran magmatic arc intruded the carbonatic rocks of the Italva group succession, along the active margin of the Oriental terrane.
b) the syn-collisional stage (ca. 595-565 Ma) was represented by collision between the Occidental and Oriental terranes. Crustal shortening took place initially through ductile thrusts and folds during the main deformation phases (D1+D2), resulting in NW-vergent transport of the tectonic domains of the Occidental terrane. S-type mylonitic granites are abundant during this stage.
c) The late-collisional stage (ca. 565-540 Ma) was characterized by oblique convergence, with pervasive deformation throughout the belt. The Serra dos Órgãos batholith and other minor metaluminous granitoid rocks are expressive examples of plutonism during this stage.
d) During the post-collisional stage (ca. 540-520 Ma), the lithospheric convergence between these two plates continued but was accomodated by discrete strike-slip ductile shear zones. Between these shear zones, shortening was achieved by steep folding of S2 foliation (D3 deformation). This stage is marked by conspicuous calc-alkaline magmatism, emplaced along subvertical shear zones.
e) The transitional stage (ca. 520-480 Ma) is characterized by transtensional shear zones temporally associated with thermal relaxation of the orogen. Stocks of calc-alkaline granitoids, mostly associated with tholeiitic rocks, are characteristic of this tectonic stage. This magmatism might be caused by melting of crustal and mantle rocks, related with high temperature conditions and uplift just after the Brasiliano collision. Probably, delamination and underplating processes resulted in the mantle contribution in the post-tectonic magmatism, that represented the transition to the Phanerozoic extensional regime. This interval also coincides with the earliest subsidence in the Paraná basin, as discussed in the following section.

4- THE PARANÁ BASIN IN GONDWANA AND DURING BREAK-UP.

4.1- Regional Setting and Mechanisms of Formation

The Paraná basin is located in the south-central parts of South America, straddling Brazil, Argentina, Paraguay and Uruguay (Fig. 8). The northern border is controlled by a NW-trending geological feature (Alto Paranaíba Arch), and towards the east, the Paraná basin is flanked by Precambrian outcrops, except in the Rio do Grande do Sul State littoral, where the Paleozoic basin reaches the present-day continental margin near Torres, in the northern portion of the Pelotas basin. The Paraná basin continues towards the Bolivian-Argentinian Chaco basin to the west, beyond the N-trending Asunción Arch near meridian 58° W, and its sediments merge with analogous sequences in the present day sub-Andean foreland domain, westwards of the Asunción Arch. Figure 9 shows a regional transect in the basin extending from the western portions in the Bra-

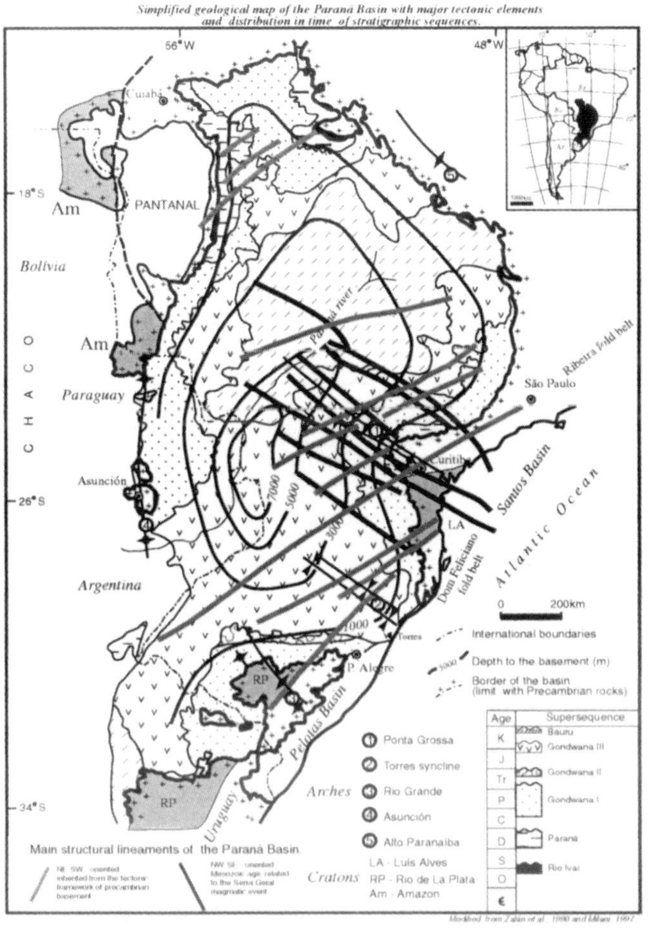

Figure 8. Simplified geological and structural map of the Paraná basin with geologic distribution of stratigraphic sequences and major tectonic elements.

zilian territory to the Precambrian outcrops in the northeast.

The Paraná basin is one of the world's largest basins in terms of depositional area, reaching more than 1,000,000 sq. km. in the Brazilian territory alone. Its sedimentary column is characterized by flat-lying siliciclastic successions, and the maximum sediment thickness is about 7,000 m (Milani et al., 1994; see Figs. 9 and 10). The present day shape of the Paraná basin is strongly controlled by erosion of the sedimentary layers along its northern and eastern borders, particularly in the region flanking the South Atlantic Ocean (between Uruguay and the São Paulo State in southeastern Brazil). Mesozoic uplift resulted in 2,500m of erosion in places (Zanotto, 1993). The Ponta Grossa Arch (Fig. 8) dissects the Paleozoic and Mesozoic basin, reflecting marked uplift and erosion along the Guapiara-Curitiba fault zone during breakup. Uplift is attributed to strike-slip processes and possibly magmatic addition of underplated rocks reflected in the Early Cretaceous diabase dike swarms that form conspicuous NW-trending lineaments in the oriental portion of the Paraná basin.

The most important tectonic lineaments in the Paraná basin (Fig. 8) form two groups, oriented NE and NW. These trends coincide with pre-oriented Precambrian grain (Zalán et al., 1990), and are inherited from Precambrian shear zones along areas of different tectonic mobility during the Precambrian orogenies, compared with the adjacent blocks. These Precambrian lineaments outcrop in the Paraná basin as isolated faults or complex shear zones for tens to hundreds of kilometers in length, and tens of kilometers in width (Campanha and Ferrari, 1984).

Soares et al. (1982) and Zalán et al. (1990) suggested that the main structural lineaments in the Paraná basin reflect basement weakness of Precambrian age, which were intermittently reactivated during breakup in the Mesozoic. Soares et al. (1982) observed that the NW-trending lineaments were intensively intruded by diabase dikes in the Early Cretaceous, whereas the NE-trending elements are notably devoid of dikes. These differences indicate that the Mesozoic dikes intruded along the trend of the Guapiara-Curitiba fault system as a result of dilation during right-lateral deformation in Neocomian times (A. Tankard, personal communication, 1998). Milani (1997a), analyzing recent geophysical data, separated the NE-SW set of lineaments, clearly related to the Precambrian history of the region, from the NW-SE ones, which are related to the Mesozoic rifting along the South Atlantic. The NW direction may be related to a possible failed arm of a rift system associated with a mantle plume in the Florianópolis region, offshore of the Luís Alves craton (Fig. 8).

The origin of the Paraná basin is still a matter of debate, typical of other intracratonic basins worldwide. These basins generally lack indicators of large-scale extension, such as rotated fault blocks. We will address two main hypotheses that attempt to explain the origin of these basins, and discuss the tectonic and sedimentary evolution of the Paraná basin within a regional framework Both hypotheses rely on compressional tectonics in the early stages of subsidence. The first hypothesis attributes initial subsidence in the Paraná basin to the closing phase of the Brasiliano orogeny, while the second relates basin formation to intermittent orogenies along the Pacific margin.

The nature of the deep structure beneath the Paraná basin may be critical to any geodynamic interpretation. One conceptual model for the underlying framework of the basin is based on the possible occurrence of a cratonic nucleus underneath the depocenter (Cordani et al., 1984). More recent geological and geophysical investigation and compilation by Petrobras E&P group showed that the depocenters of the oldest stratigraphic units (Ordovician-

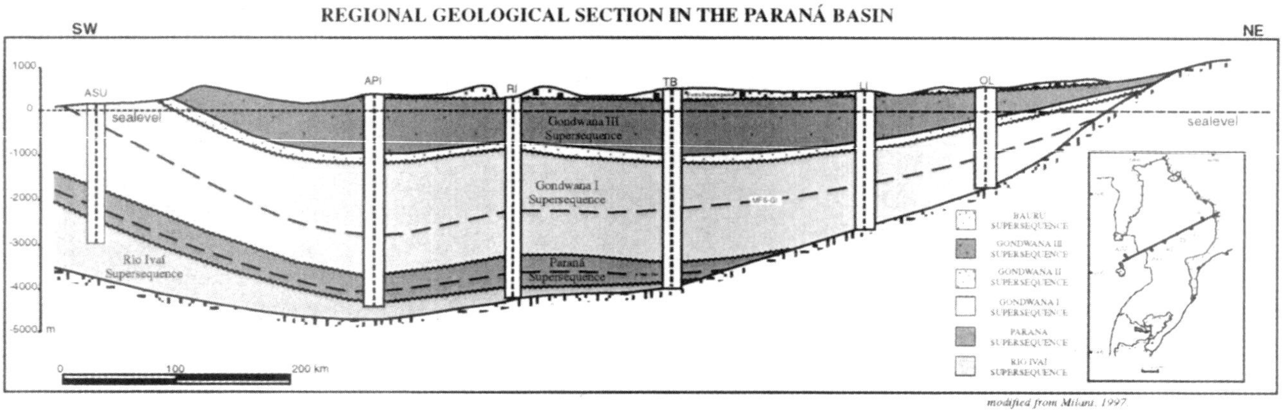

Figure 9: Schematic geological cross section in the Paraná basin showing the Precambrian flank in the northeast and the thickening of the stratigraphic sequences towards the southwest.

Silurian) are characterized by NE/SW trending structures that form a central rift (Marques et al., 1993). Several exploratory boreholes have penetrated the sedimentary cover and provided punctual information on the basement. The results of the deep drilling in the basin indicated that the basement is constituted by a number of cratonic blocks (Archean rocks) and metasediments of the Brasiliano fold belts, deformed during Precambrian orogenies (Milani, 1997a). One schematic geological section crossing the depocenter of the Paraná basin (Fig. 11) suggests that the earliest sedimentary successions (Lower Paleozoic Rio Ivaí supersequence, Fig. 10) form troughs that coincide with Brasilian belts in the basement and that the cratonic blocks have thinner covers of sedimentary units.

Cordani et al. (1984) suggested that the geological development of the Paraná basin was influenced by Precambrian structures. According to Zalán (1990), subsidence was initiated by convergence of the cratonic blocks between South America and Africa. Following the Brasiliano orogeny, release of the accumulated stress and decay of the thermal anomaly that resulted from the overthickened crust (which remained more or less fixed above hotspots in the mantle) might have been enough to trigger subsidence and accumulation of the earliest Silurian sequence (Zalán et al., 1990).

An alternative hypothesis relates the subsidence phases in the Paraná basin to compressional stresses in the western part of South America (Milani,1997a; Ramos,1998; Milani and Ramos, 1998), which resulted from almost continuous subduction and convergence along the Pacific margin. Figure 12 shows the correspondence between subsidence cycles and the stratigraphic signature of the Paraná basin supersequences to major orogenies in SW Gondwana. From the Precambrian towards the present, the Panthalassa oceanic lithosphere (now corresponding to the Nazca plate) has been subducting more or less continuously under the cratonic interior of South America. A number of catastrophic episodes are related to the lithospheric convergence and collision between sialic blocks riding on oceanic lithosphere and the cratonic blocks of the western portion of South America.

The main orogenic episodes that correspond with most intense subsidence phases in the Paraná basin (Fig. 12) are the late Ordovician – earliest Devonian (Famatinian), which includes two orogenies (Ocloyic and Precordilleran), and the latest Devonian (Chañic) and Late Permian – Early Triassic (San Rafael). The San Rafael orogeny was associated with the accretion of the Patagonian terrane, which resulted in the formation of a magmatic arch near the margin of the continent (Ramos et al., 1984). The Choiyoi Group in Argentina includes magmatic and volcaniclastic rocks with radiometric dates ranging from 275 to 250 Ma. The San Rafael was contemporaneous with the Cape orogeny of South Africa, and is also expressed in the Sierra de la Ventana of Argentina.

4.2 Tectono-Stratigraphic Evolution

The stratigraphic column of the Paraná basin (Fig. 10) records three main phases of subsidence separated by major erosional hiatuses (Milani, 1997b). These three phases of subsidence span six major, second order depositional sequences: (1) Late Ordovician Early Silurian; (2) Late Silurian-Devonian; (3) Late Carboniferous-Permian; (4) Triassic; (5), Late Jurassic-Early Cretaceous; (6) Late Cretaceous. The first four sequences are predominantly siliciclastic, and the breakup phase (Late Jurassic-Early Cretaceous) is represented by the largest flood basalt province in the world. This volcanism (Serra Geral Fm.) is more related to opening of the South Atlantic than to intra-

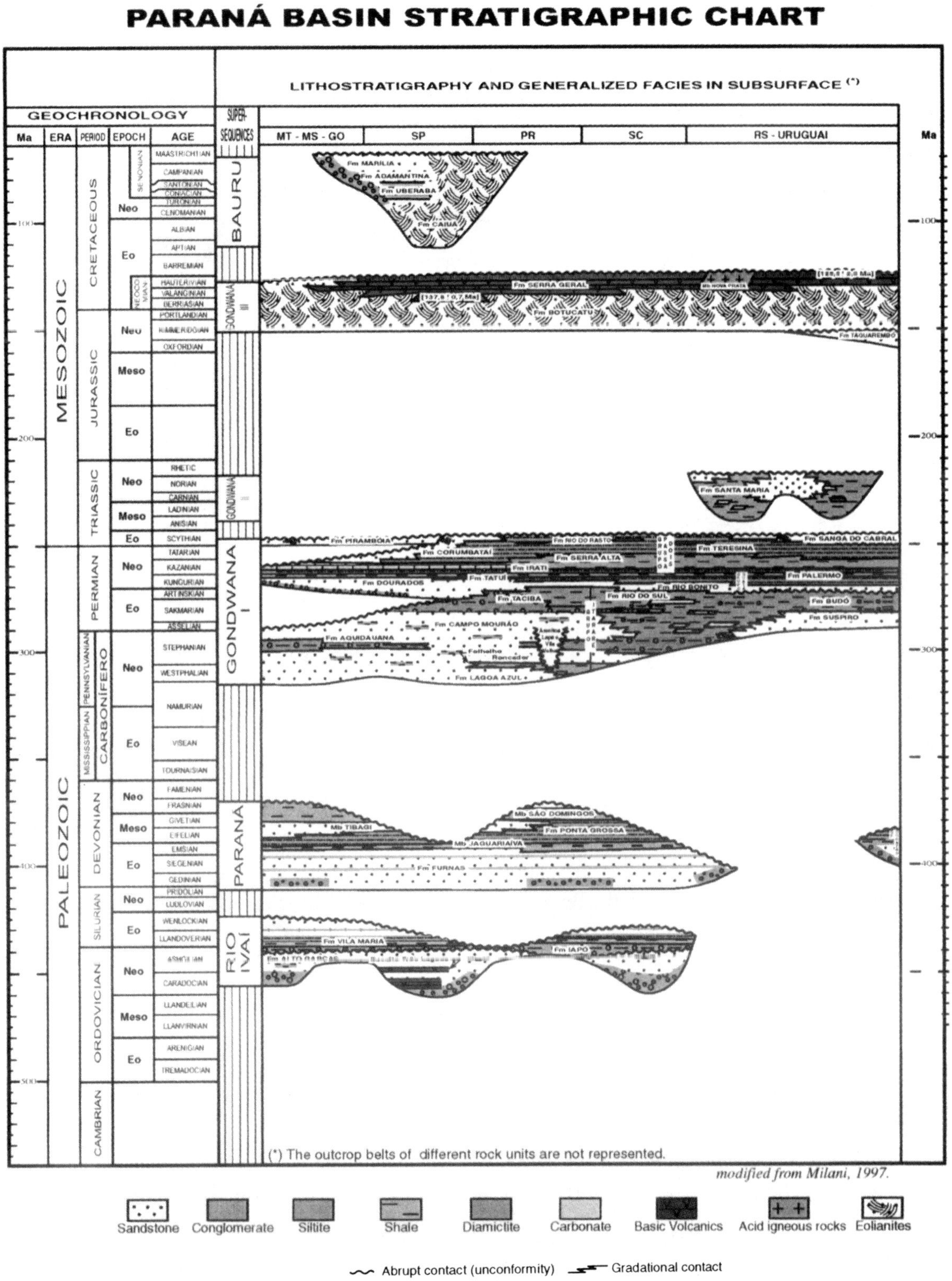

Figure 10. Litho-stratigraphic chart of the Paraná basin.

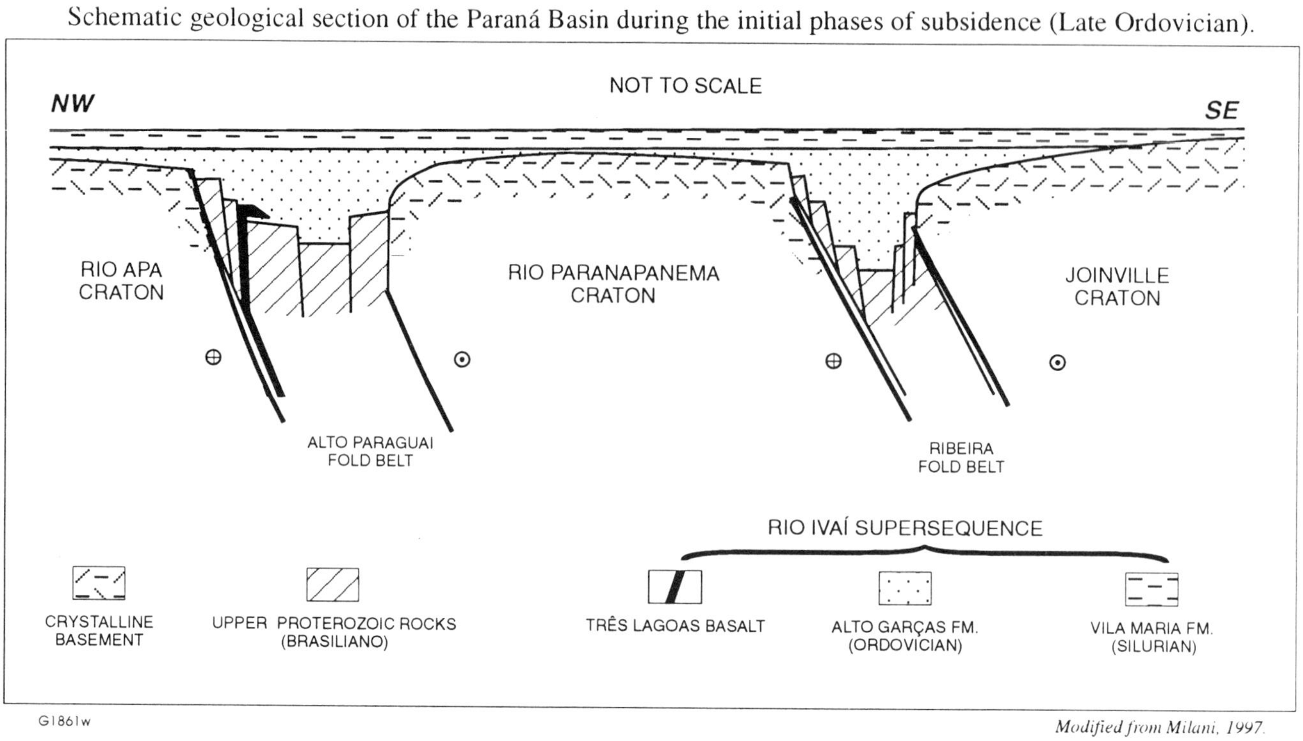

Figure 11. Schematic geological section of the Paraná basin during the inital phases of subsidence (Late Ordovician), showing the relationship between Neoproterozoic fold belts and early depocenters.

basin tectonic events. The magmatic rocks are also associated with the rift successions of the continental margin basins (e.g., Pelotas, Santos and Campos).

The stratigraphic sequences in the Paraná basin (Fig. 10) are separated by basinwide unconformities related to tectonically induced variations of base-level (Milani, 1997a). The stratigraphic signature of the Paleozoic sedimentary sequences show several onlapping and downlapping surfaces, based on analysis of sedimentary sequences in exploratory boreholes (Fig. 12). The cross-plot of the subsidence rates calculated for a number of boreholes with the eustatic sea-level curve indicates that the subsidence cycles in the Paraná basin are related to the pre-Andean orogenies in SW-Gondwana margin (e.g., increased subsidence rates in the Early Permian, coinciding with the Sanrafaelic orogeny). Maximum flooding surfaces (MFS in Fig. 12 and Fig. 13) coincide with the Ocloyic and Precordilleran orogenies. The Permian MFS is slightly delayed in time following the San Rafael orogeny.

The tectono-sedimentary framework of the SW Gondwana from Ordovician to Triassic times are shown in Figure 14 (Late Ordovician and Devonian) and Fig. 15 (Permian and Triassic). The isopach maps of the Ordovician (Rio Ivaí supersequence), Devonian (Paraná supersequence) and Late Carboníferous to Early Triassic (Gondwana I supersequence) are shown on Fig. 16.

The Late Ordovician phase of subsidence in the foreland basin coincides with the first important orogenic event in the western margin of South America. It was possibly caused by docking and accretion of the Precordillera terrane (Ramos, 1988). Figure 14-a shows the tectono-sedimentary framework of SW Gondwana in Ordovician times. The Precordillera crustal block is characterized by a thick carbonate succession rich in Olenellus trilobites dated as Cambrian, which show many similarities with the equivalent fossil assemblage in the North American Appalaches (Ramos et al., 1986). The docking of the Precordillera sialic block resulted in the Ocloyic orogeny which was contemporaneous with the Late Ordovician / Early Silurian episode of subsidence in the Paraná basin (Figs. 12 and 13).

The far-field effects of these orogenic events and consequent changes in stress field are believed to have reactivated the SW-NE trending structures as linear troughs (Fig. 16-a). These Late Ordovician Rio Ivaí depocenters (Milani et al., 1995) have a maximum sediment thickness of about 400 m, which we attribute to transtensional reactivation of pre-existing Precambrian fabrics to form small strike-slip

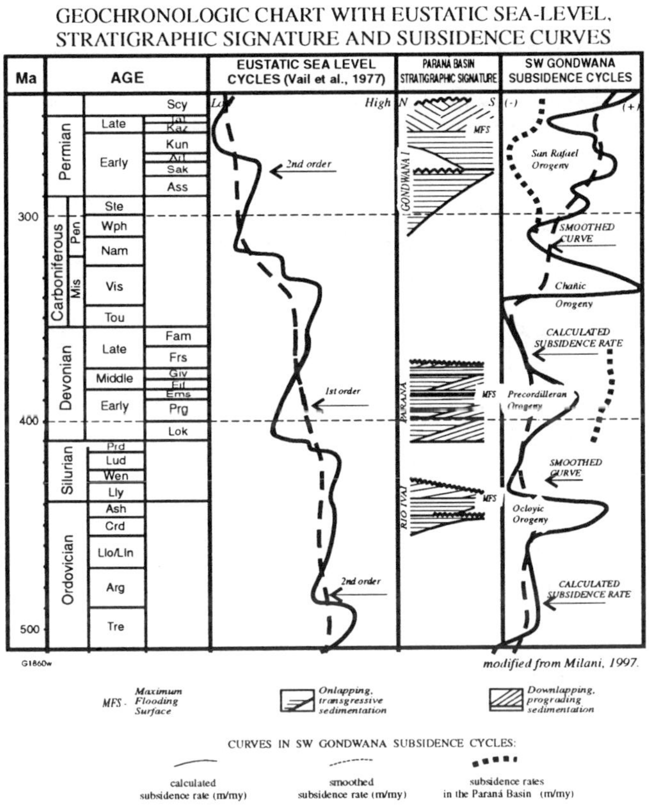

Figure 12. Geochronologic chart with eustatic sea-level, stratigraphic signature and subsidence curves of the Paraná basin.

basins. These transtensional characteristics also explain some of the recently discovered Late Ordovician extrusive basalts (Milani, 1997b). The crustal discontinuities that seem most likely to have influenced the location of the initial depocenters are the Alto Paraguai and the Ribeira fold belts of Brasiliano age (Milani, 1997; see Fig. 11). The early deposits of the Rio Ivaí Supersequence accumulated in transtensional troughs located in this province, while the cratonic regions remained positive areas, and were covered by sediments only at the end of the Ordovician (Alto Garças Fm., Upper Ordovician, around 450 Ma, Figs. 11 and 14-a).

Other important subsidence phases occurred during the Devonian (Fig. 14-b) and also coincide with accretion of sialic terranes (Chilenia). This indicates that the transpressional stresses and the flexural load (caused by the piling of accreted masses on the lithosphere of the foreland basins) may be important controls on the formation of intracratonic sedimentary basins worldwide (Stockmal and Beaumont, 1987; Milani and Ramos, 1998). The stresses responsible for these reactivations might be associated with the propagation towards the cratonic interior of far-away stresses associated with the docking and amalgamation of the sialic terranes onto the subducting margin of the South American continent.

The Precordilleran orogeny (Astrini et al., 1995) is associated with docking of the Chilenia terrane (Fig. 14-b). The sediments associated with this phase correspond to the Paraná Supersequence (Figs. 10 and 13), which superseded the Rio Ivaí supersequence. The Devonian sediments (Furnas and Ponta Grossa formations, Figs. 12 and 16-b) are characterized by an almost tabular geometry, showing only small variations in thickness for several hundreds of kilometers.

The Paraná Supersequence is characterized by a complete transgressive and regressive cycle of sedimentation, starting with coarse siliciclastics of the Lower Devonian Furnas Fm., which indicate a stable period in the basin. This was followed by fossiliferous pelitic sequences (Ponta Grossa Fm.), which indicate sealevel rise and transgression of the marine sediments into the Paraná basin. The total sediment thickness of this phase is up to 800 m in the Paraná basin, but in Argentina and Bolivia, as well as in South Africa, the the Devonian sequence may reach thicknesses greater than a few thousand meters (Gohrbandt, 1993), suggesting a communication between the gulf in the Paraná basin and the much larger open marine environments of the western Gondwana foreland (Fig. 16-b).

Lower rates of sedimentation in the Early Carboniferous are attributed to widespread glaciation in Gondwana (Caputo and Crowell, 1985). Erosion was widespread, dissecting previously deposited sedimentary layers. Above this unconformity, the most important phase of subsidence in the Paraná basin (Gondwana I) resulted in accumulation of a thick (up to 2,500 m, Fig. 16-c) sedimentary succession in diverse environments of deposition. In the Early Permian, the sedimentary basin expanded considerably, reaching the Rio Grande do Sul State in the Torres syncline (Fig. 8). This phase coincided with structural modification in the substratum of the basin. The Rio Bonito Fm. is characterized by progradation of deltaic lobes that exhibit a retrograde geometry (from south to north), coinciding with the sense of onlap of different stratigraphic markers (Fig. 12). The Guatá Group is characterized by a transgressive sequence culminating in a maximum flooding surface identified by a radioactive marker (Milani, 1997).

While the southern parts of the basin were characterized by subsidence and sedimentation in the Early Permian, the northern parts were subject to a strong erosional event. Tilting of the basement toward the south was probably related to collision of the Patagonian terrane (Fig. 15-a) and loading of the lithosphere by crustal flakes during the San Rafael orogeny (Ramos, 1988). This orogeny also formed

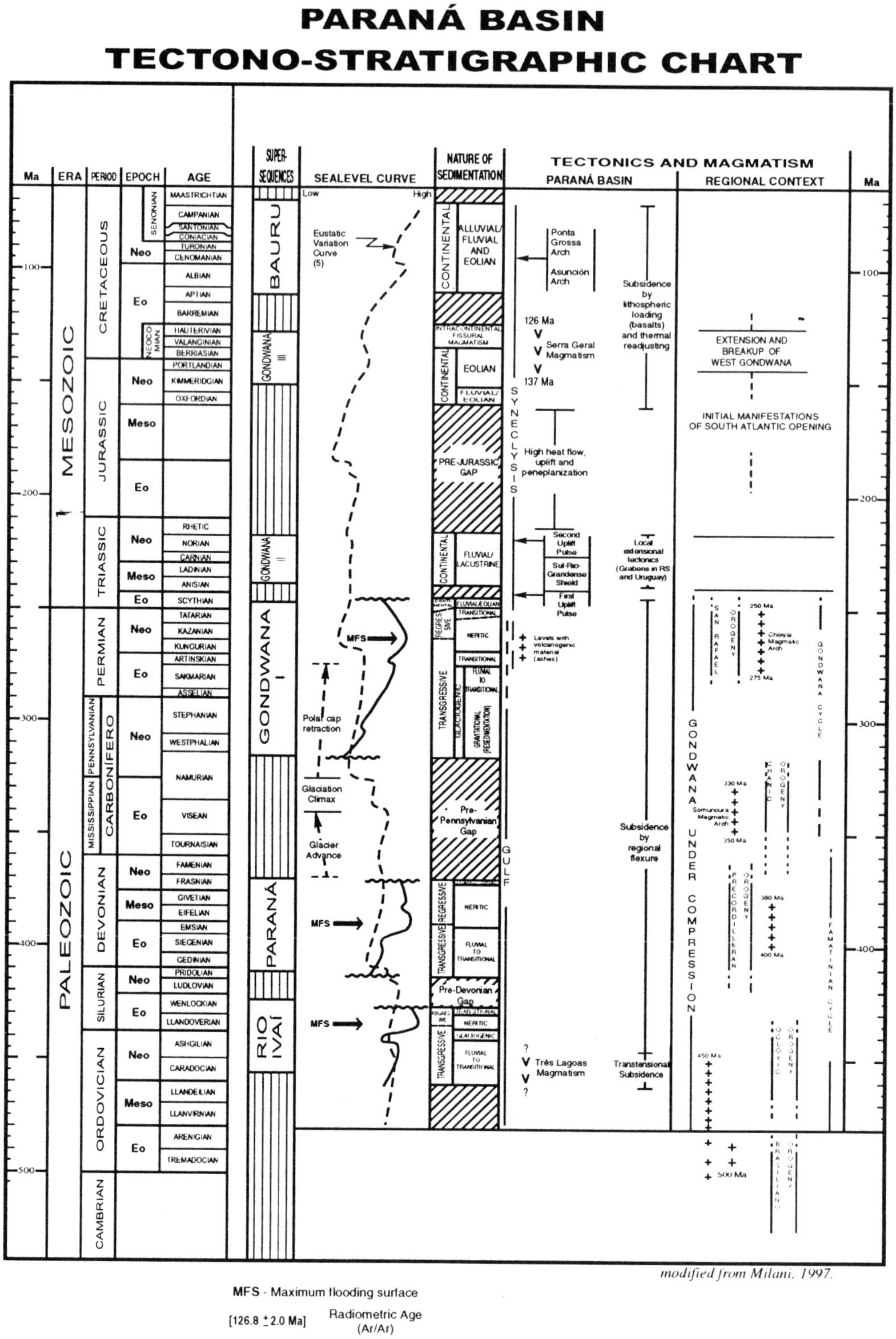

Figure 13. Tectono-stratigraphic chart of the Paraná basin.

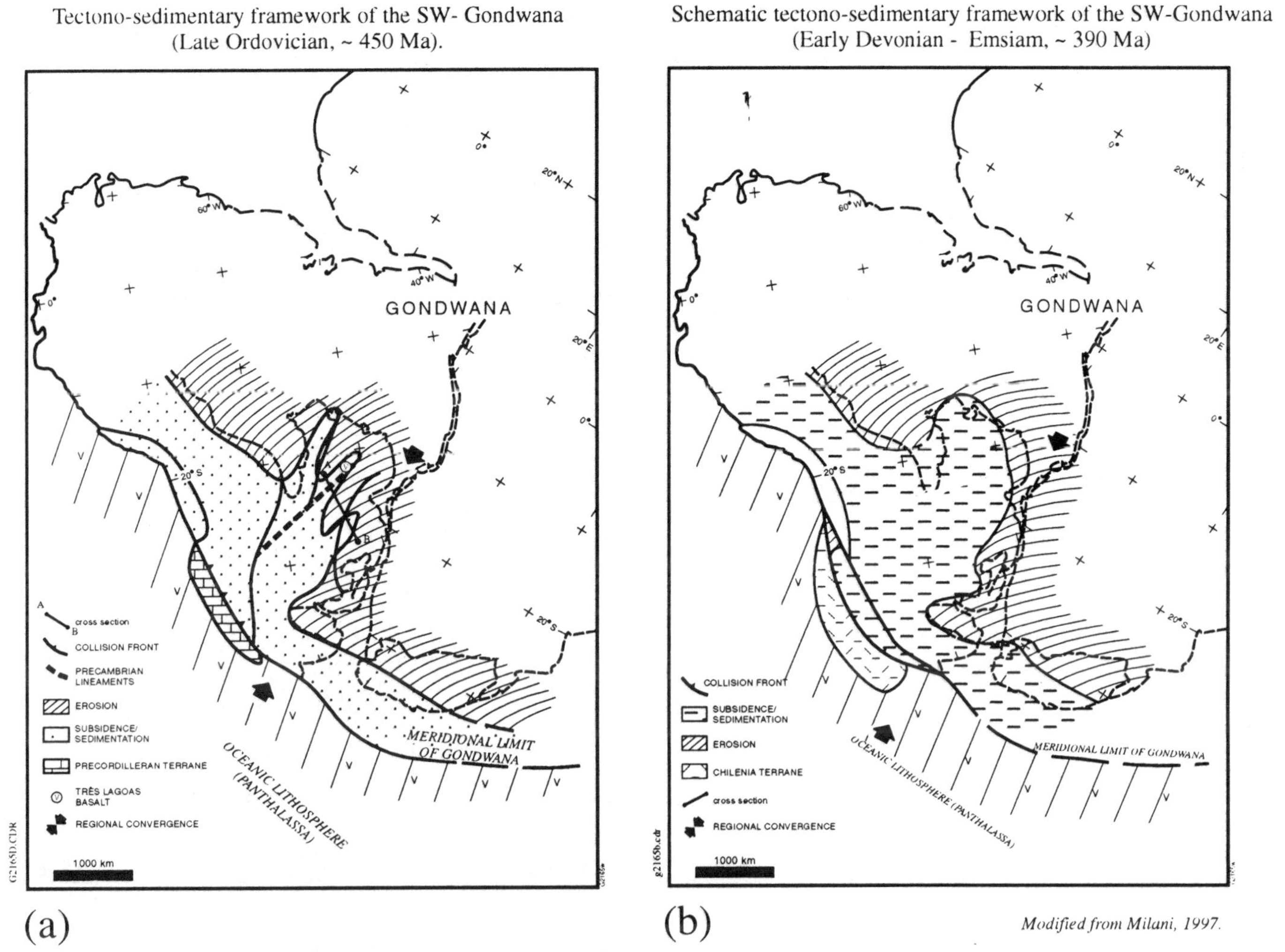

Figure 14. Tectono-stratigraphic framework of the SW-Gondwana, in the Late Ordovician (Fig. 14-a) and in the Early Devonian (Fig. 14-b).
g2165bd.cdr

a magmatic arch which had some influence on sedimentation of tuffaceous material of the Rio Bonito Fm. The magmatic arch that was implanted in the Permian had a strong impact on subsequent sedimentation in the Paraná basin (Milani, 1997b). Sedimentation rates were very high in the Paraná basin during the 255 – 245 Ma interval, and more than 1,100 m of sediments (Serra Alta to Pirambóia Fm.) accumulated during the collision of Patagonia.

The Triassic is associated with generalized extensional stresses, forming a series of NW-trending grabens in Argentina and Bolivia (Lopez-Gamundi et al., 1994), and possibly, other rifts in Uruguay and Rio Grande do Sul State (Fig. 15-b). These rifts have minor sedimentary thickness in the Brazilian territory (maximum of about 300 m of Triassic infill in local grabens) and are probably related to a relaxation of the compressional stress field that prevailed through most of the Paleozoic. Most of the northern parts of the Paraná basin were subjected to regional uplift and erosion (Fig. 10). The restricted Triasssic sedimentation in the Paraná basin (Santa Maria Fm.) was characterized by a fluvial - lacustrine environment that filled depressions, and later, by a widespread sandy desert that covered the previous sites of deposition and parts of the adjacent basement. These Mesozoic sediments in the Paraná basin are included in the Gondwana II sequence (Milani, 1997b), and are restricted to the southern parts of the basin (Figs. 10 and 15-b). During the Triassic and most of the Jurassic, the rate of subsidence was extremely low,

Schematic tectono-sedimentary framework of the SW - Gondwana (Early Permian, ~ 260 Ma)

Schematic tectono-sedimentary framework of the SW-Gondwana (Middle to Late Triassic, ~ 237 - 218 Ma)

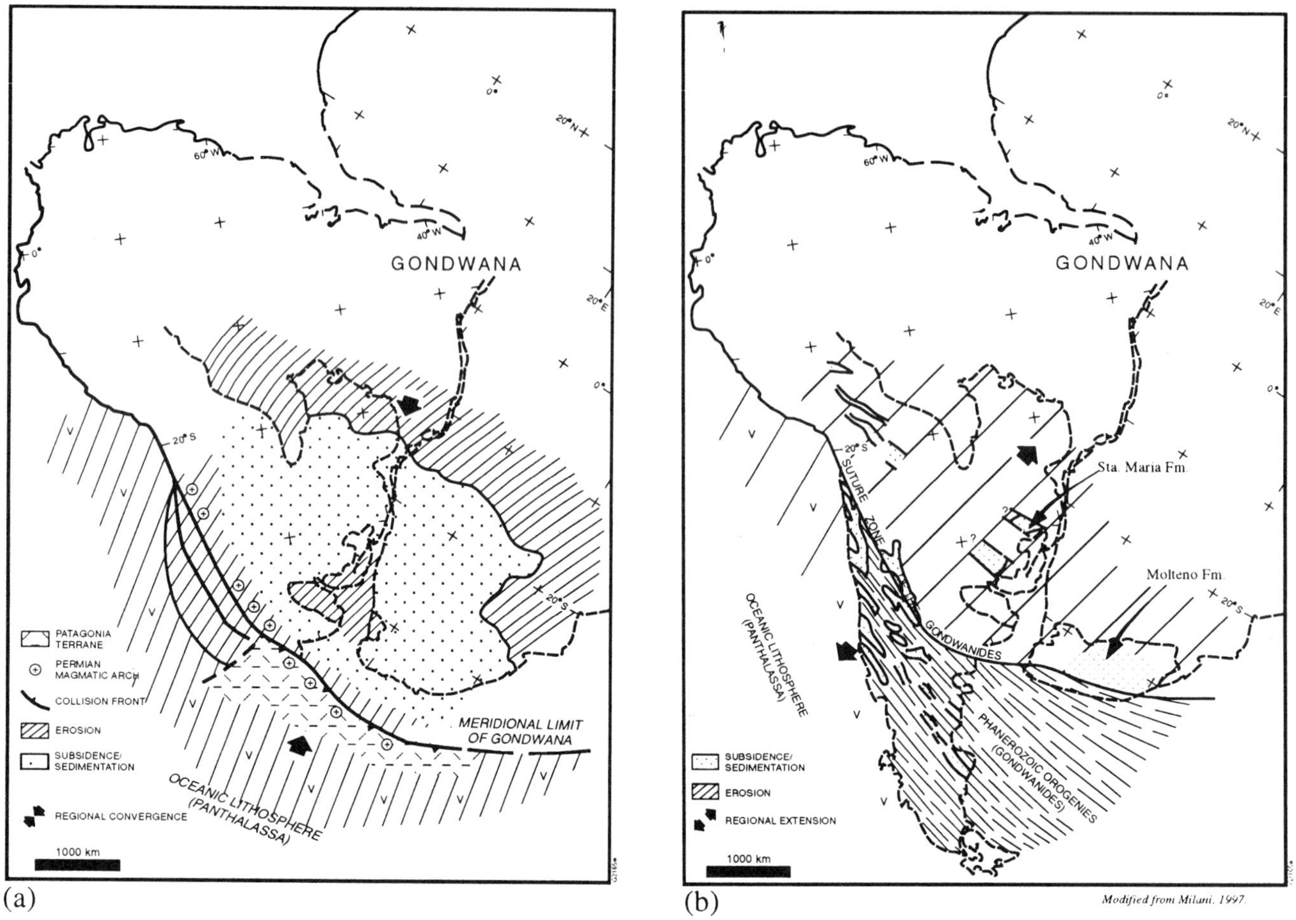

Figure 15. Tectono-stratigraphic framework of the SW-Gondwana, in the Early Permian (Fig. 15-a) and in the Triassic (Fig. 15-b).

and it did not characterize any significant pulse of tectonic subsidence in the depocenter of the basin.

The Late Triassic to Late Jurassic was a period of non-deposition in the Paraná basin, and is reflected in the largest stratigraphic gap in the tectono-stratigraphic record (Fig. 10). The fifth phase of subsidence (Gondwana III, Late Jurassic to Early Cretaceous) probably started during the Late Jurassic (Callovian – Oxfordian) and is associated with the Botucatu Fm., which corresponds to widespread deposits of unfossiliferous eolian sands that covered much of the basin, and is partially coeval with the basaltic magmatism of the Serra Geral Fm. This desert cycle was forewarning that the central part of the Gondwana craton, stable for hundreds of million years, started to suffer regional uplift. This might have been caused by thermal expansion of mantle material, probably because it remained too long above a mantle hotspot (White and McKenzie, 1989). The extensive flood basalts indicated a magmatic event that soon resulted in disruption of the western Gondwana continent and opening of the South Atlantic Ocean (Fig. 13). This pre-breakup phase was not associated with major pulses of sedimentation in the Paraná basin, and the load of the lavas may well be responsible for much of the subsidence observed in the basin.

The Serra Geral lavas are piled with ages decreasing to the top, indicating a normal piling with dike-fed basalts flooding the depositional surface. Turner et al. (1994), using Ar/Ar methods, suggested that the older lavas (137.8 +/- 0.7 Ma) were formed in the northwestern portions of the basin, and the youngest lavas (126.8 +/- 2.0 Ma) occur near the border with Uruguay, in the southeast. The magmatic layers that constitute the economic basement of the continental margin sedimentary basins also have equivalent ages (Misuzaki et al., 1988).

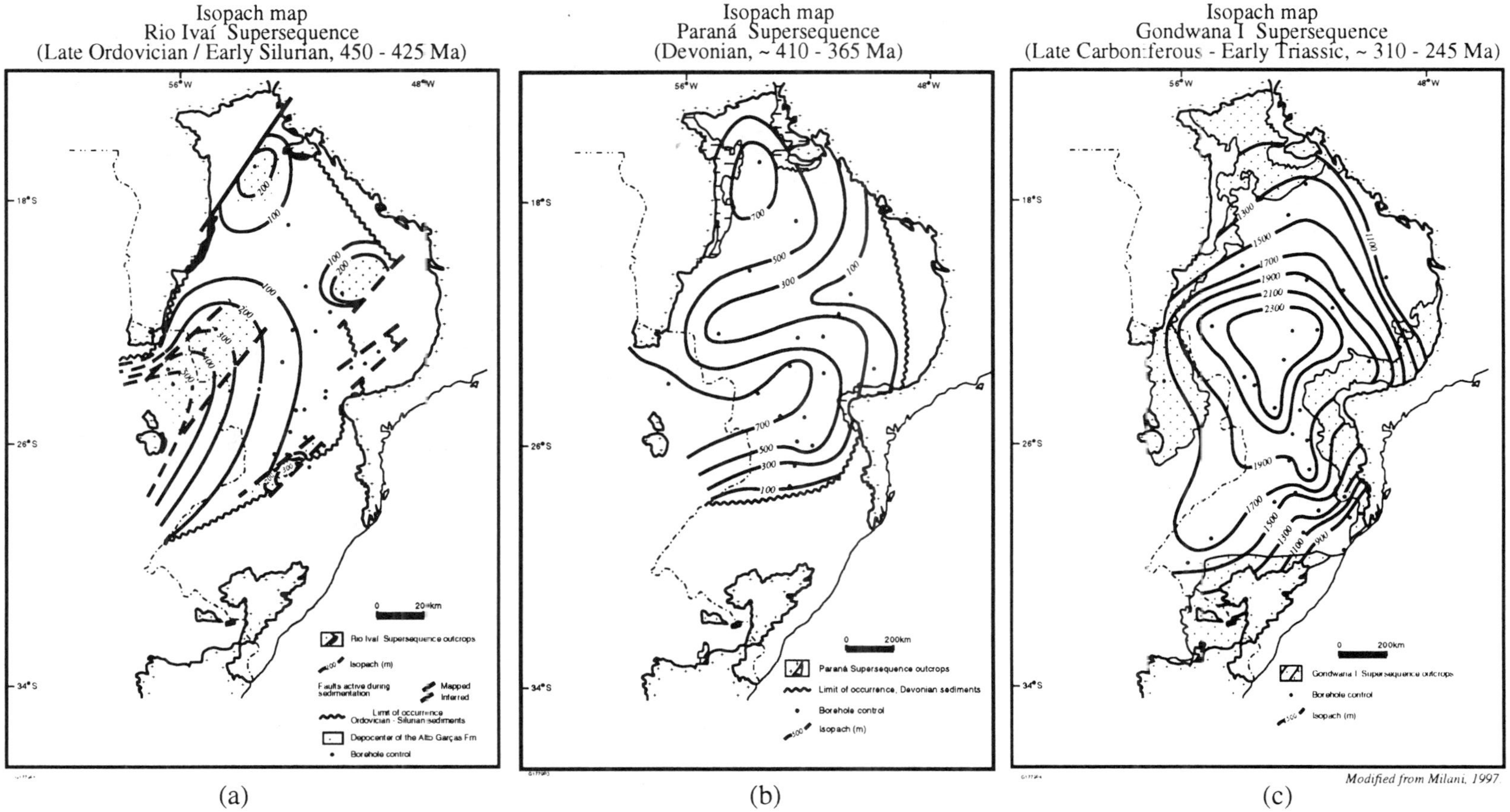

Figure 16. Isopach maps of supersequences in the Paraná basin: a) Rio Ivaí Supersequence (Late Ordovician; b) Paraná Supersequence (Devonian); and c) Gondwana I Supersequence (Late Carboniferous – Early Triassic).

The last phase of subsidence in the Paraná basin resulted in accumulation of about 300 m of Cretaceous fluvio-lacustrine sediments (Bauru Fm.), which overlie a regional unconformity separating the predominantly sandy sediments from the Serra Geral basalts (Figs. 10 and 13). The tectonic activity was subsequently diverted to the continental margin, where subsidence and sedimentation is characterized by several magmatic events, particularly in the Early Cretaceous (e.g., tholeiitic basalts of the Guaratiba Fm. in the Neocomian rift sequence of the Santos basin, Fig. 17). Recently acquired regional seismic data (e.g., Gladczenko et al., 1997) indicate that the transition from continental to oceanic crust is characterized by volcanic wedges of seaward-dipping reflectors (Fig. 18). Other pulses of igneous and tectonic activity are also registered in the Late Cretaceous and Early Tertiary (Melo et al., 1985; Mohriak and Barros, 1990).

5- THE SOUTH ATLANTIC OPENING AND THE LATE CRETACEOUS - TERTIARY TECTONISM.

The opening of the South Atlantic ocean started in the southernmost parts of the South American continent in Upper Jurassic/Early Cretaceous times and advanced towards the northeastern Brazilian margin (Rabinowitz and LaBrecque, 1979; Chang et al., 1992). This section will briefly discuss some of the main aspects of the tectonic evolution of the South Atlantic opening, with characterization of the most important tectono-magmatic episodes.

The evolution of the South Atlantic Ocean is marked by five phases with different patterns of tectonics and sedimentation (Fig. 19). The first phase (Fig. 19-I) is marked by the onset of regional lithospheric extension, which eventually led to the separation of the South American and African continents. The conceptual model for this Late Jurassic - Early Cretaceous phase admits a small asthenospheric uplift and regionally distributed thinning of the continental crust and upper mantle, with incipient faults in the upper crust controlling local depocenters associated with widespread, thin sedimentary sequences.

The next phase is characterized by increasing lithospheric stretching (cf. McKenzie, 1978), which coincides with large faults affecting the continental crust, extrusion of continental flood basalts (e.g., the Serra Geral volcanics in the Paraná basin and equivalent tholeiitic basalts in the southern offshore basins (Mizusaki et al., 1988), and formation of half-grabens (Fig. 19-II). Subsequent to this Early Cretaceous tectono-magmatic episode, the tectonic stresses were directed toward the present-day continental margin, along a series of continental rifts east of the Precambrian Ribeira fold belt. These rifts, filled with Neocomian to Barremian volcanic and siliciclastic rocks, evolved to form the present day passive margin sedimentary basins. By the end of the rifting phase, there is an increase in the lithospheric extension that is marked by large faults which rotate the rift blocks and the sedimentary layers previously deposited.

The mid-Atlantic Ridge responsible for inception of oceanic crust probably intruded the crust by the end of the rifting episodes (Chang et al., 1992). The possible mechanism for this episode involves focusing of the lithospheric stretching, previously distributed in a wider region, to a locus in the region of the mid-Atlantic Ridge (Harry and Sawyer, 1992). This phase is associated with volcanism, reactivation of large faults, and erosion of rift blocks by a regional unconformity that levels the topography. This is often designated as the Breakup Unconformity (Falvey, 1974), separating continental from transitional to marine environments of deposition. Above this unconformity and below the evaporite transitional sequence, some sedimentary basins register a substantial thickness of Aptian siliciclastic and carbonate rocks (Fig. 19-III). Both in the Brazilian and in the African margins, this sequence marks the first marine incursions and may contain hydrocarbon source rocks. The outermost rift blocks of the Santos and Campos basins pinch-out against a volcanic wedge characterized by seaward-dipping reflectors (SDR) (Souza et al., 1993, Gladczenko et al., 1997). The outburst of the SDR wedge is related to subaerial volcanism in the incipient oceanic spreading ridge (Hinz, 1981; Mutter et al., 1982), and it may be chronologically associated with the time of the breakup unconformity.

Subsequent to the continental breakup in the Early Cretaceous, most of the tectonic activity was related to the evolution of the South Atlantic Ocean and the passive margin sedimentary basins. Following the salt deposition in the Aptian (Fig. 19-IV), sedimentation becomes predominantly carbonatic. An increase in the bathymetry resulted in the deepening of the environment of deposition by the end of the Albian, with demise of the shallow water carbonates (Fig. 19-V).

The Late Cretaceous to early Tertiary was a period characterized by several tectono-magmatic events both onshore and offshore of the southeastern margin. These events include alkaline rock intrusions dating from 90 to about 50 Ma (Mizusaki and Mohriak, 1992), and the formation of several small rift basins near the continental margin of the São Paulo and Rio de Janeiro States (Fig. 2). The largest of these is the Taubaté basin, which extends for about 200 km in a NE direction, and reaches widths of about 20-30 km (Melo et al., 1985). The stratigraphy of the basin is known only superficially, because of the lack of deep exploratory boreholes. The maximum sedimentary thickness in the

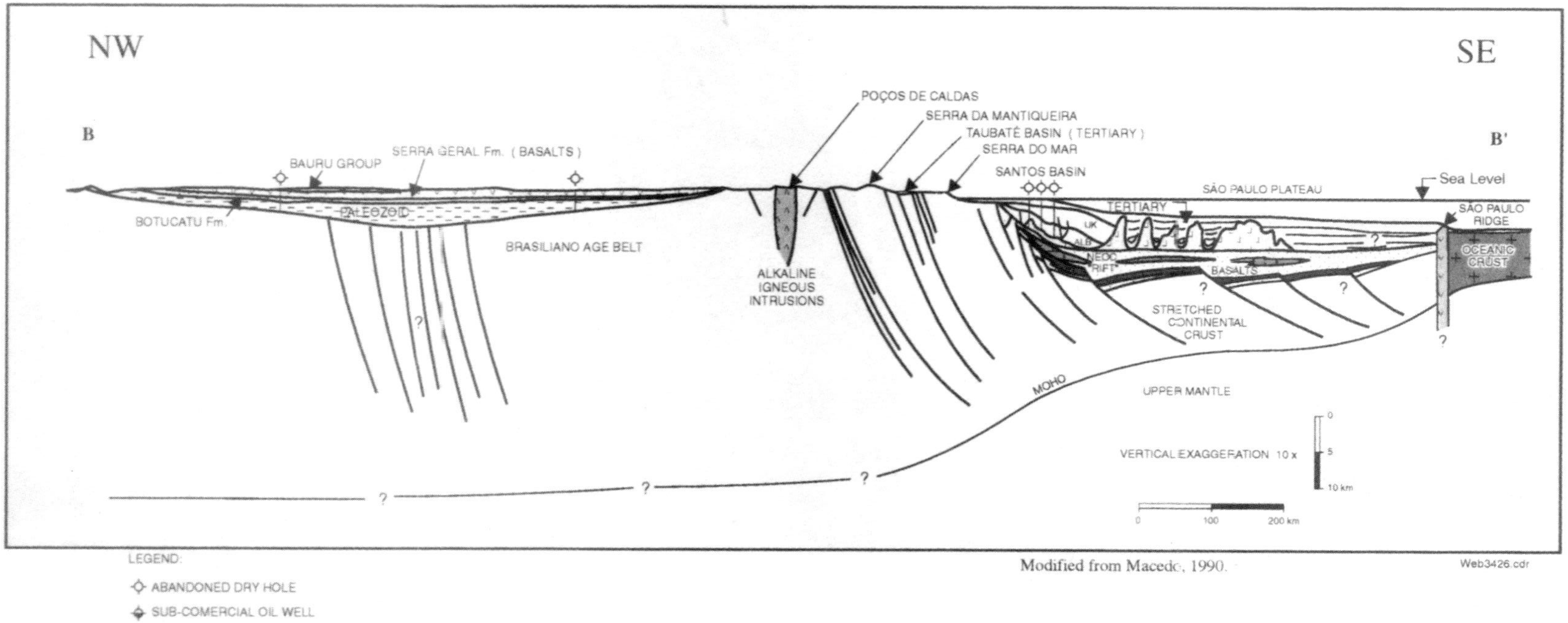

Figure 17. Schematic geological section from the Paraná basin towards the continental margin (Santos basin).

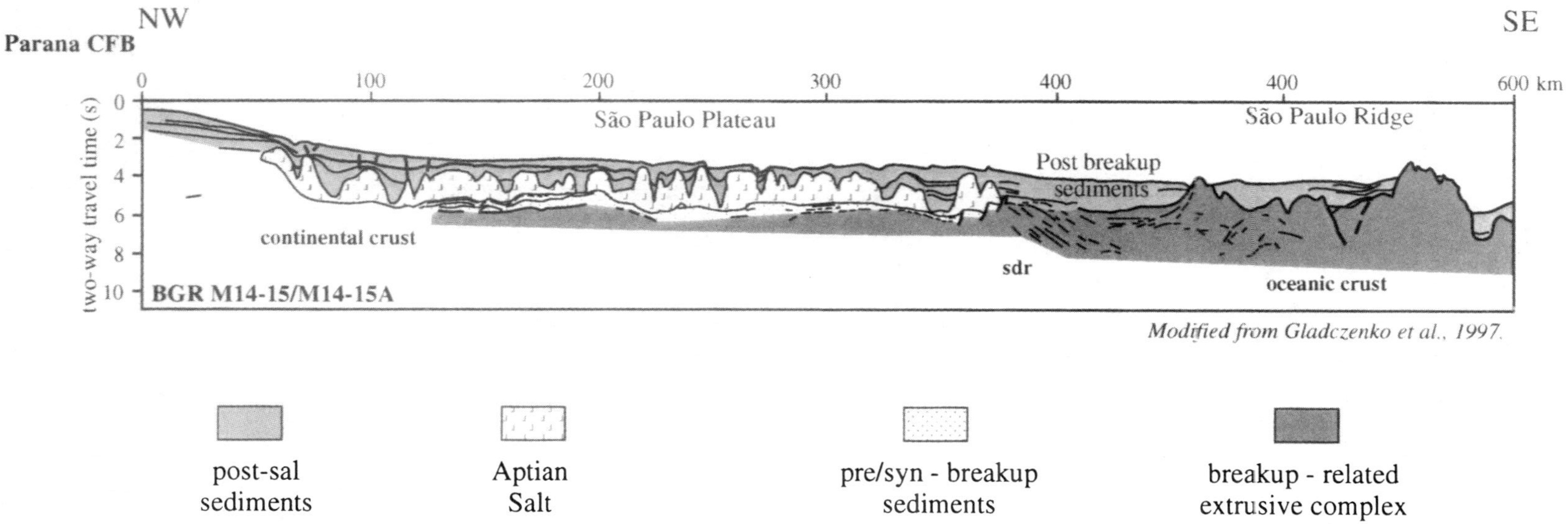

Figure 18. Schematic geoseismic section in the Santos Basin, from the platform toward the continental – oceanic crust limit. The Paraná Basin continental flood basalts occur to the west of the profile, and the transition to oceanic crust is characterized by wedges of seaward-dipping reflectors (sdr).

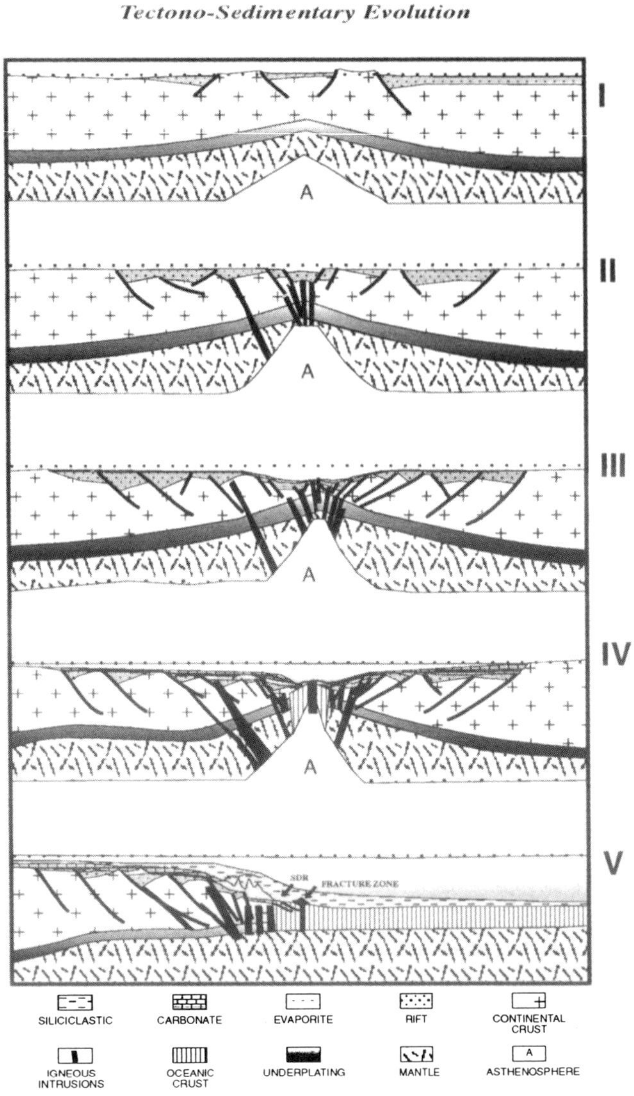

Figure 19. Schematic diagram showing sequence of events during the breakup of Gondwana and formation of the passive margin sedimentary basins in the Eastern Brazilian margin.

Taubaté basin is estimated to be about 1000 m (Marques, 1990).

The rift architecture of the Taubaté basin is characterized by a series of half-grabens separated by accommodation zones where the rift border faults change polarity (Marques, 1990). Seismic interpretation of the regional profiles in the basin indicates that the master faults were active up to the late Tertiary, and that there appears to be some transpressional component involved in the basin evolution. Zalán (1986) attributed the linear configuration of the basin and the geometry of the basin border faults to strike-slip processes. Other researchers (Chang et al., 1989; Fernandes and Chang, 1992) attributed the formation of the Taubaté basin to arching of the lithosphere and crust, resulting in small rifts.

Physical model experiments (Szatmari and Mohriak, 1995) indicate that the pattern of post-breakup tectonism observed in SE Brazil may be accounted for by compression associated with the mid-Atlantic ridge push combined with the Nazca plate subduction along the Pacific margin of South America, which resulted in a number of Andean orogenies in the Cenozoic. Tectono-magmatic episodes dated around 100 - 90 Ma in the Andes (Mirano orogeny) have corresponding tectono-magmatic episodes in the Santos and Campos basins (Mizusaki and Mohriak, 1992). The Eocene magmatism, widespread along several segments of the Brazilian Eastern Atlantic margin, is geochronologically coincident with a major event in the Andes (Incaic orogeny, Bussel, 1983). These periods of increased tectono-magmatic activity in the Late Cretaceous and Early Tertiary also correspond to intervals of sand-rich turbidite sedimentation in the deep water region, indicating rejuvenation of the source area along the Brazilian margin.

6- CONCLUSIONS

The evolution of the southeastern margin of Brazil reveals a complex geological history with numerous episodes of alternating compression and extension since at least the Lower Proterozoic. In most of these tectonic phases, a genetic link between accretional or collisional episodes and the development of sedimentary basins can be recognized. This relationship has been proposed for several regions around the world, such as the opening of the North Atlantic Ocean and the sutures of the previous Caledonian orogeny (Wilson, 1966).

Within the studied region of Brazil we can summarize some tectonic episodes that exemplify the relationship between compressional and extensional events.

- The opening of the Middle Proterozoic rifts (ca. 1.7 Ga, e.g., the São João del Rei rift) is controlled by discontinuities related to the Transamazonian orogeny (ca 2.1- 1.9 Ga). As suggested by Ribeiro (1997), the movement along the main faults at the northern border of the São João del Rei rift had a long-lived history that began during the Early Proterozoic. These master faults were probably active during different phases of rift development.
- The subsequent extensional episode that started probably around 1.0 - 0.9 Ga culminated with the opening of the Neoproterozoic ocean, and generated the Andrelândia passive margin at the border of Occidental or São Francisco continent. The subsidence in this basin was also controlled by important suture zones of the Transamazonian orogeny, e.g., the contact of the Juiz de Fora basement rocks

(Transamazonian granulites) with the reworked cratonic border.

- The above mentioned suture zone was also reactivated during the Neoproterozoic - early Paleozoic Brasiliano orogeny, concentrating the effects of convergence as a major crustal-scale thrust (Heilbron, 1995).
- The contacts between the tectonic domains of the Occidental terrane of the Ribeira belt were subsequently activated as important subvertical shear zones (e.g., the Paraíba do Sul shear zone, with the progression of the Brasiliano collision, accomodating the late oblique convergence stage of the orogeny.
- The geological development of the intracratonic Paraná basin was strongly influenced by SW/NE trending Precambrian structures and also by farfield stresses which reactivated these ancient fabrics. Two hypothesis may be postulated to explain the initial subsidence in the Paraná basin, both of them related to compressive episodes. The first hypothesis associates the initial subsidence with the latest events of the Brasiliano orogeny and an indirect control by Precambrian tectonic heritage. The second one emphasizes the direct relationship between diverse subsidence phases in the Paraná basin with Paleozoic compressional stresses along western margin of South America. Important orogenic climaxes coincided with docking and accretion of terranes that were carried by the oceanic lithosphere. These compressional pulses correlate with increased subsidence rates in the Paraná basin (Milani and Ramos, 1998).
- The breakup of Gondwana in the Late Jurassic – early Cretaceous was heralded by the massive extrusion of continental flood basalts in the Paraná basin and in the continental margin rifts. Subsequent to this Early Cretaceous tectono-magmatic episode, the tectonic stresses were focused on the present-day continental margin. The breakup of Gondwana and opening of the South Atlantic were controlled by the previous structures of the Precambrian - early Paleozoic Ribeira fold belt.
- The oceanic crust inception was marked by erosion of the outermost portions of the rift blocks and by formation of thick wedges of seaward-dipping reflectors interpreted as volcanic layers. The onshore Paraná Basin was aborted in the Mesozoic, but the offshore rifts evolved with thermal subsidence to form the present day passive margin sedimentary basins along the Brazilian eastern margin.
- The Tertiary tectonic activity onland is expressed in a series of alkaline intrusions and minor preserved extrusions, and by a number of small rift basins (e.g., Taubaté basin) that accumulated fluvial to lacustrine siliciclastic sediments. These rift basins were strongly controlled by the Brasiliano subvertical shear zones, and their origin might also be indirectly related to far-field stresses related to a combination of Cenozoic orogenies in the Andes and the ridge push at the South Atlantic spreading axis.

Acknowledgements. We thank several geoscientists from the Rio de Janeiro State University and from Petrobras E&P for very enlightening discussions on the tectonic evolution of the Eastern Brazilian Margin. We also thank Dr. B. B. Brito Neves and Dr. A. Tankard for providing thorough reviews of the first draft of the manuscript and for many helpful suggestions, and I. F. Braga for helping in the drawings.

REFERENCES

Almeida, F. F. M., Origem e Evolução da Plataforma Brasileira. *Boletim da Divisão de Geologia e. Mineralogia, Departamento Nacional da Produção Mineral,* Rio de Janeiro, 241, 1-36, 1967.

Almeida, F. F. M., Geochronological Division of the Precambrian of South America. *Revista. Brasileira de. Geociências,* 1, 1, 13-21, 1971.

Almeida, F. F. M. The system of continental rifts bordering the Santos Basin. *An. Acad. Bras. Cienc.* 48, supl, 15-26, 1976.

Almeida, F.F.M. and Y. Hasui, *O Pré-Cambriano do Brasil.* Edgard Blucher, 378 pp, 1984.

Almeida, J. C. H.; L. G. E. Silva. and C. S. Valladares, O Grupo Paraíba do Sul e rochas granitóides na região de Bananal (SP) e Rio Claro (RJ): uma proposta de formalização litoestratigráfica, in *Atas do III Simpósio de. Geologia do Sudeste,* Rio de Janeiro, SBG, pp 161-167, 1993.

Almeida, J.C.H.; M. Tupinambá, M. Heilbron and R. Trouw, Geometric and kinematic analysis at the Central Tectonic Boundary of the Ribeira Belt, Southeastern Brazil, in *Anais 39 Congresso Brasileiro Geologia*, Belo Horizonte, pp. 32, 1998.

Andreis, R. R.; A. Ribeiro. and F. V. P. Paciullo, Ciclos deposicionais no proterozóico das folhas Barbacena e Divinópolis (setor sul), 1:250.000, in *Boletim de Resumos, I Simpósio de Geologia do Sudeste*, Rio de Janeiro, pp. 97-98, 1989.

Astini, R.A., J.L. Benedetto and N.E. Vaccari, The early Paleozoic evolution of the Argentine Precordillera as a Laurentian rifted, drifted, and collided terrane - a geodynamic model. *Geological Society of America Bulletin,* 107, 253-273, 1995.

Brito Neves, B. B., De Rodínia a Gondwana. São Francisco/Congo: placa, continente/península, cráton e ante-país, in *Anais do II Simpósio sobre o Cráton do São Francisco,* Salvador, Bahia, Agosto de 1993, SBG/NBA-SE/SGM, pp. 3-5, 1993.

Brito Neves, B. B. and U. G. Cordani, Tectonic Evolution of South America during the Late Proterozoic, *Precambrian Research*, 53, 23-40, 1991.

Brito Neves, B. B.; J. M. Sá; A. A. Nilson and N. F. Botelho, A tafrogênese Estateriana nos blocos Paleoproterozóicos da América do Sul e processos subsequentes, *Geonomos,* 3, **2,** 1-21, 1995.

Brito Neves, B.B., M. Winge and M. A. Carneiro, Orogêneses precedendo e trafrogêneses sucedendo Rodínia na América do Sul, *Bol IG-USP, Série Científica,* **27,** 1-40, 1996.

Bussel, M.A., Timing of tectonic and magmatic events in the Central Andes of Peru, *Journal of the Geological Society of London*, 140, 279-286, 1983.

Campanha, G.A.C., O Lineamento Além Paraíba na área de Três Rios (RJ), *Revista Brasileira de Geociências*, 11, 3, 159-171, 1981.

Campanha, G.A.C. and A. L. Ferrari, Lineamento Além Paraíba: um exemplo de zona de cisalhamento, in *Anais 38 Congresso Brasileiro de Geologia*, Rio de Janeiro, 12, 5425-5432, 1984.

Campos Neto, M. C., A porção ocidental da Faixa Alto Rio Grande- Ensaio de Evolução Tectônica, *Unpublished Thesis, IG-USP*, 210 pp., 1992.

Campos Neto, M.C., and M.C.H. Figueiredo, The Rio Doce Orogeny, Southeastern Brazil, *Journal of South American Earth Sciences*, 8, 2, 143-162, 1995.

Caputo, M.V. and J. C. Crowell, Migration of glacial centers across Gondwana during Paleozoic Era, *GSA Bull.*, 96, 1020-1036, 1985.

Chang, H. K., C. J. Appi., C. Riccomini, J. C. Castro, M. Arai, E. L. Freitas and E.V.S. Neto, Geologia da Bacia de Taubaté, in I *Simpósio de Geologia do Sudeste, Rio de Janeiro, SBG, Boletim de Resumos*, pp. 10, 1989.

Chang, H. K., R.O. Kowsmann, A.M.F. Figueiredo and A. Bender, Tectonics and stratigraphy of the East Brazil Rift system: an overview, *Tectonophysics*, 213, 97-138, 1992.

Chrispim, S. J. and M. Tupinambá, Observações sobre o caráter transpressivo da deformação entre as zonas de cisalhamento de Guaçuí e Paraíba do Sul, in *An Anais do I Simp. Geol. Sudeste*, Rio Janeiro, pp. 120-121, 1989.

Cordani, U. G. and B. B. Brito Neves, The geologic evolution of South America during the Archean and Early Proterozoic, *Revista Brasileira de Geociências*, 12, 78-88, 1982.

Cordani, U. G.; G. Amaral, and K. Kawashita, The Precambrian Evolution of South America, *Geologische. Rundschau*, 62, 2, 309-317, 1973.

Cordani, U. G.; G. C. Melcher. and F.F.M. Almeida, Outline of Precambrian Geochronology of South America. *Canadian Journal.of Earth Sciences*, 5, 629-632, 1967.

Cordani, U.G., B. B. Brito Neves, R. A. Fuck, R. Porto, A. Thomaz Filho and F.M.B. Cunha, Estudo preliminar de integração do Pré-Cambriano com os eventos tectônicos das bacias sedimentares brasileiras, *Petrobras, Série Ciência-Técnica-Petróleo*, 15, 70 pp., Rio de Janeiro, 1984.

Correia Neto, A. V., H. Dayan, J.G. Valença and A. Cabral, Geologia e Estrutura da Zona de Cisalhamento do Rio Paraíba do Sul e Adjacências, no trecho entre Três Rios e Sapucaia (RJ), in *Atas III Simpósio de Geologia do. Sudeste*, Rio de Janeiro, SBG,pp. 194-200, 1993.

Dalziel, I.W.D., L.H. Dalla Salda and L.M. Gahagan, Paleozoic Laurentia - Gondwana interaction and the origin of the Appalachian - Andean mountain system, *Geological Society of America Bulletin*, 106, 243-252, 1994.

Delhal, J.; D. Ledent. and U.G. Cordani, Ages Pb/U, Sr/Rb et Ar/K de Formations Métamorphiques et Granitiques du Sud-Est du Brésil (Etats de Rio de Janeiro e Minas Gerais), *An. Soc .Geol. Belg.*, 92, 271-283, 1969.

Dewey, J.F. and J.M. Bird, Mountain Belts and the New Global Tectonics, *Jour. Geoph. Res.*, 75, 14, 2625-2647, 1973.

Duarte, P.D., Evolução tectônica dos ortognaisses dos Complexos Juiz de Fora e Mantiqueira na região de Juiz de Fora, M.G.: Geologia, petrologia e geoquímica. *Unpublished Thesis, Instituto de Geociências, Universidade de São Paulo*, 280 pp., 1998.

Duarte, B. P., M.C.H. Figueiredo, M. Campos Neto and M. Heilbron, Geochemistry of the Granulite Facies Orthogneisses of Juiz de Fora Complex, Central Segment of Ribeira Belt, Southeast Brazil, *Revista Brasileira de Geociências*, 27, 1, 67-82,1997.

Ebert, H., Relatório de Atividades, in R*elatório Anual do Diretor, ano 1954*, DNPM Div. Geol. Min, Rio de Janeiro, pp.79-89, 1955.

Ebert, H., Relatório de Atividades, in *Relatório Anual do Diretor*, Departamento Nacional da Produção Mineral, Divisão de Geologia e Mineralogia, Rio de Janeiro, pp. 97-107, 1957.

Ebert, H., Discordâncias pré-cambrianas em Carandaí, Minas Gerais, in *DNPM–DGM*, Bol. 183, Rio de Janeiro, 48 pp., 1958.

Ebert, H., Os Paraibides entre São João del Rei (Minas Gerais) e Itapira (São Paulo) e a bifurcação entre Araxaídes e Paraibides, *Publicação Especial, SBG-SP*, 103 pp., 1984 (in memorian).

Ebert, H. D.; Y. Hasui and J.B.S. Costa, O caráter transpressivo do Cinturão de Cisalhamento Rio Paraíba do Sul, in *Boletim III Simpósio Nacional de Estudos Tectônicos*, Rio Claro, SBG, pp. 139-141, 1991.

Ebert, H.D. and Y. Hasui, Transpressional tectonics and strain partitioning during oblique collision beteween three plates in the Precambrian of southeast Brazil, in R. E. Holdsworth, R. A. Strachan, and J.F. Dewey, (Eds.) *Continental Transpressional and Transtensional Tectonics*, Geological Society, London, Special Publications, 135, 231-252, 1998.

Ebert, H.D., M.A. Neves, Y. Hasui, J.A. Lopes, and M. Guerra, Compartimentação crustal e evolução cinemática da Província Mantiqueira através de modelagem física, in *V Simpósio Nacional de Estudos Tectônicos*, Gramado, pp. 26-28, 1995.

Falvey, D.A., The development of continental margins in plate tectonic theory, *The APEA Journal*, 14, 1, 95-106, 1974.

Fernandes, F. L. and H.K. Chang, Compartimentação e evolução tectônica da Bacia de Taubaté, in *Anais do XXXVII Congresso Brasileiro de Geologia*, São Paulo, SP, 1, 570-572, 1992.

Figueiredo, M.C.H. and M.C. Campos Neto, Geochemistry of the Rio Doce Magmatic Arc, Southeastern Brazil, *An. Acad. Bras. Cienc.*, 65, suplem, 1, 63-82, 1993.

Fonseca, A.C., Esboço Geocronológico da Região de Cabo Frio, Estado do Rio de Janeiro, *Unpublished Thesis, IG/USP, São Paulo*, 186 p., 1994.

Fonseca, A. C., U .G. Cordani. and K. Kawashita, Dados preliminares sobre a geocronologia e suas encaixantes na cidade do Rio de Janeiro. Método Rb/Sr, in *Anais do XXXIII Congresso Brasileiro de. Geologia*, Rio de Janeiro, 6, 2333-2345, 1984.

Fonseca, M.J.G, Mapa geológico do Estado do Rio de Janeiro Escala 1:400.000, Texto Explicativo. *DNPM/MME* Rio de Janeiro, 141 pp., 1998.

Fonseca, M.J.G., M. Heilbron and S.J. Chrispim, Geologia da área de Cabo Frio e Armação dos Búzios, in *Anais XXXIII Congresso Brasileiro de. Geologia*, Rio de Janeiro, SBG, 12, 5393-5424, 1984.

França, A. B. et al., Phanerozoic correlation in Southern South America, in A. J. Tankard, S. R. Suárez, and H. J. Welsinki (Eds.), *Petroleum Basins of South America*, AAPG Memoir, 6, 129-161, 1995.

Gladczenko, T.P., K. Hinz, O. Eldholm, H. Meyer, S. Neben, and J. Skogseid, South Atlantic volcanic margins, *Journal of the Geological Society of London*, 154, 465-470, 1997.

Gonçalves, M.L. and M.C.H. Figueiredo, Geoquímica dos Anfibolitos de Santana do Garambéu (MG): Implicações sobre a Evolução do Grupo Andrelândia, *Geochimica Brasiliensis*, 6, 2, 127-140, 1992.

Gohrbandt, K.H.A., Paleozoic paleogeographic and depositional developments on the central proto-Pacific margin of Gondwana: their importance to hydrocarbon accumulation, *Journal of South American Earth Sciences*, 6, 4, 267-287, 1993.

Grossi Sad, J. H. and A.L. Barbosa, A origem dos charnockitos e rochas afins da região do Médio Paraíba do Sul, estado do Rio de Janeiro, Contribuições à geologia e petrologia, *Boletim Especial da SBG*, Núcleo de Minas Gerais, 15, 281 pp.,1985.

Grossi Sad, J.H. and C.V. Dutra, Chemical composition of supracrustal rocks from Paraíba do Sul group, Rio de Janeiro State, Brazil, *Geochimica Brasiliensis*, 2, 2, 143-166, 1998.

Hasui, Y., The Mantiqueira Province: Archean Structure and Proterozoic Evolution, in *Anais International Symposium on Archean and Early Proterozoic Geological Evolution and Metallogenesis*, Salvador, BA, 1982.

Hasui, Y., C.D.R. Carneiro and A.M. Coimbra, The Ribeira Fold Belt, *Revista Brasileira de. Geociências* 5, 4, 257-266, 1975.

Heilbron, M., O limite entre as faixas de dobramento Alto Rio Grande e Ribeira na seção geotransversal Bom Jardim de Minas (MG)- Barra do Piraí (RJ), in *Anais XXXVI Congresso Brasileiro de. Geologia*, Natal, SBG, 6, 2813-2826, 1990.

Heilbron, M. Evolução tectono-metamórfica da seção Bom Jardim de Minas (MG) - Barra do Piraí (RJ). Setor Central da Faixa Ribeira, *Unpublished Thesis, IG/USP*, 268 pp., 1993.

Heilbron, M., O segmento central da Faixa Ribeira: síntese geológica e ensaio de evolução geotectônica, *Unpublished Thesis, Faculdade de Geologia, UERJ*, Rio de Janeiro, 110 pp., 1995.

Heilbron, M.; S. J. Chrispim, L.S.A. Simões and R.P. Alves, Geologia do cabo dos Búzios, *Anais da Academia Brasileira de Ciências*, 54, 3, 553-562, 1982.

Heilbron, M., M. L. Gonçalves, W. Teixeira, R. Trouw; K. Kawashita and A. Padilha, Geocronologia da área entre Lavras, São João del Rei, Lima Duarte e Caxambú, *Anais da Academia Brasileira de Ciências*, 61, 2, 177-199, 1990.

Heilbron, M., C. M. Valeriano, J. C. H Almeida and M. Tupinambá, A Megassinforma do Paraíba do Sul e sua implicação na compartimentação tectônica do setor central da Faixa Ribeira, in: *Atas 2 Simpósio de Geologia do Sudeste*, São Paulo, SBG, pp. 519-527, 1991.

Heilbron, M., F. R. Pires, and M. Bessa, Litoestratigrafia, evolução tectono-metamórfica e magmatismo do setor sudeste do Município Rio de Janeiro, in *Atas III Simpósio de Geololgia do Sudeste*, Rio de Janeiro, SBG: pp. 174-179, 1993.

Heilbron, M., C. M. Valeriano, J. C. H. Almeida, M. Tupinambá, C. S. Valladares, L. G. E. Silva, D. B. Nava, and F.B. Dios, Compartimentação tectônica e evolução geológica do segmento central da Faixa Ribeira, ao sul do Cráton do São Francisco, in *Anais II Simpósio sobre o Cráton do São Francisco*, Salvador, SBG, pp. 263-265, 1993.

Heilbron, M., R. Machado and M.C.H. Figueiredo, Lithogeochemistry of Paleoproterozoic Orthogranulites from the Bom Jardim de Minas (MG)- Vassouras (RJ) Transect, Central Ribeira belt, SE Brazil, *Revista Brasileira de Geociências*, 27, 1, 83-98, 1997.

Heilbron, M., C.M. Valeriano, C.S. Valladares and N. Machado, A orogênese brasiliana no segmento central da Faixa Ribeira, Brasil, *Revista Brasileira de Geociências*, 25, 4, 249-266, 1995.

Heilbron, M, B.P. Duarte, and J.R. Nogueira, The Juiz de Fora Complex of the Central Ribeira belt, SE Brazil: a segment of Paleoproterozoic granulitic crust thusted during the Pan African Oregon, *International Geoscience Journal/ Gondwana Research*, 1, 3-4, 373-382, 1998.

Heilbron, M., M. Tupinambá, M., J.C.H. Almeida, C.M.Valeriano, C.S. Valladares and B.P. Duarte, New constraints on the tectonic organization and structural styles related to the Brasiliano collage of the central segment of the Ribeira belt, SE Brazil, in *Abstracts of the International Conference on Precambrian and Craton Tectonics/14th International Conference on Basement Tectonics*, Ouro Preto, Minas Gerais, Brasil, Extended Abstracts, pp. 15-17, 1998b.

Hinz, K., A hypothesis on terrestrial catastrophes: wedges of very thick oceanward dipping layers beneath passive continental margins, *Geologisches Jahrbuch*, E-22, 3-28, 1981.

Hoffman, P. F., Did the break-up of Laurentia turn Gondwana inside-out?, *Science*, 254, 5011, 1409-1412, 1991.

Hoffman, P. F. Relative timing of Rodinia breakup and Gondwana assembly: critical test of Laurentiocentric models of Neoproterozoic supercontinents, *Eos*, 73, 14, 364, 1992.

Howell, D.G., *Tectonics of Suspect Terranes*, Chapman and Hall (eds), London, 232 pp., 1989.

Junho, M.C.B., Granitóides brasilianos da região central do Rio de Janeiro- geoquímica preliminar, *Anais da Academia Brasileira de Ciências*, 65, 2, 162-179, 1993.

Lópes-Gamundi, O.R, I. S. Espejo, P. J. Conaghan and C. M. A. Powell, Southern South America, in J. J. Veevers and C. M. A. Powell (Eds.), Permian-Triassic Pangean basins and fold belts along the Panthalassan margin of Gondwana, Boulder, *Geological Society of America Memoir* 184, 281-329, 1994.

Macedo, J .M., Evolução tectônica da Bacia de Santos e áreas continentais adjacentes, in G.P.R. Gabaglia and E.J. Milani (Eds.), *Origem e evolução de bacias sedimentares,* Petrobrás, Rio de Janeiro, pp. 361-376, 1990.

Machado Filho, L., M. Ribeiro, S.R. Gonzales, C.A, Schenini, A.dos Santos Neto, R.C. Palmeira, J.L. Pires, W. Teixeira and H.E.F.Castro, *Geologia das folhas Rio de Janeiro (SF 23/24) escala 1:1.000.000, mapa e texto explicativo.* RADAM Brasil-Ministério das Minas e Energia, Rio de Janeiro, 780 pp., 1983.

Machado, N. and M. A. Carneiro, U-Pb evidence of Late Archaean tectonothermal activity in the southern São Francisco shield, Brazil, *Canadian Journal of Earth Sciences*, 29, 2341-2346, 1992.

Machado, N., C.M. Noce, E. A. Ladeira, and O. Belo de Oliveira, U-Pb geochronology of Archaean magmatism and Proterozoic metamorphism in the Quadrilátero Ferrífero, southern Brazil, *Geological Society of America Bulletin,*104, 1221-1227, 1992.

Machado, N., C.S. Valladares, M. Heilbron and C.M. Valeriano, U/Pb Geochronology of Central Ribeira Belt, *Precambrian Research*, 79, 347-361, 1996.

Machado, R., Evolução geológica, análise estrutural e metamórfica da região de Vassouras e Paracambi, Rio de Janeiro, *Unpublished Thesis IG-USP,* 196pp., 1984.

Machado, R., Considerações preliminares sobre a aplicação de um modelo geotectônico para os complexos Serra dos Orgãos e Paraíba do Sul (RJ), um exemplo de colisão continental, in *Anais 34 Congresso Brasileiro de Geologia,* pp. 135, 1986.

Machado, R. and M. Demange, Classificação estrutural e tectônica dos granitóides neoproterozóicos do Cinturão Paraíba do Sul no Estado do Rio de Janeiro, *Boletim. IG-USP*, Série Científica, 25, 81-96, 1994.

Marques, A., Evolução tectono-sedimentar e perspectivas exploratórias da Bacia de Taubaté, São Paulo, Brasil, *Revista Brasileira de Geociências*, 4, 3, 253-262, 1990.

Marques A., O.A. Zanotto, A.B. França, M.A.M. Astolfi and O.B. Paula, Compartimentação tectônica da Bacia do Paraná, Curitiba, *Petrobras/Nexpar Internal Report,* 87pp. 1993.

McKenzie, D., Some remarks on the development of sedimentary basins, *Earth Planet. Sci. Lett., 40,* 25-32, 1978.

Melo, M. S., C. Riccomini, Y. Hasui, F.F.M. Almeida, and A. M. Coimbra, Geologia e evolução do sistema de bacias tafrogênicas continentais do sudeste do Brasil, *Revista Brasileira de Geociências*, 15, 3, 193-201, 1985.

Milani, E.J., Evolução tectono-estratigráfica da Bacia do Paraná e seu relacionamento com a geodinâmica fanerozóica do Gondwana sul-ocidental, *Unpublished Thesis, Instituto de Geociencias, Universidade Federal do Rio Grande do Sul,* Porto Alegre, 2 vol, 1997a.

Milani, E.J., Bacia do Paraná – Carta estratigráfica, com os principais eventos tectônicos e magmáticos relacionados a sua evolução., *Petrobras, unpublished Internal Report*, Rio de Janeiro, 1997b.

Milani, E.J. and V. A. Ramos, Orogenias paleozóicas no domínio sul-ocidental do Gondwana e os ciclos de subsidência da Bacia do Paraná, *Revista Brasileira de Geociências, 28, 4,* 527-544, 1998.

Milani, E.J., A.B. França, and R.L. Schneider, Bacia do Paraná, *Boletim de Geociências da Petrobras*, 8, 69-82, 1994.

Milani, E.J., M.L. Assine, P.C. Soares and R.F. Daemon, A sequência ordovicio-siluriana da Bacia do Paraná, *Boletim de Geociências da Petrobrás,* 8, 257-273, 1995.

Mizusaki, A.M.P, A. Thomaz Filho, and J.G. Valença, Volcano-sedimentary sequence of Neocomian age in Campos Basin (Brazil), *Revista Braileira de Geociências*, 18, 247-251, 1988.

Mizusaki, A.M.P. and W. U. Mohriak, Sequências vulcano-sedimentares na região da plataforma continental de Cabo Frio, RJ, in *Anais do XXXVII Congresso Brasileiro de Geologia, Resumos Expandidos,* São Paulo, SP, 2, 468 - 469, 1992.

Mohriak, W. U. and A.Z. Barros, Novas evidências de tectonismo cenozóico na região sudeste do Brasil: o graben de Barra de São João na plataforma continental de Cabo Frio, Rio de Janeiro, *Revista Brasileira de Geociências*, 20, 1-4, 187-196, 1990.

Mutter, J.C., M. Talwani and P.L. Stoffa, Origin of seaward-dipping reflectors in oceanic crust off the Norwegian margin by "subaerial sea-floor spreading", *Geology*, 10, 353-357, 1982.

Oliveira, M. A. F., Bulk geochemistry of the Paraíba do Sul Granulitic Belt, *Revista Brasileira de Geociências*, 12, 1-3, 369-374, 1982.

Paciullo, F.V.P., A sequência deposional Andrelândia, *Unpublished Thesis, Instituto de Geociências da Universidade Federal do Rio de Janeiro,* Rio de Janeiro, 245 pp, 1997.

Paciullo, F. V. P., Orto-anfibolitos no ciclo deposicional Andrelândia- assinatura química e ambiente tectônico, in *Boletim de Resumos Expandidos, XXXVII Congresso Brasileiro de Geologia,* São Paulo, SBG, 2, pp. 28-30, 1992.

Paciullo, F. V. P., A. Ribeiro and R. Andreis, Reconstrução de uma bacia fragmentada: o caso do Ciclo Deposicional Andrelândia, in *Anais II Simpósio sobre o Cráton do São Francisco*, Salvador, Bahia, SBG, pp. 224-226, 1993.

Paciullo, F. V. P., A. Ribeiro, R. Andreis and R. Trouw, Sedimentary, igneous and thermo-tectonic events in the folded belts at the southern border of the São Francisco craton, in *Abstracts of the International Conference on Precambrian and Craton Tectonics/14th International Conference on Basement Tectonics,* Ouro Preto, Brazil, pp. 68-69, 1998.

Pedrosa Soares, A.C., P. Vidal, O.H. Leonardos and B.B. Brito Neves, Neoproterozoic oceanic remnants in eastern Brazil: futher evidence and refutation of an exclusively ensialic evolution for the Araçuaí-West Congo orogen. *Geology,* 26, 6, 519-522, 1998.

Pires, F.R.M., J.G. Valença and A. Ribeiro, Multistage generation of granite in Rio de Janeiro, Brazil, *Anais da Academia Brasileira de Ciências,* 54, 3, 563-574, 1982.

Rabinowitz, P.D. and J. LaBreque, The Mesozoic South Atlantic Ocean and evolution of its continental margins, *Journal of Geophysical Research,* 84, B11, 5973-6002, 1979.

Ramos, V.A., Late Proterozoic - Early Paleozoic of south America: a collisional history, *Episodes,* 11, 168-174, 1988.

Ramos, V.A., T.E. Jordan, R.W. Allmendinger, S.M. Kay, J.M. Cortes and M.A. Palma, Chillenia: un terreno aloctono en la evolution paleozoica de los Andes centrales, in *II: Congresso Geologico Argentino, 9, Buenos Aires, Associación Geologica Argentina*, 2, 84-106, 1984.

Ramos, V.A., T.E. Jordan, R.W. Allmendinger, C. Mpdozis, J.M. Kay, J.M. Cortés and M Palma, Paleozoic terranes of the central Argentine-Chilean Andes, *Tectonics*, 5, 855-880, 1986.

Ribeiro, A., Diques máficos proterozóicos em São João del Rei, MG, in *XXXVII Congresso Brasileiro de Geologia,. Bol. Resumos Expandidos,* São Paulo, 1, 508-509, 1992.

Ribeiro, A., Estratigrafia e paleoambientes nas sucessões metassedimentares proterozóicas das serras do Lenheiro e São

José, São João del Rei, Sul de Minas Gerais, *Unpublished Thesis, Instituto de Geociências da Universidade Federal do Rio de Janeiro*, Rio de Janeiro, 167 pp., 1997.

Ribeiro, A., F.V.P. Paciullo, R. Andreis, R. Trouw and M. Heilbron, Evolução policíclica proterozóica no sul do Cráton do São Francisco: análise da região de São João del Rei e Andrelândia, MG, in *XXXVI Congresso Brasileiro de Geologia*, Natal–RN, 6, 1990.

Ribeiro, A. Andreis, R.R., Trouw, R.A.J., Paciullo, F.V.P. and Valença, J.G., Evolução das bacias proterozóicas e termotectonismo brasiliano na margem sul do Cráton do São Francisco, *Revista Brasileira de Geociências*, 25, 4, 235-248, 1995.

Rogers, J.J.W., A history of the continents inthe past three billion years, *Journal of Geology*, 4, 1405-1412, 1996.

Rosier, G.F., A geologia da Serra do Mar, entre os picos de Maria Comprida e do Desengano (Estado do Rio de Janeiro), *DNPM-DGM*, Boletim 166, 1957.

Rosier, G.F., Pesquisas geológicas na parte oriental do Estado do Rio de Janeiro e na parte vizinha de Minas Gerais, *DNPM, Div. Geol. Min.*, Boletim 222, Rio de Janeiro, 1965.

Schmmit, R.S. and R.A.J. Trouw, Estudo preliminar dos indicadores de transporte tectônico pré-cambriano na Região dos Lagos, Rio de Janeiro, in *Atas do V Simpósio de Geologia do Sudeste*, Penedo, Rio de Janeiro, 142-144, 1997.

Schmmit, R.S., R.A.J. Trouw, and W.R. van Schmus, The characterization of a Cambrian (~ 520 Ma) tectonometamorphic event in the Costeiro Domain of the Ribeira Belt, using U/Pb in syntectonic veins, in *Simpósio Nacional de Estudos Tectônicos*, Lençóis–Ba, Extended Abstract (in press), 1999.

Silva, L.G.E, D.B. Nava, M. Heilbron and C.M.Valeriano, Geologia de detalhe da Serra da Carioca, cidade do Rio de Janeiro, in *Atas Simposio de Geologia do Sudeste*, 2, São Paulo, SBG, pp. 161-170, 1991.

Soares, P.C., P.E. Barcellos, S.M. Csordas, J.T. Mattos, M.G. Balieiro and P.R. Meneses, Lineamentos em imagens de Landsat e radar e suas implicações no conhecimento tectônico da bacia do Paraná, in *II Simpósio Brasileiro de Sensoreamento Remoto*, CNPq-Instituto de Pesquisas Espaciais, Brasília, 1982.

Souza, K. G., R.L. Fontana, J. Mascle, J.M. Macedo, W. U. Mohriak and K. Hinz, The southern Brazilian margin: an example of a South Atlantic volcanic margin, in *III International Congress of the Brazilian Geophysical Society*, Rio de Janeiro, RJ, 2, 1336-1341, 1993.

Stockmal, G.S. and C. Beaumont, Geodynamic models of convergent margin tectonics: the southern Canadian Cordillera and the Swiss Alps, in Beaumont, C. and Tankard, A.J. (Eds.), Sedimentary basins and basin-forming mechanisms. *Canadian Society of Petroleum Geologists, Memoir* 12, 393-411, 1987.

Szatmari, P. and W. U. Mohriak, Plate model of post-breakup tectono-magmatic activity in SE Brazil and the adjacent Atlantic, in *V Simpósio Nacional de Estudos Tectônicos*, Gramado, 213-214, 1995.

Tankard, A.J et alli, Structural and tectonic controls of basin evolutionin southwestern Gondwana during the Phanerozoic, in A.J.Tankard, R.Suarez and H.J. Welsink (Eds.), *Petroleum Basins of South America, AAPG Memoir*, 62, 5-52, 1985.

Teixeira, W. and M.C.H. Figueiredo, An outline of Early Proterozoic crustal evolution in the São Francisco Craton, Brazil: a review, *Precambrian Research*, 53, 1-22, 1991.

Trompette, R., Geology of Western Gondwana (2000-500 Ma), A. A. Balkema, Rotterdam, 350 pp., 1994.

Trouw, R. A. J. Evolução tectônica ao sul do Craton do São Francisco, baseada em análise metamórfica, in *Boletim de Resumos Expandidos, XXXVII Congresso Brasileiro de Geologia*, São Paulo, 1, pp. 327-328, 1992.

Trouw, R. A. J. and R.J. Pankhrust, Idades radiométricas ao sul do cráton do Sào francisco: região da Folha Barbacena, Minas Gerais, in *Anais II Simpósio sobre o Cráton do São Francisco*, Salvador, Bahia, SBG, pp. 260-263, 1993.

Trouw, R. A. J., A. Ribeiro, and F.V.P. Paciullo, Contribuição à geologia da folha Barbacena 1:250.000, in *Anais do XXXIV Congresso Brasileiro de Geologia*, Goiânia, SBG, 2, pp. 972-986, 1986.

Trouw, R. A. J., A. Ribeiro, F. V. P. Paciullo and R.R. Andreis, A passagem do mesoproterozóico para o neoproterozóico na região sul de Minas Gerais, in *IX Simpósio de Geologia de Minas Gerais*, Ouro Preto, SBG/MG, 14, pp. 10-11, 1997.

Tupinambá, M., Rochas intrusivas e metassedimentos granulíticos do Complexo Paraíba do Sul na parte setentrional da Faixa Ribeira in *Atas do III Simpósio de Geologia do Sudeste*, SBG/RJ, Rio de Janeiro, pp. 187-193, 1993a.

Tupinambá, M., Metamorfismo granulítico do Complexo Paraíba do Sul na porção setentrional da Faixa Ribeira, a leste do Cráton do São Francisco in *Anais do II Simpósio de Geologia do Cráton do S. Francisco e Faixas Marginais*, SBG/BA, Salvador, 266-268, 1993b.

Tupinambá, M., W. Teixeira and M. Heilbron, The Pan-African/Brasiliano Arc-related Magmatism at the Costeiro Domain of the Ribeira belt, Southeastern Brazil, in *Abstracts of the International Conference on Precambrian and Craton Tectonics/14th International Conference on Basement Tectonics*, Ouro Preto, MG, pp. 12-14, 1998.

Turner, S., M. Regelous, S. Kelley, C. Hawkesworth and M. Mantovani, Magmatism and continental break-up in the South Atlantic: high precision Ar/Ar geochronology, *Earth and Science Planetary Science Letters*, 121, 333-348, 1994.

Vail, P.R., R.M. Mitchum, and S. Thompson, Seismic stratigraphy and global changes of sea-level, part 4 : global cycles of relative changes of sea-level, in C.E. Payton (Ed.), *Seismic Stratigraphy-applications to hydrocarbon exploration, AAPG Memoir* 26: 83-97. 1977.

Valeriano, C.M., J.C.H. Almeida, L.S.A. Simões, H.L. Roig, B.P. Duarte and M. Heilbron, Evolução estrutural do domínio externo da Faixa Brasília no Sudoeste de Minas Gerais: registros de uma tectônica pré-brasiliana, *Revista Brasileira de Geociências*, 25, 4, 221-234, 1995.

Valeriano, C.M., L.S.A. Simões, and M. Heilbron, Estruturação da porção meridional da Faixa Brasília, sudoeste de Minas Gerais: implicações sobre a definição do limite regional do Cráton São Francisco in *Anais do II Simpósio sobre o Cráton do São Francisco. Evolução Tectônica e Metalogenética do Cráton do São Francisco*, SBG, Núcleos Bahia/Sergipe, Salvador, BA., pp. 275-276, 1993.

Valladares, C.S., Evolução geológica do Complexo Paraíba do Sul no segmento central da Faixa Ribeira com base em estudos de geoquímica e de geocronologia U-Pb, *Unpublished Thesis, IG-USP*, São Paulo, 147 pp.

Valladares, C.S., M. Heilbron, N. Machado, and C. Valeriano, Provenance of Central Ribeira Belt metasediments using 207Pb/206Pb ages on zircon by laser-ablation ICPMS in *Extended Abstracts South American Symposium on Isotope Geology,* Campos do Jordão, SP, pp. 323-325, 1997.

Valladares, C.S., M. C. Figueiredo and M. Heilbron, Geoquímica de rochas gnáissicas paleoproterozóicas do Complexo Paraíba do Sul Complex, região de Barra Mansa, Rio de Janeiro, Brasil, *Revista Brasileira de Geociências,* 27, 1, 83-98, 1997.

Valladares, C.S., M. Heilbron and M.C.H.Figueiredo, O Granito Getulândia e sua relação com a Zona de Cisalhamento Taxaquara, in *Anais do V Simpósio Nacioanl de Estudos Tectônicos* Gramado, RS, pp. 217-218, 1995.

Van Schums, W. R., M. E. Bickford and K. Condie, Early Proterozoic evolution, in Redde, J.C., M.E. Bickford, R.S. Houston, P.K. Link, D.W. Rankin, P.K. Sims, W.R. Van Schmus (Eds), *Precambrian: conterminous* U.S. Boulder, Geological Society of America, The Geology of North America, C-2, 270-281, 1993.

White, R.S. and D.P McKenzie, Magmatism at rift zones: the generation of volcanic continental margins and flood basalts, *Journal of Geophysical Research,* 94, 7685-7730, 1989.

Wiedemann, C. M. Estado atual do conhecimento sobre a evolução do Cinturão Móvel Brasiliano no Estado do Espírito Santo, in *Boletim de Resumos I Simpósio de Geologia do Sudeste,* Rio de Janeiro, SBG, pp. 95-96, 1989.

Wiedemann, C. M., Early Paleozoic, late-to post-collisional magmatic arc of the coastal mobile belt, in the state of Espírito Santo, eastern Brazil, *Anais da Academia brasileira de Ciências,* 65, supl 1, 162-181, 1993.

Wilson, J.T., Did the Atlantic close and then re-open?, *Nature,* 211, 676–681, 1966.

Zalán, P.V., A tectônica transcorrente na exploração de petróleo: uma revisão, *Revista Brasileira de Geociências,* 16, 3, 245-257, 1986.

Zalán, P.V., S. Wolff; M.A.M. Astolfi, I. S. Vieira, J.C.J. Conceição, V.T. Appi, E.V.S. Neto, J.R.Cerqueira and A. Marques, The Paraná Basin, Brazil, in Leighton, M.W., D.R. Kolata, D.F. Oltz, J.J., Eidel (Eds.), *Interior cratonic basins.* AAPG Memoir, 51, 681-708, 1990.

Zanotto, O.A., Erosão pós-Cretáceo na Bacia do Paraná, com base em dados de reflectância da vitrinita, in *Resumos Simpósio Sul-Brasileiro de Geologia, 5,* SBG, Curitiba, 58, 1993.

Zimbres, E., K. Kawashita and W.R. Van Schmus, Evidências de um núcleo transamazônico na região de Cabo Frio, RJ e sua conexão com o Cráton de Angola, in *Anais XXXVI Congresso Brasileiro de Geologia,* Natal, RN, SBG, 6, pp. 2735-2743, 1990.

Cláudio M. Valeriano, *UERJ - Rio de Janeiro State University, Faculty of Geology, Rio de Janeiro - RJ*

Edison J. Milani, *PETROBRAS - Petróleo Brasileiro S.A. – E&P, Rio de Janeiro – RJ*

Julio Almeida, *UERJ - Rio de Janeiro State University, Faculty of Geology, Rio de Janeiro - RJ*

Miguel Tupinambá, *UERJ - Rio de Janeiro State University, Faculty of Geology, Rio de Janeiro - RJ*

Monica Heilbron, *UERJ - Rio de Janeiro State University, Faculty of Geology, Rio de Janeiro - RJ*

Webster U. Mohriak , *PETROBRAS - Petróleo Brasileiro S.A. – E&P, Rio de Janeiro – RJ*

The Mosaic of Terranes in Central Europe as seen by Deep Reflection Studies

Rolf Meissner

Institut für Geophysik, Universität Kiel, Germany

In central Europe, W of the Teisseyre Tornquist zone, the collision and collapse of continental terranes are the dominant processes of crustal evolution. Terrane collisions started in the Late Proterozoic in the Baltic Shield, continued with the Caledonian docking of East-Avalonia to Baltica, continued further with the Variscan terrane collisions, which apparently were the strongest and longest orogeny, followed by a total collapse. In the S the Alpine orogeny, caused by the terrane Adriatica, is still in full action. Considering the young nature of Western and Central Europe and the long periods in the Mesozoic when it was severely flooded, the whole area may be considered a continental margin. Seismic reflection studies were one of several investigations to decipher collisions and collapse by means of observing special reflectivity patterns. Compressional structures like indenter tectonics generally show strong and continuous reflectors, supposed to represent prominent shear-and thrust zones around the indenter, provided they were not destroyed by subsequent (hot) thermo-rheological processes. For instance, Variscan collapse and the formation of basins have created rather thin crusts. They show abundant thick seismic lamellae in the lower crust and are predominantly sialic (with some mafic-ultramafic splinters); they show only thin lamellae if affected by massive mafic intrusions in the lower crust and have created only small impedance contrasts. The very thin crusts of many passive continental margins do not show massive lamellae either, because temperature did never reach the necessary high values for a ductile creation of the lamellae by an ordering process. Survival or destruction of compressional features in old cratons depend on the post-orogenic thermo-rheological development. Strong heat pulses, from a delamination process or a plume in an extensional or collapse regime, have destroyed all deep signs of the former compressive regime in the lower crust and in the upper mantle and generated a rejuvenation of lower crust and Moho.

1. INTRODUCTION

The central Europe region (Fig. 1) is characterized by several geomorphological provinces related to the mosaic of terranes accreted during geological times spanning from the Proterozoic to the present day. The tectonic configuration of shields, orogenic fold belts, massifs, rifts and basins (Fig. 2) is associated with formation of several continental margins throughout the geological history. Figure 2 shows a schematic tectonic map of Central Europe, Figure 3 shows the early classical division of Central Europe terranes, and Figure 4 shows the new view of the most important terranes in central Europe.

Atlantic Rifts and Continental Margins
Geophysical Monograph 115

Figure 1. Shaded topographic map of the Central Europe region showing main geomorphological provinces.

Within the framework of continental plates Europe had a late start. Only Europe`s north, i.e Scotland, belonging to Laurentia, and the Fennoscanian countries, belonging to Baltica, are parts of really old cratons [*Berthelsen*,1992]. Baltica shows decreasing ages of tectonic units from Archean in the NE to Late Proterozoic in the SW, this part considered to belong to central Europe. The rest of western and central Europe accumulated in the Paleozoic (Meso-Europe, see [*Stille*,1924] and Figures 3 and 4). During the Caledonian collision between Laurentia and Baltica at about 420 to 400 Ma, a small continent, the terrane Avalonia, arrived from the S and took part in the collision [*McKerrow et al.*,1991; *Torsvik* and *Trench*, 1992]. It helped closing the Iapetus and the Tornquist oceans. East Avalonia formed the southern part of the British Isles, the southern North Sea and parts of Denmark and northern Germany. More continental terranes were split from the northern rim of Gondwana, e.g. Armorica, arriving about 20 Ma after the Caledonian collision and accreting to Avalonia, forming Western and Central Europe, and Bohemia in the SE (Bavaria and Czechia) [*Franke and Onken*, 1995]. The accretion of Armorica started the Variscan orogeny, which lasted from about 380 to 300 Ma in Western and Central Europe. All terranes from Gondwana had a continental affinity and formed the area of France, Germany, and W - Poland. Wegener`s supercontinent Pangaea was established with the accretion of Africa`s NW to the SW part of Europe. There were large or small oceans with continental margins in between all these terrane collisions, e.g., the Tornquist Ocean between East Avalonia, and SW - Baltica or the Lizzard -

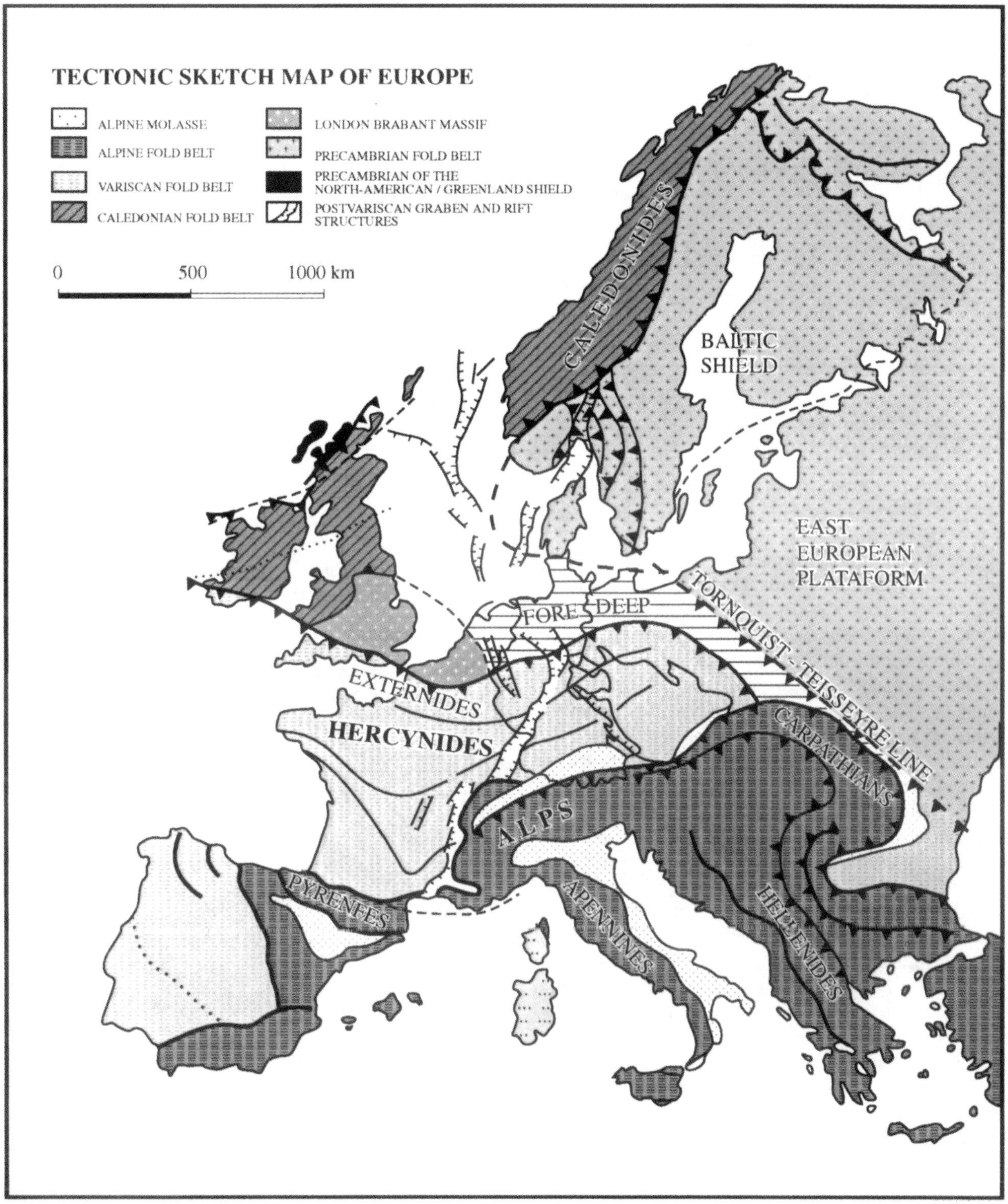

Figure 2. Schematic tectonic map of Central Europe showing the Precambrian shields, platforms, fold belts, rift structures, and main collisional fronts. Modified from Meissner and Bortfeld, 1990.

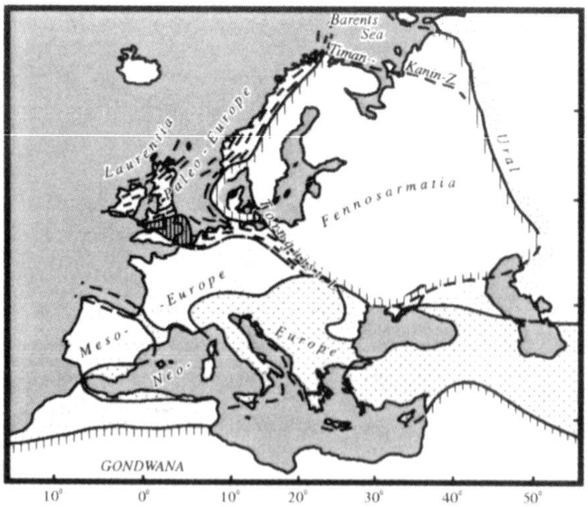

Figure 3. The classical division of Europe into three age provinces; modified after Stille,1924.

Giessen Ocean between Avalonia and Armorica. While most of the oceanic lithosphere was subducted, slices and nappes of the former continental margins are still observed today [*Franke et al.*,1990]. But also the Mesozoic development of Central and Western Europe, with its enormous marine sedimentation in various epochs, justifies to transfer the expression of "continental margin" to this area.

The collapse of the Variscan mountain belts, associated with much regional extension, had already started before 300 Ma, with the formation of rifts, grabens, extensional and strike-slip features. In the Permian, the westward drift of Africa transferred more extensional strain and movements to Europe.The collapse and the Permian extension were the last great tectono-thermal events in Central Europe. They must have been responsible for the present day structure and reflectivity of the upper and lower crust [*Meissner and Sadoviak*,1992]. The collapse was accompanied by massive granitic volcanism in the Variscan internides, indicating that the lithosphere in the center of the Variscides was extremely hot. Later, in the Permo-Carboniferous, mafic magmas from the Oslo Graben in the N to Upper Italy in the S are observed [*Matte*,1991; *Blundell et al.*,1992]. This huge magmatic outbreak in the North Sea and the North German-Polish area was not only limited to the surface area. Based on high velocities in the lower crust, massive mafic intrusion have to be assumed to have entered the stretched crust, similar to the observation at volcanic passive margins [*Bayer et el.,1995]*. Whether this period of volcanism and magmatism was the consequence of a plume, or of a build-up of excessive heat below the large continental mass of Pangaea, is still an open question. But cooling of the intrusions must have been the main reason for the long-time subsidence of the North Sea and the North German Basin (NGB). Some subsidence episodes, lasting for nearly 300 Ma, have created ideal conditions for generation and accumulation of various economically important hydrocarbon reserves [*Bayer et al.*,1995].

Europe had to wait until the Late Cretaceous to split apart as divergent margins, when the North Atlantic opened at about 65 Ma and created a large passive margin in the NW segment of the Atlantic, between Greenland and Norway [*Ziegler*,1990]. In the S, another terrane (Adriatica) arrived and accreted to Central Europe, resulting in an orogen responsible for the Alpine and surrounding collisional belts. During the Alpine orogeny, large parts of the former Penninic sediments and adjacent continental margins were migrated, compressed, faulted and folded. Today, large parts of these terranes are observed as thin and thick nappes, dominating the Alpine mountain belts. Beginning in the Tertiary, and lasting until the present day, the compressional tectonic stress from the S formed a few extensional structures and rifts in Central Europe, such as the Rhinegraben or the Egergraben, which may be classified as "impactogene" basins. Especially strong but short phases of compression were transferred as far as the Baltic Sea, where they generated inversion (flower- or harpoon-) structures in sedimentary basins SW of Bornholm [*Deeks and Thomas*,1995].

In this paper, central Europe will be the main topic of our analysis, showing that the accretion and subsidence patterns

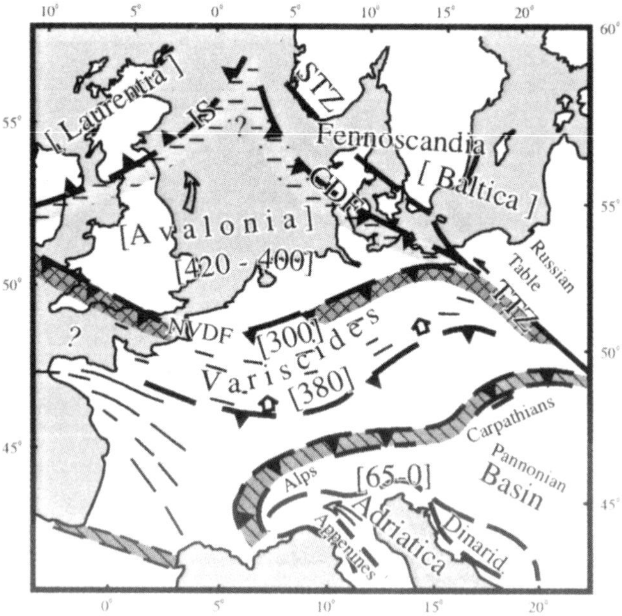

Figure 4. The new view of terrane accumulation in central Europe, slightly modified after Berthelsen, 1992.

of the various terranes, as well as their behaviour after extension and dispersal, may be revealed by geophysical (mainly seismic) investigations. It will be shown that the rigid crystalline upper crust preserves the memory of old collisional structures in form of fault and shear zones, whereas the lower crust in warm areas has been rejuvenated during the collapse processes, showing a dense seismic "lamellae" structure which is not found in any other level of the lithosphere [*Meissner*,1996]. Most listric faults from the upper crust seem to flatten with depth and turn into the horizontal direction in the lower crust. The uppermost mantle, on the other hand, except for areas with a very strong heat pulse during its evolution, is often rigid again and shows a concentration of reflectivity on single faults or shear zones, similar to the reflectivity in the upper crust. It is evident from these observations that the boundary between the crust and the mantle, the "Moho", really is one of the most prominent boundary of our planet earth (another one is the core-mantle boundary). It is definitely a boundary in fundamental parameters such as elasticity, density, rheology, petrology and mineralogy.

All observed reflections in the lithosphere seem to be created (or destroyed) by thermo-rheological processes. Figure 5 shows the DEKORP reflection network in Germany and the SW-part of the BABEL line A in the Baltic Sea [*DEKORP-BASIN Research Group* ,1998, 1999; *BABEL Working Group*,1991,1993]. The various tectonic units and their special reflectivity patterns will be discussed from N to S, i.e. from the SW part of the Proterozoic Baltic Shield, across the North German Basin, across the Paleozoic Variscides and into the Tertiary Alpine Belt.

2. THE SW - RIM OF BALTICA

It has been mentioned that northern Denmark and the Scandinavian countries belong to the old craton Baltica, which is Archaean in the very NE, getting younger toward the SW. In the Late Proterozoic province north of Bornholm, an accretion of various terranes has taken place. This is clearly seen on the BABEL line A [*Abramovitz et al.*, 1997] and on three more marine DEKORP lines [*Meissner and Krawczyk.*, in prep.]. There is a huge increase in number, amplitude, and thickness of reflectivity in the whole crust, the reflections diverging, thickening the crust toward the NE. Terrane boundaries are seen by especially strong and continuous reflections, which probably represent especially strong shear zones, as suggested by *Potter* [1989] and *Beaumont et* al. [1994]. The compressional stresses seem to be connected with the southern extension of the N-S running Protogene (or Mylonite) Zone in southern Sweden, belonging to the Late Proterozoic Grenvillian-Dalsladiam belt [*Berthelsen*, 1992]. Figure 6 shows some of the reflections with the assumed terrane boundaries indicated. This accretion seems to be the latest tectonic event in the Proterozoic. Regarding the old reflectivity and crustal structure, it seems important to note that no later thermal pulse (from a subduction, delamination, or plume) is observed in this area, which apparently has preserved the last tectonic and reflectivity pattern.

The SW end of Proterozoic Baltica is farther to the W and SW. It is marked by the Caledonian Deformation Front (CDF), a rather complex continent-continent collision, created between 420 and 400 Ma. The Caledonian fold belt (Fig. 2) is the long collisional zone between Laurentia (with Scotland and Greenland) and Baltica, where the Iapetus ocean was subducted and the Scotish and Norwegian Caledonides were formed [*McKerrow et al.*,1991; *Torsvik and Trench,*1991]. The southern part (southern England, North Sea, southern Denmark, Baltic Sea and northernmost Germany) belonged to the third partner in the Caledonian collision, the terrane (Eastern-) Avalonia. It was split from Gondwana about 490 Ma ago and docked between 425 and 400 Ma. This resulted in a partially mild collision south of the colliding continents Laurentia and Baltica. Around the British Isles, the subduction of the Iapetus is clearly observed by six BIRPS profiles [*Klemperer and Matthews,1987*]. But also in the southern Baltic Sea, where a smaller ocean (the Tornquist Sea) disappeared, the collisional structures are clearly observed. It was a rather oblique collision with a bivergent structure, i.e. the Avalonian nappes thrust onto Baltica (to the NE), the oceanic subduction (plus some Avalonian delamination (?) dipping to the NE, as seen by several reflections in the uppermost mantle. Figure 7 shows one of the lines (D 3 in the Baltic Sea, see Fig. 3) crossing the CDF. In the whole area a mild extension without a heat pulse followed, and especially in the rigid upper crust and the uppermost mantle, the reflections could survive until today. The lower crust shows some lamination, but not as strong as in areas of pronounced extension and heating typically observed elsewhere. Both the reflections in the upper crust (identified as nappes near the suface) and the reflections in the uppermost mantle (supposed to be subduction-related) seem to be decoupling in the lower crust. It should be mentioned that within central Europe, the North Sea and the Baltic Sea are the only areas where reflections from the mantle are observed. Further S, toward the center of the Variscides, where there are many signs of strong heating processes, no reflections in the mantle (i.e. no signs of former subduction or delamination) are observed, in spite of various research programmes that aimed at investigating these features [*Bartelsen et al.*,1982].

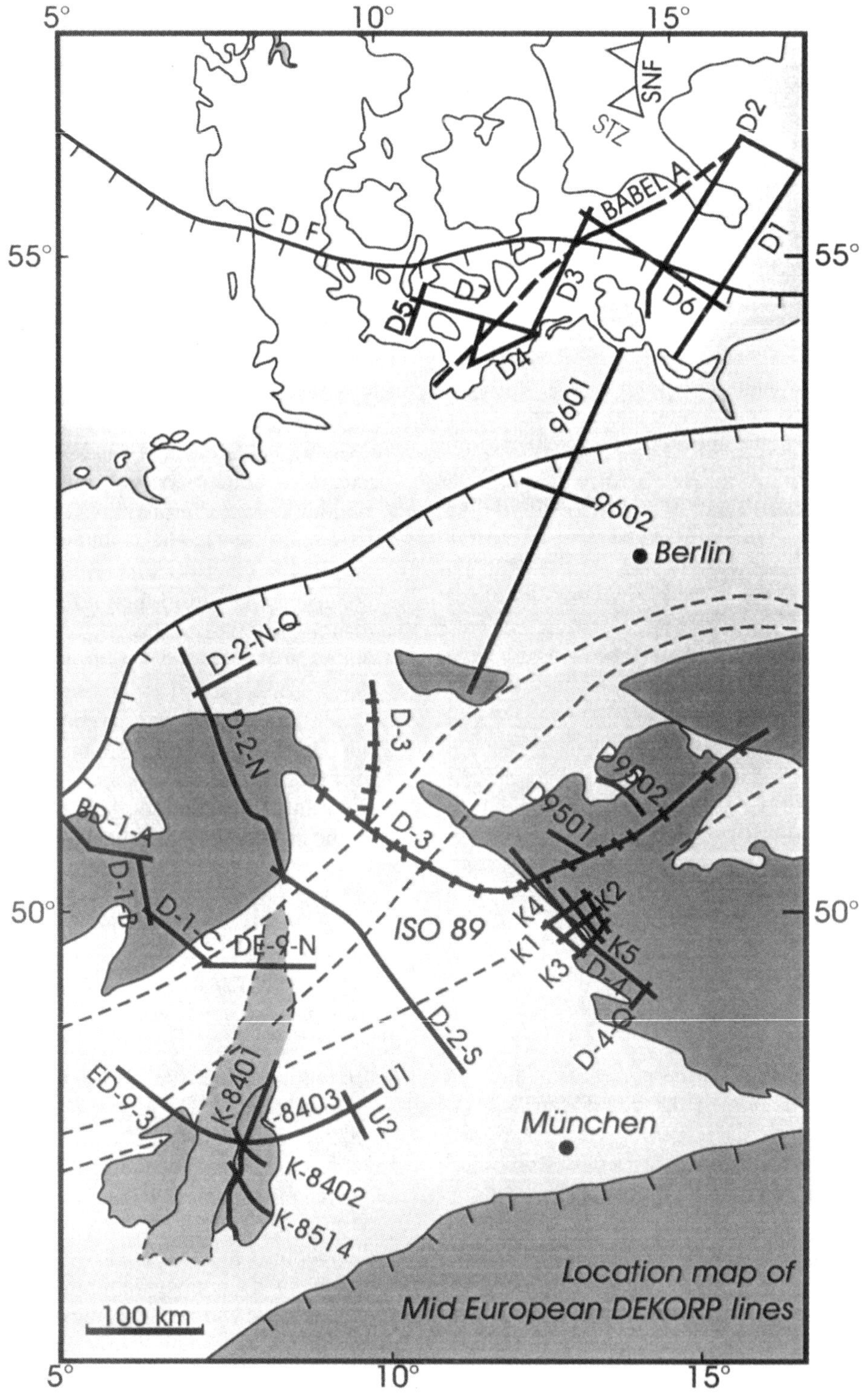

Figure 5. The DEKORP reflection network in Germany and BABEL line A in the Baltic Sea. The Rhenish Massif is located in the central-western part of the map (dark grey shading), and the NE-trending Rhine Graben, slightly to the south, is depicted in lighter tones of grey.

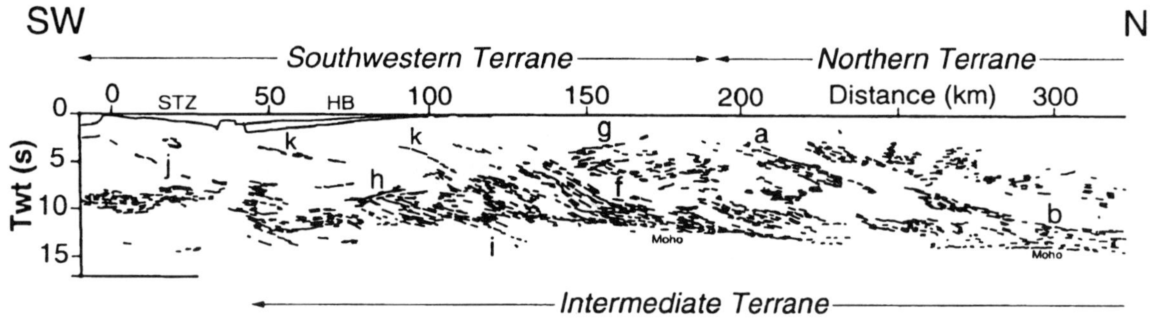

Figure 6. Reflection signals across a Late Proterozoic terrane collision NE of Bornholm; after Abramovitz el al., 1997.

3. THE NORTH GERMAN BASIN (NGB)

Although a large number of commercial reflection lines cover the NGB, only a few reflections from lower crust or Moho were observed. Subsidence history and tectonic setting is certainly determined by a polyphase evolution. After the massive mafic intrusions in the Permo-Carboniferous, especially in the eastern part of the basin, strong subsidence started, its rate decreasing over time [*Bayer et al.*,1995]. This was most probably a response to thermal recovery. Various subbasins developed, some short-time inversions interrupted the general subsidence, and signs of extension remained small [*DEKORP-BASIN Research Group*,1998, 1999]. In order to reveal the complex evolution and the present structure of the crust, a long SW-NE running DEKORP profile was acquired in spring 1996, including the marine legs in the Baltic Sea (example, line D-3 shown in Figure 7).

An automatic line drawing of this survey is plotted in Figure 8 [*DEKORP-BASIN Research Group*, 1999]. Figure 9 shows a schematic geological section based on a composite seismic transect extending from Harz towards the Baltic Sea, and Figure 10 shows a geological section across the Northeast German Basin. These profiles show that the Zechstein horizon is the most prominent reflector in the sediments and can be followed below the whole basin, marking the subsidence (since Zechstein times) very nicely. Some less reflective sedimentary layers are observed below Zechstein, giving a total sediment thickness of about 10 km. Strong, sometimes diffuse mid-crustal reflectors and high velocities in the lower crust are certainly connected to the mentioned mafic intrusions [*DEKORP-BASIN Research Group,1999]*. It was a surprise to see the Moho to be remarkably flat and rather shallow around a depth of about 29 km over the whole profile, only interrupted and partly deepening toward the NE, i.e.approaching the CDF and the Baltic rim, and to the SW, entering the Variscan externides. No Moho updoming or antiroots are observed, as usually found below grabens and small basins, compensating the low-density sediments isostatically. The Moho deepens to about 31 km only toward the NE. In the whole basin the tendency toward regional isostasy is probably provided by the massive, high-density intrusions in the lower crust, although a strong local Bouguer gravity high in the middle of the basin shows that isostasy is not complete.

Compared to the reflectivity of the lower crust in the Variscan internides to the south, or to most other extensional areas or basins, the reflecting lamellae on top of the Moho are rather thin in the NGB [*Meissner* and *Rabbel*,1999]. This may be an effect of a rather pure mafic lower crust, where the originally hot mafic magmas could not develop a differential layering with contrasting impedance. It should, however, be noted that volcanism and intrusions did not occur everywhere in the NGB. There is firm evidence that the W - part of the NGB has a thick lamella-like reflecting lower crust [*Dohr et al.*,1983] with "sialic" velocities [*Rabbel et al.*,1994]. As

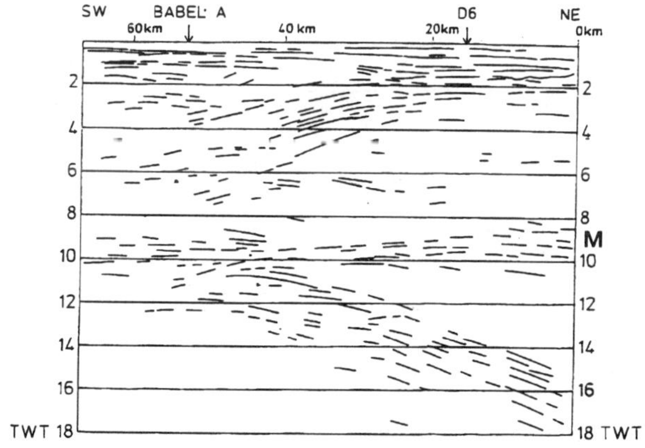

Figure 7. Example of the marine DEKORP reflection line D-3 in the Baltic Sea, crossing the Caledonian Deformation Front (CDF); from unpublished DEKORP records. M = Moho (around 10 s TWT). Northeast-diping reflectors (below the Moho) and southwest-dipping reflectors (above the Moho) are also imaged in the profile.

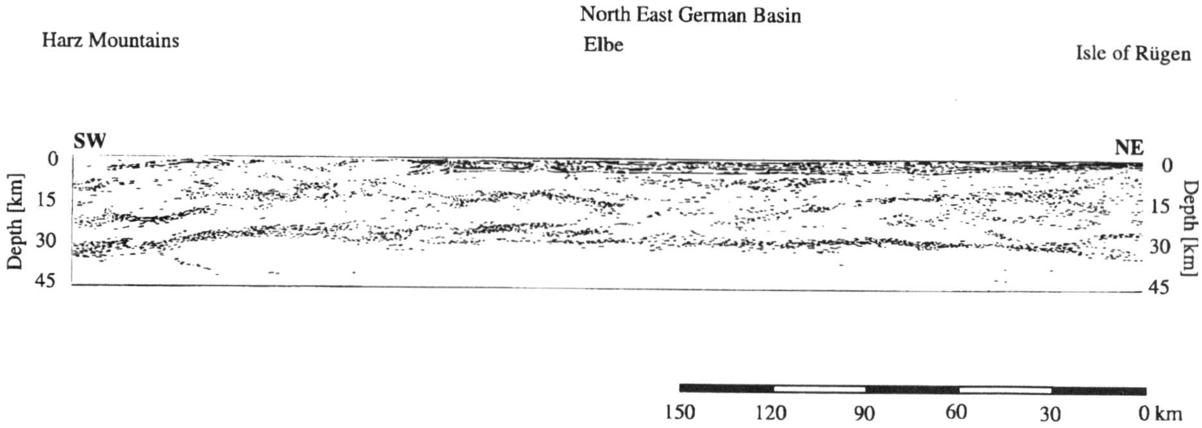

Figure 8. Line drawing of DEKORP's 96/1 profile through the North German Basin (NGB); from DEKORP-BASIN Research Group, 1998.

seen on most other basins worldwide, the thickness of the crystalline crust is reduced by about 1/3 of its original thickness, a phenomenon which is hard to explain and will again be mentioned under "discussions". The sedimentary depocenters below the Zechstein salt seem to coaxially coincide with the Triassic / Jurassic depocenters (Fig. 6-C). The southwestern extremity of the profile is characterized by compressional structures associated with the Variscides.

4. THE VARISCIDES

Between about 380 and 300 Ma large parts of central and western Europe, northwestern Africa, and the NE rim of North America were affected by the Variscan collision [*Ziegler*,1990; *Matte*,1991]. Today it is widely accepted that the Variscan orogeny in Central Europe is the result of a multiple collision of terranes with continental affinity, originally split apart from the northern rim of Gondwana in Ordovician times [*Gebauer et al.*,1989; *Tait et al.*,1996]. All continents, including Gondwana, had drifted together in Late Carboniferous to form one huge continental landmass, Pangaea [*Wegener*,1924]. This supercontinent remained more or less intact for more than 100 Ma. The African part of Gondwana had docked to the European landmass in the SW, leaving the large Tethys Ocean still juxtaposed to the S of Europe. The collisional phases might have been similar to

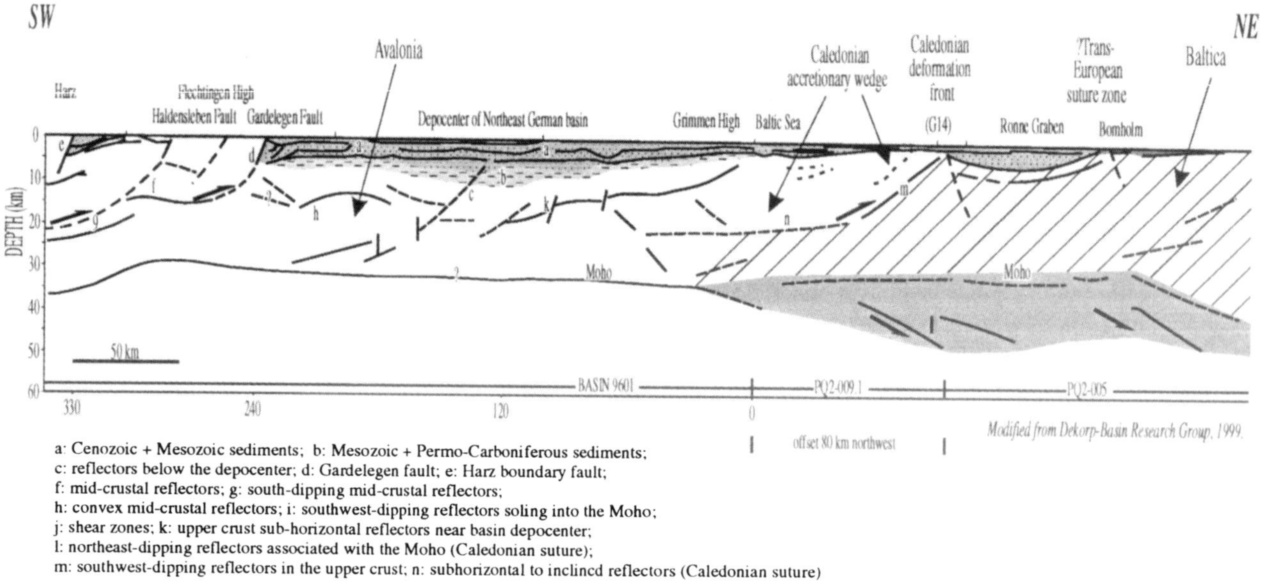

Figure 9. Crustal- scale geological section based on a composite seismic transect along the North German Basin (NGB); modified from DEKORP-BASIN Research Group, 1999.

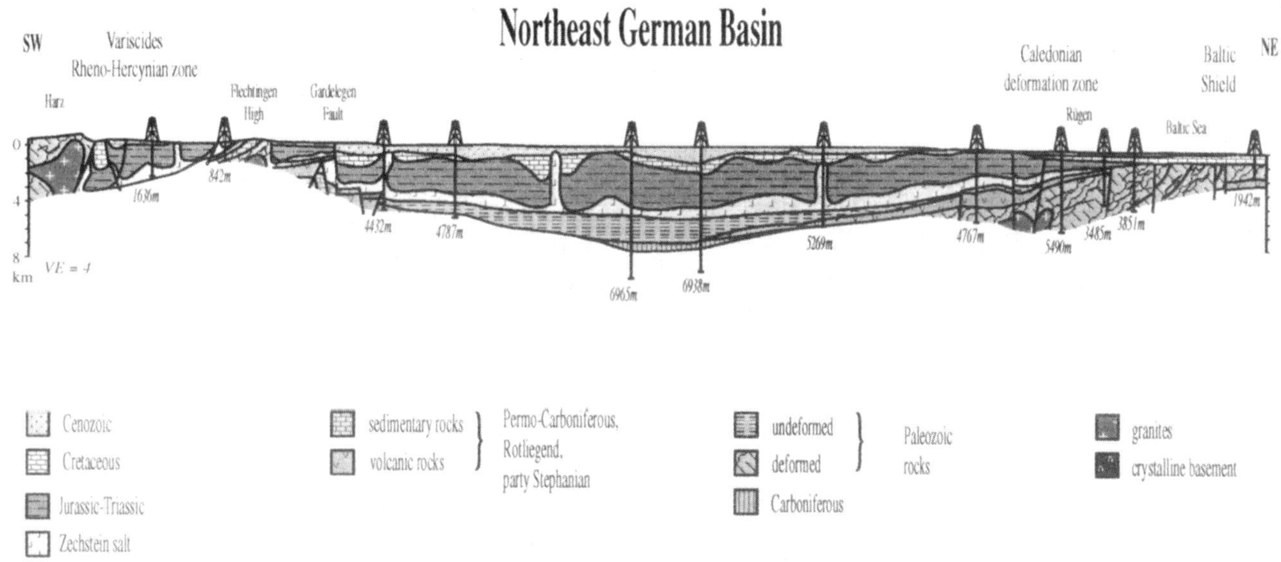

Figure 10. Schematic geological section along the North German Basin (NGB); modified from DEKORP-BASIN Research Group, 1999.

today' s continent-continent collisions, but a major collapse phase with regional extension, strike slip motion and widespread mafic magmatism, with the erosion of the mountain belts and the disappearence of crustal roots, heralded the end of compressional and the beginning of extensional stress patterns [*Meissner and Sadowiak*,1992]. It is probable that the beginning of extension was initiated by the movement of Gondwana to the W in Early Permian, but the first indication of rifting was in the Oslo Graben (*Matte,1991)*.The development of an enormous heat bulge under the huge Pangaean landmass is also hypothesized to have initiated extensional processes, possibly supported by a delamination process toward the end of the compressional phase [*Meissner and Mooney*,1998]. The extensional processes accelerated in the Permian, leading to a strong separation phase in the Jurassic and Cretaceous.

Figure 11 shows the Variscan terranes in Germany with the location of the following examples indicated. Terrane boundaries are crossed 14 times by DEKORP lines. The boundaries are often observed as strongly dipping, curved reflectors, restricted to the rigid upper crust and truncated by subhorizontal lamellae (Figure 12). Especially in the *Variscan internides*, i.e. the southern Saxothuringian and the Moldanubian, the lamellae of the lower crust are especially dense and strong, as seen in Figures 13, 14, and 15 [*Meissner and Bortfeld*,1990; *Bartelsen et al.*,1982].

The Variscan internides, which apparently suffered from a strong and long collapse period, and most probably was involved in a delamination process, have developed a large heat peak and a total rejuvenation of the lower crust and the uppermost mantle. This is the area of the most extensive development of lower crust lamellae worldwide. Former collisional roots and mantle reflections disappeared; extensional, mostly subhorizontal, stretching of inhomogeneous crustal (and some mantle) material finally led to the appearance of a shallow crust with its strongly reflecting lamellae. Only in young basins around the S and SW part of the British Isles [*BIRPS and ECORS,1986]* and in some extensional areas in the Western US [*Prussen,1991]*are there equally strong lamellae. The Variscan interior is also characterized by abundant syn-and post-orogenic granitic volcanism, one of the strongest concentration regionally, showing that the crust in this area was once extremely hot.

Other features of the Variscan internides are low (sialic) average velocities for more than 2/3 of the crust and a flat and shallow Moho (*Mooney and Meissner, 1992)*. The lower crust, which during compression must have had a mighty root, seems to be completely rejuvenated, the lamellae being created by some ordering process, partly by massive shear movement, the lower crust either delaminated or pushed outwards, the rigid upper crust subsiding. In the "discussion", a brief comparison with other extensional areas will be made.

The northern *Variscan externides* have a completely different reflectivity pattern. Reflectors, concentrating on various bands, some dipping and merging with outcrops of known surface faults, some diving into the upper mantle, especially toward the north, fill the whole crust. This

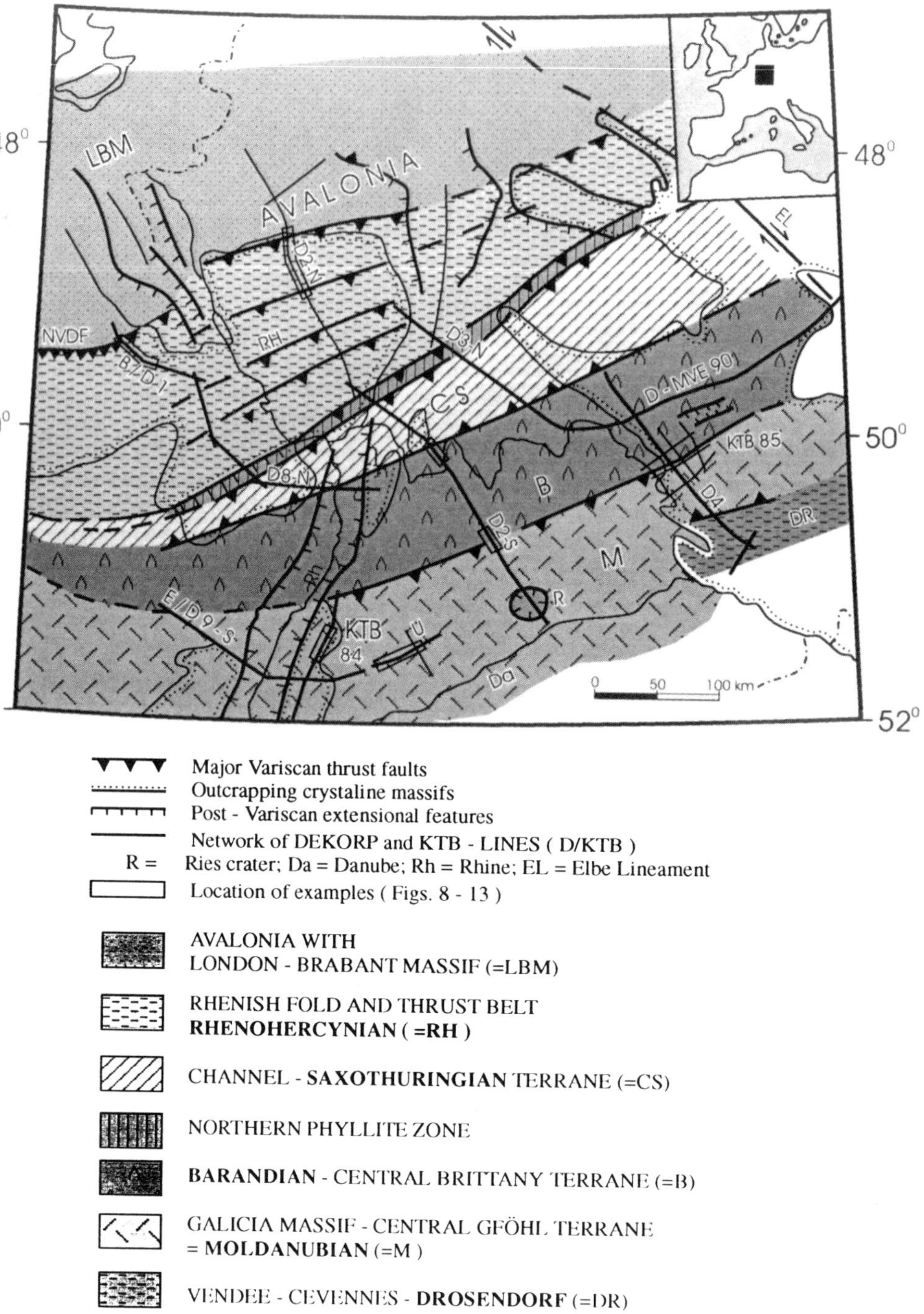

Figure 11. The DEKORP lines through theVariscan terranes.; after Franke et al. 1989. The location of the DEKORP examples in figures 8 - 13 are indicated.. LBM = London-Brabant Massif, RH = Rhenohercynian, CS= Crystalline High, B = Barandian, ST = Saxothurigian, MD = Moldanumbian.

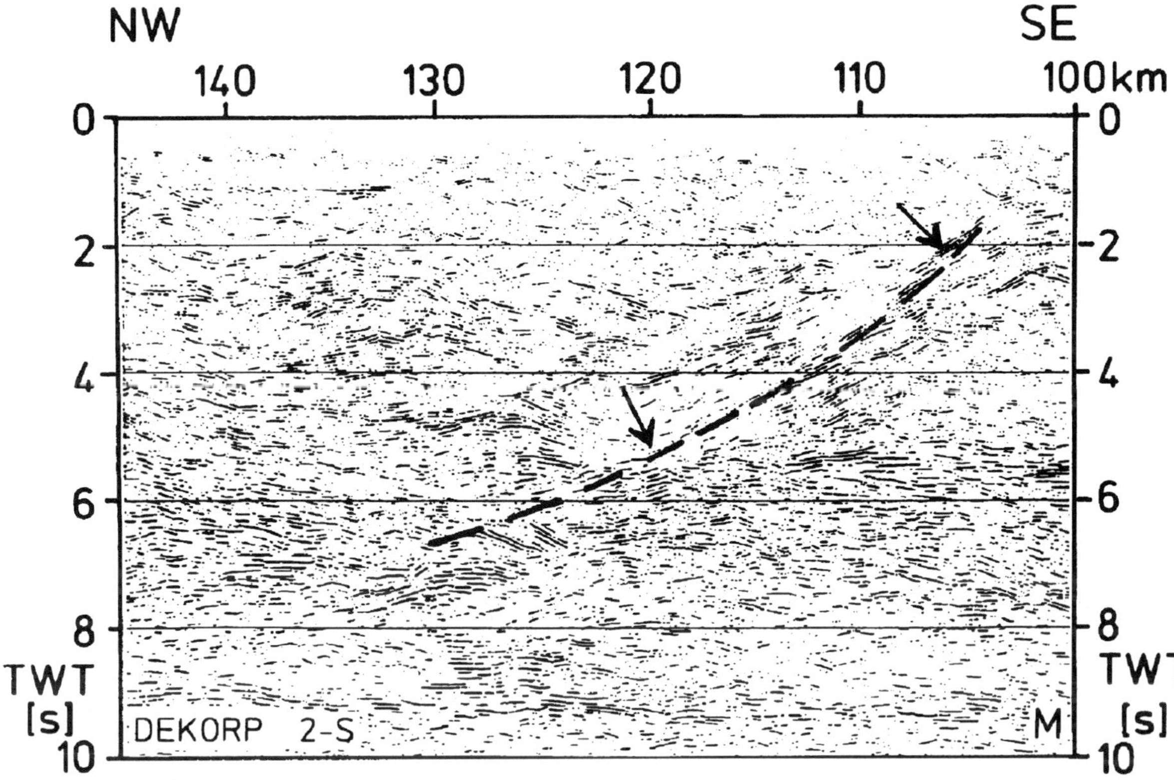

Figure 12. Example of a boundary fault inside the Saxothuringian (ST) terrane. Note the curved shape and its truncation by the lower crust lamellae around 6 s TWT; after Meissner, 1996.

concentration of strong reflections on large lineaments must represent thrust and shear zones, similar to the dense pattern of reflections of the Proterozoic terrane collision in SW Baltica. In the Rhenohercynian (RH) there were apparently several collisional phases, from the subduction of the Rheic ocean between Avalonia and Armorica and later from the compressional stresses of the late Variscan terrane collision. It should be noted that in the externides the observation of old thrusts and shear zones in the whole crust indicates that the crust never suffered from any strong post-orogenic heat pulse which would have destroyed the old compressional signature. Figure 16 shows the seismic profile DEKORP 2-N through the eastern part of the RH [*Franke et al.*,1990]. Here, the whole crust is full of thrust-and shear-zones (example, Satanella thrust in the north, and the Siegen thrust in the south). This area on the shoulders of Avalonia seems to have been reworked extensively by the Variscan collision. A similar reflectivity is observed along the western profile through the RH [*DEKORP Research Group*,1991] (profile Dekorp 1-A, Figure 17). In addition we observe at shallow depth the North Variscan Deformation Front (NVDF), a thin skinned overthrust, which can be followed for about 80 km to the SE, down to a depth of 15 km along the DEKORP 1-A profile. The track of the NVDF can be clearly observed for about 2000 km to the W by a number of deep seismic reflection surveys. It crosses France, Belgium, and the Irish Sea, and seems to be present even W of Ireland. In its course it assumes a certain curvature [*Meissner and Sadowiak*,1992]. In the Celtic Sea basins, S of Ireland, the NVDF is observed on top of strong lower crust lamellae (Figure 18) and here, clearly reveals the sequence of tectonic events: compression during the NVDF overthrust at about 300 Ma., crustal thinning and extension with the formation of lamellae during the Jurassic and Cretaceous extensional and separation phases [*Reston,1987*]. The NVDF and other Variscan boundaries have an arcuate shape, probably mainly derived from accretion and indentations of non-linear terrane boundaries; however, some of the curved lineation seem to stem from translations and rotations of upper crust material on top of a weak (fluid-like) lower crust. There are ample

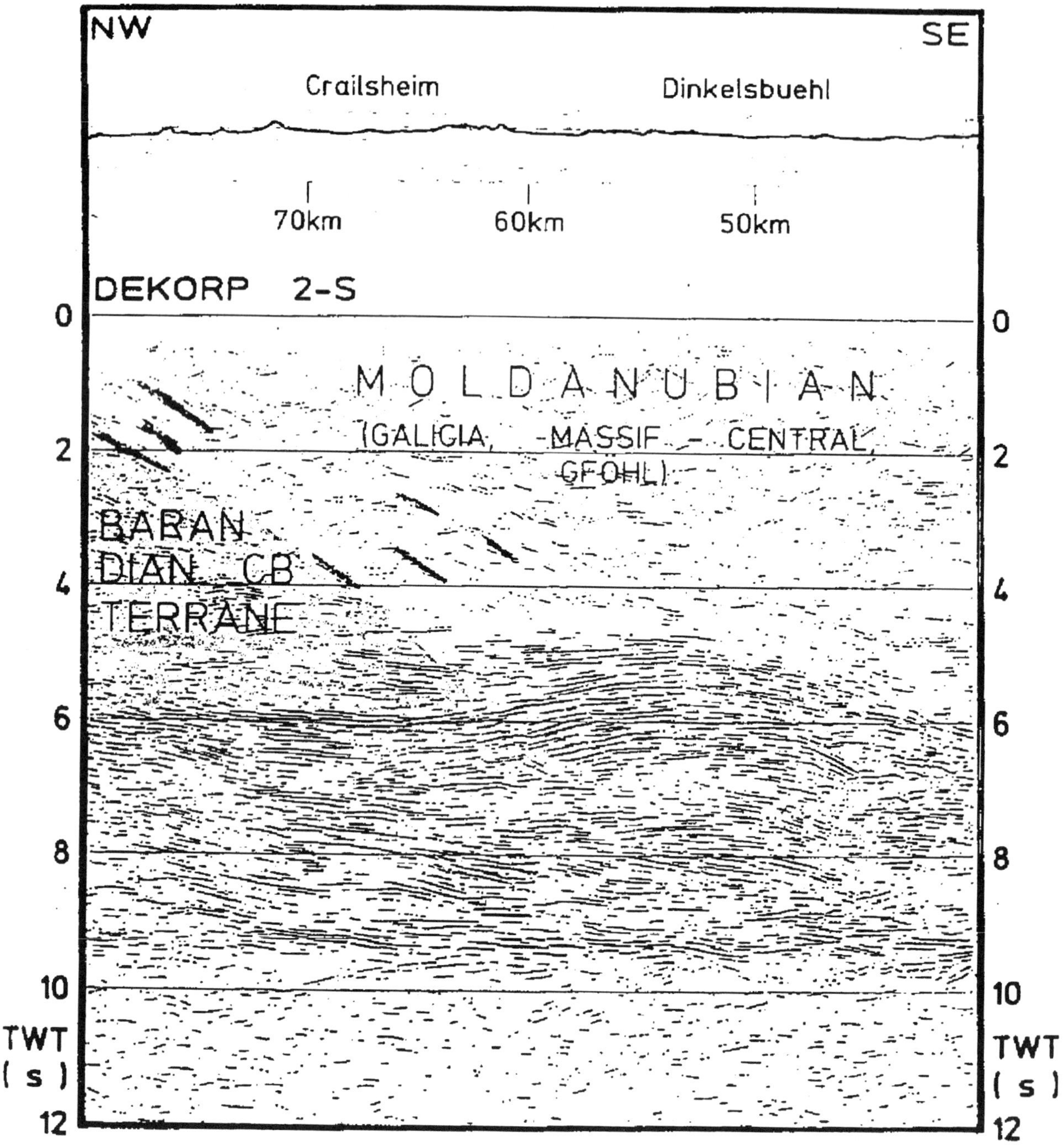

Figure 13. Automatic line drawing from a section of profile DEKORP 2-S in the Variscan internides; from DEKORP Atlas 1990 (Meissner and Bortfeld,eds.). Note strong lamellae in the lower crust and indications of a boundary fault in the upper crust.

lamellae in the lower crust of central and western Europe, especially in the extensional basins [*Meissner and Tanner*,1993].

5. THE ALPINE BELT

The Alps, south of the DEKORP network, and partly belonging to southern Europe, will not be treated in detail. The Alps have been studied by multinational and interdisciplinary working groups for more than a century [*Frei et al.*,1989; *Buness and Giese*,1990]. Toward the end of the Cretaceous, after about 100 Ma, convergence between Africa and Europe started again. The (African) terrane Adriatica became increasingly independent, rotated about 35° anticlockwise and collided with Europe. Tethyan shelf areas were included in the orogeny; they partly developed

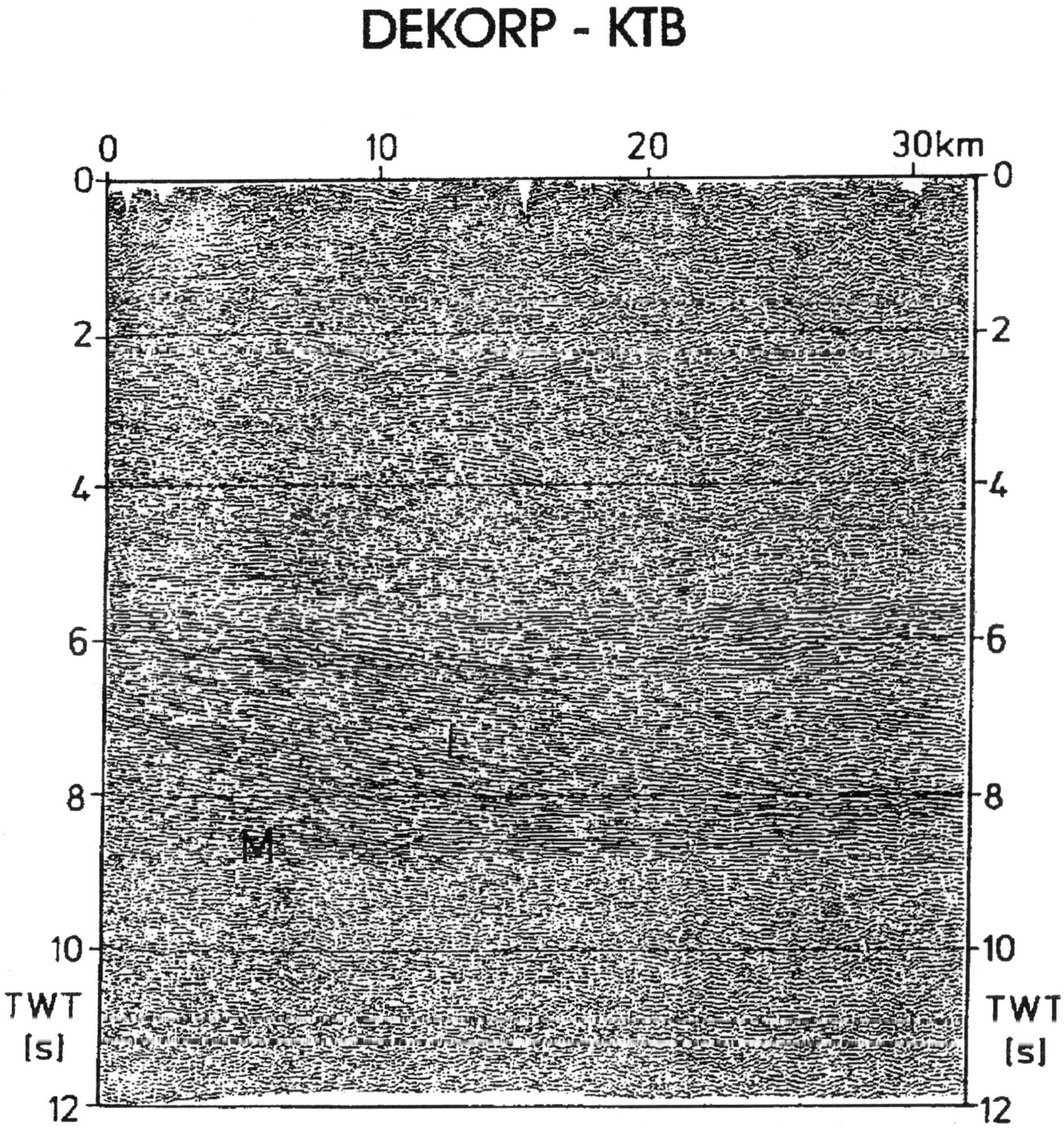

Figure 14. Automatic line drawing of a Black-Forest line in the Variscan internides; from DEKORP Atlas 1990 (Meissner and Bortfeld,eds.). The base of the strong lamellae in the lower crust corresponds to the Moho around 9 s TWT.

elongated basins which accumulated the eroded material and were later compressed and incorporated in the processes of crustal thickening and shortening. Former continental margins make up most of the Alpine nappes [*Heitzmann et al.*,1991].

A large number of refraction profiles, fan shooting, and wide-angle reflection surveys have been carried out for more than 25 years [*Buness and Giese*,1990]. These works have provided the velocity structure in the Alpine Belt nicely, including the subsuface of the Ivrea body with its strong gravity maximum, underlain by very high (mafic-ultramafic) velocities. Structural details were revealed by a great effort from the Swiss Working Group [*Pfiffner et al.*,1990; *Heitzmann et al.*,1991] and by the French-Italian ECORS-CROP working Group [*Mugnier et al.*,1991]. The northern part of the Alpine reflection lines generally show the south-dipping European Moho and lower crust down to about 55 km depth (Fig. 15). This downward dip, which seems to

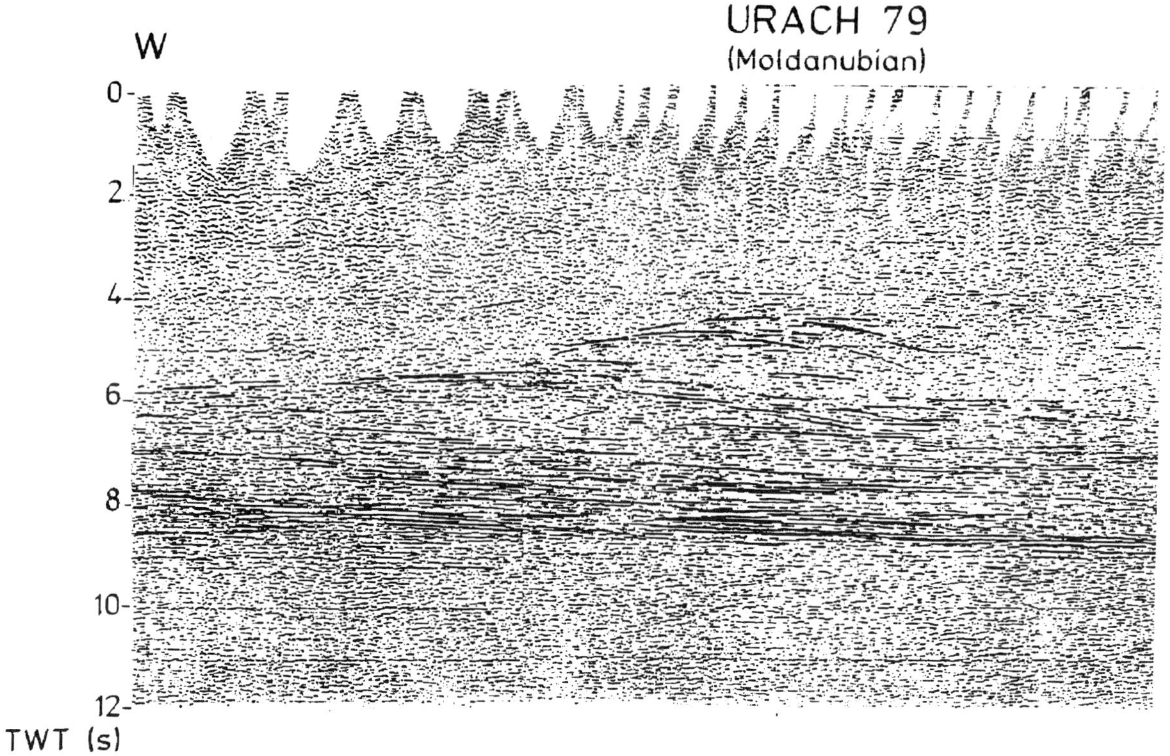

Figure 15. The 1979 Urach line 1, a pre-DEKORP profile across the Urach geothermal anomaly; after Bartelsen et al., 1982. Strong lamellae occurs rather shallow in the crust (6- 9 s TWT).

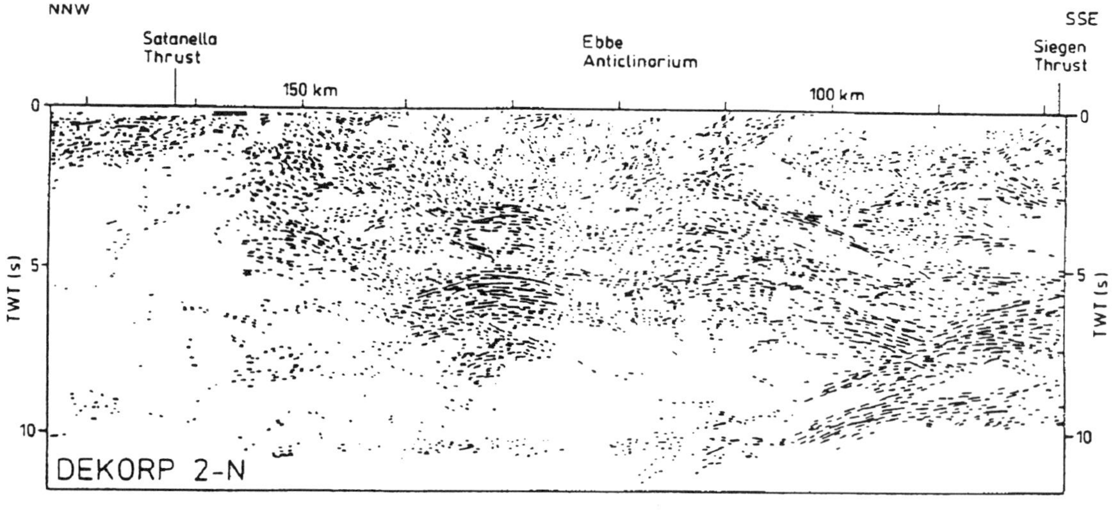

Figure 16. Line drawing of DEKORP 2-N through the Variscan externides (Rhenish Massif,RM), a Caledonian (Avalonian) massif, reworked by Variscan collisions; after DEKORP Research Group, 1991.

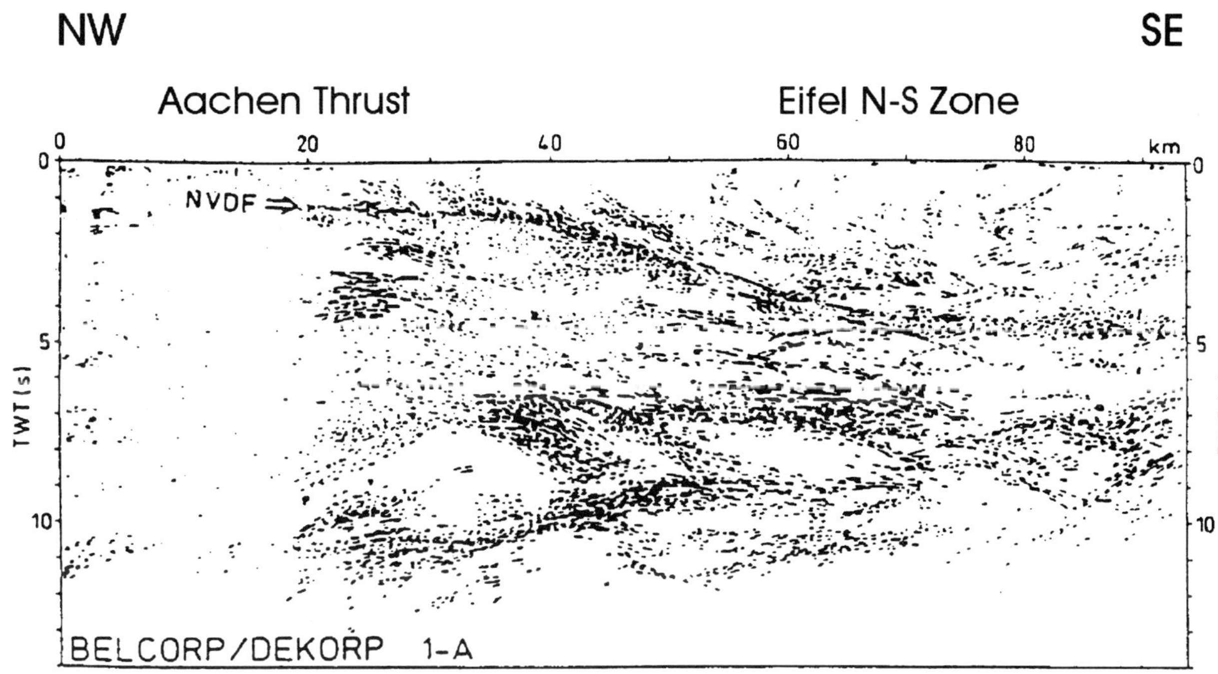

Figure 17. Line drawing of DEKORP1-A through the W- part of the Rhenish Massif; after DEKORP Research Group 1991. Note North Variscan Deformation Front (NVDF, ~ 300 Ma) (arrows), starting from a depth of about 15 km in the S, extending for about 2 000 km to the W.

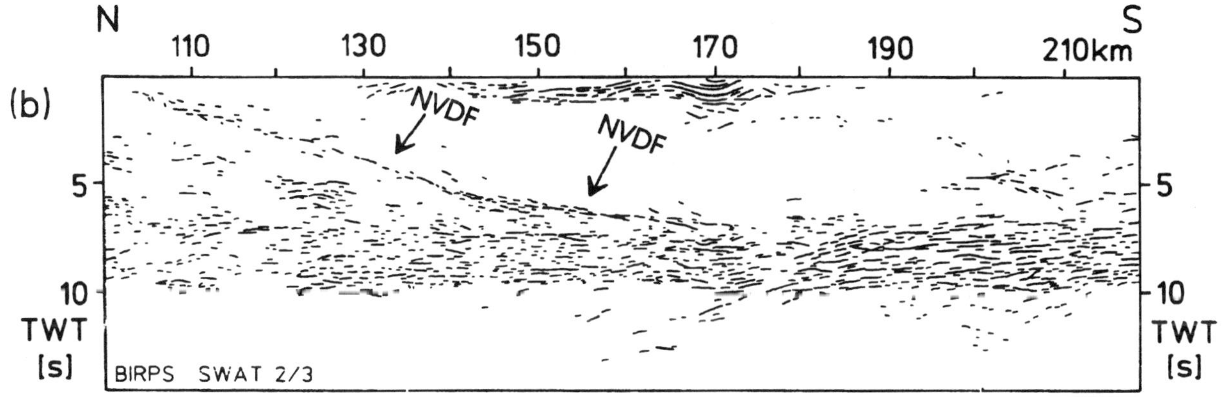

Figure 18. Line drawing of the North Variscan Deformation Front (NVDF) in the Celtic Sea Basin,S of Ireland, from BIRPS recordings; after Cheadle et al.,1987. Note NVDF originating from the top of the lower crust lamellae, and the Moho around 10 s TWT.

extend to greater depth, according to some seismological data, most probably represents the onset of delamination [*Meissner and Mooney*,1998]. Crustal shortening is not only restricted to the stacking of nappes near the surface or to the crustal thickening at depth. At mid-crustal depth, the (strong) Adriatic mantle plus some lower crust (Figure 19) intrude the weak European crust [*Pfiffner*,1990]. This indentation or "interfingering" in form of "crocodile" structures [*Meissner*,1989] is clearly observed and may play a very important role for compressional crustal shortening in general [*Beaumont et al,.1994]*. Another consequence of the compressional stress and the thickening and weakening of the

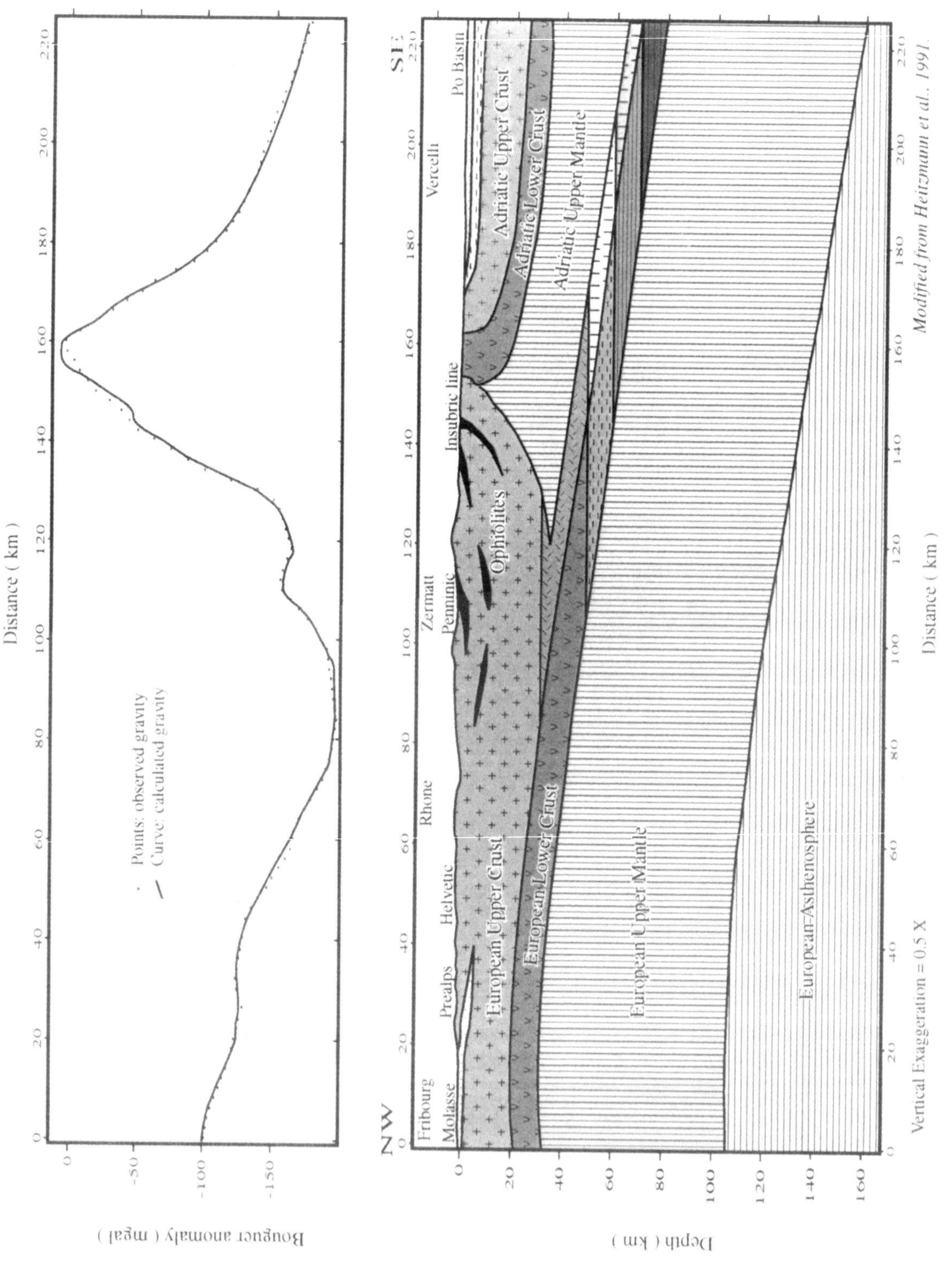

Figure 19. Conceptual tectonic model of the Swiss western Alpine profile, showing the Adriatic indenter; after seismic and gravity models from Heizsmann et al.,1991. The Adriatic indenter is also observed by other reflection profiles from Swiss and a French-Italian working groups.

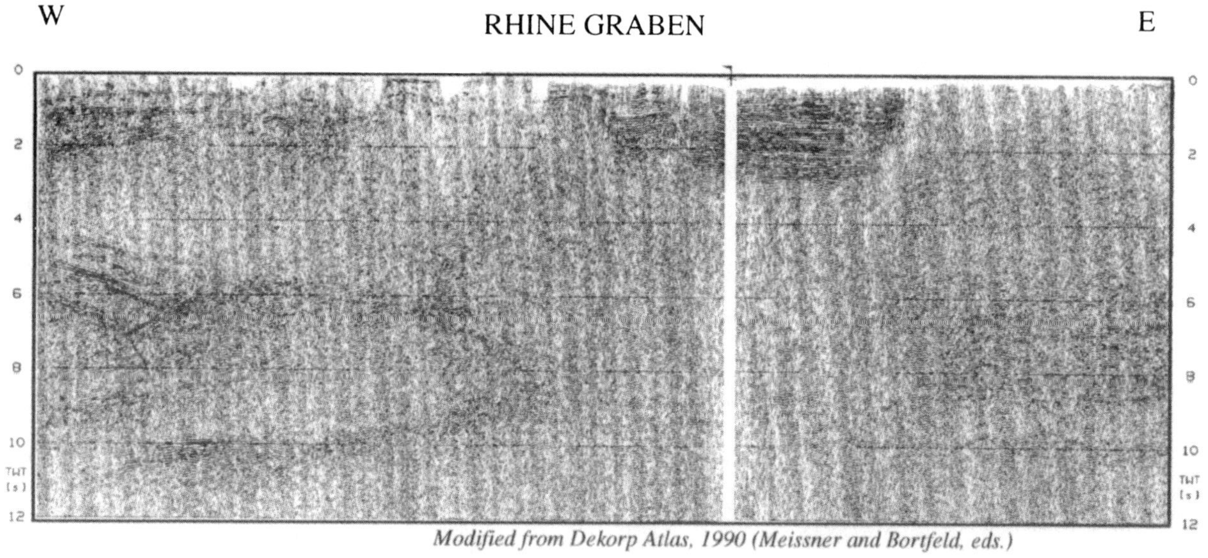

Figure 20. Deep seismic profile in the Rhine Graben, from DEKORP Atlas, 1990 (Meissner and Bortfeld, eds.)

crust is lateral escape, as studied for a number of high mountain belts [*Meissner and Mooney*,1998]. In case of the Alps, the escape is directed to the E, i.e. into the direction of the (weak and warm) Pannonian Basin.

Another consequence of the northward motion of the African plate relative to Europe is the extensional stresses developed perperdincularly to the motion vector. The east-west-directed escape vectors are associated with stresses responsible for the development of a number of Tertiary rifts, such as the Rhine Graben (Fig. 20 and Fig. 21).

6. DISCUSSION

From the comparison of certain reflectivity patterns in Central Europe with those in other countries with similar or contrasting geological and tectonic evolution, some general consequences emerged. Early ideas about the reflectivity were confirmed, whereas others were rejected on the basis of new data. As an example, before Lithoprobe in Canada and BABEL in northern Europe, the deep crusts of old shields and platforms were thought to be non-reflective, mainly based on the early COCORP data from the US. But the modern reflection surveys revealed strong crustal reflectivity, mostly in the whole crust in Proterozoic and Archean areas. These reflectors are not the long and strong lamellae of the lower crust, as found in Paleozoic and Mesozoic extensional areas, but still show much reflectivity, often concentrated on strong reflecting bands, probably representing shear zones. Even signs of very old subductions and sutures are observed in the uppermost mantle [*Warner and McGeary,1987; Mooney and Meissner*,1992]. These old cratons are rather cold and brittle up to the present day, as seen from their low heatflow values. The observed sporadic reflections in the uppermost mantle are supposed to be faults or shear zones of former sutures [*Snyder and Flack*,1990]. In such areas no strong heat pulse ever took place after the compressional period. This would have destroyed the old compressional features, similar to the rejuvenation of the lithosphere in the Variscan internides. Continuously cold environments seem to be the pre-condition for the preservation of mantle reflectors. [*Meissner*,1996]. On the other hand, cooling could have destroyed any former lamellae by transforming an early ductile environment of the lower crust into a brittle regime, and some stresses might be responsible for deforming or breaking long reflectors into smaller reflectors. It should also be mentioned that modern processing techniques more and more reveal some reflectivity in areas where earlier a "transparent" upper crust or upper mantle were reported [*Steentoft and Rabbel*,1995].

It is well known that reflectivity needs impedance contrasts ($\rho \cdot V$). In addition, deep reflectors must be rather plane in order to produce correlatable reflectors at the surface. The impedance contrast may originate from contrasting rock units,e.g. (1) sialic versus mafic material (or ultramafic splinters), (2) melt versus solid rocks, a transient phenomenon, (3) various metasediments with contrasting facies and anisotropy [*Rabbel et al.*,1998]. Reflection coefficients have been assessed or modelled by various groups (see for instance [*Warner and McGeary*,1987]), for a small crustal thickness with average spread factors and

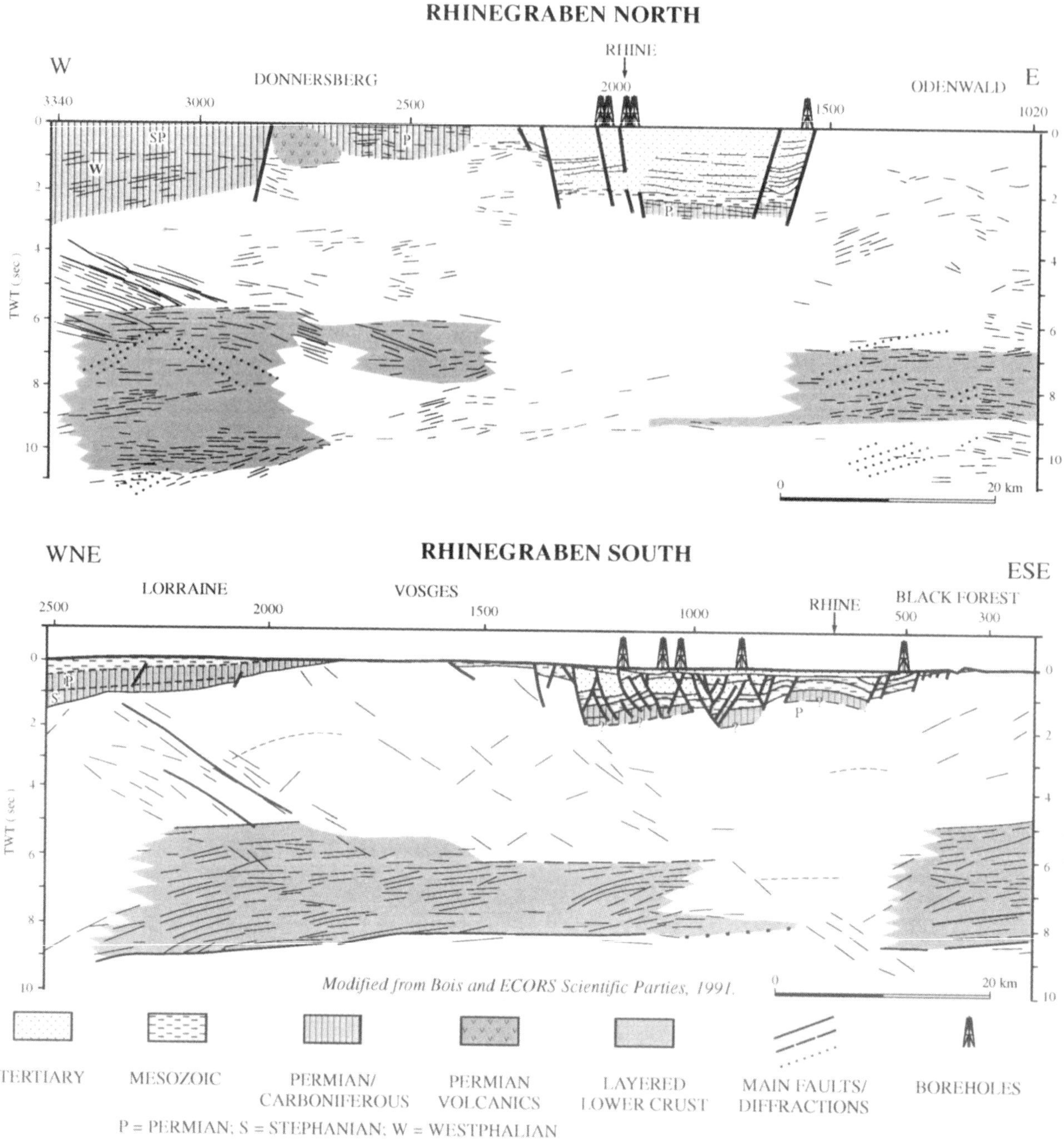

Figure 21. Schematic line interpretations of seismic profiles in the Rhine Graben (from Bois and ECORS Scientific Parties, 1991)

absorption. After *Warner*,[1990], the reflection coefficient R in the lower crust must be at least 0.05 and only seldom it is greater than 0.2. Figure 22 shows some values of the reflection coefficient R for vertical incidence versus the difference in velocity (and related density).

$$R = (1 - M)/(1+M)$$

$$\text{with } M = (\rho_{n-1} \cdot V_{n-1})/(\rho_n \cdot V_n);\ V_{n-1} - V_n = \Delta V$$

Please note that the reflection coefficients between 0.05 and 0.2 are more easily reached by low (sialic) velocities (e.g. 6.0 - 6.5 km/s or 5.5 - 6.5 km/s) than for higher (mafic) velocities. This preference for average sialic velocities in the lamellae of the lower crust is confirmed by observations. All strong lamellae, observed by DEKORP and pre-DEKORP lines through the Variscan internides have low average seismic velocities (Figure 23). This does not mean that mafic or ultramafic splinters in the lower crust are negligible. They might be responsible for especially strong lamellae (like in Urach, Figure 15), but the average velocity as a kind of matrix velocity is certainly more sialic than mafic. In those areas where simultaneous velocity observations by wide-angle studies were made (not BIRPS), the average velocity for various strong lamellae have all resulted in average velocities between 5.5 and 6.7 km/s. Figure 24 shows an example of near-vertical and wide-angle studies from a COCORP survey in the extensional part of the Cordillera [Prussen,1991].

The request for plane reflectors is more difficult to fulfill than that of impedance contrast. It is certainly not by pure chance that we see the lamellae only in the lower crust, i.e. in the zone of lowest crustal viscosity [*Rutter*,1998; *Burrow and Cloeting*,1997] and not in any other part of the lithosphere. This suggest that creep or flow processes (and certainly no rupture) play a major role for generating plane reflectors in the lower crust. It is well known that the lower crust is the weakest part of the lithosphere [*Rutter and Brodie*,1992]. It is especially weak in hot, mostly extensional, areas. Any subhorizontal tectonic stress will generate plane, subhorizontal flow and creep structures. The most important ingredient for the creation of lamellae must be this ordering process inside a weak matrix.

An important consequence of the laminated and plane layering of the lower crust is anisotropy. In a recent study [*Meissner et al., in preparation]* the connection between the formation of anisotropy and that of lamellae is explained by the same flow or creep processes. Observation of anisotropy in the lower crust is not easy and requires special

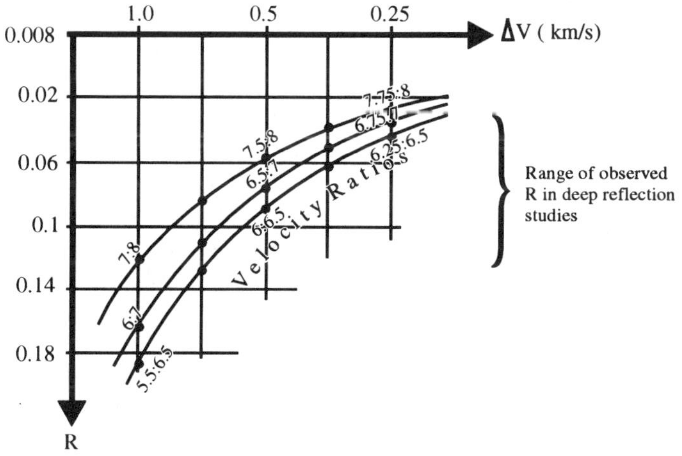

Figure 22. Reflection coefficient R as a function of the velocity contrast □V between crustal boundaries; after Meissner and Tanner, 1993. Note that big reflection coefficients are more easily reached by sialic velocity contrasts.

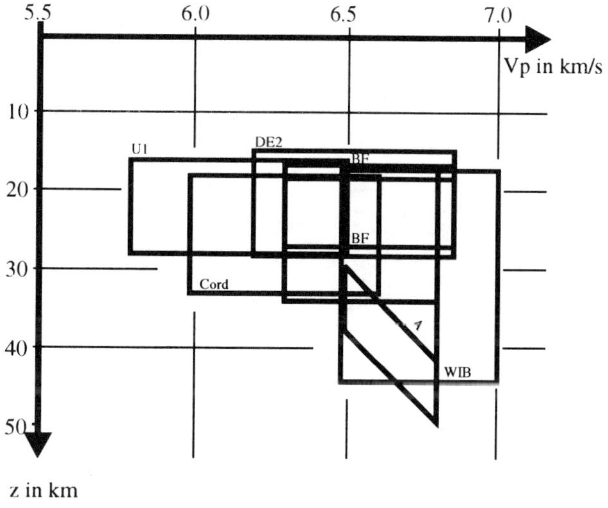

Figure 23. Average velocities of strong lamellae, compiled from coincident near vertical and wide-angle studies.

experiments, because only few raypaths will run considerable distances in the lower crust. A special wide-angle shear-wave experiment in the Urach area (Figure 15) with its strong lamellae has revealed an intrinsic, hexagonal shear-wave anisotropy [*Rabbel and Lüschen*,1996]. A theoretical approach [*Werner and Shapiro*,1998] shows that thin multilayering and an intrinsic anisotropy may both contribute to a more general anisotropic behaviour of the lower crust. Because in these studies the horizontal velocities are 3 - 7 % higher than the vertical velocities, one may state that all known velocitiy values of the lower crust, which are generally based on (subhorizontal) P_MP waves, provide the upper limit of true average velocities of laminated lower crusts. This means that the true matrix velocities are even lower and provides arguments for a sialic (upper and lower) crust in the Variscides.

Although nearly all Paleozoic and Mesozoic extensional basins have lamellae in their lower crust, there are exceptions, e.g. the North Sea and the North German Basin (NGB) [*DEKORP-BASIN Research Group,*1999]. These areas belong to the terrane Avalonia, and it seems that the interior of Avalonia is only poorly reflective in general, as also seen from reflection-poor zones in southern England and Belgium [*DEKORP Research Group,*1991]. The crust in various rims of Avalonia, i.e. in the western Baltic Sea (Figure 7) or the Rhenish Massif (Figures 12 and 13) is highly reflective, although reflectivity is concentrating on shear zones, which must have been generated by high plate-boundary stresses. These mainly compressive stresses were probably derived from the Variscan amalgamation of terranes and subduction of oceans between them. The interior of

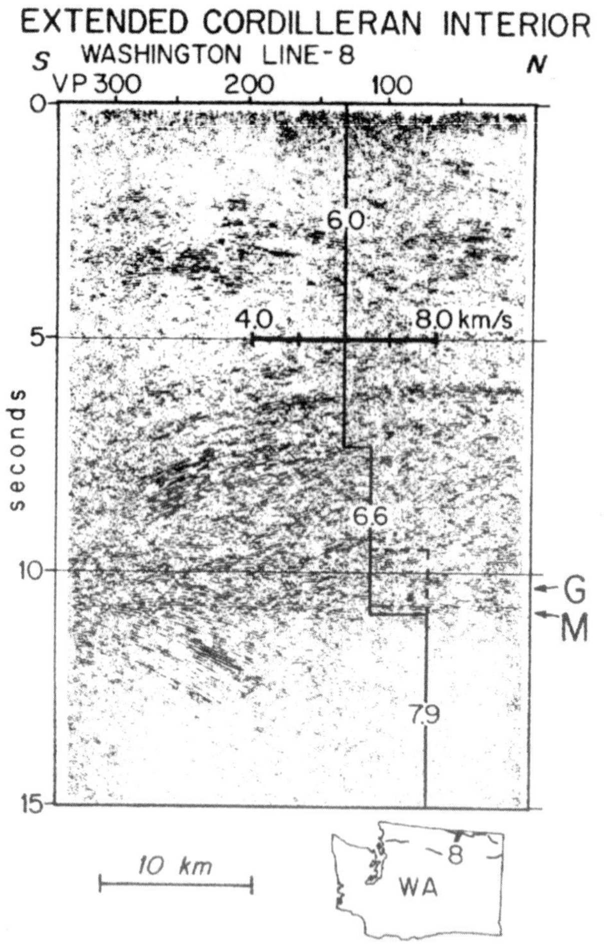

Figure 24. Example from a coincident near-vertical and wide-angle investigation from a COCORP survey through the extensional Cordilleran interior, after Prussen (1991). Note onset of lamellae on top of 6.6 km/s layer (Conrad discontinuity ?). M corresponds to Moho reflector at 11 s TWT, with seismic refraction velocity drastically increasing to 7.9 km/s. G correspond to a lower crust reflector slightly above the Moho.

Avalonia was apparently rather free of stresses. When the subsidence of the North Sea and the NGB started it was probably initiated by massive Permo-Carboniferous magmatism, not by extensional tectonics. As mentioned before, the intruding mafic melts have often replaced the former (Avalonian) lower and middle crust. They were so uniform, that only 2 or 4 km of lamellation were created.

The reduction of the normal thickness of the crystalline crust from 30 to 20 km in many basins and margins is still a mystery. The classical McKenzie stretching model might apply [*McKenzie*, 1978], as in many other basins, but evidence for a massive extension in the NGB is missing. If parts of the former middle and lower crust were expelled to the sides (the only plausible possibility), all its traces were destroyed by the massive intrusions. The missing Moho updoming today is compatible with the second phase of the McKenzie model [*McKenzie*, 1978] i.e., by an isostatic compensation by transporting hot asthenospheric material into the crust, deepening a former shallow Moho again. Today, the rejuvenated Moho is completely flat and continuous, although at some locations only very weekly reflecting.The unusually high velocities in the lower crust, as seen along many passive volcanic margins, must have reduced the impedance contrasts and produced small reflection coefficients at the Moho. The required contrasts for generating strong and ample lamellae are missing (Figure 23) in the NGB and in the North Sea, explaining also the absence of pronounced lamellae in other volcanic margins and basins. In the limited thickness of many rifted shelf areas temperatures were never high enough to create ductile creep and ordering structures.

The main basin-forming process, hence, seems to be thermal relaxation of the massive intrusions caused by an underlying plume or otherwise updoming astenosphere. This process is also part of the McKenzie model. Subsidence of the NGB continued until today, sometimes interrupted by inversions, e.g. in the Upper Cretaceous, sometimes accelerating with the forming of Mesozoic halfgrabens [*Bayer et al.*,1995]. With the opening of the North Atlantic at 65 Ma, which was preceded by a large number of continental rifts, the North Sea and the NGB became new passive margins.

In contrast to the very limited lower crustal lamellae in the NGB, lamellae in the interior of the Variscan belts are generally 12 to 15 km thick. This applies also to the Paris Basin, to the various Celtic Sea basins S of Ireland [*Matthews and the BIRPS Group*, 1990] and to several other basins outside Europe. The Variscan internides with their massive granitic volcanism, their mainly sialic velocities in more than two thirds of the crust and even inside the strong lamellae, are the result of enormous changes during the collapse period. The former crustal roots disappeared completely, most probably by a delamination process, because an over-thickened crusts at the end of a compressional phase is especially sensible toward delamination when several instabilities develop in the lithosphere [*Meissner and Mooney*,1998]. During the post-orogenic extensional period, huge strike-slip faults developed, probably induced by the westward movement of Gondwana in the south, and many rifts, grabens, and halfgrabens were formed during this period. Other extensional grabens, like the Rhinegraben and the Egergraben, were probably the result of renewed compressive stress from the South (with an E-W extension), during the formation of the Alpine belt in the Late Cretaceous and the Tertiary.

North of the Variscan internides, the externides show a completely different pattern of reflectivity. The Variscan externides belong to the southern rim of Avalonia, which came again under plate boundary stresses and massive deformation during the Variscan amalgamation of new terranes. Large and strong thrust and shear zones are observed in the whole crust, an indication that here no thermal pulse has destroyed or modified the compressive reflecting signature.The whole crust stayed cold and rigid after the southern terrane accretion.

7. CONCLUSIONS

Crossing central Europe from N to S we observe various structures and various remnants of former collisions, with terranes generally getting younger towards the S. There are many indications that, in addition to the necessary compressive stress, the thermo-rheological structure of the colliding units has played a dominant role for crustal thickening and collapse. The Proterozoic terrane collision NE of Bornholm still shows an enormous crustal thickening (from SW to NE), and some of the strongest crustal reflectivities worldwide. It is distributed in the whole (cold) crust along diverging reflectors, and it is supposed to consist of thrust-and shear zones (Figure 6) around one or several indenters, slightly similar to the strong reflectivity in the Rhenish Massif (Figure 16). The (hard) indenter from the NE apparently penetrated the slightly younger (and weaker) lower crust of the Baltic rim. The Caledonian Deformation Front (CDF) in the SW Baltic Sea is the next collisional structure. It shows a clear bivergent collision (Figure 7). Also here the lower crust shows (weak) decoupling properties, dividing the SW-dipping reflectors (the thrusts in the upper crust) from the NE-dipping reflectors in the upper mantle. The older (and colder) rim of Baltica penetrated the younger (and weaker) lower crust of Avalonia, a process similar to the compressional wedge models of *Beaumont et al.*[1994].

In contrast to these collisional structures, which were not destroyed by a later heat pulse, the Variscan orogeny suffered from a strong syn-and post-collisional collapse, having destroyed all signs of former crustal thickening, of indenters or shear zones in the lower crust and upper mantle. Only the (rigid) upper crust shows indications of thrusts and terrane boundaries. The lower crust developed the thick and highly reflecting seismic lamellae by post-orogenic shear and ordering processes. Here, the strong heat pulse was probably generated by a delamination of uppermost mantle and an over-thickened crustal root. The subsiding units mobilized the asthenosphere leading to the plume-like heating of the Variscan internides. The Alpine belt has not yet collapsed, and we can observe crustal thickening and shortening in situ, checking the thermo-rheological concept. Here the hard and rather shallow Adriatic mantle plus some mafic lower crust from the S act as a massif indenter which penetrates the weak and thick European crust. While nappes from the former continental margins dominate the surface structure of the Alpine belt, the lower and middle crust is suffering from the indentation and reacts by uplift and crustal thickening. The indenters` boundaries, i.e. strong shear zones, are seismically well observed by strong and continuous reflectors. Traces of continental margins are involved in all collisional processes; they may act as lubricating material and are often observed as nappes, the best evidence certainly provided by the youngest orogeny, the Alpine belt.

Seventy percent of today's continental margins on both sides of the North Atlantic are volcanic and influenced by the Tertiary Icelandic plume. Crustal thickness is sometimes only 20 km in the non-volcanic sections, but slightly thicker for the areas with massive magmatic intrusions, similar to the influence of Mesozoic magmatism in the North German and the Danish basins. In today's European continental margins only very thin seismic lamellae are observed in the lower crust. This can be explained by the formation processes.The volcanic margins were formed by massive intrusions of homogeneous, dense, and high-impedance mafic magmas. The rifted margins have a very small crustal thickness which has not reached the necessary temperature for generating creep processes. Compared to old shield areas in Canada, Fennoscandia or Russia, crustal thickness in Western and Central Europe is rather small and generally more sialic. Mesozoic collapse and extensional processes seem to be responsible for these observations. These processes also produced large areas of subsidence and basin developments which allowed many marine transgressions. All these Mesozoic and Cenozoic developments seem to justify the term "continental margin" for most parts of central, western, and even southern, Europe.

Acknowledgements. Thanks are due to my colleagues from the DEKORP Research Group, from the seismology branch of the USGS in Menlo Park, and the Institute of Geoscience in Kiel for helpful corrections and to Dr. Webster U. Mohriak for his steady encouragement. Sincere thanks are also due to J. Gallart and P. Szatmari for critically reading the first draft of the manuscript.

REFERENCES

Abramovitz, T., Berthelsen, A., and Thybo, H., Proterozoic sutures and terranes in the southwestern Baltic Shield, interpreted from BABEL deep seismic data, *Tectonophysics,* 270, pp. 259-278, 1997.

BABEL Working Group, Deep seismic images from the Tornquist Zone beneath the Southern Baltic Sea, *Geophys. Res. Lett.*, 16, pp. 1091-1094, 1991.

BABEL Working Group, Deep seismic reflection/refracton interpretation of crustal structure along BABEL profiles A and B in the southern Baltic Sea, *Geophys. J. Int.*, 112, pp. 325-343, 1993.

Bartelsen, H., Lueschen, E., Krey, Th., Meissner, R., and Schmoll, R., The combined seismic reflectin - refraction investigation of the Urach geothermal anomaly, in *The Urach Geothermal Project,* edited by R. Hänel, pp 247-262, Schweizerbart, Stuttgart, 1982.

Bayer, U., Scheck, M., and Hauf, Y., Structural evolution of the Northeast German Basin during Early Permian to Keuper-Volumetric analysis of sediment accumulation, *Terra Nostra,* 95/8, pp. 1-80, 1995.

Beaumont,C.,Fullsack,P.,Hamilton,J., Styles of crustal deformation in compressional orogens caused by subduction of the underlying lithosphere, *Tectonophysics 232,*pp. 119-132.

Berthelsen, A., Precambrian to Variscan Europe, in *The European Geotraverse,* edited by D. Blundell, R. Freeman and S. Mueller, pp. 153-163, Cambridge University Press, 1992.

BIRPS and ECORS, Deep seismic reflection profiling between England, Ireland, and France (SWAT), *J. Geol. Soc. London,* 143, pp. 45-52, 1986.

Blundell, D., Freeman, R., and Mueller, S., eds., A continent revealed, in *The European Geotraverse,* pp. 1-272, Cambridge University Press, 1992.

Bois, C. and ECORS Scientific Parties, 1991. Post-orogenic evolution of the European crust studied from Ecors deep seismic profiles. In: R. Meissner, L. Brown, H.J. Durbaum, W. Franke, K. Fuchs and F. Seifert (eds.), *Continental Lithosphere: deep seismic reflections*, AGU Geodynamic Series Volume 22, p. 59-68.

Buness, H., and Giese, P., A crustal section through the northwestern Adriatic plate, in *The European Geotraverse,* edited by R. Freeman, P. Giese, and S. Mueller, pp. 245-254, ESF, Strasbourg, France, 1990.

Burow, E., and Cloeting, S., Erosion and Rift Dynamics: new thermo-mechanical aspects of post rift evolution of extensional basins, *Earth and Planet Sci.Lett.*, pp. 7-26, 1997.

Cheadle, M.J., McGeary, S., Warner, M.R., and Matthews, D.H., Extensional structures of the western UK continental shelf, in *Continental Extensional Tectonics,* Geol. Soc. London Spec. Publ. 28, pp. 445-465, 1987.

Deeks, N.R., and Thomas, S.A., Basin inversion in a strike slip regime: the Tornquist Zone, southern Baltic Sea, in *Basin Inversion,* Geol. Soc. Spec. Publ. 88, edited by B. Buchanan and C. Buchanan, pp. 319-338, 1995.

DEKORP *Research* Group, Results from DEKORP 1 deep seismic reflection studies in the western part of the Rheinish Massif, *Geophys. J. Int.*, 106, pp. 203-227, 1991.

DEKORP-BASIN *Research* Group, The North German Basin and its development, submitted to *Pure Appl. Geophys.*, 1999.

Dekorp – Basin Research Group, Deep crustal structures of the Northeastern German Basin: new Dekorp-Basin ' 96 deep-profiling results. *Geology*, v. 27, p. 55-58, 1999.

Dohr, G., Lukic, P. and Bachmann, G., Deep crustal reflections in the Northwest German Basin, in *Seismic Expression of Crustal Styles,* Am. Assoc. Pet. Geol. Studies, Geol. Ser. 15, edited by A.W. Bally, pp. 1-5, 1983.

Franke, W., and 15 co-authors, Crustal structure of the Rhenish Massif: Results of deep seismic reflection lines DEKORP 2-North and DEKORP 2-North-Q, *Geol. Rundschau*, 79, pp. 523-566, 1990.

Franke, W., and Oncken, O., Zur prädevonischen Geschichte des Rhenohercynischen Beckens, Nova Acta Leopoldina, NF 71, Nr. 291, pp. 53-72, 1995.

Frei, W., Heitzmann, P., Lehrer, P., Mueller, S., Olivier, R., Ofiffner, A., Steck, A., and Valasek, P., Geotraverses across the Swiss Alps, *Nature*, 340, pp, 544-548, 1989.

Gebauer, D., Williams, J., Compston, W. and Grünenfeldes, The development of the central European continental crust since the early archean base on conventional ion-microprobe dating, *Tectonophysics,* 157, pp. 81-97, 1989.

Heitzmann, P., Frei, W., Lehrer, P., and Valasek, P., Crustal indentation in the Alps - an overview of reflection seismic profiling in Switzerland, in *Continental Lithosphere*, AGU Geodyn. Ser. 22, pp. 161-176, AGU, Washington DC, 1991.

Klemperer, S.L. and Mathews, D.H., Iapetus suture located beneath the North Sea by BIRPS deep seismic reflection profiling, *Geology*, 15, pp. 195-198, 1987.

Matte, P., Accretionary history and crustal evolution for the Variscan belt of Europe, *Tectonophysics*, 126, pp. 329-374, 1991.

Matthews, D.H. and the BIRPS Group, Progress in BIRPS deep seismic profiling around the British Isles, *Tectonophysics*, 173, pp. 387-396, 1990.

McKerrow, W.S., Dewey, J.F. and Scotese, C.R., The Ordovician and Silurian development of the Iapetus Ocean, *Spec. Papers in Peleontology*, 44, pp. 165-178, 1991.

McKenzie, D., Some remarks on the development of sedimentary basins, *Earth Planet Science Lett.*, 40, pp. 25-36,1978.

Meissner, R., Faults and folds, fact and fiction, *Tectonophysics*, 264, pp. 279-293, 1996.

Meissner, R., Rupture, creep, lamellae and crocodiles: happenings in the continental crust, *Terra Nova*, 1:89, pp. 17-28, 1989.

Meissner, R., and Bortfeld, R.K., eds., *The DEKORP Atlas*, Springer Verlag, Berlin, seismic sections 1-80, 1990.

Meissner, R., and Mooney, W.D., Weakness of the lower continental crust: a condition for delamination, uplift and escape, *Tectonophysics*, 296, pp 47-60, 1998.

Meissner, R., and Rabbel, W., Nature of crustal reflectivity along the DEKORP profiles in Germany as compared to reflection patterns in different tectonic units, accepted for special DEKORP volume, *Pure and Applied Geophysics* ,1999.

Meissner, R., and Sadowiak, P., The terrane concept and its manifestation by deep reflection studies in the Variscides, *Terra Nova*, 5, pp. 598-607, 1992.

Meissner, R., and Tanner, B., From collision to collapse: phases of lithospheric evolution as monitored by seismic records, *Phys. Earth Planet Interiors*, 79, pp. 75-86, 1993.

Mooney, W.D. and Meissner, R., Multi-genetic origin of crustal reflectivity: a review of seismic reflection profiling of the continental lower crust and Moho, in *Continental Lower Crust*, edited by D.M. Fountain, R. Arculus and R.W. Kay, pp. 45-79, Elsevier, Amsterdam, 1992.

Mugnier, J.L., Guellec, S., Menard, G., Roure, F., Tardy, M,. and Vialon, P., A crustal scale balances cross section through the external Alps, deduced from the ECORA profile, in *Deep Structure Structure of the Alps,* edited by F. Roure, P. Heitzmann and R. Polino, Mem. Soc. Geol. France, 156, pp. 203-216, 1990.

Pfiffner, O. A., Crustal shortening of the Alps along the EGT profile, *6th ed. EDT Workshop,* edited by Freeman and St. Mueller, pp. 255-262, ESF, Strasbourg, France, 1991.

Potter, C. J., The deep structure of American accreted terranes from deep seismic reflection data, in *Deep Structure and Past Kinematics of Accreted Terranes,* edited by J.W. Hillhouse, pp. 203-208, 1989.

Prussen, E., The reflection Moho along the COCORP Northwest U.S. Transect, in *Continental Lithosphere: Deep Seismic Reflections,* AGU Geodyn. Ser. 22, pp. 315-322, AGU, Washington DC, 1991.

Rabbel, W., Förste, K., Schulze, A., Bittner, R., Röhl, J., and Reichert, J. C., A high velocity layer in the lower crust of the North German Basin, *Terra Nova,* 7, pp. 327-337, 1994.

Rabbel, W., and Lüschen, E., Shear wave anisotropy of laminated lower crust at the Urach geothermal anomaly, *Tectonophysics,* 264, pp. 209-233, 1996.

Rabbel,W.,Siegesmund,S.,Weiss,T., Pohl,M., Bohlen,T., Shear wave anisotropy of laminated lower crust beneath Urach (SW Germany): a comparison with xenoliths and with exposed lower crustal sections, *Tectonophysics,298,*pp 337-356,1998.

Reston, T. J., Spatial interference, reflection character and the structure of the lower crust under extension. Results from 2-D seismic modelling, *Am. Geophys. Ser. B**, pp. 339-348, 1987.

Rutter, E.H., On the relationship between the formation of shear zones and the form of the flow law for rocks, submitted to *Tectonophysics,* 1997.

Rutter, E.H. and Brodie, K.H., The rheology of the lower continental crust, in *Continental Lower Crust,* Development in Geodynamics Ser., edited by D. Fountain, R. Arculus, R. Kay, pp. 201-268, Elsevier, 1992.

Snyder, D.B. and Flack, C.A., A Caledonian age for reflections within the mantle lithosphere north and west of Scotland, *Tectonics*, 9, pp. 439-482, 1990.

Steentoft, H., and Rabbel, W., Conflicting dips and the transparency of the upper crust, *Terra Nostra,* 95(18), pp. 127-130, 1995.

Stille, H., Grundfragen der vergleichenden Tektonik, Gebr. Bornträger, Berlin, 1924.

Tait, J.A., Bachtadse, V., Soffel, H.C., and Franke, W., Paleomagnetic constraints on the evolution of the European Variscan foldbelt, *Geologia Bavaria,* 101, pp. 221-232, Munich, 1996.

Torsvik, T.H., and Trench, A. , The Ordovician history of the Iapetus Ocean: new paleomagnetic constraints, *J.Geol.Soc. London*, 148, pp.423-425, 1991.

Warner, M., Basalt, water or shear zones in the lower continental crust, in *Seismic Probing of Continents and their Margin,* edited by J.M. Leven, D.M. Finlayson, C. Wright, J.C. Dooley, and B.L.N. Kennett, *Tectonophysics,* 173, pp. 163-174, 1990.

Warner, M. and McGeary, S., Seismic reflection coefficients from mantle fault zones, in *Geophys. J. R. Astron. Soc.**, 89, pp. 223-230, 1987.

Warner, M. R., Absolute reflection coefficients from deep seismic reflections, *Tectonophysics,* 173, pp. 15-23, 1990.

Wegener, A., Die Entstehung der Kontinente und Ozeane, 4th ed., *Sammlung Wissenschaft,* Bd. 66, Braunschweig, 1924.

Werner, U., and Shapiro, S.A., Intrinsic anisotropy and thin multilayering - two anisotropy effects combined, *Geophys. J. Int.,* 132, pp. 363-373, 1998.

Ziegler P.A., Geological atlas of western and central Europe, 2nd ed., Shell Int. Petroleum, Maatschappij, B. V., Geol. Soc. London, Elsevier, Amsterdam, 1990.

Deep Structure of North West Europe from Deep Seismic Profiling: The Link Between Basement Tectonics and Basin Development

Richard W. England

Department of Earth Sciences, Bullard Laboratories, University of Cambridge, UK.

BIRPS has acquired an extensive grid of offshore deep seismic reflection profiles over the North Sea and continental margins of North West Europe. These data can be used to map the deep expression of the major rift structures and relate the shallow rift structure to faults penetrating deep into the crust. In many cases these deep structures have an extended history of reactivation. On the North Atlantic margins it is possible to demonstrate that some of the Mesozoic basins are developed as a result of extension on normal faults that were first active during Vendian and Riphean rifting events, preceding the opening of the Iapetus Ocean. There is considerable variation in the style of rifting along the strike of the North Sea rift system. In the southern North Sea the structure of the Central Graben varies with seismic properties and thickness of the basement. Where it crosses the Mid-North-Sea – Ringkøbing-Fyn High the graben is narrow and formed by a displacement on a single large fault, in contrast to the wide rift formed by displacement on a number of smaller faults in adjacent areas. The broad distribution of sediments deposited during the thermal subsidence which followed rifting reflects deformation within the mantle lithosphere and does not reflect variations in the style of deformation of the basement. This suggests that the variations in style of faulting within the basement are due to variations in the properties of the crust and not the mantle lithosphere.

1. INTRODUCTION

Unequivocal examples of renewed movement on existing structures during deformation of the continental crust are rare, yet this process, otherwise referred to as reactivation, is a recurring theme in studies describing the geology of sedimentary basins. Most examples of reactivation describe movement on shallow structures but by implication the process must involve deformation of the deep crust and the mantle lithosphere. Deep seismic profiling is the principal method of mapping the downward continuation of near surface structures to the lower crust and mantle. The distribution and preservation of sediments within sediment-ary basins and continental margins is strongly dependent upon the distribution of strain throughout the lithosphere. Consequently it is important to understand the role of existing structures and properties of the crust and mantle in order to better understand the structure and evolution of sedimentary basins.

During the last sixteen years the BIRPS programme has built up an extensive grid of marine deep seismic reflection profiles over the shallow continental shelf of northwest Europe [*Klemperer and Hobbs*, 1992] (Fig. 1). The quality of the data is uniformly high and while development of acquisition and processing methods have resulted in a gradual improvement in data quality, older sections compare favourably with those acquired most recently. Since the data are generally processed in the same way, direct comparisons between different datasets is possible. To date, not all the data have been migrated and depth converted. The acquisition of a limited number of wide-angle/refraction datasets coincident with some of the profiles have provided good constraints on crustal thickness and velocity structure [*Barton and Wood*, 1984; *Singh et al.*, 1998], but interpretation of most of the profiles can only be based on reflection data [*Chadwick and Pharaoh*, 1999]. The acquisition of this grid of data now makes it possible to obtain a picture of the variations in basement geology across the continental shelf. Although still very widely spaced, compared with grids of data used

Atlantic Rifts and Continental Margins
Geophysical Monograph 115

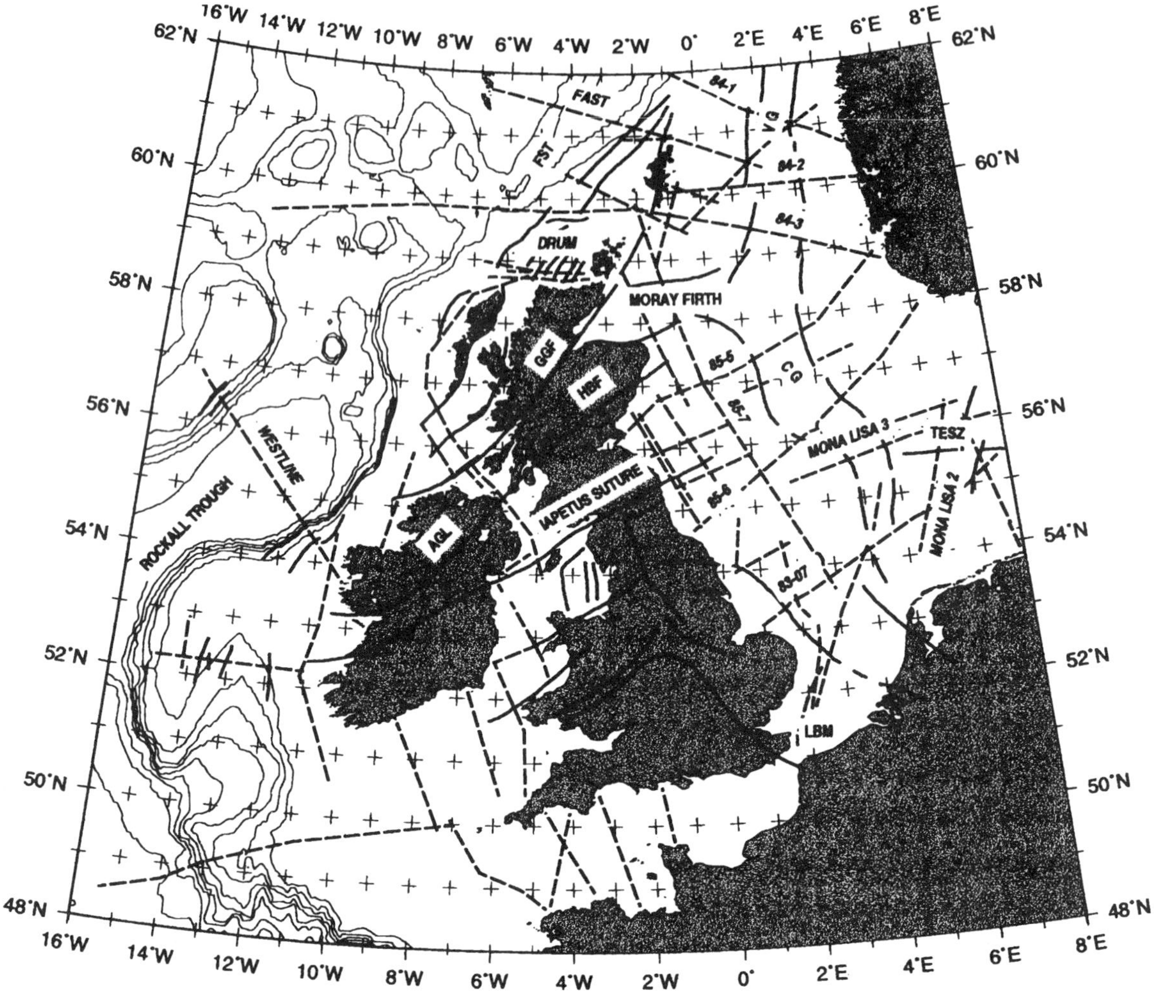

Figure 1. Map of the continental shelf of NW Europe, showing location of the BIRPS and other deep seismic profiles as dashed lines. Land areas are shown grey and bathymetry of offshore areas is contoured at 500 m intervals. Those profiles shown in Figs. 6, 7, 9, and 10 are labelled in itallics. Thick solid lines indicate the strike of major basement structure. AGL-Antrim galway line, CG-Central Graben, FST-Faroe–Shetland Trough, GGF-Great Glen Fault, HBF-Highland Boundary Fault, LBM-London Brabant Massif, TESZ-Trans-European Suture Zone, VG-Viking Graben.

by the hydrocarbons industry for regional stratigraphic studies, the lines were in many cases located to provide representative images of the deep subsurface structure in the regions they crossed. Other tie lines constrain the location of changes in structure. Thus, the grid of data provides a representative view of variations in basement structure and a broad view of the geometry of the overlying sedimentary basins. This contribution examines the role of the deep structure in defining the development of sedimentary basins and the continental margin of northwestern Europe. While this subject has been addressed in many previous studies of BIRPS data, such studies have mainly focussed on single datasets [e.g. SWAT: *Cheadle et al.,* 1987; MOBIL: *Blundell et al.,* 1991]. This study brings together observations from a number of datasets to provide a broad picture of the factors involved.

2. BASEMENT REACTIVATION AND THE DISTRIBUTION OF STRAIN WITHIN THE LITHOSPHERE

While it is generally accepted that basement reactivation of older structures is an important process in the formation of extensional sedimentary basins, *Holdsworth et al.* [1997] and *Needham and Morgan* [1997] point out that many cited examples of fault or basement reactivation rely on the orientation of basin forming and basement structures being similar. This is different from actual evidence of the same fault moving on a number of occasions to repeatedly produce accommodation space for sediments; good clear examples of which are infrequent and difficult to prove. Those studies in which reactivation of existing basement structures can be demonstrated unequivocally are based on field evidence.

The study of deep seismic reflection profiles contributes a different perspective to the issue of reactivation. Movement on faults at the near surface requires there to be deformation at greater depths in the crust and mantle lithosphere, if volume is to be conserved during deformation. The deep seismic reflection profiles provide evidence of the geometric relationship between deep and near surface structures. This relationship has been, and continues to be, used to infer that deep structures influence subsequent crustal evolution. In some cases this inference is supported because it can be demonstrated that deep reflections are coplanar with faults cropping out at the near surface which have a demonstrable history of movement on a number of separate occasions over a long period of time [*Stein*, 1988; *Stein and Blundell*, 1990; *Butler et al.*, 1995]. However, it is not immediately clear from deep seismic reflection profiling how strains are distributed through the lithosphere. While in many cases reflections exist which have geometries similar to major faults or shear zones in the crust and mantle lithosphere along which discrete deformation might occur, there are other areas described as 'reflective lower crust' in which strains of a range of magnitudes might be dissipated [*Reston,* 1990]. During rifting, horizontal strains must be distributed throughout the lithosphere such that the total strain within the crust must equal the total strain within the mantle lithosphere, if no volume change occurs. Positive and negative horizontal gradients in strain can be accommodated over decreasing or increasingly large areas with changing depth – the 'depth dependent stretching model' [*Rowley and Sahagian,* 1986]. *Kusznir and Matthews,* [1988] interpreted lower crustal reflectivity in terms of a ductile lower crustal layer into which shallow faults detached and strains within the crust and mantle lithosphere were, at least locally, decoupled. Theoretical studies (e.g. *Kusznir and Park,* [1987]) and inverse modelling of subsidence curves to derive strain rate data [*Newman and White,* 1997] indicate that the upper mantle is the strongest layer within the lithosphere. In the context of basement reactivation this concept is critical since it implies that the crust will not deform until significant strains within the mantle have begun to accumulate. As a result, it is probable that when the crust begins to fail it will do so at its weakest point, i.e. along existing weak faults. In order to keep pace with strains accumulating within the mantle lithosphere, initial strain rates may be high and then decrease with time. Consequently, crustal strains should really be considered in the context of deformation of the mantle lithosphere [*Snyder et al.*, 1997], rather than reflectors in the mantle being interpreted in terms of crustal deformation, which is the commonly adopted approach. However, as noted above, the mantle reflectors are not accessible to methods permitting unequivocal interpretation, and hence it is impossible to confirm that they are faults. In the case of sedimentary basins the best indicators of patterns of strain within the mantle lithosphere, on a broad scale, are variations in thickness of sediments accumulated during post-rift thermal subsidence resulting from thermal relaxation of the deformed mantle lithosphere [*White and McKenzie*, 1988]. Ironically, this means that study of the deformation of the upper mantle requires an examination of the near surface geology.

3. BASEMENT STRUCTURE OF THE UK AND SURROUNDING CONTINENTAL SHELF

Meissner et al. [1986] produced a map for the UK, Ireland and adjacent continental shelf areas showing the depth to the Moho in km below the surface, determined by interpretation and depth conversion of the limited amount of BIRPS data then available. More recently, *Chadwick and Pharaoh* [1998] have produced a revised set of maps from the data contained in the BIRPS Atlas [*Klemperer and Hobbs*, 1992] and a variety of data from onshore UK. *Meissner et al.* [1986] noted that the Moho around the UK was relatively flat, with no remnant roots from the Caledonian and Variscan orogenic belts. They noted that the only variations from this pattern occur beneath the major extensional sedimentary basins, where the Moho is at a shallower level (Figs. 2 and 3). These observations are still valid following further extensive deep reflection and wide-angle/refraction profiling [*Klemperer and Hobbs,* 1992; and BIRPS work in progress]. *Chadwick and Pharaoh*, [1998] confirm that regions of thin and thick crust relate to those areas which have or have not been extended by Mesozoic rifting respectively, and that present day land areas occur where the crust is thickest.

A further set of maps is presented here which incorporate interpretations of all the BIRPS data for the UK, Ireland and surrounding areas (Figs. 2-4, 11 and 12). These maps were produced by digitising the interpreted deep reflection data in twtt at 1 km intervals and plotting and contouring the digitised data on a 15 minute (Lat. and Lon.) grid using GMT software [*Wessel and Smith*, 1991]. The 15 minute grid spacing was selected after testing to minimise the misfit between the observed and contoured data after regular gridding of the irregularly spaced seismic data. The depth converted versions of the maps (Figs. 3, 11 and 12) were produced by assuming an average velocity for the basement of 6 km/s and 2.4 km/s for the sediments, which are typical values for the crust of NW Europe used in migrations and depth conversions of BIRPS data [*Klemperer and Hobbs*, 1992].

In this case the top of basement is defined as the unconformity above rocks deformed by the Caledonian (400 Ma) or Variscan (300 Ma) orogenies, whichever is present. In the North Sea this generally corresponds to the base of the lower Permian Rotliegend group. This group has been proven by drilling across most of the southern North Sea and a claystone at the top of the group forms a prominent reflection at the base of the Zechstein salt which can be traced over much of the North Sea basin. The base of the Rotliegend/top basement is poorly defined on seismic data but the well data shows that the Rotliegend rarely exceeds 500 m in thickness [*Glennie*, 1990]. Consequently a reasonable approximation to the location of the top basement can be made from the top Rotliegend reflector. North and west of Britain the unconformity is more complex and can only be defined as the top of Pre-Permian strata. In more simplistic terms the top basement defined here separates rocks affected by significant

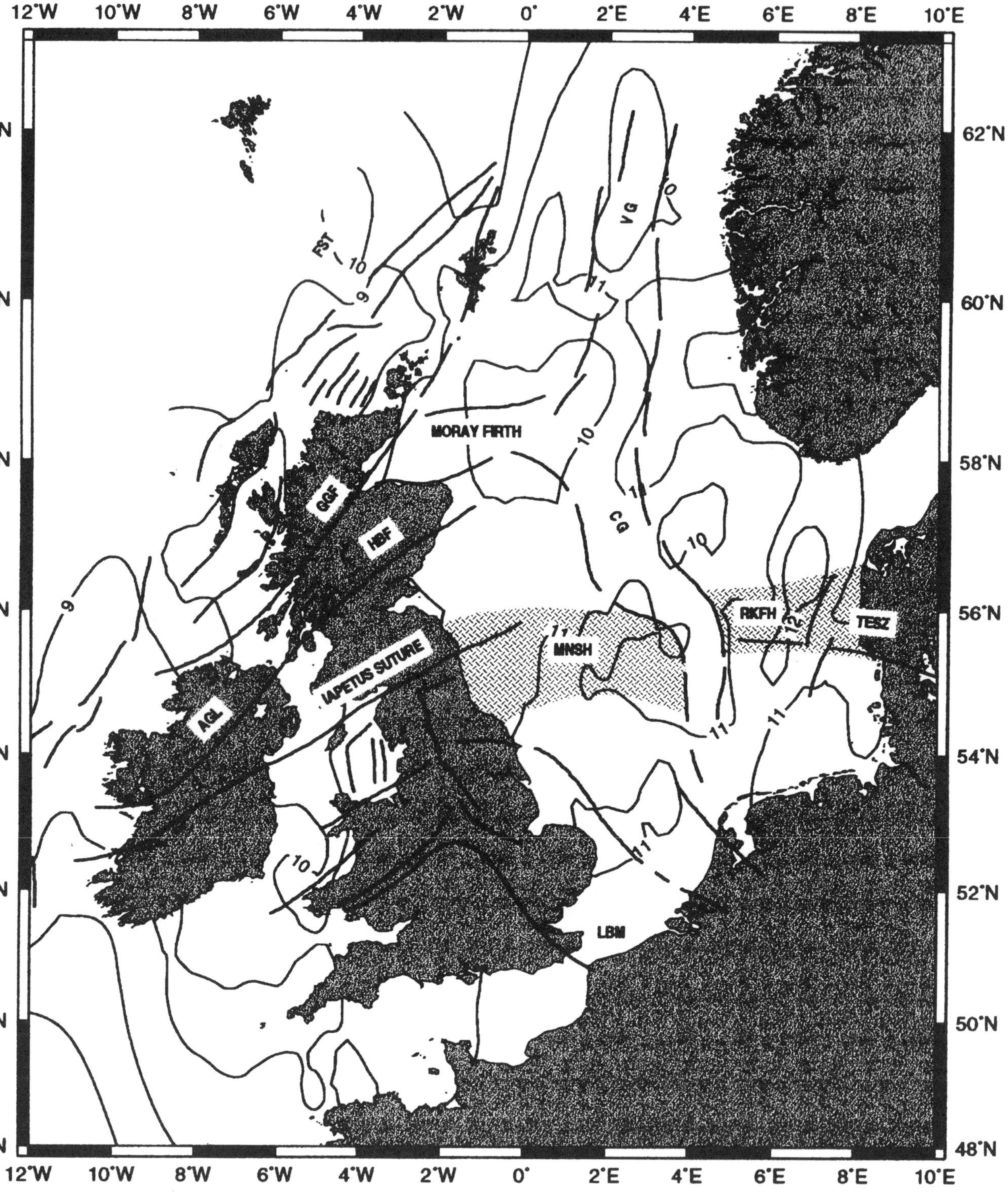

Figure 2. Contour map of two way travel time to Moho derived from digitised interpretations of the BIRPS deep reflection profiles (see Fig. 6 for examples). Contours are in seconds. Thick solid lines indicate the strike of major basement structure. Stippled region crossing the North Sea is the Mid-North Sea – Ringkøbing-Fyn high. MNSH- Mid-North Sea High, RKFH- Ringkøbing-Fyn High. Other labels as Figure 1.

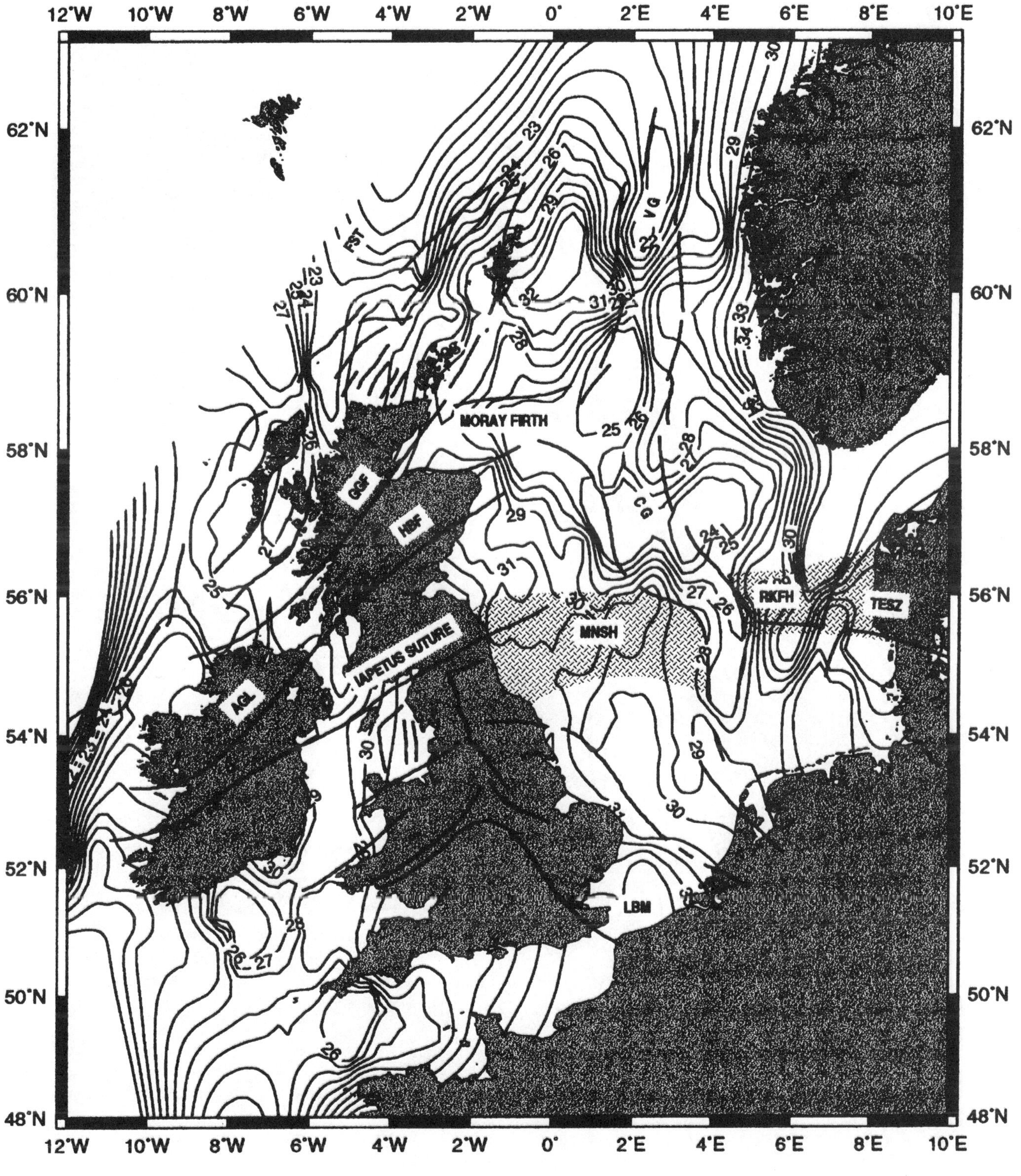

Figure 3. Contour map of depth to Moho derived from digitised interpretations of the BIRPS deep reflection profiles. Contours are in km. Thick solid lines indicate the strike of major basement structure. Stippled region crossing the North Sea is the Mid-North Sea – Ringkøbing-Fyn high. Other labels as Figure 2.

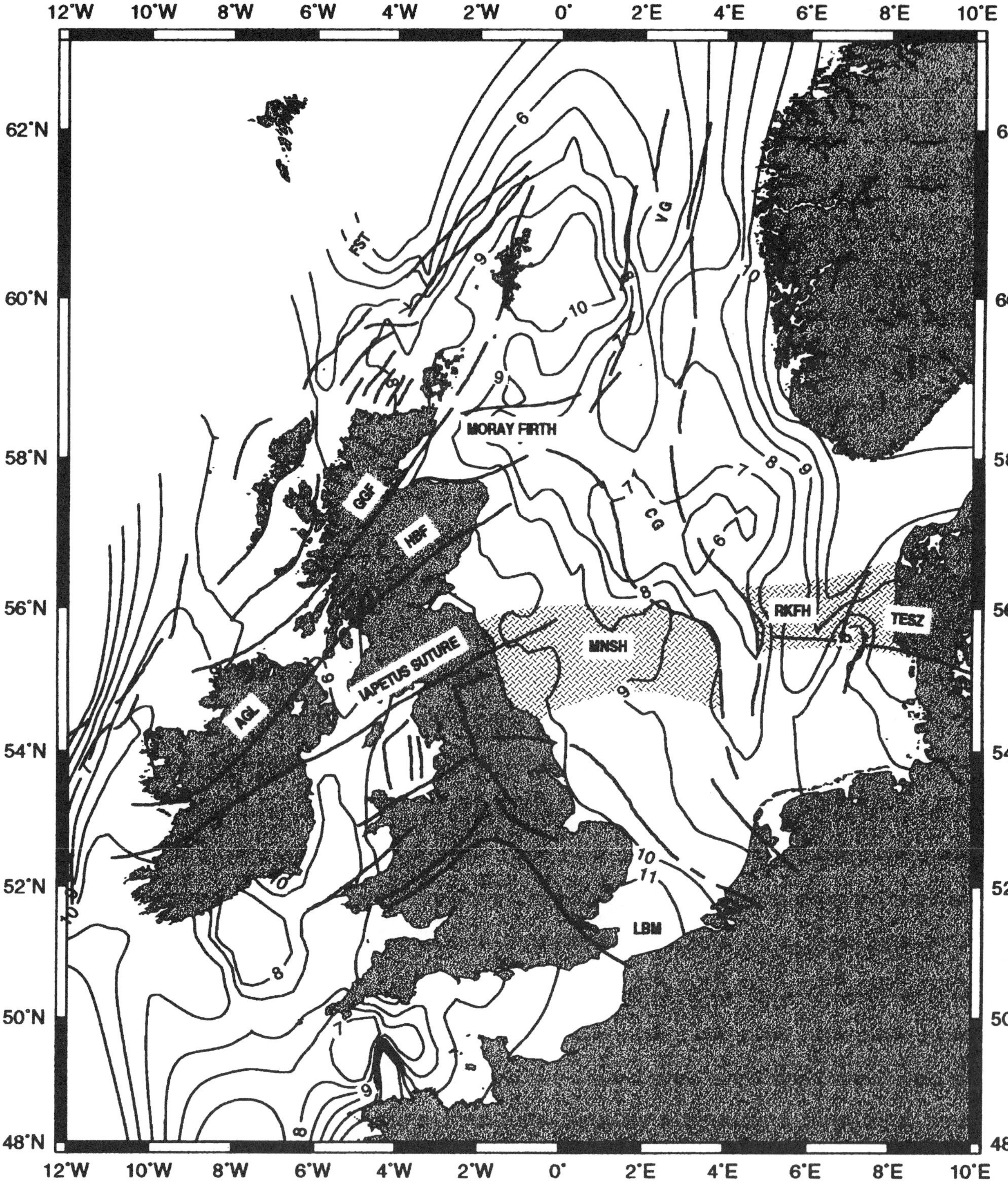

Figure 4. Contour map of basement thickness in two way travel time derived from digitised interpretations of the BIRPS deep reflection profiles. Contours are in seconds. Thick solid lines indicate the strike of major basement structure. Stippled region crossing the North Sea is the Mid-North Sea – Ringkøbing-Fyn high. Other labels as Figure 2.

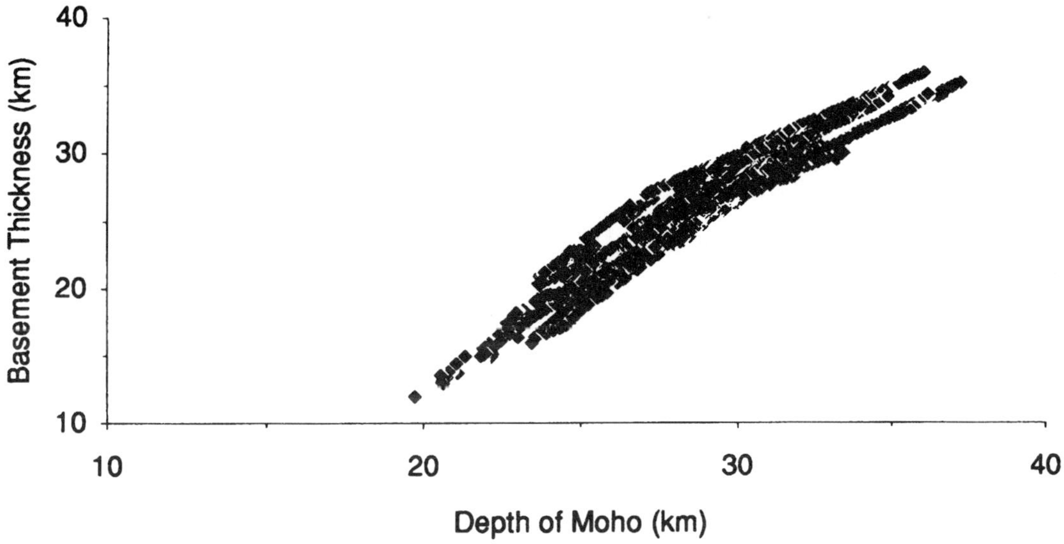

Figure 5. Basement thickness in km plotted against depth to Moho in km. Data derived from digitised cross sections.

compressional deformation which lead to crustal thickening from rocks which accumulated as a result of crustal extension and thinning.

These digitally produced maps show a more complex picture of variation in crustal structure than those previously produced by *Meissner et al.* [1986] and *Chadwick and Pharaoh* [1998]. Beneath the North Sea the Moho is generally at a shallower level under the rift than under areas of crust unaffected by Mesozoic rifting. However, there are other areas where the Moho is also shallow, such that the location of the rift is not clearly reflected in the topography on the Moho. This difference in appearance of the maps may in part be related to these being contoured using digital processing rather than by hand as in the previous approaches of *Meissner et al.* [1986] and *Chadwick and Pharaoh* [1998]. An improved velocity model for the depth conversion of the deep seismic data, based upon and including all lateral variations in seismic velocity within the crust may affect the contour pattern on the depth converted data (Fig. 3). This work is in progress. On a long wavelength scale there is up to 8 km of topography on the Moho between the Shetland Platform and the North Viking Graben, and up to 6 km of opposite variation in the topography on the top of the basement. The map showing the thickness of the basement (Fig. 4), measured between the top basement and the Moho, indicates that the crust is thickest (approximately 34 km) beneath the London Brabant Massif and the Shetland Platform, and thinnest beneath the North Viking Graben. *Barton and Wood* [1984] noted that the free air gravity anomaly pattern for the UK continental shelf contained only short wavelength anomalies, and hence that the crust is approximately in isostatic equilibrium. This is confirmed by the correlation between basement thickness and depth to Moho (Fig. 5) and the interpretations of the seismic data (Fig. 6) which show that the thick accumulations of sediments are compensated by updoming of the mantle (Airy isostasy) [*Holliger and Klemperer*, 1990]. *Chadwick and Pharaoh* [1998] noted local variations in crustal thickness were indicated by short wavelength variations in twtt to Moho. Such variations must be uncompensated or compensation must be achieved by lateral variations in the average density of the crust, or Pratt isostatic compensation. This is supported by the limited amounts of potential field modelling of the whole crust away from the main basinal areas, such as at the edge of the London-Brabant massif [*Lee et al.*, 1993] and over the Midland Microcraton [*Busby et al.*, 1993].

4. BASEMENT EVOLUTION NORTH AND WEST OF SCOTLAND

4.1. Evidence for Fault Reactivation During Basin Development

Basement evolution to the north and west of Scotland is a good example of where reactivation of basement faults is demonstrably involved in the evolution of the crust, and in particular, Mesozoic sedimentary basins. Field based studies integrated with interpretations of offshore seismic data have demonstrated that the Riphean age Torridonian rocks resting on the Lewisian basement of the Caledonian foreland of northwest Scotland were deposited in a series of fault bounded troughs [*Stewart*, 1982; *Stein*, 1988; *O'Neill and England,* 1994; *Soper and England,* 1995 and references therein]. The BIRPS' DRUM and GRID profiles, commercial offshore seismic data, drilling and onshore mapping shows that the Torridonian rocks in the Flannan, Minch and Hebrides basins are overlain by a sequence of Triassic through to middle Jurassic rocks [*Cheadle et al.*, 1987; *Fyfe et al.*, 1993; *England et al.*, 1993]. Rapid lateral changes in facies and thickness suggests that deposition of these Triassic sediments is fault controlled [*Steel,* 1971]. The field and seismic data show

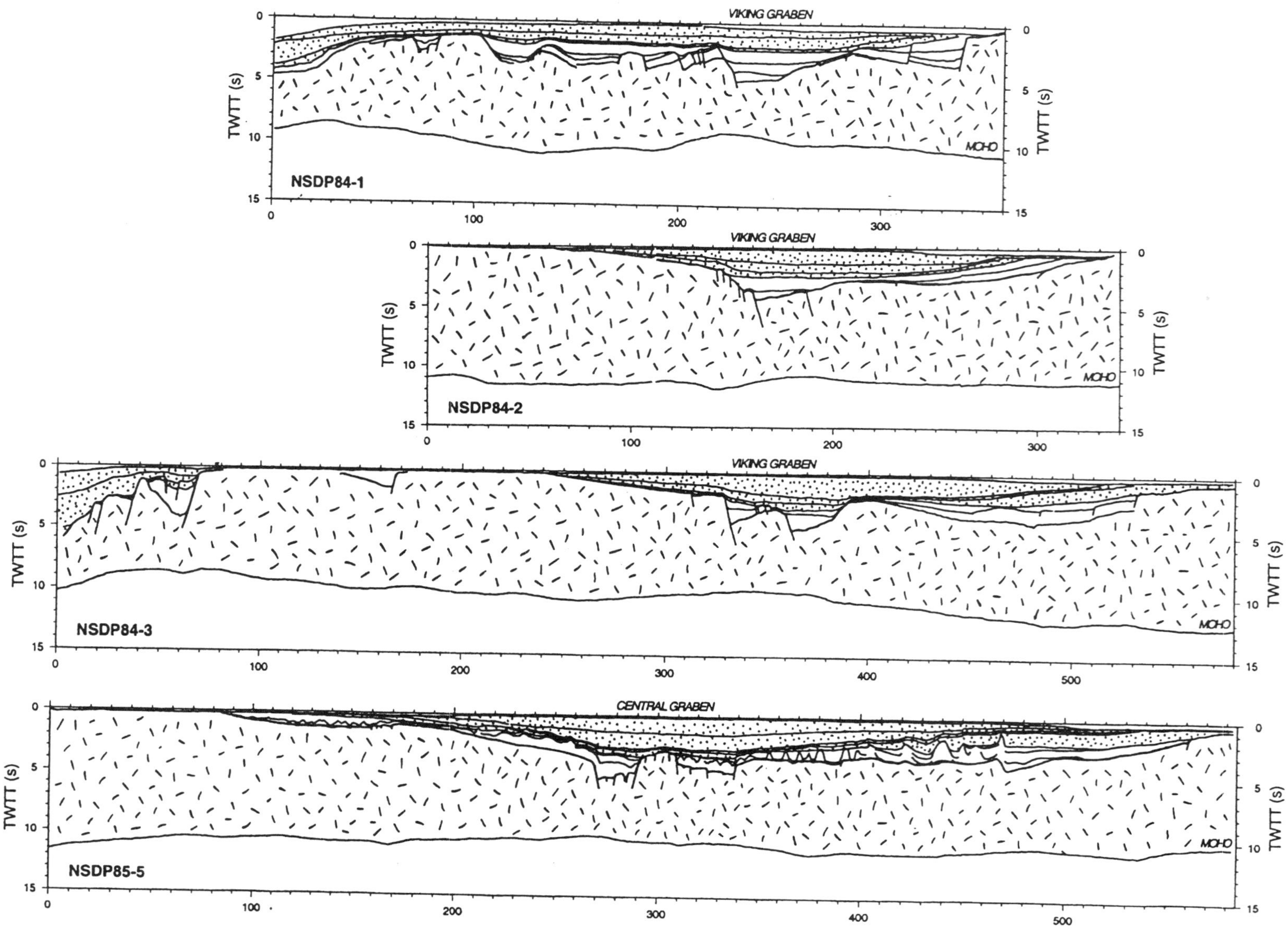

Figure 6a. Principal deep seismic reflection profiles crossing and parallel to the North Sea rift system. Major structural and stratigraphic horizons are digitised at 1 km intervals from interpreted unmigrated data. Vertical scale is two times horizontal where seismic velocity in the basement equals 6 km/s. Scale on x axis is in km. Dotted regions are post-rift sediments, stippled areas are Pre-Permian basement (below top Rotliegend). For line locations see Fig. 1. Profiles NSDP84-1 through SNST83-07 are arranged in order from north to south.Profiles NSDP85-7 and MONA LISA2 are parallel to the Central Graben. MNSH-Mid North Sea high, RKFH-Ringkøbing-Fyn high.

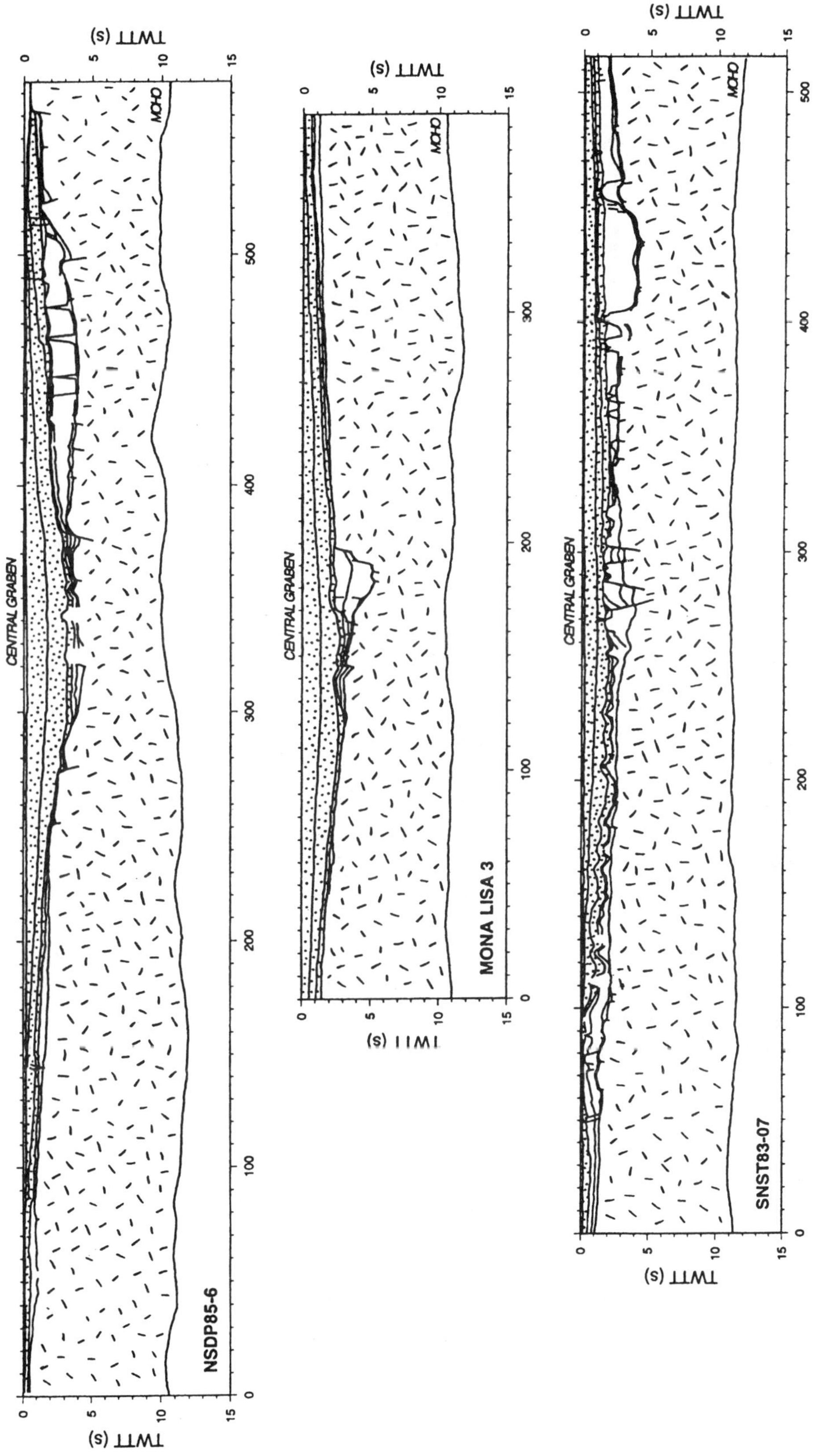

Figure 6a (Continued)

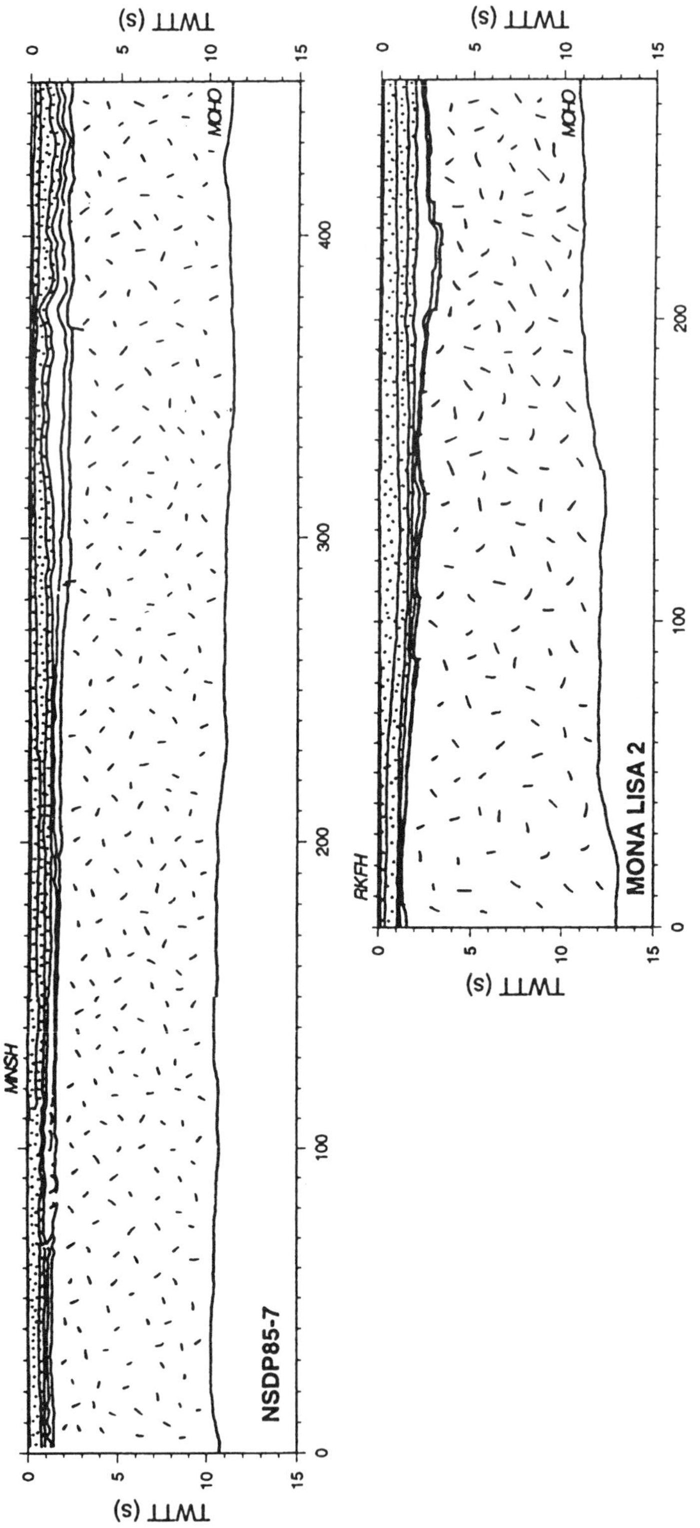

Figure 6a (Continued)

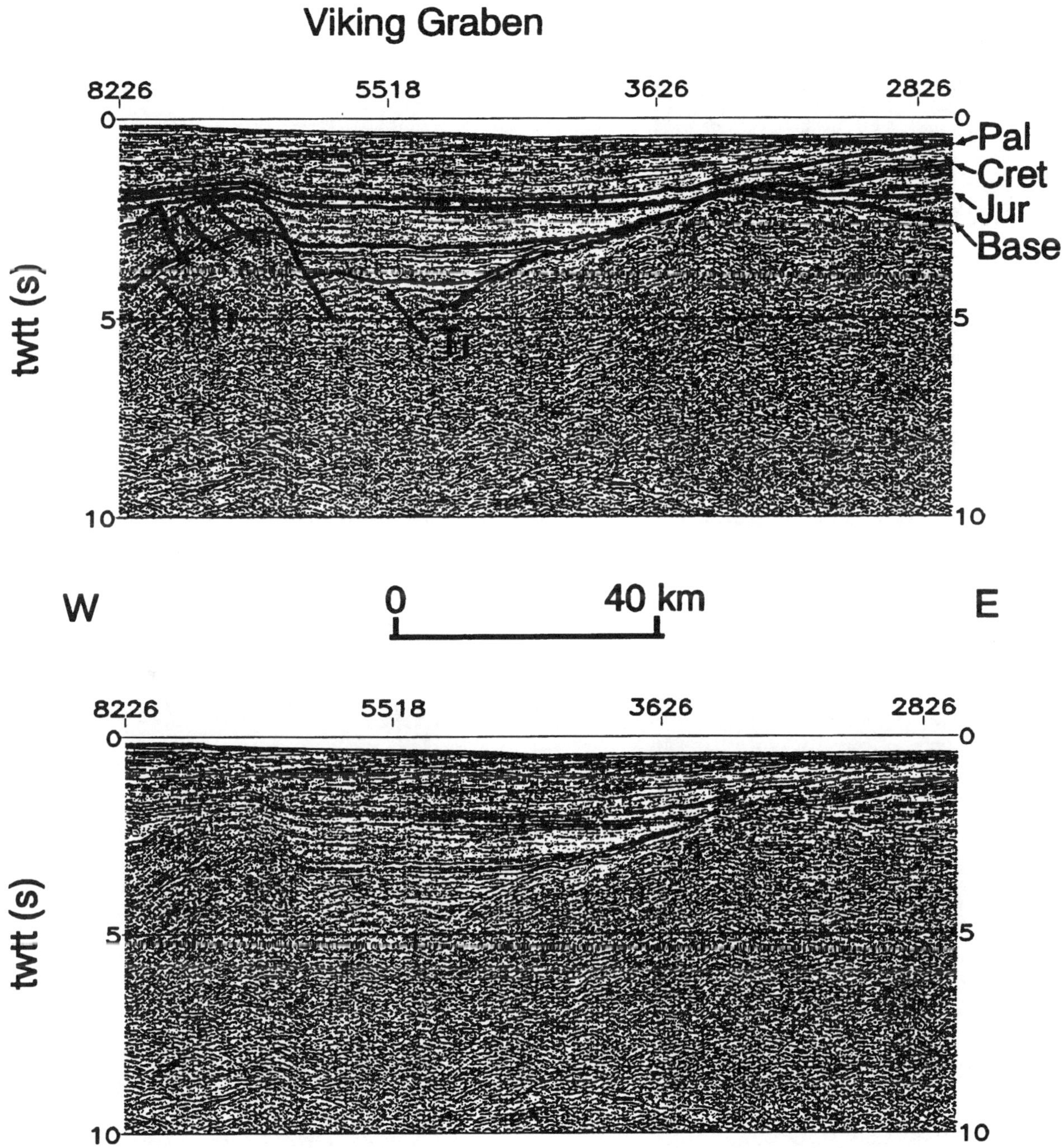

Figure 6b. Part of profile NSDP84-1 between km 210 and km 300 (See Figure 6a for location) shown interpreted above and uninterpreted below. Base-top basement, Cret-top Cretaceous, Jur-top Jurassic, Pal-top Paleocene, Tr-Triassic sedimentary rocks. Thick lines are faults. Shot point numbers along top of profile.

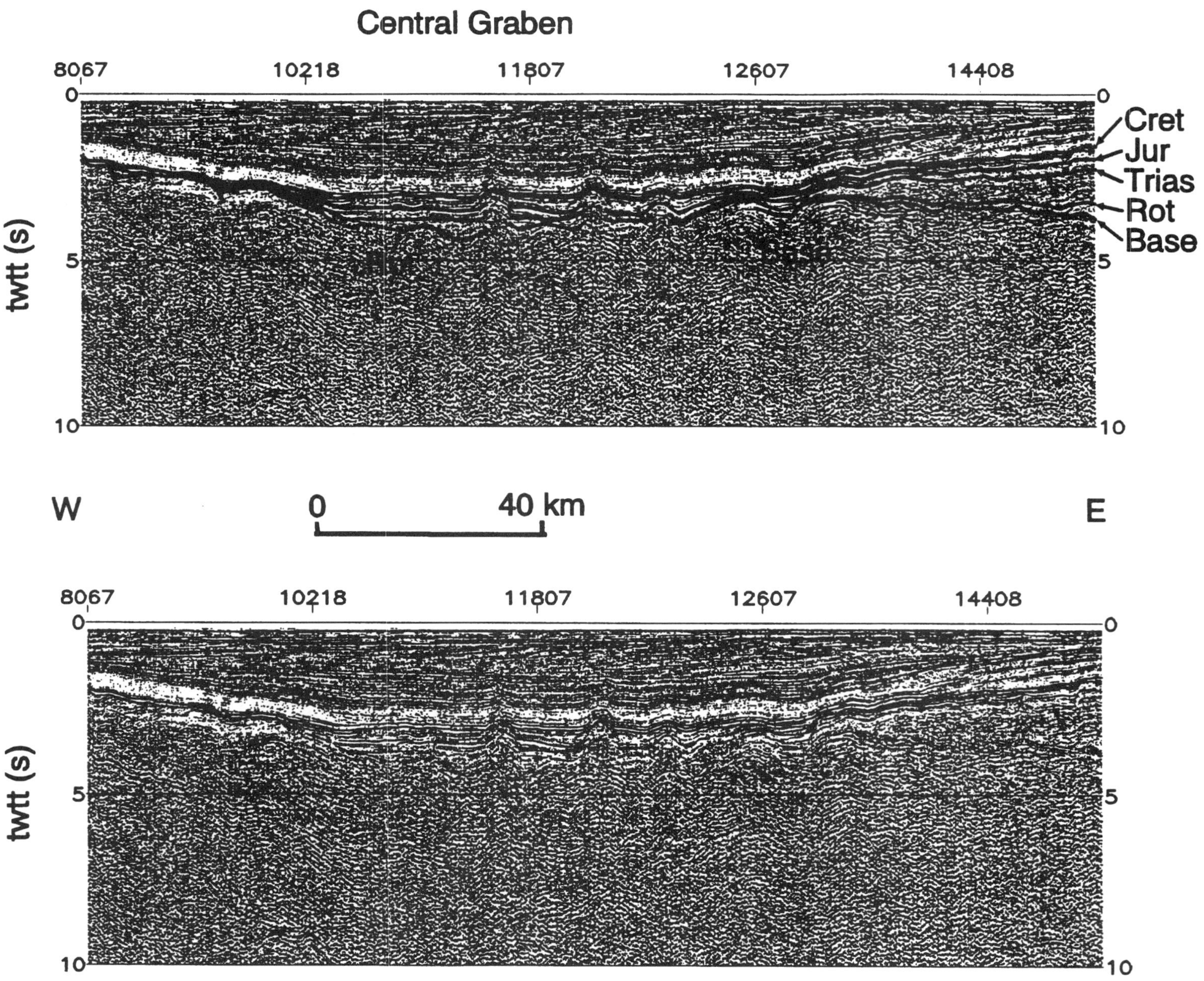

Figure 6c. Part of profile NSDP85-6 between km 240 and km 425 (See Figure 6a for location) shown interpreted above and uninterpreted below. Base-top basement, Cret-top Cretaceous, Jur-top Jurassic, Rot-top Rotliegende, Trias-top Triassic. Thick lines are faults. Shot point numbers along top of profile.

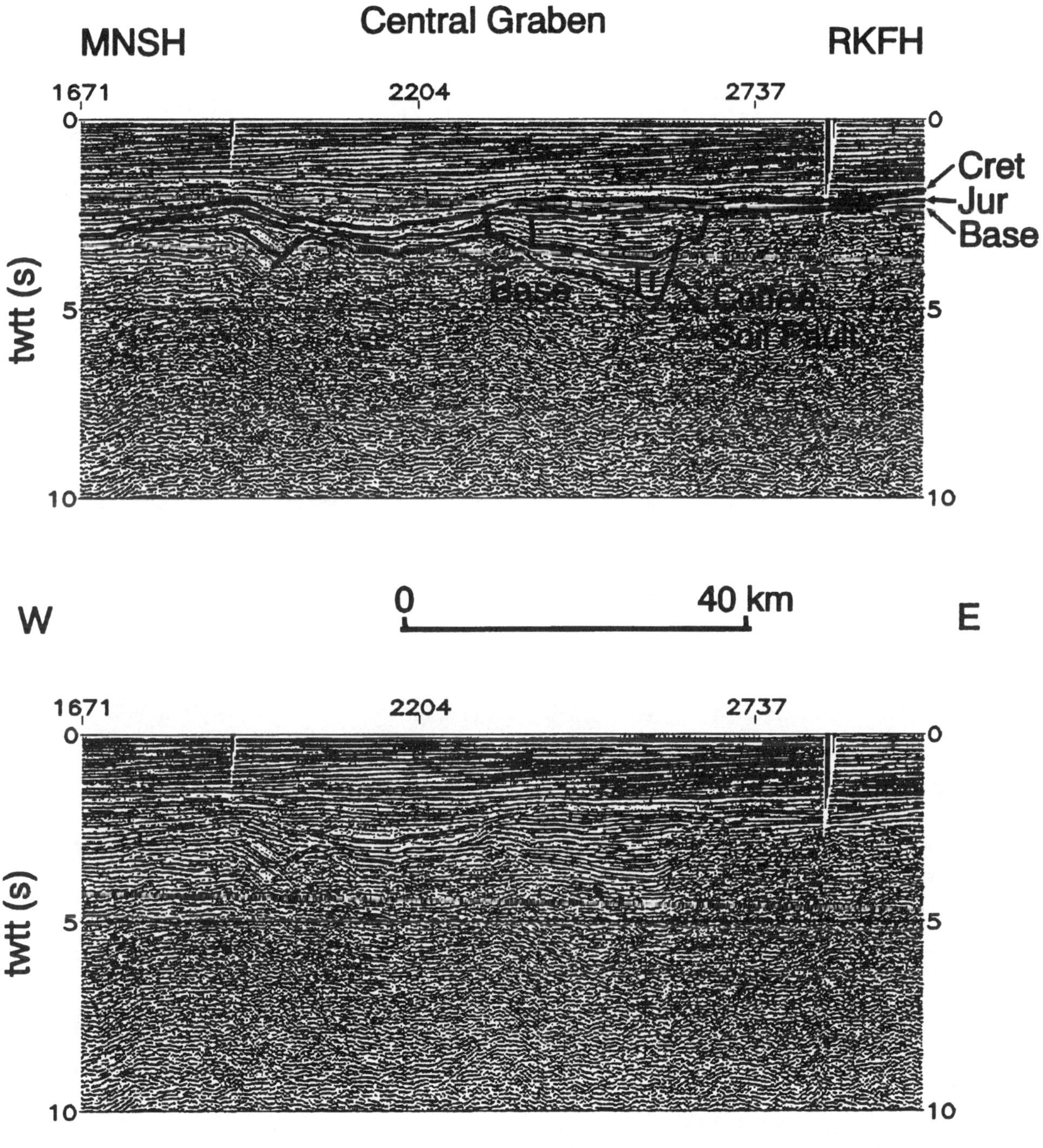

Figure 6d. Part of profile MONA LISA 3 between km 110 and km 230 (See Figure 6a for location) shown interpreted above and uninterpreted below. Base-top basement, Cret-top Cretaceous, Jur-top Jurassic, MNSH-Mid–North Sea high, RKFH-Ringkøbing-Fyn high, Tr-Triassic sedimentary rocks. Thick lines are faults. Shot point numbers along top of profile.

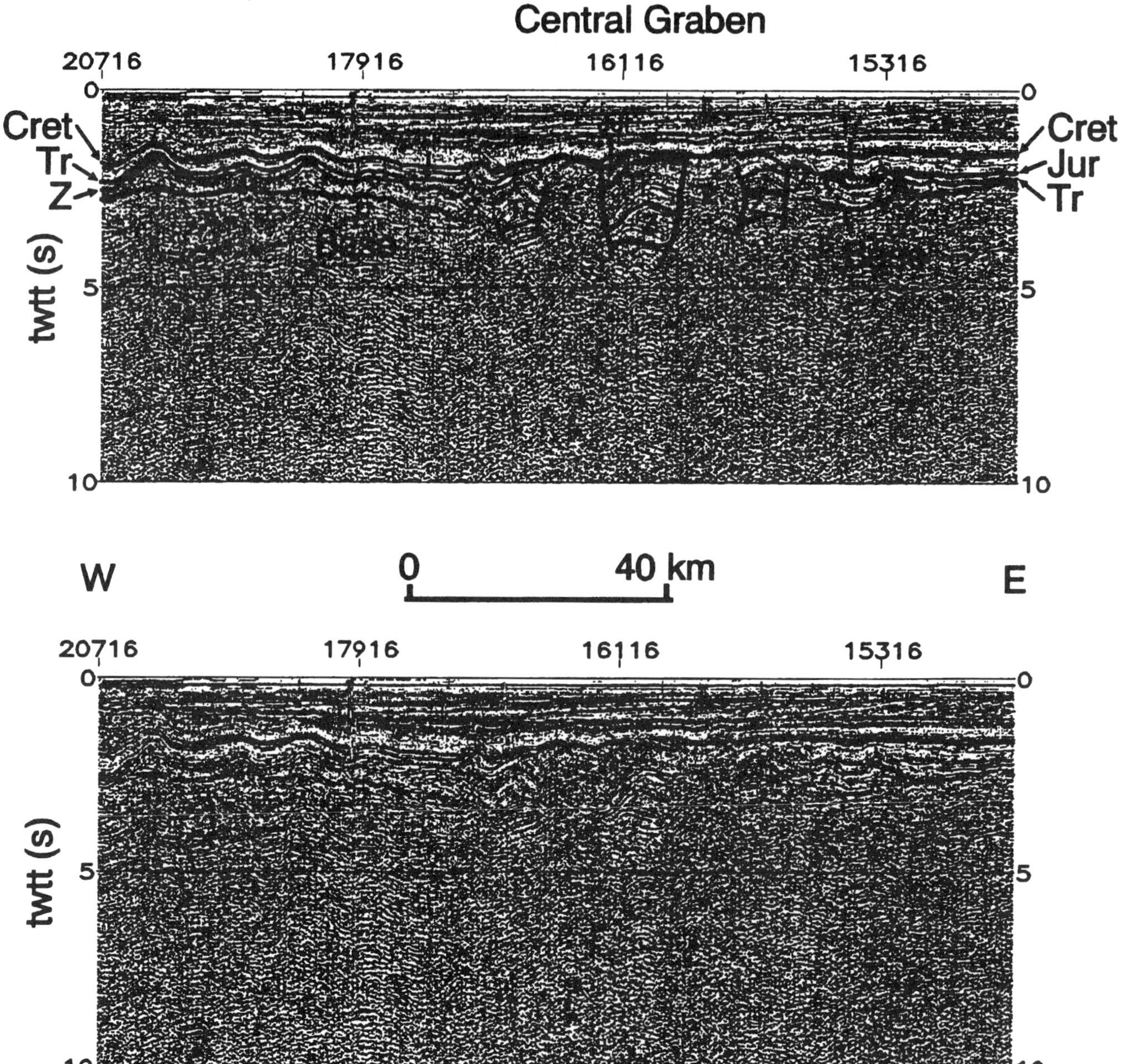

Figure 6e. Part of profile SNST83-07 between km 195 and km 335 (See Figure 6a for location) shown interpreted above and uninterpreted below. Base-top basement, Cret-top Cretaceous, Jur-top Jurassic, Tr-top Triassic, Z-Zechstein Salt. Thick lines are faults. Shot point numbers along top of profile.

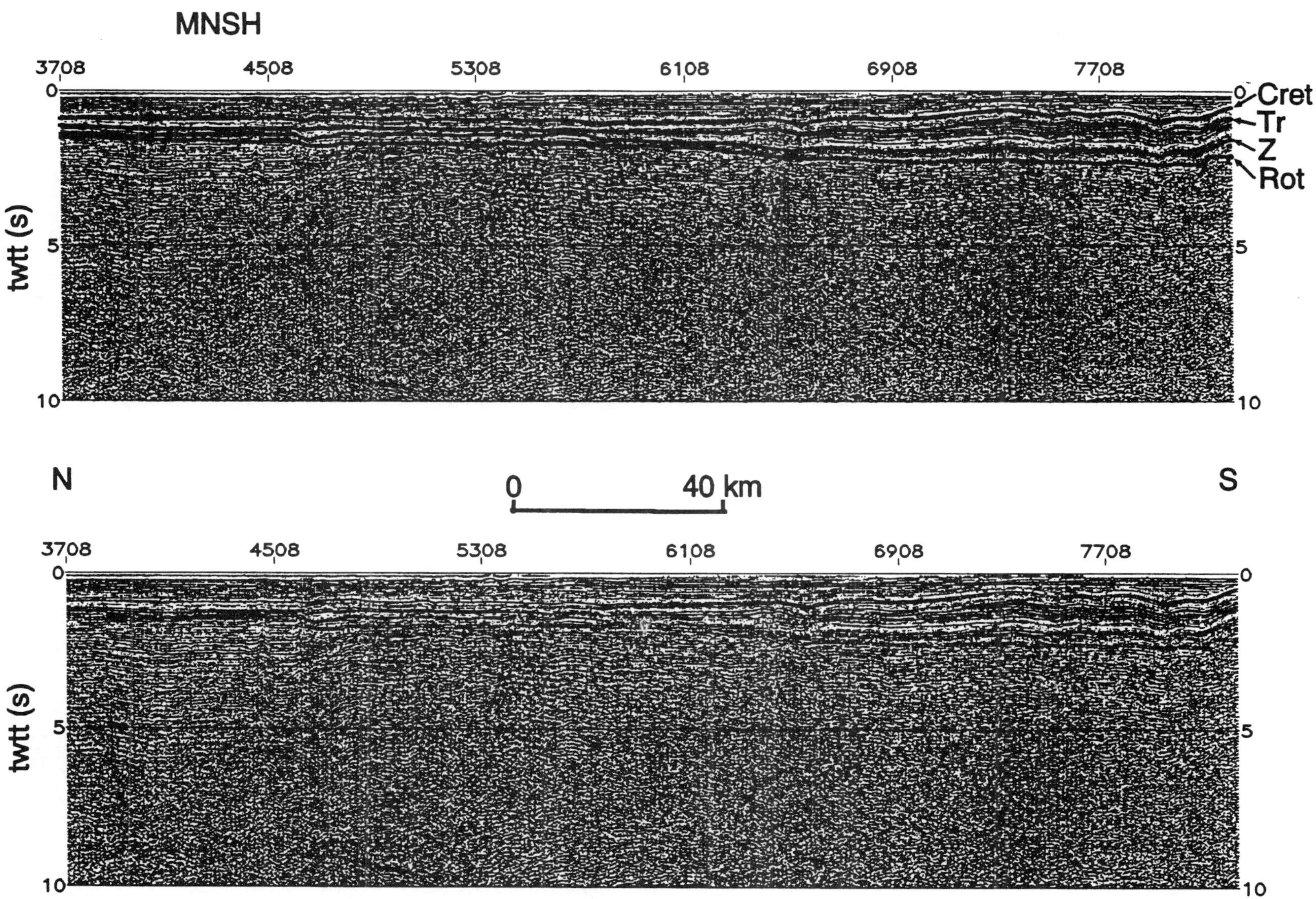

Figure 6f. Part of profile NSDP85-7 between km 150 and km 375 (See Figure 6a for location) shown interpreted above and uninterpreted below. Cret-top Cretaceous, MNSH-Mid-North Sea high, Tr-top Triassic, Rot-top Rotliegende, Z-Zechstein Salt. Thick lines are faults. Shot point numbers along top of profile.

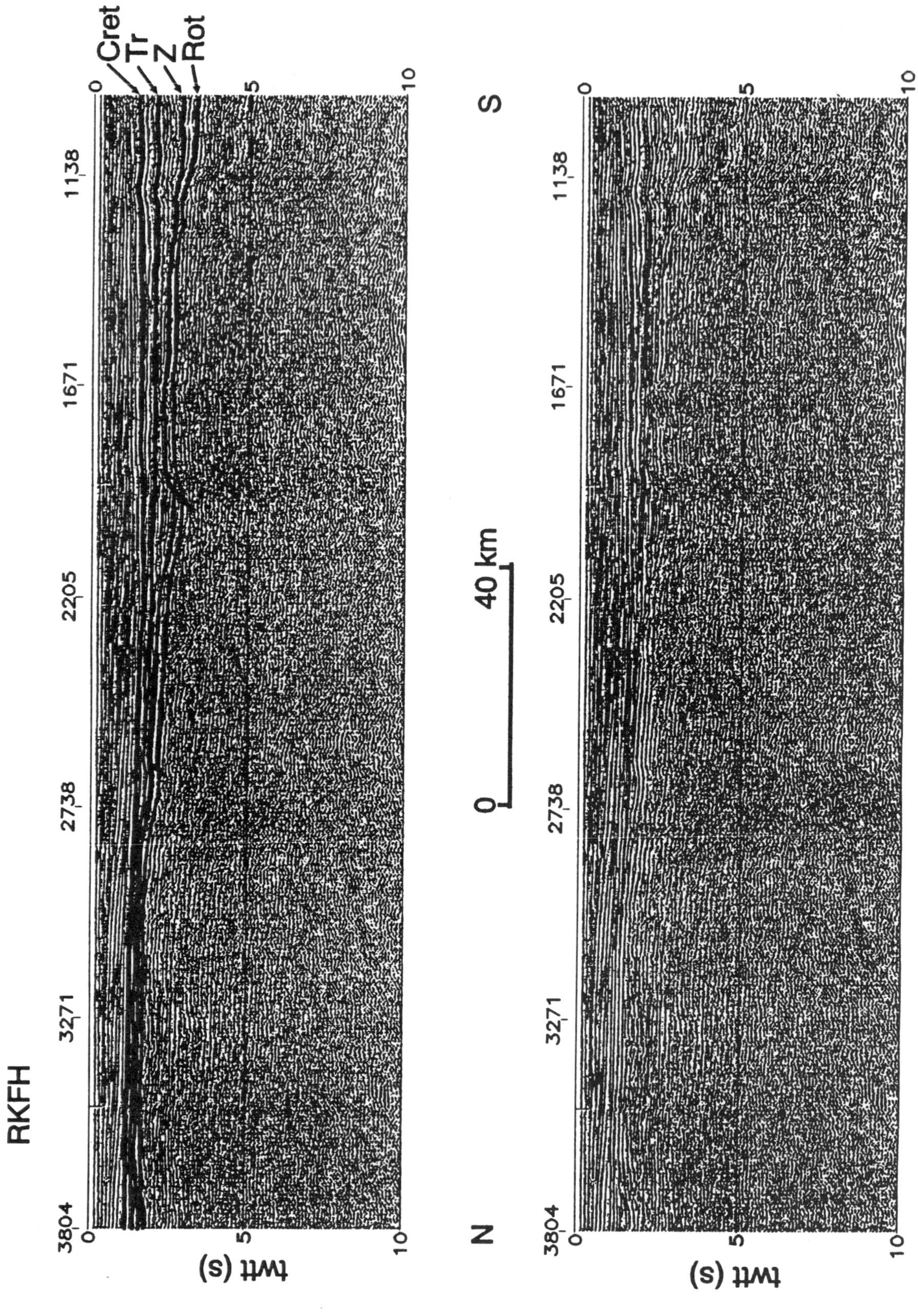

Figure 6g. Part of profile MONA LISA 2 between km 0 and km 210 (See Figure 6a for location) shown interpreted above and uninterpreted below. Cret-top Cretaceous, Tr-top Triassic, RKFH-Ringkobing-Fyn high, Rot-top Rotliegende, Z-Zechstein Salt. Thick lines are faults. Shot point numbers along top of profile.

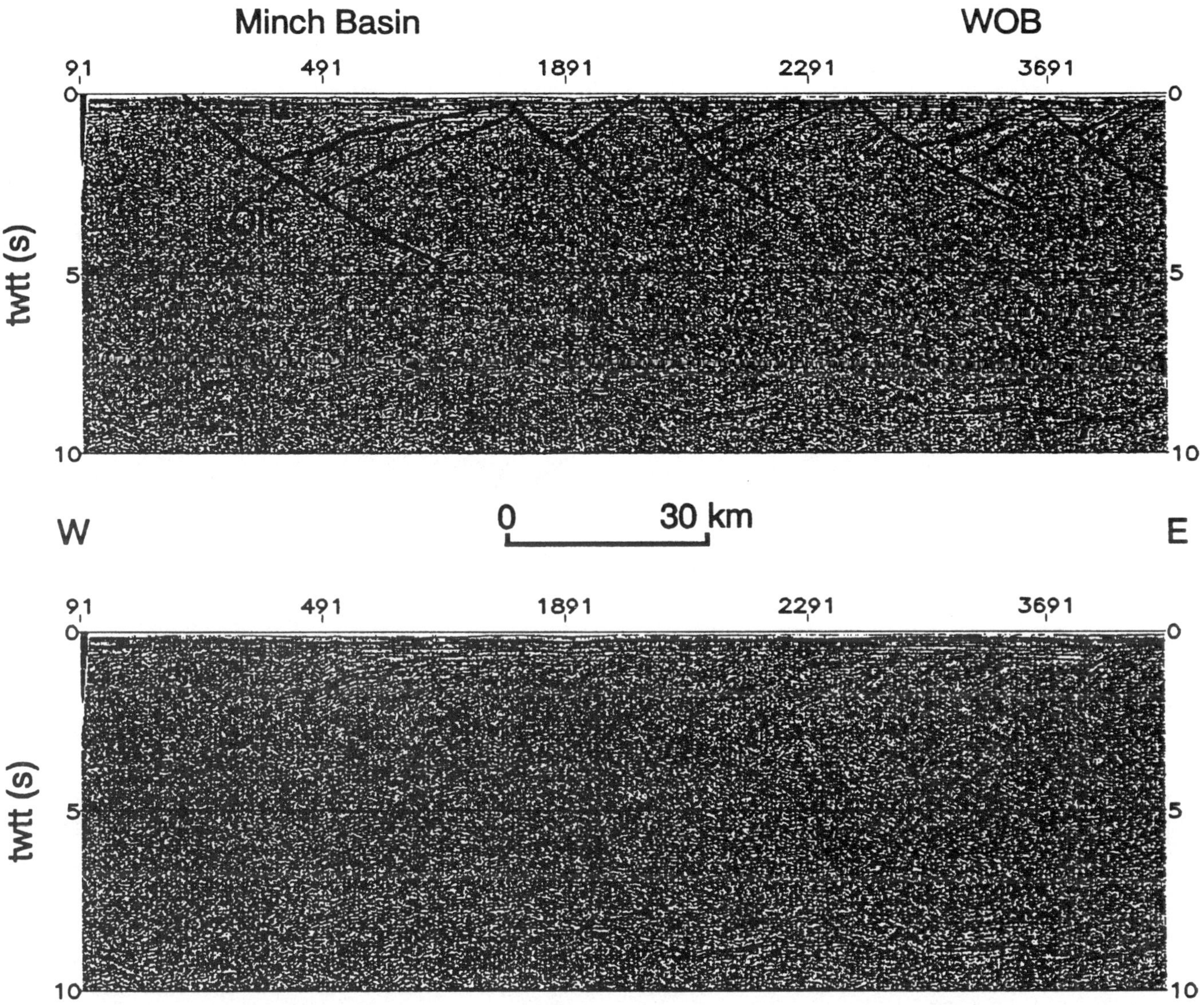

Figure 7. The BIRPS' DRUM profile, interpreted above, uninterpreted below. D-Devonian sedimentary rocks, M-Mesozoic sedimentary rocks, OIF-Outer Isles Fault, T-Torridonian sedimentary rocks, WOB-West Orkney Basin. Major normal faults are shown as thick lines, stratigraphic boundaries as thin lines. Shot point numbers along top of profile.

thickening of the Triassic sequences toward faults (Fig. 7). These faults are the same as those that bound the Torridonian rocks. Integrated seismic interpretation and gravity modelling [*O'Neill and England*, 1994] shows that the tectonic subsidence resulting from the movement on the Minch Fault and the deposition of the Mesozoic rocks in the Sea of the Hebrides Basin is less than that of the preserved thickness of the Torridonian rocks (Fig. 8). This indicates that while Mesozoic fault movement is partially responsible for the preservation of the Torridonian rocks it cannot be responsible for their entire preserved thickness. The remainder must have been deposited in a fault bounded basin.

4.1. Basin Development on the UK and Irish Continental Margin

Sedimentary basins imaged by the BIRPS GRID data between the north coast of Scotland and the Wyville-Thomson Ridge, have the same structure and stratigraphy as the Sea of the Hebrides Basin (Figs. 1 and 7). Two distinct wedges of reflectors diverging toward eastward dipping fault plane reflections bounding half graben can been seen on these profiles. Further to the north, the BIRPS deep seismic profiles (UNST, NSDP84-1 and NSDP84-3) show that the edge of the Shetland platform is bounded by west dipping normal faults. However, further

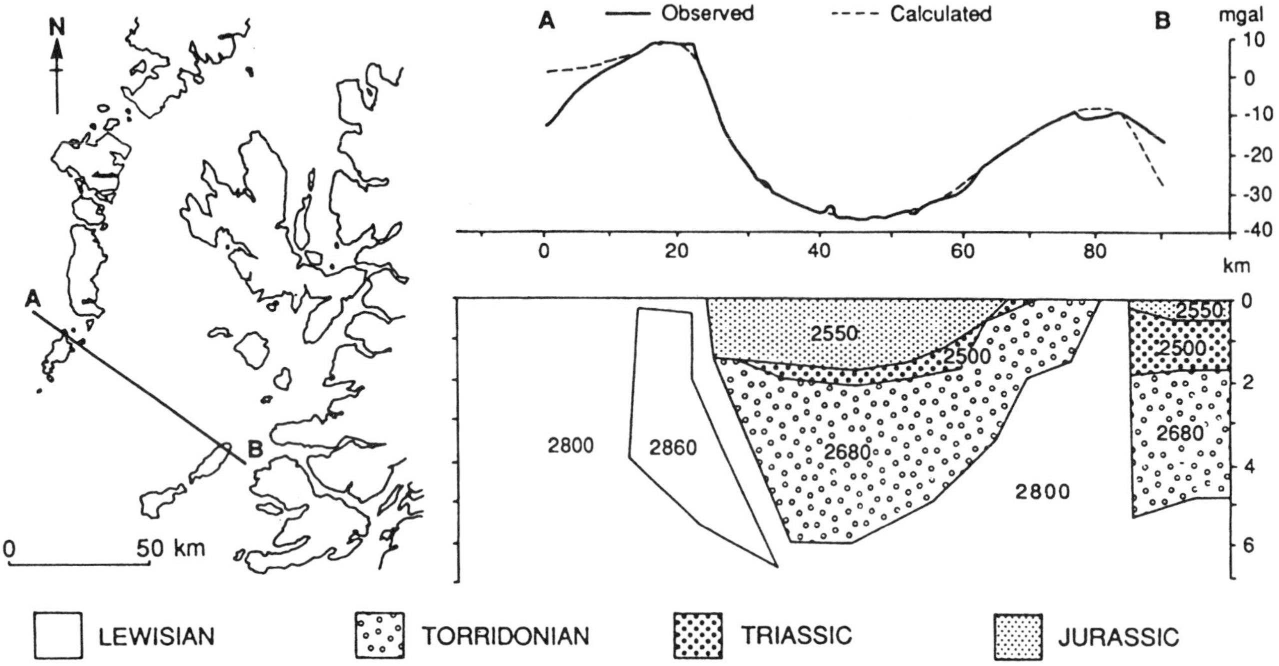

Figure 8. Interpretation of Gravity data across the Sea of the Hebrides Basin, (after *O'Neill and England* [1993]).

to the west in the centre of the Faroe - Shetland Trough some commercial and the BIRPS FAST (FAroe-Shetland Trough) seismic reflection profiles show steeply east dipping reflections which suggest that the deep structure of the tough is dominated by east dipping faults [*Mudge and Rashid*, 1987; *Duindam and Van Hoorn*, 1987] (Fig. 9). The resolution of the seismic data beneath the top Cretaceous in this area is generally not good enough, due to scattering of seismic energy by early Tertiary sills intruded into the Cretaceous section, to resolve packages of reflectors beneath the late-Jurassic–early-Cretaceous syn-rift sequence which are stacked against the reflections interpreted as faults. Sub-Cretaceous rocks in the centre of the Faroe-Shetland trough have not been penetrated directly by drilling. Consequently, it is not possible to resolve the question of whether the east dipping faults, along which late-Jurassic–early-Cretaceous extension occurred, had previously been active as the edges of Triassic, Devonian or Torridonian basins. However, on the basis of the observations to the south (in the Sea of the Hebrides and West Orkney Basins) it is probable that these east facing normal faults are major basement structures first active during the Riphean. The majority of the faults bounding the Torridonian and Mesozoic sequences appear from seismic data to be east dipping, although *Butler* [1997] noted that in NW Scotland Torridonian rocks involved in thrusting beneath the Moine thrust are bounded by both east and west facing faulted contacts with Lewisian rocks.

Across the Shetland platform, over the Rona ridge and within the eastern part of the West Orkney Basin a sequence of terrestrial Devonian rocks is preserved in a series of graben or half graben structures [*Enfield and Coward*, 1987; *Mudge and Rashid*, 1987; *Seranne*, 1992; *Coward,* 1993]. It has been argued that these Devonian rocks were deposited in a series of transtensional basins which formed during collapse of the Caledonian orogen [*McClay et al*., 1986; *Seranne,* 1992; *Coward*, 1993]. In the West Orkney basin the Devonian rocks thicken toward the same faults which bound the western edge of the Torridonian rocks, again suggesting reactivation of faults which originally formed during the Riphean. The full extent and distribution of Devonian rocks in basins to the north and west of the UK is unknown due to the thickness of the Mesozoic sequences preserved in the basinal areas. Consequently, the influence of Devonian transtension on the evolution of basins in these areas cannot be assessed. *Enfield and Coward* [1987] and *Coward* [1993] argue that minor compressional or inversion structures observed onshore at the edges of the West Orkney basin are related to late - Carboniferous transpression during the Variscan orogeny which may have resulted in the local erosion of parts of the Devonian and possible overlying Carboniferous sequences.

4.2 Deep Basins west of Britain

Extension on east facing faults is a common feature of the deep basins to the west of Britain, suggesting that the structure of the basement established by Riphean and Vendian rifting on the margin of Laurentia has had a considerable influence on Palaeozoic and Mesozoic basin development. The structure of the Rockall Trough to the west of Ireland appears to be dominated by east facing

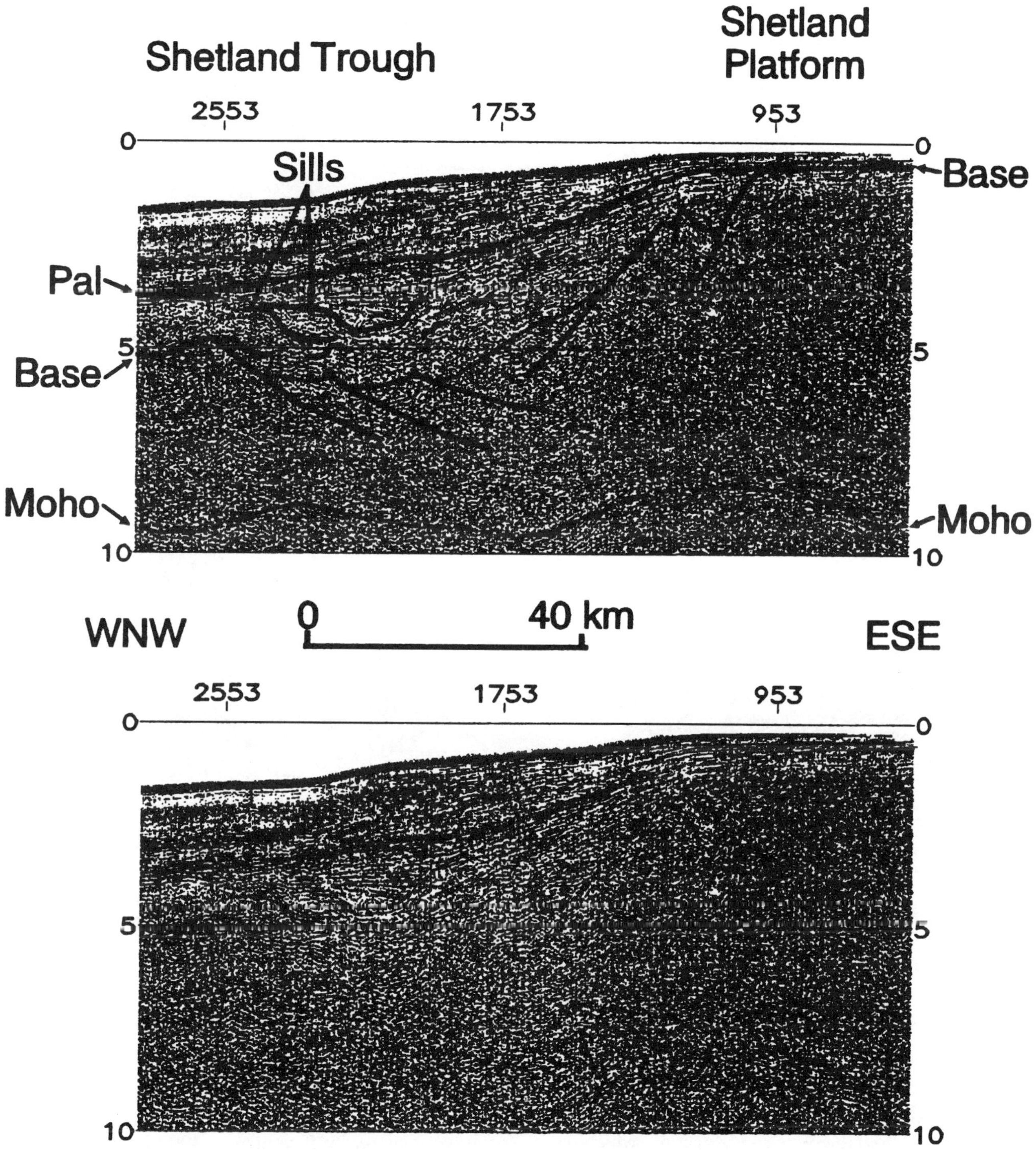

Figure 9. Part of the FAST profile crossing the Faroe-Shetland Trough and Shetland platfrom. Base-top basement, Pal-top Paleocene, Major normal faults are shown as thick lines, stratigraphic boundaries as thin lines. Shot point numbers along top of profile.

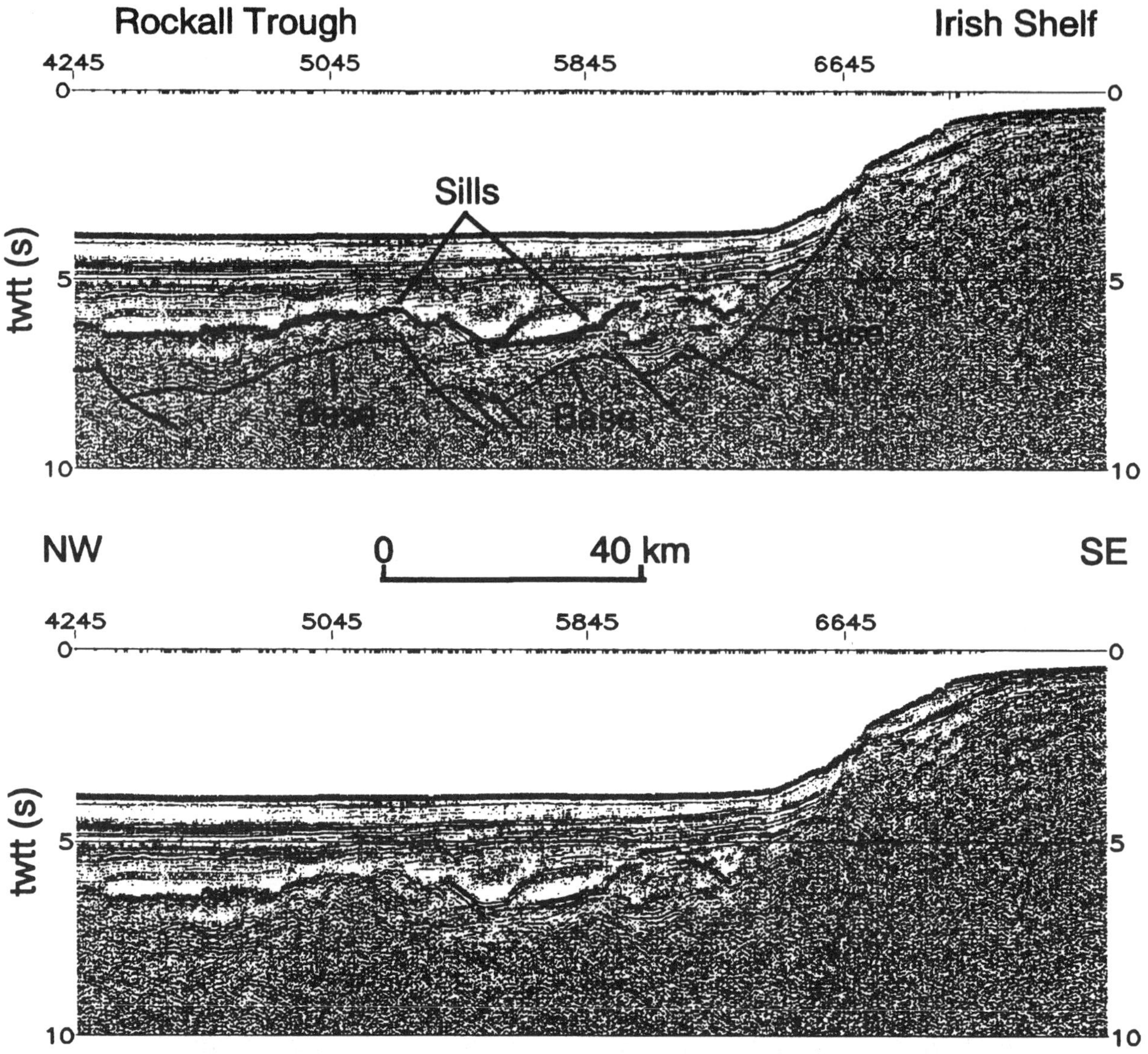

Figure 10. Part of the BIRPS' WESTLINE profile from the SE margin of the Rockall Trough and Irish shelf. Base-top basement. Major normal faults are shown as thick lines, stratigraphic boundaries as thin lines. Shot point numbers along top of profile.

normal faults across its whole width [*England and Hobbs*, 1997] (Fig. 10). In the northern part of the trough the Merlin MP-1 profile has been interpreted as showing ?Torridonian and Triassic rocks in half graben bounded by east facing normal faults [*Stein and Blundell*, 1990]. Most of these faults and the structural trends associated with them have a N-S to NE-SW strike direction which is typically referred to as Caledonian. However, since it can be demonstrated that the Caledonian compressional structures in NW Scotland and Ireland developed by inversion of normal faults active during the Riphean and Vendian [*Soper and Anderton*, 1984] this trend may be better referred to as Riphean (Fig. 1). These trends dominate the structure of Scotland and Ireland as far south as the Highland Boundary Fault, which is thought to mark the southern edge of the continental margin along which Laurentia split at c. 600 Ma to form the Iapetus Ocean [*Ryan et al.*, 1995]. To the south of the Highland Boundary Fault major faults and the strike of a cleavage produced during Acadian (early Devonian, c. 390 Ma) folding of the Silurian rocks of the Lake District, Northern England and Wales are wrapped around the edge of the Midland Craton which collided with the Iapetus Suture zone during the end Caledonian - Acadian orogenic event [*Soper et al.*, 1987]. On this basis these structures may be more correctly referred to as 'Caledonian' (Fig. 1). Tectono-stratigraphic studies of the Carboniferous basins in central and northern England suggest that these 'Caledonian' structures were reactivated as marginal normal faults to lower Carboniferous basins and were later reactivated as steep reverse faults by Variscan compressional deformation [*Fraser and Gawthorpe*, 1990; *Leeder and Hardman*, 1990].

4.3 Basin Development on the East Greenland Continental Margin

A set of observations similar to those that can be made in NW Scotland can be made in West Greenland, where there is a well documented history of repeated rifting events in the Riphean and Vendian, associated with rifting of Laurentia and opening of the Iapetus Ocean [*Soper*, 1994; *Soper and England*, 1995]. These rifting events resulted in the formation of a number of major east facing normal faults, which were partially reactivated as thrusts during the closure of the Iapetus ocean and development of the Caledonian orogen [*Soper and Higgins*, 1993]. Following the Caledonian orogeny, a series of sedimentary basins of late Palaeozoic and Mesozoic age developed along what is now the Greenland continental margin [*Larsen and Bengaard*, 1991; *Hartz et al.*, 1997]. These basins appear to have formed against the same east facing faults which controlled basin development in the Riphean and Vendian. However, reactivation of existing faults is difficult to demonstrate in this region because the post Caledonian basins completely overstep and obscure the basement. However, the orientation and spacing of the faults controlling the late Palaeozoic and Mesozoic are identical to those controlling the Precambrian basins. The timing of these crustal extension events can be broadly correlated with events in NW Scotland, on the basis of a similar tectonostratigraphy. The present day Atlantic margins north of c. 56°N run parallel to these faults and it is possible that the long history of movement on these faults resulted in zones of weakness along which the crust ruptured during separation of Greenland and Europe in the early Tertiary.

5. THE DEEP STRUCTURE OF THE NORTH SEA RIFT SYSTEM

5.1. Previous Work

Much of the existing work on, and interpretations of, the deep seismic profiles crossing the North Sea rift system have concentrated on the debate over whether the rift formed by pure or simple shear [*Barton and Wood*, 1984; *Gibbs*, 1989; *Klemperer and White*, 1989; *White*, 1989]. It is now generally accepted on the basis of the studies of the velocity structure of the crust [*Barton and Wood*, 1984], and the distribution and variation in thickness of the post-rift sediments [*White,* 1989] that the North Sea rift system formed by pure shear of the lithosphere. This section focuses on the large scale variations in the geometry of the rift system and attempts to relate variations in rift structure to the deep basement structure.

There are nine deep seismic reflection profiles crossing the North Sea orthogonal to the rift system (Fig. 6). These profiles show the general variation in structure of the rift system along its axis. In addition a number of profiles run parallel to the axis of the trough so that the relationship between variations in basement structure running orthogonally to the rift and variations in rift structure can be assessed (Fig. 6). The Central and Viking grabens do not have a simple relationship to major structural trends in the basement. The strike of major Caledonian structures, such as the Iapetus Suture Zone, mapped offshore using deep seismic data [*Klemperer et al.*, 1990] is nearly orthogonal to the strike of the graben (Fig. 1). Only in the southern North Sea does the graben run sub-parallel to the edge of the London-Brabant Massif and the Caledonian-Acadian structures which parallel the edge of the Midland Micro-craton. The MOBIL deep seismic profiles [*Blundell et al.*, 1991] which cross this region show no evidence for or against these basement features playing a direct role in the formation of the central graben. Similarly, the northern parts of the Viking graben may be considered to run parallel to the Iapetus suture zone between Laurentia and Baltica, but there is no evidence that the graben is superimposed directly on the suture. The suture is not imaged on deep seismic data in this part of the North Sea. Similarly the rift is orientated at a high angle to the WNW-ESE trending Variscan structures which cross the North German basin and the southern part of the Anglo-Brabant Massif. Consequently, it is difficult to argue that the Mesozoic North Sea rift developed by reactivation of existing structures within the basement. None of the profiles crossing the North Sea show unequivocal evidence of reactivation of pre-existing faults similar to that seen north and west of Britain and described above.

5.2. Variations in Rift Structure

The deep seismic profiles reveal distinct changes in the form of the rift along its length (Fig. 6). The most notable changes are seen in the width of the central trough. The North and South Viking grabens are distinct half graben structures with well defined fault bounded margins (NSDP84-1, km 210 - 280; NSDP84-2, km 140 - 200; and NSDP84-3, km 325 - 390; Fig. 6). This contrasts with the Dutch Central graben region where the graben is very wide, its edges are poorly defined and the basement is cut by a number of faults with small displacement (SNST83-07, km 230 - 330; Fig. 6). However, immediately to the north of this region where the rift crosses the Mid-North Sea – Ringkøbing-Fyn High the graben becomes a very narrow and well defined half graben bounded by the Coffee Soil Fault (MONA LISA 3, km 160 - 200; Figs. 6 and MONA LISA 4 not shown here). The rift widens to the north of the high (NSDP85-5, km 280 - 375; and NSDP85-6, km 270 - 335; Fig. 6) where it is divided into two distinct grabens (the East and West Central grabens) which are bounded by and formed from faults with small displacements.

On the top basement map derived from the deep reflection data (Fig. 11) the depth contours on the top of the Mid-North Sea – Ringkøbing-Fyn High extend toward the Central graben, defining the High as a region of shallow basement. Permian sediments are almost absent from the High (Fig. 6). Seismic refraction/wide-angle data

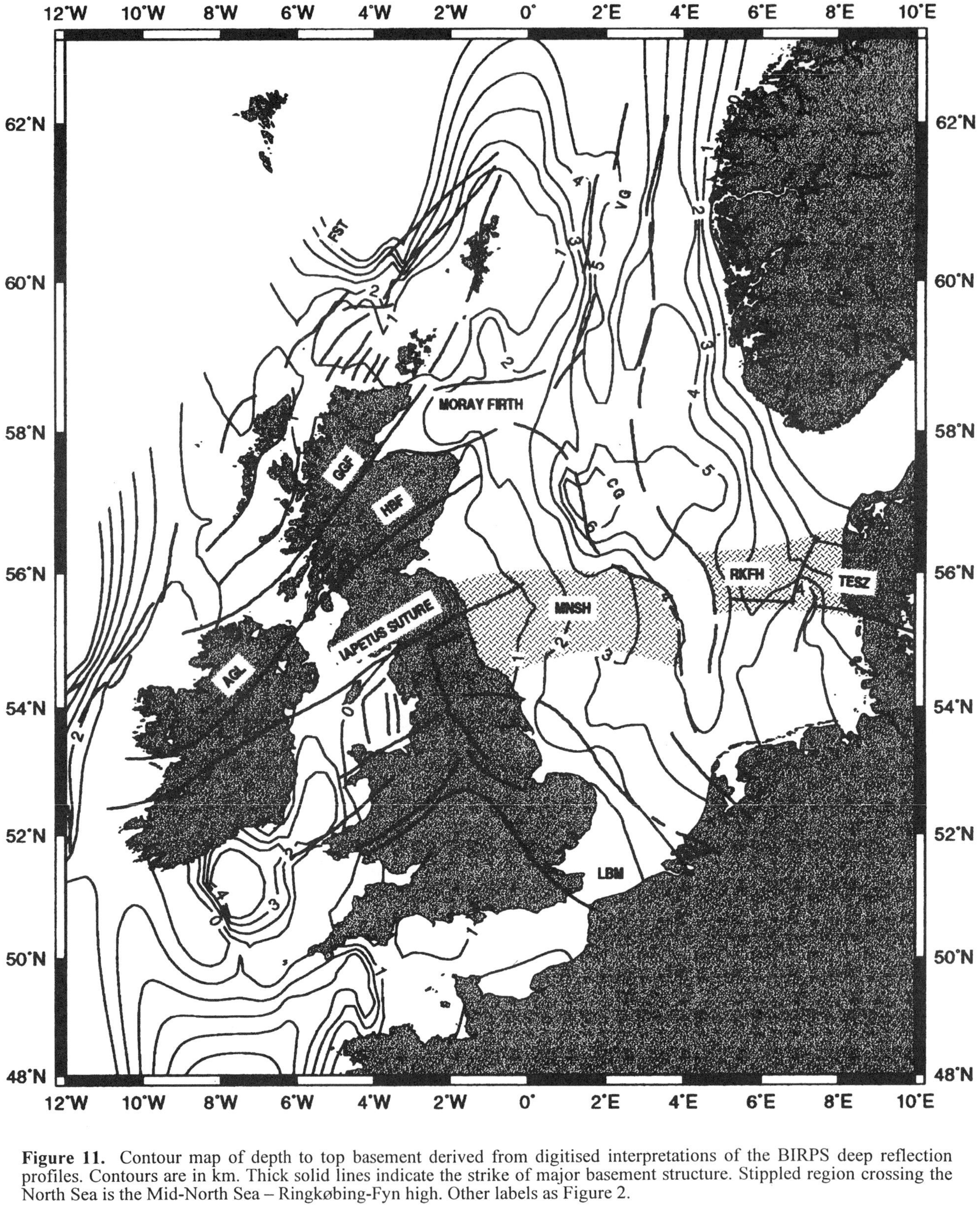

Figure 11. Contour map of depth to top basement derived from digitised interpretations of the BIRPS deep reflection profiles. Contours are in km. Thick solid lines indicate the strike of major basement structure. Stippled region crossing the North Sea is the Mid-North Sea – Ringkøbing-Fyn high. Other labels as Figure 2.

confirm that the high is formed from continental crust which is thicker that its surroundings [*MONA LISA Working Group*, 1998, *Singh et al.*, 1998, *Abramovitz et al.*, in press]. This region of thickened crust lies along the strike of the Trans-European or Tornquist suture between Eastern Avalonia and Baltica, which closed during the late Ordovician [*McKerrow et al.*, 1991] and the Iapetus suture, which closed during the late Ordovician or early Silurian, with continued shortening into the early Devonian [*Soper et al.*, 1992].

The narrowing of the Central graben where it cuts the Mid-North Sea – Ringkøbing-Fyn high (Figs. 1 and 6) may indicate that either the type or thickness of the basement exerts an influence on the style of basin formation. Before this possibility is developed it is necessary to consider alternative controls on the style of rifting.

It is possible that the changes in rift geometry are dependent upon variations in the thickness of sediments from the first (Triassic) rifting event which affected the North Sea. A thick, insulating layer of sediment would trap heat within the basement. The theoretical model of *Buck* [1991] suggests that wide rifts form from relatively warm crust (thin lithosphere), because the buoyancy force counter-acting rifting produced by dense mantle lithosphere rising into the rift is only effective if the whole lithosphere is initially relatively thin.

Permian and Triassic successions are thinner across the Mid-North Sea High than in adjacent areas [*Cameron et al.*, 1992; *Gatliff et al.*, 1994] and hence, following the model of *Buck* [1991], the regions north and south of the High, covered by thicker sediments, would behave as warm crust forming a wide rift. The High would behave as cold crust and a narrow rift would form in this region. However, the absence of any change in the thickness of the thermal subsidence sediments between the High and the margins of the central and southern parts of the graben, which were deposited after the second rift phase, would appear to exclude variations in lithospheric thickness prior to rifting. Since c. 80 Ma elapsed between the two rifting events the thermal anomaly from the first rift phase would have largely decayed. However, if Middle Jurassic doming beneath the central North Sea was caused by a thermal anomaly in the mantle [*Underhill and Partington*, 1993] the effective thickness of the mantle lithosphere may have been reduced on a regional scale by removal of its thermal boundary layer, irrespective of crustal structure, immediately prior to the second rift event. This inferred thermal event was much closer to the rift event and consequently the thickness distribution of the Cenozoic sediments reflecting changes in lithosphere thickness during rifting (Fig. 12) would be expected to reflect the shape of the Middle Jurassic thermal anomaly, as well as the anomaly imposed by Late Jurassic crustal extension. This thermal event might have had the effect of overprinting any effects due to variations in the thickness of the mantle beneath the High prior to rifting. However, the map showing the base of the Paleogene sediments (which equates to the thickness of the thermal subsidence related sediments) shows only a depression parallel to the rift system and shows no evidence for additional subsidence beneath the Outer Moray Firth - Central graben - South Viking graben triple junction resulting from the roughly circular thermal anomaly mapped from the Middle Jurassic unconformity by *Underhill and Partington* [1993]. Consequently, it must be concluded that there is little evidence of any control on the style of rifting in the North Sea from existing thermal anomalies or variations in the thermal structure of the crust or the mantle lithosphere.

It is possible that the widening of the graben to the south of the Mid-North Sea – Ringkøbing-Fyn High can be explained by reactivation of Acadian structures parallel to the edge of the Midland Microcraton, distributing strain over a wide region. However, this explanation cannot be used to account for the widening of the Central graben into the East and West Central grabens to the north of the Mid-North Sea – Ringkøbing-Fyn high, where known Caledonian basement structures have a strike orthogonal to the Central graben [*Klemperer et al.*, 1990]. Isopach maps, the NSDP85-7 profile (Fig. 6) which runs approximately parallel to the Central graben and the MONA LISA 2 profile (Fig. 6) which crosses the edge of the Ringkøbing-Fyn high parallel to the Central graben show no noticeable change in the width, thickness or symmetry of the Cenozoic post-rift sediments over the Mid-North Sea – Ringkøbing-Fyn high compared with surrounding regions (Fig. 6). Since this phase of sedimentation records thermal recovery of strains sustained by the mantle lithosphere during rifting, the lack of variation in the post-rift sedimentary package suggests that stretching within the mantle was uniform and hence variations in the syn-rift structure are related entirely to variations in the properties of the crust.

5.3. Effect of Crustal Structure on Style of Rifting

The NSDP85-7 and MONA LISA 2 profiles (Figs. 6) show that at present the basement is slightly thicker beneath the Mid-North Sea – Ringkøbing-Fyn High and the top Rotliegend reflector is at a shallower level over the High (Fig. 6). Permian (Zechstein) and Mesozoic sediments thin or are absent over the High suggesting that it remained as a positive feature during this period. These two observations suggest that the High may have been caused by the crust beneath it having properties distinct from the surrounding crust during the Permian and Mesozoic, which have survived to the present. There is no strong free air gravity anomaly associated with this feature which suggests that it is isostatically compensated. Modelling of wide-angle/refraction data from beneath the Mid-North Sea high [*Singh et al.*, 1998] and across the southern edge of the Ringkøbing-Fyn High [*Abramovitz et al.*, in press] confirms that the crust beneath the High is thicker than that surrounding it and suggests that the lower crust has a higher velocity (c. 7.2 km/s) than to the north and south of the High (c. 6.8 km/s). *Abramovitz et al.* [in press]. show that at the southern edge of the Ringkøbing-Fyn High the velocity structure of the crust changes from two layer to the south to three layer to the north, which corresponds to the edge of the Baltic plate [*BABEL Working Group*, 1993]. These observations indicate that the properties of the lower crust beneath the High contrast with those of the surrounding crust. The distinct properties and structure of the Mid-North Sea – Ringkøbing-Fyn High may be coincidental with the change in the width of

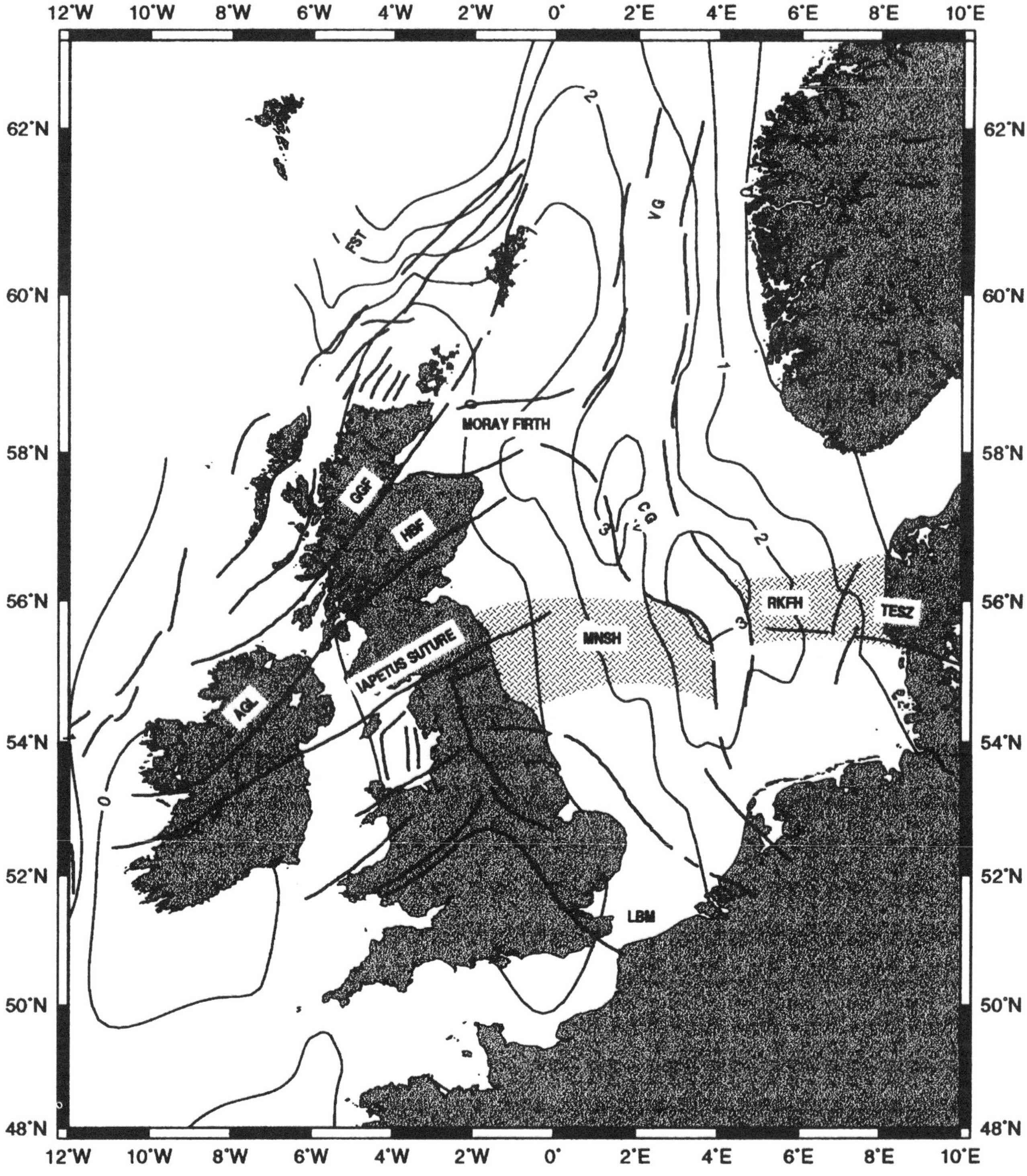

Figure 12. Contour map of depth to base Paleogene derived from digitised interpretations of the BIRPS deep reflection profiles. Contours are in km. Thick solid lines indicate the strike of major basement structure. Stippled region crossing the North Sea is the Mid-North Sea – Ringkøbing-Fyn high. Other labels as Figure 2.

the Central Graben, but the correlation between changes in crustal structure and changes in rift structure strongly suggests that it is the properties of the crust which have influenced the style of rifting.

The higher velocity of the basement beneath the Mid-North Sea – Ringkøbing-Fyn High suggests that the lower crust is formed from mafic rocks, either gneiss or basaltic underplated material. Since the high is adjacent to the Iapetus and Trans European Fault/Tornquist suture zones it is possible that it formed as a result of the closure of these two oceans during the Silurian/early Devonian and Ordovician/Silurian respectively. However, other crustal roots associated with Caledonian deformation have been removed. The seismic data indicate that the high formed before the Permian Salt basins which it divides; which dates the high at pre-Permian. It is therefore possible that the high formed as the result of underplating [*Smith*, 1992] associated with the extensive Permo-Carboniferous volcanism and intrusive activity which affected the region. The linear track of the high may record the trace of a plume beneath the continental lithosphere during this time.

The effect of stronger crust on the style of rifting would be to delay the time at which the crust failed. The lower crust would still be weaker than the upper mantle immediately beneath and faulting would only begin once the mantle began to deform. As the mantle began to deform weaker crust surrounding the high would fail gradually forming small faults since it could only accumulate low stresses before failure, but large stresses would accumulate in the stronger basement beneath the high. When the basement beneath the high eventually failed large magnitude earthquakes would occur and large faults would form, which is consistent with the presence of the large Coffee Soil fault bounding the eastern side of the graben across the high.

6. CONCLUSIONS

There is a distinct difference between patterns of Mesozoic basin development to the east and the west of the UK. North and west of Britain a case can be made, based on field data and structural and stratigraphic correlation, for basin development occurring through movement on generally east dipping large faults traversing the basement. In the West Orkney and Sea of the Hebrides Basins it can be demonstrated that such faults have been intermittently active as normal faults since the late Precambrian and that they have controlled basin geometry during the Precambrian, late Palaeozoic and the Mesozoic. It is probable that similar structures also controlled Mesozoic basin development to the west of Shetland and in the Faroe-Shetland Trough, but it is not possible to unequivocally demonstrate this with the data that has been acquired to date. Since their formation in the Riphean, these major faults appear to have remained as weak surfaces in the lithosphere, moving whenever the crust was exposed to favourable stress directions as the relatively strong underlying mantle lithosphere began to deform.

To the East of the UK, Mesozoic stretching resulted in the development of the North Sea Rift system. In contrast to the west of the UK, there is no strong evidence for the orientation of the North Sea rift system being influenced by existing structures. The North Sea rift strikes at a high angle to Caledonian and Variscan structures, except in the region of the southern gas basin where faults active during crustal extension are parallel to late Caledonian/Acadian structures which developed along the edge of the Midland Microcraton (Fig. 1). Variations in the structure of the rift along its strike correspond with variations in the velocity structure and thickness of the basement. Where it traverses the Mid-North Sea – Ringkøbing-Fyn High the rift narrows considerably and its formation is controlled by a single large fault (the Coffee Soil Fault). The smoothly varying distribution of Cenozoic sediments deposited after significant faulting ceased suggests that the mantle lithosphere deforms in a smoothly varying manner, decreasing away from the rift axis. Since the Cenozoic sediments are distributed over a wider area than the fault bounded graben, stretching of the mantle must have occurred over a broader area than in the crust, indicating that strains must decrease with depth in the mantle lithosphere. Some of this depth-dependent stretching may be accommodated by ductile flow in the weak lower crust, but it is unlikely that strain gradients due to this this process are large, since the axes of maximum sediment accumulation in the syn- and post-rift phases of basin development are coincident. If there is significant ductile deformation of the lower crust it is more probably to accommodate local variations in strain due to syn-rift deformation of crust of varying properties. Consequently, it is concluded that variations in rift structure are not related to variations in the thickness or properties of the underlying mantle lithosphere, but to variations in properties of the crust.

Acknowledgements. This paper was considerably improved following the comments of Derek Blundell, Alastair Beach, Dave Roberts and an anonymous referee, and discussion with numerous colleagues. Digitising of the BIRPS data used in the contour maps and cross sections was by Matthew Grainger. BIRPS was funded by the NERC and the BIRPS Industrial Associates (Amerada Hess Ltd., Arco British Ltd., BP Exploration Operating Co. Ltd., British Gas Exploration and Production Ltd., Conoco (UK) Ltd., Lasmo North Sea plc., Mobil North Sea Ltd., Shell (UK) Ltd., Statoil (UK) Ltd.). Cambridge Earth Sciences publication number 5634.

REFERENCES

Abramovitz, T., and H. Thybo, Seismic images of Caledonian, Lithosphere-scale collision structures in the southeastern North Sea along MONA LISA profile 2, *Tectonophysics*, in press, 1999.

BABEL Working Group, Deep seismic reflection/refraction interpretation of BABEL profiles A and B in the southern Baltic Sea, *Geophys. J. Int.*, *112*, 325-343, 1993.

Barton, P., and R. Wood, Tectonic evolution of the North Sea basin: crustal stretching and subsidence, *Geophys. J. R. Astron. Soc.*, *79*, 987-1022, 1984.

Blundell, D. J., R. W. Hobbs, S. L. Klemperer, R. Scott-Robinson, R. E. Long, T. E. West, and E. Duin, Crustal structure of the central and southern North Sea from BIRPS deep seismic reflection profiling, *J. Geol. Soc. London*, *148*, 445-457, 1991.

Buck, W. R., Modes of continental lithospheric extension, *J. Geophys. Res.*, 96, 20161-20178, 1991.

Busby, J. P., G. S. Kimbell, and T. C. Pharaoh, Integrated geophysical/geological modelling of the Caledonian and Precambrian basement of southern Britain, *Geol. Mag., 130*, 593-604, 1993.

Butler, C. A., R. E. Holdsworth, and R. Strachan, Evidence for Caledonian sinistral strike-slip motion and associated fault zone weakening, Outer Hebrides Fault Zone, NW Scotland, *J. Geol. Soc. London, 152*, 743-746, 1995.

Butler, R. W. H., Late Proterozoic rift faults and basement cover relationships within the Ben More thrust sheet, NW Scotland, *J. Geol. Soc. London, 154*, 761-764, 1997.

Cameron, T. D. J., A. Crosby, P. S. Balson, D. H. Jeffery, G. K. Lott, J. Bulat, and D. J. Harrison, *United Kingdom offshore regional report: the geology of the southern North Sea,*, HMSO, London, 1992.

Chadwick, R. A., and T. C. Pharaoh, The seismic reflection Moho beneath the United Kingdom and adjacent areas, *Tectonophysics, 299*, 255-279, 1998.

Cheadle, M. J., S. McGeary, M. R. Warner, and D. H. Matthews, Extensional structures on the western UK continental shelf: a review of evidence from deep seismic profiling, in *Continental Extensional tectonics*, edited by M. P. Coward, J. F. Dewey, and P. L. Hancock, *Geol. Soc. London Special Publ., 28*, 445-465, 1987.

Coward, M. P., The effect of Late Caledonian and Variscan continental escape tectonics on basement structure, Paleozoic basin kinematics and subsequent Mesozoic basin development in NW Europe, in *Petroleum geology of Northwest Europe, Proceedings of the 4th conference*, edited by J. R. Parker, Geological Society, London., 1095-1108, 1993.

Duindam, P., and B. Van Hoorn, Structural evolution of the West Shetland continental margin, in *Petroleum Geology of NW Europe*, edited by J. Brooks and K. W. Glennie, Graham and Trotman, London, 765-773, 1987.

Enfield, M. A., and M. P. Coward, The structure of the West Orkney Basin, northern Scotland, *J. Geol. Soc. London, 144*, 871-884, 1987.

England, R. W., R. W. H. Butler, and D. H. W. Hutton, The role of Paleocene magmatism in the Tertiary evolution of basins on the NW seaboard, in *Petroleum Geology of Northwest Europe: Proceedings of the 4th Conference*, edited by J. R. Parker, Geological Society, London, 97-105, 1993.

England, R. W., and R. W. Hobbs, The structure of the Rockall Trough imaged by deep seismic reflection profiling. *J. Geol. Soc., London, 154*, 497-502, 1997.

Fraser, A. J., and R. L. Gawthorpe, Tectono-stratigraphic development and hydrocarbon habitat of the Carboniferous in northern England, in *Tectonic events responsible for Britain's oil and gas reserves,* edited by R. P. F. Hardman, and J. Brooks, *Geol. Soc. London Special Publ., 55*, 49-86 1990.

Gatliff, R. W., P. C. Richards, K. Smith, C. C. Graham, M. McCormac, N. J. P. Smith, D. Long, T. D. J. Cameron, D. Evans, A. G. Stevenson, J. Bulat, and D. J. Ritchie, *United Kingdom offshore regional report: the geology of the central North Sea*, HMSO, London, 1994.

Gibbs, A. D., A model for linked basin development, around the British Isles, in *Extensional tectonics and stratigraphy of the North Atlantic Margins*, edited by A. J. Tankard, and H. R. Balkwill, H. R., *A. A. P. G. Memoir, 46*, 501-509, 1989.

Glennie, K. W., Lower Permian – Rotliegend, in *Introduction to the Petroleum Geology of the North Sea, 3rd edition*, edited by K. W. Glennie, Blackwell Scientific Publications, Oxford, 120-152, 1990.

Fyfe, J. A., D. Long, and D. Evans, *United Kingdom offshore regional report: the geology of the Malin-Hebrides Sea area,*. HMSO London, 1993.

Hartz, E. H., T. H. Torsvik, and A. Andresen, Carboniferous age for the East Greenland "Devonian" basin: Paleomagnetic and isotopic constraints on age, stratigraphy, and plate reconstructions, *Geology, 25*, 675-678, 1997.

Holdsworth, R. E., C. A. Butler, and A. M. Roberts, The recognition of reactivation during continental deformation, *J. Geol. Soc., London, 154*, 73-78, 1997.

Holliger, K., and S. L. Klemperer, Gravity and deep seismic reflection profiles across the North Sea rift, in *Tectonic evolution of the North Sea rifts*, edited by D. J. Blundell, and A, D. Gibbs, Oxford University Press, Oxford, 82-100, 1990.

Klemperer, S. L., and N. White, Coaxial stretching or lithospheric simple shear in the North Sea? Evidence from deep seismic profiling and subsidence, in *Extensional tectonics and stratigraphy of the North Atlantic Margins*, edited by A. J. Tankard, and H. R. Balkwill, H. R., *A. A. P. G. Memoir, 46*, 511-522, 1989.

Klemperer, S. L., and R. W. Hobbs, *The BIRPS Atlas, deep seismic reflection profiles around the British Isles*, Cambridge University Press, Cambridge UK, 1992.

Klemperer, S. L., R. W. Hobbs, and B. Freeman, Dating the source of lower crustal reflectivity using BIRPS deep seismic profiles across the Iapetus suture, *Tectonophysics, 173*, 445-454, 1990.

Kusznir, N. J., and D. Matthews, Deep seismic reflections and the deformational mechanisms of the continental lithosphere, *J. Petrology (Special Lithosphere issue)*, 63-87, 1988.

Kusznir, N. J., and R. G. Park, The extensional strength of the continental lithosphere: its dependence on geothermal gradient, and crustal composition and thickness, in *Continental Extensional tectonics*, edited by M. P. Coward, J. F. Dewey, and P. L. Hancock, *Geol. Soc. London Special Publ., 28*, 35-52, 1987.

Larsen, P-H., and H-J. Bengaard, Devonian basin initiation in East Greenland: a result of sinistral wrench faulting and Caledonian extensional collapse, *J. Geol. Soc. London, 148*, 355-368, 1991.

Lee, M. K., T. C. Pharaoh, J. P. Williamson, C. A. Green, and W. De Vos, Evidence on the deep structure of the Anglo-Brabant Massif from gravity and magnetic data, *Geol. Mag., 130*, 575-582, 1993.

Leeder, M. R. and R. P. F. Hardman, Carboniferous geology of the southern North Sea basin and controls on hydrocarbon prospectivity, in *Tectonic events responsible for Britain's oil and gas reserves*, edited by R. P. F. Hardman, and J. Brooks, *Geol. Soc. London Special Publ., 55*, 87-106, 1990.

McClay, K. R., M. G. Norton, P. Coney, and G. H. Davis, Collapse of the Caledonian orogen and the Old Red Sandstone, *Nature, 323*, 147-149, 1986.

McKerrow, W. S., J. F. Dewey, and C. F. Scotese, The Ordovician and Silurian development of the Iapetus ocean, *Special Papers in Palaeontology, 44*, Palaeontological Society, London, 165-178, 1991.

Meissner, R., D. Matthews, and Th. Wever, The "Moho" in and around Great Britain, *Annales Geophysicae, 4B*, 659-64, 1986.

MONA LISA Working Group, Closure of the Tornquist sea: constraints from MONA LISA deep seismic reflection data, *Geology, 25*, 1071-1074, 1997.

Mudge, D. C., and B. Rashid, The geology of the Faeroe Basin area, in *Petroleum Geology of NW Europe*, edited by J. Brooks and K. W. Glennie, Graham and Trotman, London, 751-763, 1987.

Needham, T., and R. Morgan, The East Irish Sea and adjacent basins: new faults or old? *J. Geol. Soc. London, 154*, 145-150, 1997.

Newman, R., and N. White, Rheology of the lithosphere inferred from sedimentary basins, *Nature, 385*, 621-624, 1997.

O'Neill, P. S., and R. W. England, The structure of the Sea of the Hebrides Basin, (NW Scotland): An integrated gravity and seismic model, *Scot. J. Geol., 30*, 1-9, 1994.

Reston, T., Shear in the lower crust during extension: not so pure and simple, *Tectonophysics, 173*, 175-183, 1990.
Rowley, D., and D. B. Sahagian, Depth-dependent stretching: a different approach, *Geology, 14*, 32-35, 1986.
Ryan, P. D., N. J. Soper, D. B. Snyder, R. W. England, and D. H. W. Hutton, The Antrim-Galway line: A resolution of the Highland Border Fault enigma of the British Caledonides, *Geol. Mag., 132*, 171-184, 1995.
Seranne, M., Devonian extensional tectonics versus Carboniferous inversion in the northern Orcadian basin, *J. Geol. Soc. London, 149*, 27-37, 1992.
Singh, S. C., P. Hague, and M. McCaughey, Study of the crystalline crust from a 2-ship normal incidence and wide-angle experiment, *Tectonophysics, 286*, 79-91, 1998.
Snyder, D. B., R. W. England, and J. H. McBride, Linkage between mantle and crustal structures and its bearing on inherited structures in northwestern Scotland, *J. Geol. Soc. London, 154*, 79-83, 1997.
Smith, K., Carboniferous magmatism in the Iapetus convergence zone: evidence from deep seismic reflection profiles, *J. Geol. Soc. London, 149*, 907-914, 1992.
Soper, N. J., Neoproterozoic sedimentation on the northeast margin of Laurentia and the opening of Iapetus, *Geol. Mag., 131*, 291-299, 1994.
Soper, N. J., and R. Anderton, Did the Dalradian slides originate as extensional faults? *Nature, 307*, 357-360, 1984.
Soper, N. J., and R. W. England, Vendian and Riphean rifting in NW Scotland, *J. Geol. Soc. London, 152*, 11-14, 1995.
Soper, N. J., R. W. England, D. B. Snyder, and P. D. Ryan, The Iapetus Suture in England, Scotland and eastern Ireland: a reconciliation of geological and deep seismic data, *J. Geol. Soc. London, 152*, 697-700, 1992.
Soper, N. J., and A. K. Higgins, Basement-cover relationships in the East Greenland Caledonides: evidence from the Eleonore Bay Supergroup at Ardencaple Fjord, *Trans. R. Soc. Edinburgh: Earth. Sci., 84*, 103-115, 1993.
Soper, N. J., B. C. Webb, and N. H. Woodcock, Late Caledonian (Acadian) transpression in north-west England: timing, geometry and geotectonic significance, *Proc. Yorks. Geol. Soc., 46*, 175-192, 1987.
Steel, R. J., New Red Sandstone movement on the Minch Fault, *Nature, Phys. Sci., 234*, 158-159, 1971.
Stein, A. M., Basement controls upon basin development within the Caledonian foreland, N. W. Scotland, *Basin Res., 1*, 107-119, 1988.
Stein, A. M., and D. J. Blundell, Geological inheritance and crustal dynamics on the northwest Scottish continental shelf, *Tectonophysics, 173*, 455-467, 1990.
Stewart, A. D., Late Proterozoic rifting in NW Scotland: the genesis of the 'Torridonian', *J. Geol. Soc. London, 139*, 413-420, 1982.
Underhill, J. R., and M. A. Partington, Jurassic thermal doming and deflation in the North Sea: implications of the sequence stratigraphic evidence, in *Petroleum geology of Northwest Europe, Proceedings of the 4th conference*, edited by J. R. Parker, Geological Society, London, 337-345, 1993.
Wessel, P., and W. H. F. Smith, Free software helps map and display data. *Eos Trans. AGU, 72*, 441, 445-446, 1991.
White, N., Nature of lithospheric extension in the North Sea, *Geology, 17*, 111-114, 1989.
White, N., and D. McKenzie, Formation of the "Steers Head" geometry of sedimentary basins by differential stretching of the crust and mantle, *Geology, 16*, 250-253, 1988.

R. W. England, Department of Earth Sciences, Bullard Laboratories, Madingley Rise, Madingley Road, Cambridge, CB3 0EZ, UK. E-mail england@esc.cam.ac.uk.

Architecture of the Continental Margin in the Gulf of Guinea as Revealed by Reprocessed Deep-Imaging Seismic Data

Bruce R. Rosendahl and Henrike Groschel-Becker

Division of Marine Geology and Geophysics, Rosenstiel School of Marine and Atmospheric Science University of Miami, Miami, Florida

Post-stack reprocessing of the PROBE seismic data set has resulted in a refined tectonic model of the continental margin between Cameroon and Southern Gabon. The margin is divided into the North Douala and Gabon Basins by the Kribi Fracture Zone, a 75 km-wide transform fault that trends NE-SW and intersects the coastline between 2-3° N. North of the Kribi Fracture Zone, oceanic crust extends essentially to the coastline and floors the narrow margin off Cameroon. Little or no offshore rift margin exists in the North Douala Basin. Reflection Moho beneath oceanic crust is relatively weak but continuous throughout the North Douala Basin, except under the aprons of the Cameroon Volcanic Line. Oceanic crustal thickness is extremely uniform in the North Douala Basin, averaging about 1.75 s in two-way travel time. A distinct NE-SW trending structural grain at both basement and reflection Moho levels occurs within the Kribi Fracture Zone. Much of the margin off Equatorial Guinea is influenced or controlled by fracture zone tectonics. South of the Kribi Fracture Zone, oceanic crust is offset about 350 km to the south-west, resulting in the broad rift margin off Gabon. Extension here has largely progressed without the complications (or simplifications) introduced by fracture zone tectonics. Rifting of the margin seems to involve brittle, upper continental crustal sheets overlying a series of ductile lower sheets or lenses that generally dip easterly to southeasterly. The ductile sheets may unroof toward outer highs, which mark the boundary between oceanic and continental lith-osphere. Strong reflection events are often associated with the ductile shear zones that define the lower sheets, but a single, continuous event representing reflection Moho cannot be delineated. We suggest the concept of a regional reflection Moho beneath rifted continental crust is invalid in this area. The tectonic and rheological models presented here may be useful in framing hydrocarbon exploration tactics and strategies along this passive margin. In a broader context, elucidation of margin architecture may be a cost-effective way to approach frontier passive margins in the next century.

1. INTRODUCTION

The handling of modern multichannel seismic reflection (MCS) data in clastic sedimentary basins has almost become standardized, and there now exist quasi-formal sets of rules to guide the collection, processing, and interpretation of such data. The remaining uncertainties principally involve second-order issues related to geometries and resolution. For example, different processors may be able to image the same normal fault structure with differing degrees of clarity, and different interpreters may disagree on the precise geometry of the fault. However, the existence of the normal fault structure (the first-order issue) is usually not in doubt.

Atlantic Rifts and Continental Margins
Geophysical Monograph 115

The "rules" become less strict where sediments are mobile (e.g., evaporites) or acoustically contorted or structureless (e.g., some carbonate bank environments), but at least data handling guidelines exist. In the above cases, the rules or guidelines have benefited enormously by high data densities, even three-dimensional seismic grids in many cases, and the ability to ground-truth the data with well information.

The situation described above contrasts with the state of MCS data handling in the deeper earth, below the sedimentary cover and within the crust and upper mantle. Here, rules or guidelines basically do not exist. There are four inter-related reasons:

1. The lack of lucid hydrocarbon incentives and the costs of "deep-imaging" MCS data have combined to limit the amount of such data available at reasonable mapping scales. Hence, the interpreter is usually faced with trying to make sense of a few widely-spaced lines that often cross widely varying geological terrains.
2. The nature of the deep-imaging end product tends to be sensitive to the original acquisitional parameters and subsequent processing routines. Thus, the interpreter is often unsure if the appearance, disappearance, or change in some acoustic parameter is geological or methodological in origin. The necessarily large source arrays and long receiver offsets used to collect deep-imaging data compound these difficulties by introducing large Fresnel zones and out-of-plane reflection events.
3. The creation of acoustic impedance contrasts in the deeper earth are presumably dominated by igneous and metamorphic processes, which theoretically can produce bizarre reflection patterns and geometries. Not surprisingly, such properties as the strength, continuity, and character of sub-basement reflections tend to be variable and ambiguous even in the best of cases. Hence, the interpreter faces considerable uncertainty in ascertaining the origin and nature of events, even where such events appear to be real reflection phenomena.
4. With a few exceptions, there is essentially no well information to directly calibrate and ground-truth deep-imaging MCS data. This problem is exacerbated by the progressive diminution in stacking velocity control with increasing depth. Thus, there is generally little information to constrain an interpreter's imagination.

The circumstances described above have apparently led academic funding agency personnel to question the cost effectiveness of deep-imaging MCS studies, as evidenced by the infrequency of such studies and the demise of certain programs (e.g., COCORP). Moreover, the exploration industry has been reticent to pick up the slack because there is a collective perception that if it cannot be reached with the drill stem, then it cannot be of much exploration value.

The treatment of deep-imaging MCS data across rifted continental margins has been subject to all of the problems and criticisms expressed above. In fact, passive margins have a reputation of being notoriously difficult to image distinctly and consistently below the break-up unconformity. In this paper we undertake an evaluation and appraisal of the utility of deep-imaging seismic data across rifted margins, in the context of the recently reprocessed PROBE data across the West African passive margin [*Rosendahl*, 1997]. What can be learned from such data, what are the interpretational limits, and what are the principal remaining uncertainties or ambiguities? Does the understanding of this margin derived from the reproceesed PROBE data justify the overall costs and efforts, and hence justify a re-evaluation of the utility of such information from both academic and explorational perspectives?

2. BACKGROUND

It has long been known that the crust must thin toward the oceanic sectors of passive margins. Accordingly, geophysical studies of passive margins have tended to focus on improved definition of Moho, especially in a spatial sense, and better resolution of the mechanical and rheological details of the thinning process. The various methodological limitations of magnetic, gravimetric, seismic refraction, and wide-angle reflection techniques opened the door for deep-imaging MCS transects of passive margins, which in theory could provide continuous profiles across the margins, including the imaging of reflection Moho (RM). In practice the definition of the structural details and the identification of RM proved elusive for the reasons discussed above. Still, enough successful imaging instances were accumulated to lend credence to the belief that RM was at least a proxy of refraction Moho, and that it may be a mappable event.

The PROBE seismic program on the West African passive margin was specifically designed to image and ultimately map the deep structure of a large segment of a rift margin spanning several hydrocarbon provinces. A grid of nearly 8000 line-km of MCS data were acquired from the region 4^0 N to 5^0 S in the Gulf of Guinea. Line coverage spanned the margin from Northern Cameroon to Southern Gabon and traversed the transition zone between continental and oceanic lithosphere. Most of the data were shot with a 7600 cu. in. airgun array and recorded on a 7100 m-long receiving cable in a 60-fold format. Almost all of the data were recorded to at least 20s of two-way travel time (TWTT).

The initial results of the PROBE Study were widely reported, beginning with *Rosendahl et al.* [1991,1992] and continuing with *Meyers* [1995], *Meyers et al.* [1996a,b], and

Groschel-Becker [1996]. One of the most intriguing elements of the dataset is the occurrence of strong, coherent, and widespread reflection events 6-11s deep in TWTT. Similar deep events in the PROBE study area also had been noted in a separate MCS data set by *Wannesson et al.* [1991].

The characteristics of these deep events have been described in similar ways by all of the above workers.

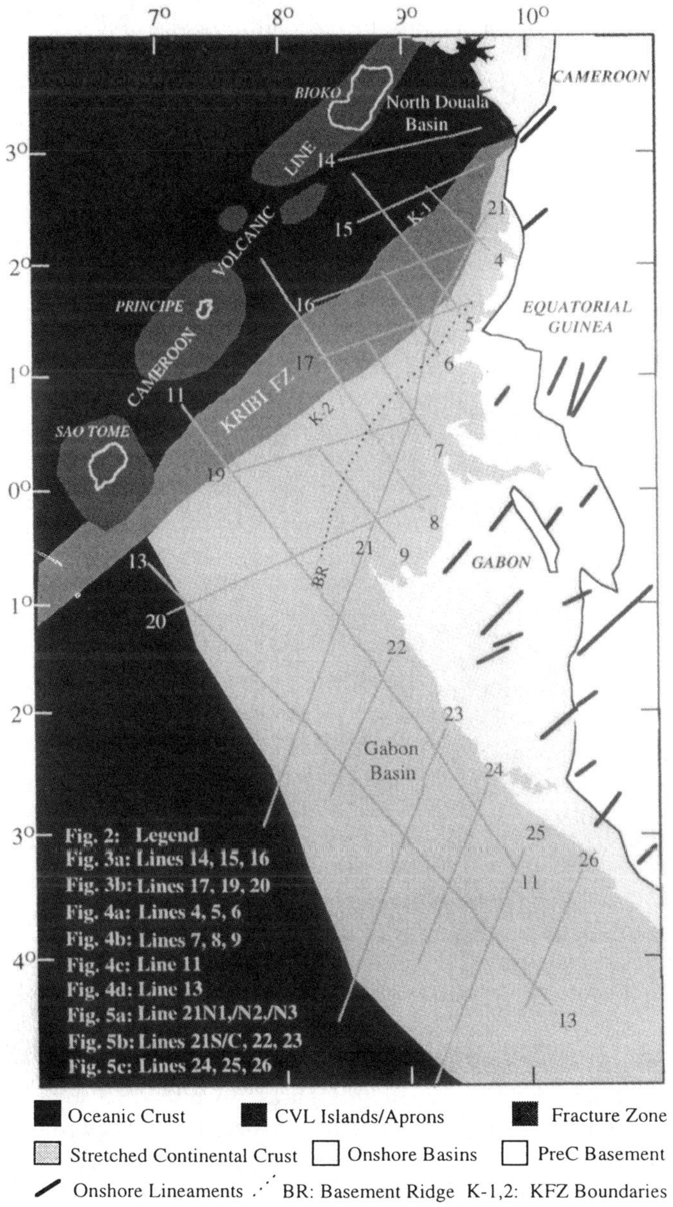

Figure 1. Locations of PROBE seismic lines used in this study superimposed on the tectonic framework of the West African margin between 4° N and 5° S. Onshore lineaments are from *Groschel-Becker* [1996] and *Meyers* [1995].

However, disagreement arose concerning the origin of the deep reflectors, specifically which set of reflector packages corresponded to reflection Moho (RM). For example, *Rosendahl et al.* [1991, 1992] generally placed RM at or near the upper level of the deepest reflector package, whereas *Wannesson et al.* [1991] differentiated the deep package into mid- to lower crustal reflections (typically 6-7 s deep) and RM (typically 10-11 s deep). Further disagreement arose within the PROBE group of workers concerning the origin of the RM acoustic impedance contrast and its relationship to the overlying crust. *Rosendahl et al.* [1992] argued that RM is a dynamic interface superimposed on a tilt-block fabric in which the main rift faults dip landward and continue through RM without offsetting it. *Meyers* [1995], *Meyers et al.* [1996a,b], and *Groschel-Becker* [1996] depict a rift fabric in which the crustal faults mainly dip seaward. *Meyers* [1995] generally terminated the faults at RM, whereas *Groschel-Becker* [1996] continued major crustal faults, associated with fracture zones, through RM. All of the PROBE workers noted the occurrence of major offsets in RM correlated with generally NE-SW trending fracture zones. However, *Rosendahl et al.* [1992] argued for "few such instances", whereas *Meyers* [1995], *Meyers et al.* [1996a,b], and *Groschel-Becker* [1996] mapped many such events. The exact trends and positions of the fracture zones also differed between the authors.

These interpretational differences were troubling and ultimately raised serious questions about what the PROBE MCS data really showed. Since early 1996, the PROBE seismic data have been subjected to systematic, post-stack reprocessing to resolve (or better define) some of the ambiguities raised in these earlier interpretations. The major problems concern the identification and positioning of fracture zones and crustal boundaries; the identification, significance and meaning of RM; the nature of the coupling of the crust with apparent RM; and the orientation and geometry of faults penetrating the crystalline crust. Although the reprocessing continues, the quality and quantity of the reprocessed PROBE data permit us to address some of these issues in a more sophisticated fashion than has previously been possible.

3. LINE DRAWING INTERPRETATIONS

The locations of key margin-crossing seismic lines are shown in Figure 1, superimposed on the generalized margin architecture inferred from this study. These key lines (or line segments in many cases) are divided into three line sets on the basis of trackline orientations. Line drawing interpretations of the reprocessed line sets are shown in Figures 3-5. A legend is provided in Figure 2. These line

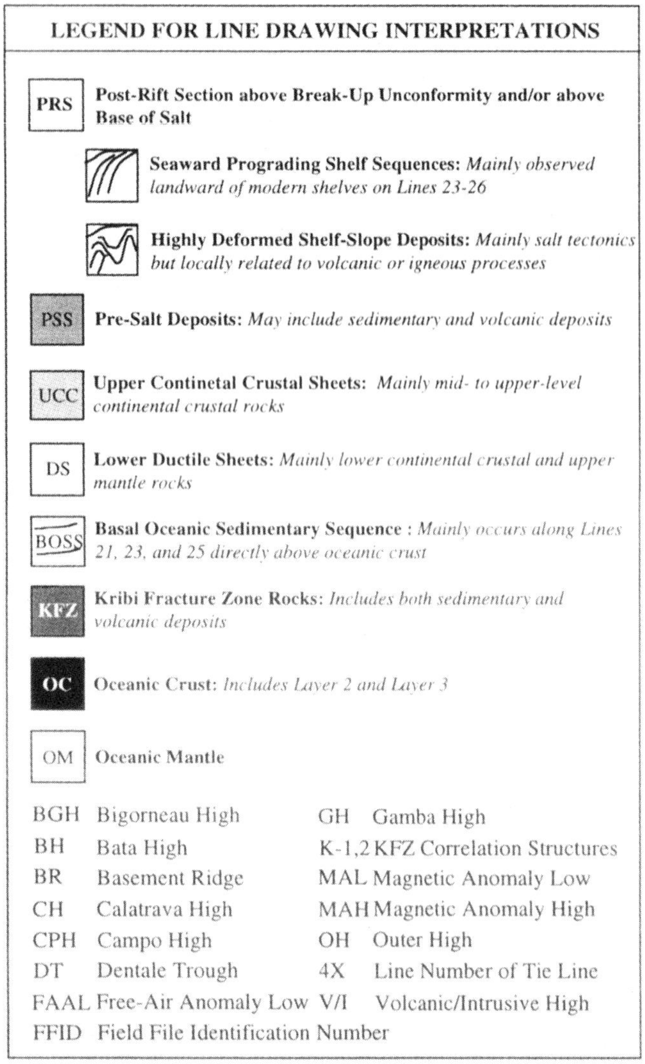

Figure 2. Legend for line drawing interpretations shown in Figures 3-5.

drawings represent a total of about 4500 km of data, or more than half of the PROBE data base. The remainder of the PROBE data are chiefly located within oceanic crust and do not add much to the results discussed below. We anticipate these line drawings will be reinterpreted and re-mapped by those seeking alternatives to the margin architecture presented in Figure 1. The groupings, arrangements, orientations, and alignments of the line drawing sets have been chosen to facilitate future mapping ease.

Considerable care has been exercised to produce line drawings that are reliable renditions of the actual data. In most cases, multiply-reprocessed data sections with various gains and filters have been used. This is a distinctive advantage of the line drawing presentation format. All crossing lines have been carefully tied and in most cases the ties of the significant reflectors are excellent. The high quality of the ties at the deeper depths imaged is surprising considering the projected diameters of the Fresnel zones and the possibility for out-of-plane reflections (e.g., sideswipe). Evidently, the source array employed was more directional than anticipated.

The least interpretative components of the line drawings are the actual tracings of the major reflections and reflector patterns. The categorization of reflector patterns and packages into units and crustal types (see Legend, Figure 2) represents a higher level of interpretation. We are very confident in our identification of oceanic crust, somewhat less confident concerning the continental units. The largest difficulty in constructing the line drawings was the identification of the post-rift/syn-rift/pre-rift boundaries along continental segments of lines, especially where salt was present. In many instances, it proved impossible to honestly define these boundaries.

Fault interpretations are not explicitly shown on the line drawings. In addition to cluttering the panels and obscuring reflectors, such interpretations impart too much bias and degrade future usefulness of these drawings. We have elected to let reflector geometries and packages "speak for themselves" regarding the existence, placement, and orientation of faults. Care also has been taken to avoid inadvertent directional aliasing during construction of the line drawings.

The decision to present the data in the form of line drawings derives from our desire to display as much of the relevant, reprocessed data as possible in the space allotted. We readily admit this presentation format is not ideal, but the alternatives are inconsistent with the intent of this paper. To compensate for the shortcomings inherent in our selection of a line drawing presentation format, we offer reasonable access to the actual reprocessed PROBE data to academic researchers, at the cost of reproduction and shipping. Unless otherwise arranged, distribution will be in SEG Y-formatted 8 mm cassette tapes covering 2000 FFIDs, or 100 km segments.

4. RESULTS

The line drawing panels in Figures 3-5 provide a loose, three-dimensional perspective of the margin architecture in this region. In plan view, the generalized architecture is deceptively simple. The main elements are shown in Figure 1 and discussed below. In actuality, there is a large amount of information contained in the line drawings that cannot be depicted in Figure 1, indicating that the margin structure is far more complex than implied by the plan view in Figure 1.

Many of these structural details cannot be mapped using the PROBE lines because the wavelength of change is much shorter than the spacing between lines. Accordingly, most of the detail contained in the line drawings is not useful to us in a mapping context, but the details still have conceptual value, as discussed below.

4.1. Oceanic Crust

The character of oceanic crust (OC) in the North Douala Basin is shown by Lines 14, 15, and 16 (Figure 3a), Lines 4, 5, and 8 (Figure 4a), and Line 11 (Figure 4c). Here, the basement is usually clearly imaged and RM is relatively weak but very continuous, except under the Cameroon volcanic aprons where RM is largely absent [*Meyers and Rosendahl,* 1991]. Oceanic RM is generally a simple reflector couplet that closely parallels the basement surface. The correlation between the basement horizon and RM is so close that the thickness of OC (layer 2 + layer 3) in TWTT does not vary by more than about 5% in the area north of the Kribi Fracture Zone (KFZ). The mean thickness in TWTT is 1.75 s. In nominal depth, oceanic crust in the North Douala Basin is 5.63 km thick, ±0.25 km. [Nominal depth is defined as the depth computed using an assumed velocity structure of 3.0 km/s for the sedimentary succession, 6.5 km/s for material designated as crust, and 8.1 km/s for the mantle]. The thickness values in time and nominal depth are somewhat less than the Atlantic mean, but well within published values for "normal oceanic crust" in the Atlantic Ocean. Considering the variability in reported thicknesses of MAR-generated oceanic crust, the uniformity is rather noteworthy. It is this uniformity in thickness, combined with the reflection character of basement and RM, that define OC in the study area. The nearly constant thickness of OC leads to the conclusion that either oceanic crustal genesis in the North Douala Basin was an extremely uniform process, or RM is coupled to basement in some dynamic way that creates or maintains an apparent constant oceanic crustal thickness. Either conclusion is possible and neither is mutually exclusive.

The oceanic crust along Line 14 (Figures 1 and 3a) is mapped to the eastern end of the profile. Here, the landward edge of oceanic crust may actually occur onshore. Hence, the narrow continental shelf in the North Douala Basin (north of 3° N) is essentially constructed on oceanic crust. Although OC appears to "climb" toward the east in the time sections of Lines 14 and 15 (Figure 3a), it actually maintains a relatively uniform total nominal depth of 14-16 km.

Within the oceanic crustal sector of the North Douala Basin, there is a clear relationship between loading and total depth to the top of OC and RM. The oceanic basement and RM are depressed (or elevated) in direct relation to the mass of the overburden. Line 5 NNW of element K-1 (Figures 1 and 4a) shows this relationship best. A plot of pressure

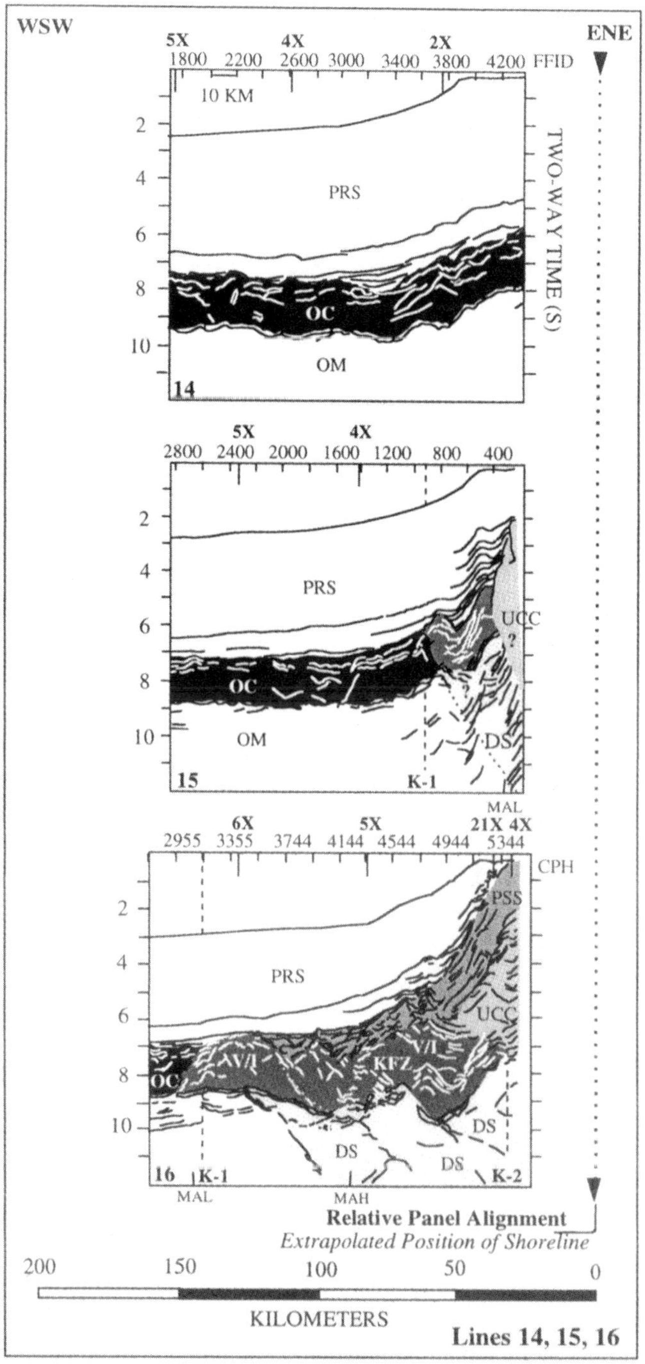

Figure 3a. Line drawing interpretations of ENE-WSW trending Lines 14, 15, and 16. See Figure 2 for legend, Figure 1 for line locations and correlations to mapped structures.

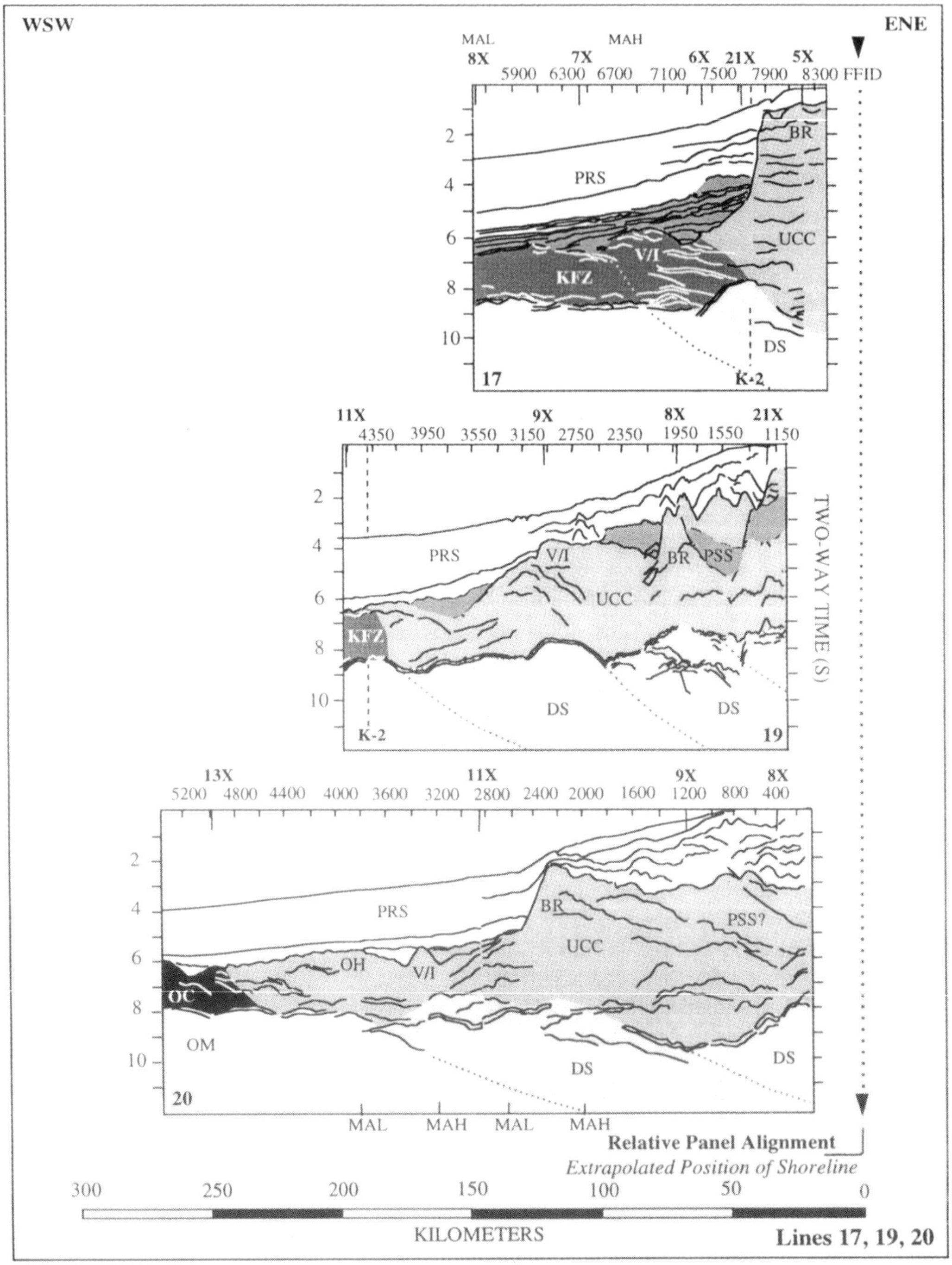

Figure 3b. Line drawing interpretations of ENE-WSW trending Lines 17, 19, and 20. See Figure 2 for legend, Figure 1 for line locations and correlations to mapped structures.

versus total depth to RM (or total depth to basement) for this sector of Line 5 would reveal a tight, sub-linear trend. Hence, the thickness of OC, using the apparent basement horizon and RM as references, does not change as a function of loading but the absolute depths of both horizons are directly related to the mass of the overburden. This suggests that the top of OC and RM are mechanically coupled, and both respond dynamically to loading.

Oceanic crust in the Gabon Basin is imaged along the seaward ends of Line 20 (Figure 3b), Line 13 (Figure 4d),

Lines 21 S/C and 23 (Figure 5b), and Line 25 (Figure 5c). However, the line coverage is limited to about 50 km or less for each of these lines. Although the average thickness of OC here remains about 1.75 s TWTT, there is a larger range in apparent thickness than in the North Douala Basin. The line drawings also suggest that OC is lumpier here than OC north of the KFZ. This may relate to the occurrence of outer highs (OH), as discussed below. It is noteworthy that the appearance of oceanic crust closely coincides with the onset of the "Basal Oceanic Sedimentary Sequence" (BOSS). This is a distinctive sequence of high-amplitude, parallel, and largely flat-lying sedimentary reflectors that infill and smooth oceanic basement just seaward of the OH.

Oceanic crust in the entire study area tends to display discontinuous, choppy reflectivity at upper levels and random, slight reflectivity at deeper levels (e.g., the OC portions of Lines 5, 8, 14, 15, and 16). Although it is tempting to correlate the former reflectivity pattern with Layer 2 and the later with Layer 3, both vertical and lateral variability is quite large and we do not, at present, give much credence to these patterns. Occasional, obliquely-trending reflectors are observed dissecting OC. However, there are fewer than about a dozen instances in which oceanic basement offsets can be confidently connected to offsets or disruptions of RM, or sub-RM events, implying the existence of through-going faults. An example of apparent through-going faulting in OC occurs between FFID 350 and 700 on Line 8 (Figure 4b); this also may be a minor fracture zone [*Groschel-Becker,* 1996]. Another example near the outer high (OH) occurs between FFID 200 and 1000 on Line 23W (Figure 5b). After careful scrutiny of all such events, we conclude that oceanic crust in the Gulf of Guinea is ruptured throughout its entire extent mainly within fracture zone terrains, or where OC joins rifted crust. Elsewhere, the mechanical breakage of the entire OC seems to be a relatively rare phenomenon, at least in the context of our seismic imagery.

4.2. Kribi Fracture Zone

The Kribi Fracture Zone (KFZ) is a 350-km transform offset crossed in whole or part by 11 different PROBE lines in the three different trackline orientations. The boundary between normal oceanic crust and the northwestern edge of the KFZ is denoted by K-1 in both Figures 1 and 3-5. The southeastern edge of the KFZ is placed at element K-2, which is structurally sensible on most crossings. Nonetheless, it is conceivable that the southeastern edge extends inboard to the "Basement Ridge" feature labeled BR in Figure 1. We return to this matter in a later discussion.

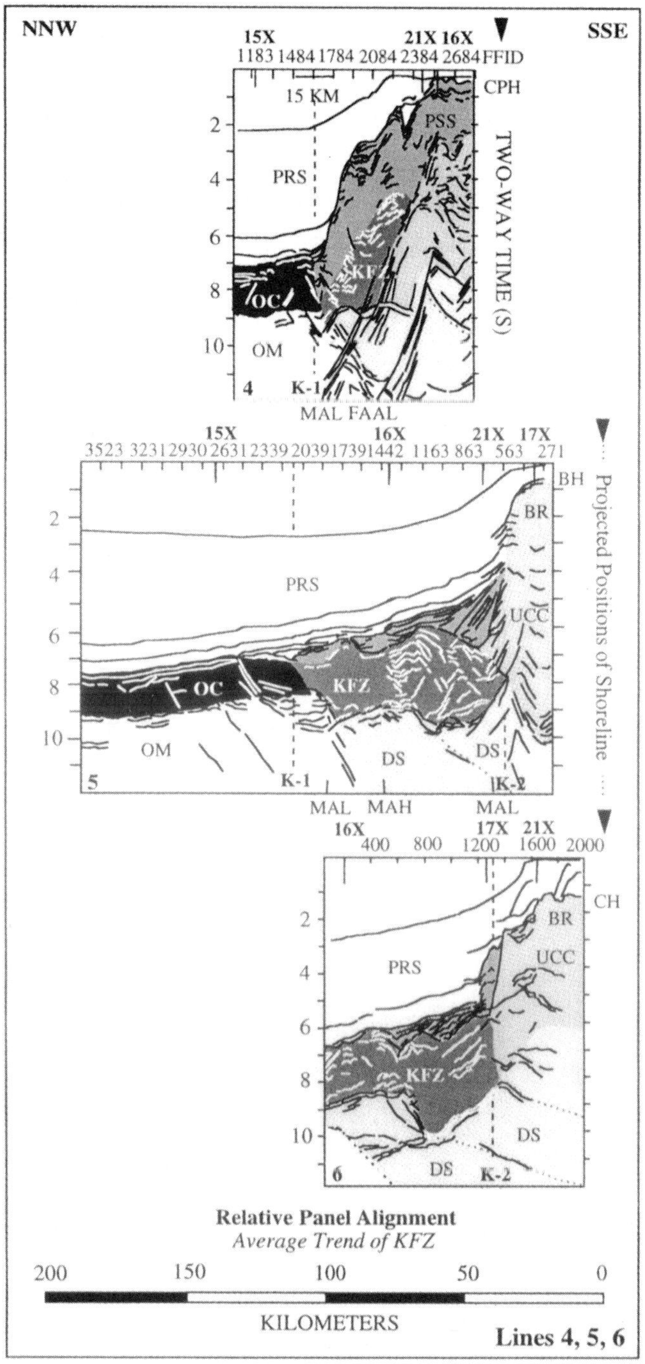

Figure 4a. Line drawing interpretations of NW-SE trending Lines 4, 5, and 6. See Figure 2 for legend, Figure 1 for line locations and correlations to mapped structures.

Crust within the KFZ is somewhat thicker in TWTT than oceanic crust immediately to the northwest of K-1. The reflectivity of apparent RM in the KFZ also is noticeably greater than RM at the base of normal OC, especially in the

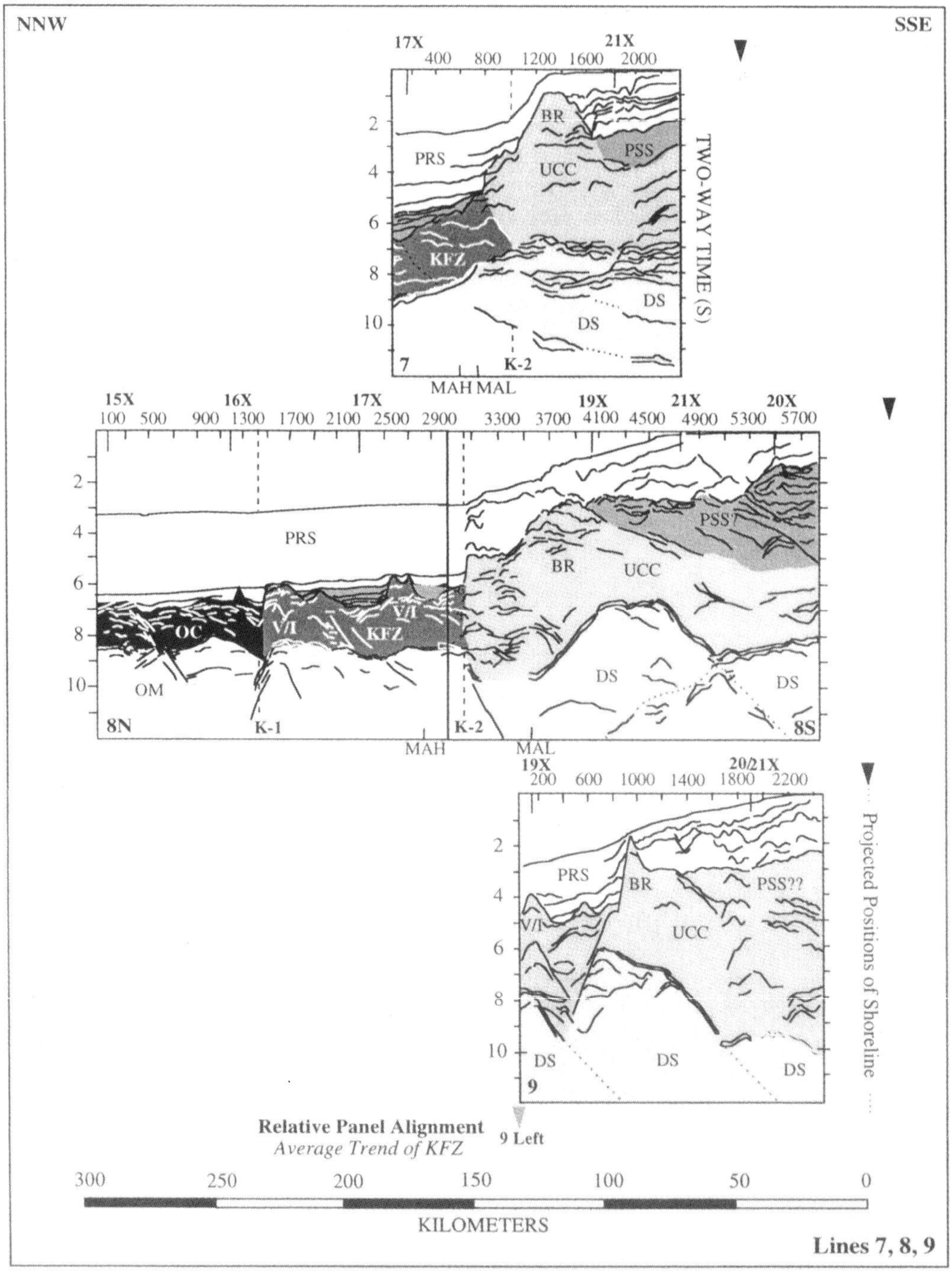

Figure 4b. Line drawing interpretations of NW-SE trending Lines 7, 8, and 9. See Figure 2 for legend, Figure 1 for line locations and correlations to mapped structures.

region of line crossings 5 and 16 (Figure 1). We suspect that fluid penetration has enhanced RM reflectivity here, via hydration of lower crustal or upper mantle rocks. Features interpreted as volcanic or intrusive highs (V/I) are common within the fracture zone, and help define its existence. We presume the KFZ contains ridge-crest and fracture zone rocks intermixed with slivers of continental rocks spalled off during creation of the transform boundary.

The internal structure of the KFZ varies considerably along its length. On Line 4 (Figure 4a), the fracture zone

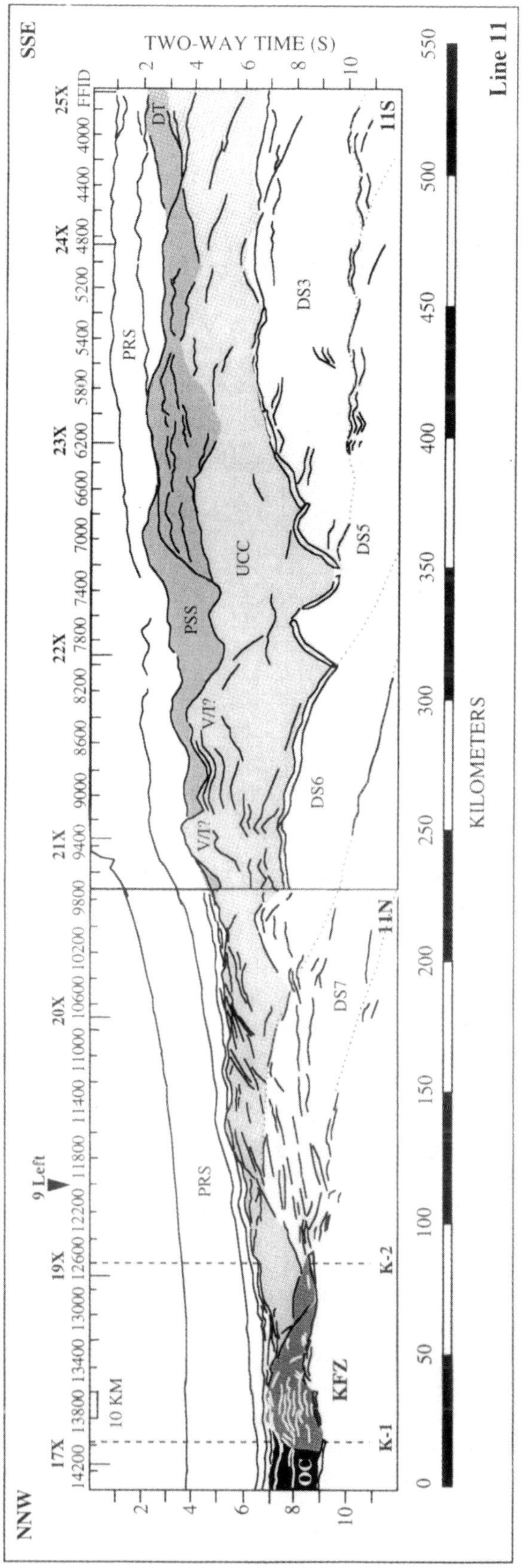

Figure 4c. Line drawing interpretation of NW-SE trending Line 11. See Figure 2 for legend, Figure 1 for line locations and correlations to mapped structures.

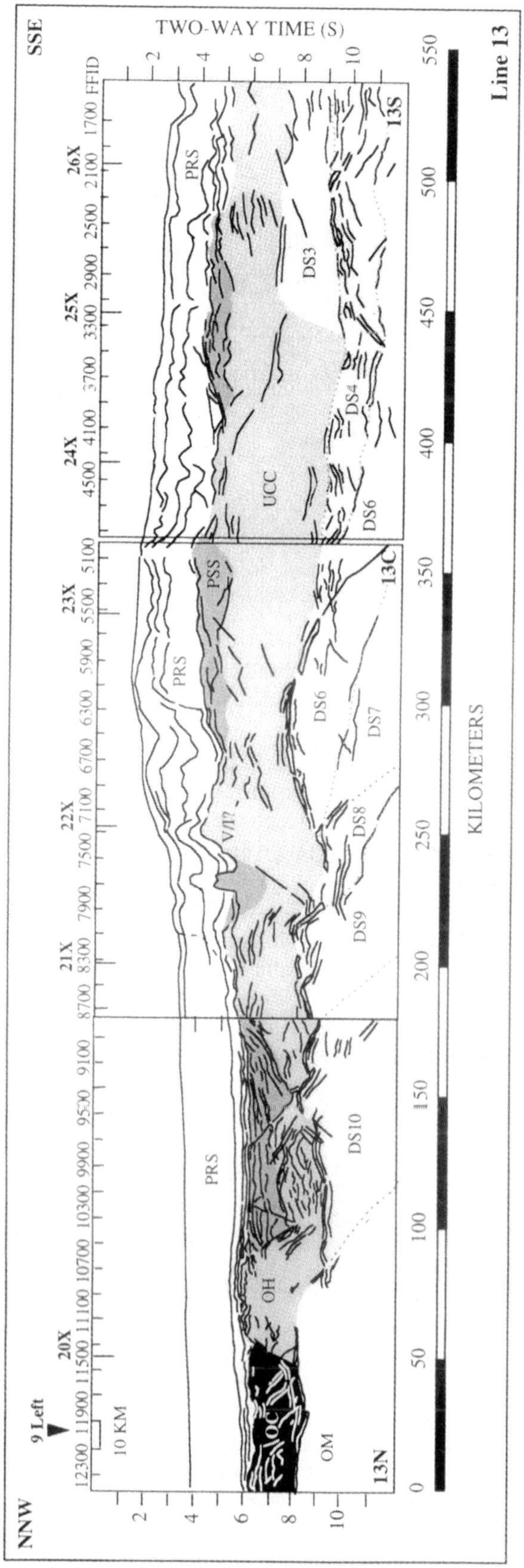

Figure 4d. Line drawing interpretation of NW-SE trending Line 13. See Figure 2 for legend, Figure 1 for line locations and correlations to mapped structures.

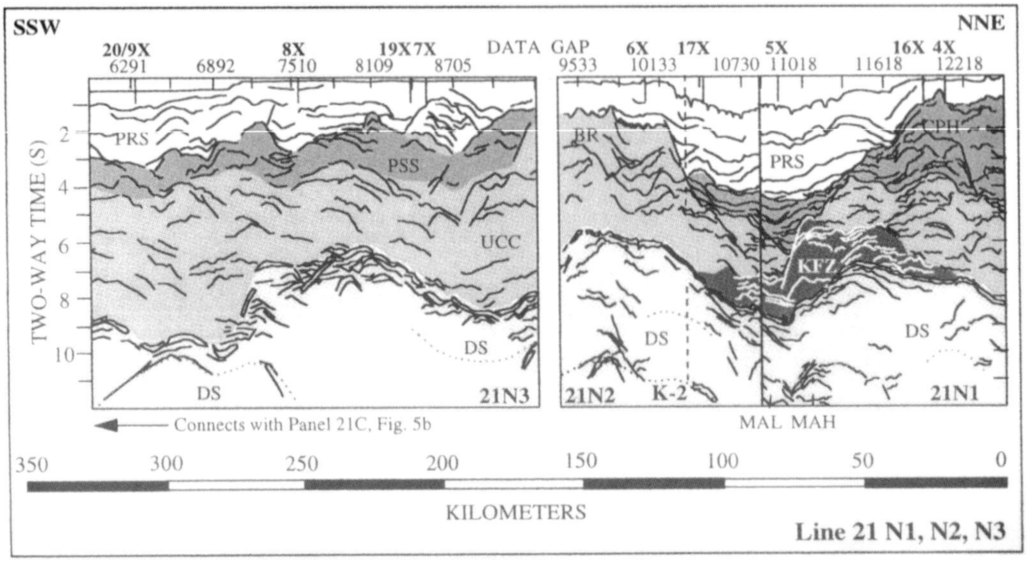

Figure 5a. Line drawing interpretation of NNE-SSW trending Line 21N1/N2/N3. Line connects with Line 21S/C in Figure 5b. See Figure 2 for legend, Figure 1 for line locations and correlations to mapped structures.

essentially juxtaposes oceanic and continental crust across a narrow zone in a rather spectacular, almost textbook-like fashion. The KFZ is broader and more complex along Line 5 (Figure 4a), but the expression is quite consistent with that shown along Line 16 (Figure 3a), and along Line 6 to a somewhat lesser degree. Between Lines 6 and 8 (Figures 4a and 4b), the KFZ becomes relatively subdued at the level of apparent RM. The KFZ is very subdued at intersection with Line 11 (Figure 4c). Lines 7 and 17 offer relatively little information because Line 7 terminates short of crossing the KFZ and Line 17 is perhaps the worst line in the PROBE data collection in terms of reflector definition (see discussion below). Between Lines 5 and 8, the KFZ has a mapped width of about 75 km, which is consistent with a fracture zone showing an offset of 350 km.

The complexity and lateral changes within and along the KFZ are not unexpected. Major oceanic fracture zones are generally complicated features and transform continental margins even more so [e.g., *Mascle et. al.,* 1996]. Although we do not believe the existence of this fracture zone is subject to much doubt, as evidenced by the radical crustal thickness differences displayed by Lines 14 and 15 compared to parallel Lines 16-19 to the south, the changes in the expression of the KFZ along strike make it difficult to map an exact trend. In our analyses this trend is defined by elements K-1 and K-2 on Lines 4, 5, 6, 8 and 16. Based upon three iterations of mapping various KFZ structures, we believe the overall trend of the KFZ shown in Figure 1 is good to within about 5° as far westward as Line 8.

The expression of the KFZ depicted here indicates that much of Line 16 (Figures 1 and 3a) traverses the KFZ at a highly oblique orientation. This is an important realization because Line 16 is a high-quality profile with many interesting structures, including tilt blocks, apparent landward-dipping faults, and asymmetric sedimentary wedges. There has been a tendency by PROBE workers, the senior author in particular, to showcase Line 16 as the type margin transect of the North Douala Basin. However, if the structures shown on Line 16 relate dominantly to fracture zone tectonics, as now believed, then it is misleading to typecast this line as the rift margin style here. This example points out the risks associated with acquiring (or indivually interpreting) one or two regional seismic transect lines without a precise *a priori* knowledge of the overall margin architecture. In fact, there is no "type transect" across the KFZ and virtually no offshore rift margin in the North Douala Basin to typecast anyway.

Some of the strongest, clearest, and most coherent basement and Moho reflections in the entire PROBE study area are observed along Line 16 within the KFZ. In fact, Line 16 displays the structural elements of the KFZ in notably better detail than the orthogonal crossings of Lines 4-9 (Figures 4a and 4b), which tend to squeeze structural detail. This may seem counter-intuitive, given that Line 16 is a highly oblique line relative to the structural trends and ought to be susceptible to various reflection-distorting phenomena. However, there are sensible explanations for this phenomenon, relating to the number of imaging points per unit length of structural slope. There may be a lesson here for future deep-imaging seismic programs in

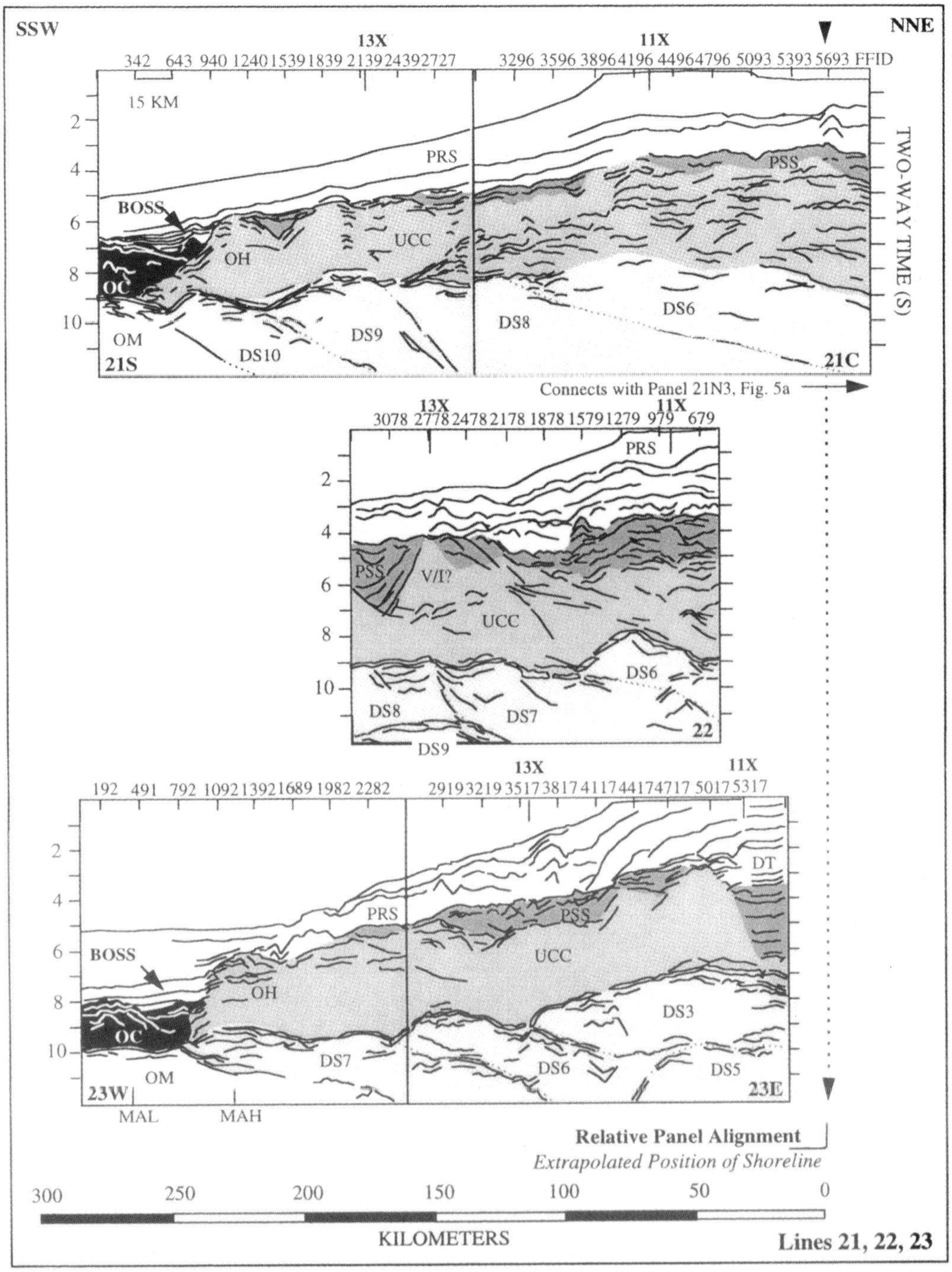

Figure 5b. Line drawing interpretations of NNE-SSW trending Lines 21S/C, 22, and 23. Line 21S/C connects with Line 21N1/N2/N3 in Figure 5a. See Figure 2 for legend, Figure 1 for line locations and correlations to mapped structures.

structurally complex terrains: Lines oriented oblique to structural trends can produce better detail than those oriented orthogonally, as long as the source array is reasonably directional.

Line 17 (Figures 1 and 3b) parallels Line 16 and also traverses the KFZ obliquely. The quality of Line 17 is relatively poor, with few definitive structures and a very weak or non-existent apparent RM. The differences between

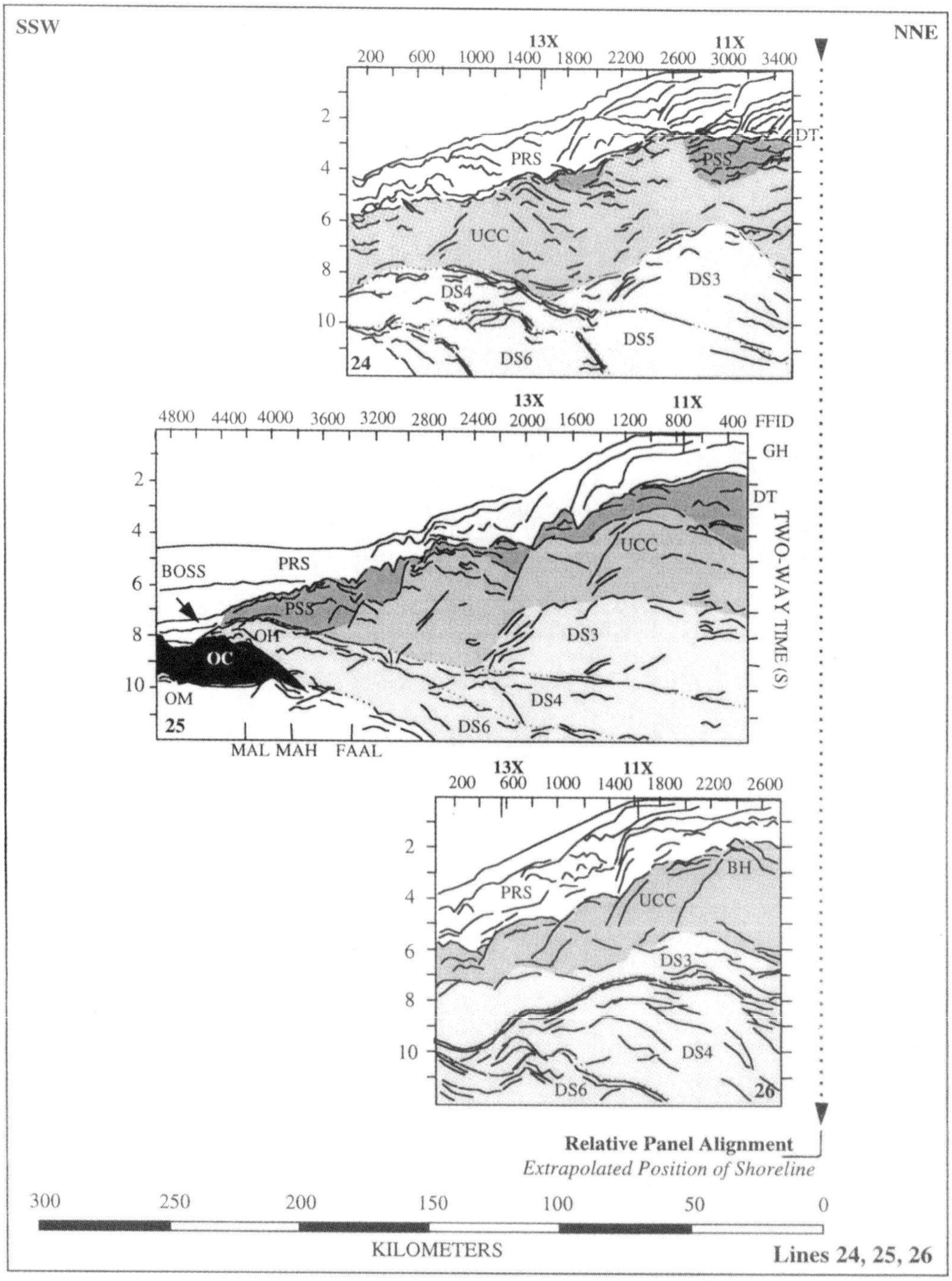

Figure 5c. Line drawing interpretations of NNE-SSW trending Lines 24, 25, and 26. See Figure 2 for legend, Figure 1 for line locations and correlations to mapped structures.

these sister lines were initially attributed to geological factors. We now ascribe them to poor initial processing of Line 17, meaning that it could be worthwhile to re-stack Line 17 (and probably other lines).

It is likely that the KFZ is the eastern end of the Ascension Fracture Zone, as mapped three decades ago by *Burke* [1969] on the basis of bathymetry. This corresponds to the 30° and 31° synthetic flow-lines of *Francheteau and Le Pichon* [1972]. The KFZ mapped here includes the Kribi, Bata, and Ascension Fracture Zones of both *Meyers*

[1995] and *Groschel-Becker* [1996]. The KFZ equivalent at the conjugate Brazilian margin may be the Pernambuco-Paraiba Fracture Zone of *Mohriak et. al.* [1995], meaning the Pernambuco-Paraiba Basin is the conjugate equivalent of the Douala Basin north of the KFZ. It also is possible that the Maceio Fracture Zone and Alagoas Basin are the conjugate equivalents. Work in progress should resolve this issue.

4.3. The Rift Margin

It is possible, even likely, that a number of small-offset fracture zones intersect the rift margin south of the KFZ, but no major-offset fracture zones are indicated in the reprocessed data. Here, the reprocessed seismic sections appear to be relatively free of fracture zone tectonics, although the occurrence of salt south of the KFZ adds new complexity to interpretations. This means that all lines south of the KFZ image the rift margin where the plates have separated in between major-offset fracture zones. We consider this to be the "cleanest" part of the rift margin in the PROBE study area, where the data offer the best possibility of constraining the extensional rifting processes that produced this margin.

Perhaps the most remarkable aspect of the originally processed PROBE data was the widespread occurrence of an *apparent* RM flooring what appeared to be oceanic, transitional, and continental crust. Similar events at similar depths at similar positions had been observed at other margins, but perhaps not to the aerial extent shown in the PROBE data. The reprocessing of the PROBE data seemingly improved the resolution of deep reflectors and the apparent coherency of RM-like events. This further reinforced the sentiment that RM here must represent true Moho, or some proxy thereof. By identifying and mapping RM, we felt it should be possible to tightly constrain the crust-mantle boundary, and thus the mechanism of rifting at this margin.

We made three separate attempts to "pick" and map a laterally coherent, continuous RM event throughout the reprocessed PROBE data, extending from the base of incontrovertible oceanic crust to the shoreward-ends of lines. This proved to be a relatively simple task within the oceanic crustal sectors, and even with the KFZ. We now have little doubt that RM beneath OC is a close and reliable proxy of seismic refraction Moho. However, this task of mapping a contiguous RM event southeastward of the KFZ and landward of the OH proved to be extremely frustrating. It may be useful to other seismic interpreters to list some of the problems we encountered. There are long segments of lines (100 km or more) over which a consistently strong and reliable *apparent* RM can be identified. The area between Lines 20 and 24 (Figure 1) is the prime example. Conversely, the area traversed by Line 20 and the strip of terrain along the northeastern ends of Lines 24-25 (Figure 1) display almost no coherent reflection events that could be equated to RM. Some of the spatial differences in reflectivity and coherence undoubtedly derive from ray-focusing (or de-focusing) through salt and variations in the quality of the initial stacks, but the sub-regional variations in apparent RM reflectivity and lateral coherence seem to be legitimate geophysical characteristics of the study area. Multiple possible RM events appear on some lines. Line 26 (Figures 1 and 5c) is a good example. Here, strong and laterally coherent reflectors occur between and within the units labeled DS3 to DS6. Any of these reflectors, or none of them, could be RM. Events picked as RM on one line sometimes mis-tied with RM on crossing lines, as long as consistent interpretational criteria were employed (i.e., reflectivity and coherency). Yet, almost always *apparent* RM on one line could be tied to a reliable, albeit lesser reflection event (or package) on another line. Reasonable RM-computed Moho depths along one line segment became unreasonable computed Moho depths at other positions along the same line, or along some tied line. In some places, *apparent* RM relief bobbed up and down by up to 12 km over lateral distances of 80 km, or less.

In the final analysis we could not generate a consistent, reliable, honest set of RM interpretations in the sector southeastward of the KFZ and landward of the OH using the premise of a singular, regional RM event. The situation was not improved by using the approach of *Wannesson et al.* [1991], and to a lesser extent *Meyers* [1995], in which the deeper reflection package was subdivided into mid- to lower- crustal reflections and Moho reflections. Essentially, the same problems re-occurred, except in duplicate (or multiplicate).

Further evaluation of these difficulties led to two conclusions: 1) The premise that a laterally continuous RM reflection event, or package, existed in our study area was probably flawed, and; 2) Attempts to force a laterally continuous, singular RM interpretation across the data set would inevitably result in misinterpretations, no matter how inadvertent. Accordingly, we defocused attention on mapping RM as such and concentrated instead on defining and tying reflection packages. This resulted in the line drawing interpretations in Figures 3-5. We believe these interpretations may represent a useful conceptual breakthrough in understanding the fundamental architecture of this margin.

The overall picture depicted in Figures 3-5 consists of one or two upper continental crustal sheets overlying a series of lower ductile sheets arranged in a loose imbricate, or shingled fashion. Extension in the upper continental sheets

(UCC) is generally accommodated by brittle deformation (i.e. faulting), with the caveats described below. Extension in the lower ductile sheets (DS) is accommodated by ductile shear within and between the sheets.

The rheological behavior of this rift margin is dictated by the foundation, in this case the lower ductile sheets. In the region of the Gabon Basin south of the KFZ about 10 ductile sheets have been identified and tied with a reasonable degree of confidence (Lines 21-26, Figures 5b and 5c, and the corresponding parts of Lines 11 and 13, Figures 4c and 4d). These ductile sheets generally dip easterly to southeasterly. Sheet numbering reflects apparent stratigraphic order; the higher the number, the lower the apparent stratigraphic level. The actual number of ductile sheets is arbitrary in the sense that perhaps twice as many discrete sheets could have been identified on a given line (but not tied with any degree of confidence). Not every ductile sheet occurs along each line. These sheets are markedly three-dimensional and might be considered as lenses, so compilation of an apparent stratigraphic order requires good line ties. In the northeastern sector of the Gabon Basin near the KFZ, we cannot specify sheet identities because the imaging quality of the critical tie-lines (Lines 19 and 20) is relatively poor. The dip of the ductile sheets in this sector may be more variable than in the southern sector.

The deformational behavior of the upper and lower sheets (or lenses) is quite variable from line to line. The lines shown in Figure 5 probably illustrate this best. On Line 26 sheets 3, 4 and 6 are strongly reflective and very highly deformed, almost to the point of apparent fluidity. The boundary between sheets 3 and 4 is a dominant reflection event west of FFID 1600, although the reflections between sheets 4 and 6 are almost as strong. No dominant reflections occur landward of FFID 1600. The overlying upper crustal sheet appears to be block-faulted in the classical sense (seaward-dipping faults). About 90 km north along Line 25, sheets 4 and 6 appear to bound the OC and may create an outer high (see below). The upper crustal sheet here is again block-faulted, with the apparent faults dipping seaward. All of the deep reflections here are weak, particularly toward the landward end of the line. On Line 24 the dominant reflectors mark the boundary between the upper crustal sheet and ductile sheets 3 and 4. Both landward and seaward-dipping faults may occur in the upper crustal sheets. The former create the Dentale Trough landward of FFID 2600. On Line 23 the upper crustal sheet is markedly non-reflective, but a very strong reflection event occurs between the upper crustal sheet and ductile sheets 3, 6, and 7. Again, the Dentale Trough is bounded by a landward-dipping fault in the upper crustal sheet. Line 22 seems to show a high degree of variability in the ductile behavior of sheets 6-9, but a very strong reflector occurs between these sheets and the overlying upper crustal sheet. Farther north along Line 21, the ductile sheets (6-10) are uniformly inclined seaward, like dominos tilted in the same direction. The brittle-ductile transition remains a very strong reflector package, especially on Panel 21S. This spatial variability also is shown on Lines 11 and 13 (Figure 4), which provides a sense of the three-dimensionality of these sheets. In fact, without Lines 11 and 13 it would have been impossible to tie Lines 21-26 together. This sort of variability occurs throughout the study area south of the KFZ.

Large and often highly reflective humps, bumps, or noses along the tops of the ductile sheets occur throughout the Gabon Basin. These are especially evident on the NW-SE trending lines (Figure 4) and the NNE-SSW trending Line 21 shown in Figures 5a and 5b. On Lines 6-9 and Line 21 just south of the KFZ, the relief on these noses can exceed 3 s in differential height (or 12 km in nominal depth) over lateral distances of 80 km. The relief is retained when time sections are converted to depth as long as reasonable velocity models are used, but the forms of the noses are usually altered. This means the relief is not an artifact of presenting the data in TWTT, although the true forms in depth may be less ridge-like than implied by the line drawings. In nominal depth, the form of these noses ranges from a distinct high to an inflection in the brittle-ductile boundary. Along the northern half of Line 21 (Figure 5a), these noses form a crude sinusoidal form with a wavelength of about 100 km and an amplitude of about 3 s. The resulting triplet of highs and lows was initially thought to be RM ridges and troughs that crossed the line at a high angle. The sinusoidal form is deceiving and actually derives from Line 21 crossing and recrossing the same nose at a low angle. Line 11 may suffer from a similar illusory problem.

In the northern part of the Gabon Basin these noses are loosely correlative with a basement ridge or high (BR) in the upper crustal sheet. The ridge form is best developed on Lines 7-9 (Figure 4b) and on Lines 19 and 20 (Figure 3b). Farther east the BR becomes the continental "wall" of the KFZ. On Line 20 the BR is the Loiret Volcanic Plug [*Reyre,* 1984: *Teisserenc and Villemin,* 1989], suggesting an igneous origin for the BR, at least where its form is ridge-like. Overall, the BR has a trend of about N30°E, which also is the trend of the Cameroon Volcanic Line (CVL*).* *Teisserenc and Villemin* [1989] and *Meyers* [1995] relate the Loiret Volcanic Plug to the CVL and assign an Early Miocene age to the intrusion. Since the Loiret Plug is clearly a component of our BR, the resulting inference is that the BR is a parallel, coeval structure to the CVL. We believe it is more likely that the BR is an igneous feature loosely related to the transform margin. In addition to the BR, there are numerous other locations where the upper crustal sheets seem to show evidence of volcanic or intrusive activity (labeled V/I in Figures 3-5).

Lines 20, 21, 23, and 25 span the entire offshore rift margin and conclusively reach OC without crossing any major, mappable fracture zones. In all cases, the transition from continental lithosphere to OC is marked by an apparent unroofing of lower ductile sheets. In two cases, the ductile sheets are modeled as impinging on OC; in the other two cases, an upper crustal sheet bounds OC. These differences may be entirely interpretational.

All four of these lines show outer highs (OH) at the juncture with OC. A modest magnetic anomaly is associated with the OC-OH transition. The anomaly has an amplitude of 200 nt along Line 25 and about 100 nt along Line 23. The anomaly presumably marks a change in remnant magnetization or magnetic susceptibility. A 70 mgal free air gravity low occurs at the inside edge of the OH on Line 25. Significant free air anomalies do not occur elsewhere in association with the OH.

Comparable outer highs have been imaged at other passive margins [e.g., *Austin and Uchupi,* 1982; *Beslier et. al.,* 1993; *Keen and deVoogd,* 1988; *Lehner and DeRuiter,* 1977; *Mohriak et. al.,* 1995; *Montadert et. al.,* 1979; *Mutter et. al.,* 1982; *Schuepbach and Vail,* 1980] and variously interpreted as uplifted blocks of continental crust, volcanic structures of oceanic crustal origin, mafic volcanic blocks, shear zones, fracture zones, serpentinite, and serpentinized peridotite. *Meyers* [1995] interpreted these same highs as faulted, mafic volcanic blocks intermixed with slivers of continental crust. Recent drilling along the Iberian margin [in *Whitmarsh et. al.,* 1996] suggests an ultramafic origin, specifically ridges of peridotite, serpentinite, and lower continental rocks. The apparent unroofing of the shingled lower ductile sheets toward the OC boundary is quite consistent with the Iberian drilling results. If ductile lower crustal or upper mantle rocks are emplaced at upper crustal levels at the outer highs, the consequent loss in shear strength may explain how (or where) seafloor spreading commences.

It is somewhat troubling that the seismic transition from continental lithosphere to OC expressed along lines 20, 21, 23, and 25 is superficially not too different from that observed along Lines 16 and 17, which cross the KFZ obliquely. Without the interpretational benefits provided by the NW-SE trending line set (Figure 4), we would be hard-pressed to distinguish oblique crossings of this fracture zone from "clean" crossings of the rift margin, except in regards to the relative amounts of relief.

The rift margin interpretations offered above are rheological in nature. They are not based upon a mapped RM and they contain few assumptions about what is crust and upper mantle, beyond the association of the upper crustal sheets with brittle continental crust. We are comfortable with the overall notion of brittle upper crustal sheets or lenses, but this simple interpretation may not be entirely accurate. There are many instances in which landward and seaward-dipping upper crustal reflection events, consistent with faults, offset apparent basement and then disappear within the upper sheets. Often, the projections of the inferred faults coincide with relief on the top surfaces of the ductile sheets. However, there is not a single instance in which fault plane continuity can be unequivocally demonstrated between the top surface of the upper crustal sheets and the ductile lower sheets, in spite of the apparent high quality and quantity of the deep-imaging data. In summary, a mechanical linkage between the uppermost crust and the lower ductile sheets or lenses cannot be demonstrated at this passive margin. Either the upper crustal sheets and ductile lower sheets are not mechanically linked here, or such linkages are not imaged very well.

On the issue of imaging, we are impressed that our reprocessed data (and thus the line drawing interpretations) tend to show clusters of organized, dipping reflector packages at or near the upper surfaces of both the upper crustal sheets and the lower ductile sheets, with few such packages in the intervening space. Line 23 (Figure 5b) may be the best example, but all the southerly lines show this phenomenon to some degree. With a few notable exceptions, it can be stated that the lower parts of the upper crustal sheets are largely free of coherent reflections. One exception is the tongue of crustal reflectors shown on Line 21 landward of FFID 2400 (Figures 5a and 5b), which seems to depict a highly reflective, discontinuously laminated upper crustal sheet. The expression continues north at least to the KFZ. Another exception is the package of sub-horizontal reflectors that occur in the upper crustal sheets and ductile sheets 6 and 7 on Line 11N (Figure 4c). These observations could suggest that the rheology of the upper crustal sheets is not uniform. However, we are suspicious that these exceptions may arise from side-swipe generated by out-of-plane faults.

5. CONCLUDING COMMENTS

We believe the line drawing interpretations in Figures 3-5 portray a 75 km-wide transform rift margin in the north, with an offset of more than 350 km across the Kribi Fracture Zone. North of the KFZ, oceanic crust extends essentially to the coast of Cameroon. South of the KFZ, a broad rifted continental margin exists landward of an outer high, which marks the basment boundary with OC. This part of the continental margin seems to have evolved by ductile shear within and between ductile sheets or lenses, below a crustal carapace that deforms more rigidly.

Progressively greater stretching and unroofing of continental lithosphere toward the OH eventually gave rise to seafloor spreading, with the consequent development of oceanic lithosphere. Depending on the reader's perspective, the model could be termed "imperfect pure shear" or perhaps "ductile multiple shear". The implications of this model are broad and perhaps provocative.

5.1. On the Meaning of Moho

We believe that RM is a close proxy of the crust-mantle boundary within oceanically-generated lithosphere. In our study area, this occurs NW of the KFZ and west of the outer highs. Landward of these boundaries, we have challenged the precept that a single and laterally coherent, continuous RM exists beneath the rifted portion of this margin. We do not think it does. Instead, we suggest that shingled sheets or lenses of ductilely deformed rocks are emplaced in the lower continental crust and upper mantle beneath the stretched continental portions of this margin. The transitions between the upper crustal sheets and lower ductile sheets are often highly reflective and laterally coherent over many tens of kilometers, but invariably the dominant reflector package eventually switches back and forth between different ductile shear zones, or disappears entirely in some areas. This is why we abandoned the notion that a single and laterally coherent RM could be mapped beneath the rift margin proper.

It might be significant that the reflectivity of the brittle-ductile transition tends to be generally greater toward the higher numbered ductile sheets, which is sensible if the higher numbered sheets represent lower ductile sheet levels, thus resulting in larger acoustic impedance contrasts. It would be useful to know to what extent the ductile sheets have unroofed during their development. In reference to Figure 5, do the higher-numbered sheets represent deeper levels that have been systematically unroofed during stretching? This is tantamount to asking how much detachment has occurred between the ductile sheets. Drilling transects across the OH here could provide some answers.

This brings us to the question of what classic Moho (essentially the seismic refraction Moho) means beneath this rifted continental margin. What is the relationship between seismic refraction Moho and the complicated, lensoid interfaces between the upper crustal sheets and lower ductile sheets? Unfortunately, deep velocity information in our area is limited to the results of 10 expanding spread profiles from the inner, southern part of the study area, presented in *Wannesson et al.* [1991]. These results seem to suggest that refraction Moho across the rifted continental parts of this margin corresponds to some highly smoothed horizon that very loosely correlates to the upper levels of the ductile sheets. Stated another way, the limited deep velocity information from the area is consistent with the rheological model presented herein. We cannot be more definitive at present.

Could the complicated, lensoid interfaces between the upper crustal sheets and lower ductile sheets represent petrologic Moho, replete with the rheological changes that should be manifested across this boundary? To indirectly test this proposition we have produced a line drawing of a nominal depth section that composites all of Line 22 with the SSW half of Line 23 (Figure 6). The composite is projected at true scale approximately parallel to the trend of the KFZ. The composite represents a best-case scenario, in the sense that the reflector packages that define the interfaces between the upper crustal sheet and lower ductile sheets are almost unequivocal across the two segments of the composite. The depth composite shows a sensible seaward thinning of the crust if the petrologic Moho is placed at the interface between the upper crustal sheets and lower ductile sheets. The long wavelength component of thinning (or stretching) follows a nearly linear trend and is approximately 1 km/17 line km of horizontal distance, or about 6%/line km. A comparable rate of stretching in nominal depth has been computed for Line 21S/C as far landward as FFID 4500, again projected parallel to the KFZ.

5.2. On Continental Rift Analogs

Passive margins must have been born as continental rifts. Some workers envision starting models akin to the East African rifts, where the basinal architecture is dominated by narrow, well-defined rift zones consisting of asymmetric half-grabens that tend to alternate basinal polarities across accommodation zones. Others envision more diffuse rifting models derived from studies of such locations as the Basin and Range province of the U.S., replete with low-angle detachments. Still others prefer to think in terms of upper plate-lower plate models. Regardless of the starting perspective, it is fashionable to expect that remnants or scars of earlier rifting stages should be preserved in the roots of passive margins. This is arguably the case along some passive margins (e.g., the Triassic basins in the offshore region of the Gulf of Maine).

In the context of the rheological model presented here, the remnants of the early rifting history along this margin would be associated with the upper crustal sheets and overlying syn-rift deposits. It could be argued that the apparent seaward-dipping faulting of the upper crustal sheets along Lines 25 and 26 gives way to the opposite or mixed orientations farther north, implying the presence of an accommodation zone or transfer fault in the vicinity of Line

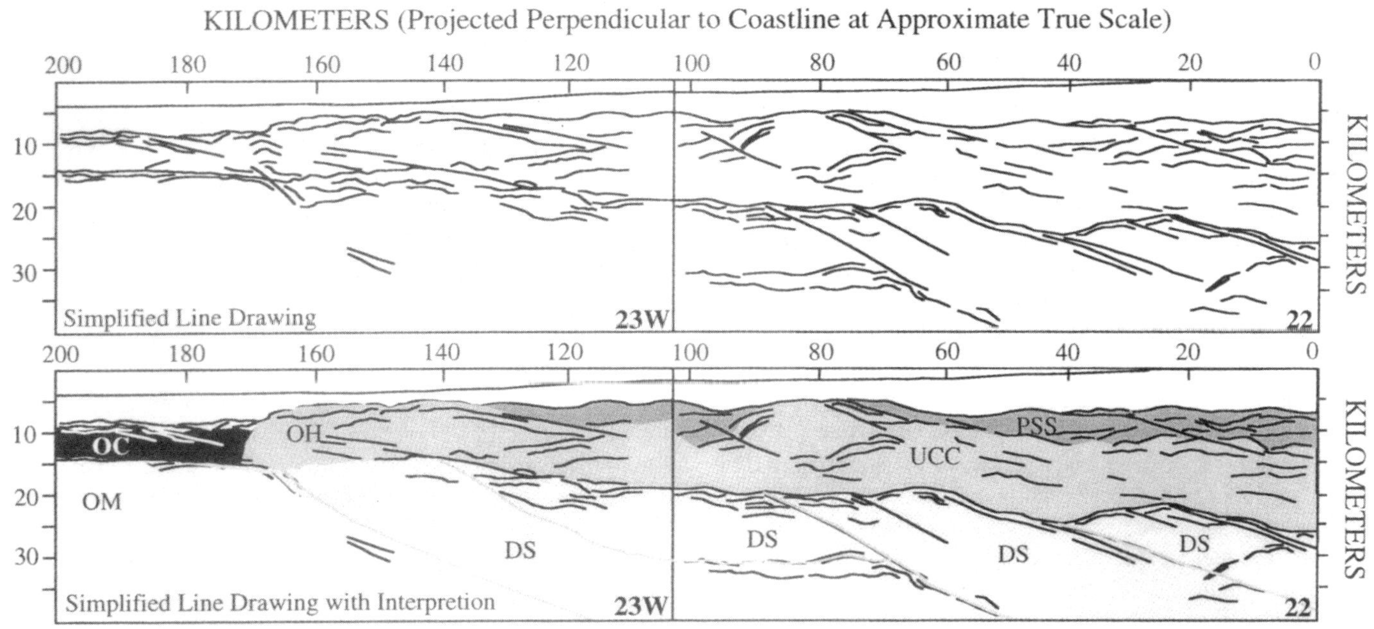

Figure 6. Line drawing composite of Lines 22 and 23W, in nominal depth. Depth conversions based upon V = 3.0 Km/s for the post-rift section (PRS); V = 6.5 Km/s for crustal units; and V = 8.1 Km/s for ductile sheets. See Figure 1 for line locations.

24. The occurrence of the Dentale Trough along the landward ends of Lines 23-26 also could be evidence of the earlier rifting history. Beyond these examples, there is scanty structural evidence for early rift remnants in our data set. If this margin began separating as a Lake Tanganyika-type continental rift, as one example, then the structural manifestations of the earlier rifting phases either have been largely obliterated by subsequent events, or they occur inboard of the PROBE lines and/or at the conjugate position off Brazil. Both *Teisserenc and Villemin* [1989] and *Dailly* [this volume] have noted the occurrence of possible rift structures landward of our profiles. Even so, the paucity of well-defined rift remnants in our data set raises the troubling prospect that continental rifts such as East Africa may not be the precursors of successful rift margins, at least not in the offshore sectors of this margin. Another precautionary flag should be hoisted. The most continental rift-like structures in the PROBE area (tilt blocks, asymmetric sedimentary wedges situated on basement, etc.) occur within the KFZ, especially along Line 16. Although we are now certain that this geology is related to transform fault tectonics, it would be quite easy to mistake it for the remnants of continental rift basins if only Line 16 were available. Here again we raise the warning that isolated, deep-imaging MCS transects can lead to serious misinterpretations of the actual regional geology.

In a way this mode of thinking -- the need for precursor rift analogs -- misses the main point of this manuscript. Given the rheological model presented here, deformational events in the upper crustal sheets may not be directly, or mechanically linked to those in the lower ductile sheets, which basically govern or control the overall rifting process. In a sense the upper crustal sheets are rafted along, stretched, and broken in response to the ductile deformation below. It is difficult to envision how this process could lead to much coherent preservation of those events which occurred in the upper-most skin during early stages of rifting, and even if it did we wonder what relevance it would bear on the fundamental processes. Although localized deformational styles in the upper crustal sheets are very important in hydrocarbon exploration (see below), they may tell us little about the fundamental evolution of this passive margin. For example, it could be argued that whether the upper crust ruptures with seaward or landward-dipping faults that are linear or listric depends more on the local geology than on the more uniform and plastic behavior of the ductile sheets.

5.3. On Costs Versus Benefits: The Exploration Perspective

The 4500 km of data interpretations shown herein represent one of the most extensive deep-imaging seismic collections available for any passive margin segment. Yet, the collection is arguably an order of magnitude too small to unravel the detailed architecture of this margin. The average spacing of the PROBE lines is on the order of 80-100 km;

the wavelength of significant structural changes at and below basement levels seldom exceeds this range and is often much less, even at the level of the ductile sheets. We estimate that a grid spacing on the order of 10 km would be necessary to map the deep structure satisfactorily.

Including only direct expenses, over one million dollars (US) has been spent acquiring, processing, and interpreting this data collection, and the job is not finished yet (see below). Ten million dollars would be necessary to do the job satisfactorily in an intellectual sense. The margin segment studied here is about 1000 km long, so the full cost would be on the order of $10,000/km of margin length.

These should be sobering figures to academic planners of future deep-imaging seismic programs on passive margins. We caution against seeking shortcuts by shooting a few deep, regional transects. Which two or three lines in Figures 3-5 would lead to the tectonic plan presented in Figure 1, or the rheological model offered herein? It could be argued that the financial limitations imposed on academia to acquire regional seismic transects instead of grids is one of the reasons why passive margin architecture is so poorly constrained. In a reconnaissance seismic mode, it also makes little sense to acquire grids along margin segments spanning less than about 300 km of margin length.

The costs cited above are not scary numbers in the petroleum industry. The cost to thoroughly understand the deep structure of this margin is equivalent to that of one or two, deep-water dry holes. In this context it is useful to consider the explorational value of our analyses.

Figure 1 concisely explains why almost all of the major offshore drilling successes on this margin have occurred south of the KFZ. If the knowledge proffered in Figures 1-6 had been available at the outset of exploration activities on this margin, the history of exploration might have been significantly different...and perhaps more cost effective. For example, it now seems futile to seek out offshore synrift plays in the North Douala Basin (north of the KFZ) because almost all the rift geology must be located onshore in Cameroon in a narrow band, or at the conjugate position off Brazil. This is not to say that the offshore area north of the KFZ is unprospective, but it means that potential sources and targets lie entirely within the post-rift section.

Figure 1 also suggests that the Niger Delta is constructed almost entirely on oceanic crust. This means that the subsidence history of the delta follows the "oceanic" relationship noted earlier for differential loading along Line 5. The reprocessed PROBE results support our earlier contentions that the Cameroon Volcanic Line is built entirely on oceanic crust [e.g. *Meyers and Rosendahl,* 1991]. Hence, the oil seeps on São Tomé-Principe must originate in a truly oceanic crustal setting even though the sources of the oils (presumably windows of North Douala Basin sediments incorporated into the volcanic edifices) may be similar to those observed on the West African shelf.

Now consider the case of the eastern terminus of the KFZ, where it abuts continental crust at depths potentially within range of the drill stem (e.g., around 2° N). The setting is a continent-bounded transform fault. The flooring is presumably fractured and intruded oceanic crust, with local fragments of continental basement. The depositional arenas are narrow troughs flanked by volcanic/plutonic ridges. The fill is mainly continental clastic sediments. Salt may have been locally present. The initial thermal regime was high but variable and hydrothermal activity would have been prevalent early on. This is not an environment in which standard exploration strategies can be employed with much hope for success, as *Dailly and Ong* [1997] and *Dailly* [this volume] have noted in their studies of offshore Equatorial Guinea. These sorts of unconventional settings could play a larger exploration role in the coming century if they can be properly exploited. This begins with a proper understanding of tectonic provenance. It would not have been possible to arrive at these results if we had been limited to conventional industry seismic data in the Gulf of Guinea.

Exactly how our rheological model aids the explorationist is questionable, but we offer the following thoughts. There is usually a crude correlation between the large-scale relief on the lower ductile sheets and major basement topography, albeit the relationship is neither simple nor direct. Along many parts of this margin, the ductile sheets are much better imaged than the basement/syn-rift or syn-rift/post-rift boundaries. In fact, details of syn-rift basin geology are often poorly resolved in the PROBE data in the continental sector. It is conceivable that the existence of major basins in frontier areas could be deduced by first mapping the topography of the ductile sheets, rather than trying to resolve the shallower basinal details from the outset. Acquisitional programs then could be designed that specifically address the basinal details. At the very least, we believe that the ductile shear zones ought to be one of the imaging targets of reconnaissance acquisitional programs in frontier areas. As we learn more about the relationships between basinal development and the underlying margin architecture, this type of information might well represent the "edge" that innovative companies could employ in the next century.

Do the results justify the costs? In the context of hydrocarbon exploration, we believe that PROBE-type studies of passive margins could prove cost-effective, at least over the long run. The question of whether or not the results justify the costs to academia is overshadowed by the glaring realities that funding agencies face. This much seems certain: If a systematic understanding of the deep seismic structure of passive margins is to happen, the task will have to be largely underwritten by the exploration industry. In the interim, much would be gained if industry seismic planners extended recording times and cable lengths to PROBE-type values, and expanded line lengths to cover more deepwater marginal areas.

Much more needs to be done with the PROBE data. Ongoing work includes: 1) conversion of time into nominal depth in a uniform, line-tied fashion; 2) computation and removal of the effects of loading to produce corrected nominal depth sections; 3) construction of complete Bouguer gravity models back-stripped to the brittle-ductile boundary; and 4) integration of these results into a conjugate margin analysis incorporating data from offshore Brazil. The later is especially important. Can a similar rheological model be deduced in the Brazilian deep-imaging data acquired by Petrobras? If so, in what general direction do the ductile sheets dip? Opposite or the same direction? We expect the ongoing work will go a long way toward resolving many of the questions expressed or implied herein, and undoubtedly raise a host of new issues to guide future research on the architecture and evolution of this segment of South Atlantic passive margin.

Acknowledgments. We thank the governments of Cameroon, Equatorial Guinea, Gabon, and São Tomé -Príncipe for permitting this field work to be done. We thank Agip, Amoco, Ark Geophysical, Geco, and Jebco for support and assistance. This manuscript has benefited from the laboratory efforts of Zan-Dong Ding and Jason Meyers. Reviews of this manuscript have been kindly provided by Albert Bally, Nick Cameron, Steve Henry, and Phil Nelson.

REFERENCES

Austin, J.A., and E. Uchupi, Continental-oceanic crustal transition off Southwest Africa, *AAPG Bull, 66,* 1328-1347, 1982.

Beslier, M.O., M. Ask, and G. Boillot, Ocean-continent boundary in the Iberian Abyssal Plain from multichannel seismic data, *Tectonophysics, 218,* 383-393, 1993.

Burke, K., Seismic areas of the Guinea coast where Atlantic fracture zone reach Africa, *Nature, 222,* 655-657, 1969.

Dailly, P., and E. Ong, Geology and petroleum potential of the Rio Muni basin-offshore equatorial Guinea (abs.), *5th International Congress of the Brazilian Geophysical Society, Vol. of Abstracts,* Rio de Janeiro, 1997.

Francheteau, J., and X. Le Pichon, Marginal fracture zones as structural framework of continental margins in South Atlantic Ocean, *AAPG Bull., 56,* 991-1007, 1972.

Groschel-Becker, H.M., Formational processes of oceanic crust at sedimented spreading centers: Perspectives from the West African continental margin and Middle Valley, Juan de Fuca Ridge, *Ph.D. dissertation,* University of Miami, Miami, FL, 328 pp., 1996.

Keen, C.E., and B. deVoogd, The continent-ocean boundary at the rifted margin off eastern Canada: New results from deep seismic reflection studies, *Tectonics, 7,* 101-124, 1988.

Lehner, P., and P.A.C DeRuiter, Structural history of the Atlantic margin of Africa, *AAPG Bull., 61,* 961-981, 1977.

Mascle, J., G.P. Lohmann, and P. Clift, *Proc. ODP, Init. Repts., 159,* College Station, TX (Ocean Drilling Program), 1996.

Meyers, J.B., Rifted continental margin architecture off West Africa, as revealed by deep-penetrating multi-channel seismic reflection and potential field data, *Ph.D. dissertation,* University of Miami, Miami, FL, 234 pp., 1995.

Meyers, J.B., and B.R. Rosendahl, Seismic reflection character of the Cameroon volcanic line: Evidence for uplifted oceanic crust, *Geology, 19,* 1072-1076, 1991.

Meyers, J.B., B.R. Rosendahl, H. Groschel-Becker, J.A. Austin, and P.A. Rona, Deep-penetrating MCS imaging of the rift-to-drift transition, offshore Doula and North Gabon Basins, West Africa, *Mar. Petrol. Geol.,* 13, 793-836, 1996.

Mohriak, W.U., J.H.L. Rabelo, R.D. de Matos, and M.C. de Barros, Deep seismic reflection profiling of sedimentary basins offshore Brazil: Geological objectives and preliminary results in the Sergipe Basin, *J. Geodynamics, 20,* 4, 515-539, 1995.

Montadert, L., D.G. Roberts, D. DeCharpal, and P. Guennoc, Rifting and subsidence of the northern continental margin of the Bay of Biscay. *Init. Rept. DSDP, 48,* 1025-1060, 1979.

Mutter, J.C., M. Talwani, and P.L. Stoffa, Origin of seaward-dipping reflectors in oceanic crust off the Norwegian margin by "subaerial seafloor spreading", *Geology 10,* 353-357, 1982.

Reyre, D., Caracteres petroles et evolution geologique d'une marge passive. Le cas du bassin bas Congo-Gabon, *Bull. Centres Rech. Explor.-Prod. Elf Aquitaine, 8,* 303-332, 1984.

Rosendahl, B.R., The seismic roots of the West African passive margin: Northern Cameroon to southern Gabon (abs.), *5th International Congress of the Brazilian Geophysical Society, Vol. of Abstracts,* Rio de Janeiro, 1997.

Rosendahl, B.R., H. Groschel-Becker, J. Meyers, and K. Kaczmarick, Deep seismic reflection study of a passive margin, southeastern Gulf of Guinea, *Geology, 19,* 291-295, 1991.

Rosendahl, B.R., J. Meyers, H. Groschel-Becker, and D. Scott, Nature of the transition from continental to oceanic crust and the meaning of reflection Moho, *Geology, 40,* 721-724, 1992.

Schuepbach, M.A., and P.R. Vail, Evolution of outer highs on divergent continental margins, in *Studies in Geophysics; Continental Tectonics, Natl. Acad. Sci., Washington, D.C., 50-61,* 1980.

Teisserenc, P., and J. Villemin, Sedimentary basin of Gabon-Geology and oil systems, in *Divergent/Passive Margin Basins, AAPG Memoir No. 48,* edited by J.D. Edwards and P.A. Santogrossi, Tulsa, OK, 117-199, 1989.

Wannesson, J., J.C. Icart, and J. Ravart, Structure and evolution of adjoining segments of the west African margin determined from deep seismic profiling, *in Continental Lithosphere: Deep Seismic Reflections,* edited by R. Meissner and others, pp. 275-289, 1991.

Whitmarsh, R.B., Sawyer, D.S., Klaus, A., and Masson, D.G. (Eds.), *Proc. ODP, Sci. Results, 149,* College Station, TX (Ocean Drilling Program), 1996.

H. Groschel-Becker and B.R. Rosendahl, Division of Marine Geology and Geophysics, Rosenstiel School of Marine and Atmospheric Science, 4600 Rickenbacker Causeway, Miami FL 33149

Tectonic and Stratigraphic Development of the Rio Muni Basin, Equatorial Guinea: the Role of Transform Zones in Atlantic Basin Evolution

Paul Dailly

Triton Energy, Dallas, Texas

The Rio Muni basin underlies the continental shelf of the West African Republic of Equatorial Guinea, located between Gabon and Cameroon. The basin is located above a series of northeast-southwest trending oceanic transform zones and their continental extensions. These constitute the boundary between the Equatorial Atlantic margin and the West African salt basin. New seismic data extending from the shelf into the deep water region permit new structural and stratigraphic observations, and interpretations of the evolution of this part of the Atlantic margin. The tectonostratigraphic evolution of the basin is divided into a number of separate phases; Early Cretaceous to Mid-Aptian rifting; Late Aptian-Cenomanian postrift onlap and regional subsidence; Early Senonian transpression and uplift of the shelfal area. This uplift is associated with margin wide gravity tectonics, the development of both extensional and compressional, thin skinned structures and the reactivation of the transform zones). Uplift may have been driven by ridge reorientation in the central and south Atlantic. Finally, during Late Cretaceous to Tertiary time, a thick passive margin wedge developed, dominated by slope fan deposits. The deposition of this sequence was punctuated by renewed uplift and renewed gravity tectonics. The Rio Muni Basin is one of the least explored portions of the West African margin and has considerable potential for hydrocarbon exploration.

INTRODUCTION

The Rio Muni Basin underlies the continental shelf of the West African Republic of Equatorial Guinea, located between Gabon and Cameroon (Fig. 1). Despite the considerable amount of exploration and the discovery of a number of oil and gas fields on the continental shelves of these adjacent countries, the Rio Muni Basin is relatively poorly explored, with only 6 wells drilled in an area of more than 6000 km^2. The basin is located at the northern end of the oil-prolific West African salt basin, but to date it has received little attention from oil companies, in part due to the narrow width of its shelf and the failure of past exploration campaigns.

In 1997 Triton Energy acquired 2 exploration blocks occupying an area of 5270 km^2 covering both the shelf and the deep-water portions of the Rio Muni continental margin. Reprocessed and newly acquired seismic data have been integrated with the six exploration wells drilled in the basin to produce a comprehensive description of the geological evolution of this portion of the West African Margin and of its relationship to deep crustal tectonics.

DATABASE

Over 6000 km of seismic have been acquired in the basin, ranging in vintage from early seventies to late nineties (1997). This resulted in a grid which varies from 10x5 km in deepwater to 2x2 km in the shelf areas.

Atlantic Rifts and Continental Margins
Geophysical Monograph 115

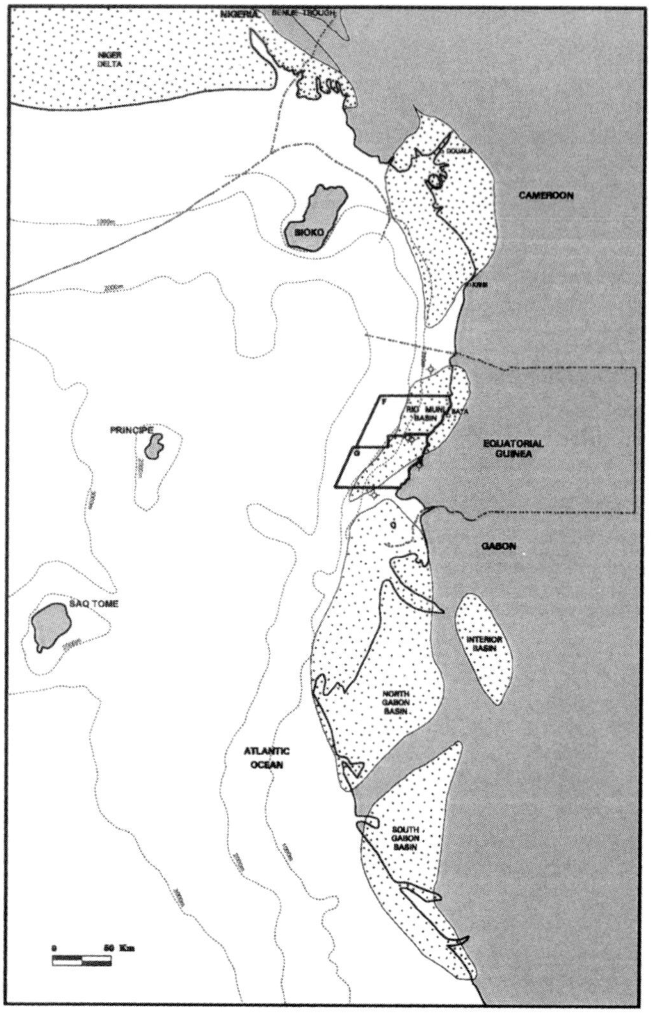

Figure 1. Location Map of the Gulf of Guinea showing basins and location of blocks F and G.

Spanish Gulf drilled the first offshore exploration well (Rio Muni 1) in 1968. This was followed by the drilling of a further four wells throughout the late seventies and eighties: Rio Muni A-1, N'Dote 1 (located onshore), Benito 1 and Matondo 1. The last well to be drilled was Eviondo East 1, operated by Elf, in 1991. The location of wells and seismic lines used in the paper are shown in Figure 2.

TECTONIC SETTING

Prior to the development of the south Atlantic, west central Africa was situated adjacent to the northern Brazilian basins *[Binks and Fairhead, 1992, Chang et al, 1992, De Matos, 1992, Klitgord and Schouten, 1986, Meyers et al, 1996a, Meyers et al 1996b]*. The Rio Muni area is thought to have been part of the Sergipe-Alagoas, Douala, North Gabon rift branch of the proto-Atlantic (Fig. 3), [*Davison 1997, DeMatos 1992*] and has a broadly similar tectonostratigraphic history to these adjacent basins.

The breakup of Pangea was developed diachronously from south to north in the South Atlantic by Late Jurassic / Early Cretaceous [Rabinowitz and LaBrecque, 1979]. Along the length of the northeastern Brazil / Western African margin, Neocomian to Aptian rifting resulted in the development of a series of thick syn-rift successions. This early rift phase was followed by Late Aptian transitional evaporitic to postrift siliciclastic sedimentation. Subsequently, following the development of oceanic crust, the stratigraphic sequences are characterized by deposition of a thick passive margin successions ranging from Albian carbonates to present-day siliclastic sequences. Although the exact ages of onset of these phases vary from one basin to the next, each has megasequences broadly corresponding to these tectonostratigraphic subdivisions. In a gross sense, the tectonostratigraphy of the Rio Muni Basin can be subdivided in this way [*Ross and Hempstead 1993*] and forms a starting point for understanding the detailed basin history.

The present day tectonic setting of the Gulf of Guinea is shown on the satellite gravity image in Plate 1. Major gravity features are associated with the Niger and Ogooue Deltas, the Congo Fan and with the Miocene aged Cameroon volcanic line, which constitutes a series of volcanic plugs intruded into both the passive margin and the Cretaceous-aged oceanic crust [*Meyers et al 1991*].

South of the Ogooue delta, a prominent NW-SE trending gravity low is associated with the Gabon basin. A clear boundary exists between the relatively quiescent, Cretaceous-aged oceanic crust adjacent to this feature and that located outboard of the Equatorial Guinea and Cameroon area.

The junction of these two oceanic provinces is the Ascension transform zone, which extends beneath the shelf of the Rio Muni Basin at its southern end (Fig. 2). North of this transform zone the gravity signature is dominated by ENE trending Atlantic transforms, which are typical of the easternmost part of the Equatorial Atlantic.

The major tectonic elements associated with basin formation in the Rio Muni area are shown in Figure 4. This part of the African margin sits at the junction between the Congo Craton to the south and the Pan African foldbelt, a Proterozoic orogen, to the north. The junction between these two terranes, exposed onshore, constitutes a complex suture zone that trends NE-SW [*Choubert et al 1968*]. This suture is contiguous with a series of NE trending structures that underlie the Rio Muni basin, and these in turn are broadly coincident with a series of major fracture zones identified by the GECO and PROBE data [*Meyers et al, 1996a, Meyers et al 1996b*]. The tectonic history of these structures is critical to understanding the structural and

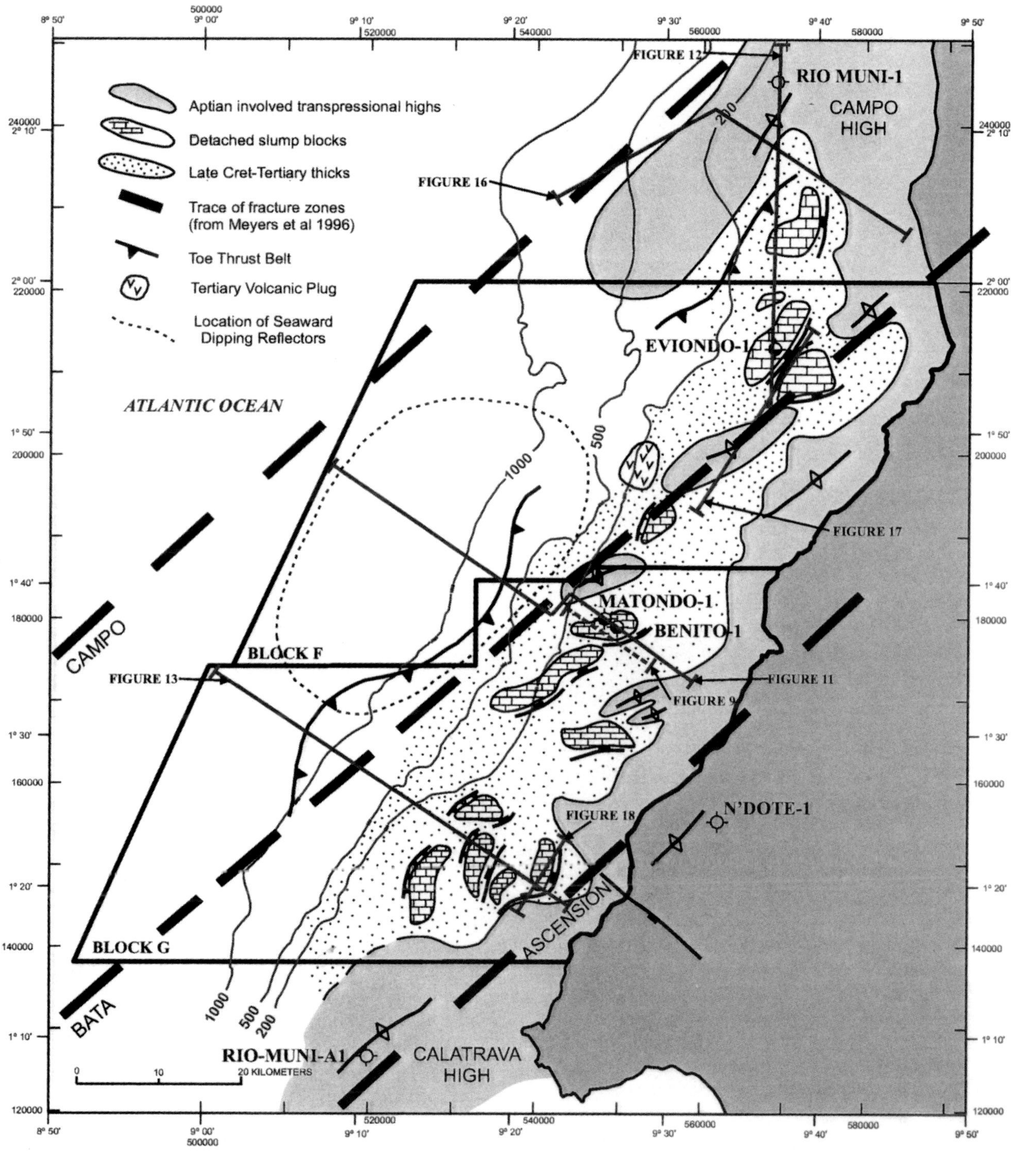

Figure 2. Geological map showing configuration of the Rio Muni Basin (structural map at near Top Albian horizon). Note the relationship of intrabasinal structural highs to transform zone trends (identified from the PROBE dataset in Meyers et al 1996b), location of extensional rafts in intervening lows, and position of toe thrust structures. The location of Tertiary volcanic plug and region with wedges of interpreted seaward-dipping reflectors are also shown.

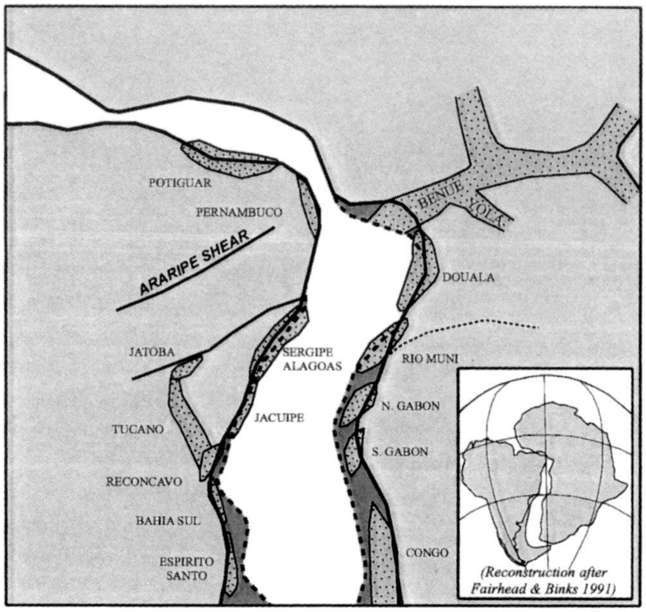

Figure 3. Cretaceous rift basin distribution in the northern part of the South Atlantic superimposed on an Aptian (115 Ma) reconstruction (modified from Binks and Fairhead 1991).

stratigraphic evolution of the Rio Muni shelf from Aptian through to Recent.

The major fracture zones of importance in the Rio Muni area are the Kribi, Campo, Bata and Ascension fracture zones (Figures 4 and 5). The Cretaceous–Tertiary shelf depocentre is bounded by the Campo and Ascension zones which trend NE –SW and separate it from the much larger north Gabon and Kribi-Douala basins.

Two major changes in the geometry of the continental margin are accommodated across the Ascension and Bata shear zones. First, a major reduction in the width of the continental shelf occurs, with the Gabon shelf to the south being relatively wide and associated with major salt basin development [*Meyers et al, 1996a*], while the shelf to the north, founded on Pan African foldbelt, is relatively narrow and has much more localised salt deposition representing the northern end of the 'Aptian salt basin' [*Meyers et al, 1996a*].

Secondly, to the south, the north Gabon basin has a strong NW-SE extensional fault grain, which is sub-parallel to the basement trend within the West Congo foldbelt, a Proterozoic orogen exposed onshore that bounds the Congo Craton on its western side (see Fig. 4). By contrast, the Rio Muni basin is dominated by an ENE trending structural trend sub-parallel to the Pan African basement grain and to the orientation of the adjacent transforms.

From this point northwards along the Central African Atlantic Margin the importance of NE-SW structures is paramount in controlling basin history [*Genik, 1992, Guiraud and Maurin, 1992, Mascle et al, 1988*].

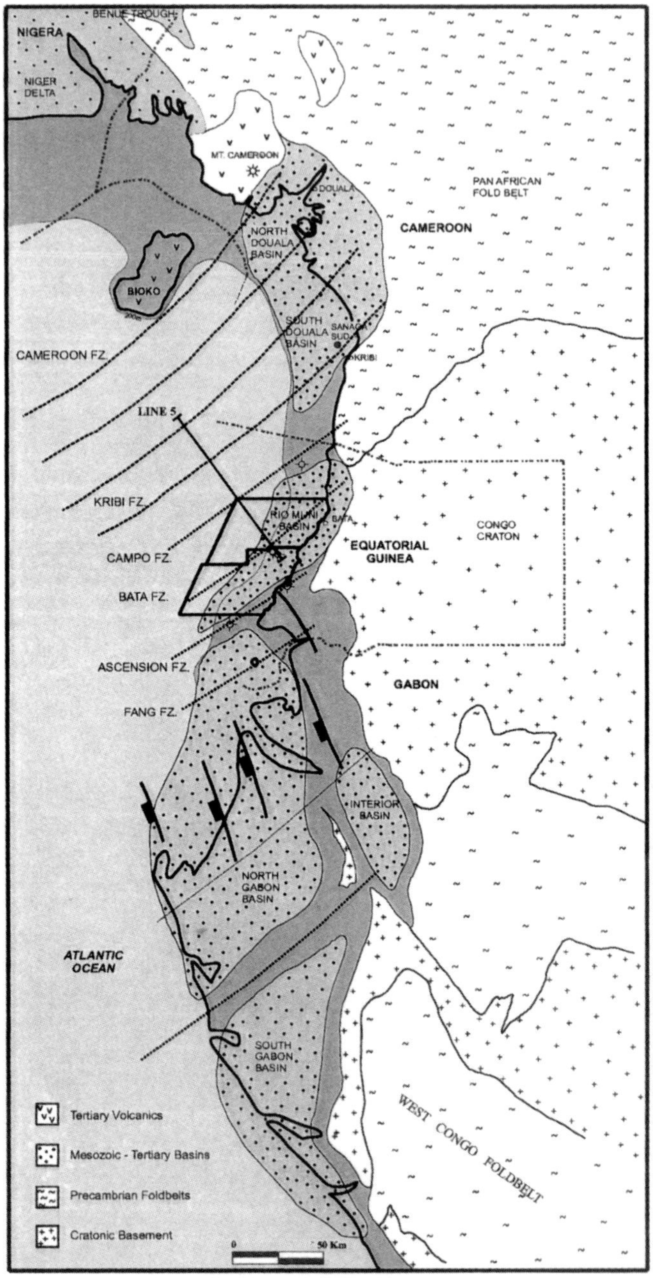

Figure 4. Map showing sedimentary basins, basement geology, main tectonic elements of West Central Africa and locations of Blocks F and G. Location of fracture zones after Meyers et al 1996b.

STRATIGRAPHY

Existing wells have penetrated sediments ranging in age from Mid Aptian to Recent and allow a basic stratigraphic

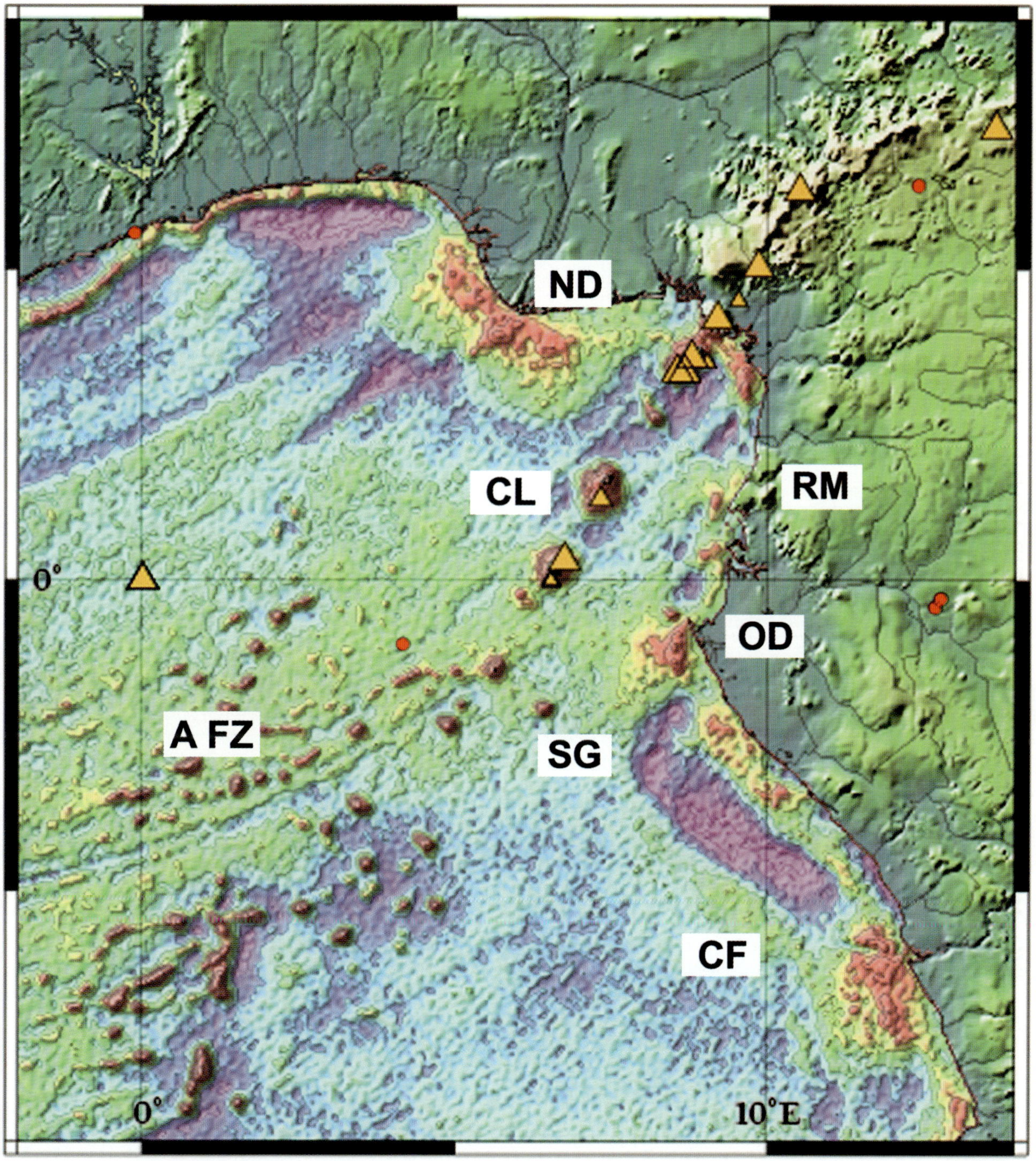

Plate 1. Satellite Gravity map of the Gulf of Guinea showing the major tectonic and sedimentary features (courtesy D. Sandwell). ND-Niger Delta, CL-Cameroon Line, RM-Rio Muni Basin, AFZ- Ascension Fracture Zone, OD- Ogooue Delta, SG-South Gabon Basin and hinge zone, CF- Congo Fan.

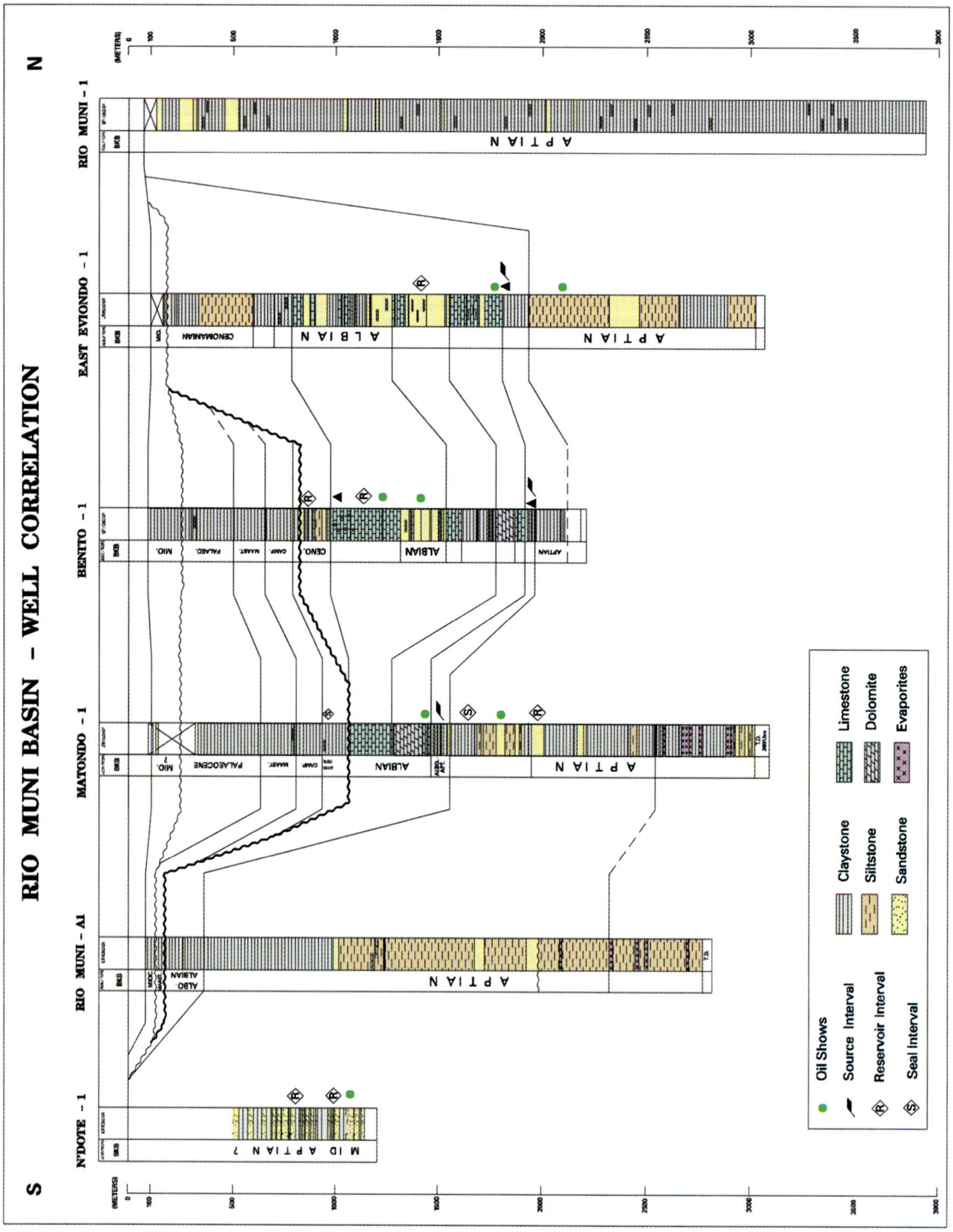

Plate 2. N-S correlation through wells in the Rio Muni basin showing penetrated stratigraphy, lithologies and hydrocarbon play elements (for well locations see Figure 2).

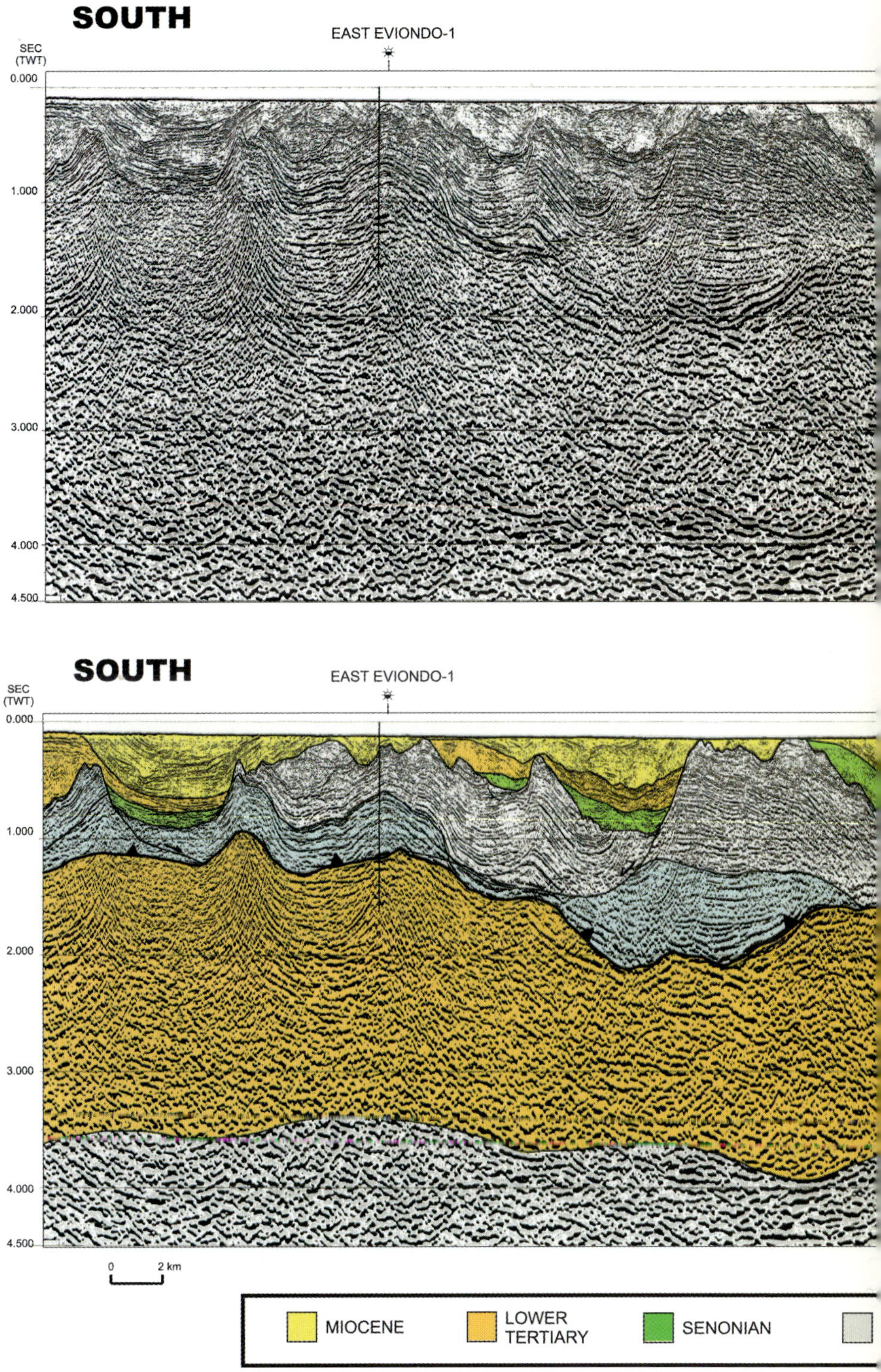

Plate 4. North-South seismic strike line showing Aptian cored high at Rio Muni 1 well locatio
–Cenomanian rafts on Apto-Albian shale detachment in the basinal area to the south. Bata ant
south of this section (see Figs. 2 and 11).

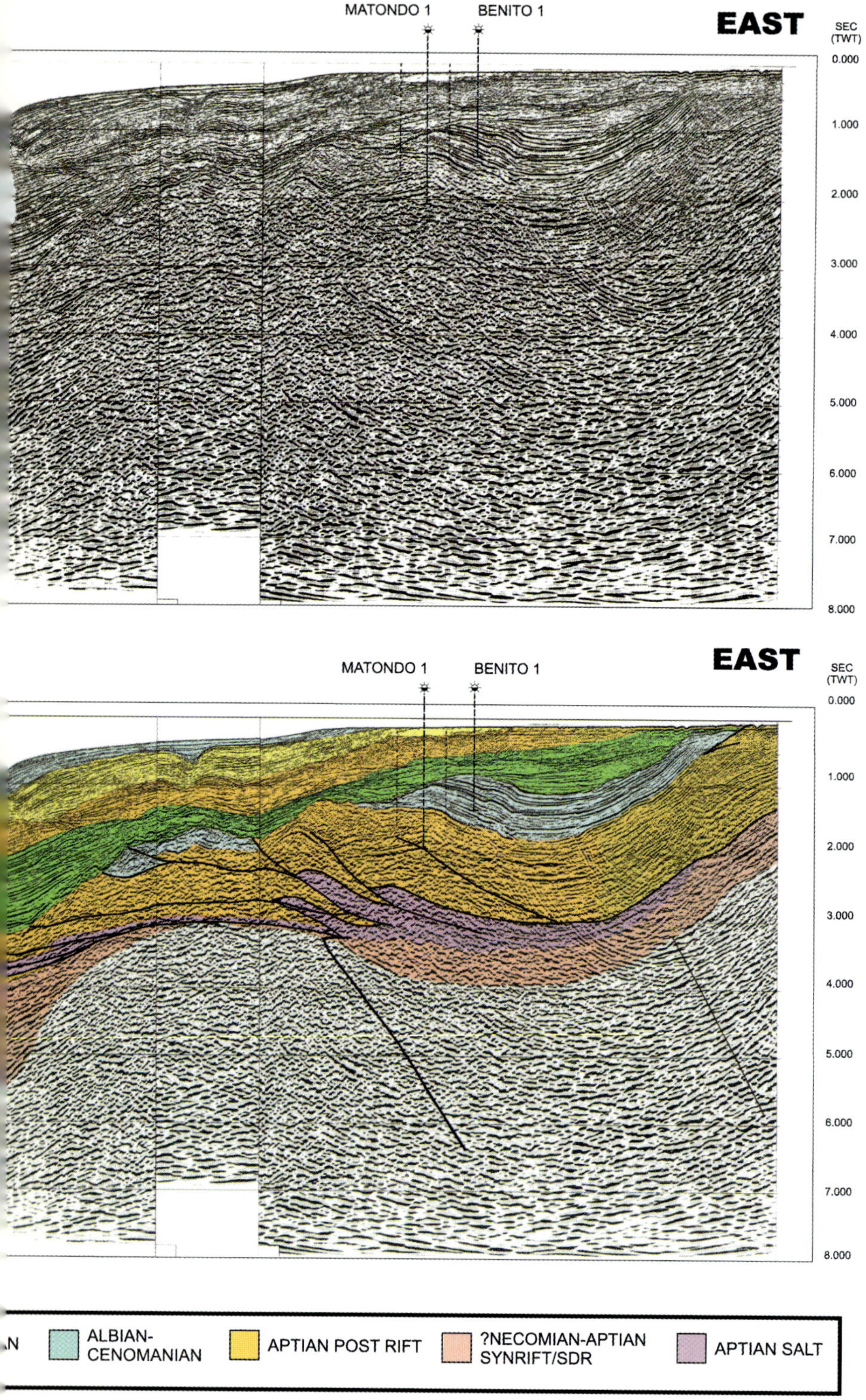

d structural high and thick, Senonian to ıe base of slope, detaching on Aptian helfal exploration wells. Note west dip- r location).

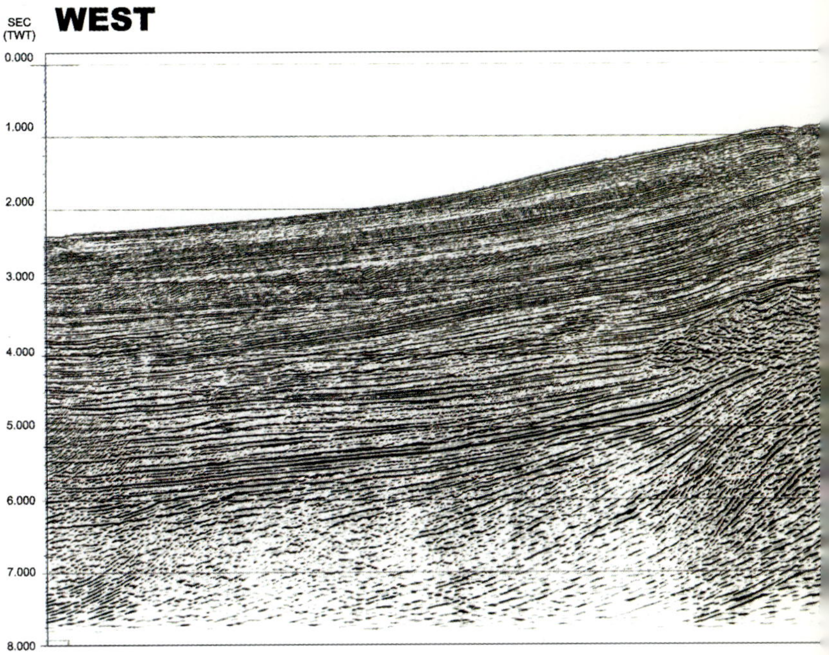

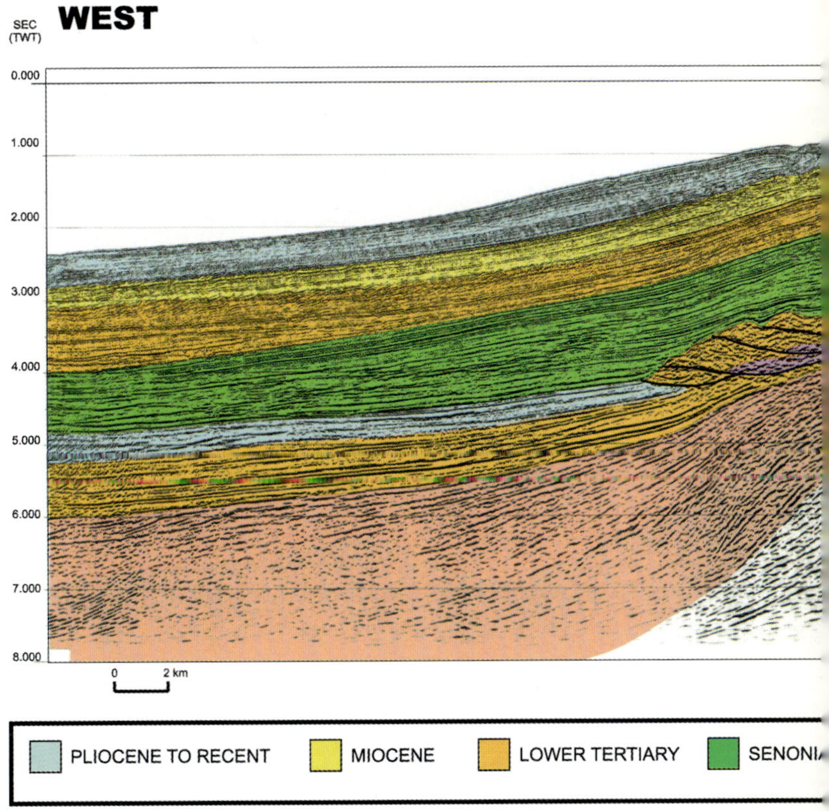

Plate 3. Regional East-West seismic line across the Rio Muni Basin showing inboa Recent deepwater sequences. A well-developed toe thrust belt is developed at t evaporites. Compressional features are interpreted to extend eastwards beneath the s ping events beneath the Aptian post-rift sequence wedging oceanwards (see Fig. 2 fc

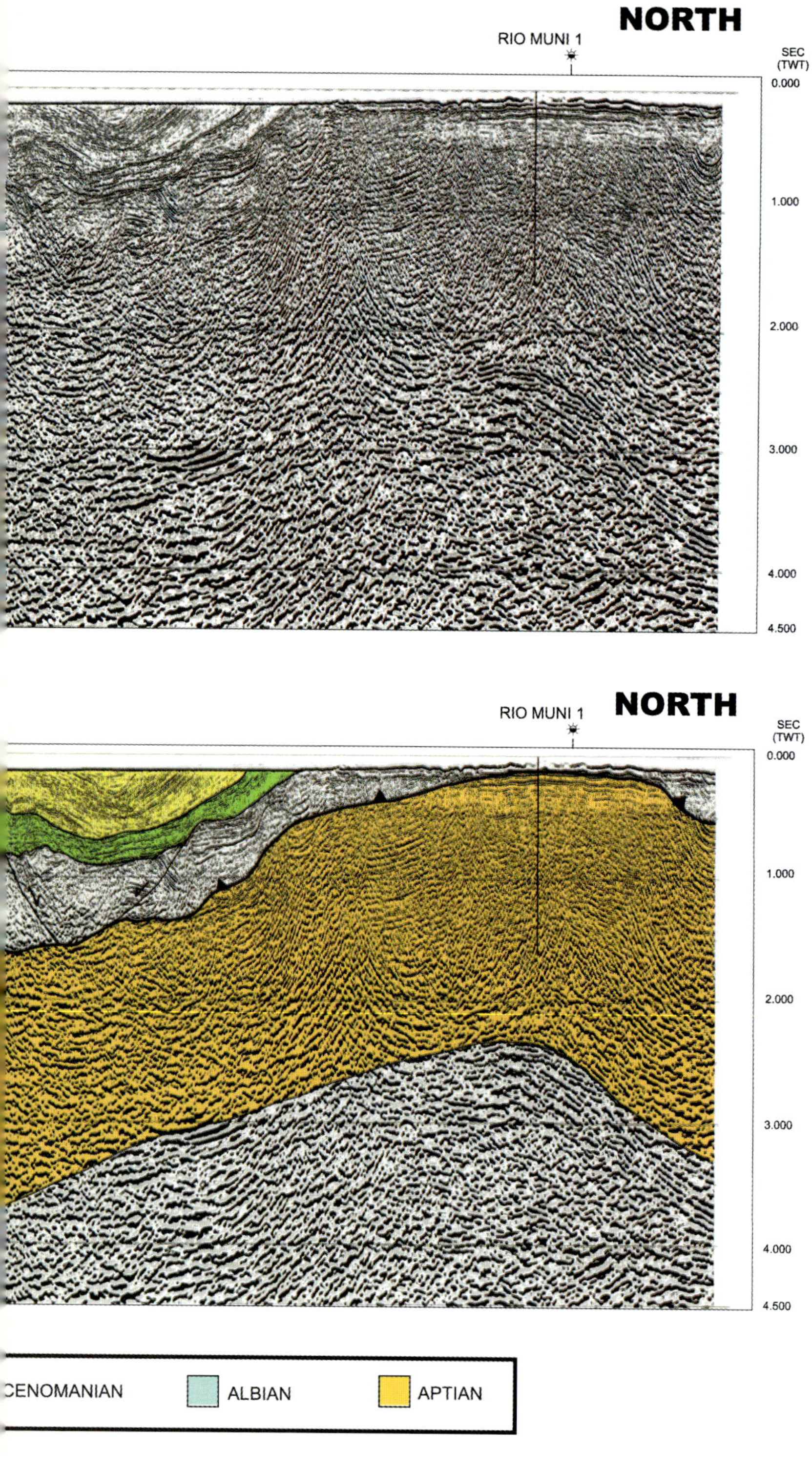

ı and development of Albian
cline is developed just to the

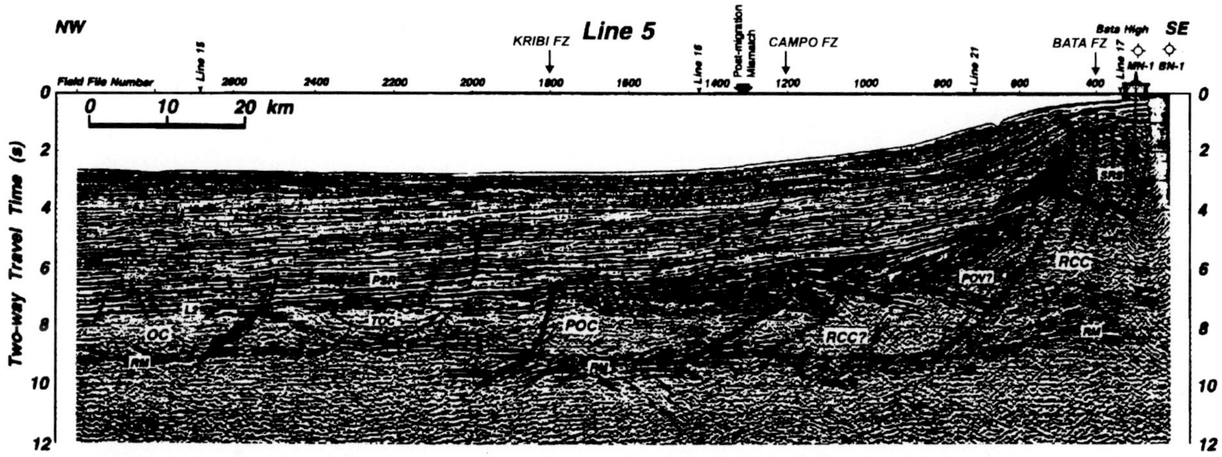

Figure 5. Seismic line 5 of the PROBE survey showing location of transform zones in the Equatorial Atlantic outboard of the Rio Muni Basin. Modified from Meyers et al 1996b (for line location, see Figure 4). SRS-syn-rift sediments, RCC-rifted continental crust, RM- reflection Moho, POV-proto oceanic volcanics, POC- proto oceanic crust, OC-oceanic crust, PSR-top pre salt reflectors, TOC-thick oceanic crust, L2- top oceanic crustal layer 2.

framework to be established within the basin. The stratigraphy is summarized in Figure 6 and Plate 2. The deepest section penetrated are a series of interbedded, fluvial sandstones and mudstones encountered in the onshore N'Dote 1 well. These are Mid-Aptian in age and may be late syn-rift sediments, although onshore seismic data is not of sufficient quality to identify their stratigraphic geometry.

Offshore, the deepest formation penetrated is a series of interbedded Upper Aptian salts, mudstones and thin sands known as the Saliferous Formation. These are best developed in the Matondo 1 well, which penetrated a thickness of 500m before reaching its total depth. These are overlain by a 1 km thick succession of Upper Aptian interbedded muds and sands known as the Matondo Formation and interpreted from core to be lacustrine turbidites [*Bottero et al 1989*]. Seismic geometries suggest that these occur within the late syn-rift to early postrift sequence. Together with the upper, penetrated portion of the Saliferous Formation, this section is underlain by thick a succession of unpenetrated syn-rift sediments and packages of seaward-dipping reflectors (sdr) in parts of the deepwater area of the basin. The sdr wedges have a possible volcanic origin, and mark the transition from continental to oceanic crust [Hinz, 1981].

The uppermost part of the Aptian section shows a transgressive character and is associated with interbedded muds, limestones and thin sands. These culminate in the deposition of organic rich mudstones with excellent source potential and which act as a detachment horizon for both extensional and compressional features in the overlying Albian carbonate successions. This transgressive package is interpreted to contain brackish to marginal marine facies [*Bottero et al 1989*]; the uppermost mudstones represent a maximum flooding surface, which is the first true marine flooding of the basin.

This transgressive package is overlain by the Albian carbonate sequence, which is 400m thick in the Matondo 1 well, but varies considerably in thickness and facies around the basin. It is dominated by oolite shoals and micritic packstones.

An interbedded sequence of marine sands and shales overlies the Albian carbonate sequence. It was penetrated in the Benito-1 well and dated as Cenomanian in age. Where sampled by exploration wells, this section is thin and shows no angular unconformity with the underlying carbonates. However, overlying this package, there is a further package which onlaps and thickens away from the structural highs on which the wells were drilled. This package attains a thickness of hundreds of metres and may extend into the Turonian (see Plate 2 and Fig. 7). These are interpreted to be early passive margin sediments.

The Aptian through Cenomanian stratigraphy is bounded by a major unconformity that cuts down dramatically through the Cenomanian and Albian sections, and is onlapped by Coniacian to Santonian mudstones, penetrated in the Matondo-1 and Benito-1 wells. This section forms part of a Late Cretaceous onlapping deep marine package, which thickens westward to over 1 km thick. A series of marine mudstone-dominated sequences with interbedded sandstones and limestones, ranging from Campanian to Recent in age, overlies the Santonian–Coniacian package. This section is subdivided by a series of seismically defined sequence boundaries (Plate 3), which amalgamate on the shelf and are represented in exploration wells by base Maastrich-

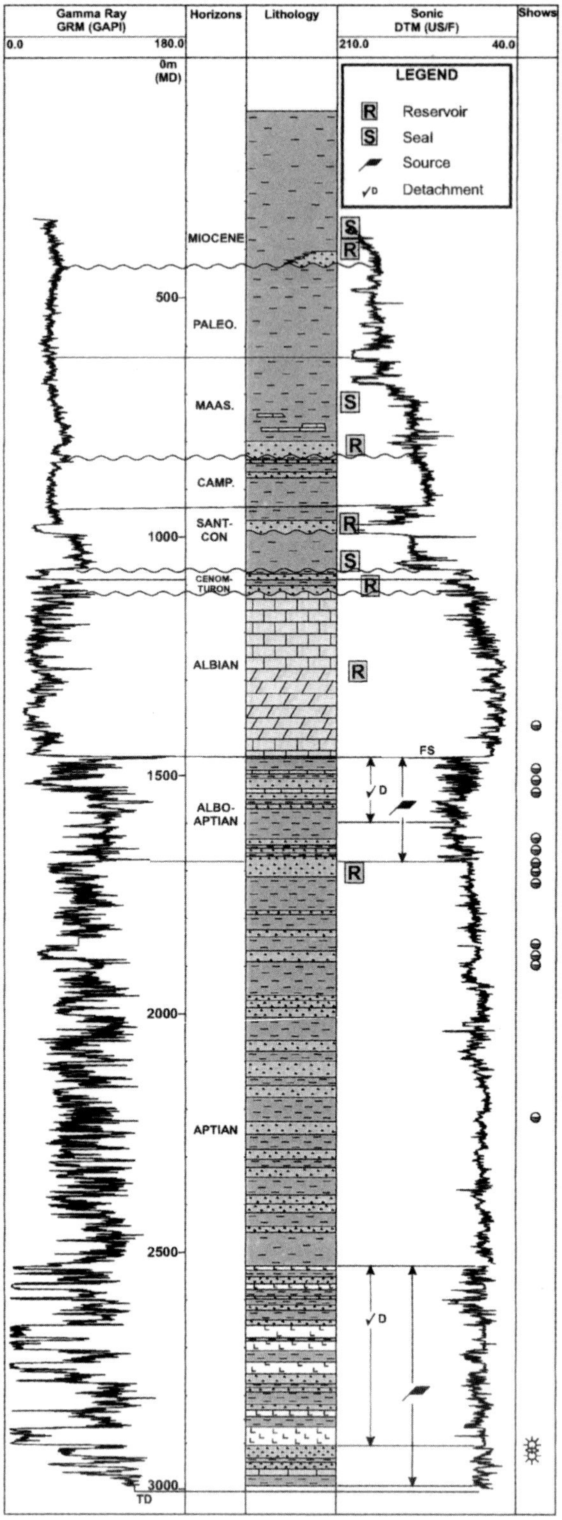

Figure 6. Lithological column showing typical shelf stratigraphy of the Rio Muni basin.

tian and intra Miocene unconformities. In the deep-water portions of the margin this section can reach up to 2 km in thickness.

CORRELATION WITH GABON AND KRIBI-DOUALA BASINS

Comparison of the Rio Muni stratigraphy with that seen in the adjacent North Gabon and Douala basins (Fig. 8), shows that although there are broad similarities in the stratigraphy of the three basins there are also major differences. The Rio Muni basin sits at the junction of two different tectonostratigraphic provinces and is the location of important stratigraphic changes. At the Upper Aptian level, the Rio Muni basin contains a thick section of lacustrine turbidites which constitute the Matondo Formation, this shows little evidence of marine influence and separates the overlying Albian carbonates from the Saliferous Formation. By contrast, the Albian aged Madiela Formation carbonates of Gabon are directly underlain by a thick Ezanga salt formation with no intervening terrestrial sequence [*Ross and Hempstead 1993*]. Thus a major phase of terrestrial sedimentation is present in the Rio Muni Basin but absent in the Gabon basin where a time equivalent sequence of salt up to 800m thick [*Teisserenc and Villemin 1990*] was deposited. This compares to individual salt beds of up to 50m in the Saliferous Formation penetrated in the Matondo 1 well (see Plate 2). Evaporite distribution is restricted to local areas in both the Rio Muni and the Kribi basins, whereas to the south in Gabon, salt deposition was developed basinwide.

The overlying Albian carbonates in the Rio Muni basin (Benito Fm.) are time equivalent to those of the Madiela Formation in the North Gabon Basin but this facies passes into interbedded marine sandstones and mudstones in the Douala- Kribi basin of Cameroon. These clastic facies range from deltaic to bathyal in environment and form the reservoir in the Sanaga Sud Field [*Pauken 1992*]. Clastic interbeds are present within the Albian of Rio Muni and these become more common from south to north (see Plate 2).

Thus, during both Upper Aptian and Albian times the Rio Muni basin occupied a zone of transition between clastic dominated deposition in the northern transform basins and carbonate/ evaporite deposition in the 'Aptian sält basin' to the south.

Deposition of the postrift megasequence was punctuated in all three basins around the Turonian-Coniacian boundary by the development of a major unconformity, termed here the Early Senonian unconformity. Subsequent deposition of Late Cretaceous to Tertiary sequences is broadly similar throughout all the basins, although thicker sequences are developed in the North Gabon basin because of the input from the Ogooue Delta.

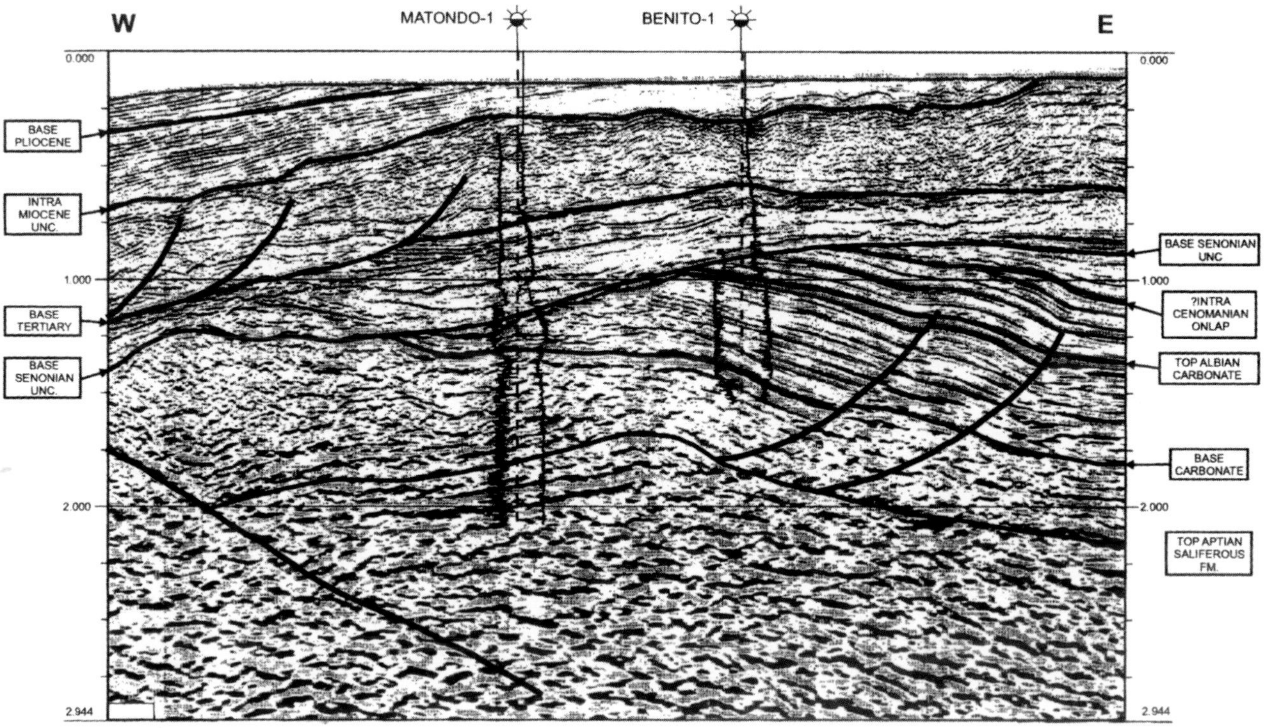

Figure 7. East-West seismic line showing the seismic expression of the main stratigraphic sequences encountered in the Benito 1 and Matondo 1 wells (see Fig. 2 for location).

SEISMIC STRATIGRAPHIC FRAMEWORK

Figure 7 shows the well to seismic tie for the Matondo 1 and Benito 1 wells and shows the main seismic markers interpreted around the basin. These are as follows:

Top Syn-rift (intra Aptian), defined by truncation beneath and onlap above, it is well developed in deepwater area. Locally it may represent the top of a seaward-dipping reflector package (Plates 3 and 5).

Top Saliferous Formation (intra Aptian), tied to the Matondo-1 and Rio Muni A-1 wells, it is associated with halokinesis on the upper slope area in the southern part of the basin. Only the upper portion of this sequence has been penetrated by exploratory boreholes, and seismic geometries suggest that the lower portion may locally contain thick salt, particularly in the southern part of the basin.

Top Albo-Aptian/Base Carbonate (Late Aptian-Early Albian), identified by a concordant-parallel, strong marker penetrated in the Matondo-1, Benito-1 and East Eviondo-1 wells, it is associated with the development of a maximum flooding surface and the base of highly reflective carbonate section. This horizon acts as detachment surface (Plate 5).

Onlap Horizon of Late Albian –Early Cenomanian age, characterized by by overlying weaker amplitude facies which thicken into the hangingwall of Albian carbonate rafts.

Early Senonian Unconformity, characterized by strong truncation below and strong onlap above, this major amalgamation surface, with up to 3 kms relief, is associated with development of major canyon systems. Onlapping stratigraphy may range from Coniacian to Miocene in age.

Base Tertiary, associated with a regional flooding surface, which acts as detachment surface over large areas of the shelf edge and slope.

Lower Tertiary (? Oligocene) Unconformity, corresponds to a major incision surface and canyon cutting event identified on seismic data and associated with basin floor deposits.

Intra-Miocene Unconformity, associated with major incision and canyon cutting events.

Base Pliocene, associated with a regional flooding surface overlain by downlapping and prograding sequences.

These surfaces are used to subdivide the stratigraphy of the basin; their geometries are summarised in Plate 6.

BASIN GEOMETRY AND TECTONOSTRATIGRAPHY

Present-Day Structure

Figure 2 shows the present day structural configuration of the basin and its relationship to deep crustal structure. To the north and south the basin is bounded by the Campo and

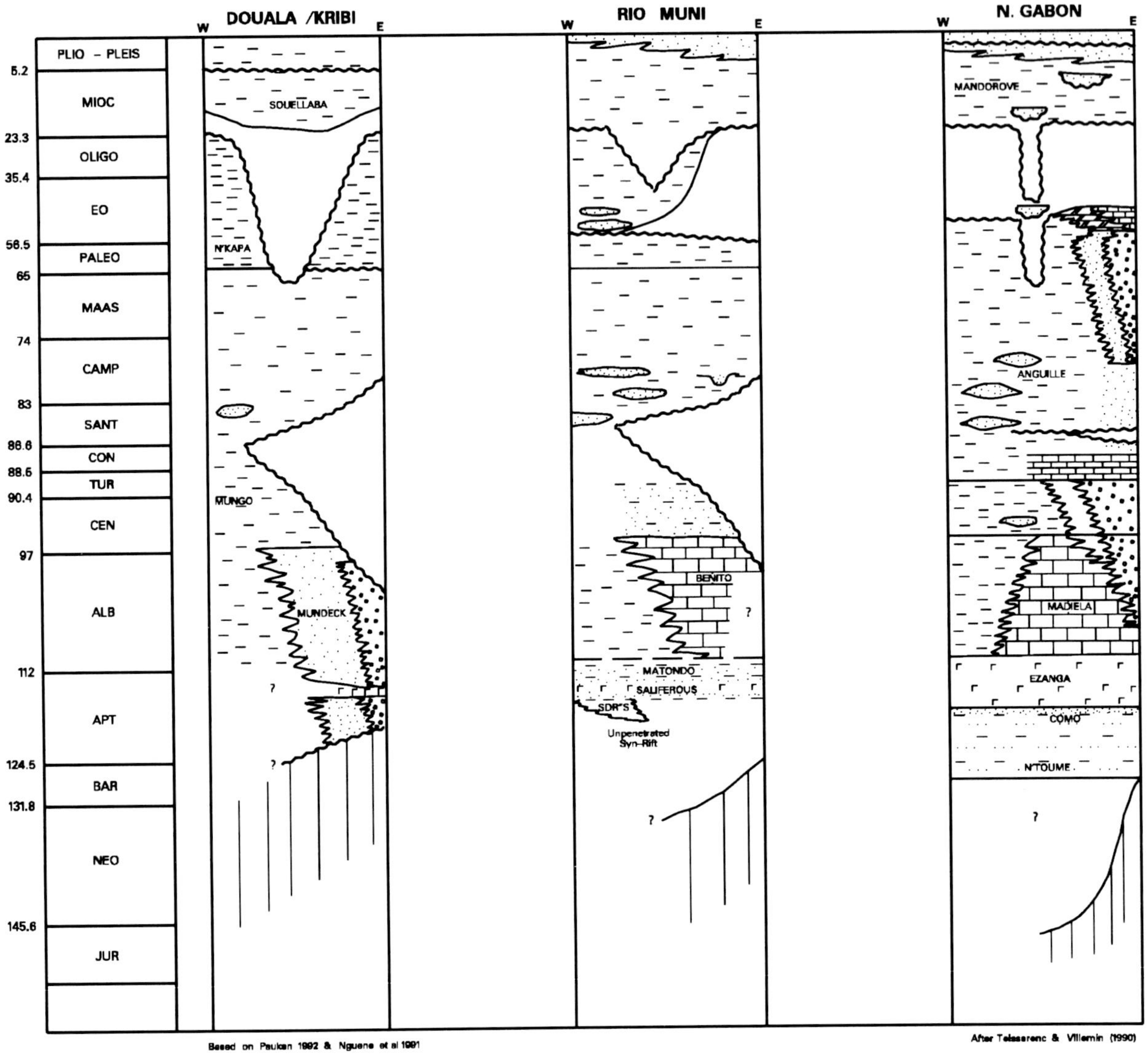

Figure 8. Comparative stratigraphy of the Douala, Rio Muni and North Gabon Basins, showing age and lithologies of major units. Douala/ Kribi column based on Pauken 1992 and Nguene et al 1992, Gabon column after Teisserenc and Villemin 1990.

Calatrava highs, which are underlain by the continental extension of the transform zones previously described. These highs trend NE and are paralleled by a series of smaller features which subdivide the basin into a series of highs and lows. These features are also oriented NE-SW and are coincident with the continental extension of the Bata transform.

The seismic line in Plate 4 is oriented from north to south and runs through the northern part of the basin in a strike orientation (see Fig. 2). It illustrates the intrabasinal highs and lows and the relationships of the major seismic packages identified. The highs consist of thick folded sections of upper Aptian post-salt clastic strata. The high at the northern end of the basin has been penetrated by the Rio Muni-1 well, which encountered 3.8 km of Aptian clastics, with no overlying Late Cretaceous or Tertiary sediment.

By contrast, the lows are dominated by rotated blocks of Albian carbonate and Cenomanian clastics detached on an underlying mudstone within the Albo-Aptian sequence. A thick Late Cretaceous to Tertiary clastic wedge overlies this succession. Thus, the present day basin appears to consist of Aptian cored highs and Albian to Tertiary aged depocentres.

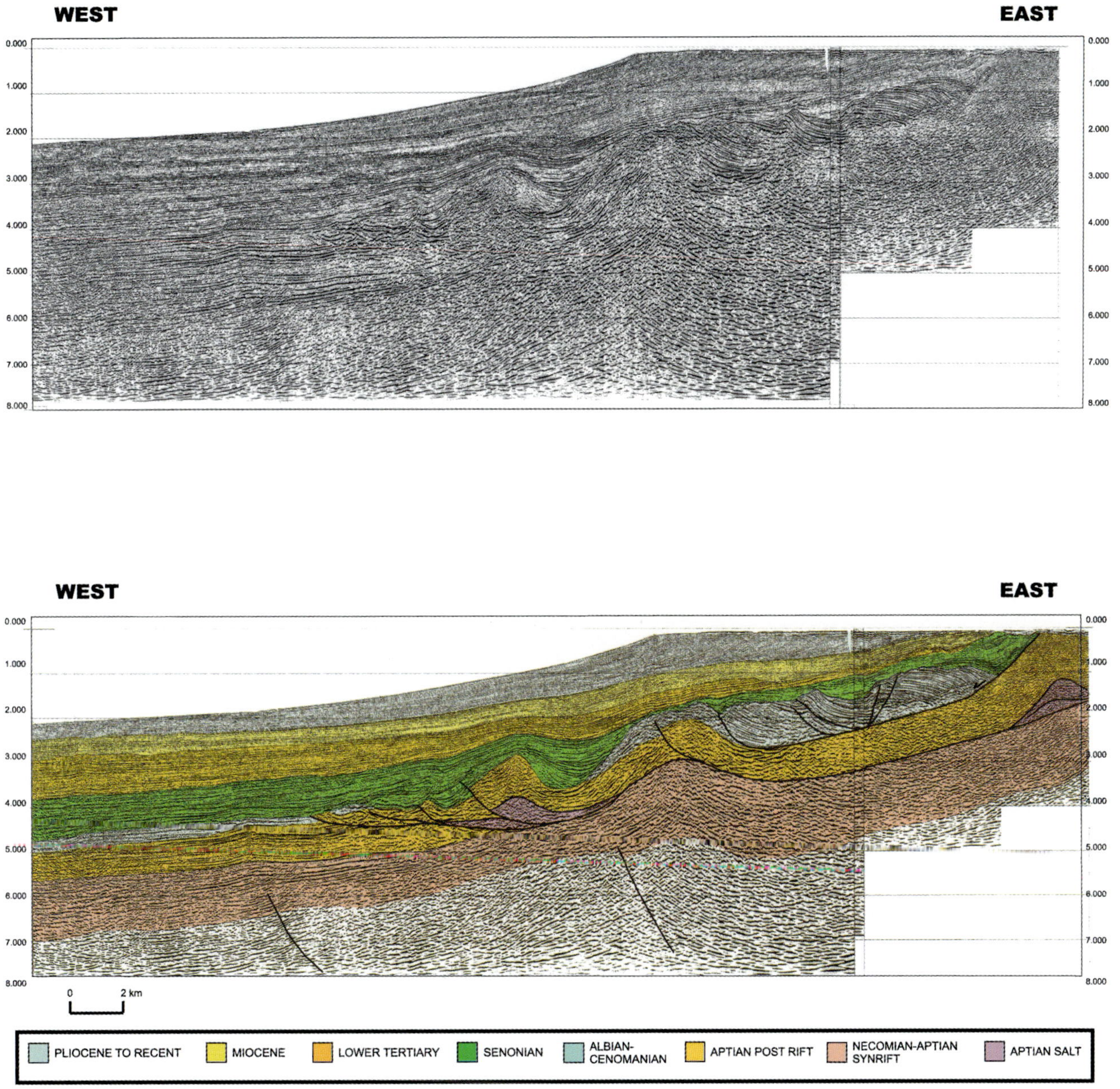

Plate 5. East –West seismic line showing relationship of Albian rafts and toe thrusts to Senonian Unconformity and to the underlying Aptian structural high, interpreted as a transpressional anticline at the Aptian level (see Fig. 2 for location).

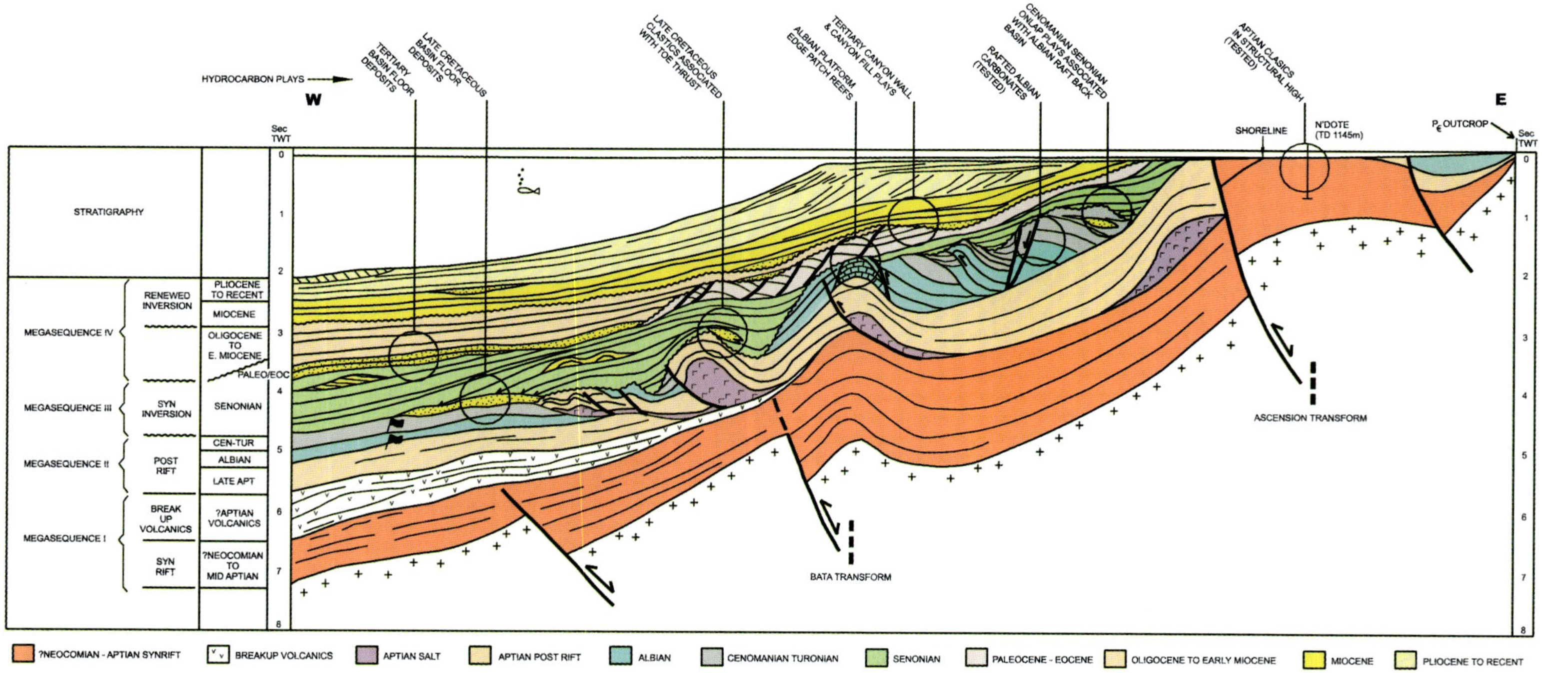

Plate 6. Geoseismic section based on the line in Plate 5 and onshore data, showing the present-day geometry of the margin: inverted extensional basins, toe-thrust development, Albian carbonate rafts, wedges of possible seaward-dipping reflectors, deepwater mounds and Miocene to Recent progradation. Hydrocarbon plays and location of potential source kitchen are also shown.

Aptian Basin Architecture

Early Senonian deformation of the shelfal area has resulted in poor seismic definition of the pre-Senonian basin geometry. Thus while it is generally accepted that a syn-rift section exists beneath the shelf [*Turner 1995, Turner 1996*], it is difficult to map the syn-rift architecture with any degree of confidence. By comparison, the deepwater area has not suffered the same degree of late deformation, and the deep basin architecture can be consistently mapped. An example of this geometry is illustrated in Plate 3, which shows strongly dipping reflectors beneath the base of the post-rift section. Wedging and thickening of these reflectors westwards and progradation into localised lows can also be seen on some lines. In the central part of the basin these reflectors dip consistently to the west and are located beneath the slope and basin areas, terminating abruptly against the structural high associated with the underlying Bata transform zone (Fig. 2). In the northern and southern parts of the basin, the dip and sense of wedging of these reflectors flip polarity from east to west. These dipping wedges are interpreted to be syn-rift in origin and are fault-bounded. Insufficient magnetic data is present to conclude whether the westward dipping packages in the central deepwater area (see Fig. 2) represent a syn-rift trough or a seaward-dipping reflector wedge characterized by offlapping volcanics associated with the final stages of the rift to drift transition [Hinz, 1981]. These wedges have been reported from a number of other basins in the north and south Atlantic [*Mutter 1985, Parson et al 1988, Larsen and Jakobsdottir 1988*], and are interpreted to be present in the conjugate Sergipe-Alagoas basin of Brazil [*Mohriak et al. 1998*]. However, while the northern and southern areas of Rio Muni have characteristics more typical of syn-rift geometries, the central area is tentatively interpreted to consist of a package of seaward-dipping volcanics located on the western side of the Bata transform. This feature may locally form the boundary between continental crust and transitional crust in this area [*Meyers 1996b*].

Middle Cretaceous Uplift and Transpression

Following establishment of the Aptian syn-rift architecture and deposition of the post-rift onlapping stratigraphy, uplift and the development of a major unconformity in the present day shelfal area terminated the subsidence-dominated regime of the Albo-Aptian.

The available vitrinite reflectance data for wells in the basin is summarised in Figure 9. The geothermal gradient and surface temperatures required to produce the observed vitrinite reflectance with depth trend in the East Eviondo 1 well are consistent with the present day geothermal gradient and surface temperature. Thus limited evidence exists to suggest that the stratigraphy penetrated in this section has been heated to higher temperatures than those present today. Therefore it is unlikely that significant net uplift of any sort has occurred at the well location [*Gibson et al 1998*] although seismic data does show some erosion and down-cutting on the western side of the structure.

By comparison, vitrinite reflectance versus depth trends for the Aptian present in the Matondo, Rio Muni 1 and N'dote 1 wells suggest that all these wells have experienced elevated palaeotemperatures and may have undergone uplift. The amount of uplift varies from a few 100m's at Matondo-1 to 1.5 km at Rio Muni 1 and up to 3 km at N'dote. In the case of Matondo 1 this is supported by AFTA analysis [*Gibson et al 1998*]. No evidence is present to suggest that the elevated palaeotemperatures at these wells are a function of localised heating effects such as hot fluids or volcanics. However, Figure 2 shows a close relationship between those wells with elevated palaeotemperatures and the location of the NE-SW trending anticlinal highs which subdivide the basin into highs and lows. Similarly, wells with Aptian vitrinite profiles suggesting limited elevation of palaeotemperatures/greater burial (East Eviondo-1 and Benito 1) are located in the intervening lows.

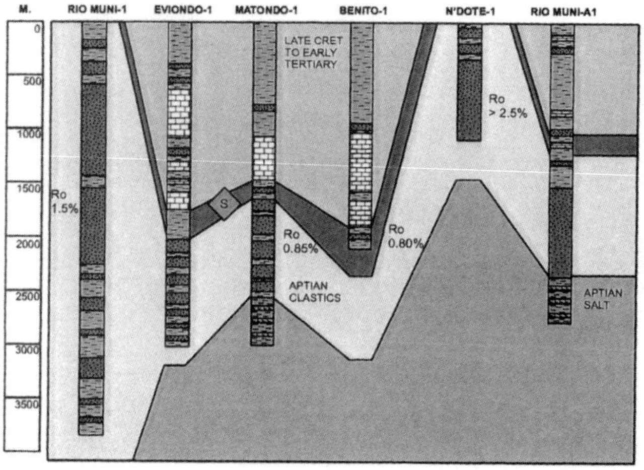

Figure 9 (a)

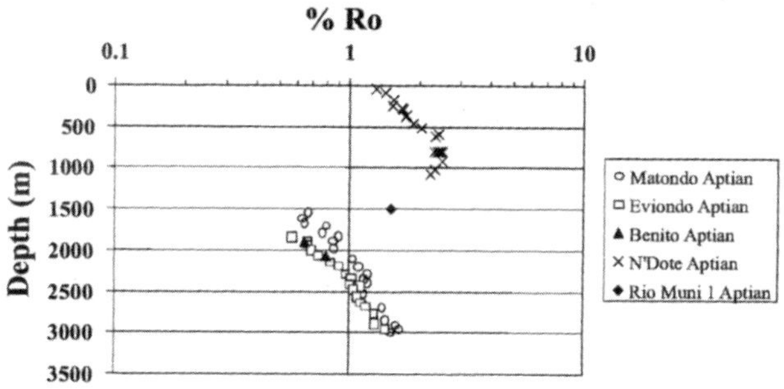

Figure 9. a) Stratigraphic section showing typical vitrinite reflectance measurements in the Aptian of the Rio Muni wells (courtesy E. Ong).
b) Plot of vitrinite reflectance data from 4 Rio Muni wells showing relative uplift of N'Dote 1, Rio Muni 1 and Matondo wells.

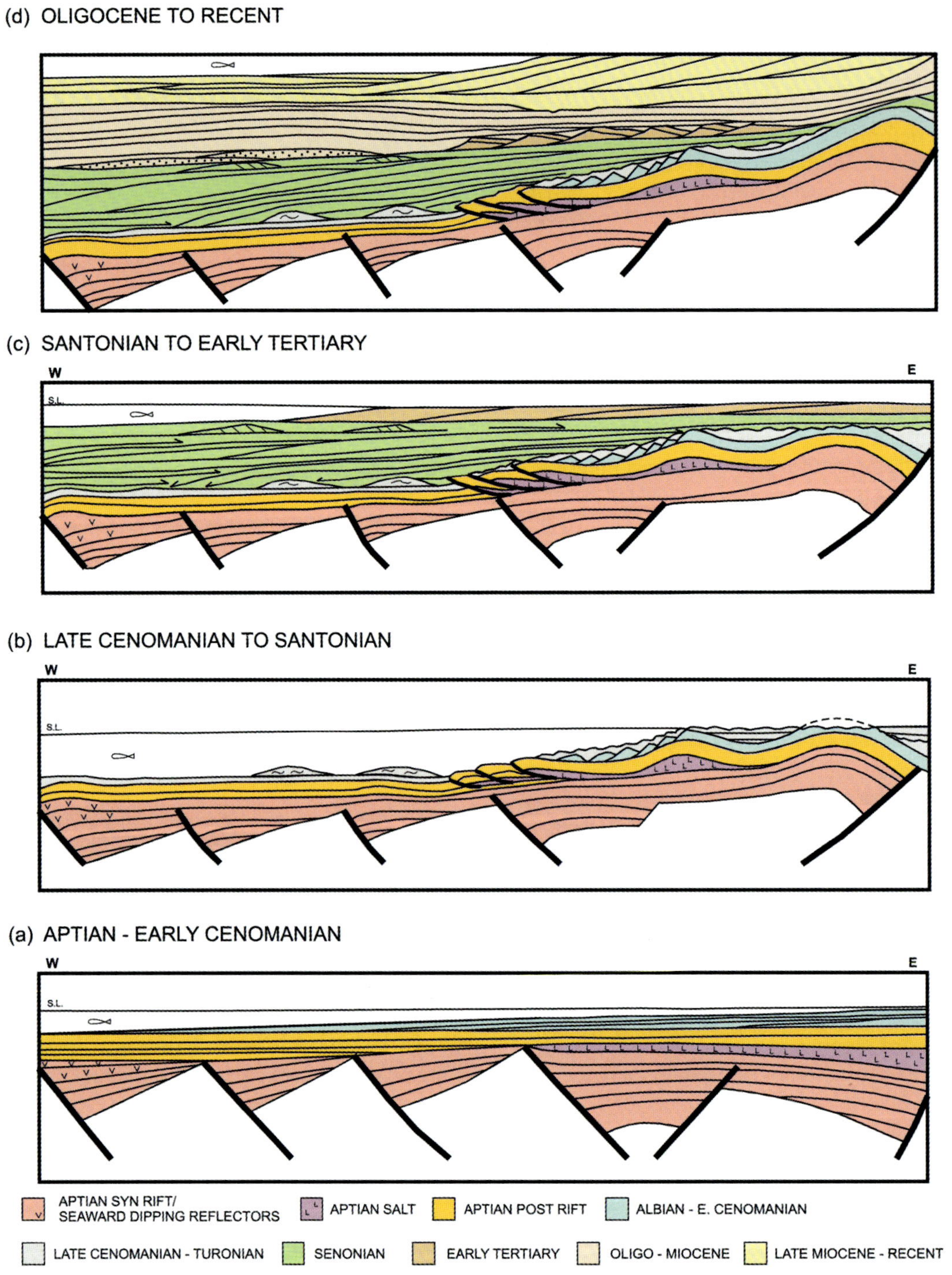

Plate 7. Series of schematic cross-sections showing main tectonostratigraphic events in evolution of the Rio Muni Basin: **a)** Syn-rift basin development with possible development of seaward-dipping reflectors in outboard basins and development of late syn-rift - early post-rift inboard salt basin, followed by gradual infill of post-rift depocentre by onlapping lacustrine fill, marine flooding and Albian-Early Cenomanian carbonate dominated shelf development. **b)** Uplift and inversion of Aptian and Albian sequences, development of anticlines above transform trends, detachment and sliding on both the Aptian salt and Apto-Albian mudstones to form deepwater toe thrusts and extensional carbonate rafts. Development of deepwater slump/fan complex and infilling of raft 'back basins' by Late Cenomanian to Santonian clastics. **c)** Infill of post-unconformity topography, development of retrograde slope fan and possible reworking of slope sediments by Late Cretaceous bottom currents forming contourite mounds. **d)** Relative base level fall results in slumping and collapse of Early Tertiary shelf-slope sediments and deposition of mounds in deep water, Miocene incision and canyon cutting events followed by Miocene to Recent shelf progradation.

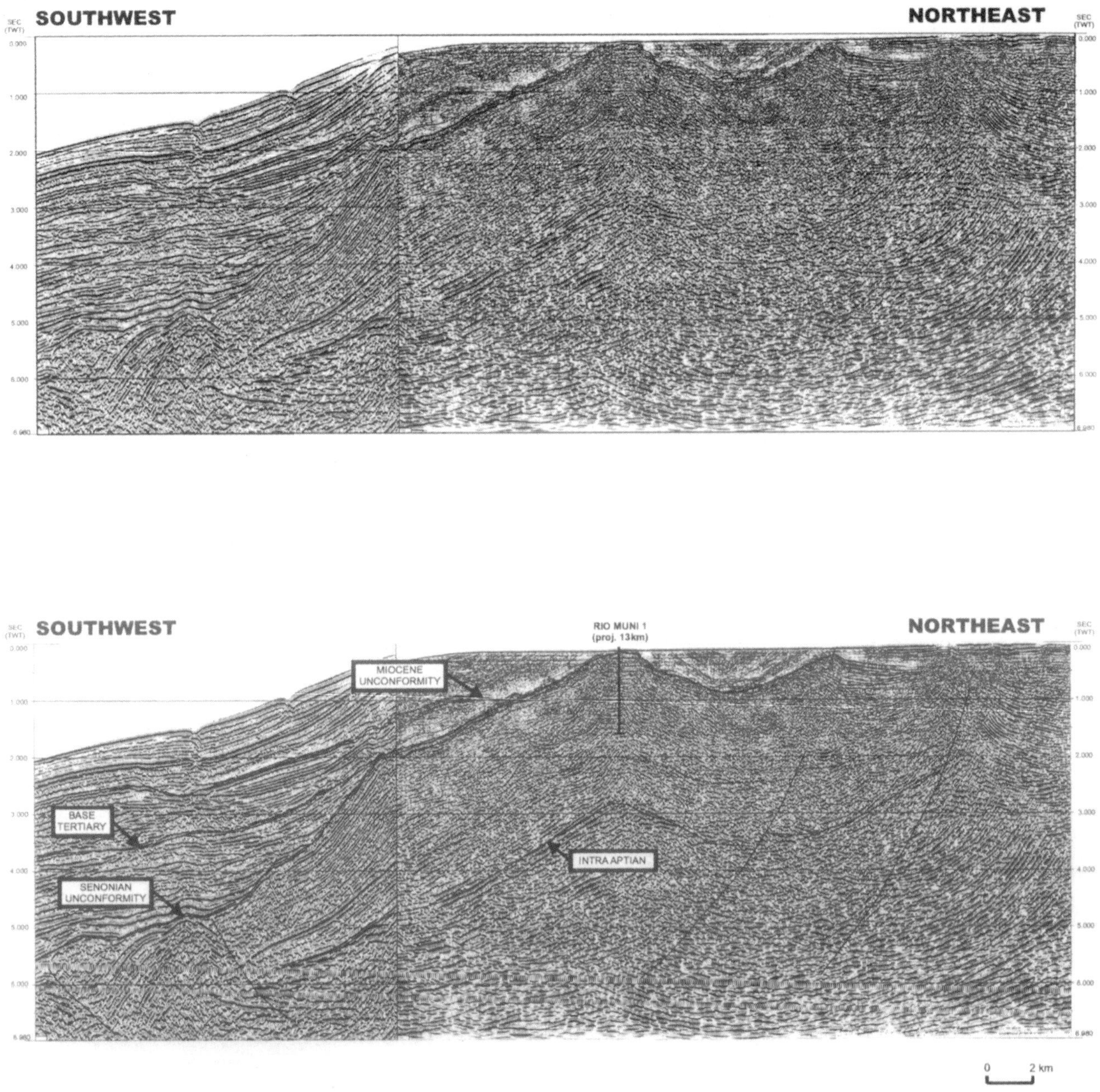

Figure 10. Composite East–West seismic line showing stratigraphic thickening of the intra-Aptian- to Senonian Unconformity package towards the crest of the Rio Muni 1 structure (see Fig. 2 for location).

The map also shows the short distances over which variation in vitrinite profiles occur, suggesting that this may be a function of intrabasinal tectonics rather than a regional phenomenon.

Figure 10 demonstrates the geometry of the structure penetrated by the Rio Muni 1 well, which encountered a thick upper Aptian section with high vitrinite reflectance values. The disposition of the Aptian section with regard to the overlying stratigraphy is shown on an East-West dip line. The well penetrated a large anticline (the Pongue Anticline) that shows evidence of thickening of the Aptian section towards the crest. This is interpreted to represent an inverted Aptian half graben. The location of this Aptian basin is broadly coincident with the location of the deep transform (see Fig. 2)

Similarly, Figure 11 shows a NE-SW trending line which runs across the center of the basin and shows a major culmination, which has folded the Aptian through Albian suc-

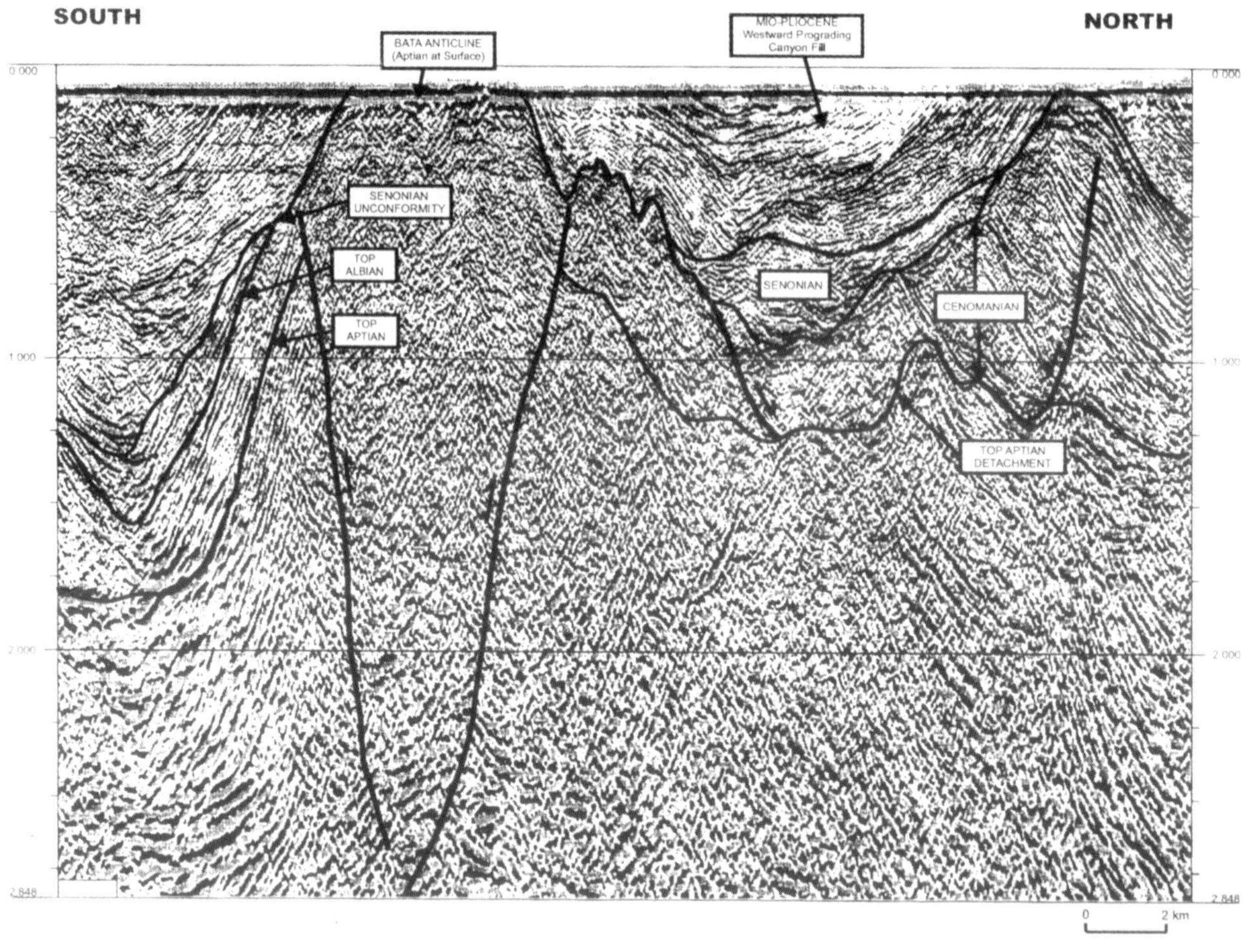

Figure 11. North –South seismic line through the Bata anticline showing folded Aptian stratigraphy onlapped by Senonian to Tertiary strata. Steep dip of Miocene to Recent section is depositional and a function of line orientation with respect to steeply dipping, westward prograding clinoforms (see Fig. 2 for location).

cession. In this area, the base of the Albian is a major detachment surface which has also undergone folding and is cut by the Early Senonian unconformity. As shown in Figure 2, when mapped out this structure forms one of a series of elongate anticlines, which directly overlies the Bata fracture zone. The southern anticline corresponds to the Calatrava high, wich was penetrated by the Rio Muni A-1 well. It is also cored by Aptian and Albo Aptian clastics, which are folded into an anticline that is truncated by the basal Senonian unconformity and onlapped by Santonian and younger stratigraphy (see Fig. 12). Thus, it appears that there are a series of anticlines, of Early Senonian age, which have uplifted the Aptian depocentres and that these are broadly coincident with the mapped positions of the transform zones, which are interpreted to extend beneath the continental margin.

Tertiary Uplift

The majority of the uplifted structures are geochronologically calibrated by the development of the Early Senonian unconformity. However, there is some evidence for renewed uplift in the Tertiary, probably associated with the development the Miocene unconformity (see Plate 6). This has been recognised in the northern part of the basin around the Pongue anticline, where Senonian to Early Tertiary stratigraphy is folded [*Ruiz 1997*]. In addition, apatite fission track analysis carried out on Aptian samples in the Matondo 1 well suggests that a second cooling event affected the stratigraphy in this well during the Oligo-Miocene, and that this was associated with several hundred metres of uplift [*Gibson et al 1998*]. These observations are consistent with the development of the unconformities in

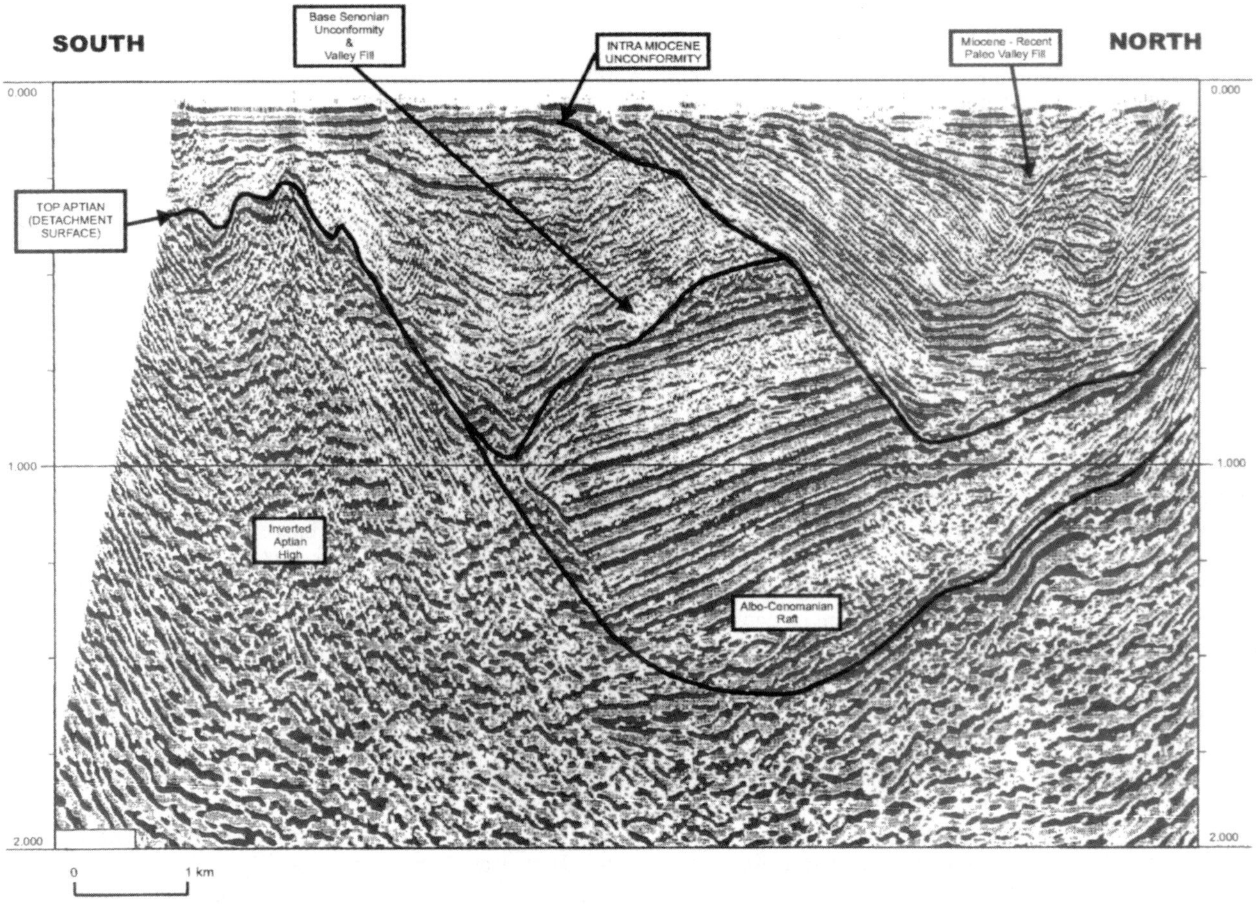

Figure 12. North –South seismic line showing the development of multiple canyon systems incising into an Aptian cored high at the southern end of the profile and a rafted Albian-Cenomanian slide block in the central part of the line (see Fig. 2 for location).

Oligocene and Miocene observed on seismic data (Plates 3,4 and 5).

Gravity Tectonics

Between the NE trending intrabasinal highs discussed above, the shelf area is dominated by a second structural style comprised largely of detached blocks of interbedded Albian- Early Cenomanian limestones and sandstones. These have been described in some detail by other workers [*Ross, 1993, Turner 1995, Turner 1996*]. An example of this is shown on the NW-SE oriented seismic line in Plate 5 which shows rotated blocks composed of Albian carbonate, which have detached on the underlying Albo- Aptian mudstone package and have slid basinwards. The figure also shows onlap of the overlying Late Cenomanian-Turonian package, dating the onset of sliding and rotation as Late Cenomanian to Turonian in age. The distribution of extensional slides can be seen in Figure 2.

Over most of the shelf area, evidence for detachment is largely at the base of the Albian carbonate, within the Albo –Aptian mudstones. Additionally, as shown in the westernmost part of the line in Plate 5, sliding may also occur at the level of the Aptian evaporites. This is particularly evident towards the outer part of the shelf and on the slope, where toe thrusts are well developed. This phenomenon is common in the south and may describe the extent of the salt section, with the toe thrust 'front' developing at the westward pinchout of the salt. The distribution of thin-skinned compressional toe thrust structures on the outer slope is shown in Figure 2. These form two broad belts, one in the south and one in the north, separated by the Bata anticlinal trend.

Onlap of Coniacian- Santonian and younger sediments

onto these thin skinned extensional and compressional structures demonstrates they were broadly contemporaneous and that this entire detached Albian to Turonian synkinematic complex is truncated by the major Early Senonian unconformity surface.

The distribution of uplift associated with this unconformity surface is shown in Figure 2 to be located for the most part on the shelf, associated with the NE trending anticlines. Beyond the shelf edge, this topography is reduced and buried beneath a thick infilling sequence associated with intense slumping and slope fan development.

The degree of uplift as indicated by the vitrinite reflectance results, and the fact that the Aptian and Albian sequences deposited prior to uplift appear to have been shallow-water facies, imply some degree of sub-aerial erosion, at least over the crests of these structures. The unconformity shows dramatic topography and is associated with major incision, canyon cutting and deposition of basin-floor lowstand deposits, prior to renewed flooding and onlap across the shelf. Figure 12 shows a NE-SW strike section illustrating incision and the development of canyons on the shelf area. The canyons cut into an Albian-Cenomanian slide block riding on a basal detachment surface. The Late Cretaceous canyon fill onlaps both the rotated slump blocks and the adjacent, Aptian cored high, indicating that both the uplifted Aptian and the Albian gravity slide developed around the same time, prior to the Early Senonian unconformity. The unconformity is interpreted to represent the culmination of the inversion event and it is likely that the uplift of the Aptian depocentres provided the topography and initial trigger for the gravity sliding of both the rafts and the toethrusts.

Post Early Senonian Unconformity Stratigraphy

Overlying the unconformity, a Late Cretaceous to Tertiary passive margin wedge developed and infilled the old topography. Where penetrated, mostly on highs, this section is mud-dominated and forms good sealing facies in the Maastrichtian to Miocene aged sediments.

Due to the location of the wells on structural highs, there are limited penetrations of the lower part of this onlapping sequence. However, seismic data extending into deep water allows description of this sequence, as shown in Plate 6.

Immediately overlying the unconformity surface and its deepwater correlative conformity, there is a lowstand complex comprised of a NW-prograding slope fan and associated mounded facies in the northwest corner of Block G.

This passes upwards into a transgressive onlapping and infilling slope fan, with well-developed east to west flowing channel levee complexes. These are thought to represent the transgressive and subsequent highstand systems. The uppermost Cretaceous sediments form a highly reflective onlapping sequence of transgressive facies, which in places are associated with the development of northeastward prograding seismic packages at the base of slope, interpreted to represent contourite deposits.

Tertiary Evolution

The upper Cretaceous trangressive package culminates in a flooding surface, representing the boundary with the overlying Tertiary section, which is up to 2 km thick. The lower part of the Tertiary section is dominated by a renewed phase of gravity tectonics associated with the development of a series of listric extensional faults which bound rotated blocks. These are interpreted to contain Palaeocene transgressive and highstand facies (Plate 6). The listric faults detach on the basal Tertiary flooding surface. Development of this fault system is thought to be associated with a combination of renewed uplift of the shelf in Oligocene to Early Miocene times, and base level fall associated with a margin-wide eustatic event [*Haq et al 1987, Rasmussen 1997*]. This resulted in collapse of the Early Tertiary shelf and reworking of the lower Tertiary sequence into seismically defined deepwater mounds. This Lower Tertiary complex is onlapped by a highly reflective lower Miocene transgressive package which in turn is overlain by two further prograding sequences of Miocene and Pliocene to Recent aged sediments.

Magnetic and seismic data demonstrates the presence of a volcanic plug in the northern part of the basin adjacent to the Bata transform zone (see Fig. 2). Onlap relationships suggest this is Miocene in age. This is consistent with the age of volcanic plugs present along the Cameroon line [*Meyers et al 1991*] and with the age of the Loiret volcanic plug associated with the South Fang Fracture Zone in the North Gabon Basin [*Meyers et al 1996a*]. A similar relationship is seen between volcanics and transforms in the Sergipe Alagoas basin [*Mohriak et al 1998*].

REGIONAL KINEMATICS

The rifting history of the south Atlantic and the relationship of the Rio Muni Basin to the northern Brazilian basins is well-established [*Chang et al 1992, Turner 1995, Davison 1997*]. Less well documented however is the distribution of mid-Cretaceous inversion along the West African margin.

The uplift described in the preceding sections culminated in the development of the Early Senonian unconformity. This event has been described from a number of other basins in Central West Africa (Fig. 13), and is associated with the development of transpressional anticlines in the Benue Trough and the Central African Rift [*Adeleye, 1975; Seiglie and Baker 1984; Fairhead and Binks, 1991; Genik 1992; Guiraud and Maurin 1992*]. North of the Rio Muni basins these structures are interpreted to result from transpres-

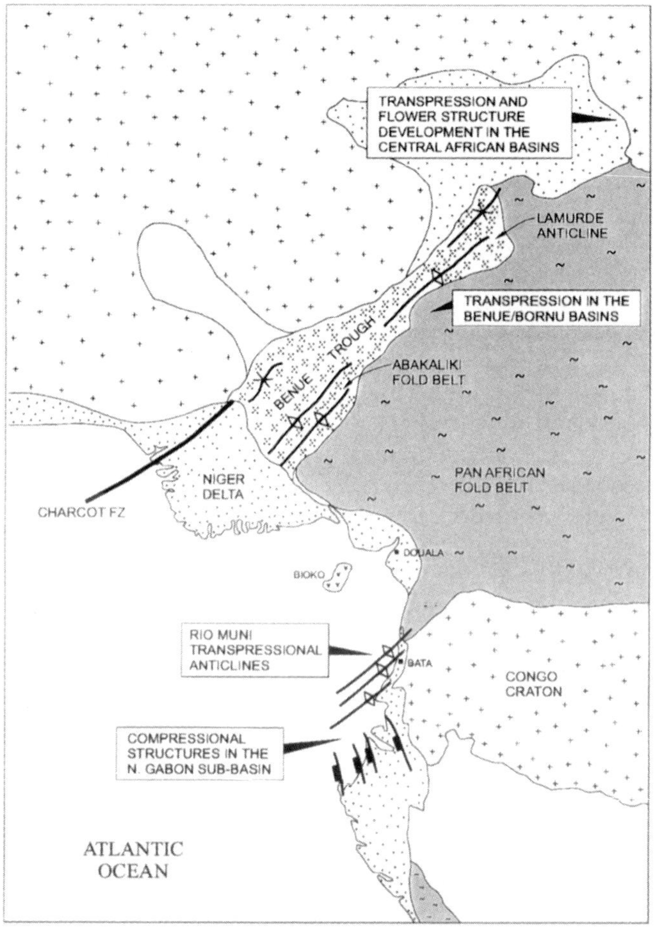

Figure 13. Map showing the distribution of transpressional structures of Early Senonian age in Central West Africa.

sional inversion of NE trending extensional faults, sub-parallel to those observed in Rio Muni.

To the south, Late Cretaceous inversion along NW-trending extensional faults is noted in the Northern Gabon sub-basin [*Teisserenc and Villemin 1990*]. A similar aged stratigraphic event is recognised as far south as the Congo Basin [*Seiglie and Baker 1984*], although no transpressional uplift is reported there.

Throughout the Equatorial African margin, this event generally occurs between upper Turonian and lower Campanian time. This period is coincident with a change in spreading direction in the opening Atlantic Ocean at Anomaly 34 (84 Ma) [*Fairhead and Binks 1991*]. This can be recognised through a change in flow line directions reflected in fracture zone geometry [*Haxby 1985, Klitgord and Schouten 1986, Fairhead 1988, Binks and Fairhead 1992*]. This is particularly well defined in the Central Atlantic (Fig. 14) but is also recognised in the South Atlantic [*Fairhead 1988*]. This event is associated with the reactivation of many of the Equatorial Atlantic fracture zones (e.g. along the Ghana-Cote D'Ivoire Margin [*Mascle et al 1997*], which can be recognised by their enhanced gravity signature away from spreading ridges [*Fairhead and Binks 1991*]. Timing of this kinematic event is broadly coincident with the age of reactivation and transpression along NE trending features in the Central West African basins. It is suggested that this event may explain the reactivation of transform zones and development of Aptian cored inversion anticlines in the Rio Muni Basin.

Thus, the tectonic evolution of the Rio Muni basin may represent southernmost manifestation of the transform-dominated tectonics of Equatorial Africa, but with a stratigraphy that has some of the components of the salt basins to the south.

HYDROCARBON POTENTIAL

Although the six wells drilled to date have not resulted in any commercial discovery of hydrocarbons [*Ross and Hempstead, 1993*], good indications of source, reservoir and seal have been encountered (Plates 2 and 6).

Source

All wells in the basin center have oil shows. The Matondo-1, Benito-1 and East Eviondo-1 wells penetrated up to 200m of Albo- Aptian aged source rock. The source facies is lacustrine -restricted marine type II/III kerogen with Total Organic Carbon values of 2-4% and Hydrogen Indices of 200-500. In addition to these Aptian post-rift source facies, Cenomanian –Turonian source rocks, encountered to the south in the Azile Formation of the Northern Gabon Basin [*Teisserenc and Villemin 1990*], may also be present downdip from the well locations. Oil and gas seeps located to the west on the volcanic islands of São Tomé and Príncipe (Fig. 1) [*Meyers and Rosendahl 1991*] support the

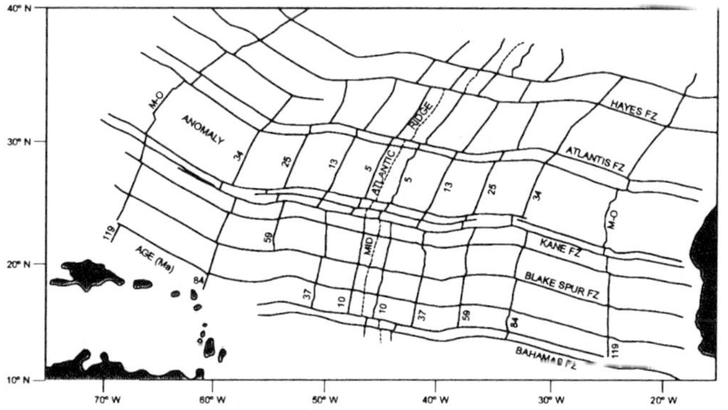

Figure 14. Map showing the distribution of magnetic anomalies and fracture zones in the Central Atlantic (modified after Fairhead and Binks 1991). Note the change in flow line directions at Anomaly 34 (84 Ma).

presence of a mature deepwater source in the post-rift succession.

Reservoir

A variety of reservoirs are present within the basin. The majority of wells drilled so far have targeted Aptian postrift sandstones on structural highs and encountered moderate to poor reservoir quality. These structures were previously interpreted to be extensional footwalls [*Ross and Hempstead 1993, Turner 1995*] but may represent inversion features, and thus the poor reservoir quality may be a function of maximum depth of burial prior to uplift. Reservoir potential remains in this play away from inverted depocentres.

Other reservoirs penetrated include Albian carbonates and Cenomanian-Turonian aged clastics which onlap rotated Albian rafts. Senonian to Tertiary aged deep-sea fans, channel levees, contourite and slump facies have been interpreted from seismic data but are not penetrated in the existing wells.

Seal

Good sealing facies have been penetrated at Albo- Aptian and Maastrichtian levels. In addition, numerous flooding surfaces can be identified on seismic data in the deep-water area, which are likely to be associated with sealing facies.

Play Types

Plate 6 shows the play types present in the basin, including those tested by the existing wells (Albian rafts and Aptian clastics in structural highs). Both of these plays depend on charge from a syn-rift source kitchen thought, by analogy to the Gabon Basin, to underlie the shelf. Drilling results have not substantiated this concept. However, a number of untested plays are located on the westernmost shelf edge, slope and deepwater areas of the basin, which could be charged from post-rift Albian and Cenomanian –Turonian coastal plain and marine source facies. These plays include Late Cretaceous clastics associated with toe thrusts which are predominantly located at the base of the slope. Late Cretaceous and Tertiary basin floor deposits result from Senonian and Oligocene relative base level falls on the shelf. Channel levee complexes developed within the Late Cretaceous slope fan and Cenomanian –Turonian clastics may be ponded behind rafted Albian carbonates. All of these plays could be charged from a kitchen area which is most likely to be developed in the western, deepwater portions of the Rio Muni Basin. These depocenters were not accessible to migration pathways that might link them to the structures in the shelf, as indicated by the results of exploratory wells drilled in the inboard parts of the Rio Muni Basin.

DISCUSSION

Plate 7 summarises the tectonic evolution of the basin. Although not fully penetrated by existing wells, seismic and regional data suggest that the Rio Muni Basin was initiated by Neocomian (?) to Aptian extensional stresses accommodated by the development of a series of half grabens. The oldest strata penetrated in the basin is the Mid Aptian of the N'Dote 1 well, which may represent late syn-rift section.

This was followed in Mid-Upper Aptian time by the deposition of a thick post-rift section of salt and siliciclastic rocks. This transitional sequence was overlain by the latest Aptian- Early Albian marine drift succession of organic mudstones, and subsequently, by the establishment of an Albian carbonate ramp over large parts of the margin. The carbonate facies is overlain by Cenomanian marine clastics, which appear from seismic data to be conformable with the carbonate section.

Deposition of this fairly typical west African stratigraphy was terminated sometime around the Late Cenomanian by the initiation of uplift associated with underlying transform reactivation, which triggered gravity sliding on both the Aptian salt and the base carbonate flooding surface. This resulted in thin-skinned extensional and compressional geometries, and the creation of an entirely new basin geometry. The culmination of this episode corresponds to the Early Senonian Unconformity, which incised Aptian through Cenomanian stratigraphy, and was then onlapped by the Coniacian-Miocene transgressive package.

Following this tectonic reorganisation the basin history has been dominated by onlapping and infilling of the resulting topography with a series of slope fans. Evidence for development of a Late Cretaceous shallow marine shelf at this time is limited. This may be due either to subsequent erosion during Tertiary base level fall or because of a narrow shelf area and rapid bypass onto the slope. Only in the Tertiary, following substantial flooding of the margin, did a broad shelf area managed to establish itself.

A Mid-Tertiary lowstand is associated with renewed uplift, extensional sliding, slumping and fan development along the transpressional zones in the northern part of the basin. This was followed by a Miocene to Recent phase dominated by rapid progradation.

CONCLUSIONS

The Rio Muni basin is located at a major tectonic boundary separating the transform-dominated, equatorial portion of the west African continental margin to the north, from the extension dominated Aptian salt basin to the south. This coincides with a major change in the exposed Precambrian geology onshore.

The tectonic history of the basin is controlled by the continental extension of a series of NE trending transform faults whose reactivation, due to change in spreading direction, has controlled the Late Cretaceous to Present basin architecture and may also have controlled the original syn-rift architecture. This in turn has determined the stratigraphic development and hydrocarbon potential.

A reduction in the thickness of the Aptian salt section occurs, decreasing from Gabon in the south across the Rio Muni Basin and into the Kribi basin in the north. Change in sedimentary facies from an Albian carbonate-prone stratigraphy in the south to a clastic-prone stratigraphy in the Douala-Kribi margin to the north was accommodated across the Rio Muni Basin.

Shear zone reactivation and uplift in the Cenomanian-Santonian of Aptian depocentres resulted in gravity sliding and the development of thin-skinned extensional rafting, which induced a compressional toe thrust belt in the deep water region, which holds considerable hydrocarbon prospectivity.

Exploration concepts based on this geological model are at an early stage but are encouraging for the hydrocarbon potential of this underexplored portion of the West African Margin.

Acknowledgments. The author would like to thank the Equatorial Guinea Ministry of Mines and Energy and Triton Energy Corporation for permission to publish. Discussion and input from Dorie McGuinness, Phil Loader, Kenny Goh, Matthew Taylor, Eddie Ong, Kevin Smith and Conrad Todd were important in formulating the ideas. Chris Coons is thanked for drafting the diagrams. Reviews by Bruce Rosendahl, Bert Bally, Webster Mohriak and John Battrick helped improved the manuscript.

REFERENCES

Adeleye, D.R., Nigerian, Late Cretaceous Stratigraphy and Palaeogeography, *American Association of Petroleum Geologists Bulletin*, V.59, No.12, pp. 2302-2313, 1975.

Binks, R.M., and J. D. Fairhead, A plate tectonic setting for Mesozoic rifts of West and Central Africa, *Tectonophysics*, V.213, pp. 141-151, 1991.

Bottero, C., J. Desneulin, JB Joubert and JP Martine, Final geological report of (the) Matondo1 well, Equatorial Guinea, *Elf Aquitaine*, 1989.

Chang, H.K., R.O. Kowsmann, A.M.F. Figueiredo, and A.A. Bender, Tectonics and stratigraphy of the East Brazil Rift system: an overview, *Tectonophysics*, V.213, pp. 97-138, 1992.

Choubert, G., Carte Tectonique Internationale de l'Afrique, 1: 5000000, *ASGA/UNESCO*., 1968.

Davison, I., Wide and narrow margins of the Brazilian South Atlantic, *Journal of the Geological Society of London*. V.154, pp. 471-476, 1997.

De Matos, R.M.D., The Northeast Brazilian Rift System, *Tectonics*, V.11, No.4, pp. 766-791, 1992.

Fairhead, J.D., Mesozoic plate tectonic reconstructions of the central South Atlantic Ocean: The role of the West and Central African rift system, *Tectonophysics*, V.155, pp. 181-191, 1988.

Fairhead, J.D., and R.M. Binks, Differential opening of the Central and South Atlantic Oceans and the opening of the West African rift system, *Tectonophysics*, V.187, pp. 191-203, 1991.

Genic, G.J., Regional framework, structural and petroleum aspects of rift basins in Niger, Chad and the Central African Republic, *Tectonophysics*, V.213, pp. 169-185, 1992.

Gibson, H., M.E. Moore, and P.G.F. Watson, Thermal history reconstruction for Matondo-1 and East Eviondo-1 in Equatorial Guinea using apatite fission track analysis and vitrinite reflectance, *Geotrack* Report No.711, 1998.

Guiraud, R., and J.C. Maurin, Early Cretaceous rifts of Western and Central Africa: an overview, *Tectonophysics,* V.213, pp. 153-168, 1992.

Haxby, W.F., Gravity Field of the World's Oceans, *Lamont-Doherty Geological Observatory*, Columbia University, Palisades, N.Y., 1985.

Haq, B.U., J. Hardenbol, and P.R. Vail, Chronology of fluctuating sea levels since the Triassic, *Science*, v.235, pp. 1156-1167, 1987.

Hinz, K., A hypothesis on terrestrial catastrophes: wedges of very thick oceanward dipping layers beneath passive continental margins, *Geologisches Jahrbuch*, E-22, p. 3-28, 1981.

Klitgord, K.D., and H. Schouten, Plate kinematics of the central Atlantic, in *The Geology of North America,* edited by P.R. Vogt and B.E. Tucholke, Vol. M, The Western North Atlantic Region, The Geological Society of America, 1986.

Larsen, H.C., and S. Jacobsdottir, Distribution, crustal properties and significance of seawards dipping sub-basement reflectors off E Greenland, in *Early Tertiary Volcanism and opening of the NE Atlantic,* edited by A.C. Morton, and L.M. Parson, Geol. Soc. London Spec. Publ., 39, pp. 95-114, 1988.

Mascle, J., P. Lohmann, and P. Clift, Development of a passive transform margin: Côte d'Ivoire-Ghana transform margin - ODP Leg 159 preliminary results, *Geo-Marine Letters*, V.17, pp. 4-11, 1997.

Meyers, J.B., and B.R. Rosendahl, Seismic reflection character of the Cameroon volcanic line: evidence for uplifted oceanic crust, *Geology*, V.19, pp. 1072-1076, 1991.

Meyers, J.B., B. R. Rosendahl, and J.A. Austin Jr., Deep-penetrating MCS images of the South Gabon Basin: implications for rift tectonics and post-breakup salt remobilization, *Basin Research*, V.8, pp. 65-84, 1996a.

Meyers, J.B., B. R. Rosendahl, H. Groschel-Becker, J.A. Austin Jr. and P.A. Rona, Deep penetrating MCS imaging of the rift-to-drift transition, offshore Douala and North Gabon basins, West Africa, *Marine and Petroleum Geology*, V.13, No.7, pp. 791-835, 1996b.

Mohriak, W.U., M. Bassetto and I.S. Vieira, Crustal architecture and tectonic evolution of the Sergipe-Alagoas and Jacuipe basins, offshore northeastern Brazil, *Tectonophysics*, V.288, pp. 199-220, 1998.

Mutter, J.C., Seaward dipping reflectors and the continent-ocean boundary at passive continental margins, *Tectonophysics* 114, pp. 117-131, 1985.

Nguene, F.R., S. Loule and C. Ngassa, Palaeoenvironments of the Douala and Kribi/Campo subbasins in Cameroon, West Africa, *Elf Aquitaine Mem. No. 13*, pp. 129-139, 1992.

Pauken, R.J., Sanaga Sud Field, Offshore Cameroon, West Africa. In: Halbouty, M, Giant Oil and Gas fields of the Decade 1978-1988, *AAPG Memoir 54*, 1992.

Parson, L.M. and ODP Leg 104 Scientific Party, Dipping reflector styles in the NE Atlantic Ocean, in *Early Tertiary Volcanism and opening of the NE Atlantic,* edited by A.C. Morton and L. M. Parson,. Geol. Soc. London Spec. Publ., 39, pp. 57-68, 1988.

Rabinowitz, P.D. and J. LaBreque, The Mesozoic South Atlantic Ocean and evolution of its continental margins, *Journal of Geophysical Research,* v.84, n.B11, pp. 5973-6002, 1979.

Rasmussen, E.S, Depositional evolution and sequence stratigraphy of the shelf and slope areas off South Gabon, West Africa, *Journal of Sedimentary Research*, Vol. 67, No.4, pp. 715-724, 1997.

Rosendahl, B.R., H. Groschel-Becker, J. Meyers, and K. Kaczmarick, Deep seismic reflection study of a passive margin, southeastern Gulf of Guinea, *Geology*, V.19, pp. 291-295, 1991.

Ross, D., and N. Hempstead,. Geology, hydrocarbon potential of Rio Muni area, Equatorial Guinea, *Oil and Gas Journal,* Aug. 30, pp. 96-100, 1993.

Ruiz, C., Tectono-stratigraphic evolution and development of raft structures in the northern Rio Muni Basin, Equatorial Guinea, *Ph. Thesis,* Univ. of London, 1997.

Seiglie, G., and M. Baker, Relative sea level changes during the Middle and Late Cretaceous from Zaire to Cameroon (central West Africa), in *Interregional unconformities and hydrocarbon accumulation*, edited by J. S. Schlee, AAPG Memoir 36, pp. 81-88, 1984.

Teisserenc, P., and J. Villemin, Sedimentary Basin of Gabon - Geology and Oil Systems, in *Divergent/Passive Margin Basins,* edited by J.D. Edwards and P.A. Santogrossi, AAPG Memoir 48, 1990.

Turner, J.P., Gravity-Driven Structures and Rift Basin Evolution: Rio Muni Basin, Offshore Equatorial West Africa, *AAPG Bulletin*, V.79, No.8 (Aug.1995), pp. 1138-1158, 1995.

Turner, J.P., Gravity-driven nappes and their relation to palaeobathymetry: examples from West Africa and Cardigan Bay, UK, in *Modern Developments in Structural Interpretation, Validation and Modelling*, Geological Society Special Publication No.99, pp. 345-362, 1996.

Paul Dailly, Triton Energy Corporation, Suite 1400, 6688 North Central Expressway, Dallas, TX

Evolution of the Angolan Passive Margin, West Africa, with Emphasis on Post-Salt Structural Styles

L. György Marton

BP Amoco Angola B.V., Luanda, Angola

Gábor C. Tari[1] and Christoph T. Lehmann

BP Amoco, Houston, Texas

In the Angolan segment of the West African passive margin gravity-driven thin-skinned deformation dominated the evolution of the post-salt succession throughout its history, connecting extensional structures in the shelf to compressive elements on the continental slope. In order to reconstruct the evolution of the various salt-related structures by analyzing the geometry of the overlying sedimentary section, regional cross-sections were constructed based on depth-converted composite seismic reflection profiles. The sequential restoration of these transects across offshore Angola clearly document the spatial/temporal transition from an eastern raft-domain to a central diapiric domain, and further westward to a compressional domain at the edge of the salt-dominated region. Comparison of several transects suggests that the development of significant outboard massive salt and allochthonous canopies is largely the function of the amount of original (and redistributed) salt and the differential load of the Tertiary sedimentary overburden. An attempt was made to quantify the distribution of extension and contraction from the shelf to the continental rise. Updip extension indeed appears to be balanced by equal amount of coeval contraction downdip on the Angolan passive margin.

INTRODUCTION

The understanding of the evolution of Mesozoic to Recent passive continental margins relies mainly on subsurface geological and geophysical data. Interestingly, in many continental margin basins, despite the large amount of data gathered, very important elements in existing models of the particular basin, such as the timing of break-up, location and nature of the continent/ocean transition, and the original geometry of salt basins remain much debated. Especially in salt basin systems, even modern exploration techniques are often challenged by the masking effect of complicated salt bodies and the frequently thick sedimentary overburden, which makes it difficult to reveal the early stages of basin history.

Based on recent work by *Marton et al.* [1998], this paper reviews some of the main tectonic events leading to the formation of the Early Cretaceous to Recent westward-facing passive margin of Angola. In this contribution we do

[1] Now at Vanco Energy Company, Houston, Texas

Atlantic Rifts and Continental Margins
Geophysical Monograph 115

not present a detailed regional overview such as the impressive books by *Emery and Uchupi* [1984] and *Uchupi* [1992]. Instead, this contribution is considered as a 'status report' for promoting further discussions among researchers of the South Atlantic region. Data in the northern/central segment of the Angolan margin, in the Lower Congo, Kwanza, and Benguela basins (Plates 1, 2, and 3) was reviewed for purposes of regional interpretation.

Recent acquisition of a high-quality regional 2D reflection seismic grid in the ultra-deep water area (i.e. water depth >1000m), complemented by older vintages of seismic data on the shelf made it possible to study the continent/ocean transition in the area. Our analysis focuses on four regional structure transects. Interpretation of composite industry reflection seismic lines of different vintages was depth-converted (Plate 4) to analyze the actual regional geometry of the margin. Interpretation of these seismic data was constrained by basinward extrapolation of wells drilled on the continental shelf, and by the results of the DSDP Leg 40, Holes 364 and 365 [*Bolli and Ryan,* 1978]. Furthermore, on regional Transect B sequential structural restoration was performed to investigate the initial basin geometry and the temporal/spatial evolution of salt-related structures along the margin. Additional regional data include satellite altimetry-based bathymetry, free-air gravity, and ship-born magnetic data.

MAJOR STRATIGRAPHIC UNITS AND MAPPING HORIZONS

Access to well control in the deep water area of offshore Angola was not available during the preparation of this work, so reliance was placed on the correlation of major seismic markers from wells on the shelf and from the DSDP wells (Plate 1). The estimated age of the stratigraphic markers and the color code for the mapped units, as well as the approximate correlation with the established formation nomenclature, are summarized in Plate 4. A simplified, horizontally layered model with constant interval velocities was used for depth conversion of the transects (Plate 4).

The deepest sedimentary unit overlying the Precambrian crystalline basement is the Neocomian syn-rift section (Plate 4). These "Bucomazi type" lacustrine sediments were deposited in half-grabens as the result of rifting between South America and Africa [e.g. *McHargue,* 1990]. The opening of the rift started from the southernmost tip of Africa/South America in the Late Jurassic (Tithonian). A northward propagating rift model for the South Atlantic is widely accepted [e.g. *Nürnberg and Müller,* 1991]. The diachronous breakup reached the Sergipe-Alagoas basin in northeastern Brazil in late Aptian to Albian times [e.g. *Chang et al.,* 1992]. Thus, along the more than 1200 km long Angolan coast a definite time-shift of the rifting age and the actual final breakup is expected.

Syn-rift sediments and structural features are not well defined on the older vintage seismic reflection data. A tentative interpretation of the syn-rift grabens was provided at the area of the Atlantic hinge zone. The Atlantic hinge zone is defined as a prominent basement high bounded by a major west-dipping normal fault zone [e.g., *Karner et al.,* 1998]. Westward from the Atlantic hinge zone (about 50-100 km from the present day shoreline) the syn-rift graben geometry is not easily recognized in the studied seismic data. Deep penetration reflection seismic data would be necessary to investigate the basement structure in this outboard domain, similar to those of *Meyers et al.* [1996].

The basement and the syn-rift section on Transects A, B and C (Plates 5, 6, and 7) is overlain by a well-defined pre-salt sedimentary package, known informally as the pre-salt wedge [e.g. *Henry et al.,* 1995]. On seismic profiles it typically exhibits a distinctive basinward-thickening character. In turn, it gradually thins and pinches out in a shelf-ward direction. Outboard the unit becomes invisible underneath the thick and complicated salt structures. The fact that this unit does not exist westward from the present day salt edge on the seismic section suggests that the unit also pinches out in a seaward direction somewhere underneath the thick outboard salt. Since the section expands dramatically from zero thickness near the shelf area – where well data are available – to more than 2000 m in the center of the basin, the exact age of the unit may only be estimated. Based on the available well data, it encompasses the Middle(?) and Upper Bucomazi Formations, overlain by the Toca and Chela/Cuvo Formations of *McHargue* [1990]. The age of the pre-salt wedge encompasses the Barremian to Early Aptian interval. The deposition of this unit is not controlled by major block faulting, and thus had to be deposited in a post-rift setting. This interpretation differs from previous models which suggest an Aptian (base-salt, e.g. *Pasley et al.,* 1988] or even younger Aptian (post-salt) cessation of continental rifting [e.g. *Brice et al.,* 1982] in this segment of the South Atlantic.

The pre-salt wedge is overlain by Aptian salt (Plate 4). The salt has greatly impacted the Albian to Present sedimentation, influencing the post-rift structural styles along the Angolan passive margin. The salt deposition oversteps the underlying pre-salt wedge limits (see e.g. Plate 6) in both landward and seaward direction. The overstep to the east simply means that by Aptian time marine transgression reached farther landward in a thermally subsiding post-rift basin. The westward overstep invokes a somewhat surprising consequence that some of the original salt had been de-

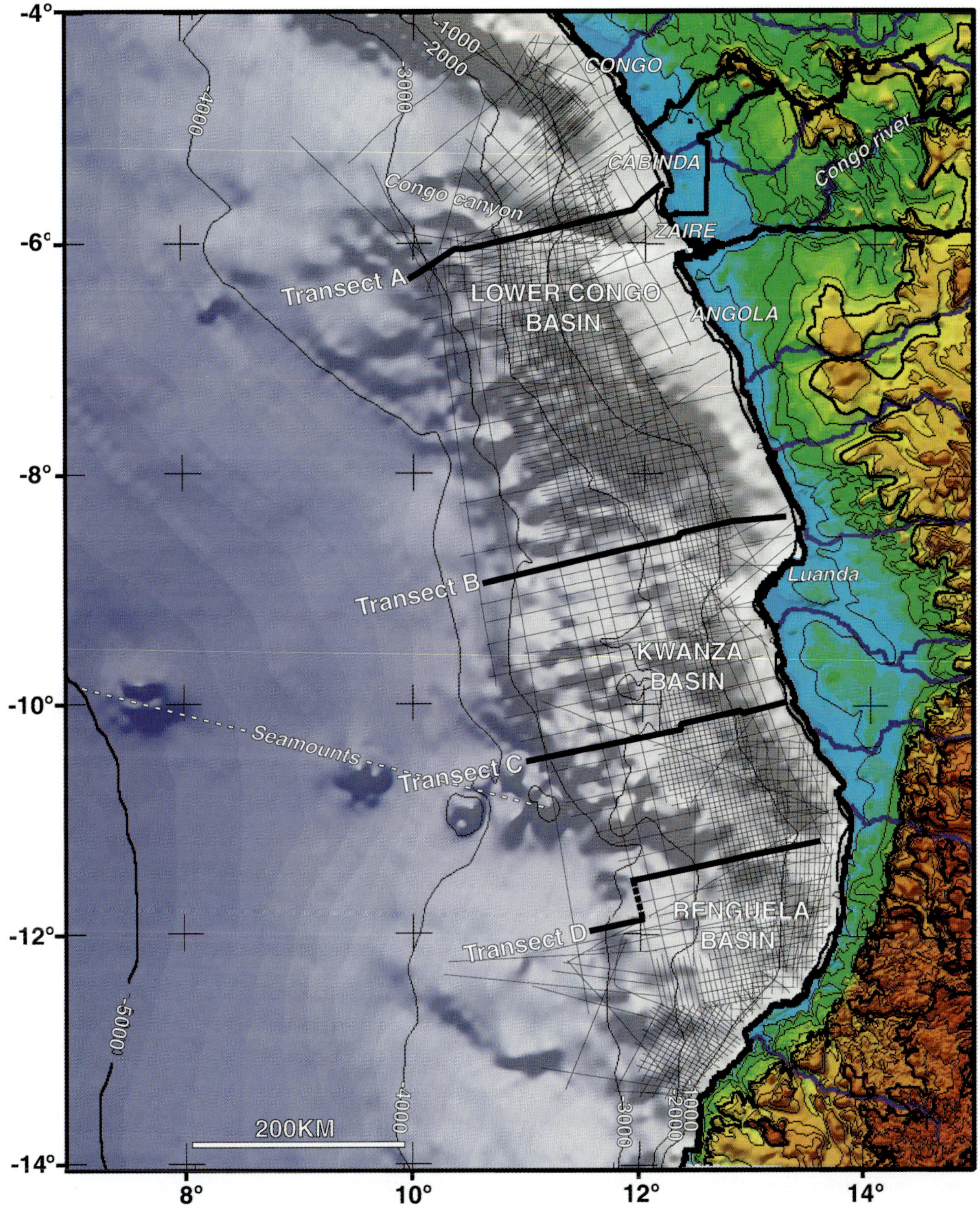

Plate 1. Bathymetry/topography of the Angolan segment of the South Atlantic region based on satellite data, adapted from *Sandwell and Smith* [1997]. Index map of reflection seismic data used in this study is superimposed. Traces of regional transects A-D are shown by thick black lines.

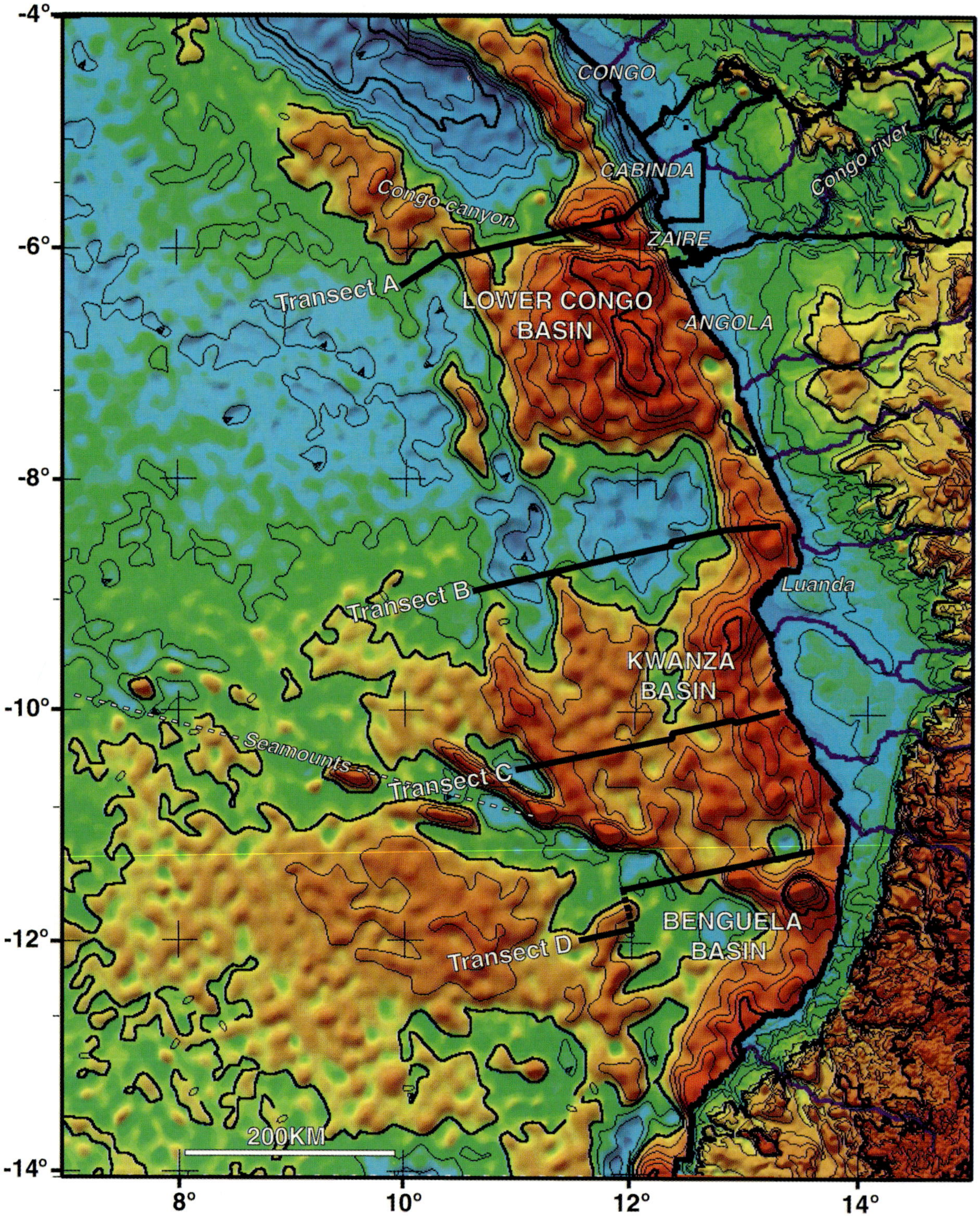

Plate 2. Free-air gravity map of the Angolan segment of the South Atlantic based on satellite data, adapted from *Sandwell and Smith* [1997]. Note that on the onshore part of the map the topography is shown.

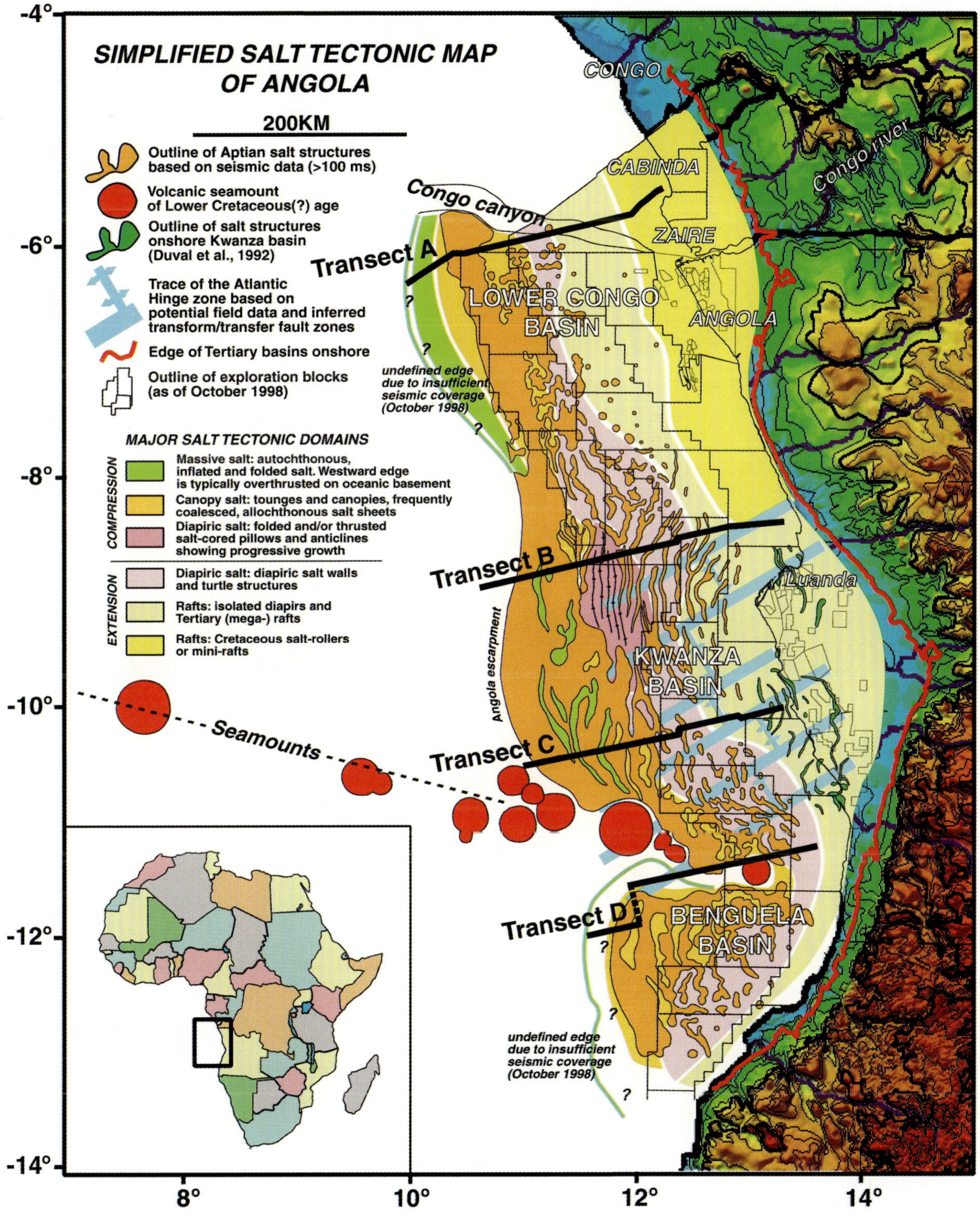

Plate 3. Simplified salt tectonic map of offshore Angola, adapted from *Tari et al.* [1998].

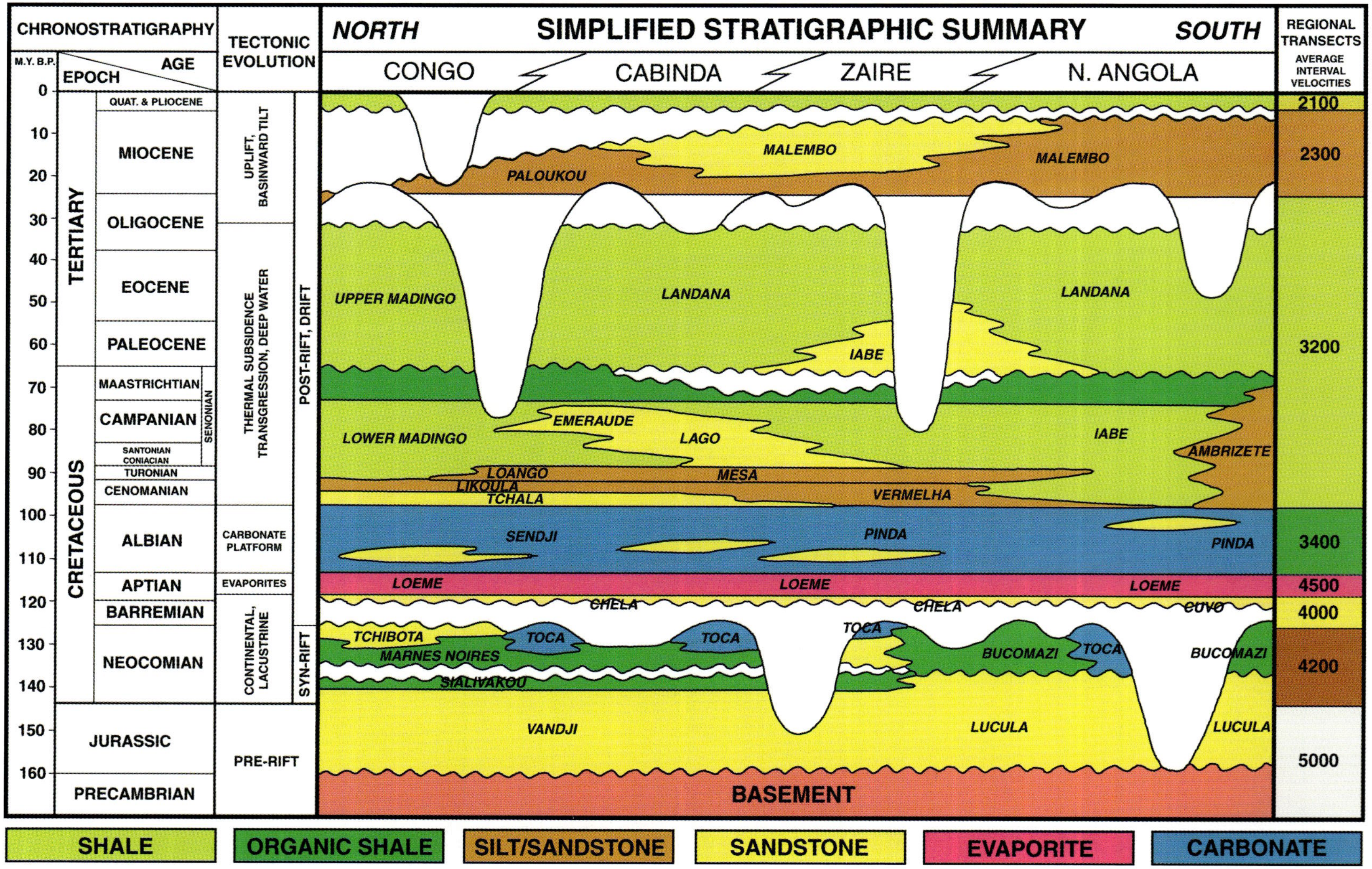

Plate 4. Simplified composite stratigraphic chart of the Congo, Cabinda, Zaire (Democratic Republic of Congo) and Angola segments of the South Atlantic, adapted from *McHargue* [1990]; *Dale et al.* [1992]; *Burwood et al.* [1995]; *Pasley et al.*[1998].

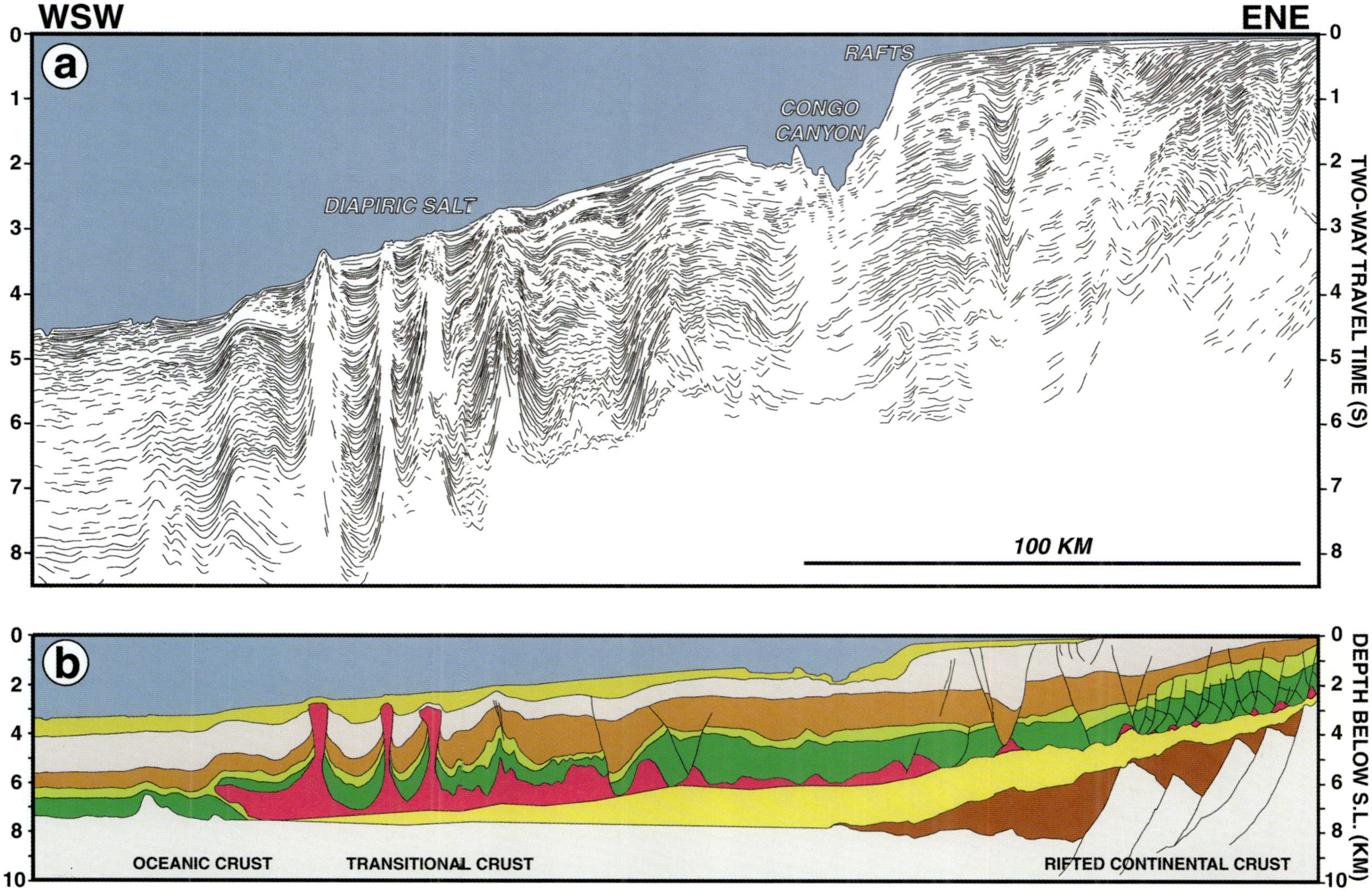

Plate 5. Transect A, Lower Congo basin. For location see Plate 1. a) Simplified line drawing interpretation of the reflection seismic data. b) Depth converted geologic section, the vertical exaggeration is 5:1. For a stratigraphic legend and the interval velocities used in the depth-conversion see Plate 4.

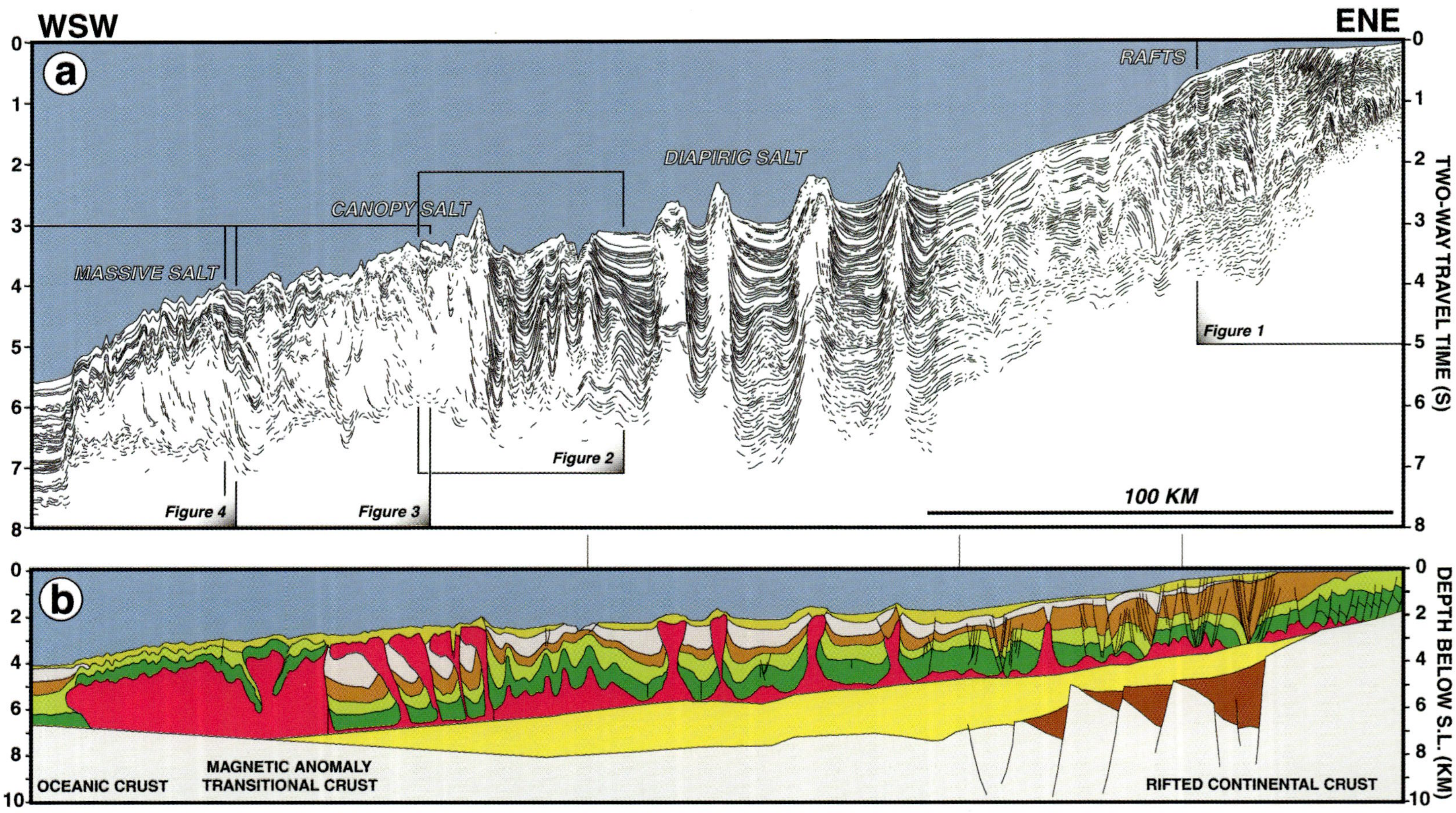

Plate 6. Transect B, Lower Congo basin. For location see Plate 1. a) Simplified line drawing interpretation of the reflection seismic data. b) Depth converted geologic section, the vertical exaggeration is 5:1. For a stratigraphic legend and the interval velocities used in the depth-conversion see Plate 4.

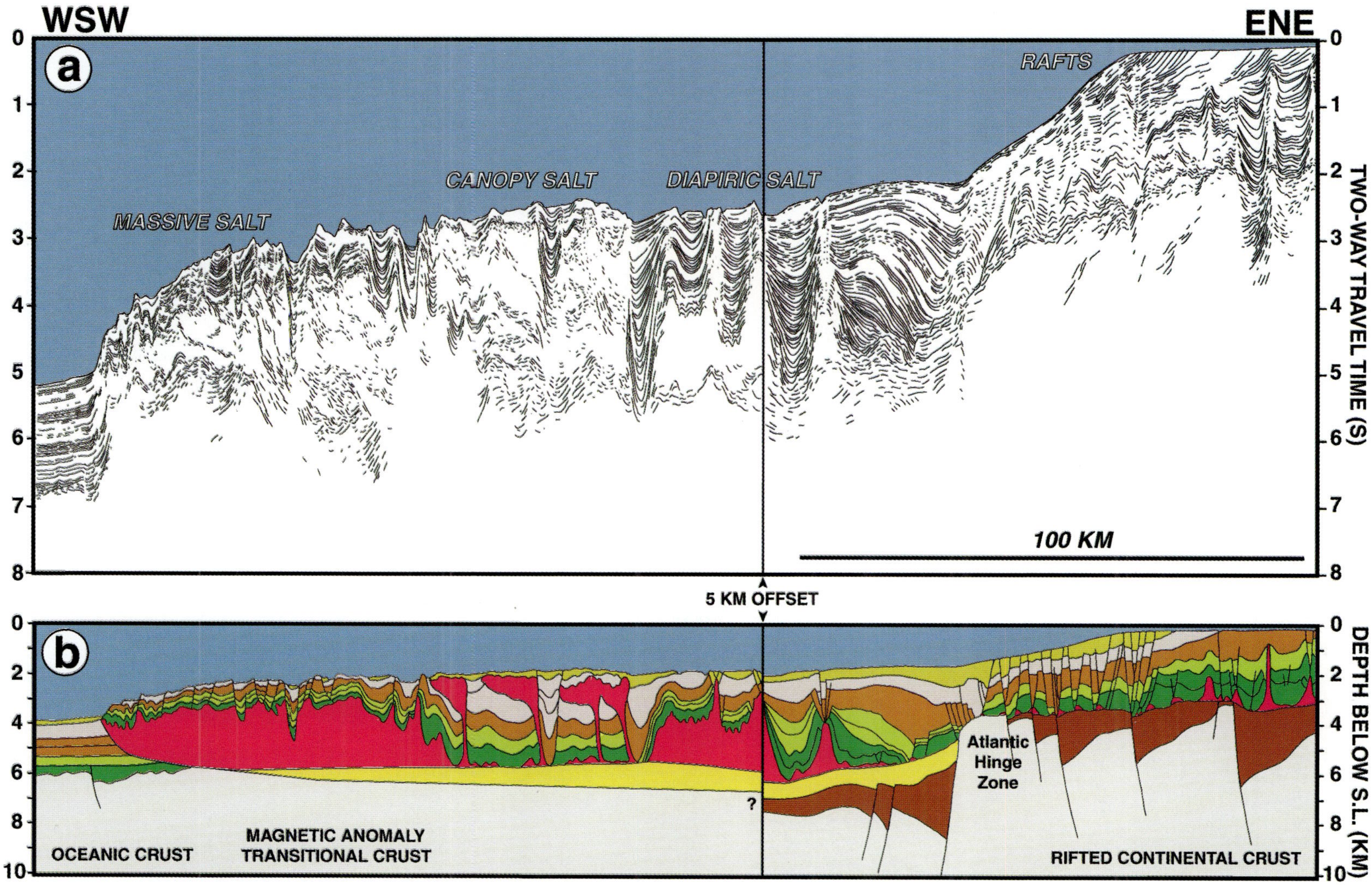

Plate 7. Transect C, Kwanza basin. For location see Plate 1. a) Simplified line drawing interpretation of the reflection seismic data. b) Depth converted geologic section, the vertical exaggeration is 5:1. For a stratigraphic legend and the interval velocities used in the depth-conversion see Plate 4.

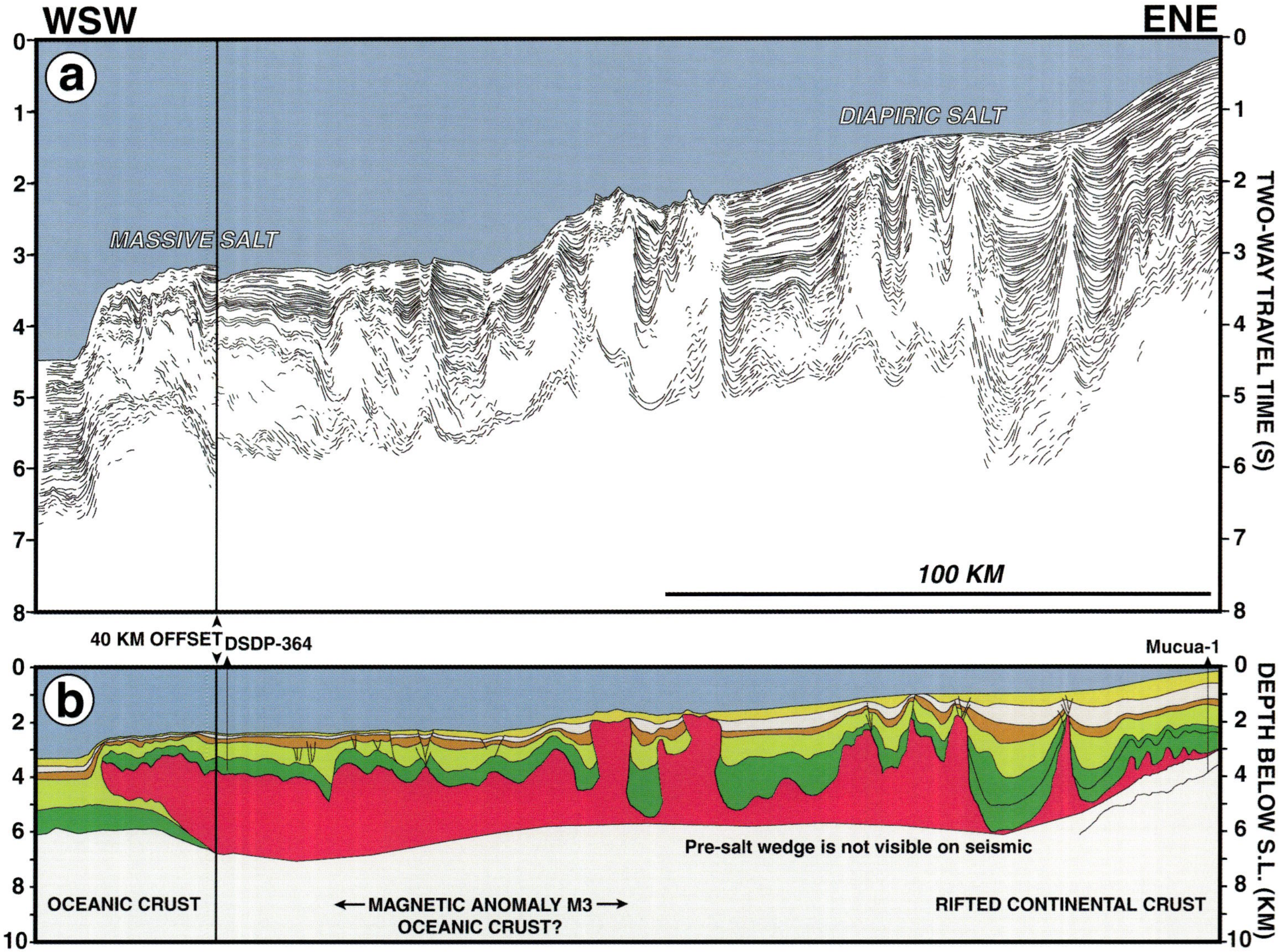

Plate 8. Transect D, Benguela/Kwanza basin. For location see Plate 1. a) Simplified line drawing interpretation of the reflection seismic data. b) Depth converted geologic section, the vertical exaggeration is 5:1. For a stratigraphic legend and the interval velocities used in the depth-conversion see Plate 4.

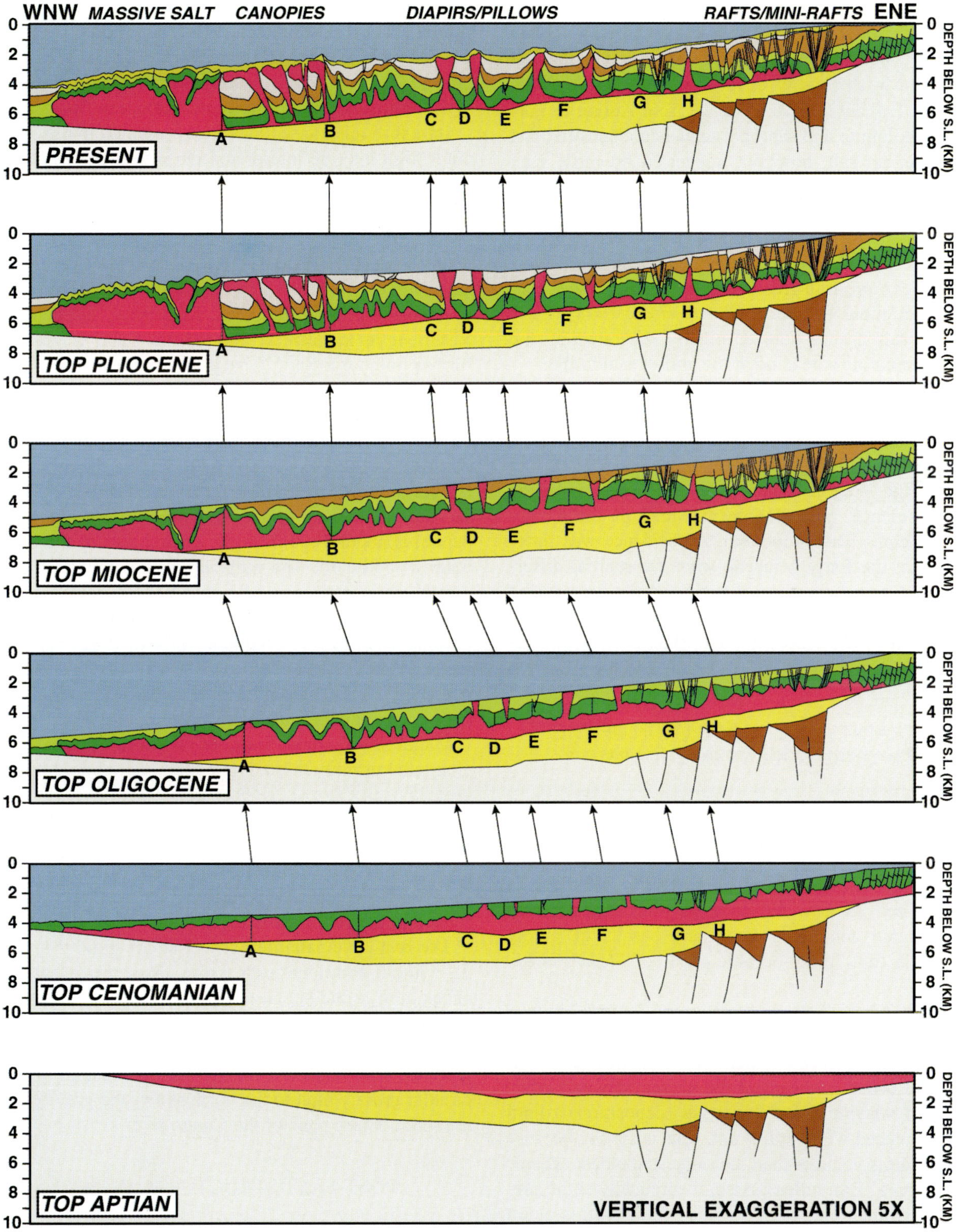

Plate 9. Sequential reconstruction of Transect B, for location see Plate 1. For a detailed explanation see text.

posited over the newly formed (Barremian to Aptian age) oceanic crust. The DSDP drilling results at Site 364 (Plate 1) were not conclusive enough to decide whether the salt was deposited in shallow or deep water setting [*Bolli and Ryan,* 1978]. Paleontological evidence from Aptian to Earliest Albian dolomite and sapropel sequence, deposited just above the Aptian salt, however, suggested no more than 500 m paleo-water depth. This implies that, initially, the Aptian oceanic crust may have formed close to the surface rather than at the 2000m theoretical depth. The Angolan and Brazilian salt basins were separated by subaerial volcanics of the proto-mid-Atlantic ridge (high) and never formed a contiguous salt basin [*Fonck et al.,* 1998]. Later on, the two salt basins were terminally separated by the continued submarine spreading at the mid-oceanic ridge.

Platform carbonates and their deep-water equivalents dominate the post-salt Cretaceous section up to the Cenomanian. The Upper Cretaceous to Tertiary section is characterized by mainly terrigeneous clastic deposition. During this period the Angolan margin experienced continued thermal subsidence, amplified by the load of the Congo sedimentary cone. This monotonous subsidence was interrupted only in the Early Miocene when substantial differential uplift exposed the former shelf and the area to the east [*Walgenwitz et al.,* 1990; *Lunde et al.,* 1992]. Erosion removed much of the Paleogene and older section and redistributed into the deeper parts of the basin to the west.

CONSTRAINTS ON THE LOCATION OF THE CONTINENT-OCEAN BOUNDARY

A self-contained model for any given passive margin should have a reasonable understanding of the location of a very important structural boundary, represented by the continent-ocean transition [e.g. *Mohriak et al.,* 1990]. It is often visualized as a wider zone of transition between heavily intruded continental crust and basalts of purely oceanic origin (c.f. "proto-oceanic crust" of *Meyers et al.,* 1996].

Evidence has been presented above to constrain the age of the final break-up in offshore Angola. If this timing of the final breakup is accepted, then the continent-ocean transition should occur landward of the present day salt edge, regardless of whether the original salt (depositional) limit has been displaced westward or not. This is simply because Aptian age salt is younger than the suggested oldest oceanic crust in the area (Barremian). Also, theoretically some pre-salt wedge sediment had to be also deposited on this oldest oceanic crust. This concept is reflected by our cross Transects A and B (Plates 5 and 6). In other words, the basement seen west from the present day salt edge is suggested to be purely oceanic [cf. *Mascle and Renard,* 1976]. Available reflection seismic data do not contradict this interpetation. In fact, on all transects (Plates 5-8), seaward from the present day salt edge, the basement reflectors show some high frequency irregularity typical to ocean floor basalts.

On a few seismic sections (not presented in this paper) an outer high may be interpreted close to the present day salt edge [cf. *Lehner and de Ruiter,* 1977]. However, on Transects A and B (Plates 5 and 6) depth conversion does not support the presence of any significant outer high.

In Transect D (Plate 8) the base salt reflector in the middle is approximately 1 km deeper than the 6 km depth for the oceanic crust to the west. In general, the depth to the basement in the central part of the basin is greater than further outboard. Simple backstripping exercise shows that the central part of Transect B (Plate 6) subsided deeper because of the extra load of sediments in this area. Measured depth to the basement is about 7600 m in the middle of the section and approximately 6700 m on the western end. However, calculated total tectonic subsidence (total observed subsidence minus sediment load induced subsidence) is about 4400 m in the central part of the basin and 5300 m outboard. The outboard value, assuming standard crustal parameters [e.g. *Sawyer,* 1986], corresponds to a β=3.7 value which is high enough to be associated with oceanic crust [*Le Pichon and Sibuet,* 1981]. The central value of 4400 m yields β of 2.6, which represents highly stretched continental lithosphere and suggests the continent-ocean boundary is located further outboard.

Magnetic anomaly maps of offshore Angola show an elongated prominent magnetic high just inboard of the present day salt edge [*Leyden et al.,* 1972]. The M3 anomaly reported from the Benguela basin [*Bolli and Ryan,* 1978] may line up with this magnetic anomaly. Due to the juxtaposition of rocks of different magnetic properties, it is reasonable to suggest that the prominent magnetic anomaly is related to the continent-ocean transition.

The available satellite-borne free-air gravity map (Plate 2), without additional processing, does not reveal a coherent pattern at the suggested continent-ocean boundary and indicates a relatively smooth transition. An elongated gravity high just offshore of Angola may reflect a trend of rift grabens, also observed on the seismic section. The semicircular gravity high to the north shows the gravity effect of the sedimentary cone of the Congo river.

POST-SALT STRUCTURAL DOMAINS

Post-salt sediments of the Angolan passive margin were deformed gradually as a response to the differential subsidence (tilting) of the basin and to the increasing sediment load. The margin provides a full spectrum of typical passive margin tectonic features and is a natural laboratory to study salt tectonic processes. A few important papers have been published on specific aspects of the salt tectonic evolution of the Angolan margins [*Brognon and Verrier,* 1966; *Verrier and Branco,* 1972; *Pautot et al.,* 1973; *Duval et al.,* 1992; *Lundin,* 1992; *Spathopoulos,* 1996]. This paper reviews the observed first-order features (Plate 3) and attempts to summarize the main evolutionary steps based on regional structure transects.

The four transects show important similarities and differences (Plates 5-8). Transects B and C exhibit the most similarities, while transects A and D differ significantly. Differences may be due to changes in salt volume, overburden volume and basement topography. The cross sectional area of salt is increasing in a monotonous fashion. For A, B, C, and D, these values are 138x, 315x, 350x, and 404×10^6 m^2, respectively. Provided that a simple relationship exists between the present-day cross sectional area of the salt and the original amount of salt deposited in the vicinity of the transects, the amount of original salt appears to decrease northward along the Angolan margin. This would imply that the southern salt basin is better developed in the sense that it subsided more, allowing for more salt accumulation during the Aptian. The thickness of the Tertiary overburden definitely decreases southward, whereas the post-salt Cretaceous section shows a subtle increase in thickness southward. The vast thickness of the Tertiary sediments (>10 km, e.g. *Raillard et al.,* 1998] in the north is due to the proximity to the Congo Cone. The exceptionally thick sedimentary overburden is responsible for the complexity and maturity of the salt structures in the Lower Congo basin [*Tari et al.,* 1998].

The two most representative structural cross sections are Transects B and C (Plates 6 and 7). From a salt tectonics point of view these sections were subdivided into four salt tectonic domains (see Plate 3, cf. *Spathopoulos* [1996]; *Jackson et al.* [1998]). From east to west these domains are: a) a raft domain, which developed to the east of the Atlantic hinge; b) a diapir domain, which developed in the central part of the basin; c) a canopy domain further outboard, which must be regarded as a transition between the diapirs to the east and the inflated salt mass to the west; and d) an massive outboard salt domain. A more detailed treatment of these domains follows below, see Figures 1-4 for seismic examples.

The eastern raft domain is well developed along most of the Angolan passive margin. An exception is Transect D to the south, where rafts were not identified on the inspected seismic line. Generally, rafts appear to be confined to the east of the Atlantic hinge zone (see Figure 1). Gravity gliding on the Aptian salt resulted in the present day distribution of salt-cored rotated fault blocks [e.g. *Spathopoulos,* 1996]. Extreme translation of blocks down the slope in many locations resulted in the total evacuation of salt (Transects A, B, and C). In some instances (Transect A about 50 km from the east) extra sediment load triggered further down-slope translation and block rotation well after total salt evacuation occurred.

Altough raft tectonics [sensu *Burollet,* 1975] has been more or less continuous since the deposition of the Albian carbonates, two main phases may be recognized: a) Albian to Cenomanian formation of "mini-rafts" or pre-rafts, characterized by small block size, close fault spacing of planar fault and high rotation of the intervening blocks; and b) mainly upper Tertiary "mega-rafts," characterized by large fault heaves, more listric type faults, larger fault spacing and large resulting block size [*Duval et al.,* 1992; *Lundin,* 1992]. The formation of the large-scale blocks was triggered by the sudden influx of Tertiary clastics and by the basinward tilt of the entire margin during the Early Miocene.

Large diapiric walls, large and very complex salt massifs, intervening turtle structures, salt pillows, and a few true cylindrical diapirs characterize the diapir domain [*Baumgartner and van Andel,* 1971; *Leyden et al.,* 1976]. The diapir domain is located at the base of the slope west of the Atlantic Hinge zone. Westward tilt of the underlying pre-salt horizon is subtle, but it is obviously sufficient to provide a transport-parallel force, required for large salt mass redistribution (see also reconstruction below). An intricate interaction between extension, due to the subtle but existing basinward tilt, and compression, due to shortening, caused by the down-the-slope translation of raft structures to the east, may have continuously changed the overall stress field in which the diapirs evolved. Most of the observed diapirs are currently compressional in nature. However, changes between extension and compression in the evolution of individual diapirs may often be observed. In this domain the diapirs are interpreted to be anchored onto the original Aptian autochthonous salt layer (Figure 2).

Perhaps the most enigmatic part of the Angolan salt basin is the canopy domain, which is located between the diapir and massive salt domains (Plate 3). Canopies here are defined as shallow allochthonous salt masses that may or may not be connected to the autochthonous salt layer. They must be very recent features as they are covered only by a thin veneer of presumably Plio-Pleistocene sediments. Seismic data deteriorate underneath the relatively shallow salt (i.e. <500m from the seafloor), making the sub-salt interpretation very difficult. Therefore it must be noted that the depicted sub-salt structures in this domain on Transects B and

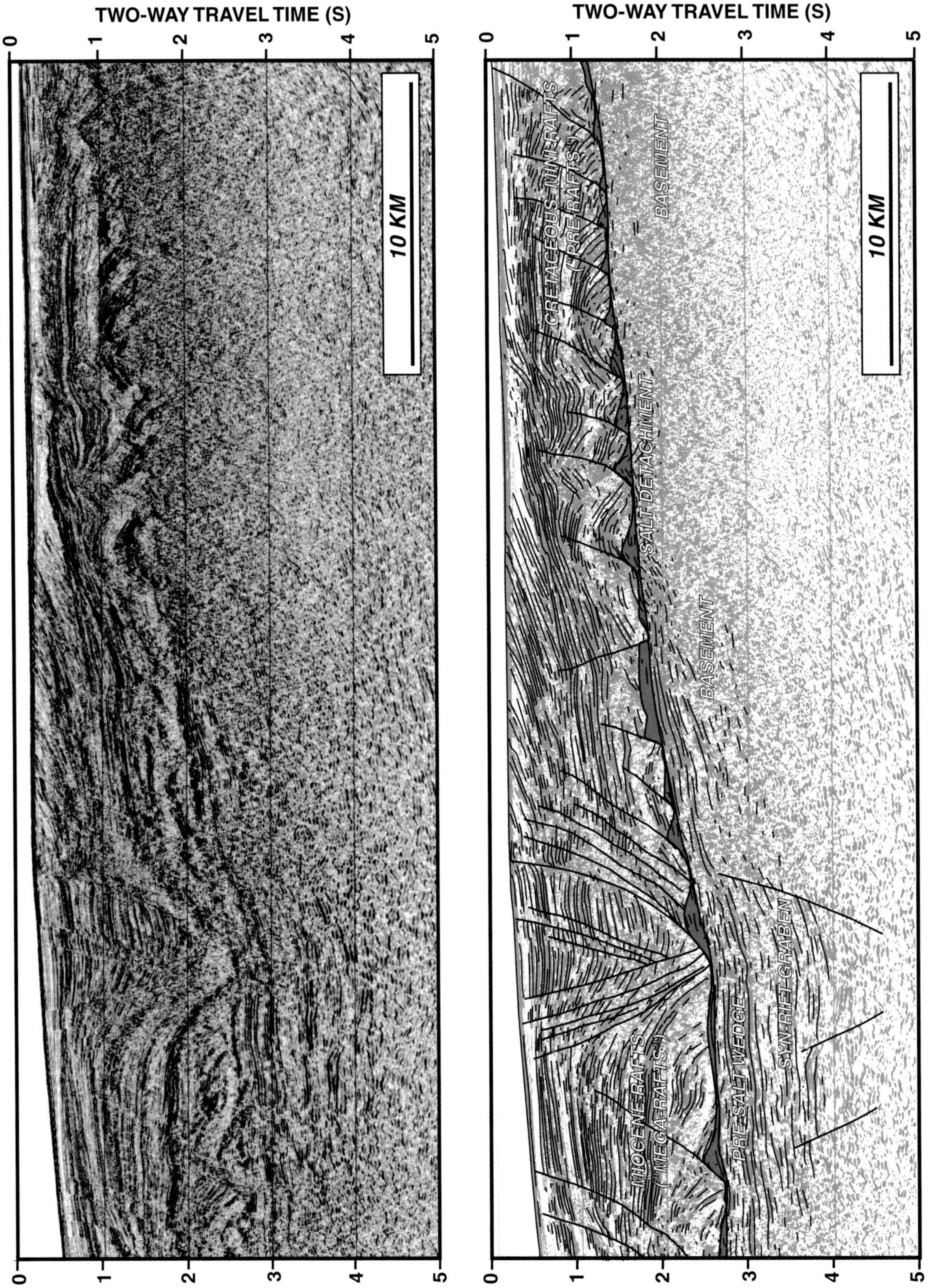

Figure 1. a) Seismic reflection example of the raft domain in Transect B, for location see Plate 6. b) Simplified structural interpretation of the data.

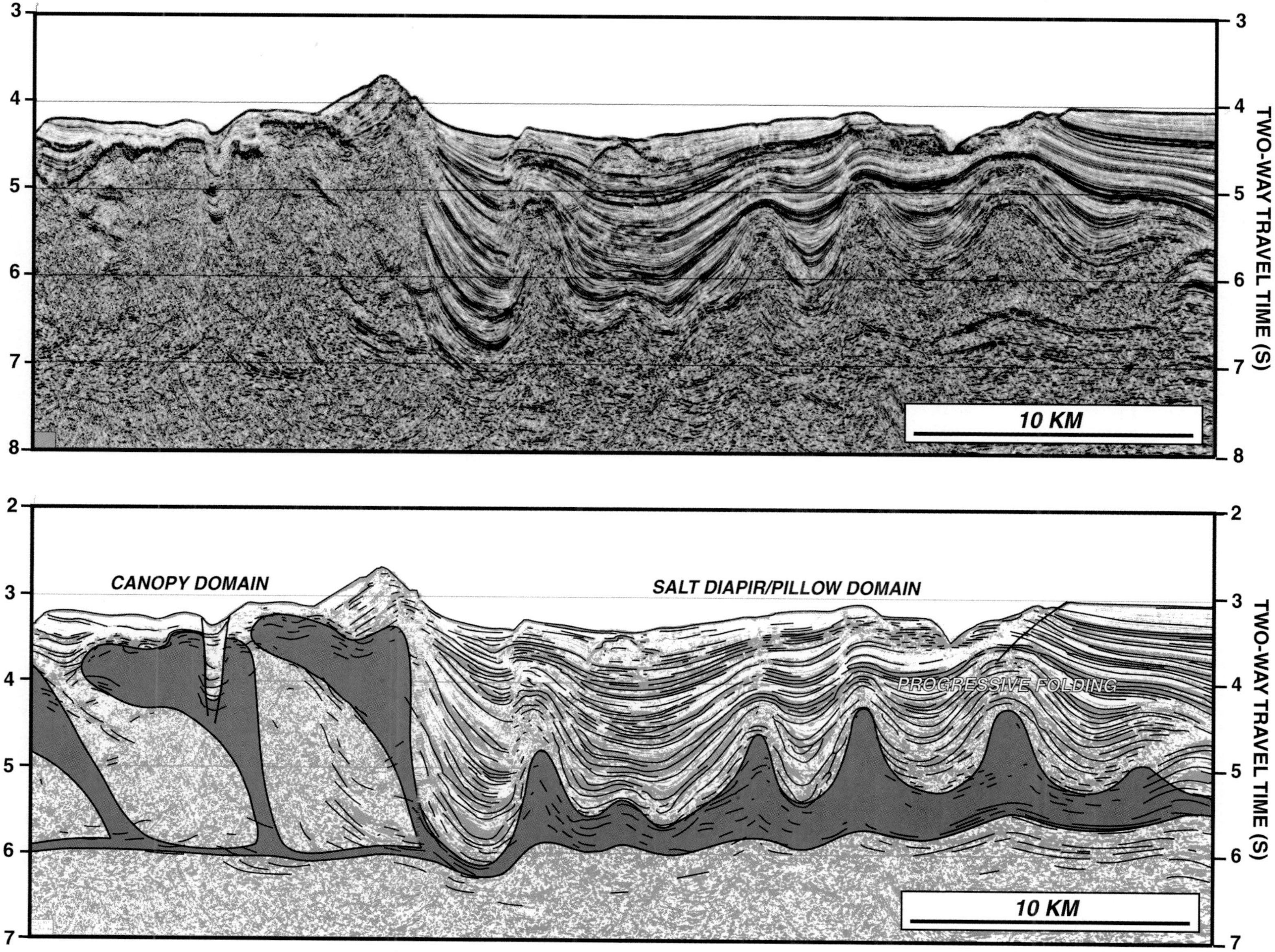

Figure 2. a) Seismic reflection example of the diapiric domain in Transect B, for location see Plate 6. b) Simplified structural interpretation of the data.

C are somewhat conceptual, and should be regarded as preliminary interpretations until more data (e.g. 3D reflection seismic) become available (Figure 3).

On Transects B and C the transition between the diapir and canopy domain is very abrupt (Plates 6 and 7). On Transect B, just east from the canopy domain, a few salt cored compressional folds are observed (see Figure 3). It may be surmised that if the rising Aptian salt breaks through these folds and escape close to the surface, than closely spaced canopies may develop. These canopies in a more advanced stage may coalesce, giving the impression of a huge, more or less coherent, allochthonous salt mass close to the seafloor. An interesting alternative interpretation of the canopy domain suggests that it is just the eastward extension of the well defined massive autochthonous salt outboard. However, the reconstruction shows (see below) that this assumption leads to an excessive amount of original salt which could not be easily restored to a prekinematic geometry.

A well-developed outer massive salt domain is interpreted on Transects B, C, and D. The massive salt domain is differentiated from the canopy domain based on the existence of a more or less undisturbed succession of Cretaceous strata above the salt (Figure 4). Stratigraphic evidence for a continuous Cretaceous strata above an approximately 3000 m thick salt massive has been provided by the correlation of drilling results at DSDP site 364 on Transect D [*Bolli and Ryan,* 1978]. The overall seismic character match of the sedimentary sequence above the salt between sections B and C establishes the correlation northward. The thick salt and the elevated Cretaceous section above indicate inflation of the autochthonous salt layer in this outboard domain. Repetition of the Cretaceous section implies a moderate amount of seaward salt flow. The degree of "overthrusting" is somewhat questionable due to the poor seismic resolution at the vicinity of the outer scarp. Restoration of section B suggests that overthrusting is relatively small, i.e on the order of a few kilometers.

On section C a number of landward dipping reflectors within the salt may be interpreted as internal thrust imbrications (Figure 4). It is mechanically difficult to imbricate the salt (and interlayered sediments) and not to imbricate the post-Aptian section above. Imbrication may occur, however, in areas of high contraction rate just in front of the Congo delta [e.g. *Raillard et al.,* 1998]. Also, recently active extensional faults observed above the massive salt domain of transect C indicate gravitational spreading (collapse) of the salt mass.

Not all the defined structural domains are well developed on all transects, but all transects have relatively more salt outboard than inboard. This suggests an overall process of salt remobilization from inboard to outboard. Transect A has the least amount of observed salt. Here no canopy domain is observed and the outboard massive salt domain is not well developed. On the other hand, on Transect D the outer massive salt domain is well developed and no canopy domain is observed. It is apparent that, to develop a significant outer massive salt domain, a large amount of original salt is necessary. The lack of a canopy domain in this latter section suggests that, to develop canopies, both a large amount of salt and significant Tertiary overburden are required.

RECONSTRUCTION OF TRANSECT B AND IMPLICATIONS FOR THE EVOLUTION OF THE ANGOLAN PASSIVE MARGIN

Regional Transect B was chosen to perform sequential restoration to validate the structural interpretation, to gain additional insight into the evolutionary steps of the above outlined structural domains, and to assess the amount of extension and contraction accommodated in the section (see Plate 9). Since the application of salt restoration techniques in the Gulf of Mexico [e.g. *Worrall and Snelson,* 1989], a general methodology was developed to restore salt-bearing formations [e.g. *Rowan,* 1994]. Starting with the present-day geometry, four intermediate restorations were created (top Pliocene, top Miocene, top Oligocene, top Cenomanian), leading to the top Aptian salt restoration.

The "salt conservation" restoration technique was chosen, which assumes that the cross sectional area of salt does not change through the restoration steps, i.e. no salt is gained or lost through time. This assumption should be valid if no significant salt dissolution occurred at the seafloor, and the amount of salt coming in and out of the plane of the section is averaged out. The assumption implies that all the deformation (extensional and compressional) is confined to the post-salt section. This assumption is reasonable for a passively subsiding continental margin that only rarely is affected by basement-involved faults during the thermal phase of subsidence. Restoration steps then attempt to restore the post-salt section to a "regional" surface. Each of these regional surfaces simulates the paleo-seafloor and represents the elevation to which beds should return in an undeformed state. Using a computer-aided two-dimensional reconstructional approach in each restoration step, a vertical-slip module (area balancing) was used to restore extensional structures, while a flexural-slip module was applied for contractional structures (line length balancing). Decompaction, isostatic correction and thermal subsidence corrections were not used for the regional analysis. Thus the restoration steps do not attempt to predict paleo-water depths.

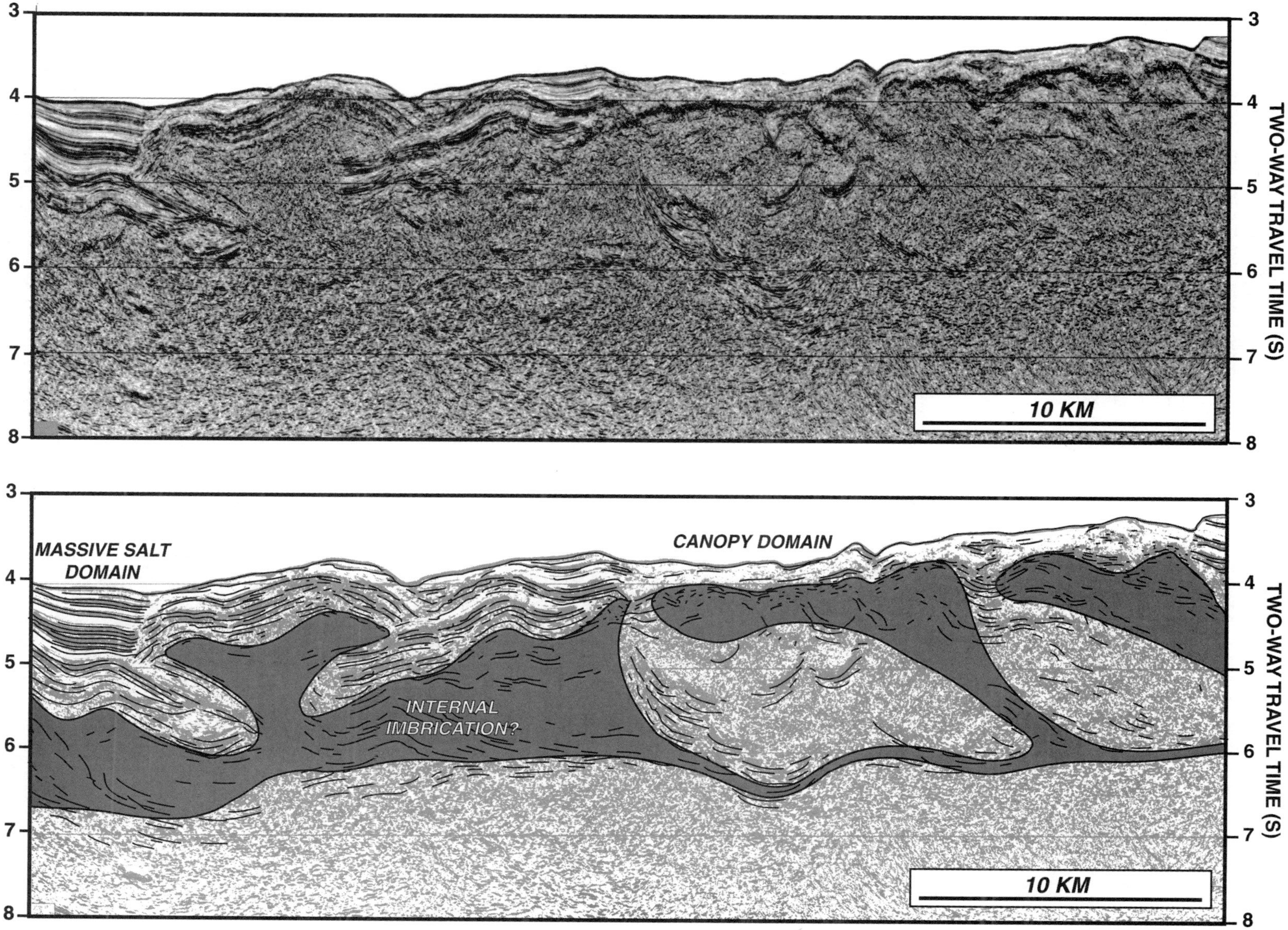

Figure 3. a) Seismic reflection example of the canopy domain in Transect B, for location see Plate 6. b) Simplified structural interpretation of the data.

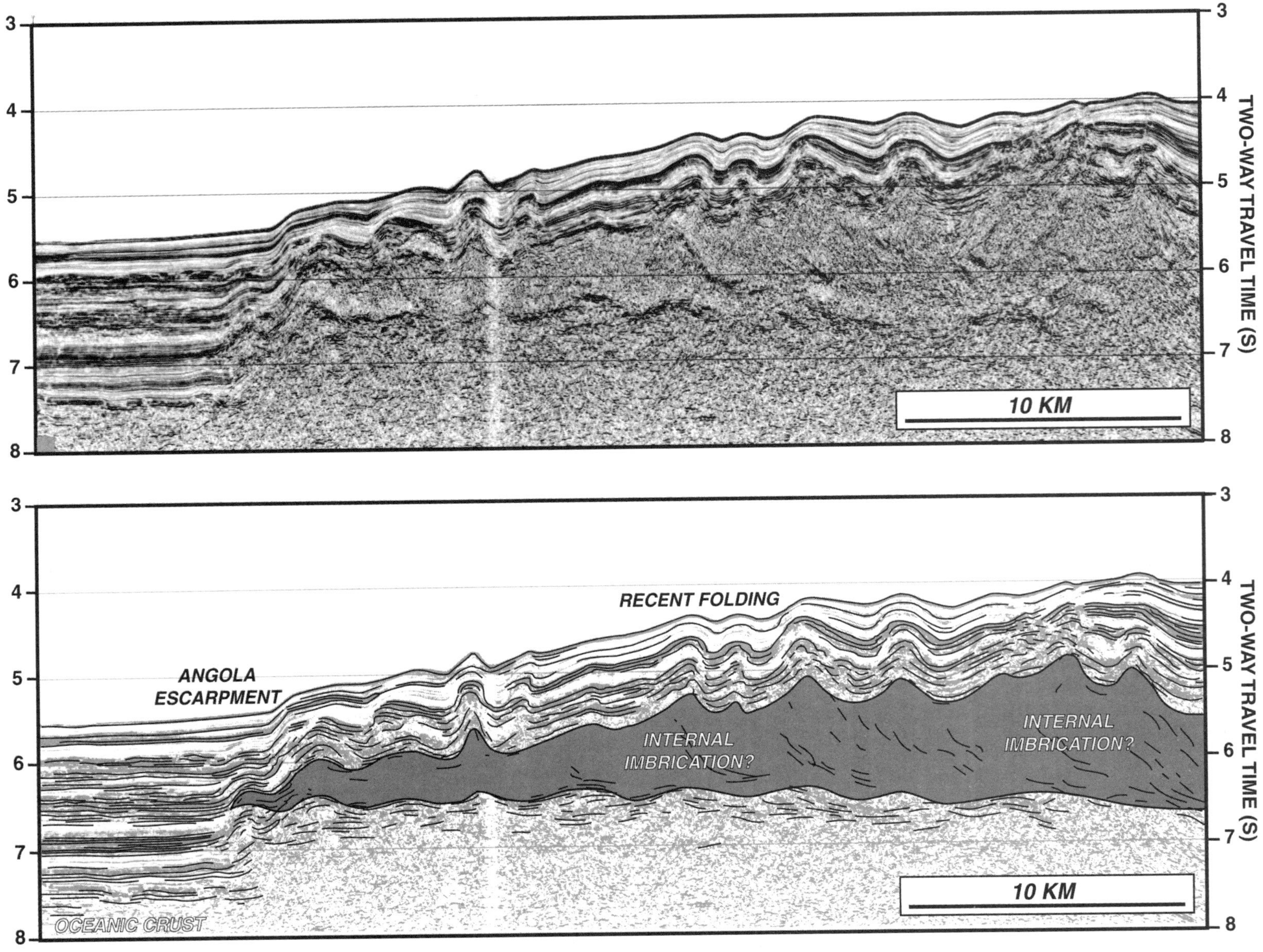

Figure 4. a) Seismic reflection example of the massive salt domain in Transect B, for location see Plate 6. b) Simplified structural interpretation of the data.

Basement was schematically restored only in the Aptian stage to show a possible configuration for the Aptian salt basin.

A discussion of the restoration results follows below, proceeding forward in time beginning with the Aptian restoration. This restoration shows the basic structural and stratigraphic relationships at the end of salt deposition. Note that the cross-sectional area of the salt is the same as in the present day section, i.e. 315×10^6 m^2. The thickest Barremian pre-salt wedge succession (2.5 km) and salt (1.4 km) were deposited in the central part of the basin due to the thermal subsidence going on since the proposed early Barremian break-up. Outboard the basement is relatively shallow as a result of the relative proximity to the thermal disturbances caused by the Aptian mid-Atlantic ridge. The model suggests that the Brazilian and Angolan salt basins were geographically separated by the Barremian to Aptian mid-oceanic ridge [cf. *Fonck et al.*, 1998].

The top Cenomanian reconstruction shows the depositional setting of the maximum 1.3 km thick Albian to Cenomanian platform section and its deep-water equivalents. Remaining fault displacements at this restoration level indicate that mini-raft formation started contemporaneously or immediately after the deposition of the Pinda carbonates. The exact amount of extension due to this mini-raft development has not been determined. Irregularities of the top salt surface reflect possible initial pillow formations and imperfections of the interpretation/restoration. Remnant salt at the location of future diapirs indicates uncertainities in the reconstruction process. At this stage marker positions from "A" to "H" were defined to refer more easily to positions on the restored sections and to discuss the amount of displacement relative to the Cenomanian positions.

The top Oligocene reconstruction shows initiation of large-scale rafting in the inboard raft domain. The measured extension for this period is 3.3 km. In the diapir domain a moderate amount of 0.8 km shortening was accommodated at the pillows between point "B" and "C." By this time the volume of the inboard salt is significantly reduced and the inflation of the outer salt masses began. About 2.8 km of shortening is accounted for in the outer massive salt and future canopy domain. Note that topography in the future canopy domain is entirely speculative.

The top Miocene reconstruction shows major changes and dramatic development of the structures. A relatively large amount of extension (7.7 km) has been accommodated along the rafted structures. Large, down-to-the-basin listric and antithetic faults developed. At the area of large extension most of the salt had been evacuated, creating touchdowns and windows [cf. *Raillard et al.*, 1998]. Most of the shortening was compensated by compressive structures in the frontal massive salt domain (6.8 km). Salt inflation in the massive salt domain was also significant in this time interval.

The top Pliocene reconstruction shows further development of the already existing Miocene structures. In the raft domain an additional 1.7 km extension was accommodated. Rim-syncline enhancement indicates ongoing diapirism. These features are fairly young and did not exist before Late Miocene-Early Pliocene times, as evidenced by thin sedimentary cover on the canopies. Significant salt evacuation from inboard and salt inflation outboard are indicated for this period.

The present day section shows the cumulative products of thin-skinned extension and contraction along the Angolan margin. Point "G" moved outward an accumulated amount of 13.1 km. This indicates all the post-mini-raft extensions accommodated at the various mega-raft structures and reactive diapir "H". The western extreme of the diapir domain (point "A") moved westward 9.7 km; and therefore a cumulative amount of 3.4 km shortening was accommodated in the diapir domain. Assuming that the western edge of the salt did not move outward significantly, a cumulative amount of 9.7 km shortening was accommodated in the combined outer massive salt and canopy domains.

CONCLUSIONS

1. The westward facing Cretaceous passive margin of Angola initiated as a northward propagating rift between South America and Africa. Associated syn-rift grabens are recognized only inboard on the currently available reflection seismic data.

2. It is estimated, based on stratigraphic and tectonic relationships, that continental rifting stopped and final breakup occurred in the Early Barremian.

3. The Barremian pre-salt wedge sediments and the overlying Aptian salt were deposited in a post-rift sag basin. The Aptian salt basin of Angola was most probably separated from the Brazilian salt basin by a subaerial proto mid-Atlantic ridge of Barremian to Aptian age.

4. Magnetic data, observed marine magnetic lineations, subsidence analysis, and stratigraphic relationships suggest that the continent-ocean boundary offshore Angola is located just eastward from the present-day salt edge.

5. Changes in cross sectional area of the salt suggest that the amount of salt increases southward from the Lower Congo to the Benguela basin. The amount of Tertiary overburden increases in an opposite direction, i.e. northward.

6. Four major salt tectonic domains are recognized along the Angolan passive margin. Due to differences in original

salt volume, sedimentary overburden, and basement configuration, every domain is not always recognizable in any given cross section.

7. Two kinds of rafts were observed in the eastern raft domain: a) mini-rafts, which developed early and involved the Albian to Cenomanian section; and b) mega-rafts which developed mainly during the Late Tertiary as a response to increased sediment load and regional basinward tilt of the margin.

8. The central diapir domain hosts a variety of salt tectonic features, including diapiric walls, pillows, turtle structures and cylindrical domes with complex evolutionary history .

9. Shallow salt with only moderate overburden characterizes the outboard canopy domain. On most sections canopies appear to be coalesced. Based on the overlying stratigraphy they are presumed to be recent, i.e. Late Miocene-Early Pliocene features.

10. The outboard massive salt domain is characterized by an inflated salt mass below and a preserved post-salt Cretaceous stratigraphic succession above. The salt gained outboard during the inflation process can be balanced by salt evacuation inboard. Recent extensional features in the carapace of the massive salt domain indicate instability and incipient gravitational spreading of the salt mass.

11. Comparison of several transects suggests that the development of significant outboard massive salt and canopy domains is largely the function of the amount of original (and redistributed) salt, as well as the amount of Tertiary sedimentary overburden.

12. Restoration of the about 300 km long transect CC' provides an evolutionary model with salt volume preserved through time. Also, distribution of extension and contraction across the margin has been quantified. A cumulative amount of 13 km extension in the raft domain was balanced by 3 km of shortening in the diapir domain and 10 km shortening in the combined canopy/massive salt domain.

Acknowledgements. We appreciate helpful discussions on Angolan geology with several colleagues, especially with Shafe Alexander, Albert Bally, Mario Brandão, Katrina Coterill, Sneha Dholakia, Jerry Dominey, Paulino Jeronimo, Tom Mason, Mark Pasley, Ralph Salomon, Ian Threadgold and Bruno Vendeville. Thanks are due to Western Geophysical for permission to show selected seismic reflection data. David Meaux of AOA Geophysics kindly provided the bathymetry and free-air gravity map. Comments on an earlier version of the manuscript by Albert Bally and Manik Talwani are gratefully acknowledged. Webster Mohriak is thanked for his editorial patience.

REFERENCES

Baumgartner, T. R. and T. H. van Andel, Diapirs of the continental margin of Angola, Africa, *Geological Society of America Bulletin*, 82, 793-802, 1971.

Bolli, H. M., W. B. F. Ryan et al., *Initial Reports of the Deep-Sea Drilling Project,* vol. 40, U.S. Goverment Printing Office, Washington D.C., 1978.

Brice, S. E., M. D. Cochran, G. Pardo, and A. D. Edwards, Tectonics and sedimentation of the South Atlantic Rift Sequence: Cabinda, Angola, in *Studies in Continental Margin Geology* edited by J.S. Watkins and C.L. Drake, pp. 5-18, American Association of Petroleum Geologists, Memoir 34, Tulsa, OK, 1982.

Burollet, P. F., Tectonique en radeau en Angola, *Bull. Soc. Géol. France*, 17, 503-504, 1975.

Burwood, R., S. M. De Witte, B. Mycke, and J. Paulet, Petroleum geochemical chaacterisation of the Lower Congo Coastal basin Bucomazi Formation, in *Petroleum Source Rocks*, edited by B.J. Katz, pp. 235-263, Elsevier, Amsterdam, 1995.

Brognon, G. P. and G. V. Verrier, Oil and geology in Cuanza basin of Angola, *American Association of Petroleum Geologists Bulletin*, 50, 108-158, 1966.

Chang, H. K., R.O. Kowsmann, A. M. F. Figueiredo, and A.A. Bender, Tectonics and stratigraphy of the East Brazil Rift system: an overview, *Tectonophysics*, 213, 97-138, 1992.

Dale, C. T., J. R. Lopes, and S. Abilio, Takula oil field and the Greater Takula area, Cabinda, Angola, in *Giant Oil and Gas Fields of the Decade 1978-1988*, edited by In: M. T. Halbouty, pp. 197-215, American Association of Petroleum Geologists, Memoir, 54, Tulsa, OK, 1992.

Duval, B., C. Cramez, and M. P. A. Jackson, Raft tectonics in the Kwanza Basin, Angola, *Marine and Petroleum Geology*, 9, 389-404, 1992.

Emery, K.O., and E. Uchupi, *The Geology of the Atlantic Ocean.* Springer, New York, 1050 p., 1984.

Emery, K.O., E. Uchupi, J. Phillips, C. Bowin, and J. Mascle, Continental margin off Western Africa: Angola to Sierra Leone, *American Association of Petroleum Geologists Bulletin*, 59, 2,209-2,265, 1975.

Fonck, J. M., C. Cramez, M. P. A. Jackson, Role of subaerial volcanic rocks and major unconformities in the creation of South Atlantic margins, in *Extended Abstracts Volume* edited by M. R. Mello and P.O. Yilmaz, pp. 38-39, American Association of Petroleum Geologists International Conference, Rio de Janeiro, Brazil, 1998.

Henry, S.G., W.B. Brumbaugh, and N. Cameron, Pre-salt source rock development on Brazil's conjugate margin: West African examples. *Proceedings of the Fourth International Congress of the Brazilian Geophysical Society*, pp. 1-3, Rio de Janeiro, Brazil, 1995.

Jackson, M.P.A., C. Cramez, and W.U. Mohriak, Salt tectonics provinces across the continental-oceanic boundary in the Lower Congo and Campos basins of the South Atlantic margins, in *Extended Abstracts Volume* edited by M. R. Mello and P.O. Yilmaz, pp. 40-41, American Association of Petroleum Geologists International Conference, Rio de Janeiro, Brazil, 1998.

Karner, G.D., N.W. Driscoll, J.P. McGinnis, W.D. Brumbaugh, N.R. Cameron, Tectonic significance of syn-rift sediment packages across the Gabon-Cabinda continental margin, *Marine and Petroleum Geology*, 14, 973-1000, 1998.

Lehner, P., and P.A.C. de Ruiter, Structural history of Atlantic margin of Africa, *American Association of Petroleum Geologists Bulletin*, 61, 961-981, 1977.

Le Pichon, X., and J.C. Sibuet, Passive margins: a model of formation, *Journal of Geophysical Research*, 86, 3,708-3,720, 1981.

Leyden, R., G. Bryan, and M. Ewing, Geophysical reconnaissance on African shelf: 2. Margin sediments from Gulf of Guinea to Walvis Ridge, *American Association of Petroleum Geologists Bulletin*, 56, 682-693, 1972.

Leyden, R., H. Asmus, S. Zembruscki, and G. Bryan, South Atlantic diapiric structures. *American Association of Petroleum Geologists Bulletin*, 60, 196-212, 1976.

Lunde, G., K. Aubert, O. Lauritzen, and E. Lorange, Tertiary uplift of the Kwanza Basin in Angola, in *Géologie Africaine: Colloque de Stratigraphie et de Paléogéographie des Bassins Sédimentaires Ouest-Africains*, edited by R. Curnelle, pp. 99-117, Recueil des Communications, 6-8 May 1991, Libreville, Gabon, 1992.

Lundin, E.R., Thin-skinned extensional tectonics on a salt detachment, northern Kwanza Basin, Angola, *Marine and Petroleum Geology*, 9, 405-411, 1992.

Marton, G., G. Tari, and C. Lehmann, Evolution of salt-related structures and their impact on the post-salt petroleum systems of the Lower Congo Basin, offshore Angola, in *Extended Abstracts Volume* edited by M. R. Mello and P.O. Yilmaz, p. 834, American Association of Petroleum Geologists International Conference, Rio de Janeiro, Brazil, 1998.

Mascle, J. and V. Renard, The marginal São Paulo Plateau, comparison with the southern Angolan Margin. *Anais da Academia Brasileira de Ciências*, 48, 179-190, 1976.

McHargue, T.R., Stratigraphic development of proto-South Atlantic rifting in Cabinda, Angola - a petroliferous lake basin, in *Lacustrine Basin Exploration: Case Studies and Modern Analogs*, edited by B.J. Katz, pp. 307-326, American Association of Petroleum Geologists, Memoir 50, Tulsa, OK, 1990.

Meyers, J.B., B.R. Rosendahl, H.Groschel-Becker, J.A. Austin, and P.A. Rona, Deep penetrating MCS imaging of the rift-to-drift transition, offshore Douala and North Gabon basins, West Africa. *Marine and Petroleum Geology*, 7, 791-835, 1996.

Mohriak, W.U., R. Hobbs, J.F. Dewey, Basin-forming processes and the deep structure of the Campos Basin, offshore Brazil. *Marine and Petroleum Geology*, 7, 94-122, 1990.

Nürnberg, D., and D. Müller, The tectonic evolution of the South Atlantic from Late Jurassic to present. *Tectonophysics*, 191, 27-53, 1991.

Pasley, M.A., E.N. Wilson, V.S. Abreu, M.G.P. Brandão, and A.S. Telles, Lower Cretaceous stratigraphy and source rock distribution in pre-salt basins of the South Atlantic: Comparison of Angola and Southern Brazil, in *Extended Abstracts Volume* edited by M. R. Mello and P.O. Yilmaz, pp. 822-823, American Association of Petroleum Geologists International Conference, Rio de Janeiro, Brazil, 1998.

Pautot, G., V. Renard, J. Daniel, and J. DuPont, Morphology, limits, origin, and age of salt layer along South Atlantic African margin. *American Association of Petroleum Geologists Bulletin*, 57, 1,658-1,671, 1973.

Raillard, S., J.J. Biteau, P. Allix, and C. Chevalier, 1998. Lower Congo Tertiary basin – offshore West Africa structural zonation and evolution, in *Extended Abstracts Volume* edited by M. R. Mello and P.O. Yilmaz, p. 364, American Association of Petroleum Geologists International Conference, Rio de Janeiro, Brazil, 1998.

Rowan, M.G., A systematic technique for the sequential restoration of salt structures. *Tectonophysics*, 228, 331-348, 1994.

Sandwell, D. and W. Smith, Marine gravity anomaly from GEOSAT and ERS-1 satellite altimetry, *Journal of Geophysical Research*, 102, 10,039-10,054, 1997.

Sawyer, D.S., Effects of basement topography on subsidence history analysis, *Earth Planet. Sci. Lett.*, 78, 427-434, 1986.

Spathopoulos, F., An insight on salt tectonics in the Angola Basin, South Atlantic, in *Salt Tectonics* edited by G. I. Alsop, D. J. Blundell, and I. Davidson, pp. 153-174, Geol. Soc. Spec. Publ., 100, The Geological Society, London, 1996.

Tari, G., J. Dominey, P. Jeronimo, and S. Alexander, Influence of salt structures on sea-floor topography and sediment dispersal, Lower Congo basin, offshore Angola, *Geol. Soc. Am. Abstracts with Programs*, 30, 364, 1998.

Uchupi, E., Angola Basin: geohistory and construction of the continental rise, in *Geologic Evolution of Atlantic Continental Rifts*, edited by C. Poag, and P.C. DeGraciansky, pp. 77-99, Van Nostrand Reinhold, New York, 1992.

Verrier, G.V. and C. Branco, Le Fosse Tertiaire et le Gisement de Quenguela-Nord (Bassin du Cuanza), *Revue de l'Institut Francaise du Pétrole*, 27: 51-72, 1972.

Walgenwitz, F., M. Pagel, A. Meyer, H. Maluski, and P. Monie, Thermo-chronological approach to reservoir diagenesis in the offshore Angola Basin: a fluid inclusion, ^{40}Ar-^{39}Ar and K-Ar investigation, *American Association of Petroleum Geologists Bulletin*, 74, 547-563, 1990.

Worrall, D.M., and S. Snelson, Evolution of the northern Gulf of Mexico, with emphasis on Cenozoic growth faulting and the role of salt, *in The Geology of North America: an Overview*, edited by A.W. Bally and A.R. Palmer, pp. 97-138, Geological Society of America, Decade of North American Geology, volume A, 1989.

G. L. Marton, BP Amoco Angola B.V., Avenida 4 de Fevereiro 95, Luanda, Angola

G. C. Tari, Vanco Energy Company, One Greenway Plaza, Houston, Texas 77046

C. T. Lehmann, 501 WestLake Park Boulevard, Houston, Texas 77079-2696

Slope Tectonics-Comparisons and Contrasts of Structural Styles of Salt and Shale Tectonics of the Northern Gulf of Mexico with Shale Tectonics of Offshore Nigeria in Gulf of Guinea

Shengyu Wu

TOTAL Exploration Production U.S.A., Inc., 909 Fannin, Suite 2200, Houston, Texas 77210

Albert W. Bally

Department of Geology and Geophysics, Rice University, P.O. Box 1892, Houston, Texas 77251

Regional structural styles on the continental slope are revealed by well-imaged deep multi-channel seismic profiles from northern Gulf of Mexico and Offshore Nigeria in Gulf of Guinea. Both provinces are situated on passive continental margins developed during the breakup of Gondwana. The northern Gulf of Mexico as well as Offshore Nigeria are underlain by complex systems of thin skinned upslope extensional fault systems, downslope contractional fold and thrust systems along with large scale salt or shale structures. The interactions between the deformation of mid-Jurassic salt and various stages and patterns of sedimentation on the continental slope are responsible for both extensional growth faults and Tertiary contractional fold and thrust systems in the northern Gulf of Mexico. The slope tectonics style of the northern Gulf of Mexico passive margin is characterized by perennial upslope extension since mid-Jurassic, associated with multi-stage autochthonous and allochthonous salt deformations along with brief periods of late Tertiary shortening in the down-dip fold belts. Offshore Nigeria is a salt-free progradational margin characterized by upslope extension, various shale structures and downslope toe-thrust systems. Gravity spreading and gliding in response to rapidly prograding Tertiary sedimentation over under-compacted shales resulted in the structural styles of the slope of Offshore Nigeria. Large upslope extensional tectonics are linked to coeval and balanced downslope contractional deformations in the form of shale ridges, massifs and diapirs, and deep basin toe thrusts. Late Miocene-Early Pliocene contractional deformation is widespread across the entire foldbelt, and with later shortening, folding migrated in a landward direction.

INTRODUCTION

On the continental slope salt and under-compacted shales under differential loading are the key elements of the complex tectonics on the otherwise stable passive margins of the northern Gulf of Mexico and Gulf of Guinea. This paper compares and contrasts salt-induced structural

Atlantic Rifts and Continental Margins
Geophysical Monograph 115

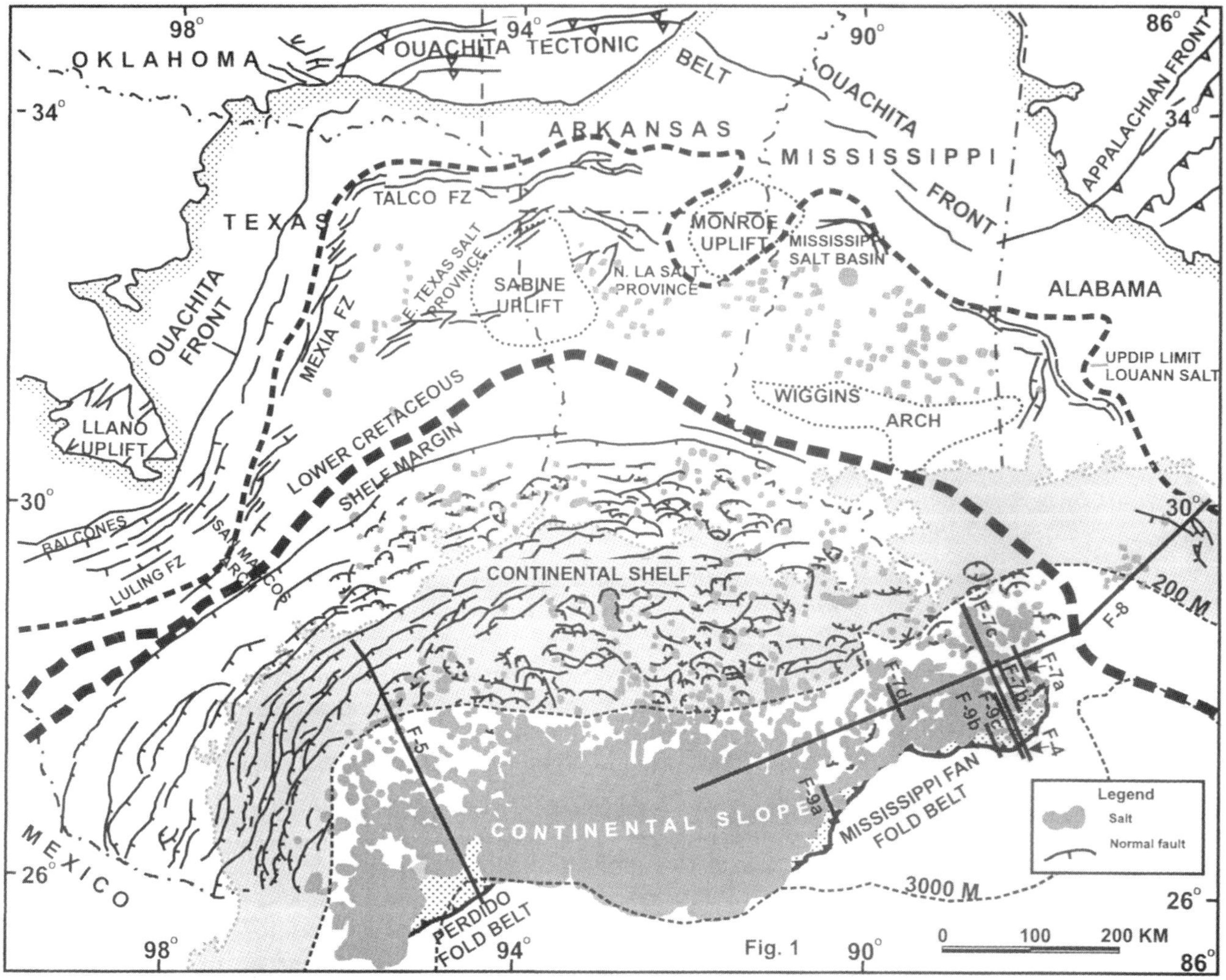

Figure 1. Tectonic setting of the northern Gulf of Mexico Basin. The tectonic features landward of Lower Cretaceous shelf margin are from *Worral and Snelson* [1987] . The tectonic features seaward of Lower cretaceous shelf margin are from *Diegel et al* [1995], *Trudgill et al* [1999] and Wu [1993].

styles of the northern Gulf of Mexico with the shale tectonic styles of the Niger Delta.

Under the same gravitational force on both continental slopes, the similarities and differences in structural styles reveal the deformation mechanisms of salt vs. shale-dominated margins.

Salt and shale structures and their associated extensional and contractional structures are discussed. *Morley and Guerin* [1996] already compared the shale and salt tectonic deformations in Niger Delta and Gulf of Mexico. The conclusions from *Morley and Guerin* [1996] are similar to this independent study. However, Morley and Guerin's work mainly emphasizes the understanding and documentation of shale tectonics deformation styles and genetic mechanisms in the Niger Delta.

The tectonic maps for the northern Gulf of Mexico and Niger Delta (Figures 1 and 3) show that both margins are characterized by upslope extensional and downslope contractional structures. The key tectonic features that differentiate the two margins are the extensive allochthonous salt structures and a narrow foldbelt over the slope of the northern Gulf of Mexico and a distinct thrust system associated with shale structures along the toe of Niger Delta slope. Changes in the Gulf of Mexico drainage system during the Tertiary (Figure 2) have significant impact on the timing of extension, salt deformation and genesis of the

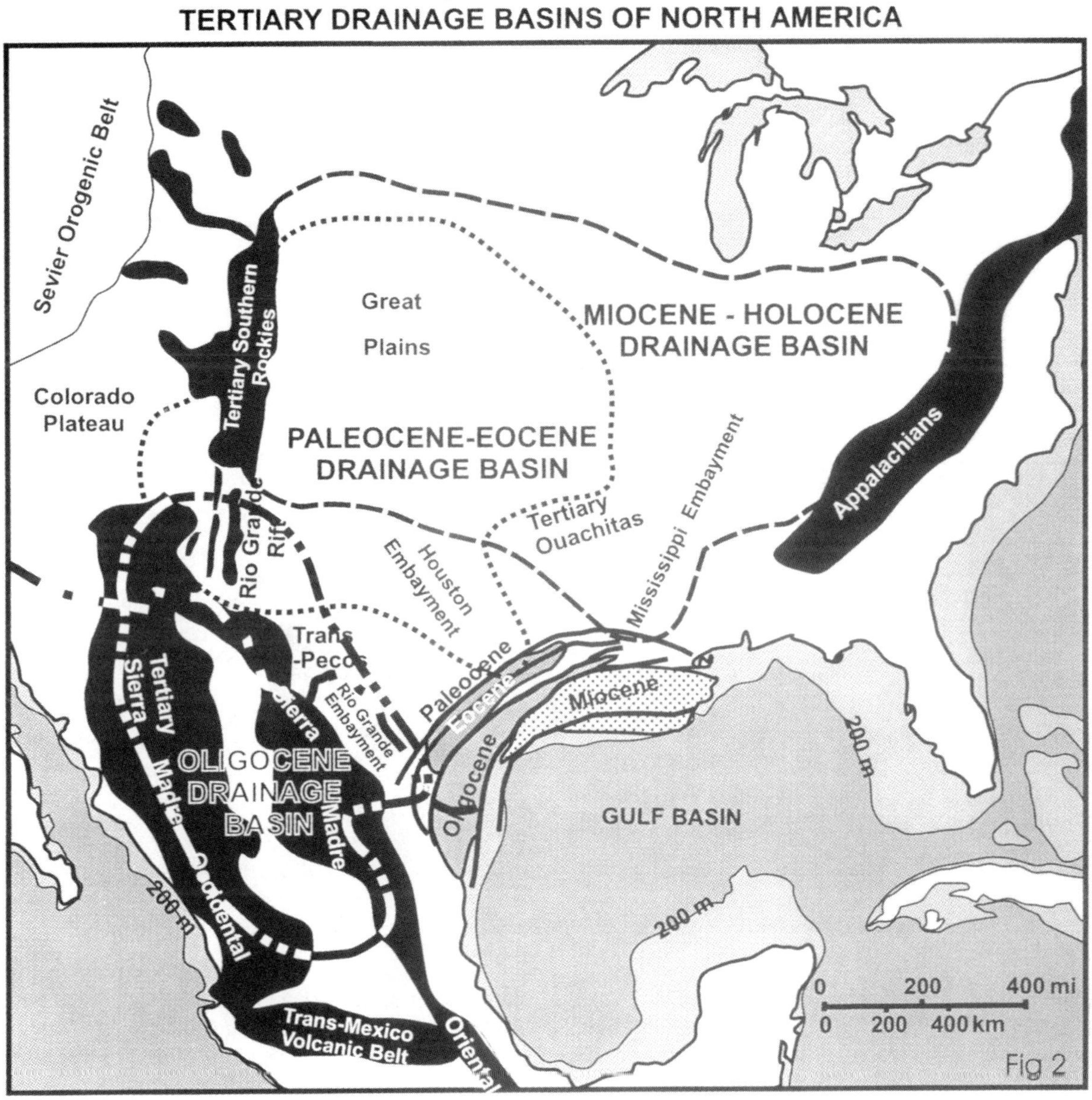

Figure 2. Tertiary depositional drainage system of the northern Gulf of Mexico [Redrawn after *Winker*, 1981].

lower slope contractional systems. In contrast, the sedimentary drainage system of the Niger Delta does not appear to have changed much during the late Tertiary. Deep multi-channel seismic profiles provide more insights of the tectonic styles of both margins.

SUMMARY OF NORTHERN GULF OF MEXICO TECTONICS

The Gulf of Mexico is one of the best known sedimentary basins dominated by salt tectonics. Many published papers have described the tectonic development of the basin [e.g. *Amery*, 1969, *Humphris*, 1978; *Buffler and Sawyer*, 1985; *Salvador*, 1987; *Worrall and Snelson*, 1989; *Wu et al*, 1990b; *Salvador*, 1991; *Sawyer et al*, 1991; *Weimer and Buffler*, 1992, *Wu*, 1993, *Peel et al*, 1995, *Diegel et al*, 1995, *Watkins et al*, 1996, *McBride*, 1998, *Trudgill et al*, 1999].

Extensive allochthonous salt structures dominate the present-day continental slope of the Gulf of Mexico. Extension on upslope growth faults of various ages is accommodated mostly by both autochthonous and allochthonous salt

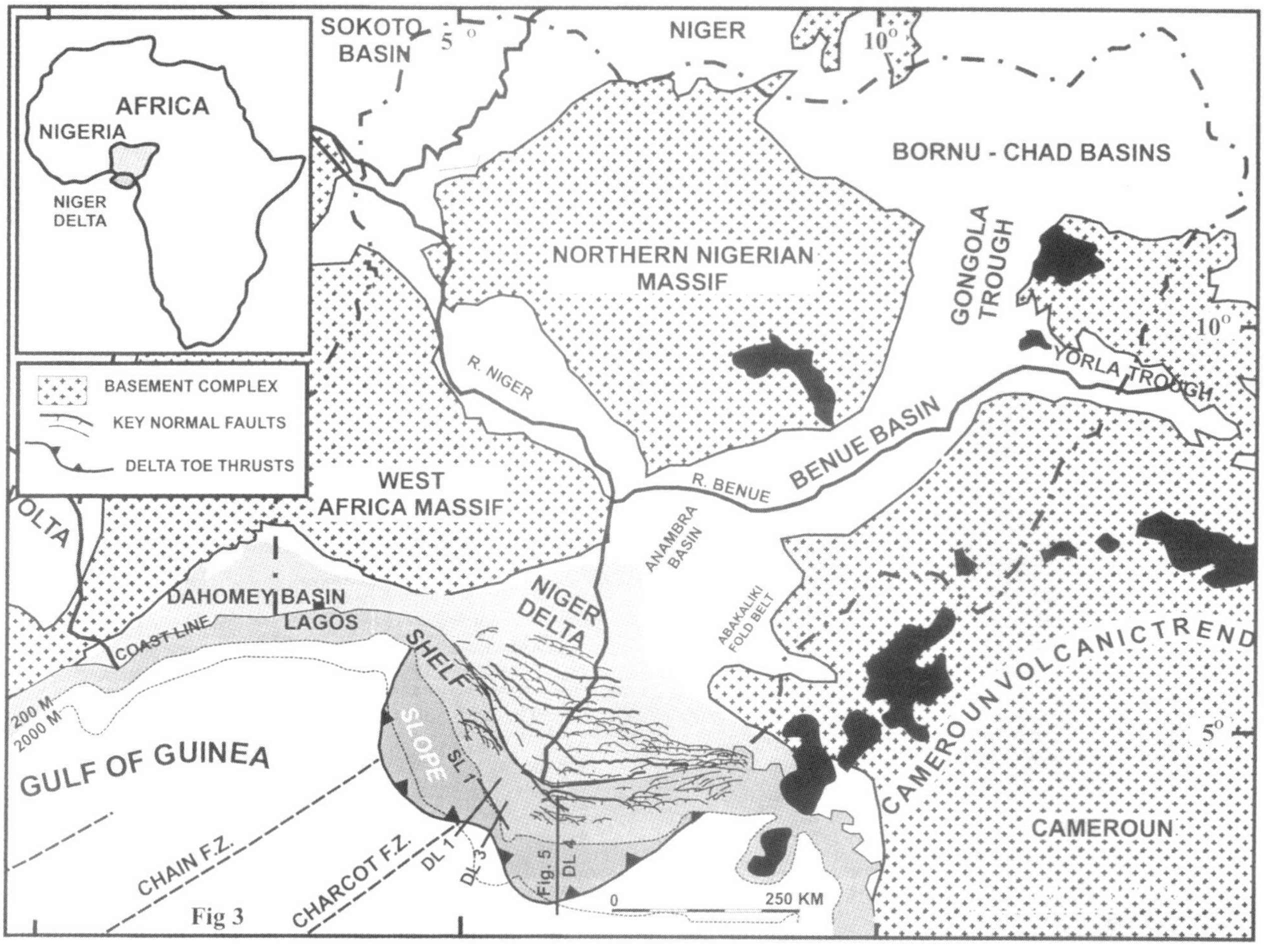

Figure 3. Tectonic setting of the Gulf of Guinea [After *Doust and Omatsola*, 1990].

deformation and partly by downslope shortening in the salt cored Mississippi Fan Fold Belt and Perdido Fold Belt, formed along the basinward edge of the mid-Jurassic salt basin.

Regional seismic line drawings across the north-central [Figure 4, *Wu*, 1993] and northwestern continental slopes [Figure 5, *Hall and Wu*, 1992, *Hall et al*, 1993] show typical structural styles of the northern Gulf of Mexico. Both Figures 4 and 5 show multi-level extensional systems, autochthonous and allochthonous salt structures and lower-slope contractional structures. Contractional episodes are diachronous across the Gulf of Mexico. They may have occurred during the Eocene-Oligocene in the Perdido Fold Belt (northwestern Gulf of Mexico, see detailed discussion by *Trudgill et al*, 1999 and *Rowan et al.,* this volume) and during the Miocene in the Mississippi Fan Fold Belt of north-central Gulf of Mexico. Much of the shortening in both foldbelts occurred during a brief contractional phase (approx. 5-10 Ma). Following the main shortening phase, continuous uplift occurred in the landward parts of the foldbelts due to their inflating salt cores.

The following discussion will mostly focus on the slope tectonics observed in the north-central Gulf of Mexico.

Salt Structures

Salt structures are distributed throughout the entire northern Gulf of Mexico slope (Figure 1, Figure 4 and Figure 5). Both autochthonous and allochthonous salt structures [e.g. *Bally*, 1981; *Worrall and Snelson*, 1989; *Wu et al*, 1990b; *Liro,* 1992, *Rowan*, 1995] are observed. These salt structures were formed as the result of interactions between sediment loading and salt deformation. We understand that the allochthonous salt evolved from the original mid-Jurassic autochthonous salt under specific conditions, involving changes thickness and drainage area of salt, density inversion, and differential loading in a slope environment.

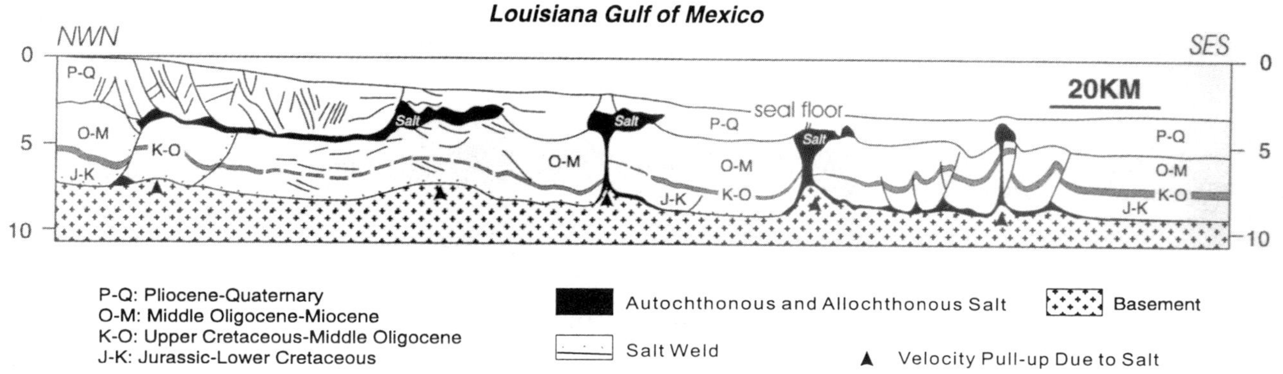

Figure 4. Regional seismic line drawing of offshore Louisiana slope showing extensional, salt and contractional structures..

The original mid-Jurassic salt bed is presently either very thin or has been removed from its original area, leaving behind salt welds [*Jackson and Cramez*, 1989]. An exception to this inference is observed in the Mississippi Fan Fold Belt and Perdido Fold Belt, where thick autochthonous salt forms the cores of the folds (Figure 4 and 5). Elsewhere in the deep Gulf of Mexico salt basin, most of the original salt has been displaced to form various allochthonous salt structures.

The timing of salt deformation is interpreted as a response to syntectonic sedimentary loading of the varying drainage systems (Figure 2), i.e. major salt-related deformation occurs in response to post-Eocene sedimentation on the northwestern Gulf of Mexico slope and post-Oligocene sedimentation on the north-central slope. Large scale allochthonous salt sheets began to form during late Eocene to mid-Oligocene in the northwestern Gulf of Mexico (Figure 5) and during the Miocene in the north-central Gulf of Mexico (Figure 4, 7b) as determined by the age of the sediments underlying and overlying the allochthonous salt bodies.

Different stages of allochthonous salt deformation are illustrated by 2D seismic profiles in Figures 7a-d. A salt stock with downslope overhang in Figure 7a shows the end of autochthonous salt deformation and the beginning of the formation of an allochthonous salt body. The primary weld is formed along the original salt bed. A more extensive allochthonous salt sheet can be formed when a large amount of original salt existed. Figure 7b shows an early stage allochthonous salt body emplaced in the sediments near the seafloor. The landward inclined salt welds in Figure 7b and 7c are interpreted as depleted large salt feeder stocks or the walls of their associated allochthonous salt sheets. The leading edge of the spreading salt sheet deforms the sea floor, forming seafloor scarps. The most prominent one is the present-day Sigsbee Escarpment [*Amery*, 1969]. The earlier allochthonous salt sheet can be deformed again by later sediment loading as shown clearly in Figure 7c where an amalgamated allochthonous salt sheet has been deformed. Part of the original allochthonous salt has been displaced forming "tertiary" salt welds. When the original allochthonous salt is thick, a second generation al-

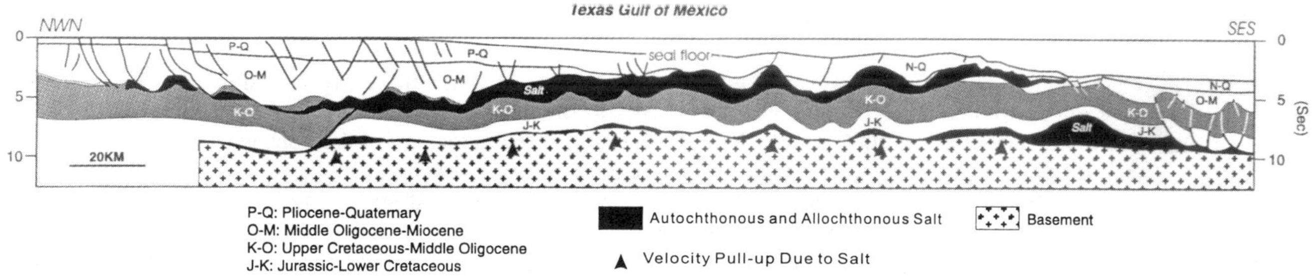

Figure 5. Regional seismic line drawing of offshore Texas slope showing extensional, salt and contractional It illustrstes structures from near the mid-Cretaceous shelf margin across the continental shelf and slope to the Perdido Fold Belt. The interpretation is based on a special 18.0-20.0 seconds experimental seismic Line GBRS-1 acquired by the joint Gulf of Mexico Regional Study program of TGS-CALIBRE Offshore Co., GECO-PRAKLA Geophysical Company, Inc. and the Department of Geology and Geophysics of Rice University sponsored by the oil industry. The stratigraphy is correlated based on well data and published sources [e.g. *Worrall and Snelson*, 1989, *Wu et al*, 1990a, *Wu*, 1993, *Feng*, 1995].

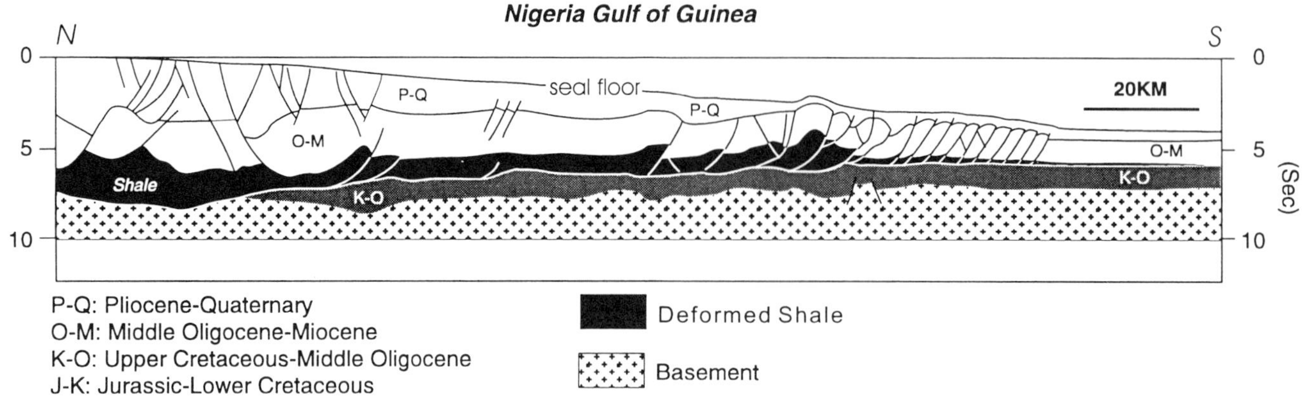

Figure 6. Regional seismic line drawing of the Gulf of Guinea Slope showing extensional, shale and contractional structures.

lochthonous salt may be formed over the first generation allochthonous salt as illustrated in Figure 7d.

A line drawing-interpretation of a regional seismic line in Figure 8 shows various stages and styles of salt deformation in north-central and northeastern Gulf of Mexico, from autochthonous salt domes and diapirs in the east, to first and second generation allochthonous salt structures basinward to the west. The various salt-related structural styles, from the edge of the mid-Jurassic salt basin toward the center of the Gulf of Mexico basin to the west, are related to changes of salt thickness, lithology, geometry, and timing of overburden sedimentation.

Extensional Systems

Extensional systems affecting the post mid-Jurassic sedimentary sections evolved with the development of various styles of salt structures. Diachronous (mid-Jurassic to present-day) and multi-level extensional systems characterize the Gulf of Mexico extensional systems (Figure 4 and 5). Most of the extensional systems are related to salt deformation. This is especially true for the north-central Gulf of Mexico.

The pre-Tertiary extensional systems are mostly located landward of mid-Cretaceous shelf margin (Figure 1). The extensional system on the present-day continental slope began to form since Eocene in the Western Gulf of Mexico (Figure 5) and in the Miocene in the Central Gulf of Mexico (Figure 4) in response to the changes of the Gulf of Mexico drainage systems during Tertiary times (Figure 2, *Winker*, 1981). These extensional systems are still active on the present-day slope.

Multi-level extensional systems and detachment surfaces are common in the northern Gulf of Mexico slope. The oldest faults sole out on top of or within the thin autochthonous salt or salt weld. Figure 7a shows a series of faults associated with the deformations of autochthonous salt. A set of pseudo-clinoforms [*Wu et al*, 1990b, *Mohriak*, 1995] below MCFS formed in response to Mesozoic clastic progradation and autochthonous salt withdrawal. Figure 7b illustrates a large down-to-the-basin growth fault soling out in autochthonous salt behind a tabular allochthonous salt sheet. The extension on this fault is largely accommodated by the deformation of salt feeders and partly translated downdip to form the Mississippi Fan Fold Belt [*Wu et al*, 1990b].

The second level of extensional faults form above an amalgamated allochthonous salt sheet as illustrated in Figure 7c. This second level of extensional system is detached from the lower level extensional system which decouples on the lower autochthonous salt.

Third or even higher order extensional systems exist in the northern Gulf of Mexico slope. They are related to second or higher order allochthonous salt as shown in Figure 7d. In this example, the extension occurs along salt-sediment interfaces for both second and third level extensional systems.

Lower Slope Contractional Fold Belts

Two distinct fold belts are observed in the lower slope of the northern Gulf of Mexico continental margin. The Perdido Fold Belt is located in the northwestern lower slope of Texas while the Mississippi Fan Fold Belt is located in the northeastern lower slope of Louisiana [*Blickwede and Queffelec*, 1988; *Buffler*, 1989; *Worrall and Snelson*, 1989; *Weimer and Buffler*, 1992, *Trudgill et al,* 1999, *Rowan*, this volume]. The southern margin of both fold belts coincide with the basinward zero isolith edge of mid-Jurassic autochthonous salt. Both fold belts are interpreted to be salt-based. Please refer to Rowan et al. [this volume] for a detailed analysis of the Perdido Fold Belt. We will show examples from the Mississippi Fan Fold Belt in the following discussion.

Toward the north (landward) of the Mississippi Fan Fold Belt there are large extensional systems of various ages and extent as well as widespread allochthonous salt structures on the upper and middle slope. To the south of the Mississippi Fan Fold Belt mostly undeformed lower-slope and abyssal plain sediments overlie a salt-free oceanic crust basement. The shortening of the fold belt occurred mostly during mid-late Miocene. Figure 9a is a seismic profile across the western end of the Mississippi Fan Fold Belt where the shortening began and ended within the mid-Miocene. A large section of the landward verging fold has been eroded during a period of non-deposition. In the central part of the Mississippi Fan Fold Belt, Figure 9b illustrates two double verging folds formed during late-mid-Miocene to late Miocene. Strong erosion also occurred during particularly the later part of the compression. Figure 9c shows the fault-propagating fold system across a full section of the Mississippi Fan. The shortening occurred in a very brief period during mid-late Miocene as indicated by the convergence of MM2 (13.8Ma)-UM (5.5 Ma) sediments toward the apexes of the anticlines. We recognize that the overall shortening in this cross section is less than 5%, far less than the total amount of extension observed in the upslope extensional systems. The episode of the shortening is very short relative to the perennial upslope extensions. Following the main shortening of the foldbelt, later salt core inflation caused slight uplift of some of the anticlines. Some salt diapirs developed from the inflating salt cores in the northern parts of the Mississippi Fan foldbelt (Figure 9c).

Trudgill et al [1999] demonstrated a similar sequence of deformation in the Perdido Foldbelt, i.e., the main shortening during early Eocene has been followed by continuous uplifts in the landward parts of the foldbelt due to inflating salt cores.

Summary of the Structural Evolution of the Gulf of Mexico in Terms of Salt Tectonics

The northeastern Gulf of Mexico post mid-Jurassic tectonic history can be summarized by Figure 12a-f [adapted from *Wu,* 1993]. The tectonic history in the Western Gulf of Mexico is similar to that of Central Gulf of Mexico but developed mostly in response to earlier Eocene-Miocene sediment influxes.

Mid-Jurassic salt was deposited on a relatively smooth regional unconformity without major faults. The salt thins toward the south and pinches out near a roughly E-W and NW-SE trending transitional-oceanic crust boundary [*Salvador*, 1987]. Four episodes of Meso-Cenozoic tectonics can be differentiated as follows:

(1) From its mid-Jurassic deposition to the mid-Cretaceous, Louann salt was deformed perennially by sediment loading on both shelf and slope. As a result, peripheral fault systems, salt rollers, salt domes, salt anticlines, salt massifs, pseudo-clinoforms, turtle structures, and local growth structures [refer to *Jackson and Talbot*, 1991 for terminology] were formed. By middle Cretaceous, much of the salt was pushed toward the basin by the sediment load and concentrated near the present-day slope. Some salt had already formed small scale allochthonous salt bodies. An extensive primary weld formed as salt was squeezed out under the Early Cretaceous platform and upper slope by clastic progradation episodes.

(2) During a significant period of sediment starvation, between mid-Cretaceous (Turonian) and mid-Oligocene time, previously formed salt structures were stabilized (Figure 12a).

(3) From mid-Oligocene to early Middle Miocene time, early salt structures were reactivated by rapid sediment loading (Figure 12b). By middle Middle Miocene time, large diapiric salt ridges, walls and stocks formed in response to increased sediment loading. The growing salt structures, driven by buoyancy and differential loading, caused the downslope displacement of the overlying Mesozoic-Middle Miocene section (Figure 12c). This lateral displacement is transmitted downslope, decoupling the Mesozoic-Middle Miocene section over the salt and toward the basinward pinchout of salt. Thus compressional strains, not accommodated by the primary salt withdrawal basins, were transmitted to and concentrated along the basinward edge of salt to form the Mississippi Fan Fold Belt. Shortening in the Mississippi Fan Fold Belt ceased by Late Miocene as the salt walls and stocks had made their way up through the overburden and began to be emplaced in the less dense sediments of the slope to form allochthonous salt structures. During and following the formation of the Mississippi Fan Fold Belt and updip from the fold belt, large primary withdrawal basins, along with landward- and seaward-dipping and local growth fault systems, were formed due to massive salt withdrawal from under the Meso-Cenozoic overburden, resulting in primary welds. Inclined and sub-vertical secondary welds were also formed as feeder stocks or walls thinned to accommodate some of the updip extension in the primary withdrawal basins.

(4) From the Late Miocene to the Present, large-scale asymmetric primary allochthonous salt sheets continued to form and advanced downslope (Figure 12e). As sedimentation continued during the primary allochthonous salt development, secondary salt withdrawal basins with associated basinward- and landward-dipping growth faults, tertiary welds, local folds and reverse faults were formed over the deformed allochthonous salt. The secondary allochthonous salt

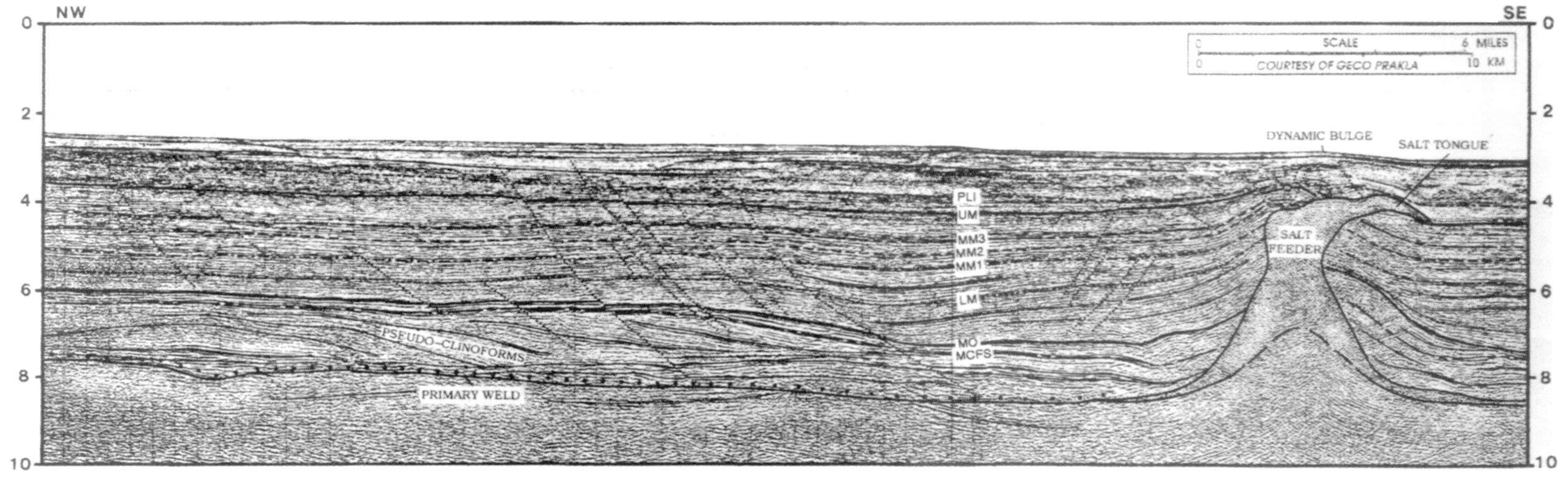

Figure 7. Seismic examples of extensional systems and related salt structures of the northern Gulf of Mexico slope. See Figure 1 for locations. The correlated horizon are PLI (1.9 Ma), UM (5.5 Ma), MM3 (10.5 Ma), MM2 (12.5 Ma), MM1 (13.8 Ma), LM (15.5Ma), MO (30 Ma), MCFS (91.5 Ma), TS (top of autochthonous salt), BS (base of autochthonous salt). Courtesy of Geco-Prakla for permission of publishing the seismic examples.

a) A profile showing a salt stock with overhang, primary salt welds, pseudo-clinoforms and extensional faults due to salt withdrawal.

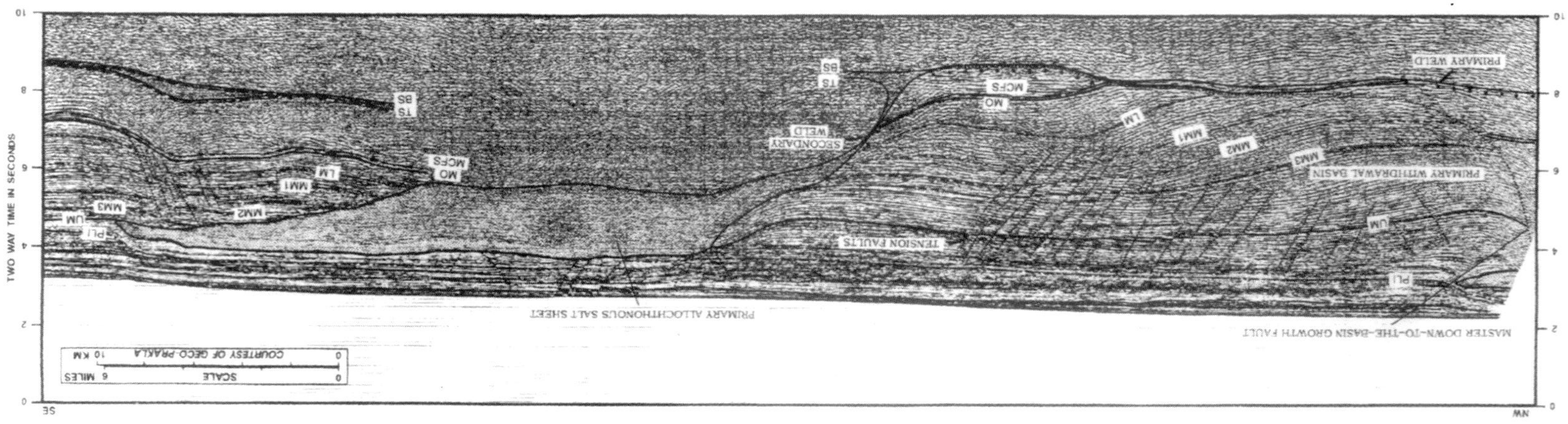

b) A profile showing a large allochthonous salt sheet, primary and secondary welds and a large growth fault behind the salt sheet.

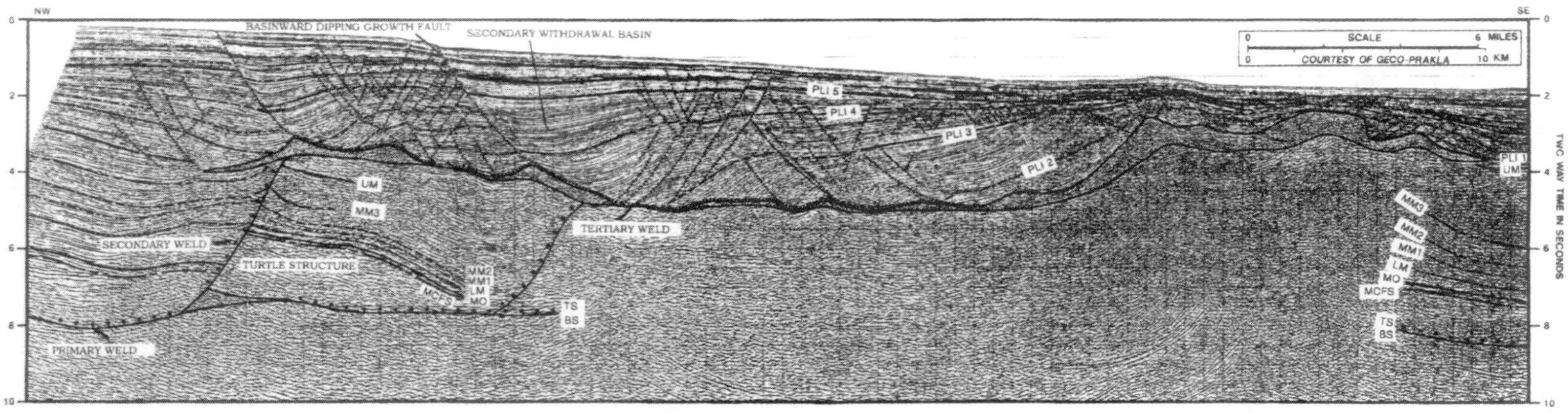

c) A profile showing a deformed allochthonous salt, primary, secondary and tertiary salt welds and a series of growth faults over the secondary allochthonous salt.

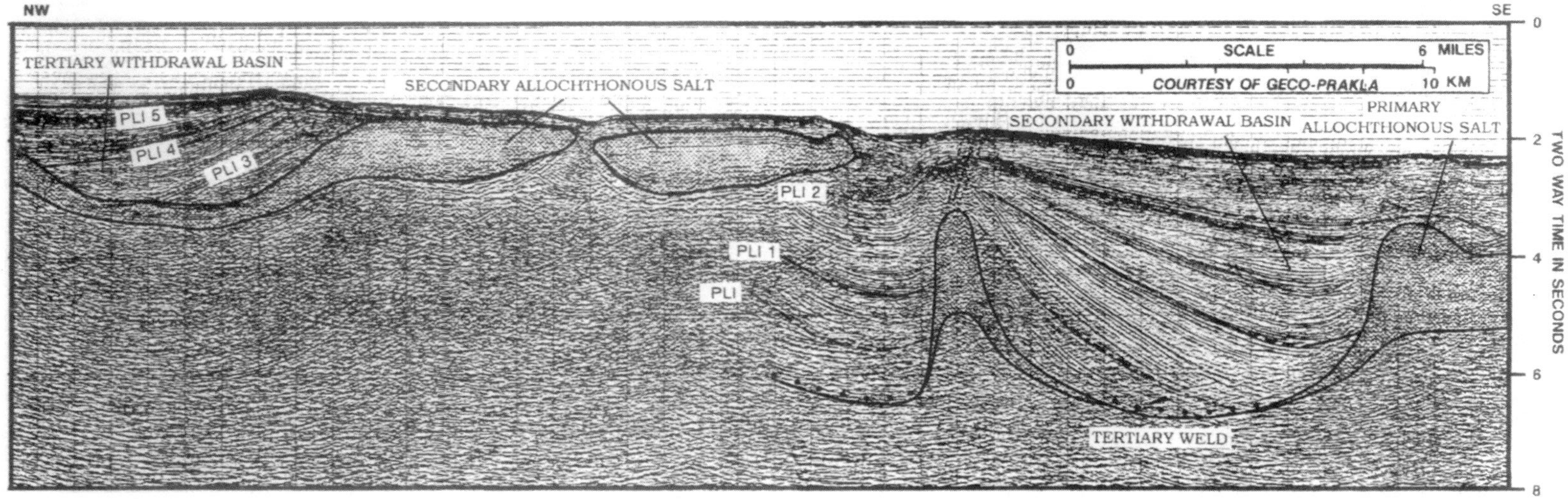

d) A profile show a deformed primary and secondary allochthonous salt and associated extensional systems

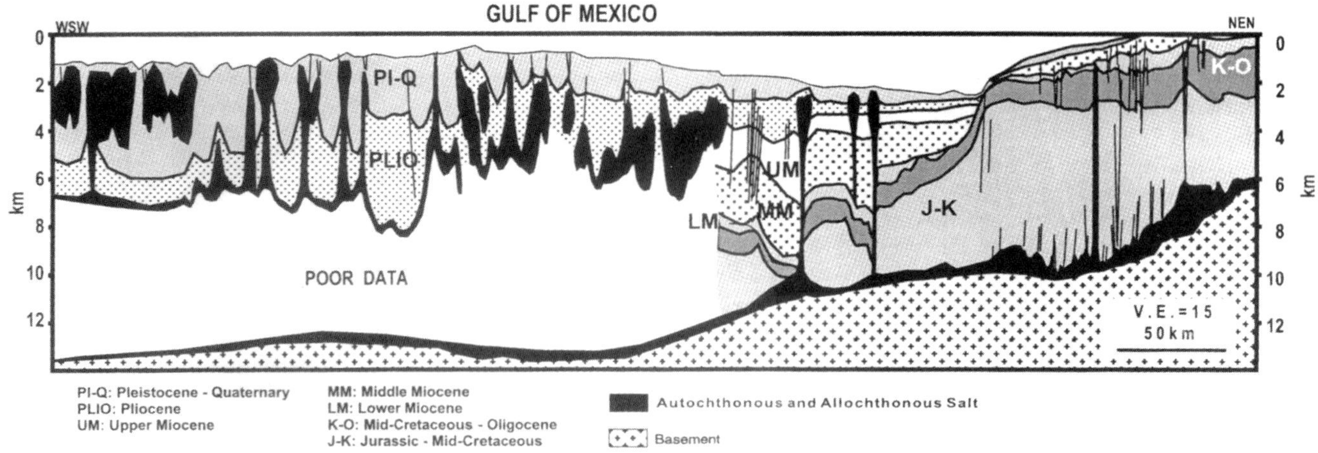

Figure 8. Regional seismic line drawing of various salt structures in the Central and Eastern Gulf of Mexico, from Destin dome in the east to the slope of Louisiana.

structures, tertiary withdrawal basins and later welds evolved as the primary allochthonous salt was deformed by the Plio-Pleistocene sedimentation in some areas. Some of the salt cores of the Mississippi Fan Fold Belt pierced through the overburden to form allochthonous salt bodies (Figure 12f). Active primary and secondary allochthonous salt formed sea floor highs. The allochthonous salt structures of various generations are widely distributed over more than sixty percent of the present-day slope of the Gulf of Mexico.

STRUCTURAL STYLES AND EVOLUTION OF THE CONTINENTAL SLOPE OFFSHORE NIGERIA

Offshore Nigeria is a "classical" shale tectonic province. The geology of the Niger Delta has been described in many papers [e.g. *Mascle et al*, 1973; *Beck and Lehner*, 1975; *Lehner and de Ruiter*, 1977; *Knox and Omatsola,* 1989; *Doust and Omatsola*, 1990, *Morley and Guerin,* 1996] .

Figure 3 [after *Knox and Omatsola*, 1989] shows the tectonic provinces of the Niger Delta and the location of the regional seismic profiles shown in this paper. The Neogene Niger Delta is underlain by thick Tertiary overpressured marine shales. Under differential loading due to rapidly deposited sediments, large growth fault-bounded depobelts formed on the shelf [Knox and Omatsola, 1989]. Shale ridges, diapirs and inter-diapir depocenters are more typical for the slope [*Morley and Guerin*, 1996]. Finally, the basinward end of the delta is marked by a distinct toe thrust system [*Doust and Omatsola*, 1990, *Morley and Guerin*, 1996].

Some of the detailed aspects of shale tectonics in this area are illustrated by several regional seismic lines (Figures 10 and 11) from a grid of data obtained by TGS-NOPEC.

Only a few time surfaces are interpreted and tied at few points along the interpreted lines. Line DL4 (Figure 10a and 11a) is interpreted based only on seismic character. The age determination of "Lower Pliocene" (4.2 Ma), "Middle Miocene" (10.5 Ma ?) is approximate and based on proprietary well information in the upper slope area and on the correlation of a TGS-CALIBRE regional sequence stratigraphic project and published sources [*Beck and Lehner* 1975; *Haq et al*, 1987; *Knox and Omatsola*, 1989; *Doust and Omatsola,* 1990]. The "Paleocene" top is only a rough estimate since no direct correlation with well information can be made across the mobile shale province in the middle slope. This surface is characterized by long wavelength and high amplitude reflectors. It can be recognized over most of the slope area.

The basement is recognizable on seismic data as a very strong and low frequency reflector below which no coherent reflections exist. The basement is thought to be Albian oceanic crust developed during the opening of the Atlantic Ocean [*Knox and Omatsola*, 1989; *Doust and Omatsola*, 1990]. For most of the area the basement underlying the delta is smooth. It dips gently landward (Figures 10a, 11a) except near the Charcot Fracture Zone where the basement is associated with a high relief block as shown in Figures 11b and 11d. This basement relief has affected the position and style of toe thrusting (Figure 11b; *Morley and Guerin*, 1996; *Knox and Omatsola,* 1989; *Mascle et al*, 1973].

From the seismic lines in Figures 10 and 11, we can see that from the shelf to the lower slope, the structural styles change from an extension-dominated style to a diapiric shale province and finally to the imbricates and toe thrusts

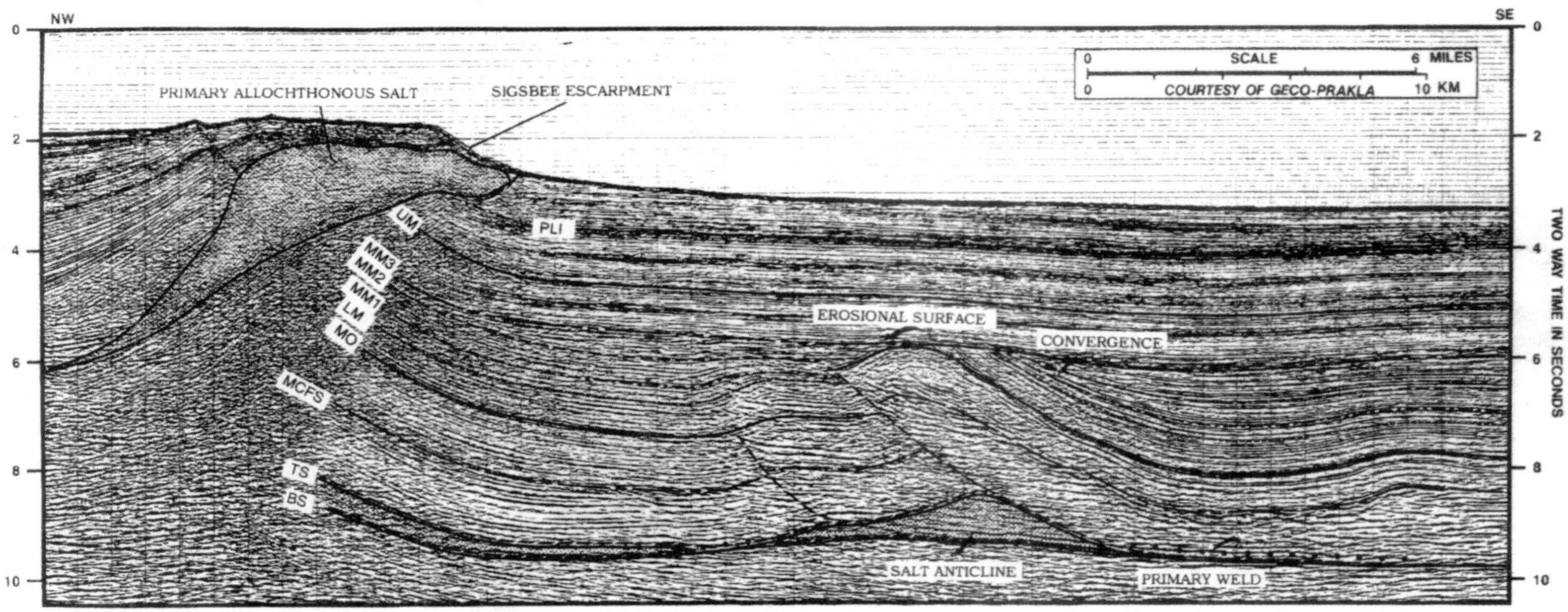

Figure 9. Interpreted seismic examples of salt cored contractional features. See Figure 1 for locations. The correlated horizon are PLI (1.9 Ma), UM (5.5 Ma), MM3 (10.5 Ma), MM2 (12.5 Ma), MM1 (13.8 Ma), LM (15.5Ma), MO (30 Ma), MCFS (91.5 Ma), TS (top of autochthonous salt), BS (base of autochthonous salt). Courtesy of Geco-Prakla for permission of publishing the seismic examples.

a) A profile from western Mississippi Fan Fold Belt

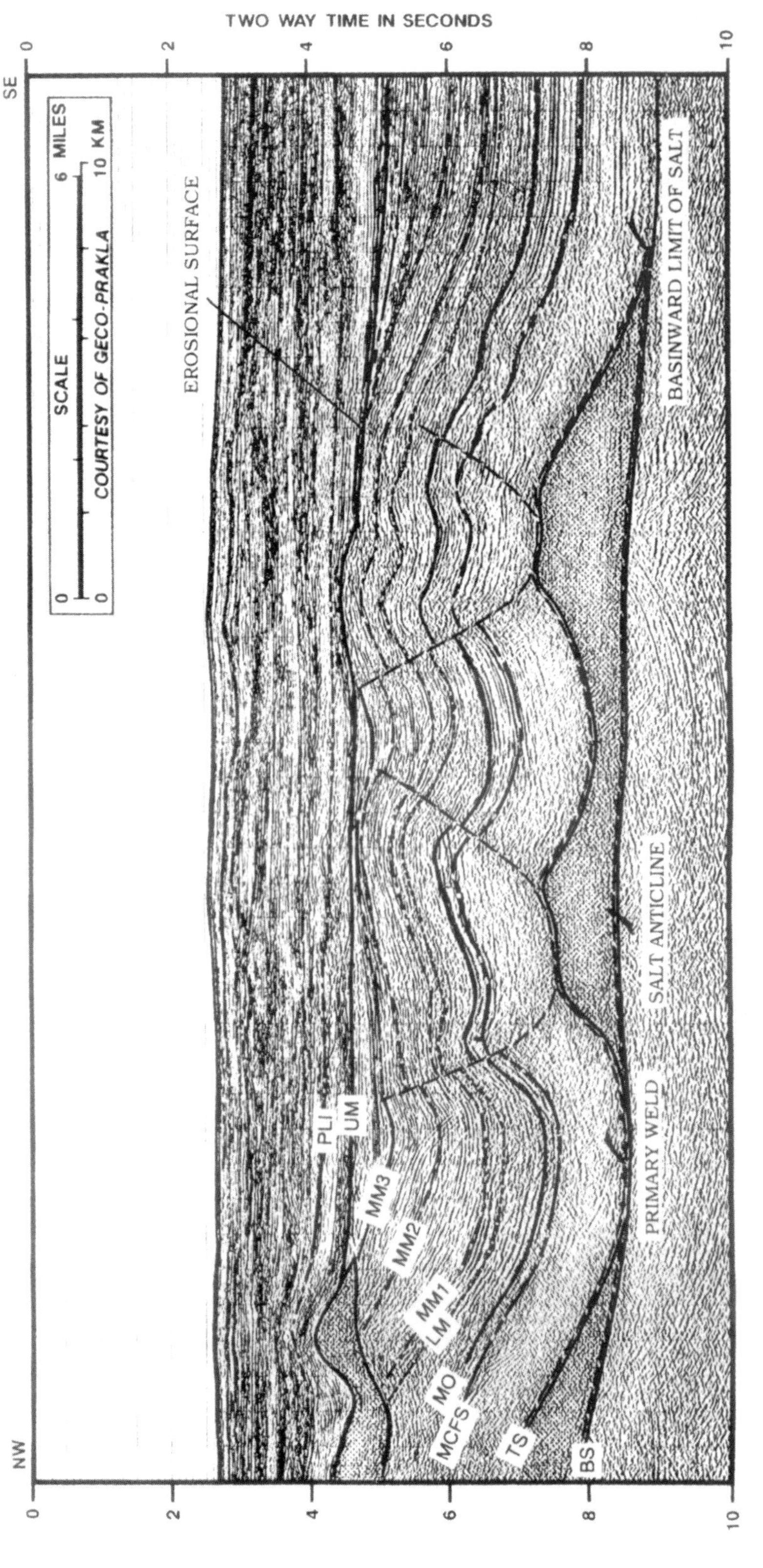

b) A profile from central Mississippi Fan Fold Belt

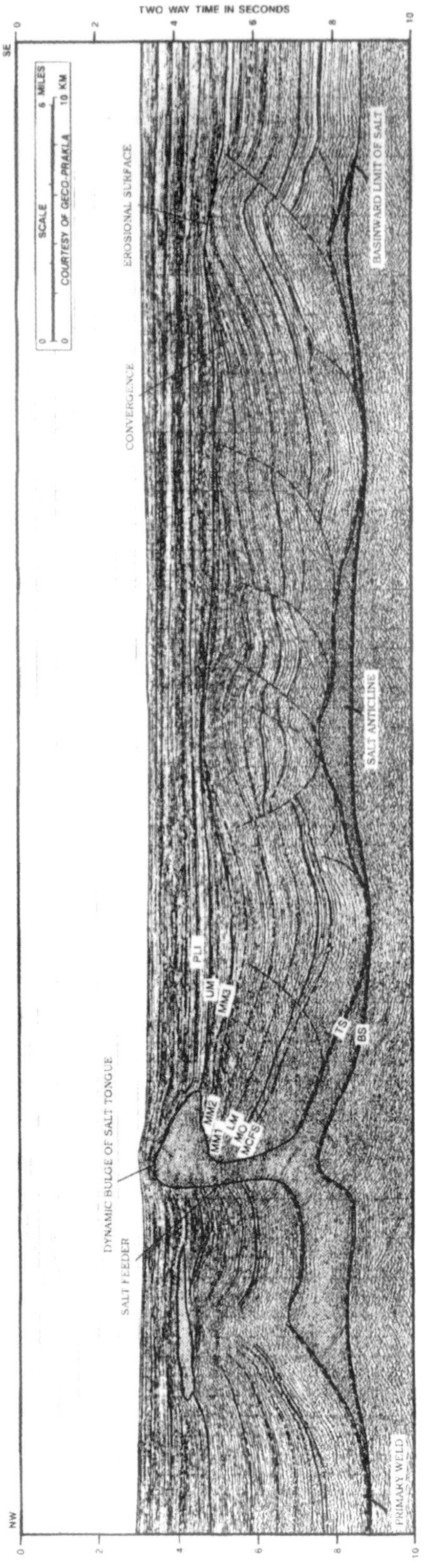

c) A profile from eastern Mississippi Fan Fold Belt

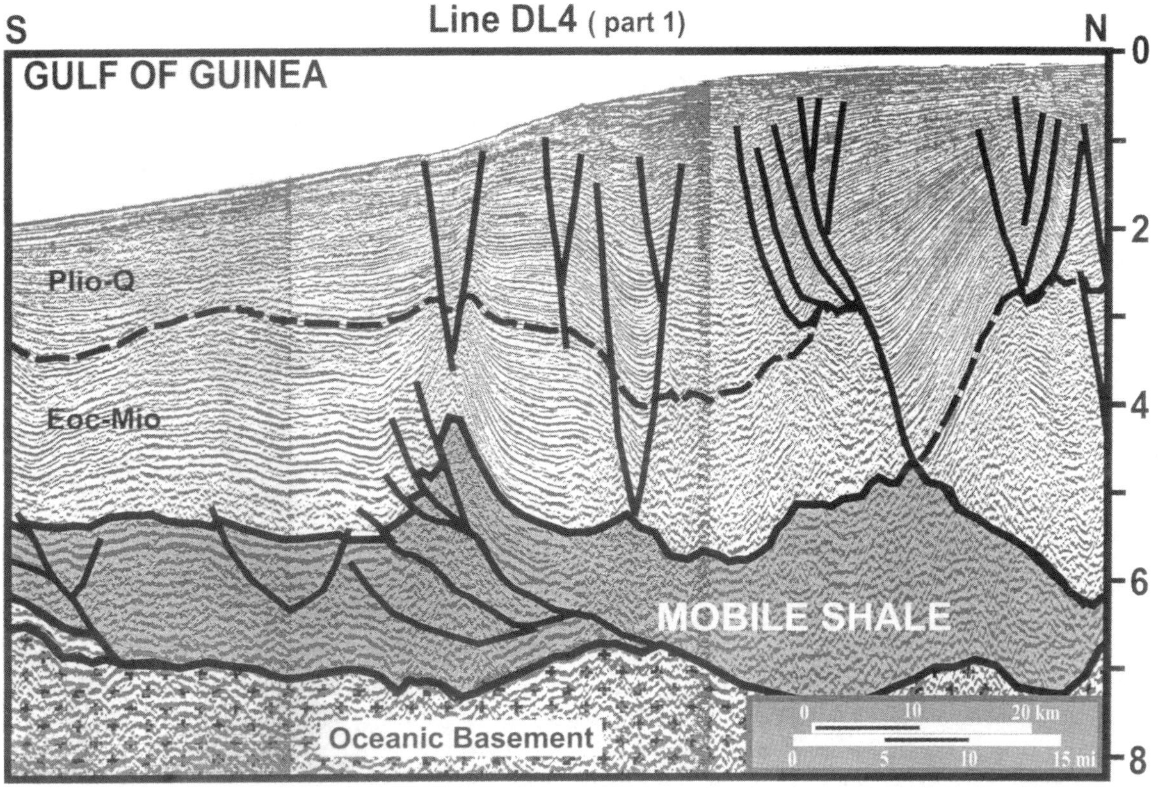

Figure 10. Seismic examples of extensional systems and related shale structures of Gulf of Guinea. See Figure 3 for locations. Courtesy of TGS-NOPEC for permission of publishing the seismic examples.

a) Counter regional fault over a shale ridge in dip line DL4 (part 1).

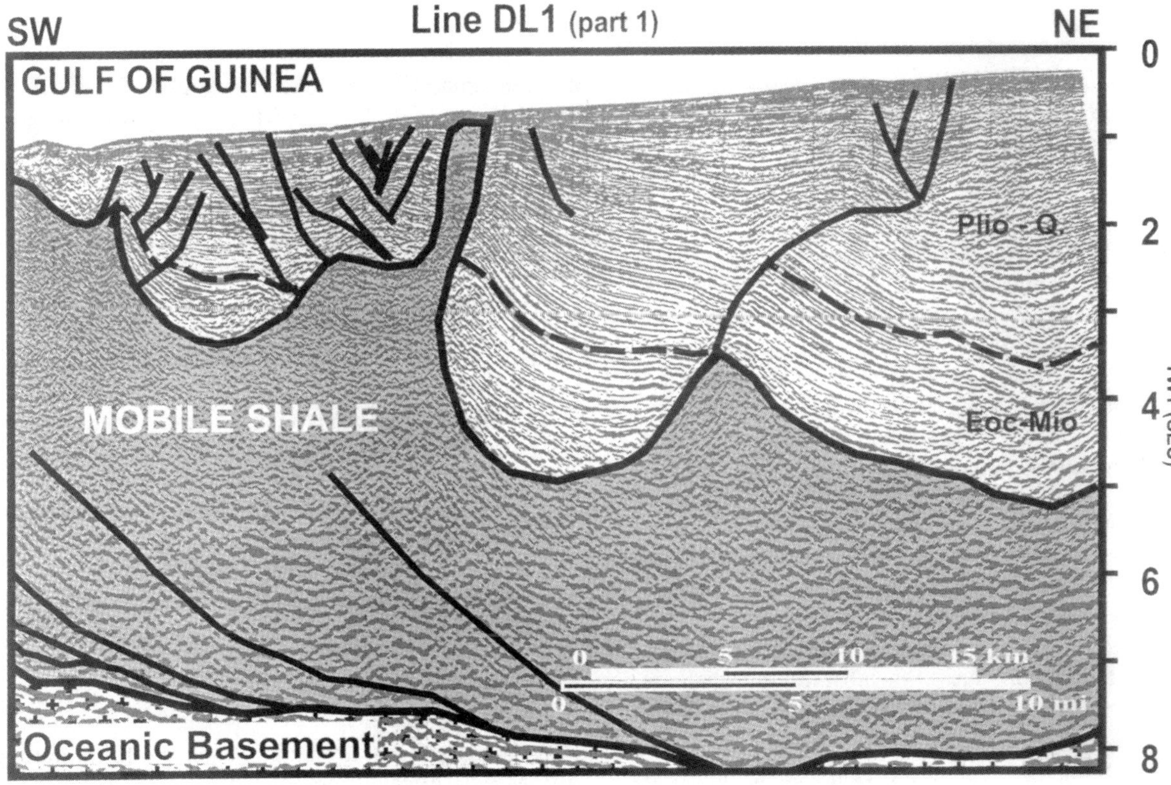

b) Extensional faults over mobile shale structures in dip line DL1 (part 1).

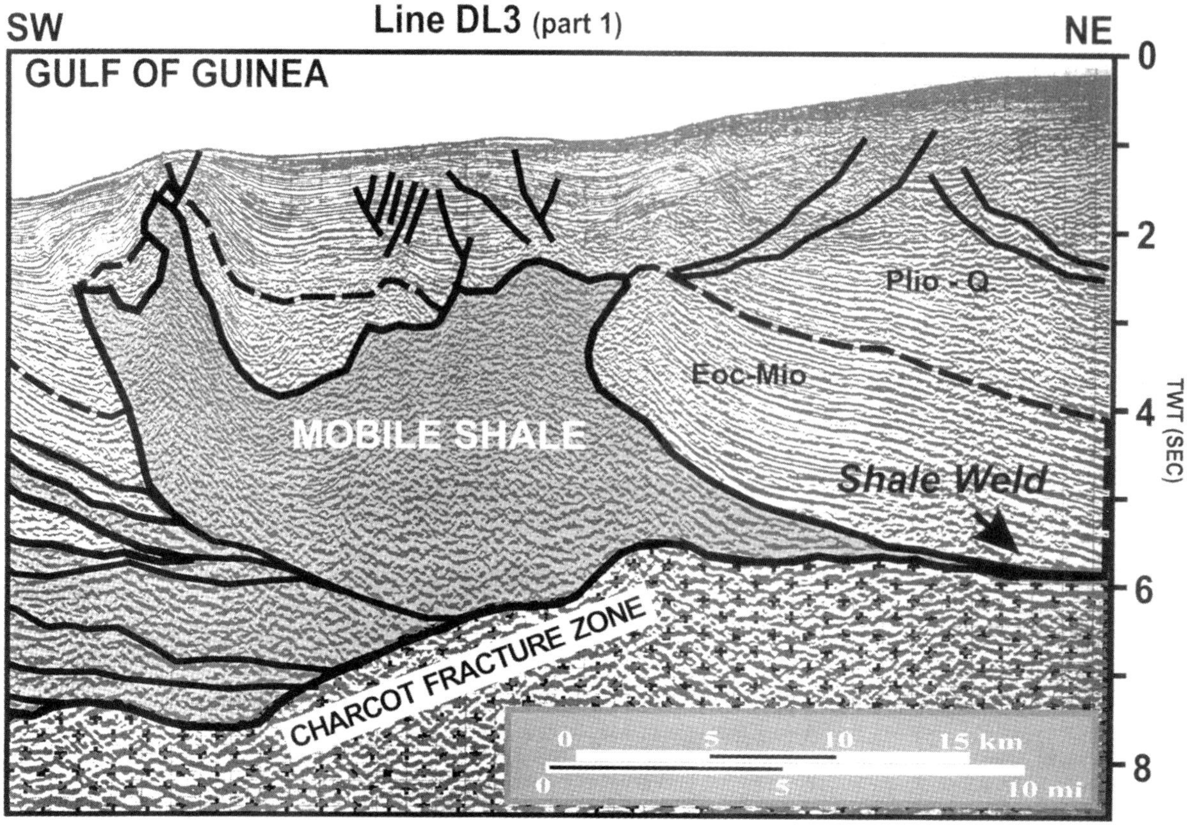

c) An example of shale weld behind a shale ridge in dip line DL3 (part 1)

of the lower slope. The seismic examples show the relationship between shale structures, extension, and contraction in the slope of offshore Nigeria.

Shale Structures

Widely distributed marine shales deposited since the late Cretaceous underlie the entire Niger Delta [e.g. *Mascle et al,* 1973; *Lehner and de Ruiter,* 1977; *Knox and Omatsola,* 1989; *Doust and Omatsola,* 1989]. Since "Paleocene" the load of the rapidly developing Niger Delta causes and deforms the overpressured shales. Overpressured shale was squeezed progressively farther seaward under and in front of the prograding delta loads [*Dailly*, 1976; *Knox and Omatsola,* 1989] forming large shale ridges and diapirs as shown in Figures 10 and 11. As sediments continued to prograde, the older shale ridges were deformed and pushed farther seaward. We can observe that some local basins are being developed on top of the shale massif (e.g. Figure 10 b) as sedimentation continues. As most of the shale is squeezed in front of the sediment load to form large shale massifs, some deformed shales remain behind and are trapped under the delta, forming individual shale ridges in between adjacent depobelts [*Knox and Omatsola*, 1989]. The shale weld is formed when sediments younger than mobile shale are in direct contact with older stratum (i.e., basement in this particular case), after the removal of mobile shale. Although not as common as salt welds, shale welds do exist. The seismic example in Figure 10c shows a shale weld where mobile shale is completely squeezed out from under the sediment loading.

Most of the deformed shales form large shale ridges. Limited shale overhangs are visible on the seaward advancing shale massifs. Seen on the seismic lines in Figure 10 and 11, the large shale massifs separate areas with normal faults from areas with toe thrusts. Most of the large shale structures are involved in a toe thrust system suggesting that large shale structures are under compression, being squeezed seaward from under the prograding delta. Line DL4 in Figure 10a and 11a does not show any large shale diapir. It is interpreted to be located along depocenters between large shale structures.

Extensional Systems

Large extensional systems began to form during the early Middle Eocene stage of the delta development [*Knox*

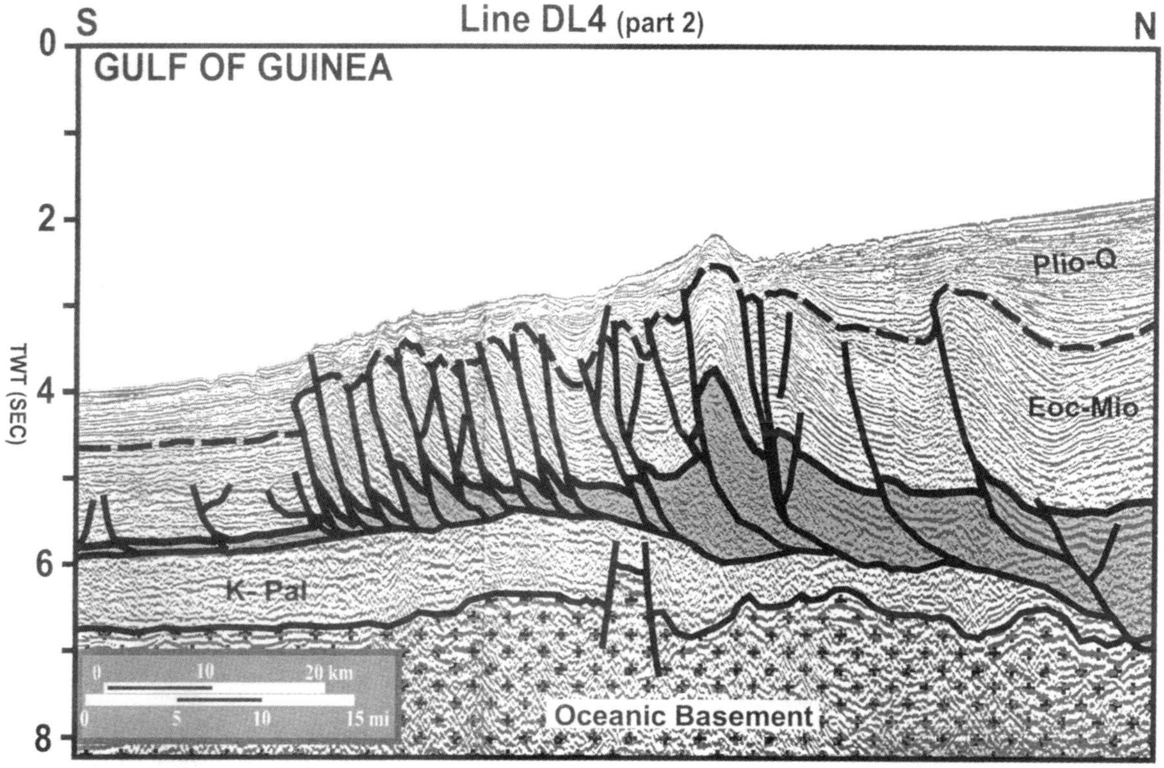

Figure 11. Seismic examples of contractional toe thrusts. See Figure 3 for locations. Courtesy of TGS-NOPEC for permission of publishing the seismic examples.

a) Toe thrusts without strong basement influence in dip line DL4 (part 2, continuation of part 1).

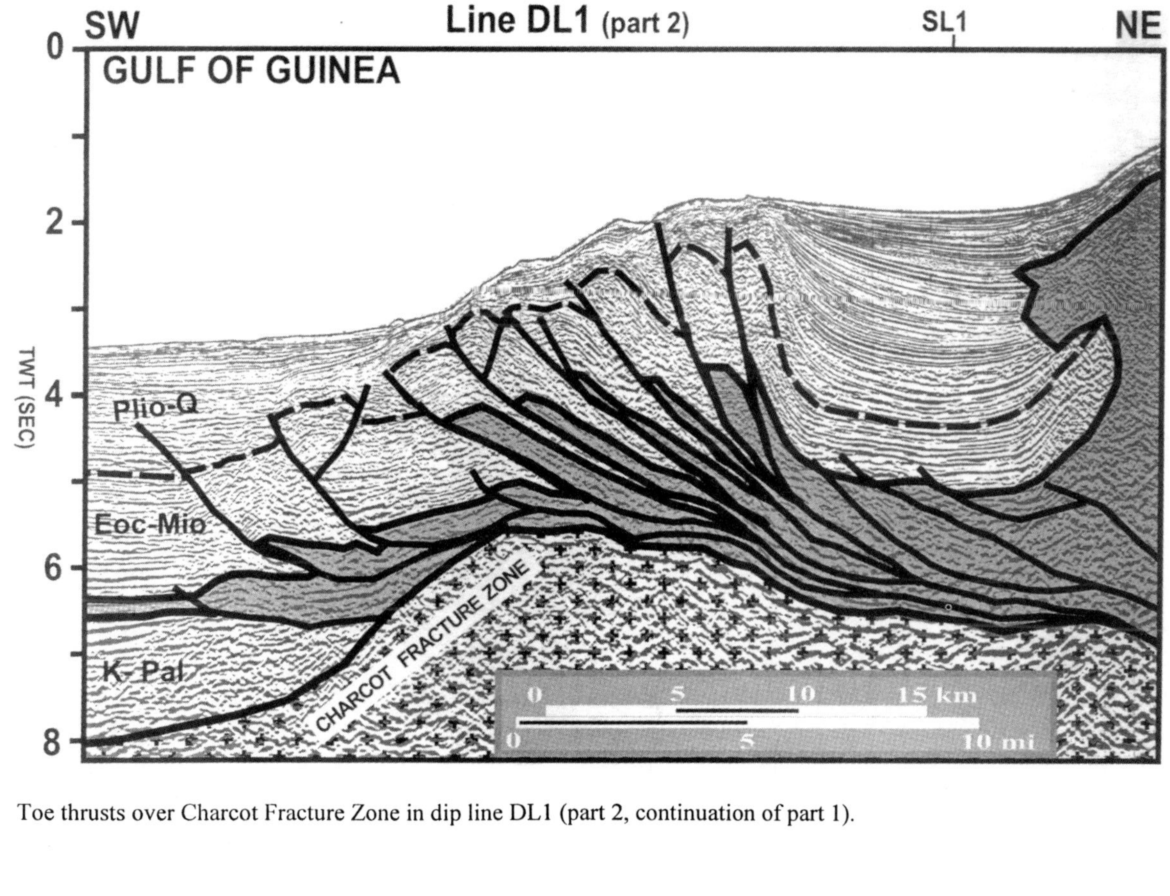

b) Toe thrusts over Charcot Fracture Zone in dip line DL1 (part 2, continuation of part 1).

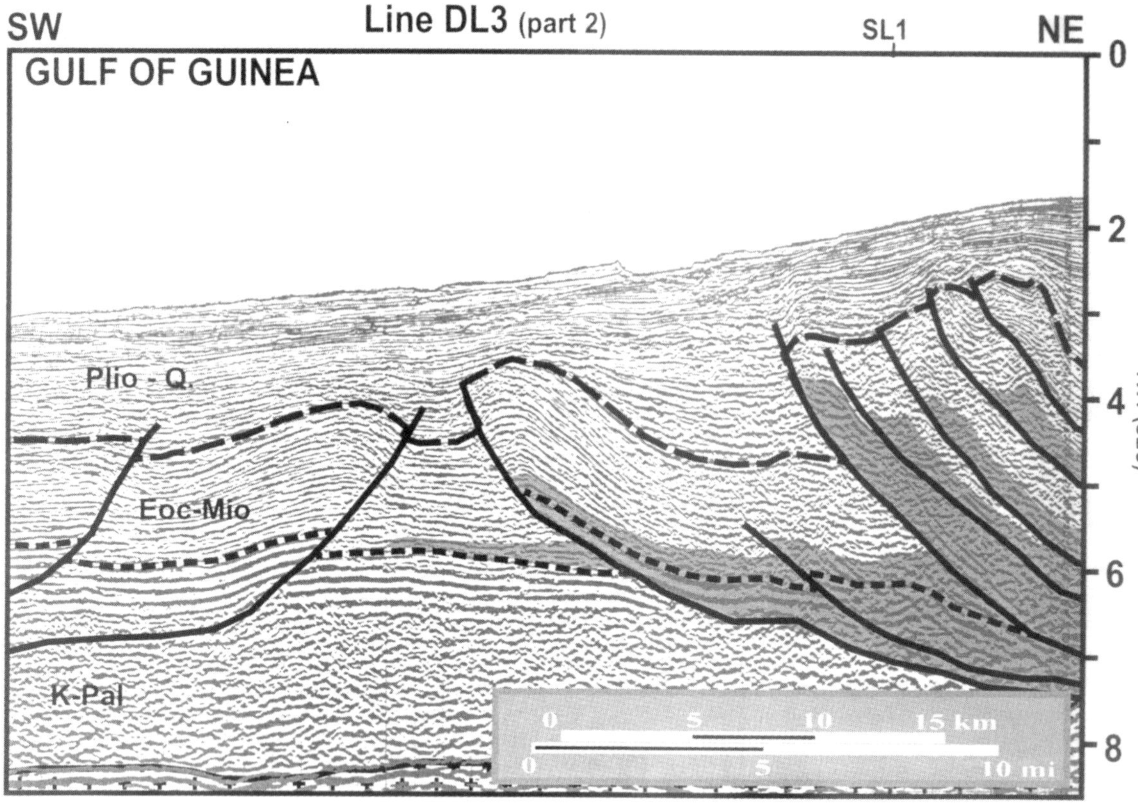

c) A seismic example showing the "out of sequence" toe thrust deformation in dip line DL3 (part 2, continuation of part 1).

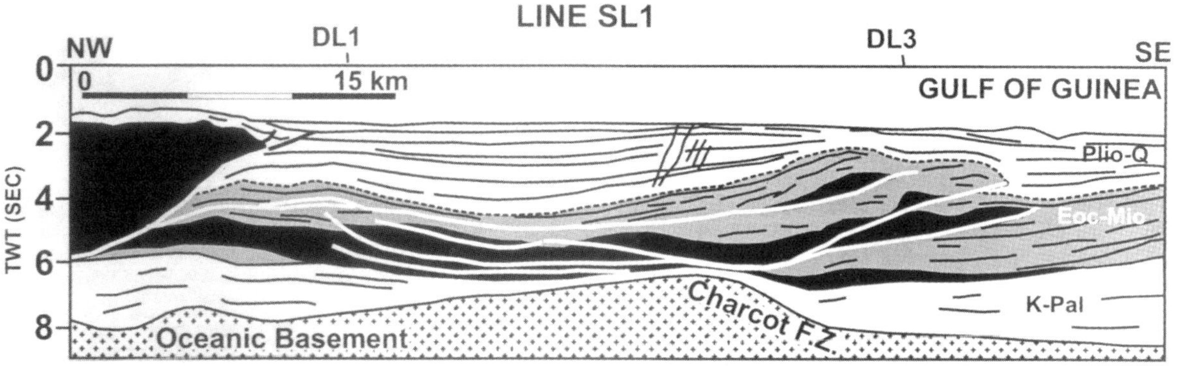

d) Seismic line drawing for strike line SL1 showing lateral ramps of the toe thrusts over the Charcot Fracture Zone.

MESO-CENOZOIC STRUCTURAL EVOLUTION OF THE NORTHEASTERN GULF OF MEXICO

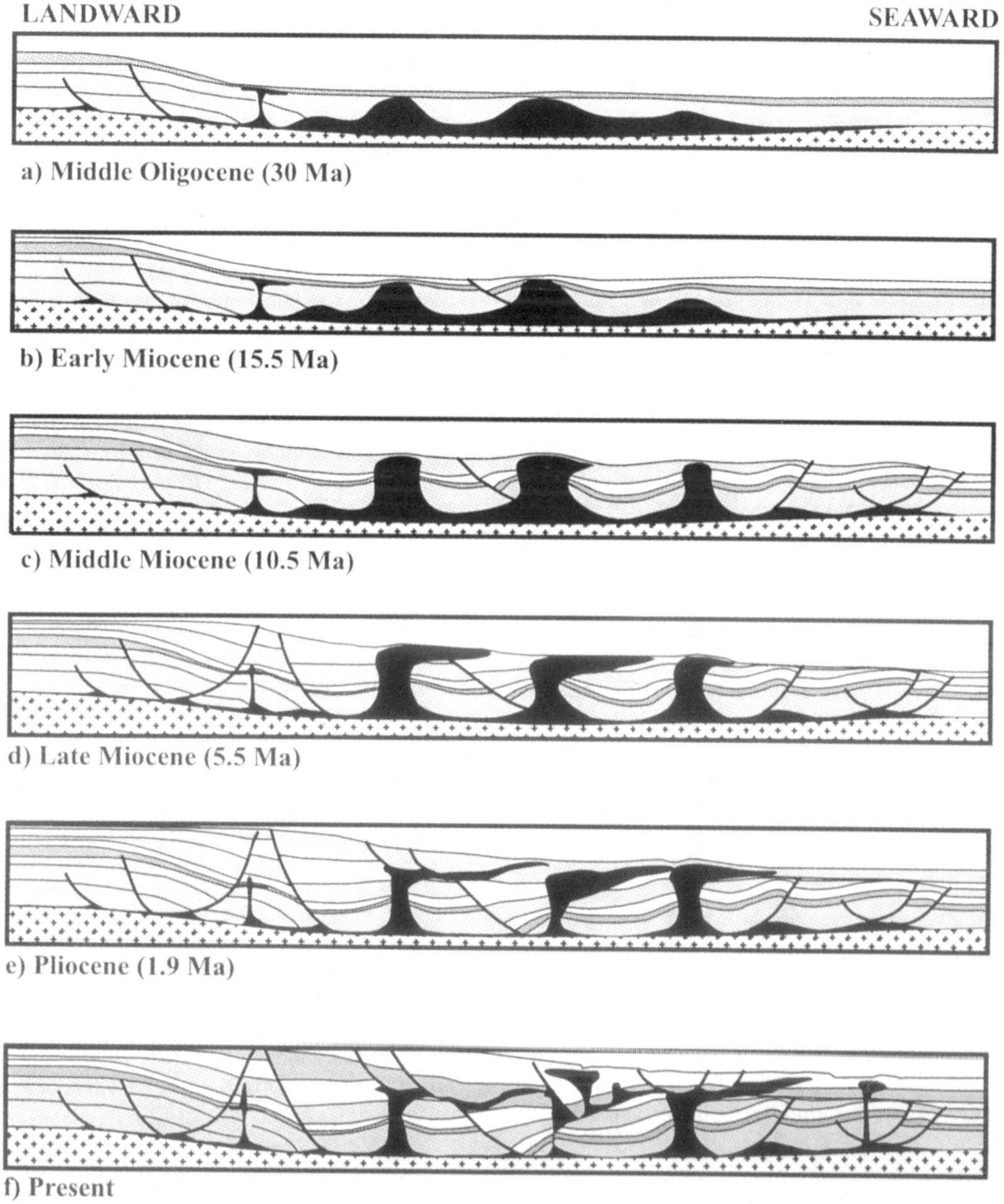

Figure 12 (a-e). Schematic tectonic evolution of offshore Louisiana

a) Middle Oligocene (30 Ma): Autochthonous salt structures stabilized beyond mid-Cretaceous platform margin.
b) Early Miocene (15.5 Ma): Reactivation of pre-existing salt due to early Micoene sediment influx.
c) Middle Miocene (10.5 Ma): Rapid Middle Miocene sediment influxes induced large rising salt ridges and walls and the formation of the Mississippi Fan Fold Belt at the basinward limit of the Jurassic salt basin.
d) Late Miocene (5.5 Ma): Formation of widespread allochthonous salt sheets and associated extensional systems due to salt withdrawal from autochthonous salt. Mississippi Fan Fold Belt ceased to develop due to salt withdrawal which accommodated the upslope extension.
e) Present: Allochthonous salt continues to deform and form supra-allochthonous salt mini-basins and secondary allochthonous salt bodies. Mult-level detachment systems correspond to multi-salt levels. The upslope loading caused some salt diapirs to form from the salt cores of the Mississippi Fan Fold Belt.

and Omatsola, 1989; *Doust and Omatsola*, 1990]. A number of growth faults merge into the mobile marine shale and affect deltaic paralic sediments. The growth faults are formed together with the deformation of the underlying marine shales in response to large loading of prograding deltaic sediments [e.g. *Beck and Lehner*, 1975; *Knox and Omatsola*, 1990]. Most of the extensional systems are located in the present-day onshore portion of the basin as well as along the continental shelf. The oldest growth fault system in the landward direction ceased to develop as most of the mobile shales were squeezed to the slope area. The growth fault systems become progressively younger seaward according to *Knox and Omatsola* [1990]. Also most of the faults are synthetic, with dips toward the sea (e.g. Figures 10b, 10c), but note that some major "conjugate" or antithetic faults have dips in the landward direction (e.g. Figure 10a).

On the slope, where most of our data set is located, we can see growth faults developed behind and on top of the shale massif. They are associated mostly with the Pliocene-Pleistocene depocenters Dominantly seaward-dipping growth faults decouple within the mobile shale. Some small-scale tension faults are also present. Figure 10d presents a present-day snapshot of the slope tectonics in the Niger Delta, where we can see that a Pliocene-Pleistocene depocenter is still developing. Landward it is bounded by a growth fault and seaward by a shale massif. The expansion of the growth fault is translated to the coeval deformation of shale massif and the contractional toe thrusts. However, some toe thrusts ceased to be active while the shale massif continued to grow.

Toe Thrusts

A spectacular toe thrust system occurs at the base of the Niger Delta slope [e.g. *Mascle*, 1973; *Beck and Lehner*, 1975; *Lehner and de Ruiter*, 1977; *Mandl and Crans*, 1981; *Knox and Omatsola*, 1989; *Doust and Omatsola*, 1990]. The toe thrusts are situated seaward of the shale massif and landward of the undeformed deep marine deposit. The seismic examples in Figures 10 and 11 show typical toe thrusts in front of the Niger Delta. The observed shortening began during the "Late Miocene" and has been active since then. The active toe thrusts deform the present-day sea floor. The toe thrust belt is decoupled within the landward portion in the deformed shale, which sometimes appears to involve the Paleogene and frequently soles out close to the top of the oceanic basement. In the seaward portion of the thrust belt, the decoupling level ramps up from the basement to near the interpreted top of the "Paleocene" surface.

The influence of basement geometry on the styles of toe thrusts can be observed in Figure 11b. The basement relief shown in Figure 11b is part of the Charcot Fracture Zone. The presence of the basement high acts as a strain boundary where the shortening is concentrated as stacked thrust imbricates. The effect of the basement relief and the lateral ramp of the thrusts are also observed on the strike line in Figure 11d. On Line DL4 in Figure 10a and 11a, the basement is slightly inclined landward following depth conversion. The shortening is spread out over an area that is 100 km wider than the toe thrust areas near the Charcot Fracture zone. An indentation of the leading edge of the toe thrust near the Charcot Fracture Zone is formed as a result of the basement relief (Figure 3). The extensional deformation on the shelf and upper slope is translated into shortening of the toe thrusts and the deformed mobile marine shales.

In a classical foreland folded belt the typical sequence of deformatin is from the inside toward the outside, i.e. toward the foreland and from top to bottom. “Out of sequence” relations are often described when thrust fault emanating from deeper and later décollement levels cut across higher and older décollement system [*Bally et al*, 1966].

In the non-orogenic folded belt of the Niger Delta the late-Miocene to present-day deformation appears to be mostly “out of sequence”, i.e., seaward folds appear to be slightly older than more landward folds (e.g. Figure 11a, b,c). This relation is puzzling in view of the seaward migration of the depocenters and their associated growth fault systems. A possible explanation may be that the seaward folds loose the overpressured section due to water escape whereas more landward shale structures maintain the higher pore pressures throughout their deformation. All in all a possible widespread slope failure [*Michalowski*, 1995] in form of gravity gliding due to rapid progradation of the Niger Delta may be responsible for the late Miocene to early Pliocene widespread toe thrust development. Following this widespread slope failure, the later deformation is focused near the landward depocenter of the prograding Niger Delta in form of gravity spreading. The leading and distal folds ceased to deform due to their distance to the depocenter. Eventually these distal folds are expected to be reactivated again when the Niger Delta continues to prograde seaward.

Summary of the Structural Evolution of Gulf of Guinea in Terms of Shale Tectonics

The seismic profiles in Figures 10 and 11 illustrate the relationship between extensional growth faults, deformed mobile shale and contractional toe thrusts on the continental slope of the Niger Delta. From onshore to the lower slope, the structural styles change from stabilized and buried extensional growth faults to active growth faults, to highly deformed shale massifs and diapirs, and toe thrusts. Various stages of extensional growth faults developed since Eocene. The growth faults sole out within the deformed marine shale. Thus extension is translated downdip through ductile marine shales into the diapiric shale massif and shortening in the toe thrust. The deformed mobile shale has been moving seaward to form large shale massifs and diapirs under the loading of prograding delta. Toe thrust systems have been active since early Miocene. The late Miocene-early Pliocene widespread shorting of the toe thrust belt may be related to a possible slope failure due to rapid development of Niger Delta. Increased shortening in the thrust belt results from the updip extensions. The uplifting of the landward side of the thrust belt is due to the inflating mobile shale in front of the prograding delta. The basement relief of the Charcot Fracture Zone causes the concentration of shortening within the toe thrust belt and the formation of lateral ramps. The tow thrust system decouples mostly within the Paleogene marine shale.

The tectonic history of offshore Nigeria since Miocene time can be summarized in Figure 13 a-d.

a) Early Miocene (~15 Ma): Toe thrusts and large shale ridges developed behind the prograding Niger Delta. Large depocenters developed over mobile shale whithdrawal basins behind large shale ridges. The toe thrusts and shale structures are interpreted to be developed by differential loading and gravity spreading.
b) Early-mid Miocene (~11 Ma): Toe thrusts and shale ridges continue to migrate seaward in front of prograding Niger Delta. More overpressured marine shale is involved in the toe thrusts and shale structures. The depocenters behind mobile shale continue to develop.
c) Late Miocene (~6 Ma): A possible failure of slope due to rapid prograding delta with large sediment influx caused widespread toe thrust development. The widespread toe thrusts and associated updip extensions are interpreted to be deformed mainly by gravity gliding.
d) Present: Leading toe thrusts, initially formed during interpreted slope failure, ceased to deform and are apparently fossilized. The "out of sequence" deformations of the toe thrusts are locally observed. The toe thrust and mobile shale deformations are concentrated in the landward part of the thrust belt near the depocenters. The deformation is interpreted to be mainly driven by differential loading and gravity spreading.

DISCUSSIONS AND CONCLUSIONS

A comparison of the structural styles documented by seismic data in the northern Gulf of Mexico and the Niger Delta reveals similarities and differences between salt and shale structures and their relations to associated extensional and contractional structures. The evolution of the slope tectonic styles in each margin is summarized in Figures 12 and 13.

Tectonic Setting

The northern Gulf of Mexico slope and the Niger Delta are both formed by deltaic deposits and situated on passive continental margins. The sedimentary patterns are strongly influenced by deformed salt and shale structures respectively. The basement is not significantly involved in the tectonics of either basin. The tectonic activities in both margins are the results of prograding clastic sediments over salt of Gulf of Mexico slope and an overpressured shale substratum in the Gulf of Guinea slope. The Pliocene-Pleistocene rate of sedimentation at the shelf margin near the central part of the Niger Delta reaches over 1.2 mm/year (4 seconds two way time in 4.2 million years at seismic velocity of 2.5 km/second). The overall slope surface was about 1.6 degrees with local variations. Similar slopes and rates of Pliocene-Pleistocene sedimentation are observed in northern Gulf of Mexico.

Salt versus Shale Structures

Salt and overpressured shale are both ductile and deform under gravity, extension and contraction. Gravity spreading under differential loading is common for salt and overpressured shale tectonics. Similar diapiric structures such as ridges, domes and large massifs are found in deformed mobile shale and salt. But the large-scale allochthonous structures are unique to salt basins, especially in the northern Gulf of Mexico slope. The strong density difference between salt and compacted and under-compacted sediments allows salt to rise higher in the section due to buoyancy forces. Salt appears to be more mobile in the Gulf of Mexico than in the overpressured shale of the Niger Delta. This may be related to the rheological differences between salt and overpressured shale. The mobility of salt does not reduce within the sediments in the process of deformation, while the overpressured shale loses its pressure after extrusion and loss of pressured fluid and hence the reduced mobility. Today in the Gulf of Mexico much of the salt has become allochthonous, whereas in the Niger Delta shale structures remain deeply rooted.

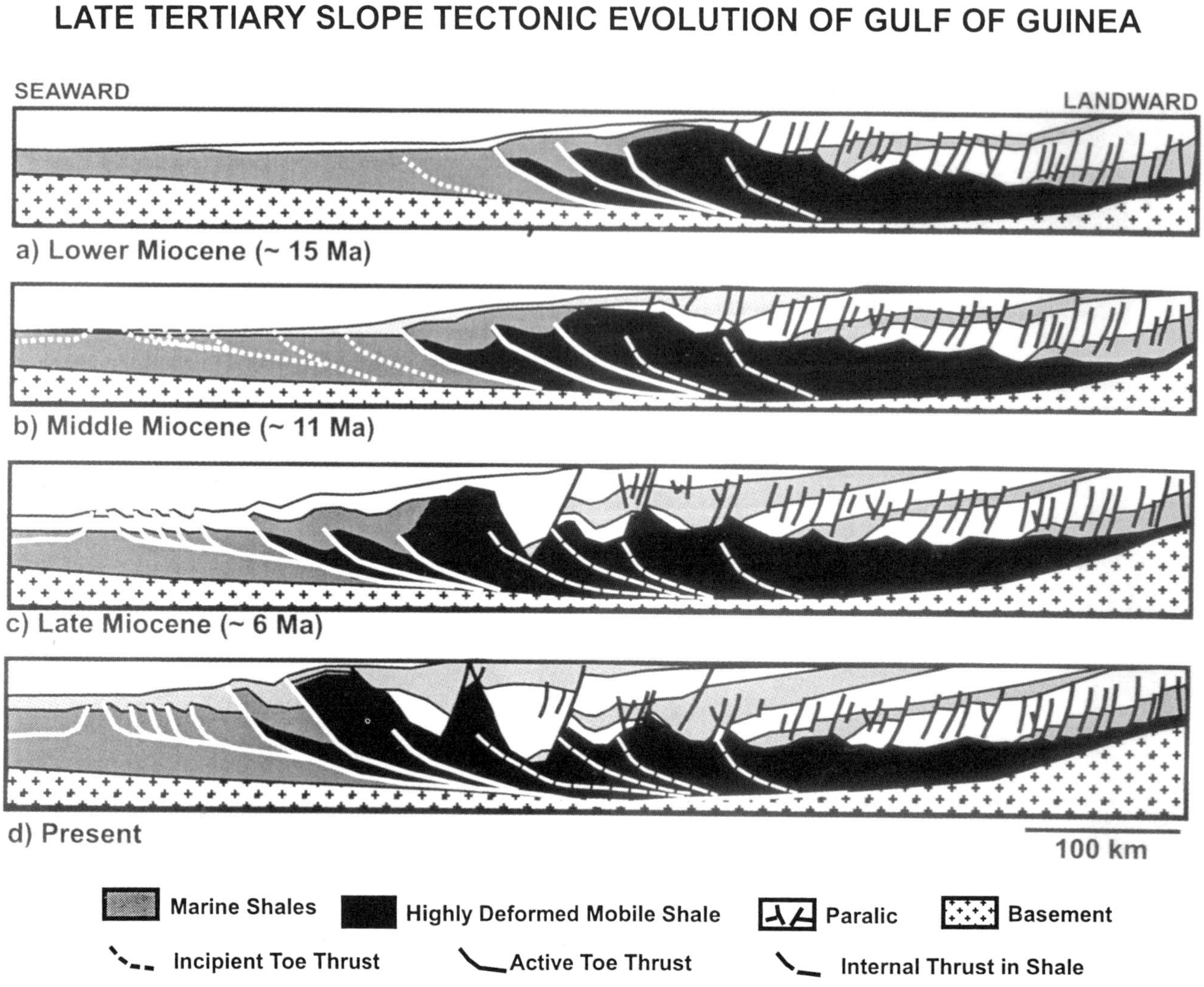

Figure 13 (a-d). Schematic tectonic evolution of Gulf of Guinea.

a) Early Miocene (~15 Ma): Toe thrusts and large shale ridges developed behind the prograding Niger Delta.

b) Early-mid Miocene (~11 Ma): Seaward migrating toe thrusts and shale ridges in front of prograding Niger Delta.

c) Late Miocene (~6 Ma): Possible failure of slope due to rapid prograding delta caused widespread toe thrust development.

d) Present: Leading toe thrust ceased to deform. The toe thrust and mobile shale deformation is concentrated in the landward part of the thrust belt near the depocenter.

Extensional Systems

Extensional systems of varying ages extend from onshore to the upper continental slopes of both margins. As seen in both the northern Gulf of Mexico slope and Gulf of Guinea slope, extensional systems are situated over and behind salt or mobile shale massifs. In the northern Gulf of Mexico, because the presence of autochthonous and multilevel allochthonous salt, different decoupling levels exist for the extensional and contractional systems. The extension of growth faults is accommodated by the withdrawal and seaward spreading of salt in both autochthonous and allochthonous levels. Little or no extension over the allochthonous salt is translated into the shortenings of Perdido Fold Belt and Mississippi Fan Fold Belt. Shale structures in the Gulf of Guinea slope are mostly deeply rooted in the Gulf of Guinea slope. There is basically one main decoupling level at the base of the deformed mobile shale. Extension in the Gulf of Guinea slope is accommodated by the deformation of mobile shales and the toe thrusts as illustrated by regional seismic profiles.

Contractional Systems

Contractional systems are present in both the lower slopes of the northern Gulf of Mexico and Gulf of Guinea. The salt-cored Perdido Fold Belt and Mississippi Fan Fold Belt are distributed along the southern depositional edge of the mid-Jurassic salt basin. They were formed briefly during late Eocene to mid-Oligocene in Perdido Fold Belt and mid-late Miocene in Mississippi Fan Fold Belt. The fold belt decouples in the autochthonous salt. The main shortening phase in both Perdido Fold Belt and Mississippi Fan Fold Belt is interpreted to correspond to the period of large-scale rising of salt walls and diapirs as feeders for the later development of allochthonous salt bodies. The fold belts ceased to develop shortly after the initial main compressional phase. Once large allochthonous salt structures began to form, the space provided by salt withdrawal from both the autochthonous level and from the feeder stocks accommodates the increased sediment influxes and extensions. Due to the prograding shelf margin, the inflating salt cores caused the uplifting of the landward parts of the fold belts after the main shortening phase. The later upslope extensions have been accommodated by the deformation of allochthonous salt and subvertical salt feeders.

In contrast, the compressional toe thrust belt in the Niger Delta was caused by upslope extension and has been deforming almost continuously since its initiation in the Miocene. A possible late-Miocene to Early Pliocene slope failure and continued subsequent progradations of Niger Delta have caused perennial shortening in the toe-thrust belt in an "out of sequence" manner. The decoupling level in the overpressured marine shale is less confined and thus differs from the limited depositional distribution of mid-Jurassic salt in lower Gulf of Mexico slope. The upslope extensions in the Gulf of Guinea slope are largely accommodated by defomations of overpressured shales in the form of shale ridges, walls and diapirs in the platform and slope, and by toe-thrust systems in the deep basin. Overall shortening in the Gulf of Guinea slope toe thrust belt is substantially greater than the shortenings in the Gulf of Mexico slope and deep basin, as quantified in the Perdido Fold Belt and Mississippi Fan Fold Belt (~5% shortening).

Acknowledgments. TOTAL generously provided financial support for the work of S. Wu in the period 1990 – 1993 during a Ph.D. program at Rice University when most of the results of this work were achieved under the direction of Dr. A.W. Bally. We especially thank Dr. Carlos Cramez for his support and encouragement. We thank GECO-PRAKLA for providing seismic data and allowing the publication of seismic data from north-central Gulf of Mexico. We thank TGS-NOPEC Geophysical Company for providing seismic from northwestern Gulf of Mexico and granting permission of publishing seismic data from the Niger Delta.

REFERENCES

Amery, G. B., Structure of Sigsbee scarp, Gulf of Mexico; *AAPG Bulletin*, v.53, p. 2480-2482, 1969.

Bally, A.W., Gordy, P. L., Stewart, Structure, seismic data and orogenic evolution of southern Canadian Rocky Mountains, *Bulletin of Canadian Petroleum Geology*, v. 14, no. 3, p.337-381, 1966.

Bally, A. W., Thoughts on the tectonics of folded belts, in K. R. McClay, and N. J. Price (eds.), *Thrust and Nappe Tectonics*, Geological Society of London, 13-32, 1981.

Beck, R. H. and P. Lehner, Oceans, new frontier in exploration. *American Association of Petroleum Geologists Bulletin*, 58: 1672-1678, 1975.

Blickwede, J. F., and T. A. Queffelec, Perdido foldbelt: a new deep-water frontier in western Gulf of Mexico (abstract). *American Association of Petroleum Geologists Bulletin*, 72: 163, 1988.

Bruce, C. H., Pressured shale and related sediment deformation-mechanism for development of regional contemporaneous faults. *American Association of Petroleum Geologists Bulletin*, 57: 878-886, 1973.

Buffler, R. T., and D. S. Sawyer, Distribution of crust and the early history, Gulf of Mexico basin. *Gulf Coast Association of Geological Societies Transactions*, XXXV: 333-344, 1985.

Dailly, G. C., A possible mechanism relating progradation, growth faulting, clay diapirism and overthrusting in a regressive sequence of sediments. *Canadian Petroleum Geologists Bulletin*, 24: 92-16, 1976.

Diegel, F.A., J.F. Karlo, D.C. Schuster, R.C. Shoup, and P.R. Tauvers, Cenozoic structural evolution and tectono-stratigraphic framework of the northern Gulf Coast continental margin, in M.P.A. Jackson, D.G. Roberts, and S. Snelson, eds., *Salt tectonics: a global perspective*: AAPG Memoir 65, p.109-151, 1995.

Doust, H., and E. M. Omatsola, Niger Delta,. in J.D. Edwards and P.A. Santogrossi, eds., *Divergent/Passive Margins*, American Association of Petroleum Geologists Memoir, No. 48, p. 201-238, 1990.

Feng, J., Post mid-Cretaceous seismic stratigraphy and depositional history, deep Gulf of Mexico, *Ph.D. dissertation*: Austin, Texas, University of Texas at Austin, 253p., 1995.

Hall, D. J., and S.Wu., Salt tectonics of the Mississippi Delta and shale tectonics of the Niger Delta: Comparisons and Contrasts (abstract) . Abstracts and Program, South-Central Section, *Geological Society of America*: 13, 1992.

Hall, D. J., B. E. Bowen, R. N. Rosen, S. Wu, and A. W. Bally, Mesozoic and early Cenozoic development of the Texas margin: a new integrated cross-section from the Cretaceous shelf edge to the Perdido foldbelt: *Gulf Coast Section of the SEPM Foundation 13th Annual Reasearch Conference Proceedings*, p.21-31, 1993.

Haq, B. U., J. Hardenbol, and P. R. Vail, Chronology of fluctuating sea levels since the Triassic. *Science*, 235: 1156-1167, 1987.

Humphris, C. C. Jr., Salt movement on continental slope, northern Gulf of Mexico, p. 69-85. In A. H. Bouma, G. T. Moore, and J. M. Coleman (eds.), Framework, Facies and Oil-Trapping Characteristics of the Upper Continental Margin. *American Association of Petroleum Geologists Studies in Geology*, No. 7, 1978.

Jackson, M. P. A., and C. Cramez, Seismic recognition of salt welds in salt tectonics regimes. *Gulf Coast Section SEPM Foundation Tenth Annual Research Conference Program and Extended and Illustrated Abstracts*: 66-71, 1989.

Jackson, M.P.A. and Talbot, C. J., A glossary of salt tectonics: *Geological Circular*, v.91-4, 44p, 1991.

Knox, G. J., and E. M. Omatsola, Development of the Cenozoic Niger Delta in terms of the 'Escalator Regression' model and impact on hydrocarbon distribution: *Proceedings KNGMG Symposium 'Coastal Lowlands, Geology and Geotechnology** : 181-202, 1989.

Lehner, P., and P. A. C. de Ruiter, Structural history of Atlantic margin of Africa. *American Association of Petroleum Geologists Bulletin*, 61: 963-981, 1977.

Liro, L. M., Distribution of shallow salt structures, lower slope of the northern Gulf of Mexico, USA. *Marine and Petroleum Geology*, 9: 433-451, 1992.

Mandl, G. and W. Crans, Gravitational gliding in deltas, p. 41-54. In K. R. McClay and N. J. Price (eds.), *Thrust and nappe tectonics*, Geological Society of London Special Publication, 9, 1981.

Mascle, J. R., B. D. Bornhold, and V. Renard, Diapiric structures off Niger delta. *American Association of Petroleum Geologists Bulletin*, 57: 1672-1678, 1973.

McBride, B. C., The evolution of allochthonous salt along a megaregional profile across the northern Gulf of Mexico Basin, *AAPG Bulletin*, v. 82, No. 5B, p.1037-1054, 1998.

Michalowski, R. L., Stability of slopes: Limit analysis, in Haneberg, W.C. and Anderson, S. A., eds., *Clay and Shale Slope Instability*: Boulder, Colorado, Geological Society of America Reviews in Engineering Geology, v. X, 1995.

Mohriak, W. U., Salt tectonics structural styles: contrasts and similarities between the south Atlantic and the Gulf of Mexico, in C. J. Travis et al, eds, *Salt, Sediment and Hydrocarbons,* Gulf Coast Section SEPM Foundation, 16th Annual Research Conference, p. 177-191, 1995.

Moretti, I., S. Wu, and A. W. Bally, Balanced Cross Section (LOCACE) to reconstruct allochthonous salt sheet Offshore Louisiana. *Marine and Petroleum Geology*, 7: 371-377, 1990.

Morley, C.K., and Guerin, G., Comparison of gravity-driven deformation styles and behaviour associated with mobile shales and salt, Tectonics, v. 15, p. 1154 – 1170, 1996.

Peel, F.J., C.J.Travis, and J.R.Hossack, Genetic structural provinces and salt tectonics of the Cenozoic offshore U.S. Gulf of Mexico: a preliminary analysis, in M.P.A. Jackson, D.G. Roberts, and S. Snelson, eds., *Salt tectonics: a global perspective*: AAPG Memoir 65, p.109-151, 1995.

Rowan, M.G., Structural styles and evolution of allochthonous salt, central Louisiana outer shelf and upper slope, in M.P.A. Jackson, D.G. Roberts, and S. Snelson, eds., *Salt tectonics: a global perspective*: AAPG Memoir 65, p.109-151, 1995.

Rowan, M.G., Three-dimensional geometry and evolution of a segmented detachment fold, Mississippi Fan foldbelt, Gulf of Mexico: *Journal of Structural Geology*, v.19, p.463-480, 1997.

Rowan, M.G. et al., this volume

Salvador, A., Late Triassic-Jurassic palaeogeography and origin of Gulf of Mexico. *American Association of Petroleum Geologists Bulletin*, 71: 419-451, 1987.

Salvador, A., Triassic-Jurassic, , in Salvador, A., ed., *The Gulf of Mexico Basin*: the geology of North America: Geological Society of America, v. J, p. 131-180, 1991.

Sawyer, D. S., Buffler, R. T. and Pilger, R., Crust of the Gulf of Mexico basin, in Salvador, A., ed., *The Gulf of Mexico Basin*: the geology of North America: Geological Society of America, v. J, p. 53-72, 1991.

Trudgill, B.D et al, The Perdido Fold Belt, Northwestern Deep Gulf of Mexico: Part 1: Structural geometry, evolution and regional implications. *AAPG Bulletin*, v.83, No. 1, p. 88 –113, 1999.

Vail, P. R.., Seismic stratigraphic interpretation using sequence stratigraphy, Part 1, p. 1-10. In A. W. Bally (ed.), *Atlas of Seismic Stratigraphy*. American Association of Petroleum Geologists Studies in Geology, No. 27, Vol.1, 1987.

Vendeville, B.C., and M.P.A. Jackson, The rise of diapir during thin skinned extension: *Marine and Petroleum Geology*, v.9, p. 331-352, 1992.

Vendeville, B.C., Remobilization of salt structures by sediment progradation: Experimental models and potential applications to the Gulf of Mexico and other continental margins (Abstract), *AAPG Hedberg Research Conference*, Galveston, Texas, September 20-24, 1998.

Watkins, J.S., Bradshaw, B. E., Huh, S., Li, L., Zhang, J., Structure and distribution of growth faults in the northern Gulf of Mexico OCS, in J. O. Jones and R.L. Freed (eds), *Structural Framework of the Northern Gulf of Mexico*, Special publication of Gulf Coast Association of Geological Society, p. 63-77, 1996.

Winker, C. D., and R. T. Buffler, Palaeogeographic evolution of early deep-water Gulf of Mexico and margins, Jurassic to Middle Cretaceous. (Comanchean), *American Association of Petroleum Geologists Bulletin*, 72: 318-346, 1988.

Worrall, D. M., and S. Snelson, Evolution of the northern Gulf of Mexico, with emphasis on Cenozoic growth faulting and the role of salt, in A. W. Bally and A. R. Palmer (eds.), *The geology of North America; An Overview*. Geological Society of America, The Geology of North American, Vol. A, p. 97-138, 1989.

Wu, S., Salt and slope tectonics offshore Louisiana: *Ph.D. dissertation*, Houston, Texas, Rice University, 251p., 1993.

Wu, S., and A. W. Bally, Tectonics of the northeastern Gulf of Mexico, *South-Central Section, Geological Society of America, Abstracts with Program*: 51, 1992.

Wu, S., P. R. Vail, and C. Cramez, Allochthonous salt, structure and stratigraphy of the northeastern Gulf of Mexico, Part I: Stratigraphy. *Marine and Petroleum Geology*, 7: 318-333, 1990 a.

Wu, S., A. W. Bally, and C. Cramez, Allochthonous salt, structure and stratigraphy of the northeastern Gulf of Mexico, Part II: Structure. *Marine and Petroleum Geology*, 7: 334-370, 1990b.

Wu, S., C., Cramez, A. W. Bally, and P. R. Vail, Evolution of allochthonous salt in Mississippi Canyon Area. Gulf Coast Section *SEPM Foundation Tenth Annual Research Conference Extended and Illustrated Abstracts*: 161-165, 1989.

Albert W. Bally, Department of Geology and Geophysics, Rice University, P.O. Box 1892, Houston, Texas 77251

Shengyu Wu, TOTAL Exploration Production U.S.A., Inc., 909 Fannin, Suite 2200, Houston, Texas 77210

e-mail: shengyu.wu@total.com

Deep-Water, Salt-Cored Foldbelts: Lessons from the Mississippi Fan and Perdido Foldbelts, Northern Gulf of Mexico

Mark G. Rowan

Rowan Structural Consulting, Boulder, Colorado

Bruce D. Trudgill

Department of Geological Sciences and Energy & Minerals Applied Research Center, University of Colorado, Boulder, Colorado

J. Carl Fiduk

Boulder, Colorado

The Mississippi Fan and Perdido foldbelts are Tertiary deep-water contractional provinces located at the basinward limit of the Jurassic Louann Salt in the northern Gulf of Mexico. Both formed at different times in response to updip sedimentary loading and extension during gravity spreading of the continental margin above the salt layer. The structural styles are characteristic of salt-cored foldbelts, consisting of linear, generally symmetric detachment folds with regular wavelengths and relatively minor, high-angle reverse faults. However, there are significant differences between the two foldbelts: (1) fold wavelengths, amplitudes, and segment lengths are all greater in the Perdido foldbelt; (2) the Mississippi Fan foldbelt is more asymmetric (basinward-vergent); (3) fold geometries are more rounded in the Mississippi Fan foldbelt and more angular in the Perdido foldbelt; (4) the fold envelope is tilted basinward in the Perdido foldbelt but is subhorizontal in the Mississippi Fan foldbelt; (5) domal structures are found in the Perdido foldbelt; and (6) antecedent folds, highly-rotated fault blocks, and deeply-rooted diapirs are present in the Mississippi Fan foldbelt but absent in the Perdido foldbelt. Variations in the thickness and mechanical stratigraphy of the folded multilayers explain the wavelengths and fold geometries. The asymmetry, fold-envelope orientation, and occurrence of domal structures are ascribed to the facies and original thickness distribution of the evaporite layer, which in turn was controlled by the underlying rift-basin geometry. Finally, preexisting salt structures from an early phase of deformation account for the short fold-segment lengths, the rotated fault blocks, and the presence of diapirs in the Mississippi Fan foldbelt.

INTRODUCTION

Salt-cored foldbelts exist in the deep-water provinces of some continental margins around the world, including the northern Gulf of Mexico [*Martin*, 1980; *Wu et al.*, 1990;

Atlantic Rifts and Continental Margins
Geophysical Monograph 115

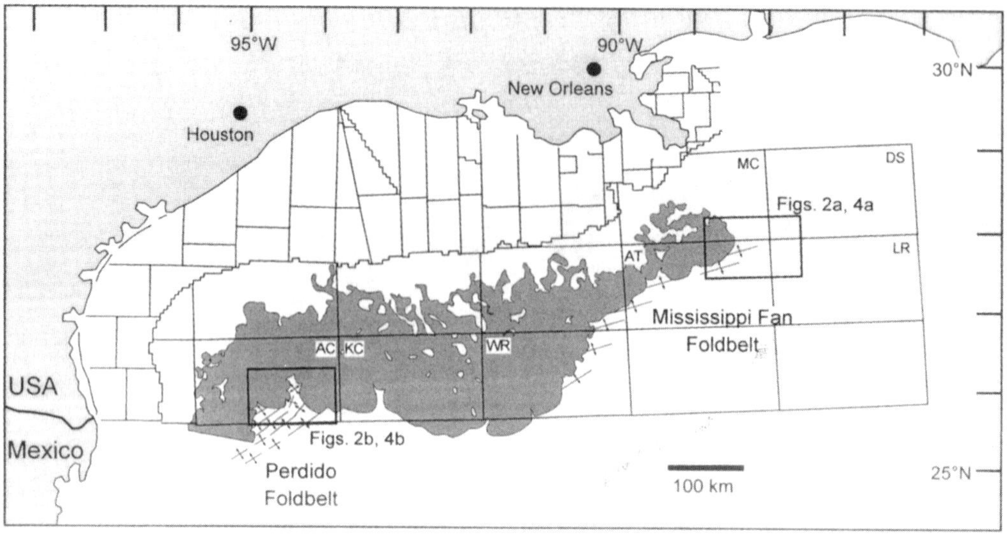

Figure 1. Map of the northern Gulf of Mexico showing the study areas within the Mississippi Fan and Perdido foldbelts just basinward of the allochthonous Sigsbee salt nappe (dark grey). AC - Alaminos Canyon; AT - Atwater Valley; DS - De Soto Canyon; KC - Keathley Canyon; LR - Lloyd Ridge; MC - Mississippi Canyon; WR - Walker Ridge.

Weimer and Buffler, 1992; *Peel et al.*, 1995; *Trudgill et al.*, 1995; 1999; *Rowan*, 1997], offshore Brazil [*Demercian et al.*, 1993; *Cobbold et al.*, 1995; *Mohriak et al.*, 1995], offshore West Africa [*Lundin*, 1992; *Duval et al.*, 1992; *Spathopoulos*, 1996; *Morley and Guerin*, 1996], the Yemeni Red Sea [*Heaton et al.*, 1995], the Eastern Mediterranean [*Sage and Letouzey*, 1990; *Letouzey et al.*, 1995], and possibly the eastern Canadian margin (based on our interpretation of salt geometries illustrated in *Keen and Potter* [1995]). The foldbelts are found at the toe of the slope or at the basinward limit of the original salt basin, and form in response to gravity gliding and/or gravity spreading above the salt detachment. They are characterized by linear, salt-cored, detachment-fold trains with regular wavelengths. The folds range from symmetric to asymmetric, with vergence either basinward or landward, and are often cut by reverse faults on one or both flanks.

The contractional folds potentially contain giant hydrocarbon accumulations, with possible traps provided by both structural closure and stratigraphic onlapping onto fold limbs. However, the foldbelts are relatively unexplored because of the great water depths. To our knowledge, they have been tested only in the northern Gulf of Mexico. In the Mississippi Fan foldbelt, four structures have been drilled, with two announced discoveries. In the Perdido foldbelt, a well was drilled in 1996 in world-record water depths (~2,320 m or 7,612 ft); it did not reach all the intended targets because of mechanical failure, but producible oil was encountered and exploration of this structure is con-tinuing.

In this paper, we compare and contrast parts of the Mississippi Fan and Perdido foldbelts in the northern Gulf of Mexico (Figure 1). The two foldbelts are distinct: although the regional setting and the stratigraphy are broadly similar, the timing of deformation is different and there are significant variations in structural style [*Weimer and Buffler*, 1992; *Peel et al.*, 1995; *Trudgill et al.*, 1999]. The timing of folding has been attributed to regional shifts of more proximal sedimentary loading and associated extensional deformation during the Tertiary [*Peel et al.*, 1995], and is not pursued here. Instead, we focus on the differences in structural style. We use 2-D seismic data (Figure 2) to illustrate typical map- and profile-view geometries from each foldbelt, and then examine numerous factors in an attempt to explain both the similarities and differences between the two foldbelts. We suggest that the variations in structural style can be attributed primarily to three factors: (1) the thickness and mechanical stratigraphy of the folded strata; (2) the original thickness and facies of the dé-collement layer; and (3) the presence or absence of an ear-lier phase of deformation. The relative lack of concrete data means that many of the conclusions are speculative. Never-theless, the results should aid in the understanding of the structure and evolution of the deep-water northern Gulf of Mexico, and hopefully will also help in the interpretation and understanding of deep-water, salt-cored foldbelts along other continental margins.

The analysis presented here is based in large part on our previous investigations of both the Mississippi Fan foldbelt [*Rowan et al.*, 1993; *Rowan*, 1997] and the Perdido foldbelt [*Trudgill et al.*, 1995; 1999; *Fiduk et al.*, 1997; 1999]. The interested reader is referred to these publications for more detailed explanations of the structure and stratigraphy of the two areas.

COMPARISON

Regional Setting

Both foldbelts are interpreted to occur just landward of the basinward depositional limit of the autochthonous Louann Salt [*Fiduk et al.*, 1995] and are believed to represent part of the down-dip, contractional component of linked systems of gravity spreading [*Blickwede and Queffelec*, 1988; *Worrall and Snelson*, 1989; *Apps et al.*, 1994; *Peel et al.*, 1995; *Diegel et al.*, 1995; *Trudgill et al.*, 1999]. Early models suggested that the margin of the original salt basin occurred near the boundary between thinned continental crust and oceanic crust [*Buffler*, 1989; *Salvador*, 1991a; 1991b; *Sawyer et al.*, 1991; *Buffler and Thomas*, 1994]. More recent models have moved the oceanic-continental boundary to the north, close to the present shelf edge [*Marton and Buffler*, 1994; *Hall*, 1995]. In the latter interpretation, much of the Louann Salt was deposited over oceanic crust and the deep-water foldbelts are underlain by oceanic crust.

We will suggest below that some of the differences in structural style between the two foldbelts are related to thickness variations in the Louann Salt, which in turn are related to the geometry of underlying basement. Northeast-southwest trending half-grabens and northwest-southeast trending transverse features have been identified beneath the décollement layer in the Perdido foldbelt [*Trudgill et al.*, 1999], and we have identified similar features on proprietary seismic data from the Mississippi Fan foldbelt. Although it is tempting to cite these as evidence for rifted continental crust, rotated fault blocks can also occur in oceanic crust [*Keen et al.*, 1991; *Keen and Potter*, 1995]. Thus, we consider the nature of crust underlying the Gulf of Mexico foldbelts enigmatic, and it is even possible that the situation is different for the two foldbelts.

Timing

The timing of fold growth in both foldbelts is defined by growth strata, with the ages derived from horizon correlations to biostratigraphic data in wells. In the Mississippi Fan foldbelt, the only well with publicly available biostrat-igraphic data (Shell AV471-1 well, Figure 2a) encountered a crestal unconformity that spanned part of the growth interval, and in the Perdido foldbelt, no biostratigraphic data have been released. Thus, published estimates of the timing of deformation are dependent on

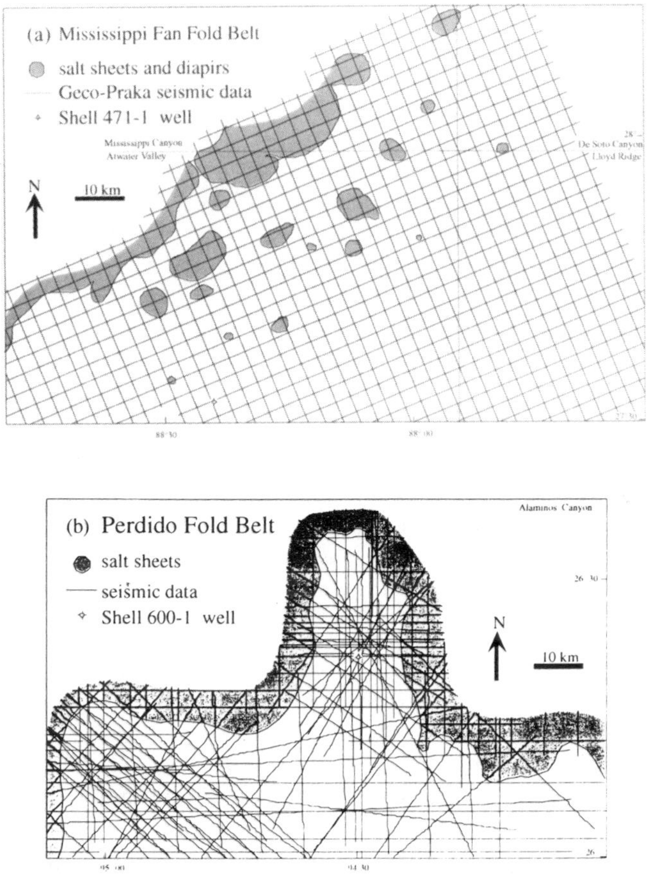

Figure 2. Database maps plotted at the same scale: (a) Mississippi Fan foldbelt - seismic data courtesy of Geco-Prakla; (b) Perdido foldbelt - seismic data courtesy of Amoco, Mobil, Shell, Texaco, and Western Geophysical. Locations shown on Figure 1.

long-range cor-relations and will almost certainly be refined as more data become available.

Folding within the Mississippi Fan foldbelt was dated as middle to upper Miocene by *Wu et al.* [1990], *Feng and Buffler* [1991], *Weimer and Buffler* [1992], and *Wu* [1993]. Subsequently, the time range was enlarged to span the late Oligocene (30 Ma) to the Pleistocene (1.35 Ma), with the main pulse of deformation occurring during the middle to late Miocene (15.5 Ma to 10.5 Ma) [*Rowan et al.*, 1994; *Rowan*, 1997]. In the Perdido foldbelt, *Peel et al.* [1995] suggested that folding occurred during the latest Oligocene to middle Miocene, whereas *Trudgill et al.* [1999] invoked a main phase during the early Oligocene (with possible continuation into the early Miocene) and a minor phase of reactivation of some folds during the middle Miocene.

Regardless of the exact ages of deformation, it is clear that the Perdido foldbelt formed before the Mississippi Fan foldbelt, although there may be some overlap in timing.

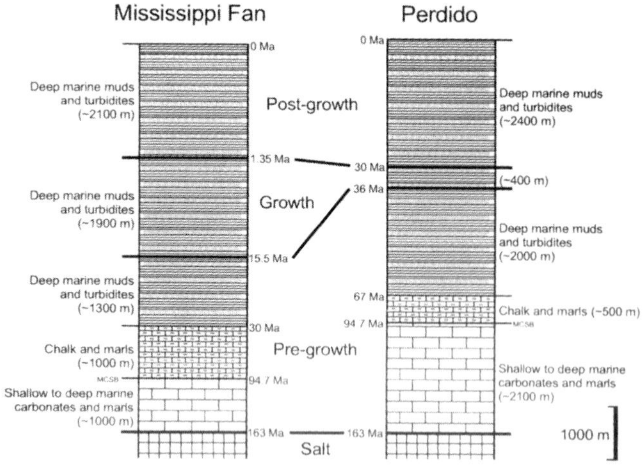

Figure 3. Simplified stratigraphic columns for the two foldbelts, plotted at the same depth scale (time scale is variable), showing the division into pre-growth, growth, and post-growth sections. Ages and lithologies taken from *Winker* [1982], *Shaub et al.* [1984], *Winker and Buffler* [1988], *Galloway* [1989], *Galloway et al.* [1991], *Feng and Buffler* [1991], *Feng* [1995]. Thicknesses are from depth conversions of seismic data presented in *Rowan* [1997] and *Trudgill et al.* [1999].

Wu et al. [1990] and *Peel et al.* [1995] suggested that folding was related to episodes of extensional faulting farther landward, which in turn were caused by sedimentary load-ing. Regional shifts in the location of Tertiary depocenters are compatible with the younging of folding to the east: the major deltas were located in south Texas, updip from the Perdido foldbelt, from the Paleocene through the Oligocene, but shifted eastward to the Louisiana shelf, updip from the Mississippi Fan foldbelt, during the Miocene [e.g., *Winker*, 1982; *Galloway*, 1989; *Worrall and Snelson*, 1989]. However, much updip loading and extension occurred prior to folding in both areas, and it is likely that some of the downslope translation of the suprasalt overburden was ac-commodated by the extrusion of allochthonous salt [*Peel et al.*, 1995; *Trudgill et al.*, 1999].

Stratigraphy

The post-salt stratigraphic history of the northern Gulf of Mexico is broadly similar in both the Texas and Lou-isiana offshore [e.g., *Winker*, 1982; *Shaub et al.*, 1984; *Winker and Buffler*, 1988; *Galloway*, 1989; *Galloway et al.*, 1991; *Feng and Buffler*, 1991; *Feng*, 1995]. Thus, the vertical stacking of lithologies is interpreted to be the same in both foldbelts, although there are differences in thick-nesses and ages (Figure 3). Above the Louann Salt is a seq-uence of carbonates (possibly shallow-marine) that changes upward into deep-marine carbonates, marls, and shales. The top of this unit is the Mid-Cretaceous Sequence Boundary (MCSB) [*Buffler*, 1991], dated at 94.7 Ma [*Feng*, 1995], which can be correlated throughout both foldbelts. Above the MCSB is a sequence of chalk and marls representing a deep-water, carbonate-dominated depositional environment that persisted in each area until the first major influx of shelf-derived Tertiary clastics. This influx occurred near the end of the Cretaceous in the Perdido foldbelt, but not until the Oligocene in the Mississippi Fan foldbelt because of the regional shift of delta systems mentioned above. The clastic section is composed of deep-marine muds and distal turb-idites. In the Perdido foldbelt, the main periods of turbidite deposition were during the late Paleocene to early Eocene, the early Oligocene, the middle Oligocene, and the early Miocene [*Fiduk et al.*, 1999]; in the Mississippi Fan fold-belt, they probably occurred during the early, middle, and late Miocene, the early Pliocene, and the Quaternary.

Although there is an overall similarity, the mechanical stratigraphy of the prekinematic section is thought to be dif-ferent in the two foldbelts based on regional correlations (Figure 3). In the Mississippi Fan foldbelt, the pre-growth section is about 3,300 m (11,000 ft) thick and consists of approximately equal proportions of the carbonate, chalk, and siliciclastic sequence. In contrast, the Perdido foldbelt pre-growth section is about 4,600 m (15,000 ft) thick, with a thin chalk sequence between equal thicknesses of the carbonates and siliciclastics. In both cases, the details of the thickness and proportion of interbedded competent and incompetent layers are unknown.

Map-View Geometry

Both the Mississippi Fan and Perdido foldbelts comprise elongate, usually linear folds with fairly consistent length and spacing (Figure 4), although there are significant differences between the two. Anticlines in the Mississippi Fan foldbelt (Figure 4a) are generally smaller and more closely spaced, with segment lengths of 5-15 km, wave-lengths of 5-10 km, and maximum amplitudes of approx-imately 2 km (Table 1). In contrast, anticlines in the Perdido foldbelt (Figure 4b) have segment lengths of 15-40 km, wavelengths of 10-15 km, and maximum amplitudes of about 2.5 km (Table 1). Most fold segments in both areas are

Table 1. Fold Dimensions

Foldbelt	Wavelength	Segment Length	Maximum Amplitude
MFFB	5-10 km	5-15 km	2 km
PFB	10-15 km	15-40 km	2.5 km

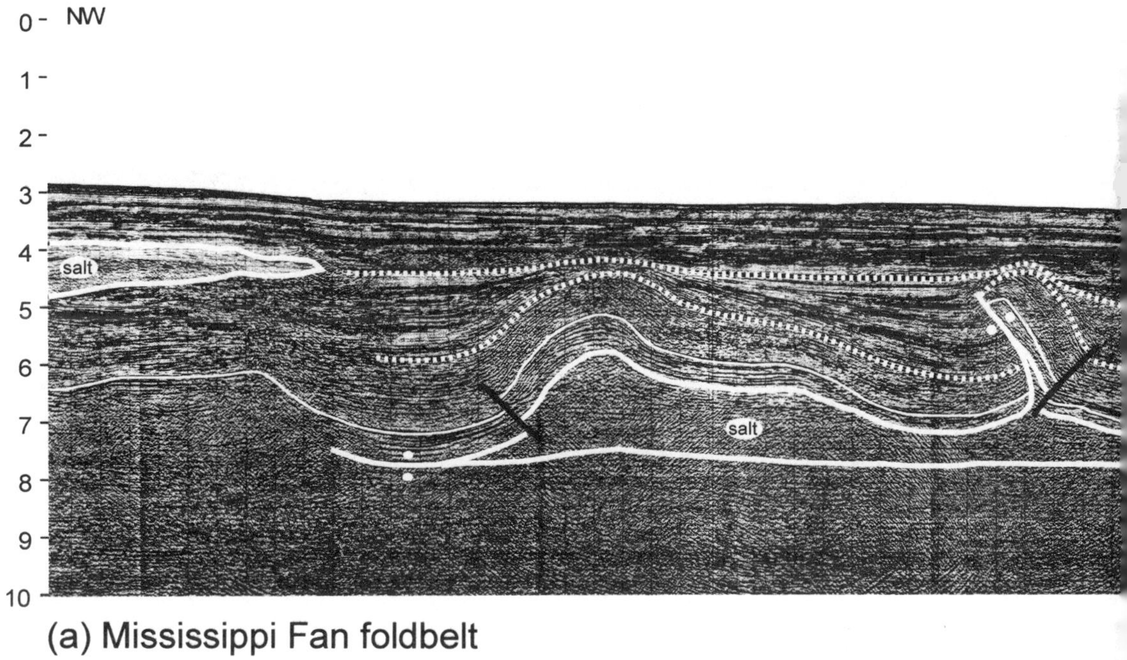

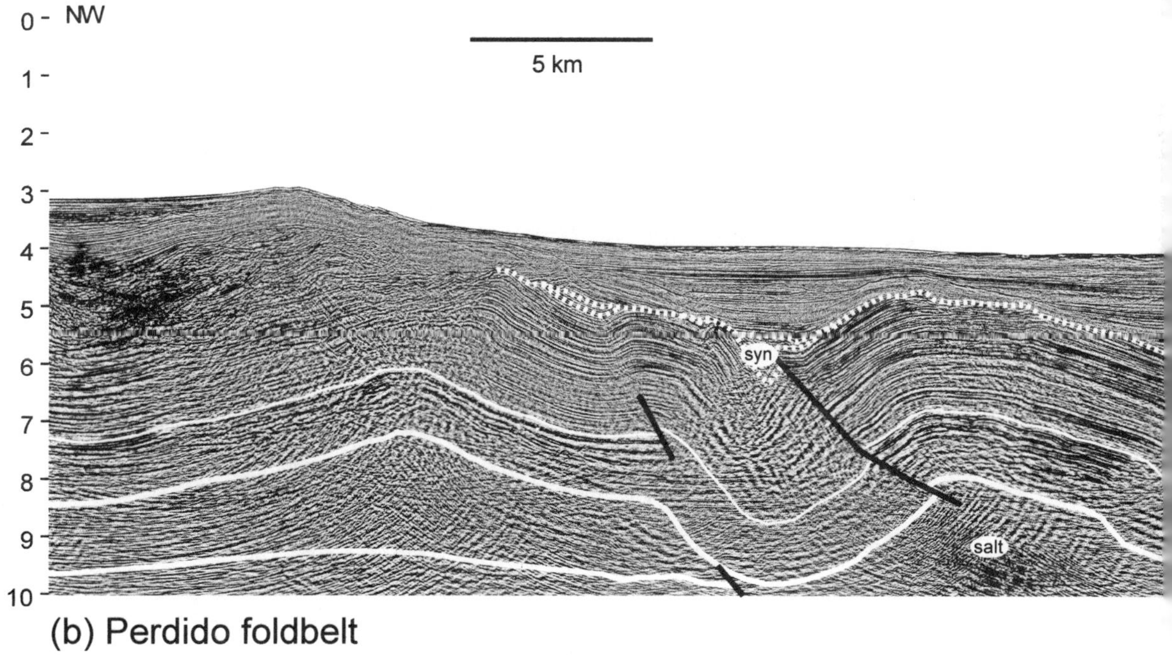

Plate 2. Interpreted versions of the seismic profiles illustrated in Figure 5: (a) Mississippi Fan foldbelt; (b) Perdido fo Horizons shown are the Mid-Cretaceous Sequence Boundary (MCSB) and those that bound the major tectonostratig units. Thinning and thickening of the pre-growth section in the northwestern two-thirds of profile (a) is due to diapir-salt withdrawal and inflation. See text for further discussion of profile geometries. Locations shown in Figure 4.

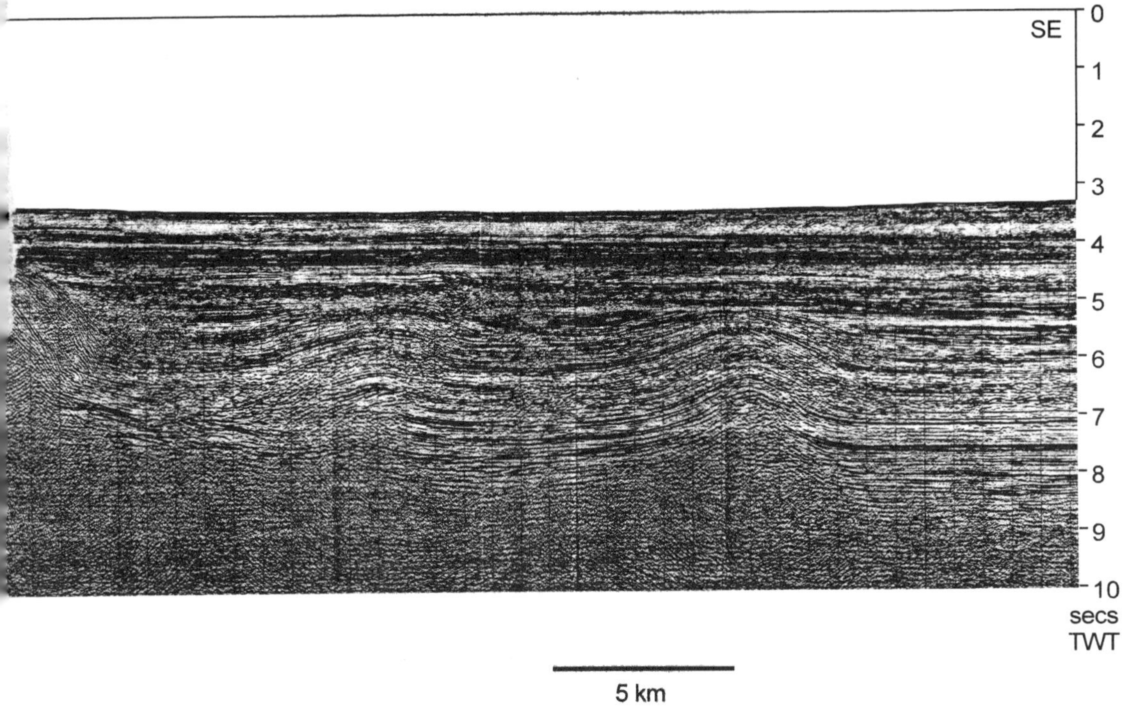

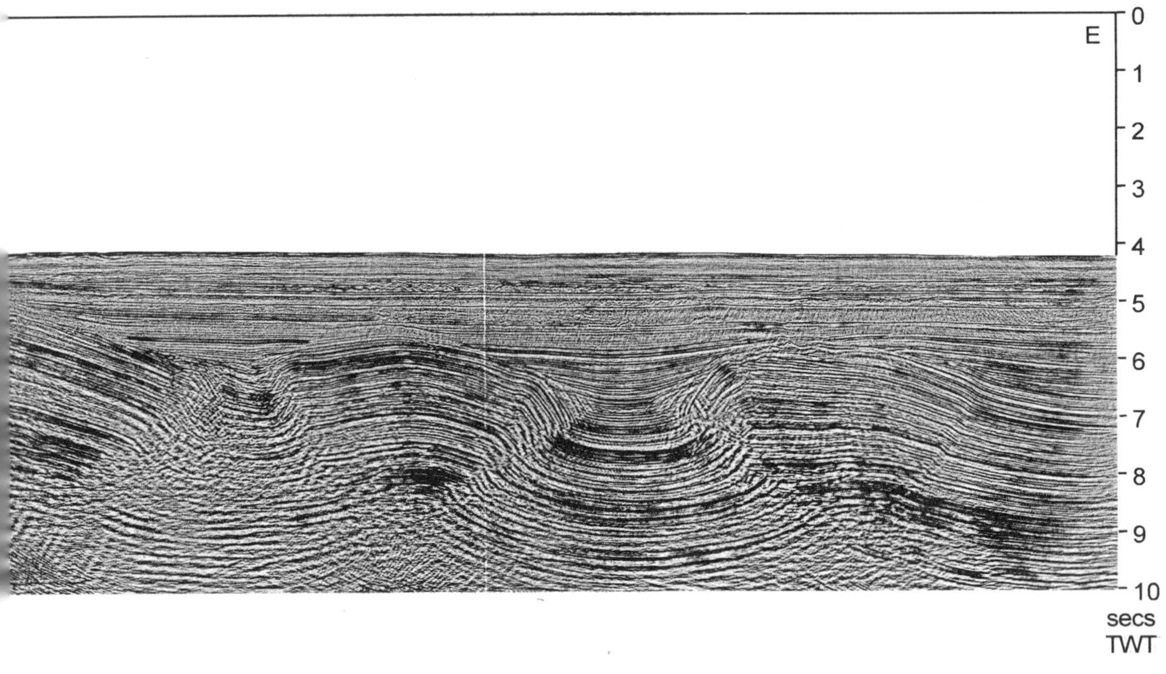

i Fan foldbelt, cour-

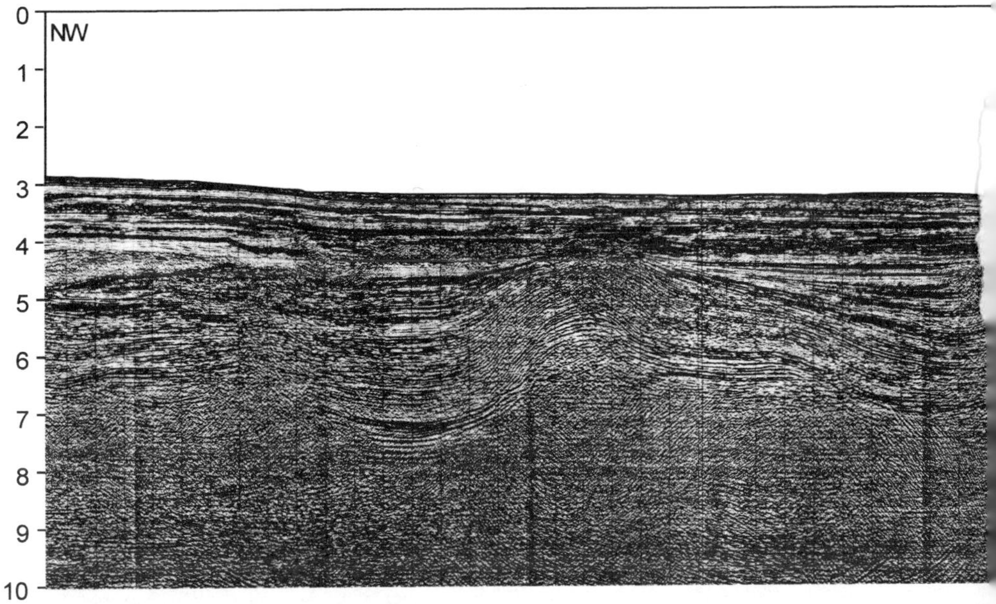

(a) Mississippi Fan foldbelt

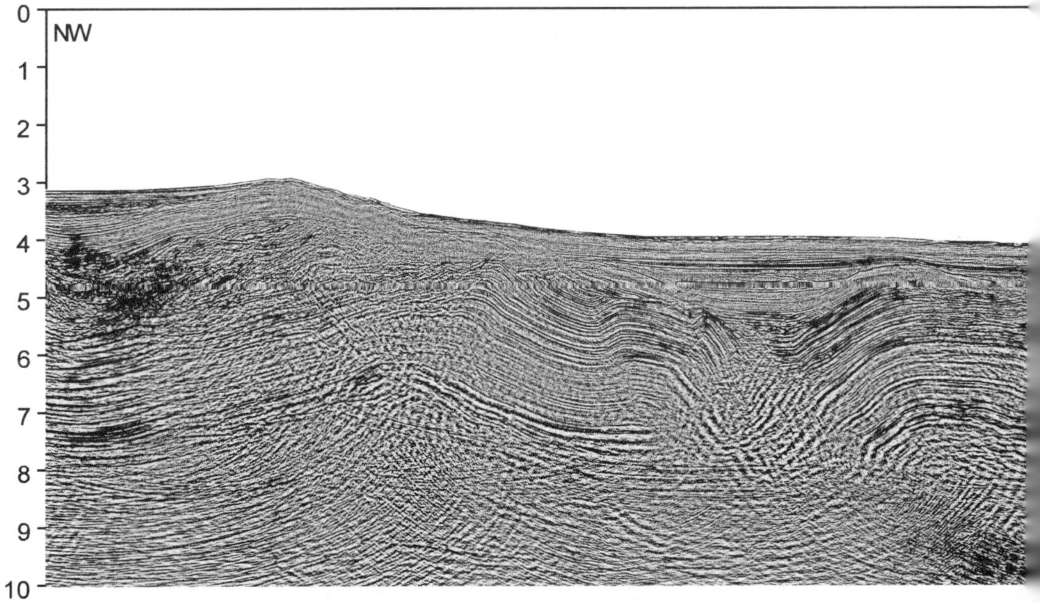

(b) Perdido foldbelt

Plate 1. Regional, uninterpreted, time-migrated seismic profiles plotted at the same scale: (a) Mississip
tesy of Geco-Prakla; (b) Perdido foldbelt, courtesy of Mobil. Locations shown in Figure 4.

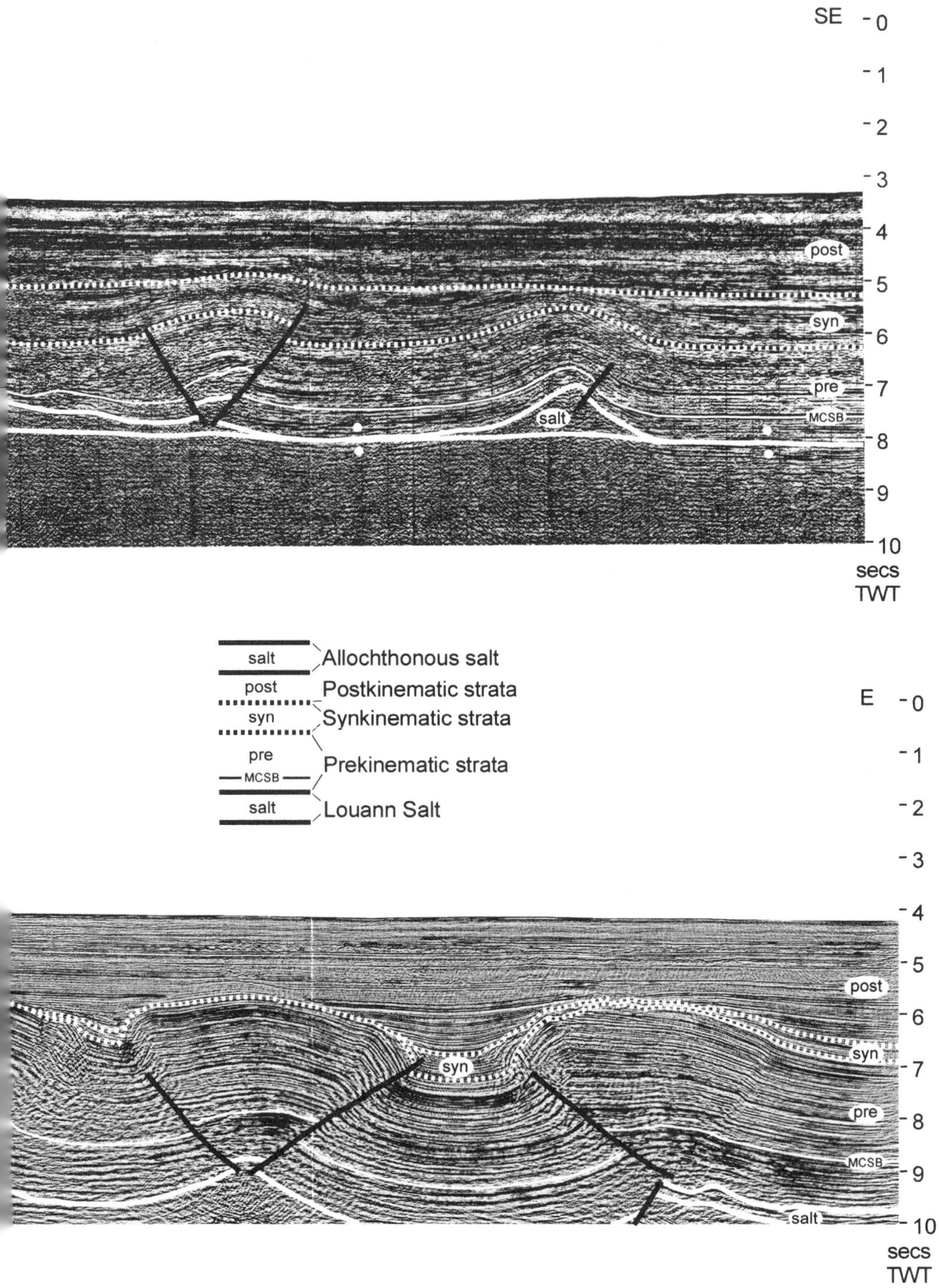

dbelt.
raphic
elated

isolated, although there are instances of apparent along-strike linkage of separate segments (indicated by curves or kinks in fold-trace geometry) in both foldbelts. The Perdido foldbelt is anomalous in that it also contains two domal, rather than elongate, anticlines (Figure 4b).

Profile-View Geometry

Two time-migrated seismic profiles at the same scale illustrate the characteristic fold geometries of each foldbelt (Plates 1 and 2). Folds are symmetric or asymmetric, with both basinward and landward vergence (Figure 4). The proportion of symmetric folds is similar for both areas (60% in the Mississippi Fan foldbelt versus 64% in the Perdido foldbelt, ignoring the domal structures). However, the asymmetric folds in the Perdido foldbelt are equally divided between basinward- and landward-vergent anticlines, whereas basinward-vergent folds comprise 77% of the asymmetric structures in the Mississippi Fan foldbelt. Thus, the Mississippi Fan foldbelt is slightly more asymmetric and basinward-vergent than the Perdido foldbelt [*Weimer and Buffler*, 1992; *Trudgill et al.*, 1999].

Anticlines in the Mississippi Fan foldbelt (Plates 1a and 2a) are generally more rounded than those in the Perdido foldbelt (Plates 1b and 2b), which can have distinct kink bands on fold limbs. Minor kink bands sometimes intersect such that anticlines can overlie synclines and vice versa [*Mount*, 1989; *Mount et al.*, 1990]. Synclines in the Mississippi Fan foldbelt are usually wide and flat-bottomed, but those in the Perdido foldbelt are often tighter, with limbs from adjacent anticlines sometimes interfering with each other. This occurs, even though the anticlines have a greater wavelength in the Perdido foldbelt (Table 1), because of the greater thickness of the folded section (Figure 3) and the greater amplitude of the Perdido folds (Table 1). One result of the tight synclines in the Perdido foldbelt is the occurrence of syncline-flank pop-up struc-tures rooted in out-of-the-syncline reverse faults (Figure 5).

Many anticlines are cut on one or both limbs by high-angle reverse faults (Plates 1 and 2). The faults tend to be slightly more asymmetric (basinward vergent) in the Mississippi Fan foldbelt, in keeping with the associated fold geometries. In the Perdido foldbelt, some of the faults cut the entire prekinematic (and synkinematic?) section, where-as others lose displacement upward into unfaulted kink bands (Plate 2b). The same may or may not be true for the Mississippi Fan foldbelt, where faulted limbs usually are not imaged as well. Fault footwalls in both foldbelts are usually folded, indicating that the folds formed first and were later cut by the faults.

An anomalous type of structure occurs occasionally in

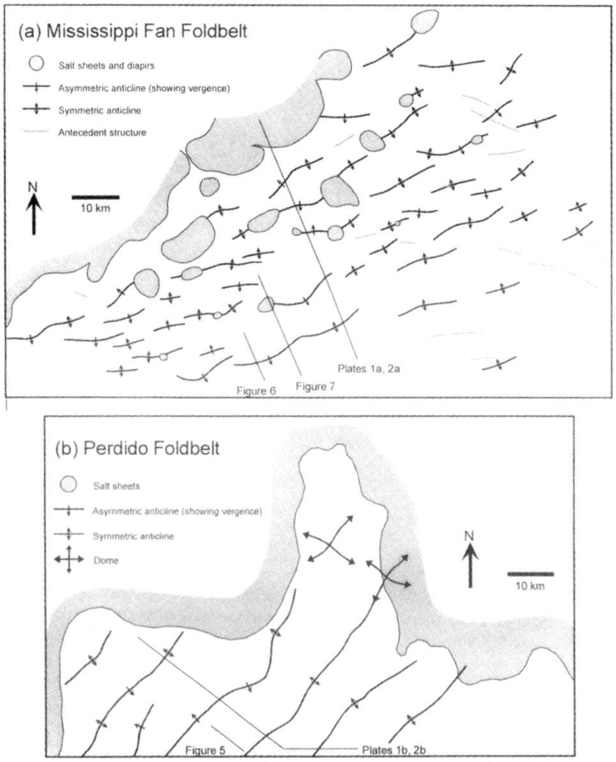

Figure 4. Structure maps plotted at the same scale: (a) Mississippi Fan foldbelt; (b) Perdido foldbelt. The maps show the outlines of allochthonous salt sheets and diapirs, the crests of symmetric and asymmetric anticlines, the orientation of antecedent folds in the Mississippi Fan foldbelt (a), and the location of illustrated seismic profiles. Note the smaller wavelengths and segment lengths of folds in the Mississippi Fan foldbelt compared to the Perdido foldbelt. Locations shown on Figure 1.

the Mississippi Fan foldbelt. These are asymmetric folds that have the appearance of highly-rotated fault blocks (cen-ter of Plates 1a and 2a). The dominant fault is actually inter-preted to be a secondary salt weld formed by the horizontal closure of a salt diapir, an interpretation supported by the presence of diapirs on either side of the illustrated fold (Figure 4a). The geometry is similar to that of a rejuvenated diapir, in which shortening affects a preexisting diapir [Vendeville and Nilsen, 1995].

The stratigraphic interval just above the Louann Salt is essentially constant thickness in the Perdido foldbelt (Plates 1b and 2b) but shows significant thickness variations in the Mississippi Fan foldbelt (Plates 1a and 2a) that were caused by Upper Jurassic to Cretaceous deformation [*Rowan*, 1997]. These older structures have strikes that curve from approximately parallel to the major folds in the center of the foldbelt to highly oblique (with WNW strikes) in the eastern part of the foldbelt (Figure 4a). These structures are more common than is shown in Figure

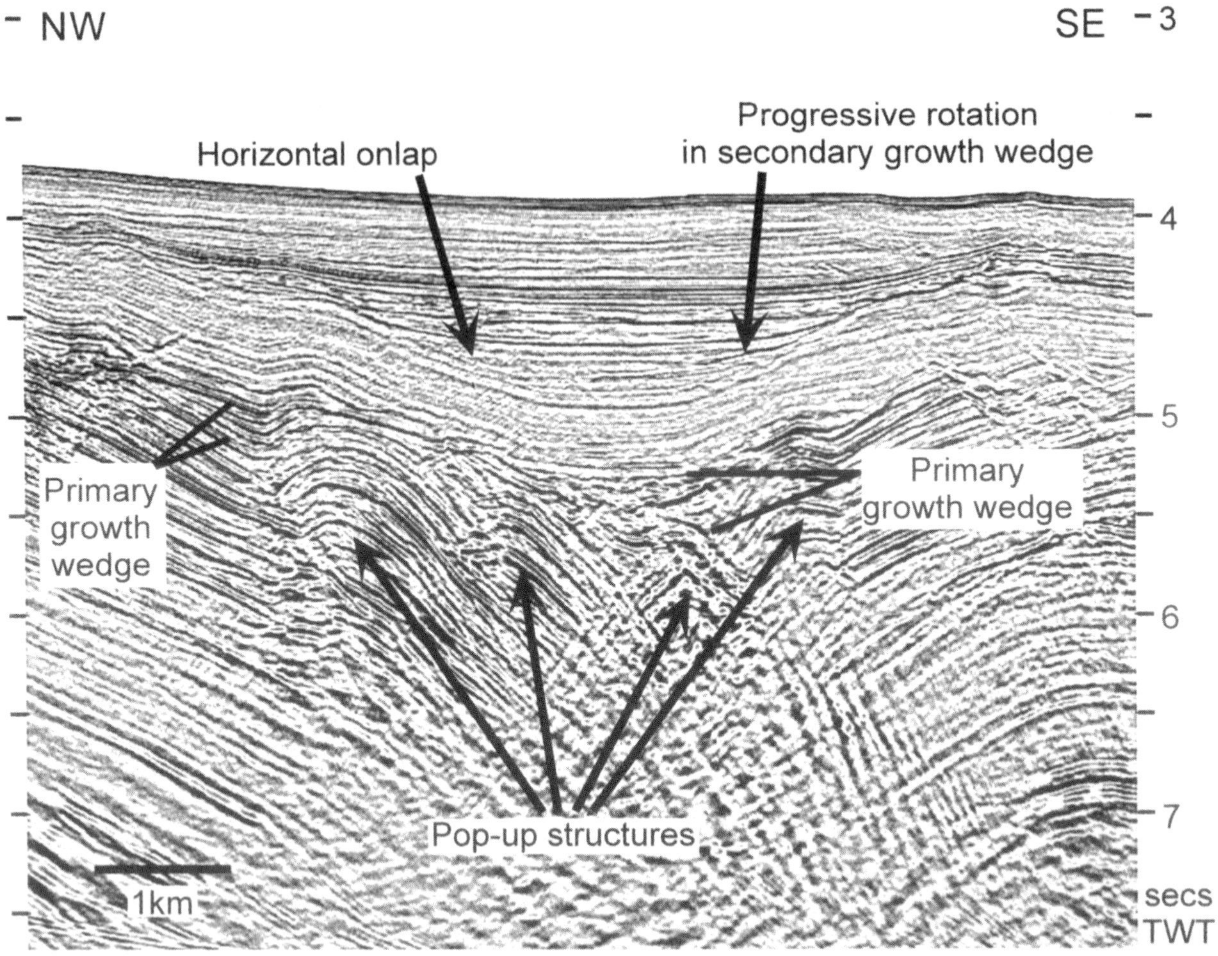

Figure 5. Time-migrated seismic profile crossing a syncline in the Perdido foldbelt. The syncline is poorly imaged, but pop-up structures rooted in out-of-the-syncline reverse faults are visible on both flanks. Stratal geometries indicate a complicated history: (1) small growth wedges record the main growth phase which was followed by erosional trun-cation; (2) low-amplitude, parallel strata mark a period of quiescence; and (3) young onlap and minor progressive rotation indicate a late phase of reactivation, with the backlimb of the southeastern anticline still active after the forelimb of the northwestern anticline had stopped growing. Note that the central pop-up structures were active only during the main growth phase but that the outer ones were active in both the main and late stages of deformation. Seismic data courtesy of Mobil. Profile location shown on Figure 4b.

4a because many of them underlie the larger, mapped folds (e.g., frontal fold in Plate 2a).

Fold-Train Geometry

In the Mississippi Fan foldbelt, the geometry of the fold train is usually regular. Synclines are generally at the same level, and although the elevation of anticline crests can vary because of strike-parallel plunge, the regional fold envelope is subhorizontal. This is not apparent in the fold train illus-trated in Plates 1a and 2a, which is misleading because the frontal folds are traversed near their terminations and are thus lower and the landward synclines are elevated because of abundant nearby salt (Figure 4a). The fold envelope in the Perdido foldbelt, in contrast, genuinely dips basinward, such that more landward synclines are elevated relative to more frontal folds (Plates 1b and 2b; seen even better in profiles illustrated in *Trudgill et al.* [1999]). This dipping envelope is not due to folding of a landward-thickening wedge of prekinematic strata because this interval has roughly constant thickness [*Fiduk et al.*, 1999] (Plate 2b). Instead, depth-converted profiles show that the more land-ward portion of the fold train is elevated with respect to the basinal regional [*Trudgill et al.*, 1999], and a decent base-salt reflection indicates that the extra material beneath the folds is thickened autochthonous salt.

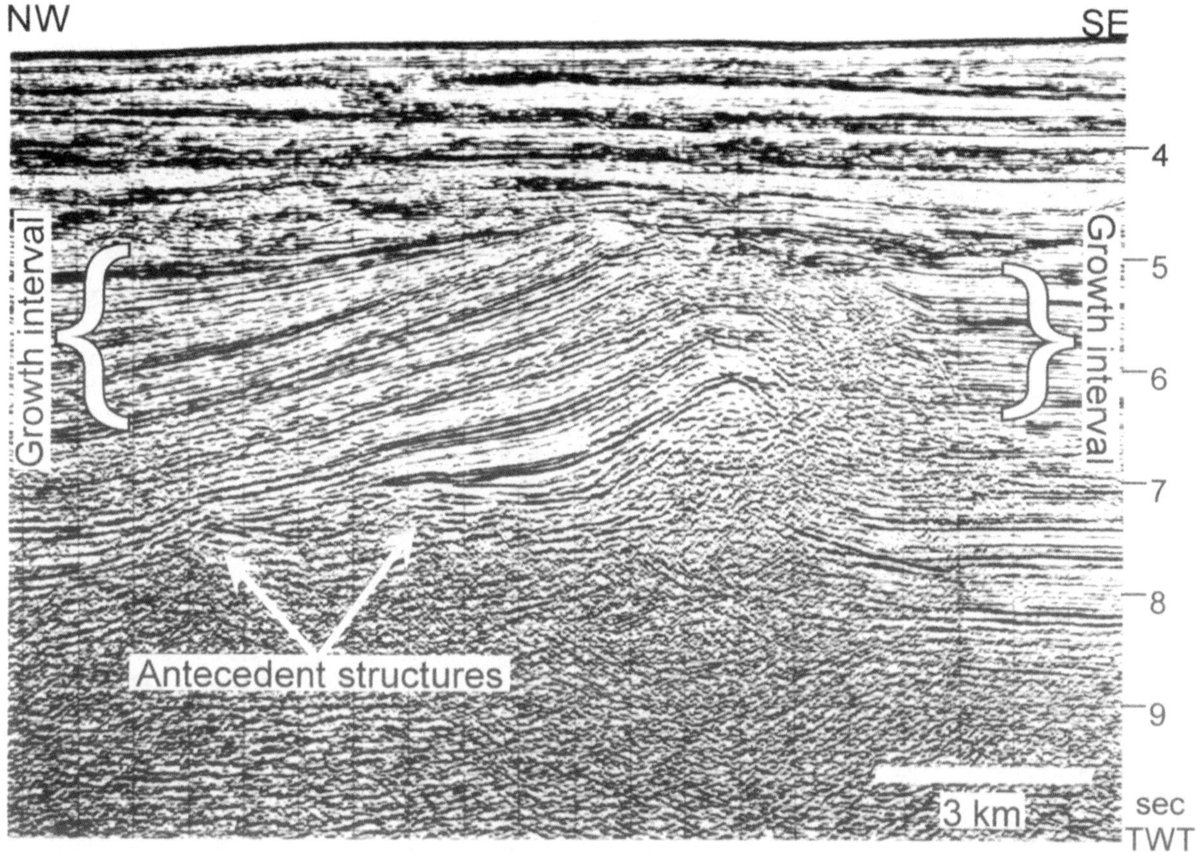

Figure 6. Time-migrated seismic profile through a frontal fold in the Mississippi Fan foldbelt showing antecedent (Upper Jurassic to Cretaceous) structures on the backlimb and the main (Mio-Pliocene) growth interval on both limbs. The youngest growth strata are thinned and progressively rotated on the backlimb but mostly onlap the forelimb [see *Rowan*, 1997]. Seismic data courtesy of Geco-Prakla. Profile location shown on Figure 4a; the fold is the same as the frontal fold on Plates 1a and 2a and Figure 7.

In addition, the syn-cline between the two frontal folds in Plates 1b and 2b is below the basinal regional, implying subsidence into orig-inally thicker salt.

Growth Strata

Folds within the Mississippi Fan foldbelt display thick (up to 1.9 km; Figure 3) synkinematic intervals on both limbs. Strata are progressively rotated (with less rotation higher in the section), thin toward the anticlines, and are often eroded over the fold crests (Plates 1a and 2a, Figure 6). In detail, the growth geometries differ on the two fold limbs: whereas all backlimb growth strata are progressively rotated and thinned, the youngest coeval units on the forelimb are constant thickness and simply onlap the forelimb (Figure 6). This geometry has been attributed to a late stage of deformation in which the forelimb locks up while the backlimb continues to be folded and uplifted [*Rowan et al.*, 1993].

In contrast, growth strata appear to be much thinner in the Perdido foldbelt, attaining a maximum of only about 500 m (Figure 3). These strata are mostly preserved within the tight synclines (where they are generally poorly imaged) between the prekinematic strata and a parallel-bedded, postkinematic interval that was deformed during a minor phase of fold reactivation (Figure 5). The younger defor-mation is marked by progressively rotated growth strata on backlimbs and subhorizontal onlap onto forelimbs (Figure 5).

The geometry of syn- and post-growth strata suggest that the two foldbelts had different bathymetric expression (Plates 1 and 2). In the Mississippi Fan foldbelt, sedimentation appears to have mostly kept up with uplift. The crests of many folds were eroded, but this erosion was accompanied by sedimentation in the synclines and on the fold flanks so that there was relatively minor bathymetric relief. In contrast, sedimentation did not keep pace with uplift in the Perdido foldbelt. There, folds remained as

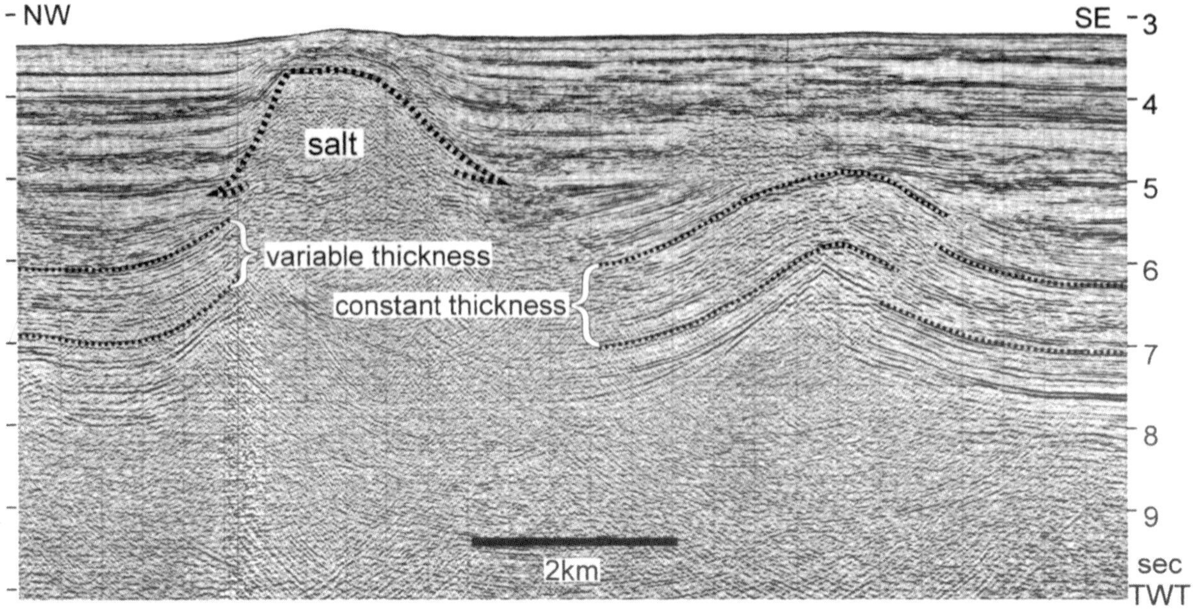

Figure 7. Time-migrated seismic profile through a frontal fold and a diapir in the Mississippi Fan foldbelt. The structure beneath the diapir crest is poorly imaged, but is along strike with another fold (Figure 4a). The indicated part of the pre-growth section is constant thickness over folds with no associated diapirs (e.g., frontal fold) but has variable thickness adjacent to diapirs, indicating long-lived diapir growth (see also Plate 2a). Seismic data courtesy of Geco-Prakla. Profile location shown on Figure 4a; the fold is the same as the frontal fold on Plates 1a and 2a and Figure 6.

large bathymetric highs for an extended period after deformation. The crests were eroded and post-kinematic strata only gradually filled in the bathymetric relief. The contrast between the two foldbelts is exemplified by the different geometries of the boundary between the growth and post-growth sections (Plates 1a and 1b).

Diapiric Salt

Both foldbelts are imaged just basinward of the allochthonous Sigsbee salt nappe and both continue beneath the allochthonous salt for unknown distances. The Perdido fold-belt contains no salt diapirs or allochthonous salt outboard of the Sigsbee Nappe, although diapirs have been reported from the southwestern continuation of the foldbelt in Mex-ican waters (J. Fox, personal communication, 1995). In con-trast, shallow diapirs are common in the Mississippi Fan foldbelt (Figure 4a). These diapirs are not part of the Sigs-bee Nappe, but are instead subvertical stocks rooted in the autochthonous Louann Salt that have spread radially to form outwardly flared salt flanks (Figure 7). There is a direct cor-relation between diapirs and folds, with the diapirs always located along anticline crests (Figure 4a). Strata around the diapirs show thinning and thickening throughout the (fold-ing) pre-growth section, in contrast to the constant thickness of part of this interval in folds where no diapirs are present (Figure 7).

Evolution

Two-dimensional structural restorations show that the two foldbelts have different histories. In the Mississippi Fan foldbelt, five stages of deformation can be defined [*Rowan et al.*, 1993; *Rowan*, 1997]: (1) an initial stage of gravity gliding during the Late Jurassic to Cretaceous that created small contractional salt pillows; (2) a period of quiescence that lasted until the middle Oligocene; (3) an early stage of slow buckle folding from 30 Ma to 15.5 Ma; (4) the main phase of more rapid shortening from 15.5 to 10.5 Ma, in which the folds grew and were cut by reverse faults; and (5) a final stage of slow fold amplification, ending around 1.3 Ma, in which the forelimbs locked up while the backlimbs were uplifted and rotated.

In contrast, there was no early (Mesozoic) phase of deformation in the Perdido foldbelt, and the Tertiary shortening can be divided into two stages [*Trudgill et al.*, 1999]: (1) the main phase of folding in the early Oligocene; and (2) a minor phase of reactivation between 15.5 Ma and 10.5 Ma that affected only some folds (Figure 5). In addition, there was a late stage (post-5.5 Ma) of local inflation of the Louann Salt that is not attributed to shortening but to load-ing effects of the allochthonous Sigsbee Nappe [*Trudgill et al.*, 1999].

Shortening

Structural restorations also quantify the amount and rates deformation. In the Mississippi Fan foldbelt, one fold accommodated 800 m of shortening, with a maximum shortening rate of 10 cm/ka [*Rowan et al.*, 1993]. Two-dimensional cuts through other folds show shortening values between 30 m and 1340 m (Table 2). In the Perdido foldbelt, four folds accommodated about 4.8 km of shortening [*Trudgill et al.*, 1999], an average of 1200 m and 20 cm/ka per fold. Measurements on individual folds yield shortening values between 290 m and 1270 m (Table 2). In comparing the rate calculations, it must be remembered that they are based on the poorly constrained age correlations of the growth strata.

ANALYSIS

Many aspects of the two foldbelts are similar, primarily because of the analogous regional settings and the presence of salt. The deformation in each case is the basinward response to updip loading and gravity-driven extensional collapse of a passive margin section underlain by salt. The amount of shortening does not balance the amount of extension (even if folds hidden beneath the Sigsbee Nappe are included), implying that much of the downslope translation was accommodated by salt and the extrusion of alloch-thonous sheets and nappes [*Peel et al.*, 1995; *Trudgill et al.*, 1999]. The folds themselves are typical of salt-cored detachment folds, with an overall symmetry, a regular wavelength, and a prevalence of folding over faulting, in contrast to the asymmetric, thrust-dominated structural styles of foldbelts without salt detachments [*Davis and Engelder*, 1985].

Despite the general similarity, there are significant differences in geometry between the two foldbelts. These are primarily in the size of the folds (amplitude, wavelength, and segment length), the asymmetry, the shape of the folds, the attitude of the fold envelope, and the presence or absence of salt diapirs. These variations in structural style cannot be explained by differences in the regional setting or the lithologies of the folded layers. The apparent increased amount of shortening in the Perdido foldbelt could explain in part the greater fold amplitudes and segment lengths, but not other aspects. Variations in shortening rate or syn-kinematic sedimentation rate may play a role, but it is uncertain just exactly what the effects would be. There-

Table 2. Shortening and Depth-to-Detachment Calculations

Fold	l_1	l_0	s	A	d
Mississippi Fan					
A1	5.68	6.39	0.71	3.70	5.21
A2	9.75	9.86	0.11	2.64	24.00
B1	9.31	9.37	0.06	2.56	42.70
B2	5.86	5.90	0.04	1.03	25.75
C1	15.24	15.48	0.24	9.09	37.88
D1	8.16	8.42	0.26	3.77	14.50
D3	6.88	7.16	0.28	4.00	14.29
E1	12.18	12.23	0.05	3.96	79.20
F1	8.38	8.41	0.03	1.99	66.30
F2	7.45	7.52	0.07	3.11	44.40
C	19.14	20.48	1.34	14.70	10.97
E	17.47	18.39	0.92	11.21	12.18
G	17.44	18.26	0.82	16.52	20.15
J	13.30	13.55	0.25	6.69	26.76
L	14.06	14.18	0.12	6.91	57.58
Perdido					
A1	4.93	5.34	0.41	1.28	3.12
A1A	5.43	5.72	0.29	1.47	5.07
A2	6.00	6.86	0.86	4.34	5.05
A3	15.14	15.49	0.35	9.17	26.20
B1	8.84	9.61	0.77	5.84	7.58
B2	8.18	9.21	1.03	7.09	6.88
B3	10.46	11.73	1.27	10.52	8.28
B4	13.81	14.17	0.36	9.77	27.14
B5	9.41	10.65	1.24	7.52	6.06

l_1 - horizontal distance between synclines; l_0 - bed length between synclines (MCSB horizon); s - shortening (=l_0-l_1); A - measured area between synclines; d - depth-to-detachment (=A/s). All values in km or km^2 (A).

fore, we focus here on three factors that are likely to have impacted the development of the foldbelts: (1) the thickness and mechanical stratigraphy of the folded multilayer; (2) the areal distribution, thickness changes, and facies variations of the décollement layer; and (3) the presence and nature of an early deformation event.

Mechanical Stratigraphy

The geometry of buckle folds is largely dependent on the mechanical stratigraphy of the layers being compressed. Buckle folds generally form when a more competent multilayer is bounded on both sides by less competent material of adequate thickness. In the case of the Gulf of Mexico fold-belts, the lower bounding incompetent layer is salt and the upper is water.

In single-layer buckle folding, the fold wavelength is a function of the layer thickness and the viscosity contrast [*Ramberg*, 1960; *Biot*, 1961]. Although other factors enter into the equation in the case of a multilayer, there is still a direct correlation between multilayer thickness and fold wavelength. This relationship is apparent in the two foldbelts: in the Mississippi Fan foldbelt, the prekinematic mul-tilayer is approximately 3300 m thick (Figure 3) and the fold wavelength is 5-10 km (Table 1); and in the Perdido foldbelt, the multilayer is about 4600 m thick (Figure 3) and the wavelength is 10-15 km (Table 1). Thus, the observed wavelengths are interpreted to be primarily a function of the thickness of the prekinematic strata. In addition, we have already stated that the greater fold amplitudes in the Perdido foldbelt may be due to the increased net shortening. The variation in segment length (Table 1), however, cannot be explained by either of these factors. Although increased shortening causes segments to link into longer folds, we have attempted to identify original segments in tabulating the data. Segment length will be addressed in more detail below.

In a buckled multilayer, some of the main controls on fold geometry are: the rheology of the layers, the thickness of each layer or packet of layers, and the mechanical properties of the layer interfaces [*Ramsay and Huber*, 1987]. Al-though this information is unknown in the two foldbelts, we infer that such factors probably explain the more rounded fold geometry in the Mississippi Fan foldbelt and the more kink-band fold geometry in the Perdido foldbelt. Kink-like geometries, with sharper hinges and more planar limbs, develop when the proportion of incompetent material within the folded multilayer is small and the competency contrast is large [*Ramsay and Huber*, 1987]. Thus, shale or marl units within the Perdido foldbelt may be thinner than in the Mississippi Fan foldbelt. Alternatively, fold geometry is often controlled by dominant units within a multilayer. The chalk layer in the Mississippi Fan foldbelt is thicker (1000 m versus 500 m) and represents a third of the total thickness rather than only a tenth in the Perdido foldbelt. If it is a massive unit with little mechanical anisotropy, it would have formed rounded folds with broad hinges that may have controlled the geometry of the entire multilayer.

Décollement Layer

Both foldbelts are interpreted to be underlain by autochthonous Louann Salt. Salt is clearly indicated in the Mississippi Fan foldbelt by the presence of deeply rooted diapirs (Figures 4a and 6), but is more questionable in the Perdido foldbelt. Nevertheless, there are numerous arguments that support a salt interpretation for the Perdido foldbelt: the stratigraphic level of the décollement layer; the geometry of the layer and overlying strata; and the reported presence of salt to the southwest in Mexican waters [*Fiduk et al.*, 1995; *Trudgill et al.*, 1999]. Seismic data also suggest that the frontal fold in each foldbelt occurs just landward of the basinward pinchout of salt [*Fiduk et al.*, 1995].

When a highly efficient lubricating layer (e.g., salt) is present, symmetric detachment folds with little faulting tend to develop. In contrast, thrust-dominated, asymmetric struc-tures form when the décollement layer is relatively com-petent (e.g., shale). These are end-member scenarios; a shale layer may be weaker because of high pore pressures, and a salt layer may be stronger because of interbedded shales, carbonates, and other evaporites. Ultimately, it is the rheology of the décollement layer, as well as its thickness and thickness distribution, that influence fold geometry. We examine each of these factors below.

Rheology. The internal stratigraphy and the associated rheologies of the décollement layer are unknown in both foldbelts. However, there may be differences between the two areas that partially explain the observed fold geometries. If the layer consists primarily of salt, then folds will be more symmetric, as is the case in the Perdido foldbelt. However, both foldbelts are located near the original depositional limit of salt [*Fiduk et al.*, 1995], where other lithologies may be intercalated and lateral facies variations are more likely. We speculate that an increase in such lithologic heterogeneity might explain the more asymmetric nature of the folds in the Mississippi Fan foldbelt.

Thickness. If the décollement layer is very thin and efficient, synclines are flat and anticlines are essentially isoclinal with vertical limbs [*Dahlstrom*, 1990] (Figure 8a). The folds are tight because there is not enough volume of the weak layer to fill the fold cores. Alternatively, a thin detachment layer leads to thrust-dominated folding rather than detachment folding [*Stewart*, 1996]. In contrast, if the décollement layer is thick, it first flows from beneath synclines into anticlinal cores and then out again as the folds tighten (Figure 8b). The result is more open folds (at least during early stages of deformation) and synclines that first subside below regional and then are raised again [*Homza and Wallace*, 1995; 1997]. The open geometry of folds in both the Mississippi Fan and Perdido foldbelts suggests that an adequate thickness of salt existed in both.

This interpretation is supported by the failure of depth-to-detachment calculations. Where synclines remain at a constant regional elevation, measurement of the excess area and amount of shortening in an anticline yields the depth to the detachment, or salt layer. Results in the Gulf of Mexico foldbelts are highly variable and much too large: from 5.2 km to 79.2 km below the MCSB in the Mississippi Fan foldbelt and from 3.1 km to 27.1 km in the Perdido foldbelt (Table 2). Seismic interpretations suggest that the detach-ment layer is actually about 1 km below the MCSB in the Mississippi Fan foldbelt and 2 km in the Perdido foldbelt (Figure 3). The wide variability and clearly incorrect values are probably due to differential subsidence of synclines, three-dimensional flow of salt in the fold cores, and an earlier deformation event (in the Mississippi Fan foldbelt – see below). Three-dimensional flow may be illustrated in the Mississippi Fan foldbelt, where folds C, E, G, J, and L (Table 2) are separate traverses through the same fold. The calculated depth to detachment ranges from 11.0 km at the culmination (fold C) to 57.6 km near the termination (fold L), all unreasonable values.

One implication of an adequate thickness of salt and the consequent open fold geometry concerns the kinematics of folding. When the décollement layer is thin, rock layers move through fold hinges as the limbs lengthen [*Dahlstrom*, 1990] (Figure 8a). When the décollement layer is thicker, this same process is possible, but it is more likely that the material does not move through the hinges and that the limbs have fixed length but variable dip through time [e.g., *Poblet and Hardy*, 1995; *Poblet and McClay*, 1996; *Homza and Wallace*, 1997] (Figure 8b). The kinematics of prog-ressively rotating fold limbs is documented by the geometry of growth strata in the Mississippi Fan foldbelt [*Rowan et al.*, 1993], but the relative lack of growth strata in the Perdido foldbelt makes such a determination difficult.

Thickness distribution. It is unlikely that the décollement layer had constant thickness across the two foldbelts. Whereas the top of the salt layer was probably subhorizontal prior to any deformation, the base may have had significant relief if the evaporites filled in any relict rift-related topography. Even where rift basins were filled in prior to evaporite formation, differential compaction of half-graben sediments and footwall basement would have resulted in thickness variations in the overlying evaporites. Thus, lateral thickness changes in the weak layer could be gradual or abrupt, depending on the underlying rift geometry and the amount of synrift clastic fill.

We have suggested before that the geometry of the fold envelope in each foldbelt might be explained by thickness variations in the original salt layer [*Trudgill et al.*, 1999]. In the Mississippi Fan foldbelt, where the fold envelope is subhorizontal, we see evidence on seismic data for only minor basement structures. Thus, it is likely that the Louann salt gradually thinned to the south to its pinchout just basinward of the frontal folds (Figure 9a). In contrast, we have mapped larger basement structures beneath and basin-ward of the Perdido foldbelt [*Trudgill et al.*, 1999; *Fiduk et al.*, 1999]. These take the form of northeast-southwest trending basement highs and half-grabens and northwest-southeast trending transverse structures. Furthermore, we interpreted a large gravity anomaly just northwest of the Perdido foldbelt (EDCON Inc., unpublished data, 1996) as thick, relict autochthonous salt in a half-graben adjacent to a basement high, an interpretation compatible with published regional cross sections [*Hall et al.*, 1993; *Peel et al.*, 1995]. Thus, we suggested that basinward salt flow in response to updip loading and exten-

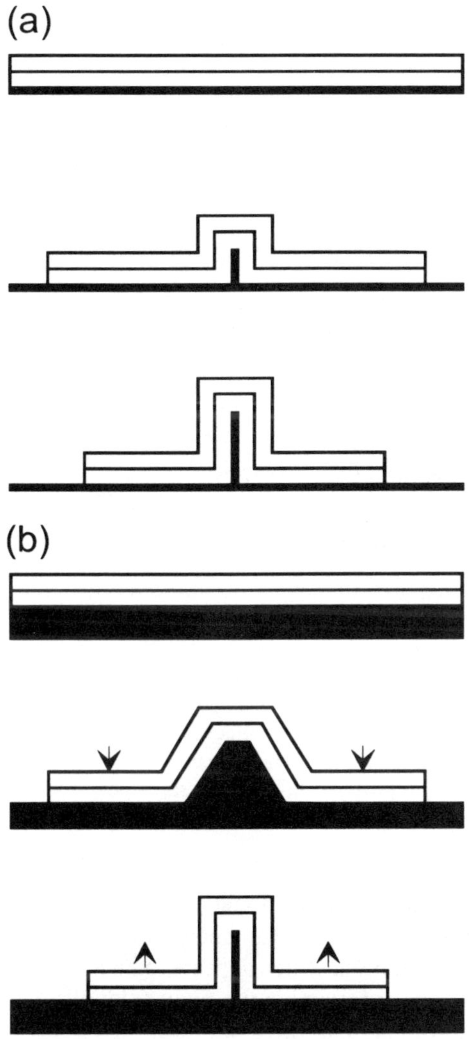

Figure 8. Kinematic models of salt-cored detachment folds: (a) thin décollement layer; (b) thick décollement layer. In (a), the salt volume is insufficient to fill the fold core, so that the fold is isoclinal and grows by limb lengthening and hinge migration [after *Dahlstrom*, 1990]. In (b), in which the fold grows by rotation of limbs with constant length, the synclines first sink as salt flows into the core, and then rise again as salt is forced out of the tightening fold [after *Homza and Wallace*, 1995; 1997; *Poblet and McClay*, 1996]. A thin décollement layer may also lead to thrust-dominated folding rather than detachment folding [*Stewart*, 1996].

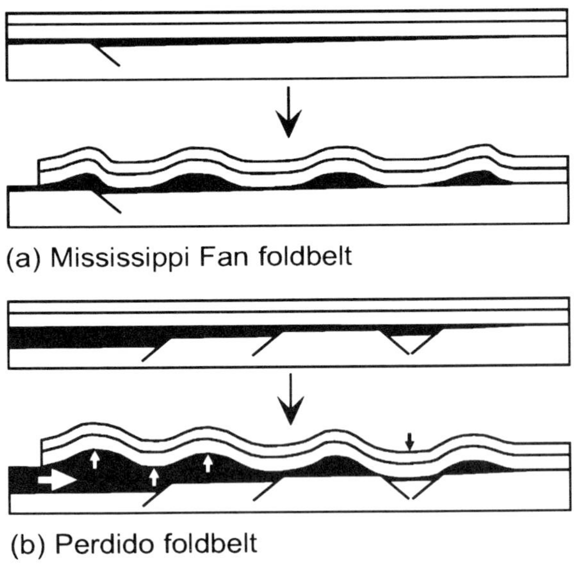

Figure 9. Models of basement influence on fold-train geometry [after *Trudgill et al.*, 1999]. In the Mississippi Fan foldbelt (a), a lack of large basement structures and a gradually thinning evaporite layer leads to a horizontal fold envelope and slightly more asymmetric folds. In the Perdido foldbelt (b), large basement structures have several effects: thicker salt in a frontal graben results in a syncline below basinal regional (small black arrow); abrupt basinward thinning of salt leads to inflation (white arrows) and a dipping fold envelope; and generally thicker salt yields more symmetric folds.

sion, combined with local loading, caused inflation of the Louann Salt where the it thinned rapidly over a basement high. This inflation, which appears to have occurred during or immediately following the main folding stage, uplifted the overburden and tilted the fold envelope basinward (Figure 9b). In addition, subsidence into thicker salt in some half-grabens could explain the present elevation of some synclines below the basinal regional level (Figure 9b).

This proposed model of varying basement geometry and original salt thickness is compatible with our suggestion that facies variability within the evaporite layer can explain the increased basinward asymmetry of folds in the Mississippi Fan foldbelt. Where the Louann Salt gradually thinned (Mississippi Fan foldbelt; Figure 9a), different evaporite facies or other lithologies may have been more common because sea-level changes during evaporite deposition would have affected a wider area, thereby leading to a greater fold asymmetry. Where the Louann thinned suddenly (Perdido foldbelt; Figure 9b), the basin margin would have been narrower and lateral facies transitions would have been more abrupt. In this case, salt may have formed a greater proportion of the evaporite layer volume, leading to more symmetric structures.

The presence of larger basement structures in the Perdido foldbelt may also explain the occurrence of domal, rather than elongate, anticlines. The two domes line up in a northwest-southeast orientation (Figure 4b) and are thought to overlie a transverse basement structure with different original thicknesses of salt on either side [*Trudgill et al.*, 1999]. Interference between the regional shortening dir-ection and an oblique component of shortening located over this structure may have caused the domal geometries.

Early Deformation

We have shown that the sequences just above the Louann Salt have variable thickness in the Mississippi Fan foldbelt but are roughly constant thickness in the Perdido foldbelt (Plates 1 and 2, Figure 6). The thickness changes in the Mississippi Fan foldbelt reflect an early deformation event. Although the early structures could have had extensional origins, their position at the basinward toe of the salt layer implies that they are more likely the result of early contraction that took place in the Late Jurassic and Cretaceous in response to basinward gravity gliding on the Louann Salt [*Rowan*, 1997]. Differential thermal subsidence resulting from Middle Jurassic rifting and Upper Jurassic oceanic spreading would have generated a basinward tilt that was later reversed by loading subsidence along the continental margin to the north. In this scenario, a three-phase deformation history is predicted for the Mississippi Fan foldbelt that is compatible with the observed geometry and reconstructed evolution (Figure 10): (1) early, small-wavelength buckling beneath a thin overburden in response to gravity gliding; (2) a period of quiescence as the basinward tilt of the décollement layer was reversed by flexural loading; and (3) large-wavelength buckling beneath a thicker overburden in response to gravity spreading of the prograding clastic margin.

The presence or absence of early salt structures, whether contractional or extensional has implications for some of the observed differences between the two foldbelts. Antecedent folds may explain variations in segment length as well as the origin and role of deeply-rooted salt diapirs.

Segment length. Most of the early salt structures (marked by thins in the overlying sequence) are located beneath the more recent folds, suggesting that they served as buckling instabilities during the initiation of Tertiary contraction [*Rowan*, 1997]. Others, however, were not reactivated and are visible beneath present-day synclines (Figures 4a and 6). The reason that not all early structures were reactivated is that the overburden thickness, and thus the fold wavelength, was greater during Tertiary deformation. The antecedent folds have trends that vary from east-northeast striking, parallel to the major folds, to west-northwest striking, highly oblique to the major folds (Figure 4a). The orientation roughly mimics that of the Cretaceous carbonate margin as it wrapped around from Texas and Louisiana to a northwest strike in Florida [*Winker and Buf-fler*, 1988], supporting a gravity gliding interpretation for the early deformation. The later shortening direction is rel-ated to failure of the Tertiary deltas to the north-northwest. The presence of antecedent folds may explain the shorter segment lengths in the Mississippi Fan foldbelt (Figure 4, Table 1). Salt-cored folds tend to be highly elongate [e.g., *Davis and Engelder*, 1985; *Harrison*, 1995], but local, smaller-wavelength, perturbations in the salt thickness (especially if they are oblique to the shortening direction) probably controlled the initial segment distribution and thus the subsequent growth and linkage [*Rowan*, 1997].

Antecedent folds appear to be absent from the Perdido foldbelt, which may explain the greater segment lengths (Figure 4, Table 1) even though the net amount of shortening in the two foldbelts is similar. The absence of antecedent folds is perplexing, in that the Perdido foldbelt, like the Mississippi Fan foldbelt, was located at the basinward limit of a detachment layer that presumably dipped basinward during the latest Jurassic and Cretaceous (Figure 10a). We suggest that early gravity gliding was accommodated farther landward through either contractional folding or by salt inflation and possibly extrusion. If early gravity gliding actually occurred, its lack of expression in the area of the Perdido foldbelt may be related to the rift-basin geometry and the associated thickness changes in the Louann Salt. For example, the large interpreted half-graben and basement high just northwest of the foldbelt may have insulated the Perdido area from any early deformation.

Diapirs. The occurrence of early deformation in the Mississippi Fan foldbelt also explains why diapirs are present there but not in the Perdido foldbelt. There are two end-member scenarios for the origin of the diapirs and their association with the folds. First, the diapirs could have formed by late piercement from the cores of anticlines (fold-first model; Figure 11a). Alternatively, the diapirs could have predated the Tertiary folding (diapir-first model; Figure 11b). In the latter case, folds would have nucleated at the diapirs (the weakest locations in the rock volume) [*Vendeville and Nilsen*, 1995], just as early structures served as buckling instabilities.

There are four lines of evidence and argument that support the diapir-first model. First, there is no geometric evidence that anticline cores were cannibalized in order to supply salt to diapirs breaking through fold crests [see *Coward and Stewart*, 1995]. Second, even though fold crests were eroded, the amount of overburden above the fold-core salt was too large in most cases to enable active diapirism [see *Schultz-Ela et al.*, 1993]. Third, there is evidence of long-lived diapir growth, in that strata that are normally constant thickness over the folds thin and thicken adjacent to the diapirs (Plate 2a, Figure 7). Finally, anomalous fold and fault geometries in the Mississippi Fan foldbelt are interpreted as representing diapirs that welded shut during shortening (Plates 1a and 2a). Also, note that the diapir in Figure 7 is widest at the top of the growth section and then narrows upward. This is compatible with the diapir-first model in that the salt would have spread laterally during shortening and then would have been gradually buried after folding ceased. However, the salt geometry is also compatible the fold-first model. In this scenario, salt would have broken through the fold crest and flowed laterally as a salt glacier, only to be buried by the rapid Pleistocene sedimentation.

If the diapirs really did exist prior to Tertiary contrac-

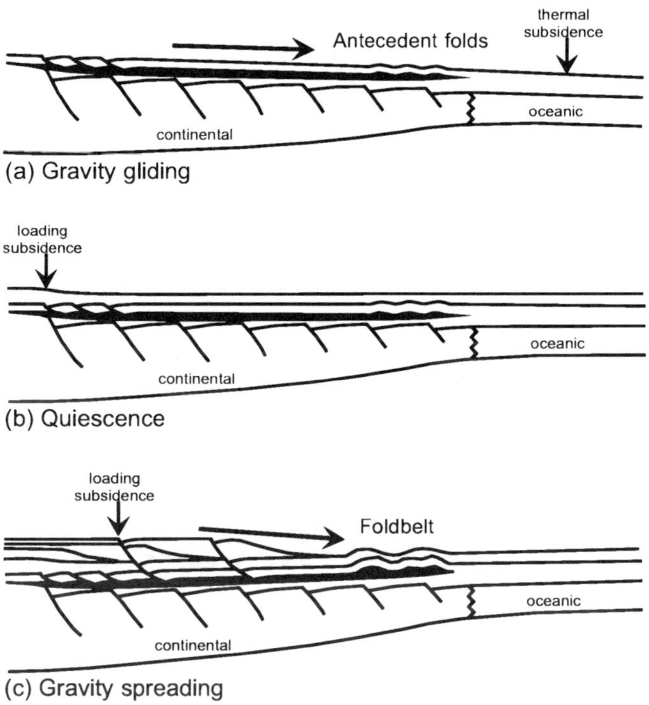

Figure 10. Three-stage model of passive margin development when a regional salt layer (black) is present: (a) greater thermal subsidence where oceanic crust cools leads to a basinward slope, gravity gliding, and the formation of antecedent folds; (b) as thermal subsidence wanes, flexural subsidence due to landward sedimentary loading reverses the sense of regional tilt, resulting in a period of quiescence; and (c) progradation of a clastic margin causes gravity spreading with contraction at the basinward toe of the salt layer, even though continued loading subsidence creates a landward-dipping décollement layer. Note the greater wavelength of folds in (c) compared to (a).

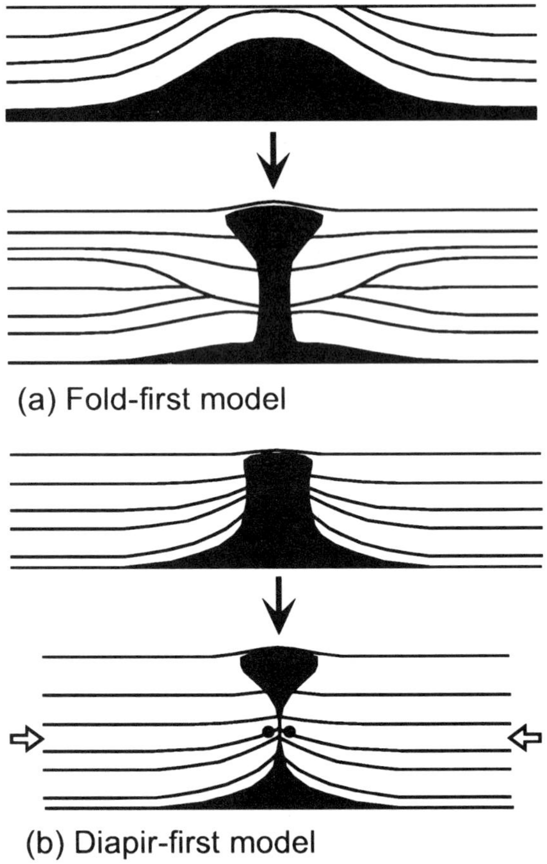

Figure 11. Alternative models for the origin of deeply-rooted diapirs in the Mississippi Fan foldbelt. In (a), a con-tractional fold is breached by active diapirism through the fold crest, and salt in the fold core is cannibalized as the diapir grows passively [after *Coward and Stewart*, 1995]. In (b), a long-lived passive diapir is squeezed during contraction, leading to diapir rejuvenation and secondary weld formation [after *Vendeville and Nilsen*, 1995]. Seismic data suggest that the diapir-first model (b) is more appropriate for the Mississippi Fan foldbelt.

tion, how were they triggered? We suggest that they were initiated during the early deformation event. Folding at that time occurred with only a thin overburden, one that would have been much easier to break through at fold crests [*Coward and Stewart*, 1995]. Once the active diapirs reached the sea floor, they would have continued growth as passive diapirs as long as there was still an adequate source of salt [*Vendeville and Jackson*, 1992]. When contraction began in the Miocene, the diapirs would have been re-juvenated, with further diapir growth and extrusion fed by a narrowing feeder while folding and reverse faulting ac-com-modated shortening adjacent to the diapirs [*Vendeville and Nilsen*, 1995]. This model also explains the lack of diapirs in the Perdido foldbelt because there was no early deformation to trigger diapirism. By the time shortening occurred, the overburden was too thick to allow piercement.

DISCUSSION AND CONCLUSIONS

The gravitational collapse of passive margins is facil-itated by the presence of salt because it is such an efficient lubricant. Updip sedimentary loading and extensional de-for-mation is balanced downdip by a combination of con-trac-tional deformation and salt extrusion, with the salt serving as the linking detachment surface. The contrac-tional defor-mation is usually accommodated in two places. Much of it occurs in deep-water foldbelts near the basinward limit of the salt layer, but it can also occur at the toe of the slope, which progrades basinward through time (*Vendeville et al.*, 1997]. In the northern Gulf of Mexico, the Mississippi Fan and Perdido deep-water foldbelts are located just landward of the autochthonous salt pinchout and just basinward of the allochthonous Sigsbee salt nappe. The existence of folds that formed at the toe of the paleo-slope is unknown be-cause of the inability to image well beneath the shallow salt.

In the parts of the foldbelts studied, both the Mississippi Fan and Perdido foldbelts have geometric features that are characteristic of salt-cored foldbelts. The folds are pre-dominantly symmetric, they have regular wavelengths, and they are often cut by relatively minor, high-angle reverse faults on one or both flanks. However, there are significant differences in structural style between the two foldbelts: (1) fold wavelengths, amplitude, and segment lengths are all greater in the Perdido foldbelt; (2) the Mississippi Fan fold-belt is more asymmetric (basinward-vergent) than the Per-dido foldbelt; (3) fold geometries are more rounded in the Mississippi Fan foldbelt and more angular in the Per-dido foldbelt; (4) domal anticlines are present in the Per-dido foldbelt but not the Mississippi Fan foldbelt; (5) the fold envelope is tilted in the Perdido foldbelt but subhori-zontal in the Mississippi Fan foldbelt; (6) growth strata are better developed in the Mississippi Fan foldbelt; and (7) ante-cedent folds, highly-rotated fault blocks, and deeply-rooted diapirs are present in the Mississippi Fan foldbelt but absent in the U.S. portion of the Perdido foldbelt.

Certainly, the mechanical stratigraphy of the pre-kinematic layers was a critical factor in determining the structural style. The greater thickness of this section in the Perdido foldbelt explains the greater wavelength (but not the greater amplitude, which is probably related to in-creased shortening). The rheologies and thicknesses of in-terlayered units within the folded multilayer may account for differences in fold style, with the kink-band geometry of the Perdido foldbelt compatible with thinner incompe-tent layers (shales and marls) and the rounded geometry of the Mississippi Fan foldbelt possibly controlled by a thicker, massive chalk unit.

Other differences between the two foldbelts cannot be explained by the mechanical stratigraphy. Facies variations within the evaporite layer may have controlled foldbelt symmetry, with a greater proportion of salt resulting in more symmetric structures and a greater proportion of other intercalated lithologies creating more asymmetric folds. Thickness variations in the evaporite layers, whether prim-ary or secondary, were also important. Original thickness variations related to the underlying basement geometry caused salt inflation, thereby creating the dipping fold envelope in the Perdido foldbelt. In contrast, the gradual thinning of salt and relatively small size of basement structures in the Mississippi Fan foldbelt resulted in a subhorizontal fold envelope. Thickness variations can also be secondary, formed by an earlier episode of shortening. Antecedent folds occur in the Mississippi Fan foldbelt but not in the Perdido foldbelt. The thickness irregularities served as buckling instabilities that controlled fold nucleation, segment length, and segment linkage. The early deformation beneath a thin overburden also triggered diapirism in the Mississippi Fan foldbelt, with the diapirs localizing later contractional strain such that shortening was accommodated by squeezing and rejuvenation of the diapirs, linked along strike to folds and reverse faults.

It is uncertain whether the foldbelts are underlain by thinned continental crust or oceanic crust. Gravity/magnetic data and seismic data suggest the presence of horsts and grabens beneath both foldbelts, with the scale of basement structures being larger beneath the Perdido foldbelt. We infer a continental affinity but cannot rule out the possibility of oceanic crust because it, too, can contain rotated fault blocks. In either case, both foldbelts, as well as deep-water foldbelts on other passive margins around the world, are important future petroleum provinces. They potentially contain giant accumulations that await evolving drilling and production technologies.

Acknowledgments. We wish to thank many individuals for their contributions to the ideas presented here, even if they do not agree with all the conclusions: Jim Dischinger, Steve Dobbs, Ray Fletcher, Bill Gafford, Peter Gale, Cindy Guu, Lyle Henage, Roy Kligfield, Bryant Korn, Roger Lowe, Randy Marrett, Wayne Orlowski, Ron Phair, Tomas Queffelec, Tim Seeley, Bob Ratliff, Geneva Roberts, and Paul Weimer. Dick Buffler, Bert Bally, and editor Webster Mohriak offered helpful suggestions to improve the manuscript. Seismic data were provided by Geco-Prakla (Abu Chowdhury), Western Geophysical (Denby Auble), Amoco, Mobil, Shell, Texaco, and the University of Texas Institute for Geophysics (Dick Buffler); gravity data were made available by EDCON Inc. (Alan Herring); and software was donated by Landmark and CogniSeis (now Paradigm). The work was funded by the EMARC Perdido industrial consortium (Amoco, Mobil, Shell, and Texaco) and the EMARC Gulf of Mexico industrial consortium (Agip, Amerada Hess, Amoco, Anadarko, BHP, BP, Burlington, CNG, Conoco, CXY, Enterprise, Exxon, Marathon, Maxus, Mobil, Occidental, Phillips, Shell, Texaco, UPR, Unocal, and Vastar). Paul Weimer is thanked for his role in both research projects.

REFERENCES

Apps, G. M., F. J. Peel, C. J. Travis, and C. A. Yielding, Structural controls on Tertiary deep water deposition in the northern Gulf of Mexico, in *Submarine Fans and Turbidite Systems: Sequence Stratigraphy, Reservoir Architecture, and Production Characteristics*, edited by P. Weimer, A. H. Bouma, and B. J. Perkins, pp. 1-7, Gulf Coast Section SEPM Foundation Fifteenth Annual Res. Conf., Program with Papers, 1994.

Biot, M. A., Theory of folding of stratified viscoelastic media and its implications in tectonics and orogenesis, *Geol. Soc. Am. Bull.*, *72*, 1595-1620, 1961.

Blickwede, J. F., and T. A. Queffelec, Perdido foldbelt: a new deep-water frontier in western Gulf of Mexico (abstract), *Bull. Am. Assoc. Pet. Geol.*, *72*, 163, 1988.

Buffler, R. T., Distribution of crust, distribution of salt, and the early evolution of the Gulf of Mexico Basin, in *Gulf of Mexico Salt Tectonics, Associated Processes and Exploration Potential*, pp. 25-27, Gulf Coast Section SEPM Foundation Tenth Annual Res. Conf., Program with Papers, 1989.

Buffler, R. T., Seismic stratigraphy of the deep Gulf of Mexico basin and adjacent margins, in *The Gulf of Mexico Basin*, edited by A. Salvador, pp. 353-387, The Geology of North America, vol. J, Geol. Soc. Am., Boulder, 1991.

Buffler, R. T., and W. A. Thomas, Crustal structure and evolution of the southeastern margin of North America and the Gulf of Mexico basin, in *Phanerozoic Evolution of North America Continent-Ocean Transitions*, edited by R. C. Speed, pp. 219-263, DNAG Continent-Ocean Transect Volume, Geol. Soc. Am., Boulder, 1994.

Cobbold, P. R., P. Szatmari, L. S. Demercian, D. Coelho, and E. A. Rossello, Seismic and experimental evidence for thin-skinned horizontal shortening by convergent radial gliding on evaporites, deep-water Santos Basin, Brazil, in *Salt Tectonics: a Global Perspective*, edited by M. P. A. Jackson, D. G. Roberts, and S. Snelson, pp. 305-321, Am. Assoc. Pet. Geol. Memoir 65, 1995.

Coward, M., and S. Stewart, Salt-influenced structures in the Mesozoic-Tertiary cover of the southern North Sea, U.K., in *Salt Tectonics: A Global Perspective*, edited by M. P. A. Jackson, D. G. Roberts, and S. Snelson, pp. 229-250, Am. Assoc. Pet. Geol. Memoir 65, 1995.

Dahlstrom, C. D. A., Geometric constraints derived from the law of conservation of volume and applied to evolutionary models for detachment folding, *Bull. Am. Assoc. Pet. Geol.*, *74*, 336-344, 1990.

Davis, D. M., and T. Engelder, The role of salt in fold-and-thrust belts, *Tectonophysics*, *119*, 67-88, 1985.

Demercian, S., P. Szatmari, and P. R. Cobbold, Style and pattern of salt diapirs due to thin-skinned gravitational gliding, Campos and Santos basins, offshore Brazil, *Tectonophysics*, *228*, 393-433, 1993.

Diegel, F. A., J. F. Karlo, D. C. Schuster, R. C. Shoup, and P. R. Tauvers, Cenozoic structural evolution and tectono-stratigraphic framework of the northern Gulf Coast continental margin, in *Salt Tectonics: A Global Perspective*, edited by M. P. A. Jackson, D. G. Roberts, and S. Snelson, pp. 109-151, Am. Assoc. Pet. Geol. Memoir 65, 1995.

Duval, B., C. Cramez, and M. P. A. Jackson, Raft tectonics in the Kwanza Basin, Angola, *Marine and Petroleum Geology*, *9*, 389-404, 1992.

Feng, J., Post mid-Cretaceous seismic stratigraphy and geologic history, deep Gulf of Mexico basin, Ph.D. thesis, Univ. of Texas, Austin, 1995.

Feng, J., and R. T. Buffler, Preliminary age determinations for new deep Gulf of Mexico basin seismic sequences, *Gulf Coast Assoc. Geol. Soc. Proceedings*, 283-289, 1991.

Fiduk, J. C., B. C. McBride, P. Weimer, M. G. Rowan, and B. D. Trudgill, Defining the basinward limits of salt deposition, northern Gulf of Mexico, in *Salt, Sediment, and Hydrocarbons*, edited by C. J. Travis et al., pp. 53-64, Gulf Coast Section SEPM Foundation Sixteenth Annual Res. Conf., Program with Papers, 1995.

Fiduk J. C., P. Weimer, B. D. Trudgill, M. G. Rowan, P. E. Gale, B. E. Korn, R. L. Phair, W. T. Gafford, G. R. Roberts, R. S. Lowe, and T. A. Queffelec, Seismic interpretation of Mesozoic-Cenozoic strata, Perdido fold belt area (Alaminos Canyon), northwestern Gulf of Mexico, *Gulf Coast Assoc. Geol. Soc. Proceedings*, 47, 159-168, 1997.

Fiduk J. C., P. Weimer, B. D. Trudgill, M. G. Rowan, P. E. Gale, B. E. Korn, R. L. Phair, G. R. Roberts, W. T. Gafford, R. S. Lowe, and T. A. Queffelec, The Perdido fold belt, northwestern deep Gulf of Mexico: Part 2. Seismic stratigraphy and petroleum systems, *Bull. Am. Assoc. Pet. Geol.*, 1999 (in press).

Galloway, W. E., Genetic stratigraphic sequences in basin analysis II: Application to northwest Gulf of Mexico Cenozoic basin, *Bull. Am. Assoc. Pet. Geol.*, *73*, 143-154, 1989.

Galloway, W. E., D. G. Bebout, W. L. Fisher, J. B. Dunlap Jr., R. Cabrera-Castro, J. E. Lugo-Rivera, and T. M. Scott, Cenozoic, in *The Gulf of Mexico Basin*, edited by A. Salvador, pp. 245-324, The Geology of North America, vol. J., Geol. Soc. Am., Boulder, 1991.

Hall, D. J., B. E. Bowen, R. N. Rosen, S. Wu, and A. W. Bally, Mesozoic and early Cenozoic development of the Texas margin: a new integrated cross-section from the Cretaceous shelf edge to the Perdido foldbelt, *Gulf Coast Section SEPM Foundation Thirteenth Annual Res. Conf. Proceedings*, 1993.

Harrison, J. C., Tectonics and kinematics of a foreland folded belt influenced by salt, arctic Canada, in *Salt Tectonics: A Global Perspective*, edited by M. P. A. Jackson, D. G. Roberts, and S. Snelson, pp. 379-412, Am. Assoc. Pet. Geol. Memoir 65, 1995.

Heaton, R. C., M. P. A. Jackson, M. Bamahmoud, and A. S. O. Nani, Superposed Neogene extension, contraction, and salt canopy emplacement in the Yemeni Red Sea, in *Salt Tectonics: A Global Perspective*, edited by M. P. A. Jackson, D. G. Roberts, and S. Snelson, pp. 333-351, Am. Assoc. Pet. Geol. Memoir 65, 1995.

Homza, T. X., and W. K. Wallace, Geometric and kinematic models for detachment folds with fixed and variable detachment depths, *J. Struct. Geol.*, *17*, 575-588, 1995.

Homza, T. X., and W. K. Wallace, Detachment folds with fixed hinges and variable detachment depth, northeastern Brooks Range, Alaska, *J. Struct. Geol.*, *19*, 337-354, 1997.

Keen, C. E., B. C. MacLean, and W. A. Kay, A deep seismic reflection profile across the Nova Scotia continental margin, offshore eastern Canada, *Can. J. Earth Sci.*, *28*, 1112-1120, 1991.

Keen, C. E., and D. P. Potter, Formation and evolution of the Nova Scotian rifted margin: evidence from deep seismic reflection data, *Tectonics*, *14*, 918-932, 1995.

Letouzey, J., B. Colletta, R. Vially, and J. C. Chermette, Evolution of salt-related structures in compressional settings, in *Salt Tectonics: A Global Perspective*, edited by M. P. A. Jackson, D. G. Roberts, and S. Snelson, pp. 41-60, Am. Assoc. Pet. Geol. Memoir 65, 1995.

Lundin, E. R., Thin-skinned extensional tectonics on a salt detach-ment, northern Kwanza Basin, Angola, *Marine and Petroleum Geology*, *9*, 405-411, 1992.

Martin, R. G., Distribution of salt structures in the Gulf of Mexico, U.S. Geol. Surv. Miscellaneous Field Studies Map MF-1213, 1980.

Marton, G., and R. T. Buffler, Jurassic reconstruction of the Gulf of Mexico basin, *Int. Geol. Rev.*, *36*, 545-586, 1994.

Mohriak, W. U., J. M. Macedo, R. T. Castellani, H. D. Rangel, A. Z. N. Barros, M. A. L. Latgé, J. A. Ricci, A. M. P. Mizusaki, P. Szatmari, L. S. Demercian, J. G. Rizzo, and J. R., Aires, Salt tectonics and structural styles in the deep-water province of the Cabo Frio region, Rio de Janeiro, Brazil, in *Salt Tectonics: A Global Perspective*, edited by M. P. A. Jackson, D. G. Roberts, and S. Snelson, pp. 273-304, Am. Assoc. Pet. Geol. Memoir 65, 1995.

Morley, C. K., and G. Guerin, Comparison of gravity-driven deformation styles and behavior associated with mobile shales and salt, *Tectonics*, *15*, 1154-1170, 1996.

Mount, V. S., State of stress in California and a seismic structural analysis of the Perdido fold belt, northeast Gulf of Mexico, Ph.D. thesis, Princeton Univ., Princeton, 1989.

Mount, V. S., J. Suppe, and S. Hook, A forward modeling strategy for balancing cross sections. *Bull.. Am. Assoc. Pet. Geol.*, *74*, 521-531, 1990.

Peel, F. J., C. J. Travis, and J. R. Hossack, Genetic structural prov-inces and salt tectonics of the Cenozoic offshore US Gulf of Mexico: a preliminary analysis, in *Salt Tectonics: A Global Perspective*, edited by M. P. A. Jackson, D. G. Roberts, and S. Snelson, pp. 153-175, Am. Assoc. Pet. Geol. Memoir 65, 1995.

Poblet, J., and S. Hardy, Reverse modelling of detachment folds: application to the Pico del Aguila anticline in the south central Pyrenees (Spain), *J. Struct. Geol.*, *17*, 1707-1724, 1995.

Poblet, J., and K. McClay, Geometry and kinematics of single-layer detachment folds. *Bull. Am. Assoc. Pet. Geol.*, *80*, 1085-1109, 1996.

Ramberg, H., Relationship between length of arc and thickness of ptygmatically folded veins, *Am. J. Sci.*, *258*, 36-46, 1960.

Ramsay, J. G., and M. I. Huber, *The Techniques of Modern Structural Geology, Volume 2: Folds and Fractures*, 700 pp., Academic Press, London, 1987.

Rowan, M.G., Structural styles and evolution of allochthonous salt, central Louisiana outer shelf and upper slope, in *Salt Tectonics: A Global Perspective*, edited by M. P. A. Jackson, D. G. Roberts, and S. Snelson, pp. 199-228, Am. Assoc. Pet. Geol. Memoir 65, 1995.

Rowan, M. G., Three-dimensional geometry and evolution of a segmented detachment fold, Mississippi Fan foldbelt, Gulf of Mexico, *J. Struct. Geol.*, *19*, 463-480, 1997.

Rowan, M. G., R. Kligfield, and P. Weimer, Processes and rates of deformation: preliminary results from the Mississippi Fan foldbelt, deep Gulf of Mexico, in *Rates of Geologic Processes*, edited by J. M. Armentrout, R. Bloch, H. C. Olson, and B. F. Perkins, pp. 209-218, Gulf Coast Section SEPM Foundation Fourteenth Annual Res. Conf., Program with Papers, 1993.

Sage, L., and J. Letouzey, Convergence of the African and Eurasian plates in the eastern Mediterranean, in *Petroleum and Tectonics in Mobile Belts*, edited by J. Letouzey, pp. 49-68, Editions Technip, Paris, 1990.

Salvador, A., Triassic-Jurassic, in *The Gulf of Mexico Basin*, edited by A. Salvador, pp. 131-180, The Geology of North America, vol. J., Geol. Soc. Am., Boulder, 1991a.

Salvador, A., Origin and development of the Gulf of Mexico Basin, in *The Gulf of Mexico Basin*, edited by A. Salvador, pp. 389-444, The Geology of North America, vol. J., Geol. Soc. Am., Boulder, 1991b.

Sawyer, D. S., R. T. Buffler, and R. H. Pilger, Jr., The crust under the Gulf of Mexico Basin, in *The Gulf of Mexico Basin*, edited by A. Salvador, pp. 31-52, The Geology of North America, vol. J., Geol. Soc. Am., Boulder, 1991.

Schultz-Ela, D. D., M. P. A. Jackson, and B. C. Vendeville, Mechanics of active salt diapirism, *Tectonophysics*, *228*, 275-312, 1993.

Shaub, F. J., R. T. Buffler, and J. G. Parsons, Seismic stratigraphic framework of deep central Gulf of Mexico basin, *Bull. Am. Assoc. Pet. Geol.*, *68*, 1790-1802, 1984.

Spathopoulos, F., An insight on salt tectonics in the Angola Basin, South Atlantic, in *Salt Tectonics*, edited by G. I. Alsop, D. J. Blundell, and I. Davison, pp. 153-174, Geological Society Special Publication No. 100, 1996.

Stewart, S. A., Influence of detachment layer thickness on style of thin-skinned shortening, *J. Struct. Geol.*, *18*, 1271-1274, 1996.

Trudgill, B. D., M. G. Rowan, P. Weimer, J. C. Fiduk, P. E. Gale, B. E. Korn, R. L. Phair, W. T. Gafford, J. B. Dischinger, G. R. Roberts, and L. F. Henage, The structural geometry and evolution of the salt-related Perdido fold belt, Alaminos Canyon, northwestern deep Gulf of Mexico, in *Salt, Sediment, and Hydrocarbons*, edited by C. J. Travis et al., pp. 275-284, Gulf Coast Section SEPM Foundation Sixteenth Annual Res. Conf., Program with Papers, 1995.

Trudgill, B. D., M. G. Rowan, J. C. Fiduk, P. Weimer, P. E. Gale, B. E. Korn, R. L. Phair, W. T. Gafford, G. R. Roberts, and S. W. Dobbs, The Perdido fold belt, northwestern deep Gulf of Mexico, Part 1: Structural geometry, evolution and regional implications, *Bull. Am. Assoc. Pet. Geol.*, *83, 88-113*, 1999.

Vendeville, B. C., and M. P. A. Jackson, The rise of diapirs during thin-skinned extension, *Marine and Petroleum Geology*, *9*, 331-353, 1992.

Vendeville, B. C., M. P. A. Jackson, and S. Raillard, Interaction between sedimentation and deformation during the structural evolution of salt-bearing divergent continental margins, talk presented at the 5th International Congress of the Brazilian Geophysical Society, Sao Paulo, Brazil, Sept. 28-Oct. 2, 1997.

Vendeville, B. C., and K. T. Nilsen, Episodic growth of salt diapirs driven by horizontal shortening, in *Salt, Sediment, and Hydrocarbons*, edited by C. J. Travis et al., pp. 285-295, Gulf Coast Section SEPM Foundation Sixteenth Annual Res. Conf., Program with Papers, 1995.

Weimer, P., and R. Buffler, Structural geology and evolution of the Mississippi Fan Fold Belt, deep Gulf of Mexico, *Bull. Am. Assoc. Pet. Geol.*, *76*, 225-251, 1992.

Winker, C. D., Cenozoic shelf margins, northwestern Gulf of Mexico, *Gulf Coast Assoc. Geol. Soc. Trans.*, *32*, 427-448, 1982.

Winker, C. D., and R. T. Buffler, Paleogeographic evolution of early deep-water Gulf of Mexico and margins, Jurassic to middle Cretaceous (Comanchean), *Bull. Am. Assoc. Pet. Geol.*, *72*, 318-346, 1988.

Worrall, D. M., and S. Snelson, Evolution of the northern Gulf of Mexico, with emphasis on Cenozoic growth faulting and the role of salt, in *The Geology of North America: an Overview*, edited by A. W. Bally, and A. R. Palmer, pp. 97-138, Geol. Soc. Am. Decade of North American Geology, vol. A, 1989.

Wu, S., Salt and slope tectonics, offshore Louisiana, Ph.D. thesis, Rice University, Houston, 1993.

Wu, S., A. W. Bally, and C. Cramez, Allochthonous salt, structure and stratigraphy of the north-eastern Gulf of Mexico, part II: structure, *Marine and Petroleum Geology*, *7*, 334-370, 1990.

J. Carl Fiduk, 4215 Evans St., Boulder, CO 80303, USA

Mark G. Rowan, Rowan Structural Consulting, 1633 D 4th St., Boulder, CO 80302, USA (e-mail: mgrowan@uswest.net)

Bruce D. Trudgill, Department of Geological Sciences, Campus Box 399, University of Colorado, Boulder, CO 80309-0399, USA (e-mail: bruce@emarc.colorado.edu)

Geological and Geophysical Interpretation of the San Julian Basin—Offshore Argentina

J.R.S. Soares F., A.P. Miranda and A.M.F. de Figueiredo

Petrobras Internacional S.A. - BRASPETRO

Seismic and potential field data were acquired in the San Julian Basin in the south offshore of Argentina. The interpretation of these geophysical data enabled the definition of the structural framework of the basin and its tectonic evolution. Tectonic phases related to rifting, thermal subsidence and passive margin were identified. A possible Late Cretaceous wrenching event strongly affected previous structures and created several features typical of this kind of deformation. Several prospects were identified within the basin and one exploratory well was drilled, but resulted dry.

1. INTRODUCTION

The San Julian Basin is situated in the South Atlantic, offshore of Argentina, and has an area of approximately 14,300 km ^{2} (Figure 1). BRASPETRO and partners started to work on a block within San Julian Basin in 1990, the CSJM-1 Block. At that time, the basin was an exploration frontier area, with only four regional seismic lines acquired within the basin, with a total length of 627 km, and no exploratory wells had been drilled.

In order to evaluate the hydrocarbon exploratory potential of the basin, the existing seismic lines were reprocessed and new geopotential, magnetic and gravity surveys were carried out. Seismic lines were acquired in two different campaigns, totaling 4,518 km of 2D seismic data.

Interpretation of the geophysical data enabled the identification of the main rift phase of Jurassic age and a subsequent phase of thermal subsidence. A possible Late Cretaceous wrenching event modified the rift phase architecture and created several structures typical of this kind of deformation; these features were very well imaged by the new seismic data. At the end of the Cretaceous, the basin was uplifted and eroded. This event was followed by a passive margin phase of subsidence that took place during the Tertiary.

Rift, sag and passive margin megasequences were interpreted on the seismic data by integration of the regional seismic grid with stratigraphic data.

This paper will present an interpretation of the different geophysical datasets and the integration of this data to describe the regional geology.

Atlantic Rifts and Continental Margins
Geophysical Monograph 115

2. PLATE TECTONICS SETTING

Recent studies on the tectonic evolution of Southwestern Gondwana (Tankard et al., 1995; Musset and Taylor, 1994) have shown that this zone is constituted by the amalgamation of several microplates around cratonic areas. The San Julian Basin is located on the northern edge of the Deseado / Malvinas microplate (Figure 2), which collided against the southern edge of the Gondwana continent by Permian-Triassic times (Figueiredo et al., 1995). During the Triassic, with the onset of the Gondwana continental breakup, these zones were subjected to an extensional period and evolved into isolated grabens with intense block-faulting and strong volcanic activity (Uliana and Biddle, 1988).

Once oceanic crust was formed in the area, and the drift phase started, the Deseado/Malvinas microplate moved ad-

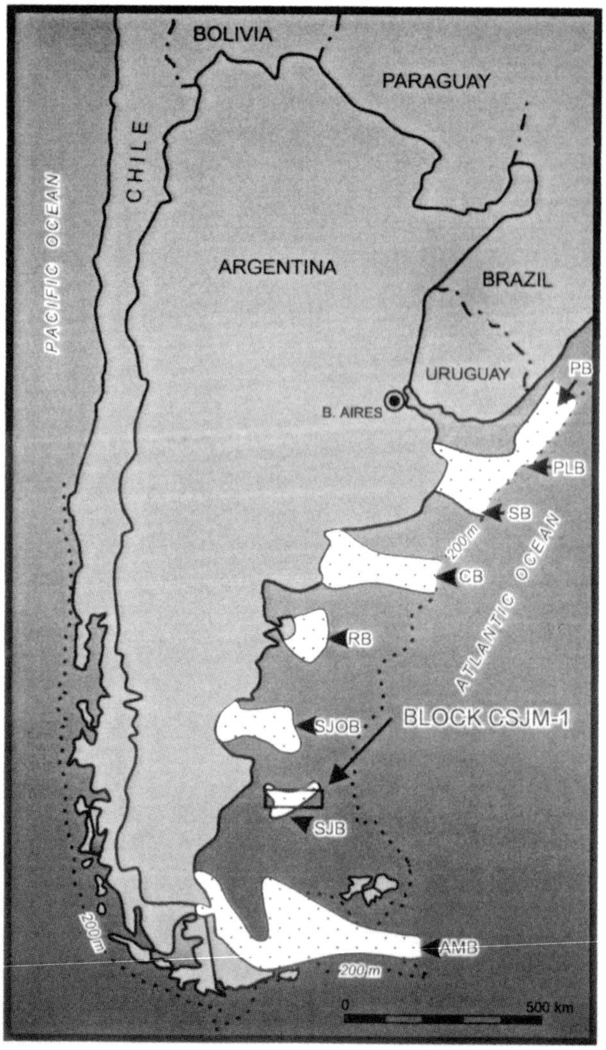

Figure 1. Situation map with South America Atlantic Offshore Basins, AMB – Austral Malvinas Basin, SJOB – San Julian Basin, SJo – San Jorge Basin, RB- Rawson Basin, CB- Colorado Basin, SB- Salado Basin, PLB – Punta del Leste Basin and PB – Pelotas Basin.

hered to the newborn South American continent, constituting a large massive plate.

The existence of an external ridge at the edge of this new continental margin from Punta del Este to San Julian area has limited the access of marine waters in the Cretaceous, preventing the flooding of this large platform until the mid-Tertiary (Uliana and Biddle, 1988; Stoakes et al., 1991).

For this reason, the thermal subsidence of the entire area was not very pronounced, and the sag phase from Early Cretaceous to Maastrichtian was characterized by tectonically more stable conditions, and massive volcanic activity.

During the sag phase, the San Julian Basin was mildly affected by regional compressional stresses, generated by the collision of the westward drifting South America continental mass with several Pacific microplates and also by compressional stresses related to the movement along the adjacent Malvinas-Agulhas fracture zone. These stresses were responsible for gentle folds and fault rearrangements, which were modified by a strong event of wrenching that occurred at the end of Cretaceous/ Paleocene time (Zalán, P.V., unpublished data).

This later wrenching event may be related to differential motion between the South America Plate and the Deseado/ Malvinas microplate, during a phase of general plate reorganization. This wrenching event was responsible for the inversion of San Julian Basin and generation of reverse faults, transpressive fold trends and even lateral displacement of blocks. Due to the strong uplift of the entire basin, a regional unconformity developed, eroding deeply into the sag sequence, with the formation of a peneplain in much of the area.

Tertiary marine sediments covered the peneplained surface. These deposits characterized the transgression of the marine waters over this large platform. This marks the onset of the passive margin phase. These passive margin sediments have a thickness ranging from 150 on the west sector to 400 meters on the east sector of the basin. Progradational packages within this sequence indicate a shelf progradation to east and northeast, sourced from the south and west.

3. GEOPHYSICAL DATA SET

Interpretation of the geophysical dataset acquired within the San Julian Basin was conducted in an integrated way. Different types of geophysical surveys were carried out with specific objectives, as shown in Table 1 below:

Interpretation of the reprocessed available seismic lines allowed a preliminary definition of the geometry and structural styles of the basin.

The geopotential survey, in which the sea surface topography is measured by satellite, was carried out and integrated with regional maps. This survey technique uses an indirect method of measuring the gravity field and is based on the assumption that the average sea surface is an equipotential surface of the earth's gravitational field (Sandwell and McAdoo, 1988). Therefore, variations of the sea surface can be related to mass variations within the upper crust of the earth. Maps obtained by this method are the satellite gravity maps (Figure 3).

The integration of the geopotential data with the reprocessed seismic data enabled definition of areas of basin development.

A new modern marine gravity and magnetic survey comprising 2766 km of gravity data and 2430 km of

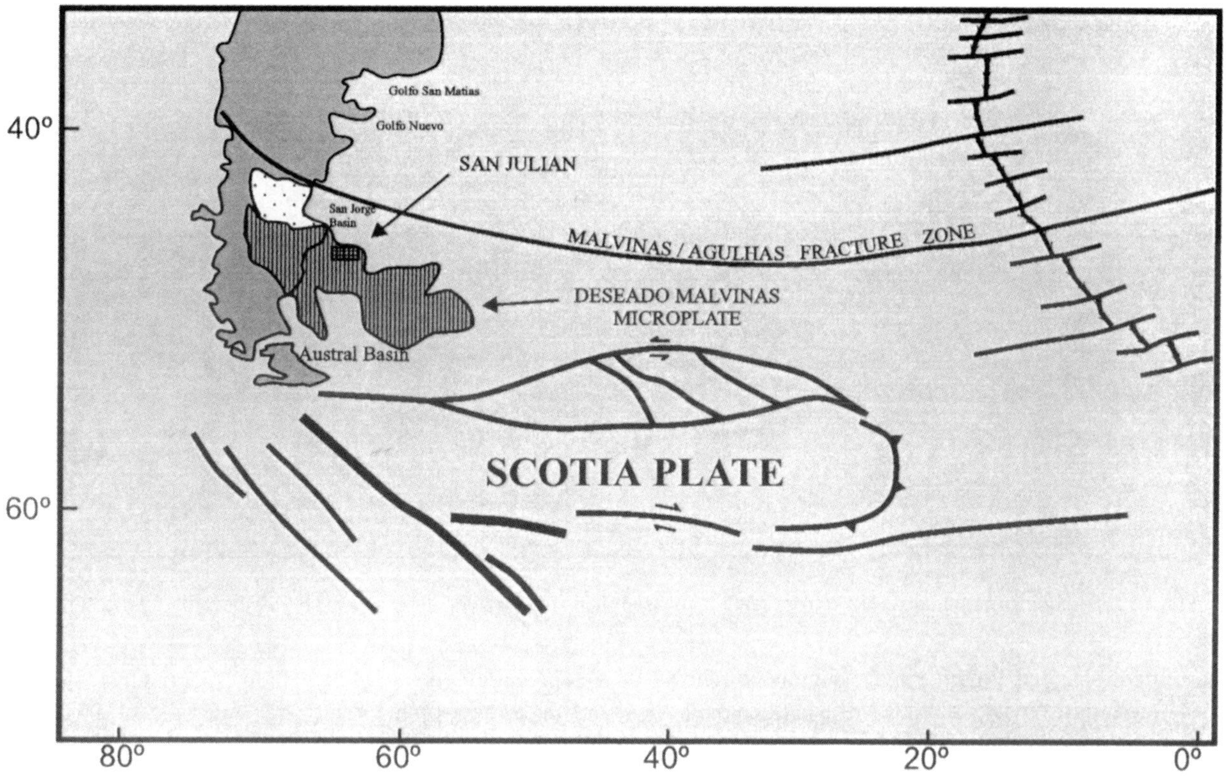

Figure 2. Plate tectonics setting, this map shows the position of the San Julian Basin in the northern part of the Deseado Malsvinas microplate.

magnetic data was acquired together with the regional seismic survey. The interpretation of the potential field data was conducted in conjunction with the interpretation of the regional seismic data (Braga, L., unpublished report).

The Bouguer anomaly map (Figure 4) shows good agreement between the position of the main depocenters interpreted from seismic and the satellite gravity maps. A major depocenter is characterized in the western sector of the block.

The sediment stripped Bouguer map (Figure 5), where the gravity effect of the sedimentary load has been removed, shows the area of maximum crustal thinning that corresponds to the main rift development and maximum heat flow.

Figure 6 shows a comparison of depocenters interpreted from potential field data with depocenters interpreted from seismic data . The coincidence of these depocenters is illustrated on Figure 6-a. The main conclusions from the comparative analysis of these depocenters are:

- The position of the main basin depocenters as depicted from gravity (Figure 6b) and geopotential (Figure 6c) data shows an agreement with the position of the depocenteres interpreted from seismic data (Figure 6d).

Table 1 – Objectives of Geophysical Surveys

Survey	Products	Objectives
Old Seismic	Reprocessed seismic dataset and preliminary basement map	Basin limits and structural style preliminarily defined
Geopotential	Satellite Gravity Maps	Defining main areas of basin development
Potential Field	Gravity and Magnetic Maps	Defining areas of basin development, modeling of basement depth and basement structures
Regional Seismic	Regional seismic dataset, structural and isopach maps and leads inventory	Define basement configuration, megasequences and identify leads
Detail Seismic	Detail seismic dataset, structural and isopach maps and prospects inventory	Detail basement structures and refined sequence stratigraphy to characterize prospects.

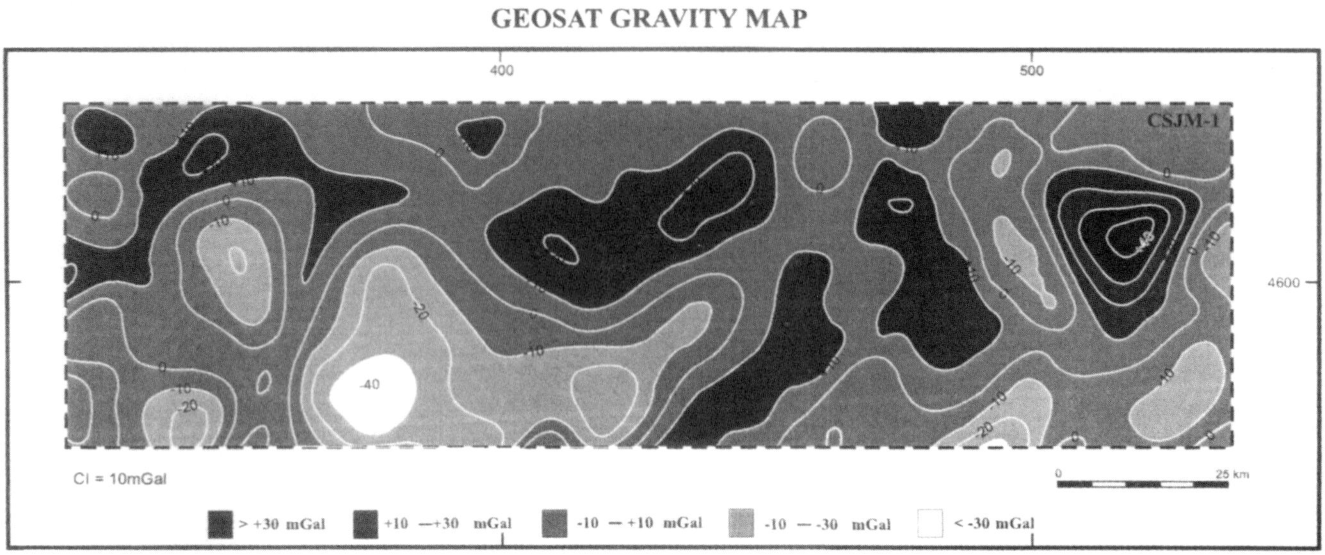

Figure 3. Satellite gravity map, derived from satellite measurements. Interpreted depocenters are indicated by light colors.

- The sediment stripped Bouguer map (Figure 6e) shows the area of maximum crustal thinning, which is coincident with the main rift depocenters.

Two seismic 2D acquisition campaigns were carried out. In the first campaign, 2818 km of regional seismic lines were acquired over the main depocenters shown on the Geopotential data. Acquisition parameters were 144 channels, shot-point and source intervals of 26,67 meters, cable length of 3840 meters, cable depth 10 meters and 32 guns at 7 meters depth.

In the second acquisition campaign, 1770 km of seismic lines were acquired over the most prospective areas, as

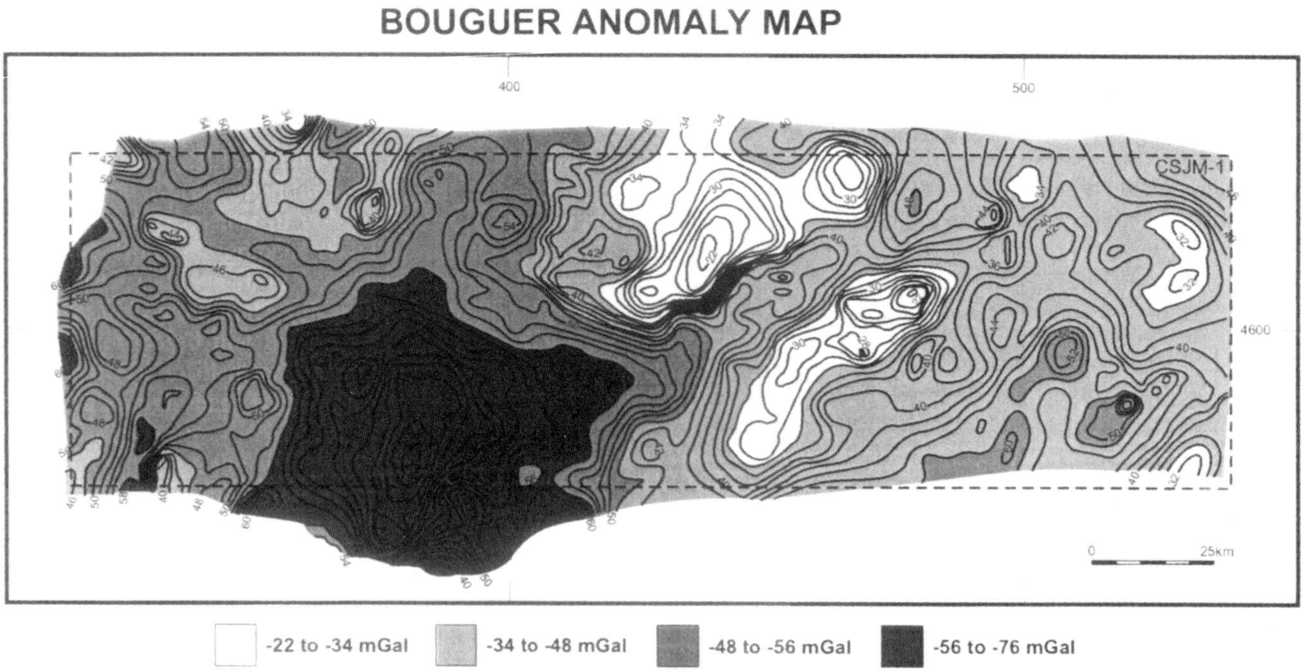

Figure 4. Bouguer anomaly map, depocenters in this map are indicated by darker colors, related to highly negative Bouguer values.

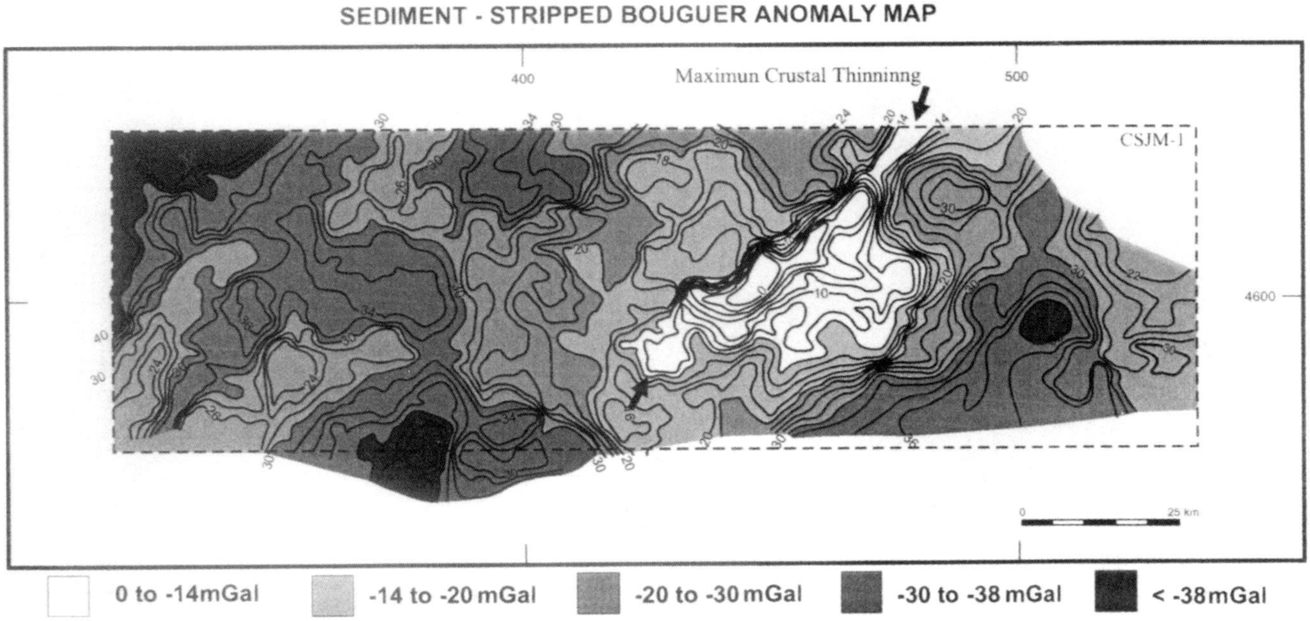

Figure 5. Sediment stripped Bouguer anomaly map, the effect of the sedimentary load on gravity has been removed. Higher values, close to zero, indicate where the crust is thin and high density mantle rocks are closer than nearby areas.

defined by the interpretation of the first phase regional seismic lines. Parameters of acquisition of these lines were 120 channels, shot-point and source intervals of 26,67 meters, cable length of 3200 meters, cable at 10 meter depth and 32 guns at 7 meters depth.

4. EXPLORATORY WELL

The San Julian ES-1well was spudded on October 22nd, 1994 and was abandonmed as a dry hole on December 15, 1994, after reaching a total depth of 2,940m. Figure 7 shows the stratigraphic column of the borehole, the depositional environments of the stratigraphic sequences, and the correlation to main tectonic events.

The well was drilled on an elongated NE trending structural high which was partially bounded by reverse transpressional strike-slip faults. The main targets were the Aptian to Turonian continental sandstones, which were thought to be correlative to the Chubut Group of the productive San Jorge basin.

The prognosed stratigraphy for the prospect San Julian es-1was prepared considering the analogies between the San Jorge and San Julian basins. However the well results showed some major differences between the predicted and the drilled sedimentary section: an older stratigraphic record exists in the San Julian basin as indicated by biostratigraphic analysis; and the Aptian to Turonian sandstones, the main reservoir-prone interval in the San Jorge basin, are absent in the San Julian basin.

Neither reservoir-prone intervals nor live oil shows were detected from samples or cores in the drilled section. Source prone intervals were drilled from 1350 to 2505 meters in tuffaceous claystone, with organic carbon content ranging from 1.5 to 5.35 % and type I or II organic matter. Hydrocarbon generation and migration models need further study to de determined. No hydrocarbons zones were detected from electric logs. However dead oil stains were reported from Middle Jurassic and Cretaceous rock samples, associated to tuffaceous sandstones and siltstones, with very low porosity.

No radiometric dating was provided from the San Julian es-1well samples, due to ditch sample contamination and the mineralogical composition of the volcanics. Biostratigraphic conclusions and age of the drilled sequences are based on palynological data, because the entire interval is barren of calcareous microfossils and nanofossils. For this reason dating of the intervals above 1,000m and below 2,500m is very tentative.

5. SEISMIC HORIZONS AND MEGASEQUENCES

Selected horizons for interpretation were correlated to megasequence boundaries (Soares Filho and Miranda, 1997), as described below.

Economic basement is interpreted as a high amplitude and discontinuous reflector, having the same seismic characteristics of basement in nearby basins, where it is represented by volcanic rocks of Jurassic age. Drilling data

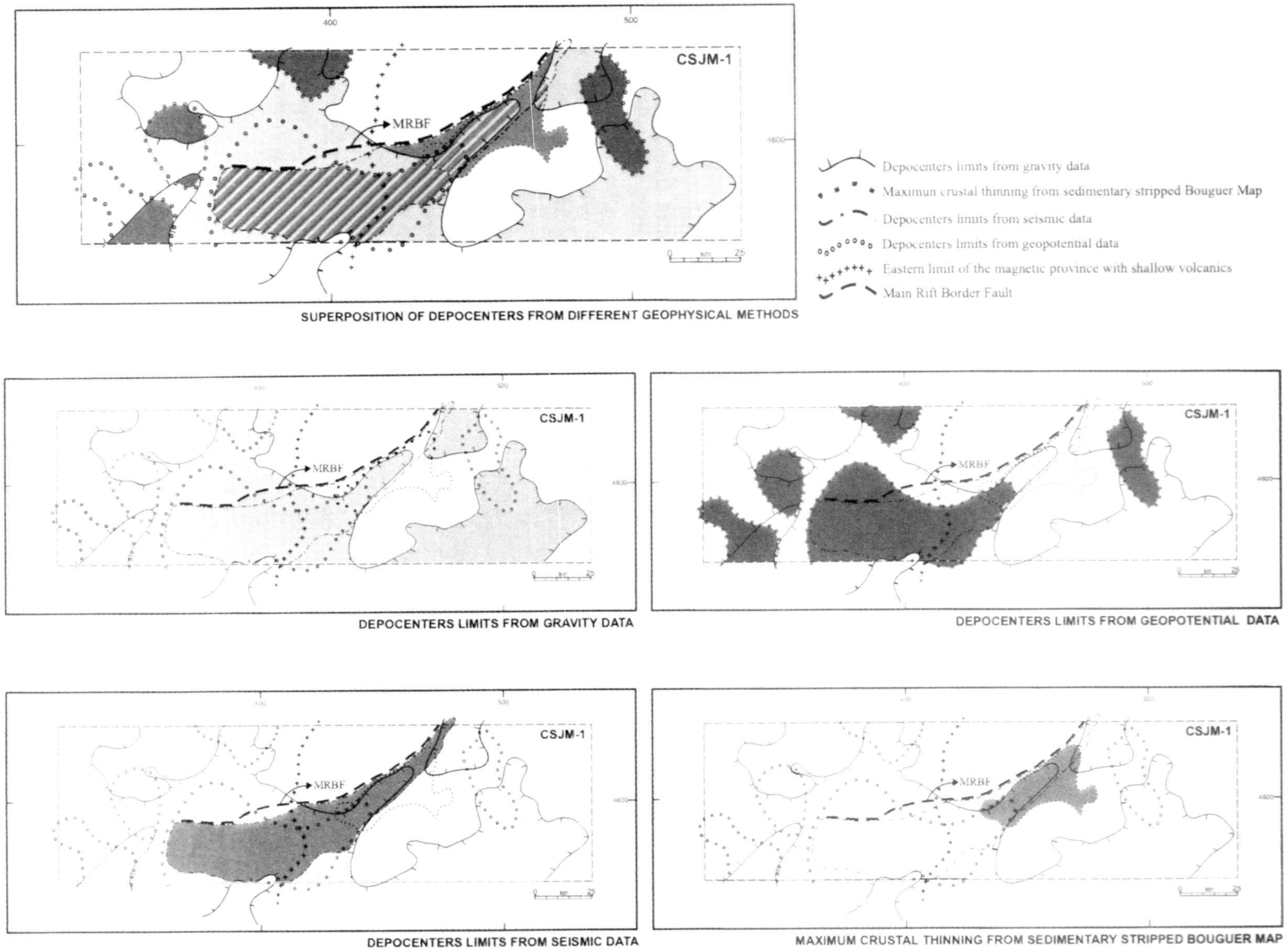

Figure 6. The interpretation of depocenters based in different geophysical methods is shown in Figure 6a. Interpretation based in gravity data (6b) and geopotential (6c) shows a good agreement regarding the position of the main basin depocenter, nearby the main rift border fault (MRBF). This main rift development area is also shown in the basin limits map based in the seismic interpretation (6d). The area of maximum crustal thinning based in the sedimentary stripped Bouguer map (6e) is in the northeastern end of this main rift development zone, also adjacent to the (MRBF).

indicate that the basement is represented by metamorphic rocks of Cambrian age.

The rift megasequence was subdivided in two parts: The upper part consists of a set of parallel reflectors, generally concordant with the base of the sag megasequence, with local truncations. The lower part is a reflection free seismic facies, which has not yet been drilled. Faults within the rift megasequence show thickening on the down thrown side.

The sag megasequence is characterized by a parallel set of continuous reflectors, with high amplitude in the lower part; and low amplitude discontinuous reflectors in the upper part. The upper part is associated with the most conspicuous angular unconformity observed in the basin, reflecting uplift, basin tilting and erosion.

Parallel reflectors (related to marine transgressions) and sigmoidal reflectors (associated with progradation) constitute the Tertiary passive margin megasequence.

6. STRUCTURAL FRAMEWORK

Based on the structural map of the basement obtained from regional interpretation of the seismic data (Figure 8), the basin may be divided in two different sectors, northeastern and western sectors.

The northeastern sector is a half-graben controlled by a main border fault running in a NE - SW direction and downthrown to the east. The basin depocenter reaches 6000 m near the main border fault. The main border fault has a horsetail termination at its southwestern end, where it splits into several smaller faults. In the northeastern sector, faults trend NE - SW, are parallel to the main fault and present normal throw.

In the western sector, structural elements are distinct. The basin is symmetric, with basement depth of 2 to 3 km and faults that strike east - west. In both sectors the faults have a strike-slip character: straight traces on profiles, positive flower structures, negative flower structures, faults merging at depth into a single vertical trace, and throw change from normal to reverse along the strike, following typical characteristics of wrench tectonics (Harding, 1985). In the northeastern sector, there is prevalence of normal throw in mapped faults, indicative of transtensional wrenching. Conversely, the western sector shows the predominance of reverse throw in the interpreted faults, reflecting the prevalence of a transpressional component of the wrenching event. This Late Cretaceous wrenching event overprinted most of the rift elements within the basin.

7. TECTONIC AND STRATIGRAPHIC EVOLUTION

The predicted stratigraphy for the prospect San Julian es-1was prepared considering the analogies between the San

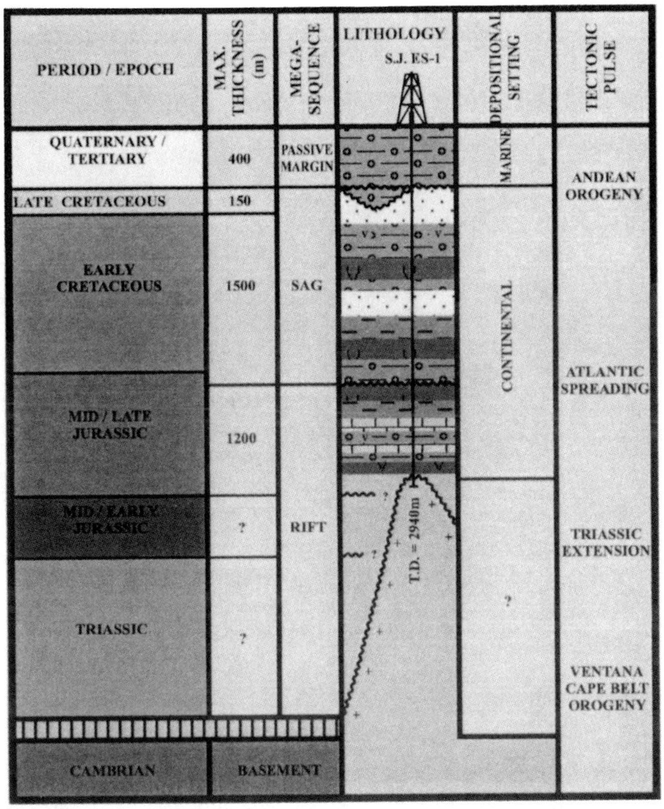

Figure 7. Stratigraphic chart, with tectonic pulse data based on data from Tankard et ali. (1995).

Jorge and San Julian basins. Since the well results proved the inaccuracy of the preliminary analogy between the San Julian and San Jorge basins, another tentative stratigraphic correlation was then estabilished with the Deseado massif. A field trip to El Tranquilo area, in the central part of the Deseado massif in southern Patagonia of Argentina, was carried out in order to examine possible facies changes and thickness of potential reservoirs in the area (Figure 9).

Several selected outcrops of Triassic to Tertiary rocks were visited. The outcropping Middle Jurassic to Cretaceous sequences are similar to the ones drilled by the San Julian es-1well, but the outcropping pre-Callovian interval, consisting of Triassic and Lower Jurassic continental clastics, was absent in the San Julian es-1well, probably because the well was drilled on a structural high.

Low grade metamorphic basement occurs in both El Tranquilo and San Julian basins. In El Tranquilo area, basement is related to the lower Paleozoic (Cambrian) La Modesta Formation, which is composed of quartzose and micaceous schists. A similar lithological description was obtained from the 6 m core recovered at the bottom of the San Julian es-1well.

In the San Julian ES-1well, basement is directly overlain

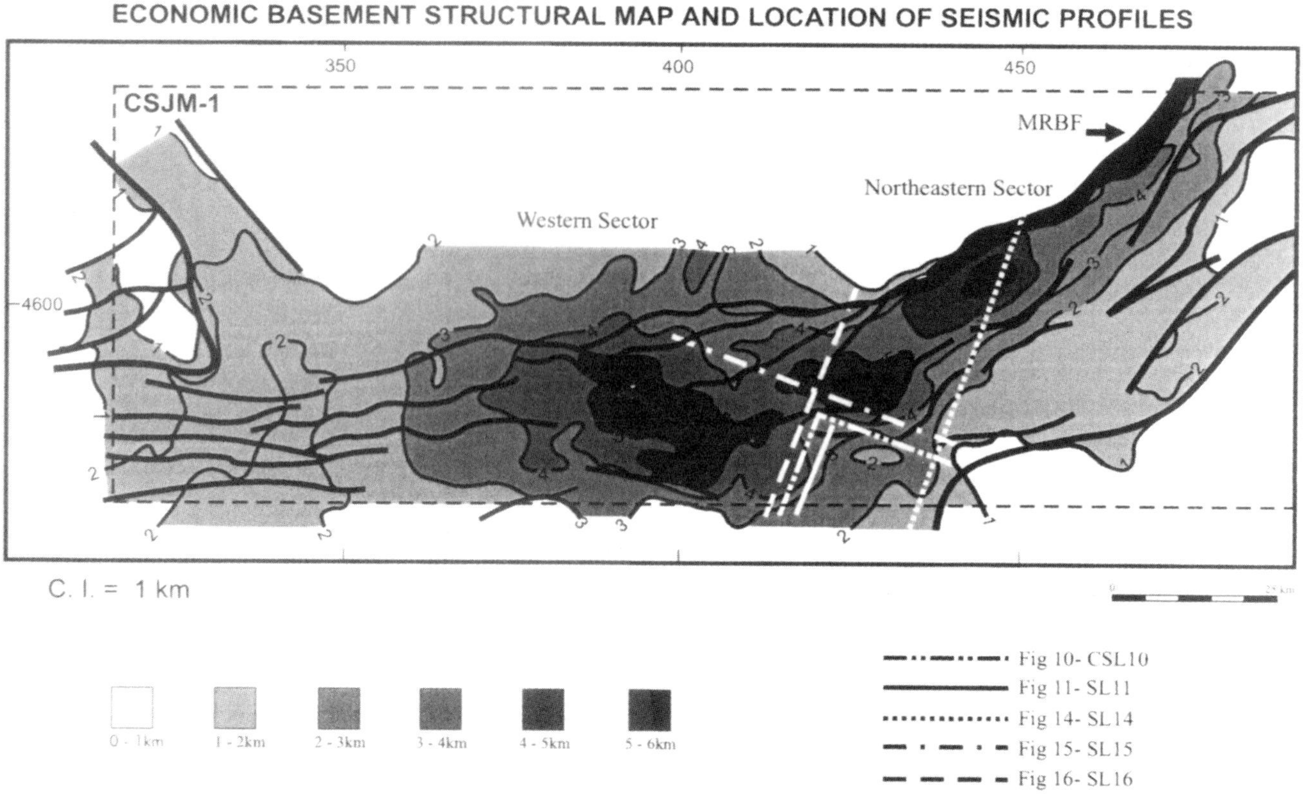

Figure 8. Economic basement structural map based on seismic data, two different compartments are shown in this map. The northeastern sector, a half-graben controlled by the main rift border fault (MRBF) and the western sector, a more symmetric basin compartment, with faults trending E/W.

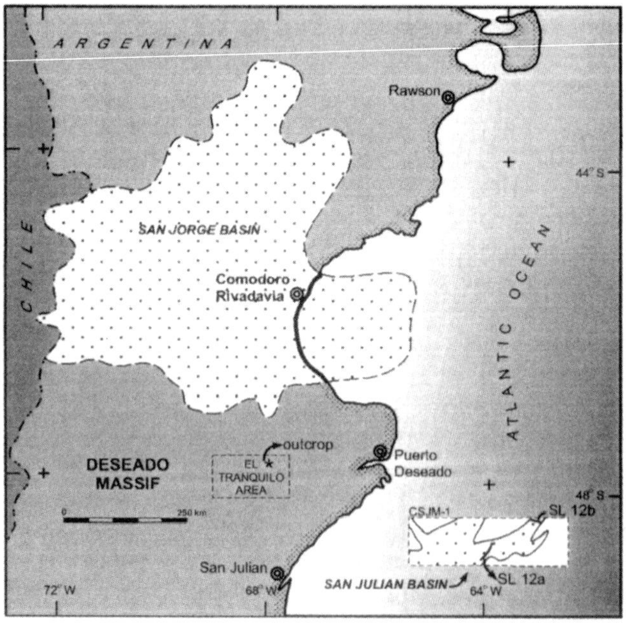

Figure 9. El Tranquilo area situation map, SL 12 – Location of seismic line shown in Figure12 and OC – Outcrop picture shown in figure 13.

by the Callovian to Oxfordian Bahia Laura Group. Seismic data suggest that a pre - Callovian interval occurs in basin depressions away from the well, thickening in the northeastward direction (Figure 10).

The depocenter of the undrilled pre - Callovian sequence in the San Julian basin is controlled by the main rift border fault (MRBF), along the northeastern graben. This major basement - related normal fault controlled the deposition of pre - Callovian sediments during rapid subsidence of the early rift phase. Rollover structures along listric faults were developed during deposition of contemporary sequences. The hydrocarbon potential of the pre-Callovian sequence remain untested by the SJ es-1well.

Callovian to Oxfordian rocks (Bahia Laura Group) outcrop in the El Tranquilo area, consisting of acidic volcanics, ignimbrites, tuffaceous sandstones, siltstones and claystones, with a maximum thickness of 1,100 meters. An equivalent Mid / Late Jurassic interval was deposited in the San Julian basin, reaching 1,200 meters in thickness. The distal lacustrine deposits of this volcanic complex are represented by tuffaceous claystones and oolitic limestones drilled by the SJ es-1well in the San Julian basin, which correlates to the late rift phase.

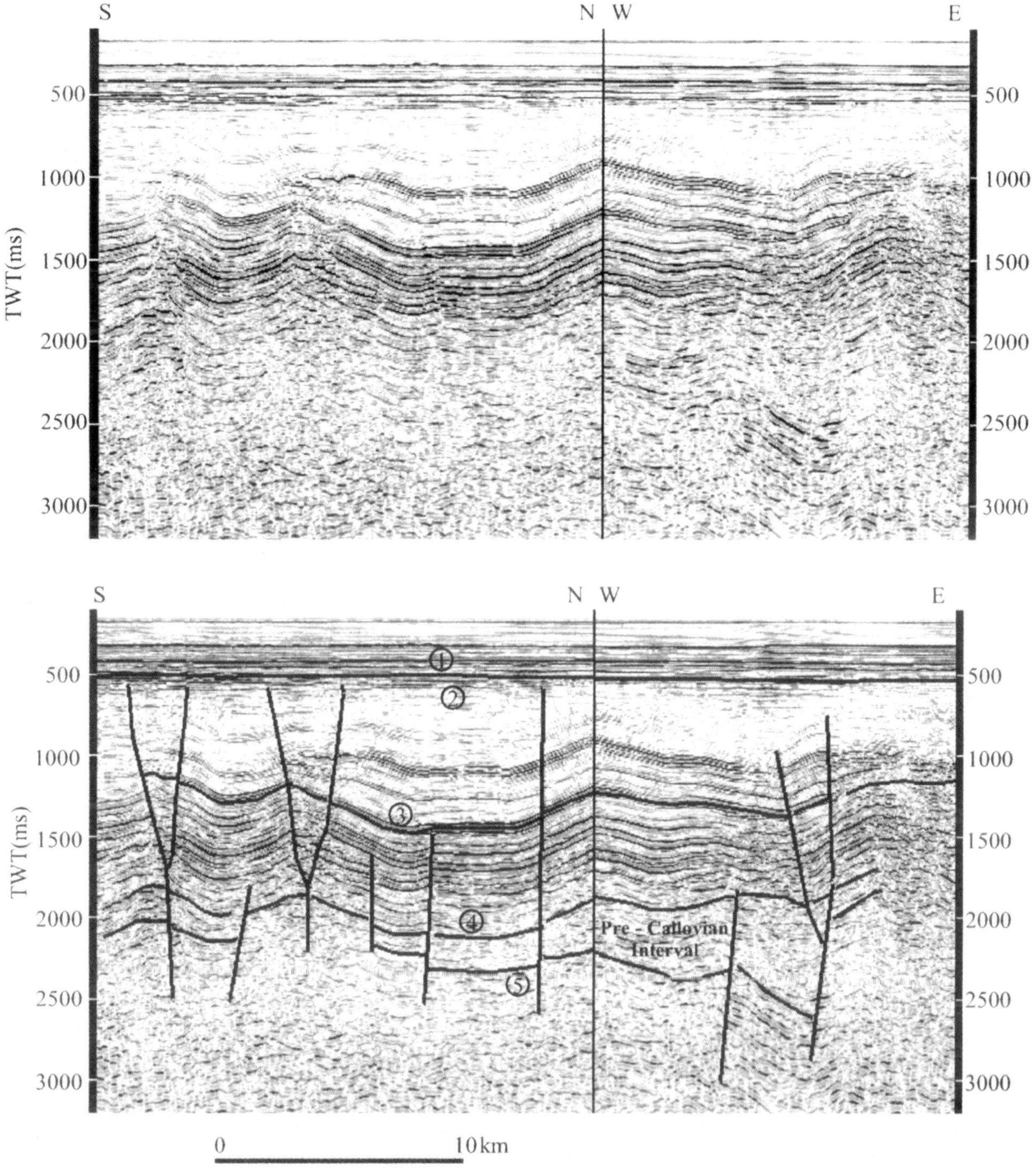

Figure 10. Composite seismic line 10 (10 – a interpreted profile, 10 –b uninterpreted profile) with identification of the pre-Callovian interval. Numbers indicate interpreted horizons: 1-Base of passive margin megasequence, 2 – Base of Maastrichtian Interval, 3 - Top of rift megasequence, 4 – Top of Pre-Callovian Interval and 3 – Economic basement. The Pre-Callovian section was not drilled by the SJ ES-1well, and remains as an untested interval regarding the hydrocarbon potential. Note that this section thickens northeastward. In San Julian well, Late Jurassic volcanics and volcaniclastics occur over the basement, although older units may occur away from the well, as indicated by seismic data. See Figure 8 for location.

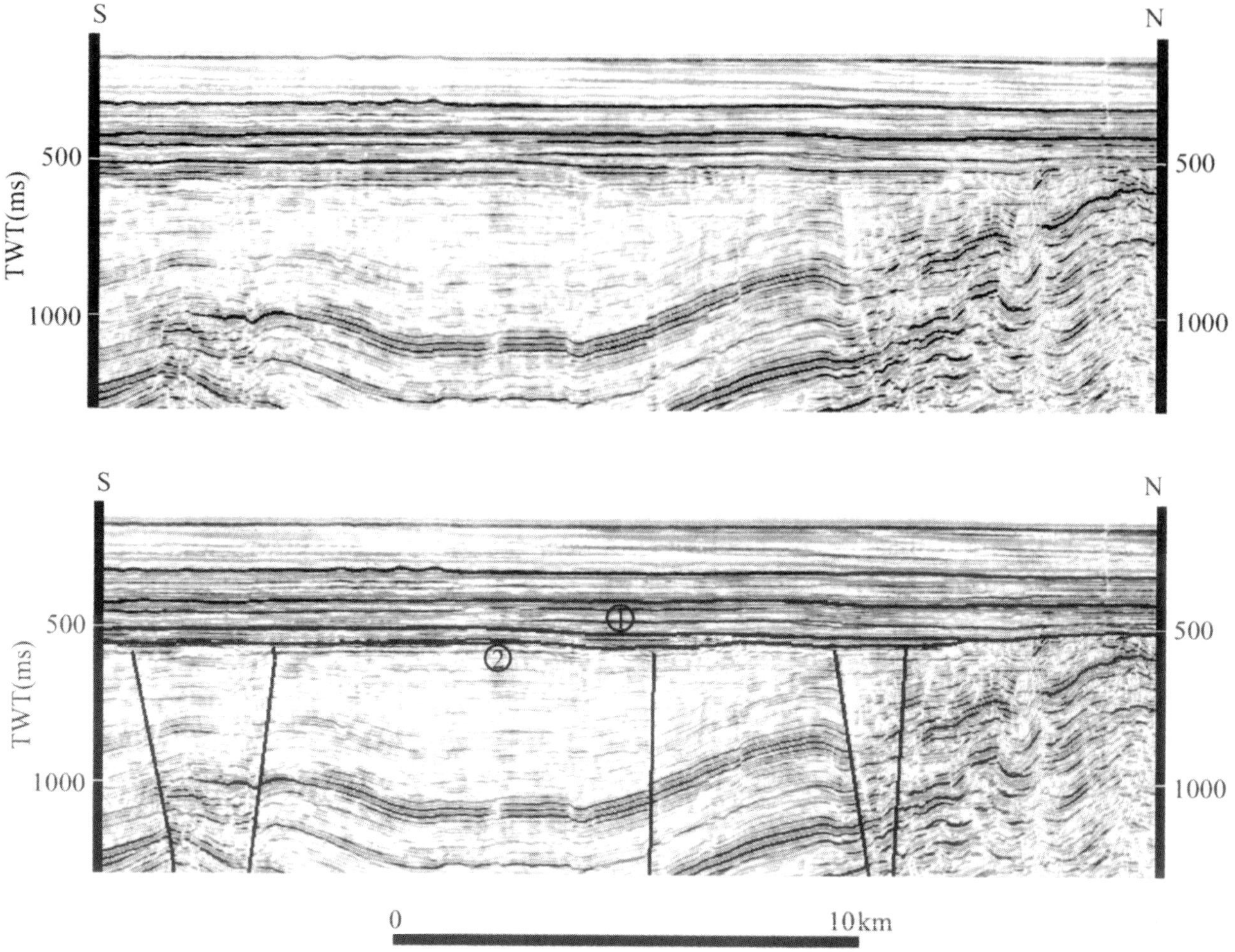

Figure 11. Interpreted (11a) and not interpreted (11b) seismic line with Maastrichtian interval. Numbers indicate interpreted horizons: 1-Base of Maastrichtian Interval, 2 – Base of passive margin megasequence. See figure 8 for location.

Near the base of this section in the San Julian well, there is a thick acidic igneous body (trachyte or rhyolite). This type of extrusive bodies is believed to be widespread. Also in the top of the rift package, oolitic calcarenites, calcretes and other paleosoils were defined, suggesting the existence of shallow ephemeral lakes at the end of this phase.

Tithonian to Neocomian rocks (Bajo Grande Fm.) outcrop in the El Tranquilo area, consisting of tuffaceous sandstones, siltstones and claystones, with a maximum thickness of 1,000 meters. An equivalent Lower Cretaceous interval was deposited in the San Julian basin, reaching 1,500 meters in thickness. Several layers of acidic volcanics were also drilled by the San Julian ES-1well within this interval. Although the Neocomian is considered to be part of the regional rifting period in southern Argentina, a distinct tectonic evolution seems to have occurred in the San Julian basin. The main border fault reduced its tectonic activity and did not have a strong depositional control. The basin depocenter shifted westward as a thermal subsidence regime was developed, characterizing a distinct sag depocenter.

Maastrichtian rocks (Laguna Palacios Fm.) outcrop in the El Tranquilo area and the stratigraphic column consists of lapilli tuffs, with a maximum thickness of about 200 meters. An equivalent Upper Cretaceous interval was anticipated in the San Julian basin and was also recognized in the well, reaching up to 150 meters in thickness with the same general composition of the previous sequences. The areal distribution of Upper Cretacous tuffs within the San Julian basin seems to be restricted to the vicinities of the sag depocenter and may be considered as a late sag unit (Figure 11).

The relative tectonic stability observed during most of the Cretaceous in the San Julian basin preceded a pronounced wrenching event that promoted marked graben inversion and lateral displacement between blocks, overprinting the

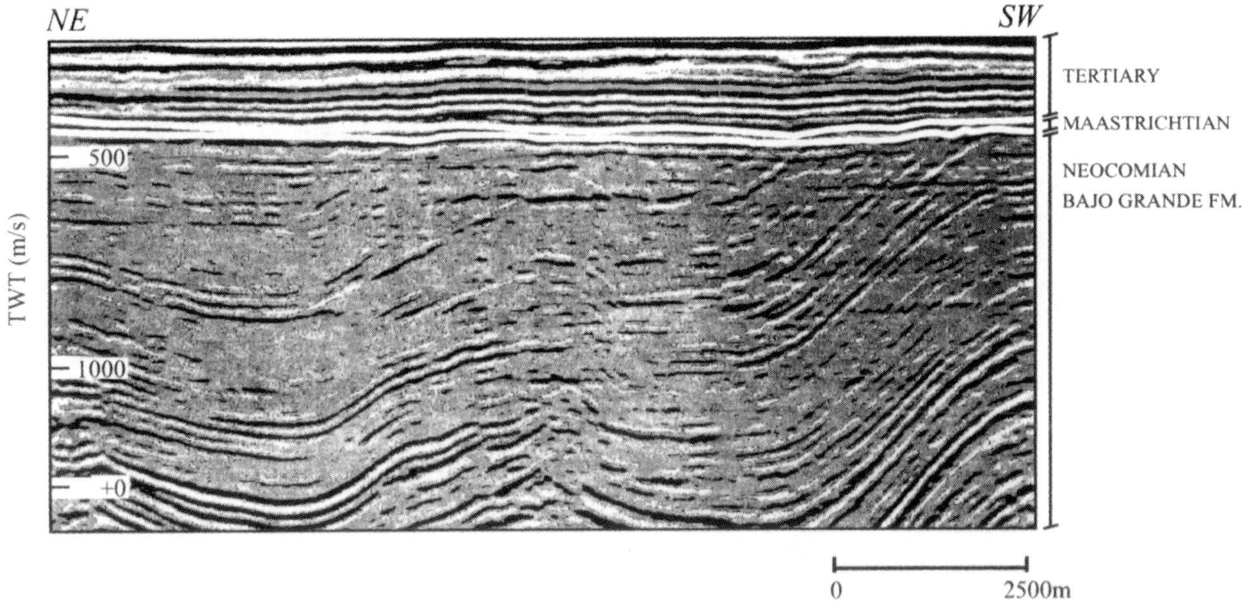

Figure 12a. Interpreted seismic profile in the San Julian basin showing the angular unconformity that separates the Tertiary plus Maastrichtian above (Laguna Palacios equivalent) from Neocomian below (Bajo Grande equivalent). See figure 9 for location.

older rift trends within most of the basin depocenters. The wrench character of this event is expressed in several positive and negative flower structures observed in San Julian seismic data. These wrench-related faults have a NE – SW trend in the northeastern sector of the basin; and they are characterized by normal throw and transtensional character. In the western sector, they have an east-west trend and reverse throw, indicating transpressional processes.

The age of this tectonic reactivation is believed to be associated with the strong angular unconformity observed on seismic data in the San Julian basin (Figure 12-a and 12-b). A similar angular unconformity was also observed in outcrops along the El Tranquilo area (Figure 13). Horizontal Maastrichtian Laguna Palacios tuffs rest directly over dipping Neocomian Bajo Grande tuffaceous siltstones, which possibly indicate an early Maastrichtian age for this tectonic event, as corroborated by seismic data (Figure12-a). The Maastrichtian is absent in the northeastern sector of the basin crossed by seismic line 12-b.

Evidence from seismic and stratigraphic analysis indicate that the first marine transgression into the San Julian basin occured during the Paleocene over a peneplained terrain. Prograding marine sequences were deposited throughout the Tertiary, reaching up to 400 meters in thickness. This event corresponds to the passive margin phase of subsidence in the basin.

8. SEISMIC EXAMPLES OF WRENCH TECTONICS STRUCTURAL STYLES

Some of the seismic lines provide remarkably good examples of wrench tectonics affecting sedimentary basins. Also tectonic provinces and structural styles found within San Julian basin are shown in these lines.

Line 14 (Figure14) shows the half-graben character of the northeastern sector of the San Julian basin. Wrench style tectonics is shown on this line by the normal throw, almost vertical faults which affect rift and sag phase sediments. Negative flower structures are shown on the seismic section and, together with the intense, almost vertical trace of normal faults, are the most typical structures of this part of the basin.

Line 15 (Figure15) in the central part of the basin shows three positive flower structures, formed during the wrenching episode that has a transpressive character in this part of the basin. The flower structure in the middle of the section is a very good example of a rift phase fault reactivated as a positive flower structure. The right and left ends of the section show the angular unconformity separating the almost horizontal reflectors of the passive margin phase and the high angle reflectors of the sag and rift phases, which were affected by the wrenching episode.

Line 16 (Figure 16), situated at the western limit of the main rift, shows other interesting examples of the wrench-

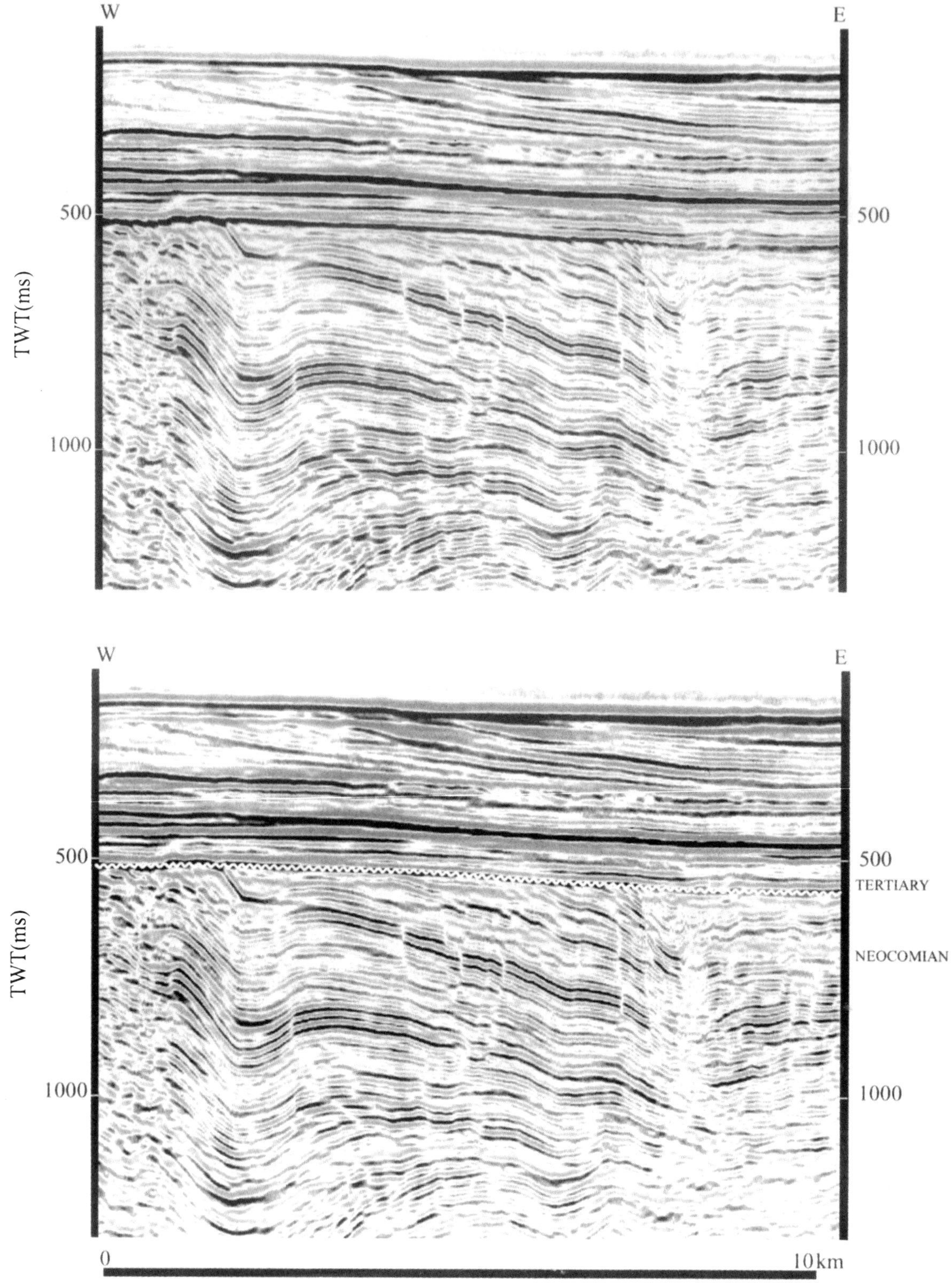

Figure 12b. Uninterpreted (12b) and interpreted (12bi) seismic profile in the San Julian basin showing the angular unconformity that separates the Tertiary from Neocomian below (Bajo Grande equivalent). See figure 9 for location.

Figure 13. Outcrop in El Tranquilo Area showing the angular unconformity that separates Laguna Palacios Formation (LP) above and Bajo Grande Formation (BG) below. In figure 13b the approximate trace of the unconformity is indicated.

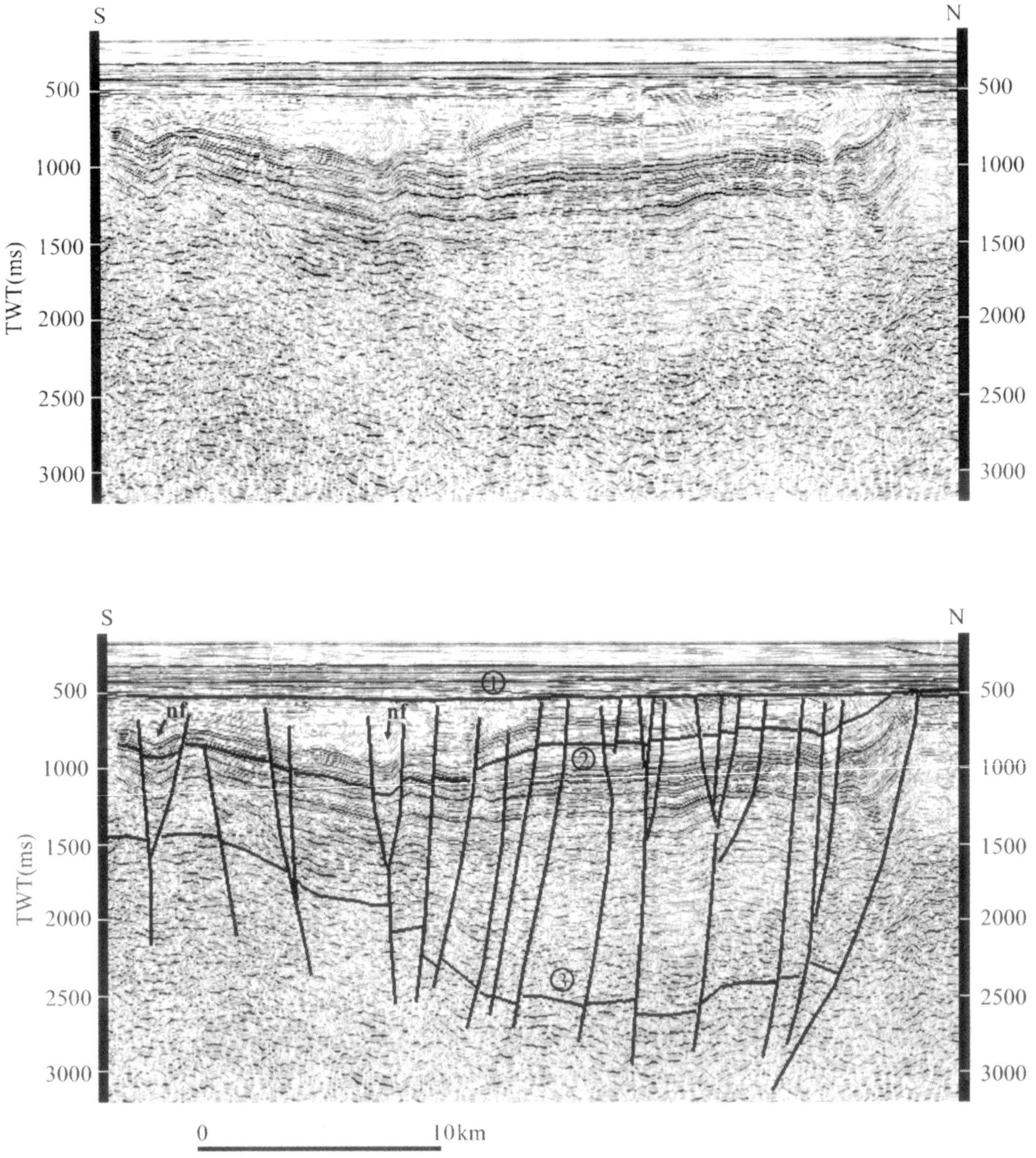

Figure 14. Interpreted (14 a) and uninterpreted (14 b) seismic line 14. Numbers indicate interpreted horizons representative of megasequence boundaries: 1-Base of passive margin megasequence, 2 – Top of rift megasequence and 3 – Economic basement. Arrows indicate interpreted negative flower (nf) structures. Several normal throw, near to vertical trace faults are represented. See Figure 8 for location.

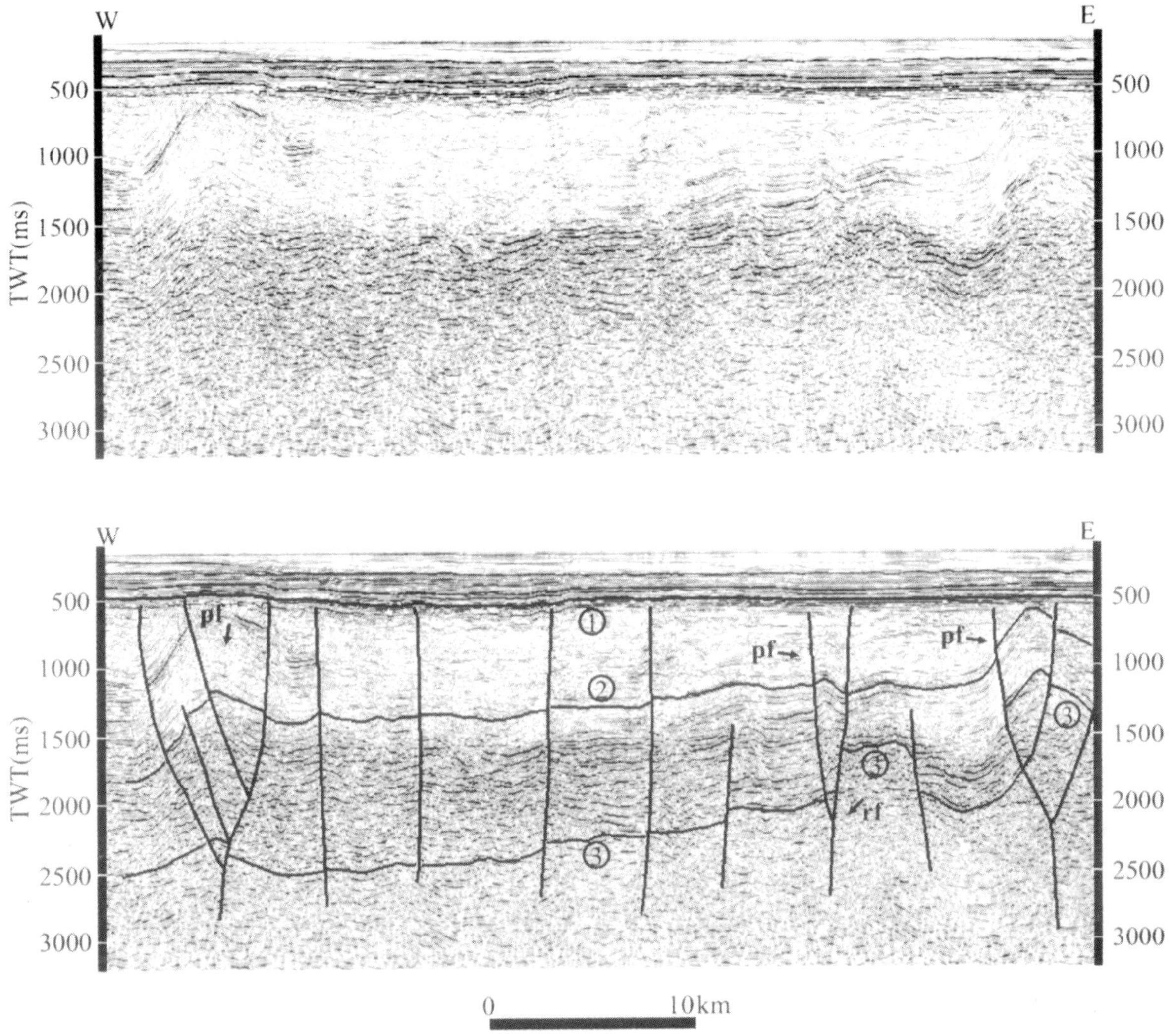

Figure 15. Interpreted (15 a) and uninterpreted (15 b) seismic line 15. Numbers indicate interpreted horizons representative of megasequence boundaries: 1-Base of passive margin megasequence, 2 – Top of rift megasequence and 3 – Economic basement. Arrows indicate interpreted positive flower (pf) structures. Several near to vertical trace faults are represented. Note rift phase fault reactivated (rf) as a part of a positive flower structure. See Figure 8 for location.

ing processes that affected the basin. The main distinguishing criteria are: 1) almost straight vertical traces of faults; (2) positive flower structures on the left part of the section; (3) negative flower structures on the right hand side of the section; and (4), faults where the throw changes from normal for deeper reflectors to reverse for shallower reflectors along the same fault trace.

9. CONCLUSIONS

Geophysical data consisting of potential field and seismic data acquired within San Julian basin were interpreted and integrated within a regional tectonic framework. Two distinct structural domains were recognized in the basin; the northeastern sector where a main rift was interpreted; and the west sector where the basin is more symmetric and characterized by sags. In both sectors, geophysical data presented evidence of wrench tectonics, transtensional in the northeastern sector and transpressional in the western sector.

The evolution of the basin through rift (Triassic to Late Jurassic), sag (Cretaceous) and passive margin (Tertiary to Holocene) phases was established based on the geophysical and geological dataset.

The Late Cretaceous wrench event that affected the basin left several remarkable examples of this kind of tectonics that are shown on the seismic data.

The drilled well presented very poor exploratory results.

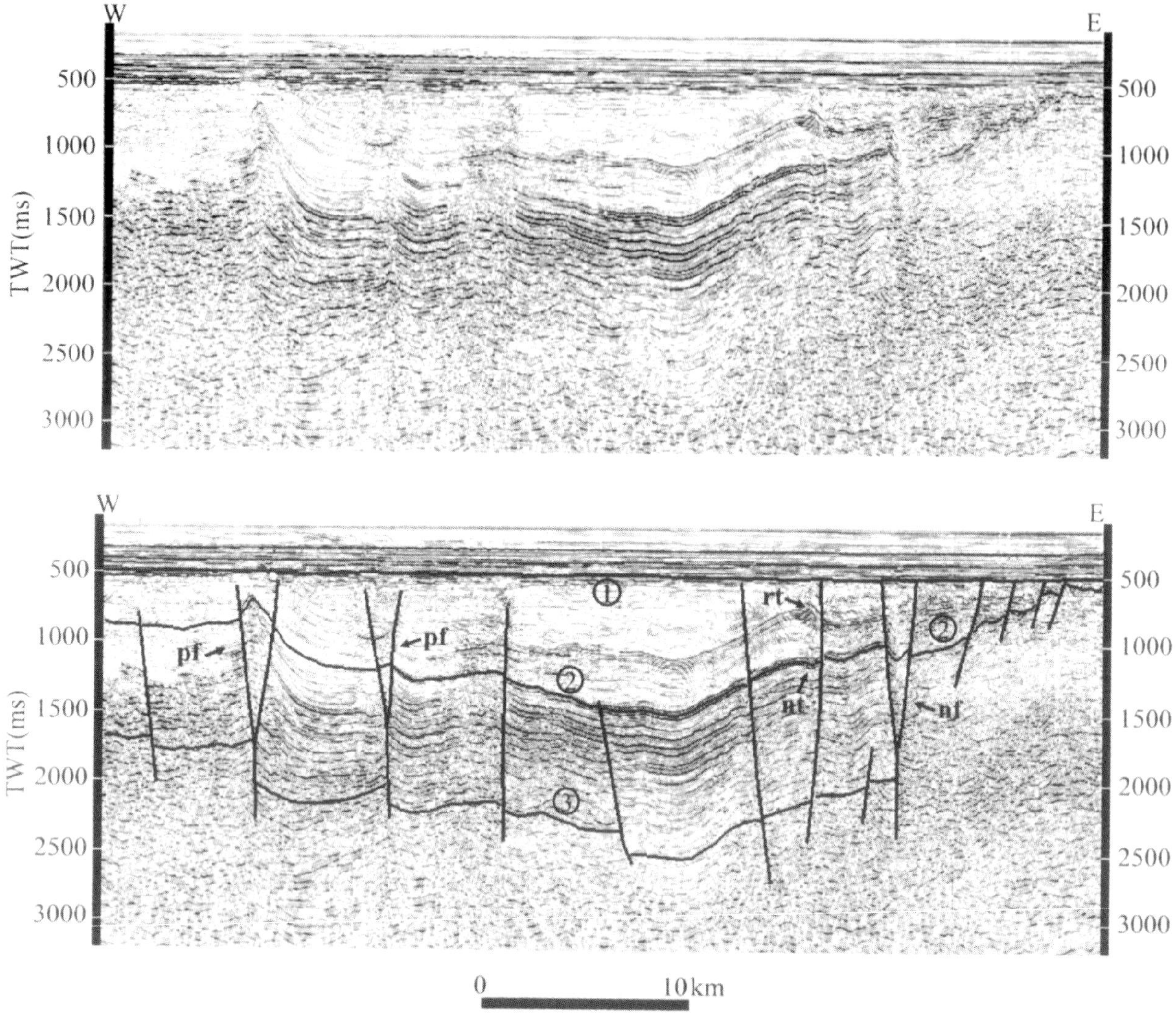

Figure 16. Interpreted (16 a) and uninterpreted (16 b) seismic line 16. Numbers indicate interpreted horizons representative of megasequence boundaries 1-Base of passive margin megasequence, 2 – Top of rift megasequence and 3 – Economic basement. Arrows indicate interpreted positive flower (pf) structures and negative flower (nf) structures. Several near to vertical trace faults are represented. Note fault with changes in the throw from normal (nt) below to reverse (rt) above. Normal faults affecting the basement are also observed on the right hand side of the seismic line. See Figure 8 for location.

However the pre-Callovian sequence that occurs in depressions away from the well was not tested, and the exploratory potential of the basin still remains to be assessed.

Acknowledgements. This paper presents results of teamwork performed by Petrobras Internacional – BRASPETRO Exploration Group. Several other BRASPETRO and PETROBRAS colleagues contributed to the analysis and conclusions presented in this paper. Among those we would like to mention Antonio Affonso Rocha, Antonio Carlos de Almeida Rocha, Fausto Machado, Idelso Piazza, Luis Braga, Paulo Ledur, Pedro Victor Zalan and Rudy Ferreira. We thank also Dr. Webster Ueipass Mohriak for his innumerous suggestions and incentive.

REFERENCES

Figueiredo, A.M. de F., A.P. de Miranda, R.F. Ferreira and P.V. Zalán (1995): *Cuenca de San Julian, Geología y Recursos Naturales de la Plataforma Continental Argentina*, V.A.Ramos y M.A. Turic Eds. p. 193-212.

Fitzgerald, M.G., R.M. Mitchum, M.A. Uliana and K.T. Biddle (1990): Evolution of the San Jorge Basin, Argentina:

American Association of Petroleum Geologists Bulletin, v.74, p. 879 – 920.

Harding, T.P. (1985): Seismic Characteristics and Identification of Negative Flower Structures, Positive Flower Structures, and Positive Structural Inversion: *American Association of Petroleum Geologists Bulletin,* v. 69, p. 582-600.

Mussett, A.E. and G.K. Taylor (1994): ^{40}Ar – ^{39}Ar ages for dykes from the Falkland Islands with implications for the break-up of southern Gondwanaland: *Journal of the Geological Society,* London, Vol. 151, p. 79-81.

Sandwell, D.T. and McAdoo, D.C., 1988. Marine gravity of the southern ocean and Antarctic Margin from Geosat. *Journal of Geophysical Research*, vol. 93, p. 10,389-10,396.

Soares Filho, J.R. de S. and A.P. de Miranda (1997): Interpretação Geofísica na Bacia de San Julian – Offshore da Argentina: 5º Congresso Internacional da Sociedade Brasileira de Geofísica – v. 1, p. 37-40.

Stoakes, F.A., C.V. Campbell, R. Cass and N. Ucha (1991): Seismic Stratigraphic Analysis of the Punta del Este basin, Offshore Uruguay, South America: *American Association of Petroleum Geologists Bulletin,* v. 75, p. 219-240.

Tankard, A. J., M.A. Uliana, H.J. Welsink, V.A Ramos, A.B.França, E.J. Milani, B.B de Brito Neves, N. Eyles, J. Skarmeta, H. Santa Ana, F. Wiens, M. Cirbián, O. López Paulsen, G.J.B. Germs, M.J. De Wit, T. Machacha and R. McG. Miller (1995): Structural and Tectonic Controls of Basin Evolution in Southwestern Gondwana During the Phanerozoic, in A.J.Tankard, R.Suárez and H.J. Welsik, *Petroleum Basins of South America: AAPG Memoir* 62, p 5-52.

Uliana, M. A. and K. T. Biddle (1988): Mezozoic-Cenozoic Paleogeographic and Geodynamic Evolution of Southern South America: Revista Brasileira de Geociências – v. 18 (2), p 172-190

Inferences Regarding Initiation of Oceanic Crust Formation From the U.S. East Coast Margin and Conjugate South Atlantic Margins

Manik Talwani

Rice University and Houston Advanced Research Center

Vitor Abreu

Unocal Corporation

The U.S. East Coast volcanic continental margin is compared and contrasted with volcanic continental margins of the South Atlantic (Pelotas and Walvis basins). Seismic reflection data acquired by the EDGE project off the US East Coast, supplemented by wide-angle reflection and refraction surveys, provided the best data across this margin. One important geophysical feature associated with this margin is the presence of a very prominent Total Intensity magnetic anomaly that lies over the Continental Slope. The East Coast Magnetic Anomaly may be associated with the seaward dipping reflectors (SDR's) along the US East Coast and this wedge is interpreted to be a part of the Initial Oceanic Crust. Migration of the seismic reflection data suggested that the strongly reflecting continental crust ended abruptly at an interface with a steep landward slope. It seems that the dipping reflectors, the underlying intrusive material, and the high velocity material lying further below (interpreted earlier as underplated material) actually constitutes the earliest crust and upper mantle generated at the time of the initial opening of the ocean. The volcanic complex of the Paraná-Etendeka continental flood basalts and the SDR's of the Pelotas and Walvis basins are the result of the rifting and subsequent breakup of the South American and African plates under an initial influence of the Tristan da Cunha hotspot. The SDR's wedges of Pelotas and Walvis basins were emplaced after continental breakup at least partially over an extended continental crust. Two major SDR's provinces characterize the South Atlantic: Santos-San Jorge (South America) and Walvis-Orange (Africa) provinces. These provinces form a broad and symmetrical volcanic complex. It is proposed that these SDR's provinces were emplaced as a subaerial oceanic ridge that would represent the initial stage of formation of the oceanic crust in most passive margins.

INTRODUCTION

While much progress has been made in resolving details of the crustal structure across passive margins, at least two important problems remain: one involves the detailed nature of the transition from continental to oceanic crust and

Atlantic Rifts and Continental Margins
Geophysical Monograph 115

the other involves the along-strike variation of passive margin structure. We believe that these need to be resolved in order to provide a firm basis for theories regarding the evolution of these margins.

In this study, the interpretation of a volcanic margin off the U.S. East Coast is compared with a complementary interpretation of two conjugate South Atlantic continental margins. The U.S. East Coast study includes deep crustal experiments and thus provides information about the detailed nature of the transition from continental to oceanic crust.

The South Atlantic study, which incorporates seismic reflection data, provides detailed comparison of two conjugate margins as well as along-strike variations along both margins.

Most theories of the evolution of passive margins start with McKenzie's elegant exposition of extension and consequent thinning of continental crust (McKenzie, 1978). Later, authors have elaborated on the extent of thinning and have generally concluded that the crust thins to a few kilometers and then breaks, leading to formation of oceanic crust and to the onset of drift (e.g., *Le Pichon and Sibuet*, 1981). A critical question is whether there is firm seismic evidence for a continental crust that has been thinned down to few kilometers. There does appear to be evidence from studies of the Northeast Atlantic passive margins that the crust thins to extreme degrees under the continental margins. One of the earliest studies that demonstrated the thinning was by *Montadert et al.* (1979), and there have been later and more elaborate studies since then. Does similar thinning occur on all passive margins? In particular, does it occur on volcanic margins?

Part of the difficulty in answering this question is how continental crust is to be identified (or even defined). The two diagnostic features normally used are the presence of listric normal faults and identification of materials by their seismic velocities. Unfortunately, it is not possible to carry out drilling to depths where the nature of the crust is most questionable. While listric normal faults clearly define faulted continental blocks at many locations, they may not do so at all locations. Seismic reflection data near the axis of Labrador Sea, where identified marine magnetic anomalies clearly indicate that the crust is oceanic, show listric normal faults (*Chalmers and Laursen,* 1995). Therefore, such faults cannot always be associated with continental crust. The LASE experiment (LASE, 1986) first called serious attention to the observation that material with velocity about 7.2 km/sec forms the lower part of the crust, interpreted by most authors as underplated material rising out of the underlying mantle, which was not designated oceanic or continental. There is the added question arising from the presence of the Seaward Dipping Reflectors (SDR's), first discovered by *Hinz* (1981). As more and more deep seismic reflection profiles have been obtained by equipment capable of penetrating to progressively greater depths, two facts about the SDR's have become clear. The first fact is that they appear to be almost ubiquitous in margins that have been explored with deep penetrating seismic reflection equipment. Secondly, they extend to great depth - in some cases approaching the region of the 7.2 km/sec material. Is there, then, space for a thinned continental crust lying below the SDR's?

In the North Atlantic EDGE experiment, vertical incidence reflection was carried out with a powerful sound source and a long (6 km) towed array, and wide angle reflection and refraction were carried out using the same sound source and a number of ocean bottom instruments. As a result, the most detailed crustal section to date across this margin was obtained.

While the results from the EDGE experiments have been reported earlier, further processing of seismic data was able to shed new light on the details of ocean continent transition. The second area is in the South Atlantic, where we present results from two conjugate (Brazil and Namibia) margins, and discuss the along strike variations along these two margins. We believe that these results will shed some light on the questions posed above.

THE US EAST COAST MARGIN

The EDGE experiment off the US East Coast combined seismic reflection and refraction experiments and extended from New Jersey across the southern part of the Baltimore Canyon trough. It provides the most detailed crustal information across the US Atlantic margin (*Sheridan et al.*, 1993; *Holbrook et al.*, 1992; *Holbrook et al.*, 1994; *Talwani et al.*, 1995; *Glover et al.*, 1997). Figure 1 locates the multichannel seismic reflection lines and the ocean bottom hydrophones and seismometers that recorded the airgun shots fired on Line 801, which extends from the mouht of the Chesapeake Bay toward the oceanic crust with bathymety exceeding 3,000 m.

The crustal structure along the profile, which was obtained by multidisciplinary integration of several geophysical methods, is summarized in Figure 2 (*Talwani et al.*, 1995). The top part of Figure 2 shows reflectors traced by Sheridan et al. (1993). A principal discovery was the presence of a characteristic wedge of seaward dipping reflectors extending down to depths approaching 25 km. The reflection time section has been converted to a depth section using velocities obtained by *Holbrook et al.*, 1994. The depth to Moho decrease seawards to about 30 km under the shelf, but there is no evidence of further thinning under the shelf. Note that this is an unmigrated stack and migration severely alters the position of the reflectors, particularly in the part of the section considered continental.

The wide-angle reflection and refraction crustal velocity section constructed by *Holbrook et al.* (1994) is shown in

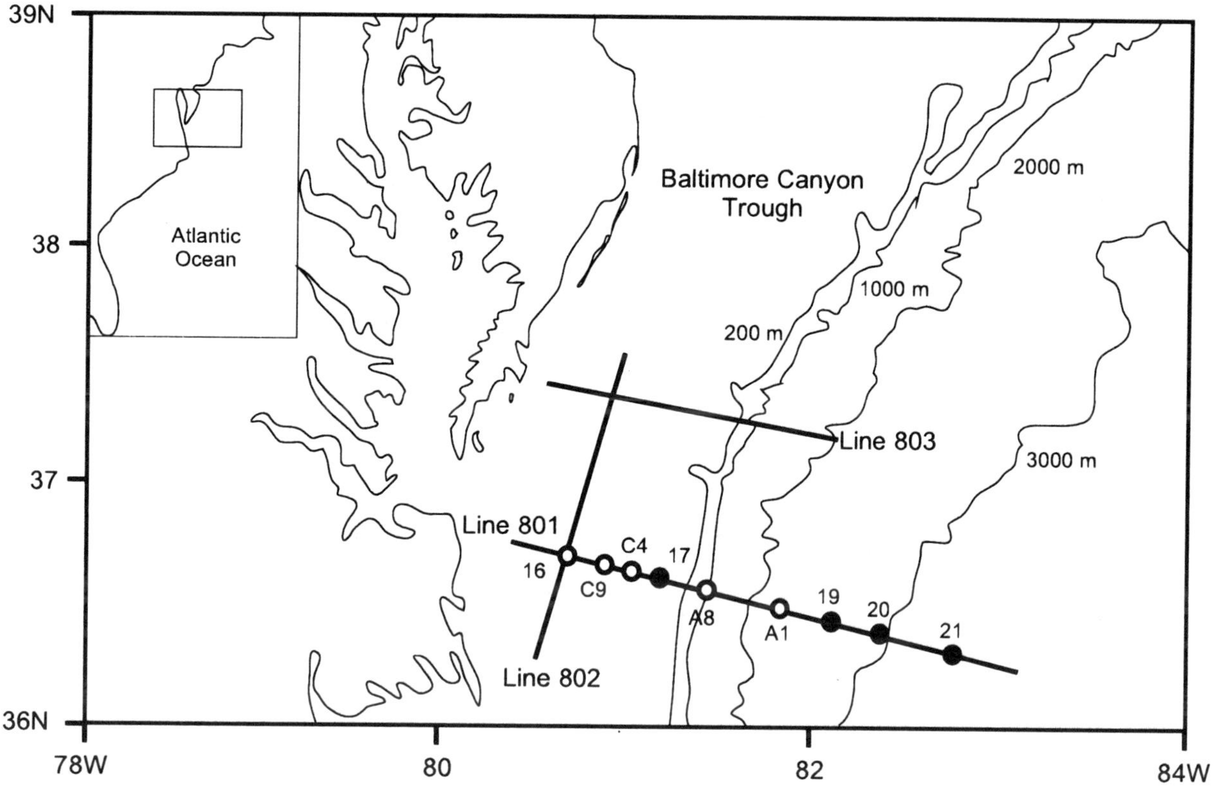

Figure 1. Location map of the EDGE seismic experiment. Vertical Incidence seismic reflection was carried out on lines 801, 802, and 803 (Sheridan et al., 1993). Airgun shots along line 801 were recorded by ocean bottom hydrophones (open circles) and ocean bottom seismometers. Wide-angle reflection and refraction data were used to obtain crustal structure (Holbrook et al., 1994). Airgun shots along line 802 were also recorded at ocean bottom seismometer #16.

Figure 2 (middle section). It was constrained at the upper part of the crust (sediments and seaward dipping reflectors) by the vertical incidence seismic reflection information. Data recorded by ocean bottom instruments provided velocities as well as information about deeper crustal structure where no reflectors had been obtained by vertical incidence reflection profiling. Thus, velocities greater than 6 km/sec were associated with the deeper parts of the seaward dipping reflectors and it was inferred that the seaward dipping reflectors continued seawards into the upper part of the normal oceanic crust. Wide-angle reflections from the base of the upper part of the crust (interface 7 in Figure 3) are clearly resolved in data recorded by the ocean bottom instruments (7R in Figures 4 and 5) and serve to establish this crustal boundary. Reflections from M, the interface that defines the base of the crust (see synthetic and observed arrivals marked M in Figures 4 and 5) are also well documented by the ocean bottom instruments. An important result is that the lower part of the crust possesses a velocity exceeding 7 km/sec. On the shelf, support for the flat part of interfaces 8 and 9 was derived from vertical incidence reflections (*Sheridan et al.,* 1993). Velocities were obtained from unreversed wide-angle reflection data observed by *Holbrook et al.* (1992), from shots on Line 802 and received at OBS #16 (Figure 1). Evidence is lacking for the nature of the seaward continuation of interfaces 8 and 9 (Figure 3).

The bottom section in Figure 2 shows that the observed magnetic anomaly can be explained by the association with SDR's. Two assumptions have been made: that the volcanic layers represented by the reflectors extend seaward (where they have not been imaged), and secondly, that the feather edge of the reflectors extends about 10 km landward (from 40 km to 30 km) from the westernmost point where it is imaged seismically.

The material lying below the deepest parts of the SDR's (from 80 to 100 km, Figure 2) has a velocity of 7.5 km/sec. Many authors interpreted this to be underplated material arising from below. At any rate, this material is not considered continental in nature because of the high velocity value. Thus, *Talwani et al.* (1995) inferred that the entire crustal material lying above Moho and underlying the

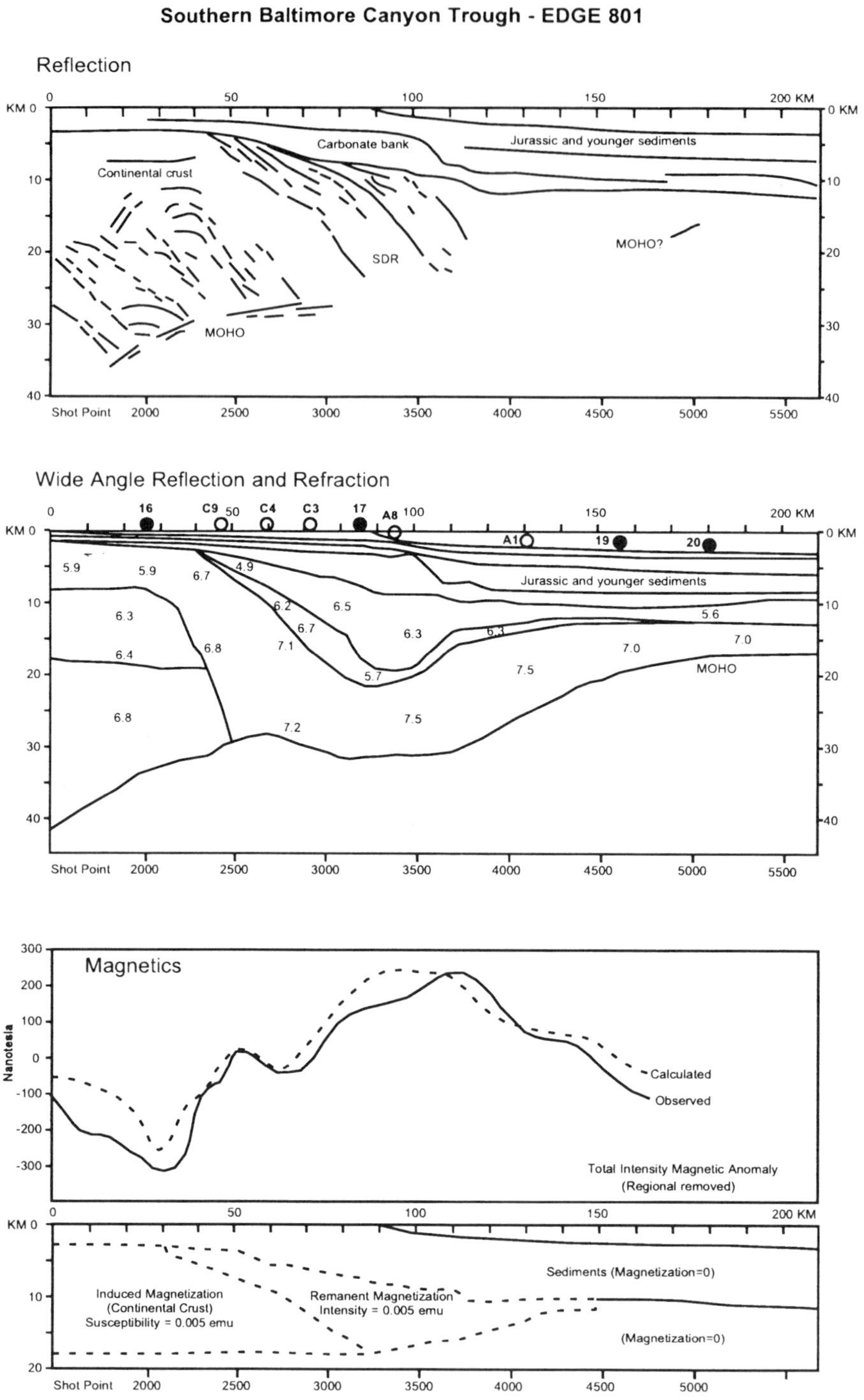

Figure 2. Top - Seismic reflection section along EDGE line 801 (Sheridan et al., 1993). Time section converted to depth section using velocities given by Holbrook et al. (1994).
Middle - Crustal section with seismic velocities along EDGE line 801 (Holbrook et al., 1994).
Bottom - East Coast Magnetic Anomaly along EDGE line 801 attributed to volcanic layers with high magnetic intensity (Talwani et al., 1995).
The seismic reflection section is unmigrated. Migration alters the geometry of the reflectors drastically. See Figures 5a and 5b for migrated time sections. The altered geometry is reflected in Figure 3.

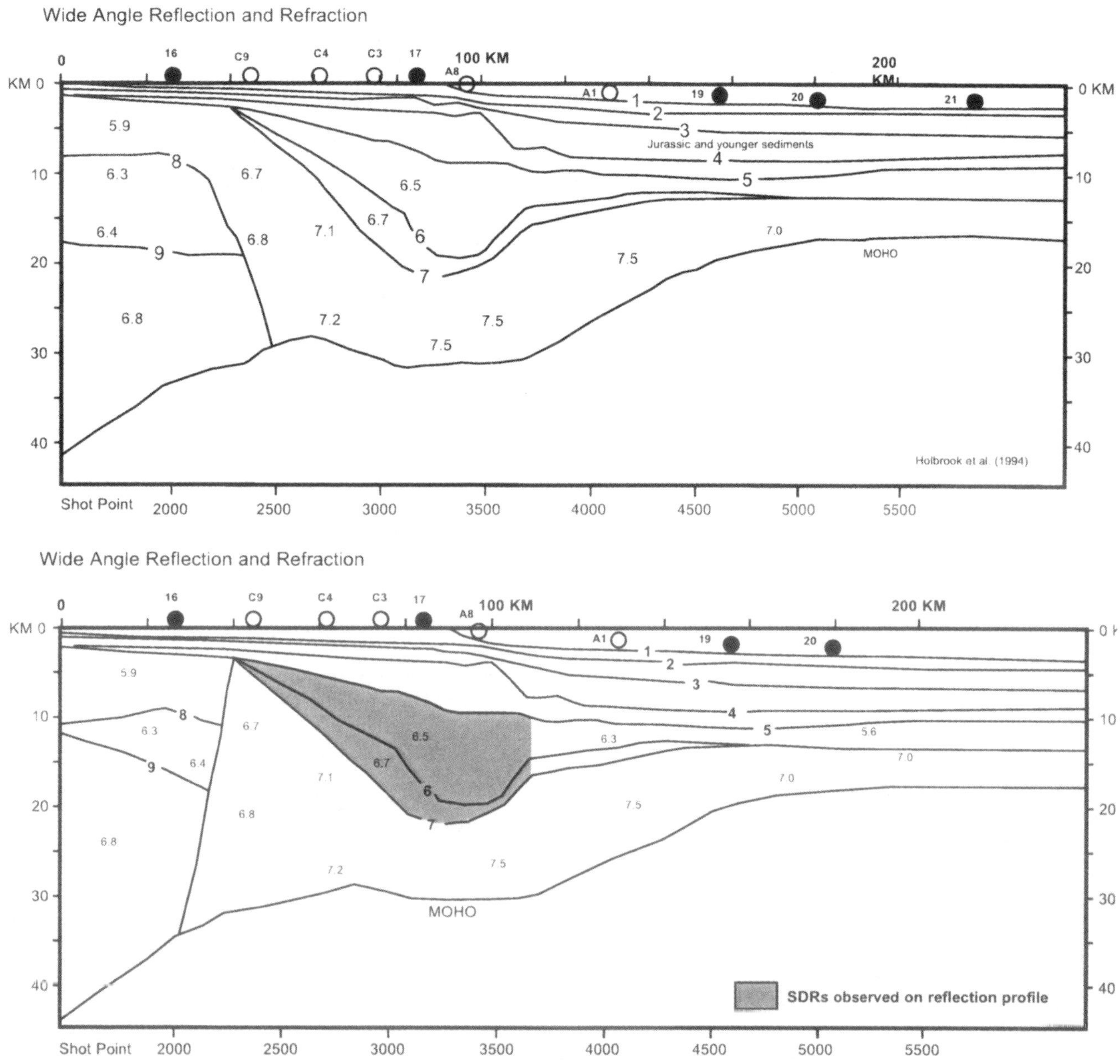

Figure 3. Top - Crustal section obtained by Holbrook et al., 1994.
Bottom - Holbrook's crustal section modified by replacing boundaries 8 and 9 by the boundaries between the domains identified in Figure 5b. These domains are truncated against the steeply dipping boundary (between the reflecting and non-reflecting crust) identified in Figure 5b. The increase in seismic velocities across the steeply dipping boundary is noteworthy.

SDR's wedge was emplaced after separation, and called it "initial oceanic crust." Their views were in contrast to the opinion held by many authors that thin continental crust lied between the SDR's and Moho. Even if we accept the idea that the SDR's constitute the extrusive part of the initial oceanic crust, the exact nature of the boundary between continental and oceanic crust and the processes operating at the time of initial opening have not been entirely understood to this moment.

DEFINITION OF OCEAN CONTINENT BOUNDARY FROM MIGRATED SEISMIC DATA

Sheridan et al. (1993) interpreted an unmigrated stack of Line 801. In a stack, diffractions can be mistaken for pri-

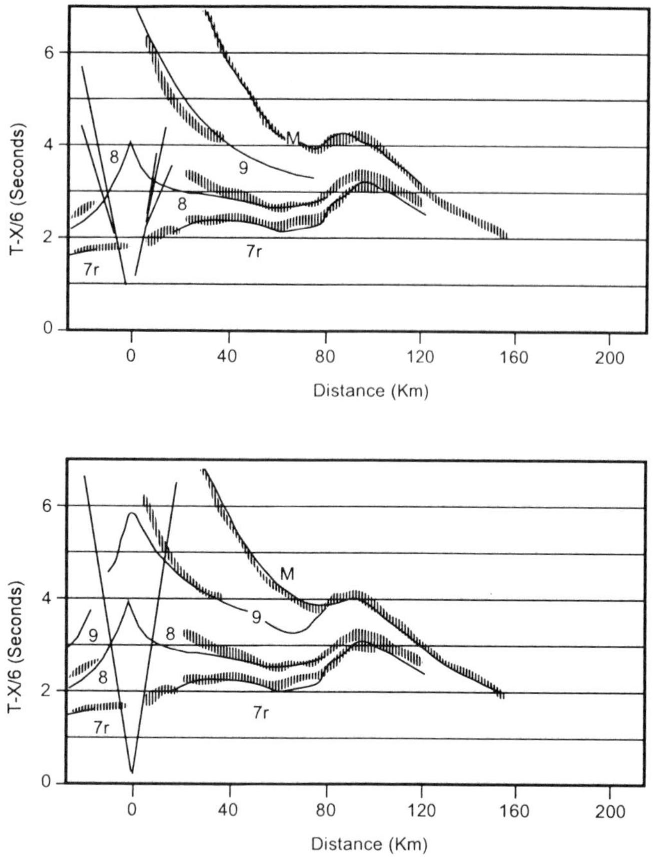

Figure 4. Top - Vertical lines represent interpreted arrivals observed on receiver gathers at ocean bottom seismometer #16. Synthetic wide-angle reflection arrivals obtained for crustal structure in Figure 3 (Top) by asymptotic ray theory are also plotted (Holbrook et al., 1994). The numbers associated with the arrivals correspond to interface numbers in Figure 3 (Top).
Bottom - Vertical lines as above. Synthetic arrivals obtained for crustal structure in Figure 3 (Bottom) by wave equation modeling using the pseudo-spectral method. The numbers associated with the arrivals correspond to interface numbers in Figure 3 (Bottom). The object of showing this figure is to demonstrate that the wide angle reflection data cannot be used to choose between the position of interfaces 8 and 9 as originally obtained by Holbrook et al. (1994) and as obtained in this paper.

mary reflectors. To resolve this issue, *Lizzaralde and Holbrook* (1997) carried out a depth migration and discovered an abrupt decrease in crustal reflectivity east of 40 km, which correlated with an increase in crustal velocity. From the decrease in reflectivity they concluded that the crust east of 40 km is more homogeneous than the crust to the west, and that the increase in velocity was due to intrusion of mafic material. They interpreted the crust below the SDR's (between 40 and 75 km) as thinned continental crust intruded by mafic material. In the unmigrated stack, *Sheridan et al.* (1993) observed coherent reflectors to 75 km and interpreted as continental crust. The migrated stack demonstrated that the reflectors in this region (40 - 75 km) were in fact part of diffractions, and migration moved them out (upwards and landwards) of the region. Therefore, the nature of the crust in this region was questionable.

In order to resolve this issue, we used a number of migration algorithms available at Houston Advanced Research Center, including prestack depth migration. However, prestack depth migration did not markedly improve the results of *Lizzaralde and Holbrook* (1997). In particular, they did not show the coherent reflectors seen in the original stack. We believe that a lack of accurate knowledge of velocities and structure, which are essential to a good depth migration, were responsible for the relatively poor images in our prestack depth migration.

A conventional post-stack time migration was attempted using DISCO software. The results were considerably better than the depth migrations. Figures 6 and 7 contrast the stack and the time migration images. As in *the Lizzaralde*

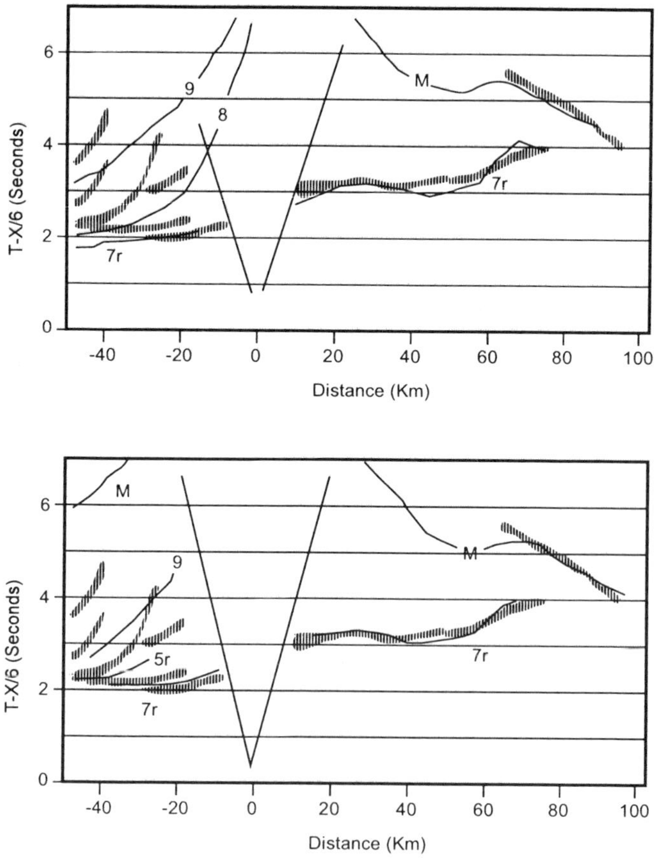

Figure 5. Caption for Figure 6 applies with the difference that the gathers and synthetic data are for ocean bottom hydrophone #C9.

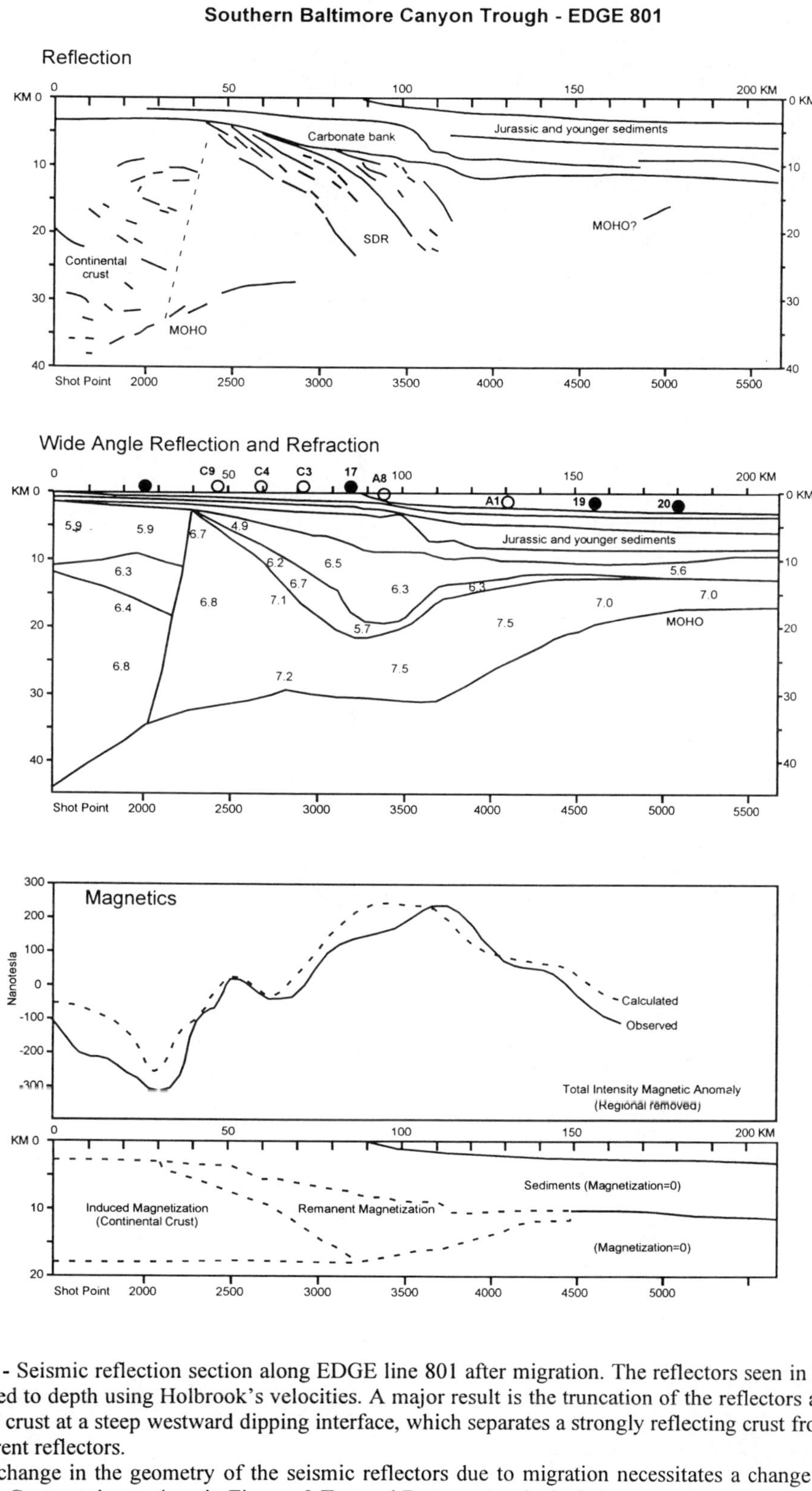

Figure 6. Top - Seismic reflection section along EDGE line 801 after migration. The reflectors seen in Figures 5a and 5b are converted to depth using Holbrook's velocities. A major result is the truncation of the reflectors associated with the continental crust at a steep westward dipping interface, which separates a strongly reflecting crust from a crust with very few coherent reflectors.

Middle - The change in the geometry of the seismic reflectors due to migration necessitates a change in Holbrook's crustal section. Contrast the sections in Figures 3 Top and Bottom. A principal change is that the domains identified in the continental crust (in Figure 5b) are truncated against the steep westward-dipping interface.

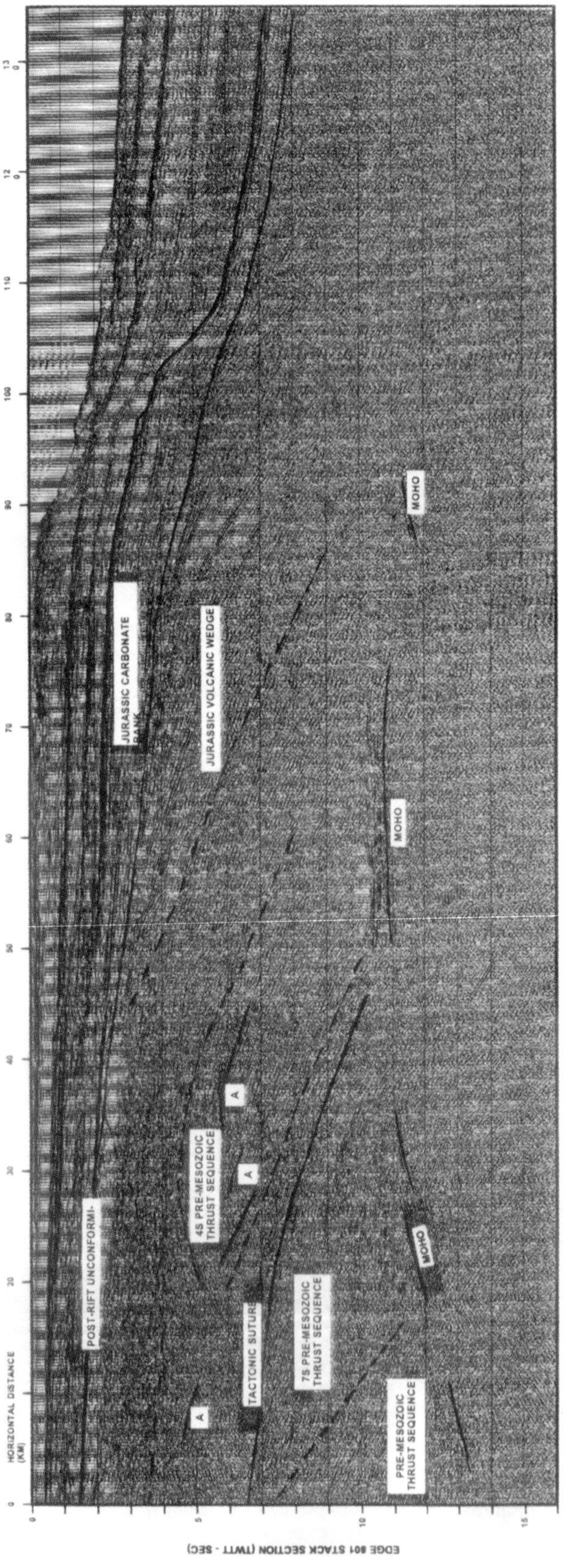

Figure 7 a. Unmigrated section of part of line 801 (After Sheridan et al. 1993)

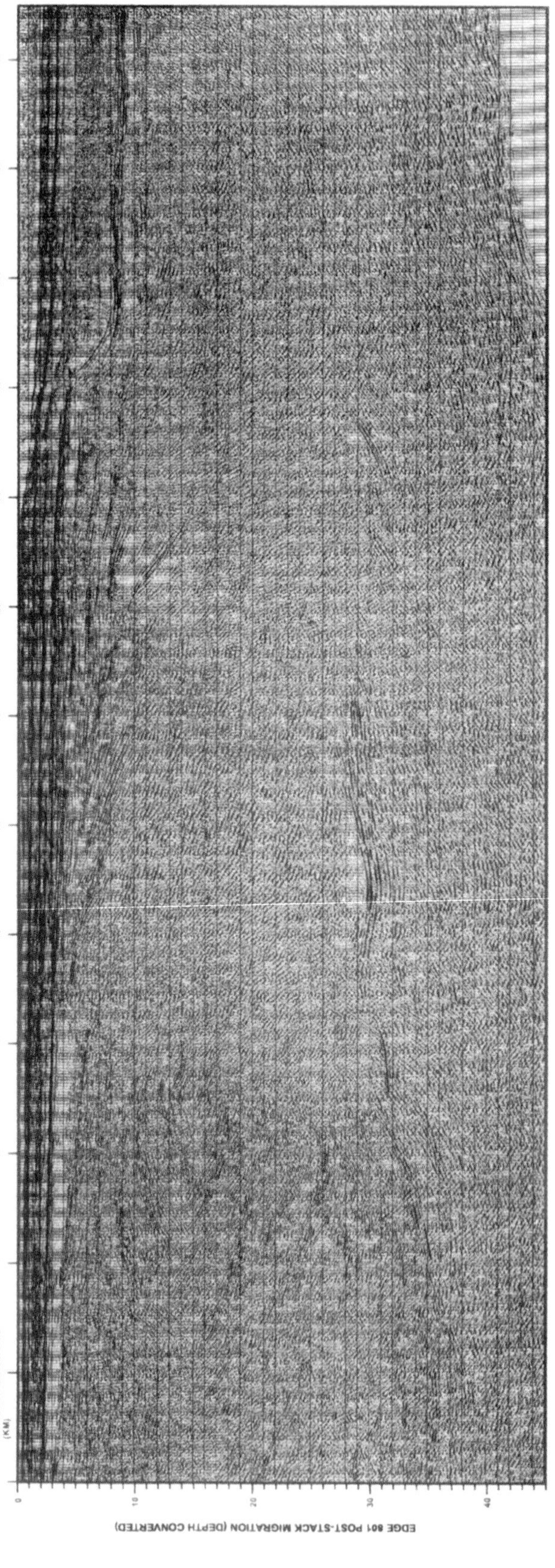

Figure 7 b. Post-stack time-migrated section of part of line 801.

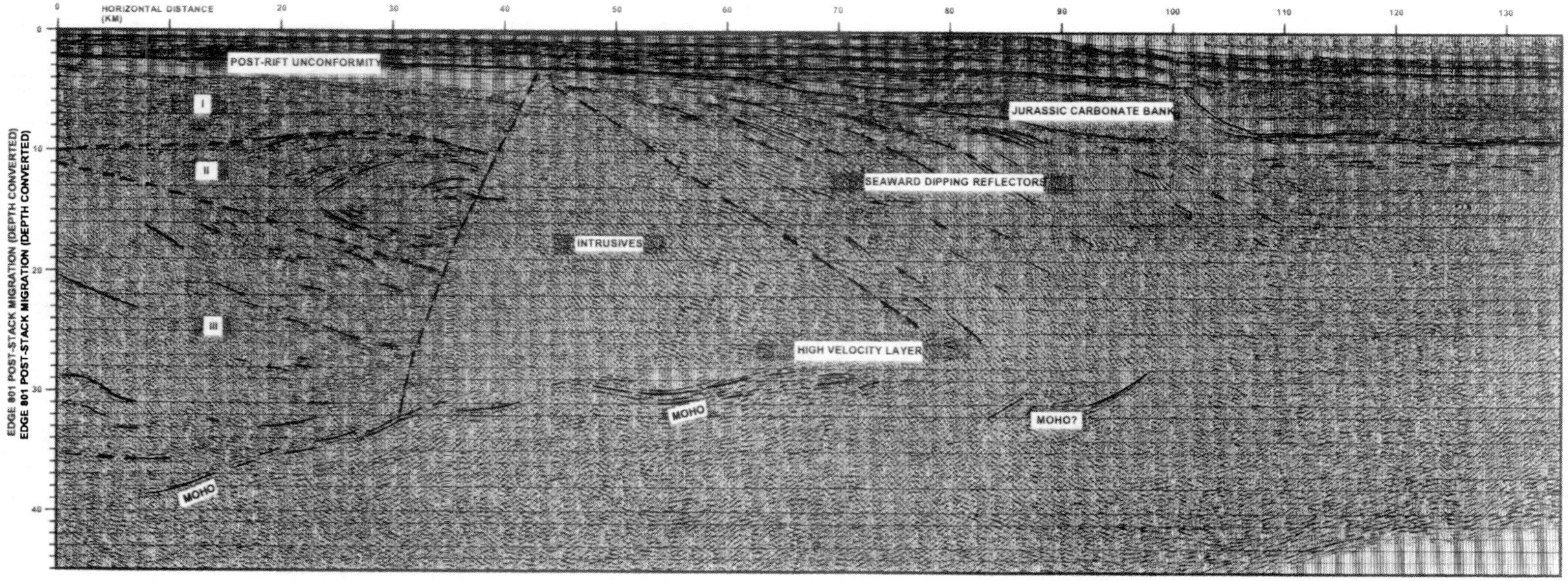

Figure 7 c. Interpreted post-stack time-migrated section of part of line 801 (Figure4b). This section contrasts strongly with the unmigrated stack presented by Sheridan et al., 1993. Migration alters the geometry of the reflectors. They move upwards and landwards and fall into three domains (I, II, and III) defined by the varying dips of the reflectors. In Zone I they are nearly horizontal. In zone II the dips vary greatly, suggesting an intensely folded original geometry. In zone III they mostly dip gently oceanwards. All the reflectors appear to truncate abruptly seawards along a steeply dipping (landwards) interface. Seaward of this interface coherent reflectors are generally lacking.

and Holbrook (1997) results, the diffraction tails moved out of the 40 km to the 75 km region. In addition, there are two other very important differences in the migrated image from the stack image. The reflectors in the continental crust appear to fall into three regions (Figure 7c) that are somewhat different from the regions described by Sheridan et al. (1993). Below the pre-rift unconformity and down to about 4 seconds (or about 10 km, using *Holbrook et al.* (1994) velocities for depth conversion) the reflectors are subhorizontal (Region I). The reflectors at the bottom part of the section in Region III seem to uniformly dip to the east (except at the very bottom). Between these two regions, there is a portion of the crust (Region II) in which the reflectors show important variations in dip. No reflectors seem to be present at the interfaces dividing the three crustal regions. These interfaces would roughly correspond to interfaces 8 and 9 of *Holbrook et al.* (1994).

The most important observation arising from the migration image is the marked change in the pattern of reflectivity that takes place along a sloping interface (from 10 seconds (34 km) at 28 km to 2 seconds (4 km) at 40 km). The reflectivity is higher on the landward side of the interface, with coherent reflectors being abruptly truncated at the interface. *Lizzaralde and Holbrook* (1997) observed the seaward decrease in reflectivity in this area. We have mentioned further above that (1) this change takes place clearly along the interface that dips landward; (2) that coherent reflectors exist landward of the interface; and that (3) there is a marked lack of coherent reflectors seaward of the interface. We believe that the existence of this interface has very important implications for the evolution of the area.

The slope of this interface contrasts with the seaward slope of the interface that terminates interfaces 8 and 9 (*Holbrook et al.*, 1994, Figure 3). *Holbrook et al.* (1994) OBS's 16 and C9 (Figure 1) are closest to the location of this interface. The receiver gathers were examined at these stations in an attempt to discriminate between the two alternatives: a landward dipping interface obtained from migrated vertical incidence reflection data (this work), and a seaward dipping interface inferred by *Holbrook et al.* (1994).

The crustal velocity section from *Holbrook et al.* (1994) is reproduced in Figure 3 (top), and is modified in the present study in Figure 3 (bottom). In compiling the synthetic receiver gathers for our modified section, we have retained the velocities obtained by *Holbrook et al.* (1994), even though this procedure is not strictly accurate. In Figures 4 and 5, we compare the synthetic receiver gathers obtained from the two models with the observations by *Holbrook et al.* (1994). However, the synthetic gathers in the two studies have been obtained in different ways. *Holbrook et al.* (1994) obtained theirs by using asymptotic ray theory, whereas we have carried out wave-equation modeling using the pseudo-spectral method (*Kosloff and Kessler,* 1990). While there are differences in the travel time curves obtained from the two models, it is evident that these models cannot resolve the question of the important interface discussed above, which terminates the crust with coherent reflectors that lie on its landward side. The new model fits the observed receiver gathers better than the earlier model of *Holbrook et al.* (1994). The depth to interface 9 below station 16, which is based largely on unreversed receiver gathers from shots on Line 802, is 2 or 3 km greater in the *Holbrook* model; but this difference is not important for the present discussion.

This brings us to the question of the nature of the crustal material lying between about 40 to 75 km along the profile. This portion of the crust directly underlies the seaward dipping reflectors. Coherent reflectors are generally absent and the crustal velocities increase seawards across the landward dipping interface. As noted earlier, coherent reflectors that characterize continental crust terminate abruptly at this interface. Earlier authors have characterized the 40 to 75 km crust as continental crust that thins seaward. *Holbrook et al.* (1994) attribute the higher velocities and the decrease in reflectivity to mafic intrusions. However, there are two important questions unanswered: what is the nature of the steeply dipping interface and what caused it, and secondly, why there is no apparent boundary between the thicker continental crust and the oceanic crust lying seaward?

An alternative interpretation of the crustal model is that the material lying seaward of the steeply dipping interface and below the seaward dipping reflectors, represents intrusive rocks emplaced contemporaneously with the overlying extrusives. Note the increase of velocity seawards across this interface. The intrusives continue further seawards as the high velocity (>7 km/sec) material that forms the base of the crust.

In this scenario, crustal extension prior to opening does not stretch the crust thinner than about 30 km, at which point it breaks with emplacement of an initial oceanic crust of almost 30 km, largely consisting of intruded material with the extruded volcanics appearing as seaward dipping reflectors on top. This scenario would explain the sharp boundary to the continental crust landward, and the absence of a sharp boundary seaward. However, an open question still remains: is this structure purely local or does it persist along strike in the entire US margin ? *Grow and Sheridan* (1988) inferred that the US East Coast is a non-volcanic margin, with rift blocks infilled by siliciclastic sediments bounded seaward by the abrupt change to an oceanic crust. The EDGE experiment demonstrated, after using a more powerful seismic source capable of penetratring deeper into the crust, that not only this segment of the margin was volcanic, but it also possessed one of the thickest sequences of SDR's (identified as consisting of volcanic material) in the whole Atlantic margin. Similarly,

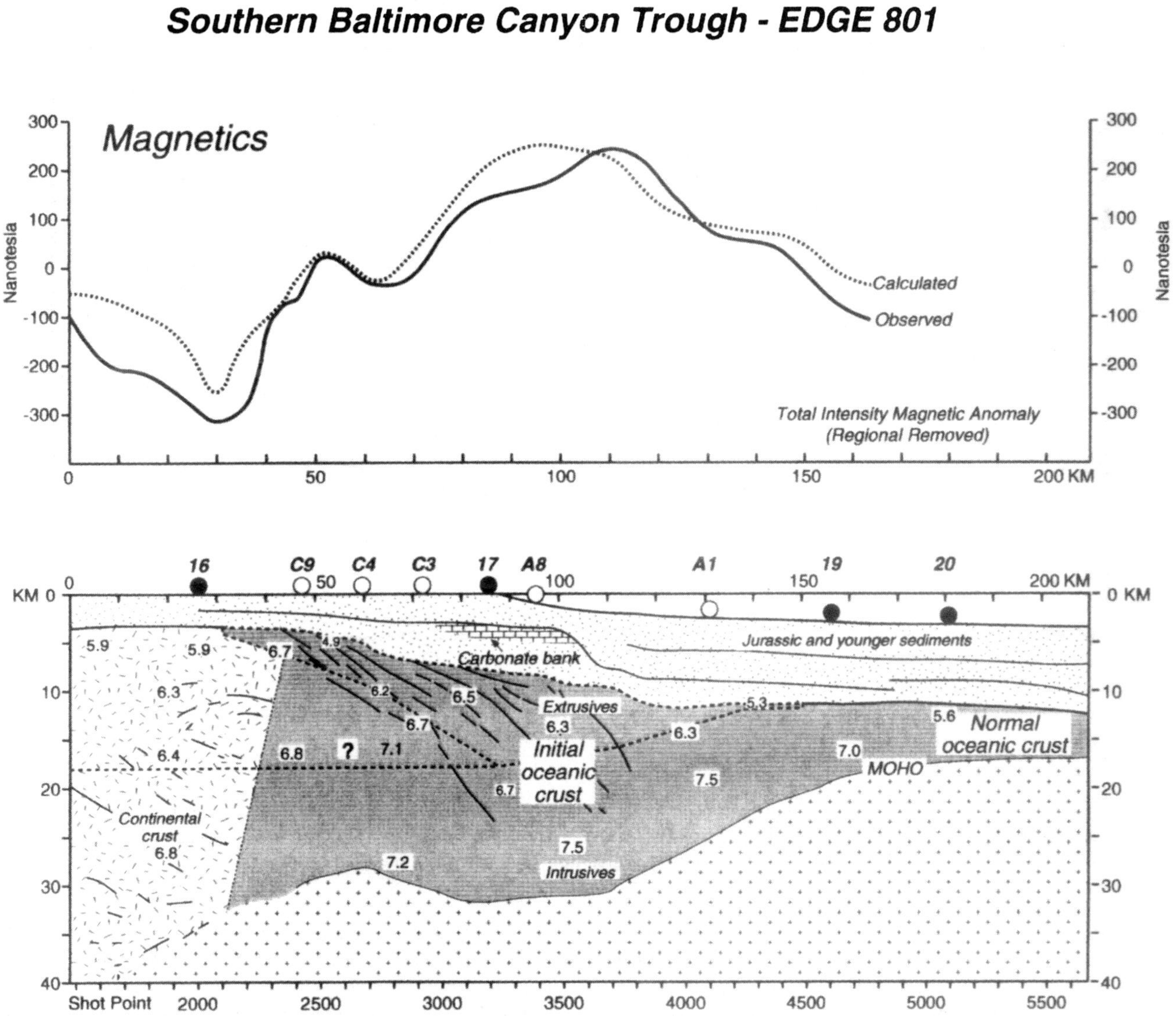

Figure 8. Crustal section along EDGE line 801. The initial oceanic crust consists of three parts: (1) Volcanic extrusives - seaward dipping reflectors, (2) underlying high velocity 7+ km/sec material (interpreted in the past as underplating), and (3) the crust lying immediately seaward of the steeply dipping boundary, which we interpreted as initially emplaced crustal intrusives.

it is quite possible that better seismic reflection data may show a similar structure across other portions of the US East Coast margin. Other examples of the change from a conceptual interpretation of a non-volcanic to a volcanic margin include the Sergipe-Alagoas and Jacuípe basins offshore the northeastern Brazilian margin (*Mohriak et al.*, 1998).

An important question is how thick the continental crust can be stretched to before it breaks. *Le Pichon and Sibuet* (1981), assuming reasonable values of temperatures in the mantle, inferred that the crust can be thinned to a few kilometers. Our results, which would conflict with their ideas, may require much higher mantle temperatures, as well as possible preexisting weaknesses in the crust, which would initiate breakup at a much earlier stage in the evolution of the margin. Clearly, our data suggests further studies are required to elucidate this problem.

It has been argued that whether or not the crust below the SDR's is mainly oceanic (with some entrained continental material) or continental (heavily intruded with mafic material) is just a matter of semantics (e.g., *White and*

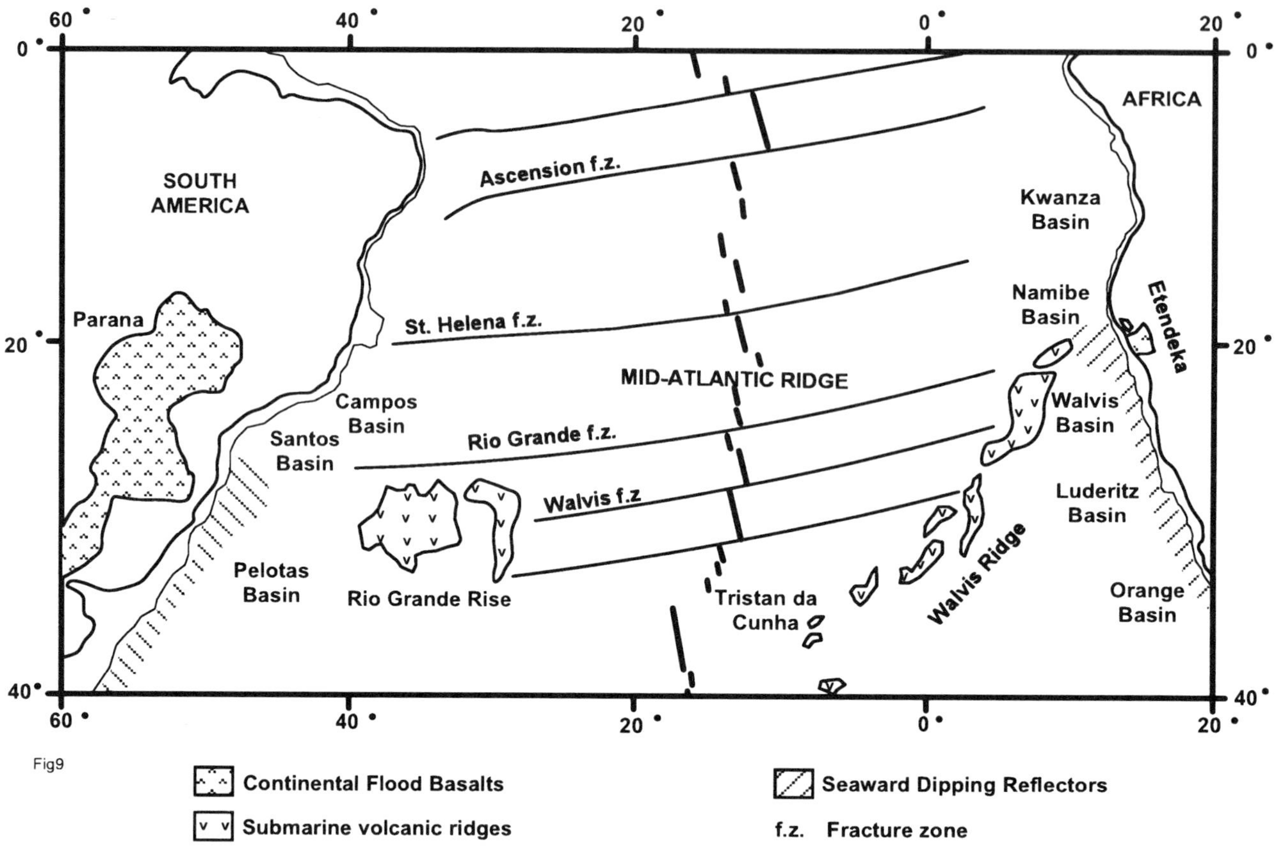

Figure 9. Present configuration of the South Atlantic showing the basins discussed in this study, the thick volcanic ridges of the Rio Grande and Walvis Ridge, and the Mid-Atlantic Ridge (Lohmann et al., 1995; Hark and Hinz, 1995; Gladczenko et al., 1997; this study). Note the conjugate positions of the Walvis and Pelotas basins.

McKenzie, 1989). Therefore, it would be more appropriate to call it "transitional crust", implying some sort of a gradual transition from continent to ocean. However, we believe that there is strong evidence that the clearly defined reflectors in the highly reflective continental crust are abruptly truncated at the steeply dipping interface. Moreover, the higher crustal velocities and the absence of coherent reflectivity seawards would indicate an abrupt change in the nature of the crust. However, the crust seaward of the steeply dipping interface differs from normal oceanic crust in its thickness and velocity structure. Nevertheless, as far as its genesis is concerned, we consider it to be a crust first generated during the early phases of seafloor spreading and therefore we prefer to call it the "Initial Oceanic Crust." These ideas are represented in Figure 8.

SOUTH ATLANTIC MARGIN

The evolution of conjugate volcanic passive margins of the South Atlantic during the early Cretaceous is discussed here by comparing the Pelotas Basin (offshore South Brazil) with the Walvis Basin in the northern portion of the Namibian margin (Southwest Africa, Figure 9). Approximately 1,500 km of seismic lines and four industry wells (drilled by Petrobras) in the Pelotas Basin, and almost 3,000 km of seismic profiles in the Walvis basin (acquired by Namcor and PGS) were analyzed, together with data from DSDP sites 362, 530, 531, and 532. The Pelotas and Walvis basins are particularly well imaged examples of volcanic margins (e.g., *Light et al.,* 1992; *Souza et al.,* 1993; *Mohriak et al.,* 1995; *Fontana,* 1996; *Gladzenko et al.,* 1997; *Abreu,* 1998). The continental counterparts of these volcanic margins (Figure 9) are the Serra Geral Formation (Paraná Basin, Brazil) and the Etendeka Formation (Walvis Basin, Namibia).

A shallow marine depositional environment characterized the Paraná Basin during most of the Paleozoic in the Brazilian portion of the basin (*Zalán et al.* 1990). During the Triassic/early Jurassic, the basin was thermally uplifted prior to the extensional processes that ultimately caused continental breakup. During the early Cretaceous, when the

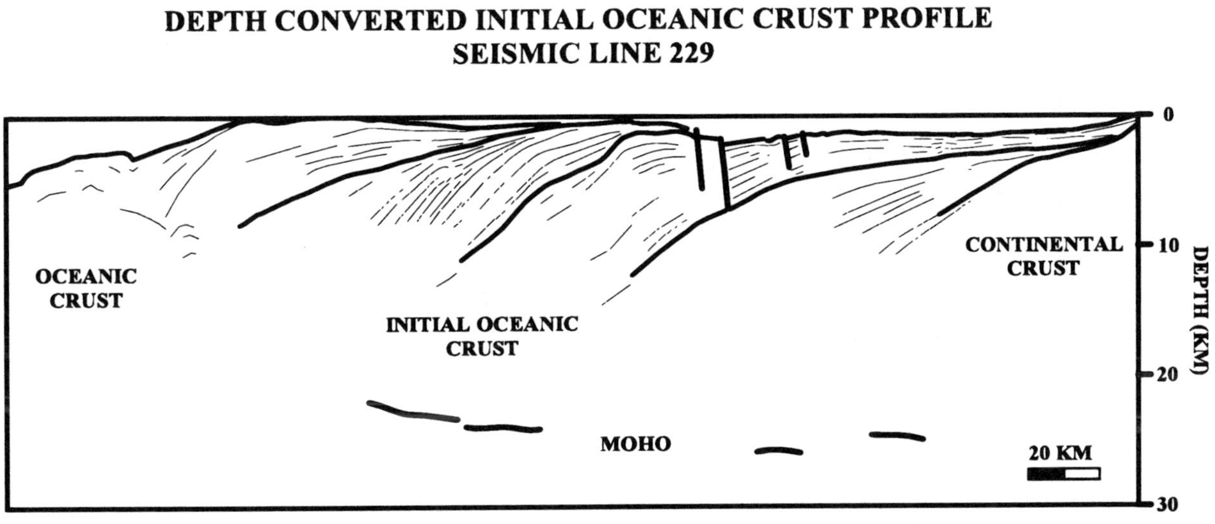

Figure 10. Seismic profile 229 in the Walvis Basin time/depth converted. For location of the Walvis Basin see Figure 9.

break-up of the continent began, thick basaltic layers were emplaced in Pelotas and Walvis basins. The offshore extension of these basalts form a thick and broad wedge of Seaward Dipping Reflectors (SDR's) that may reach 15 km in thickness (Figure 10) and are prominent features in volcanic margins. There is no evidence of any significant synrift sedimentation in the Walvis basin and in the northern portion of the Pelotas basin apart from small, isolated half-grabens in the platform (*Chang et al.,* 1992; *Abreu,* 1998). The deep water region is characterized by the seaward-dipping wedges and the transition to a pure oceanic crust. In this paper, possible mechanisms for emplacement of the SDR's are discussed.

The sedimentary basins in West African and East Brazilian margins originated during Lower Cretaceous extension that culminated with the separation of the South American and African continents. The Namibia continental margin extends from 17 to 29 degrees south and is composed of four basins (from north to south): Namibia, Walvis, Lüderitz, and Orange basins (Figure 9). The conjugate Pelotas basin lies along the southeasternmost portion of the Brazilian continental margin between 28 and 34 degrees south and covers an area of 50,000 square km (Figure 9). These basins formed as parts of a passive volcanic margin during the early Cretaceous when Africa separated from South America by a northward propagation of the rift system (*Rabinowitz and LaBrecque,* 1979). The early proximity of these rift basins to the Tristan da Cunha hotspot, with great influence on the genesis of the Paraná basalts, the Rio Grande Rise, the Walvis Ridge, and the Etendeka basalts, offer an opportunity to evaluate the importance of a hotspot in the formation of the volcanics. This is in marked contrast to the US East Coast margin, for which there is no obvious evidence of a hotspot.

During the early Cretaceous, when the break-up of the continent began, thick basaltic layers were deposited in the Walvis and Pelotas basins. Neocomian to Aptian siliciclastic rocks filled rift troughs that in the Pelotas Basin are mainly located in the platform and bounded by faults that dip landward (*Chang et al.* 1992). During Albian and Turonian times, due to thermal subsidence in the early drift stage, an extensive clastic/carbonate platform was developed along the Eastern Brazilian margin (*Ojeda,* 1982). Sedimentation from the upper Cretaceous to Tertiary is characterized by a predominance of siliciclastics, showing several cycles reflecting the interplay of subsidence and eustatic fluctuations. The sedimentary package tends to decrease in thickness over the volcanics, increasing in thickness where the SDR's are small in volume, or absent.

The formation and evolution of the Pelotas Basin and the offshore Walvis basins as volcanic margins were investigated using seismic/sequence stratigraphy on high quality reflection seismic lines (*Abreu,* 1998). SDR's sequences associated with subaerially emplaced basalt flows are clearly visible along several regional seismic profiles in the Pelotas and Walvis basins (Figures 11A and 11B and Plate 1). The data show strike as well as dip variations in the distribution and character of the SDR's. The SDR's commonly exhibit truncation and onlap features typically associated with the sequence stratigraphy of siliciclastic rocks. In the southern part of the Pelotas and Walvis basins, SDR's appear to be associated with antithetic faults. The thickness of the overlying sediment appears to correlate inversely with the extent of the volcanics. In areas where the volcanics extend downdip over greater distances, the overlying sediment thickness is reduced if compared with areas where volcanics are less widespread (Figures 11A and 11B). In the Pelotas Basin, post-volcanic subsidence in

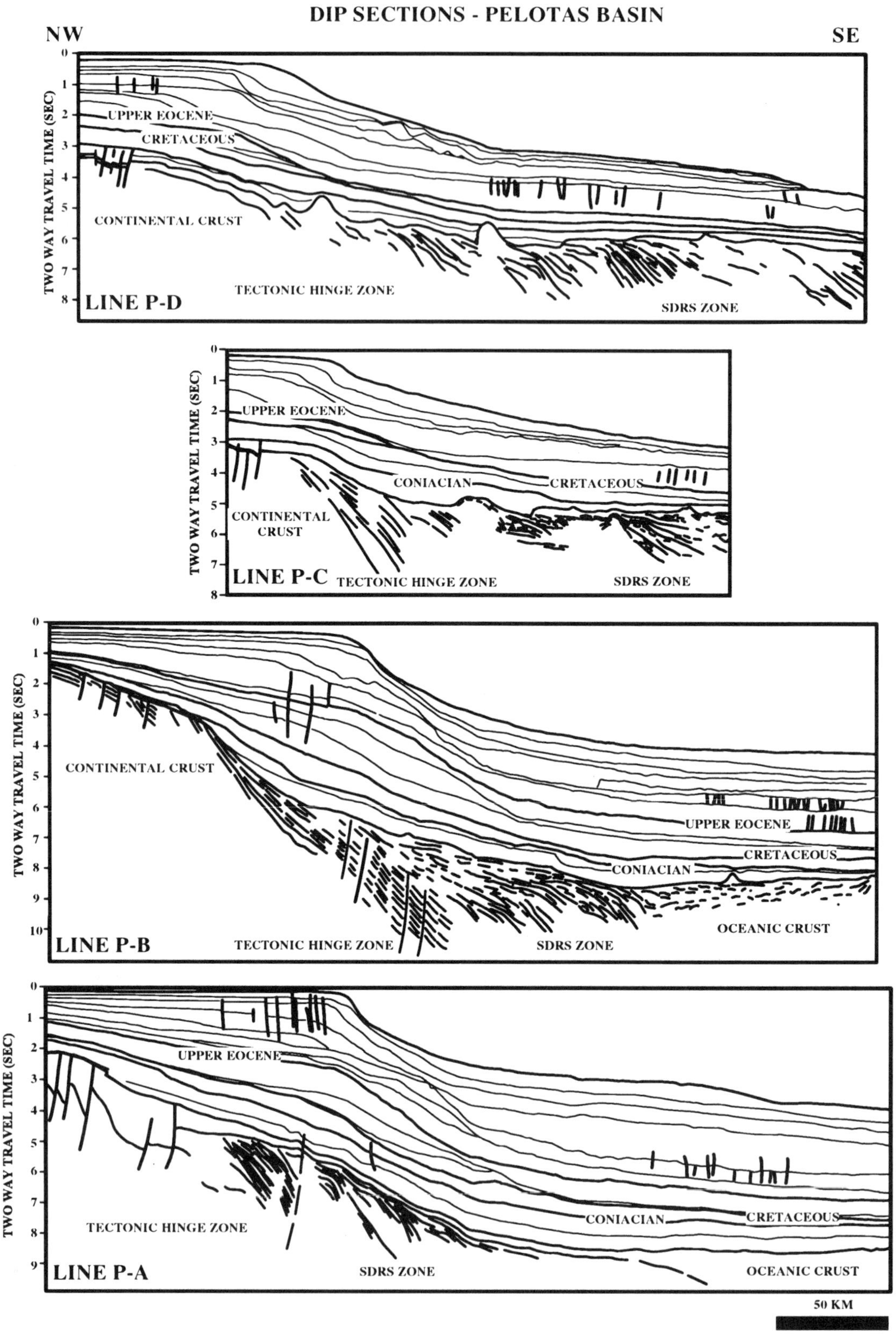

Figure 11 a. Comparison of four dip seismic profiles studied in Pelotas Basin. The top profile is northernmost.

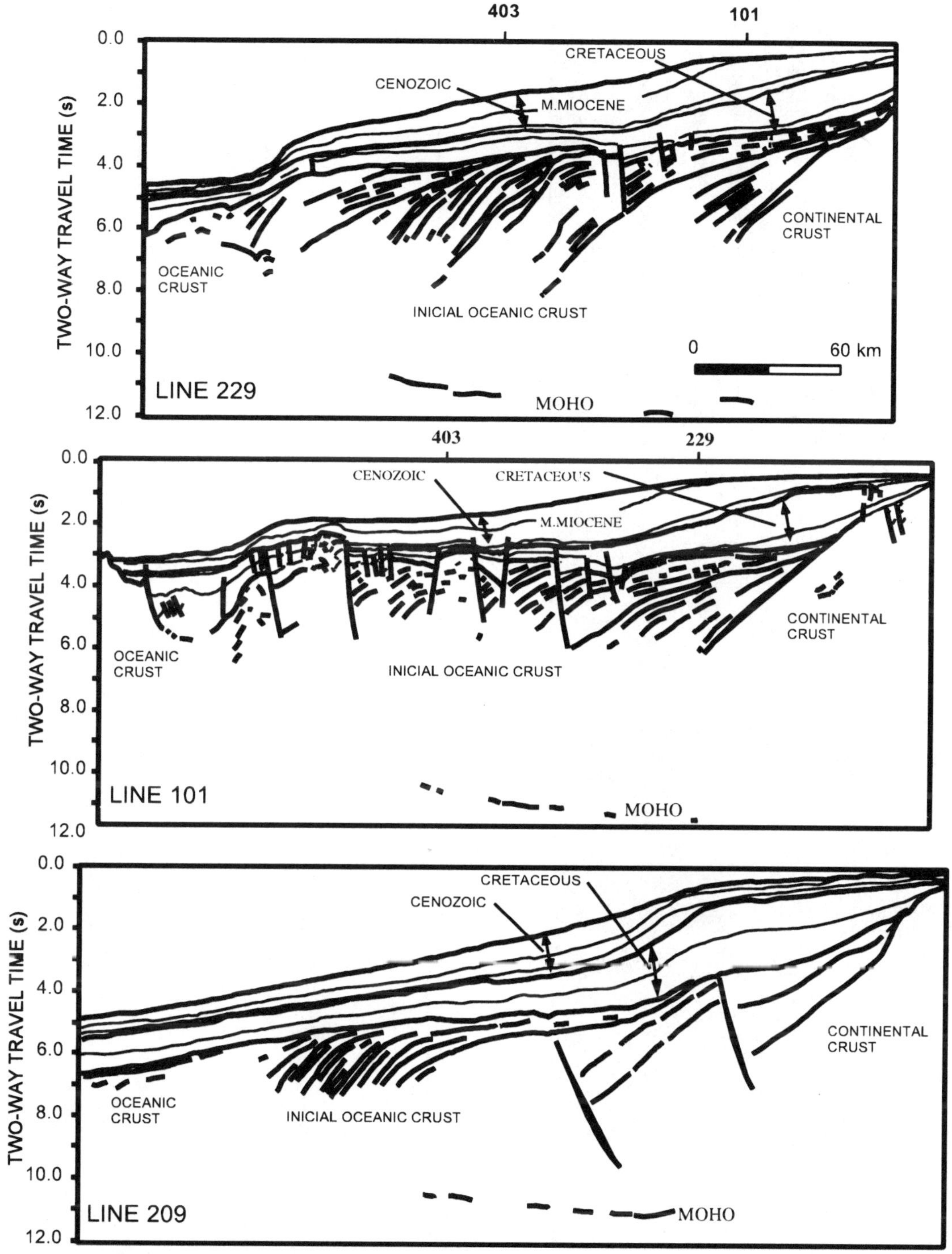

Figure 11b. Interpretation of three dip seismic profiles in the Walvis Basin. The top profile is northernmost.

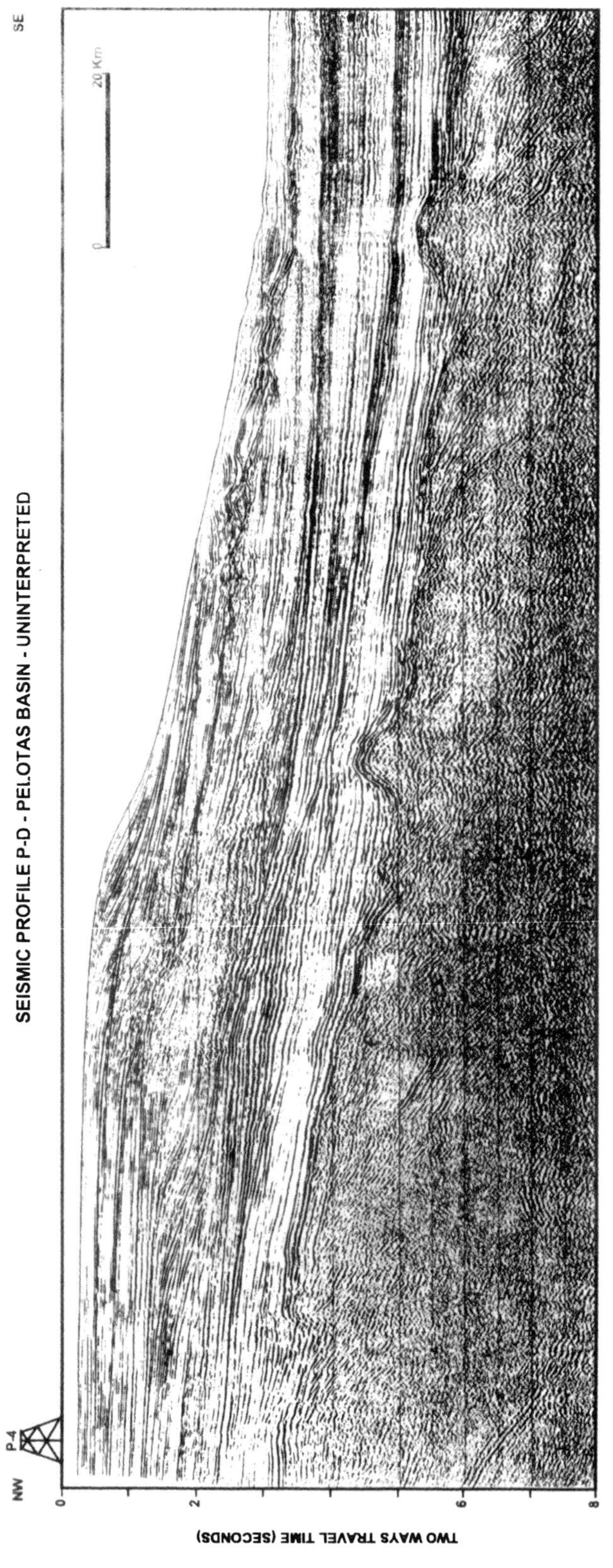
SEISMIC PROFILE P-D - PELOTAS BASIN - UNINTERPRETED
NW
SE
P-4
20 Km
TWO WAYS TRAVEL TIME (SECONDS)
0
2
6
8

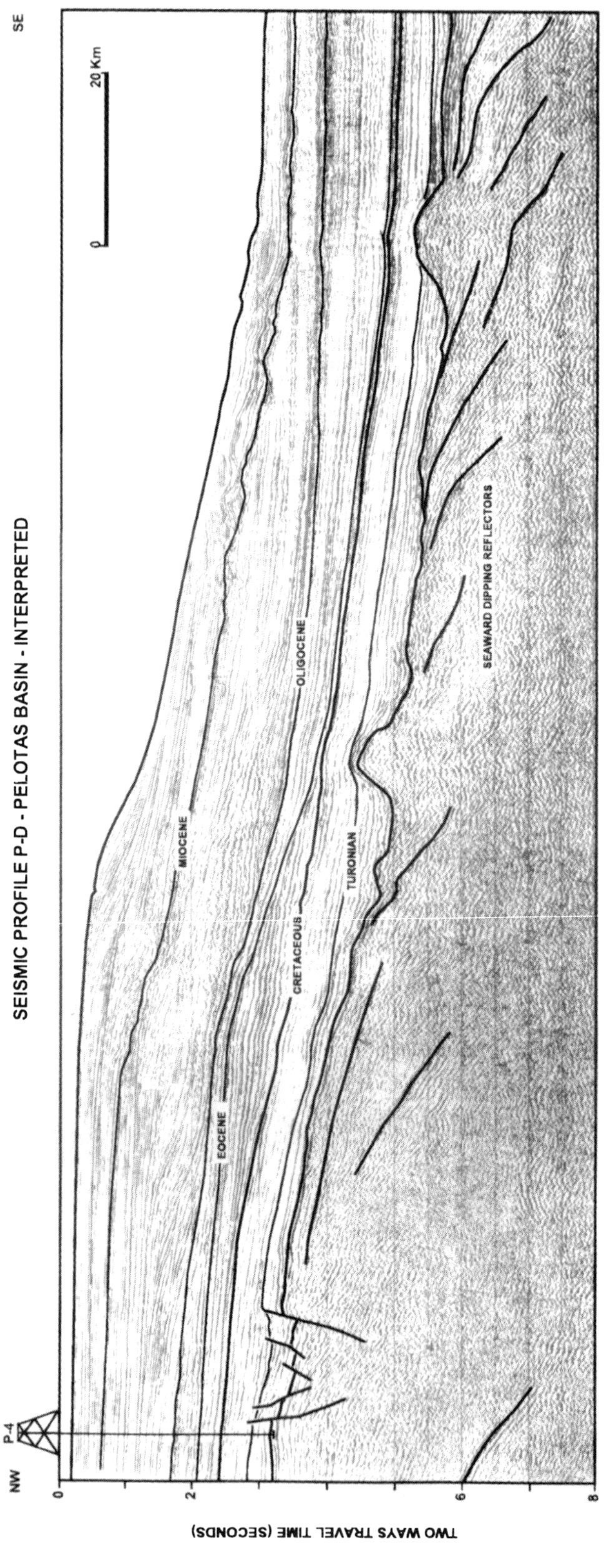
SEISMIC PROFILE P-D - PELOTAS BASIN - INTERPRETED
NW
SE
P-4
20 Km
MIOCENE
EOCENE
CRETACEOUS
OLIGOCENE
TURONIAN
SEAWARD DIPPING REFLECTORS
TWO WAYS TRAVEL TIME (SECONDS)
0
2
6
8

Plate 1

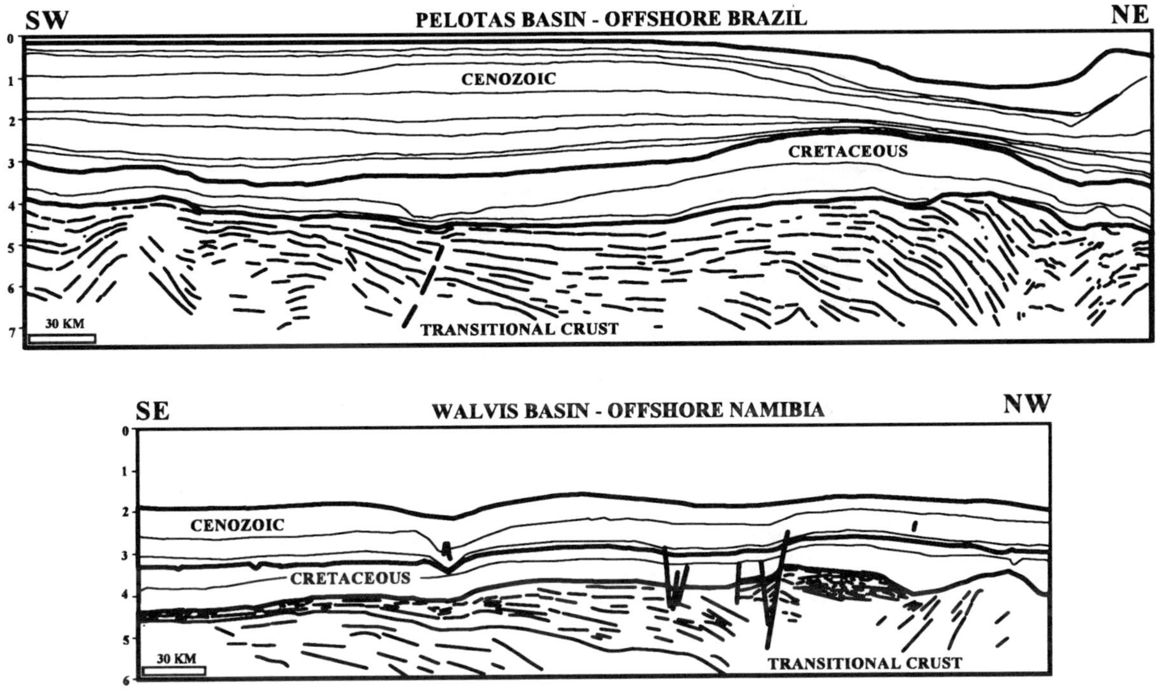

Figure 12. Strike reflection profiles along Pelotas and Walvis basins showing northerly dips as well as increased thickness of the volcanics to the north.

widespread volcanic areas (northern portion) appears to be less than in areas where the volcanics are more narrowly distributed (southern portion). In the northern part, faults do not appear to be present. However, in the southern part, antithetic normal faults are visible (Figure 11A). The SDR's dip both parallel and perpendicular to the coast of each continental margin. On strike profiles from the Pelotas and Walvis basins, the SDR's show a northerly dip component (Figure 12). Figure 13 compares the configuration of the continental flood basalt in the Paraná Basin with the Albian reconstructed SDR's in the Pelotas Basin. Note the northerly dips and the increased thickness to the north.

The SDR's show features that are comparable to stratigraphic sequences. It is possible to map and subdivide the volcanic section through the establishment of volcanic layered sequences. Like its stratigraphic counterparts, volcanic sequences are unconformity-bounded units (Figure 12). The unconformities are defined by onlap over the previous volcanic sequence and truncation below the following volcanic sequence.

In the Pelotas Basin, two distinct patterns in the SDR's are observed. The southern part of the basin shows SDR's within fault-bounded half-graben systems (Figure 11A). The internal facies are planar-divergent, resembling a growth fault in the northern part of the basin. Where faults are less obvious, the lateral extent of the volcanics is greater and mounds are visible, the SDR's show a convex upward, divergent pattern. A similar pattern is observed in the Walvis Basin (Figure 11B). Volcanic mounds occur mainly in the northern portion of the Pelotas Basin (Figure 11A). They appear to be aligned in a northwest/southeast direction, following other lineaments along the same trend in the Southeastern Brazilian margin (Souza et al., 1993).

Figure 14 shows a simple geometric reconstruction of the conjugate Walvis and Pelotas basins for the present (view from north), and Figure 15 depicts the palinplastic reconstruction for the Albian. The Moho position is interpreted from seismic evidence on the Namibian margin and extrapolated to the Brazilian margin. In seismic dip profiles from the Walvis Basin, a strong and discontinuous reflector is observed around 11 seconds (two-way travel time), and is interpreted as the Moho (*Abreu,* 1998). In the Pelotas Basin, the Moho may also be interpreted at equivalent depths in the deep water region (*Fontana,* 1996).

The updip portion of the northern Pelotas and Walvis systems consist of a gentle oceanward dipping basement monocline. Only in the upper part of the southern Pelotas Basin some obvious half-grabens are displayed on seismic profiles. Extension in the northern portion of the Pelotas and Walvis basis is not evident, although large scale antithetic faults and half-graben systems can be interpreted on seismic lines from the southern portion of these basins. The nature of the infill of these grabens beyond the platform is

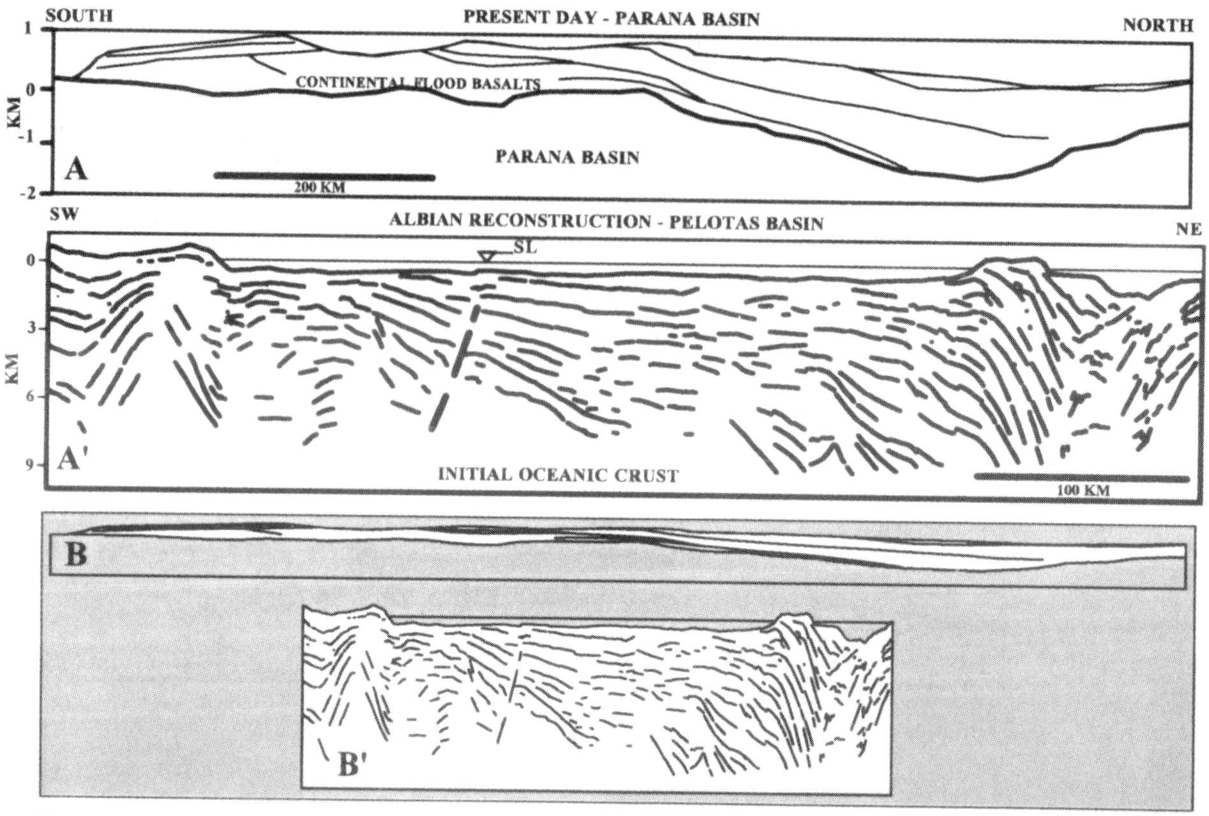

Figure 13. The configuration of the continental flood basalts in the Paraná Basin is compared with the configuration of SDR's reconstructed to the Albian in the Pelotas Basin. Note in both cases the increased thickness to the north and the northerly dips.

uncertain. Sparse drilling into the proximal grabens of Pelotas Basin recovered, in addition to Aptian to Neocomian sediments, tens of meters of volcanic rocks (*Chang et al.*, 1992). Therefore, the grabens could be infilled by pre-rift (Paraná basin) or syn-rift (fluvio-lacustrine) sedimentary rocks overlain by a thinner volcanic wedge. *Gladczenko et al.* (1997) proposed that the grabens observed in Walvis Basin would be infilled mostly by pre-rift sediments (Paleozoic), although there is no well control to confirm this hypothesis.

The presence of the half-graben system with a bare minimum of local depocenters in the platform supports the inference of an extended continental crust that could project below at least the feather edge of the SDR's. If the volcanics were deposited over an extended continental crust during early stages of emplacement before the breakup, it is possible that the half-grabens may also be filled with siliclastic rocks originated from a continental source, as indicated by the presence of downlap surfaces within the volcanic sequences.

MODEL FOR SDR's EMPLACEMENT PELOTAS AND WALVIS BASINS

Two end-member models for SDR's emplacement can perhaps be exemplified on the one hand by the work of *White and McKenzie* (1989) and on the other hand by *Holbrook et al.* (1994), and *Talwani et al.* (1995). *White and McKenzie* (1989) suggested that SDR's provinces were formed during the rift phase under the influence of a hotspot (in this work referred to as Rift Model). *Talwani et al.* (1995) emphasized that the SDR's would be emplaced during continental breakup, implying that a hotspot is not necessary for the SDR's formation (in this work referred to as Drift Model). The Rift Model requires that the volcanics would be emplaced over an extended continental crust and they would not be necessarily symmetrical in conjugate margins. Normal listric faults, occasionally present in SDR's provinces, could be responsible for the subsidence of the volcanics in the Rift Model (extension). In the Drift Model, the SDR's are emplaced as an initial oceanic crust

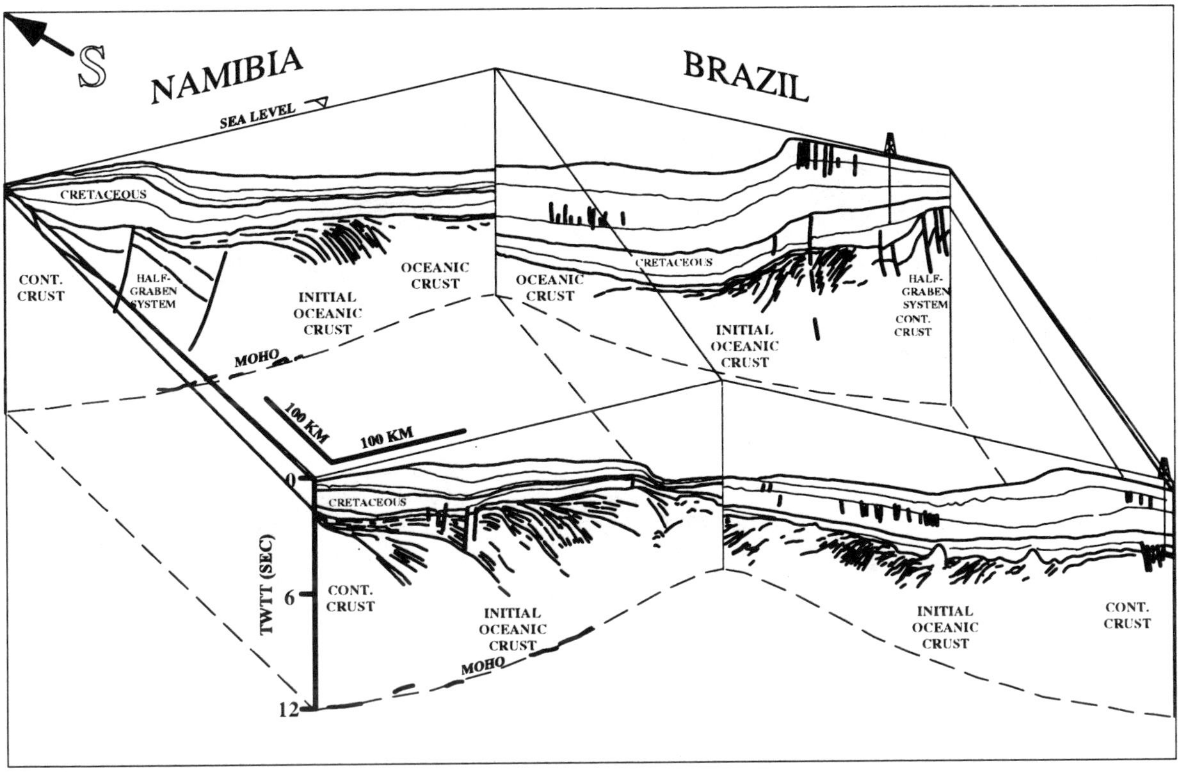

Figure 14. Three-dimensional reconstruction for the conjugate Pelotas and Walvis margins for the present. The Moho position is interpreted from seismic evidence on the Namibian margin and extrapolated for the Pelotas Basin.

and only a small portion of it would be emplaced over continental crust (feather edge), forming symmetrical wedges in conjugate margins. The reflectors would subside mostly due to loading and the SDR's provinces would be present in most passive margins.

The Paraná and Etendeka flood basalts were formed under the influence of the Tristan da Cunha hotspot before continental breakup. The SDR's are considered in this setting as an initial oceanic crust with emplacement during continental breakup and therefore slightly younger than the continental flood basalts dated around 130 Ma (*Turner et al.,* 1994). Line P-D (Pelotas Basin) shows three volcanic mounds over the SDR's which are associated with the Tristan da Cunha hotspot track (Figure 11a). These volcanic mounds are interpreted to be buried volcanic islands formed by the Tristan da Cunha hotspot as the South American plate moved to the West after breakup and SDR's emplacement. This is evidence for the location of the Tristan da Cunha hotspot below the Paraná Basin before and shortly after continental breakup.

The Pelotas and Walvis basins SDR's provinces are symmetrical. Antithetic faults and half-grabens probably mostly infilled by volcanics (Tectonic Hinge Zone in Fig. 11a) occur adjacent to a narrower SDR's wedge (around 50 km) in the southern portion of the basins. In the north, a broad SDR's wedge is present in both margins, reaching a width of more than 200 km, and half-grabens are not well-developed. This may be the most extensive suite of SDR's known worldwide.

The Pelotas Basin SDR's are part of a volcanic system (Figure 16) that extends from the northern coast of the Argentine margin (San Jorge Basin, Figure 16) to at least as far north as the southern portion of the Brazilian margin (Santos Basin). The Santos-San Jorge SDR's province is approximately 3,000 km long and 60 to 120 km wide, extending continuously from north of the Malvinas Plateau (*Lohmann et al.,* 1995) to the continental margin of Uruguay (*Hark and Hinz,* 1995; *Hinz et al.,* 1999) and Brazil (southern portion of the Santos Basin, *Souza et al.,* 1993; *Gladczenko et al.,* 1997). Likewise, the Walvis Basin SDR's are part of a province with similar magnitude, and is referred to as the Walvis-Orange SDR's Province in Figure 15. The volcanic system formed by the Paraná-Etendeka flood basalts and the Santos-San Jorge and Walvis-Orange SDR's provinces would largely exceed the 2,000 km diameter of influence of a hotspot suggested by *White and McKenzie* (1989). Although a hotspot does not appear to be related to the emplacement of the Santos-San

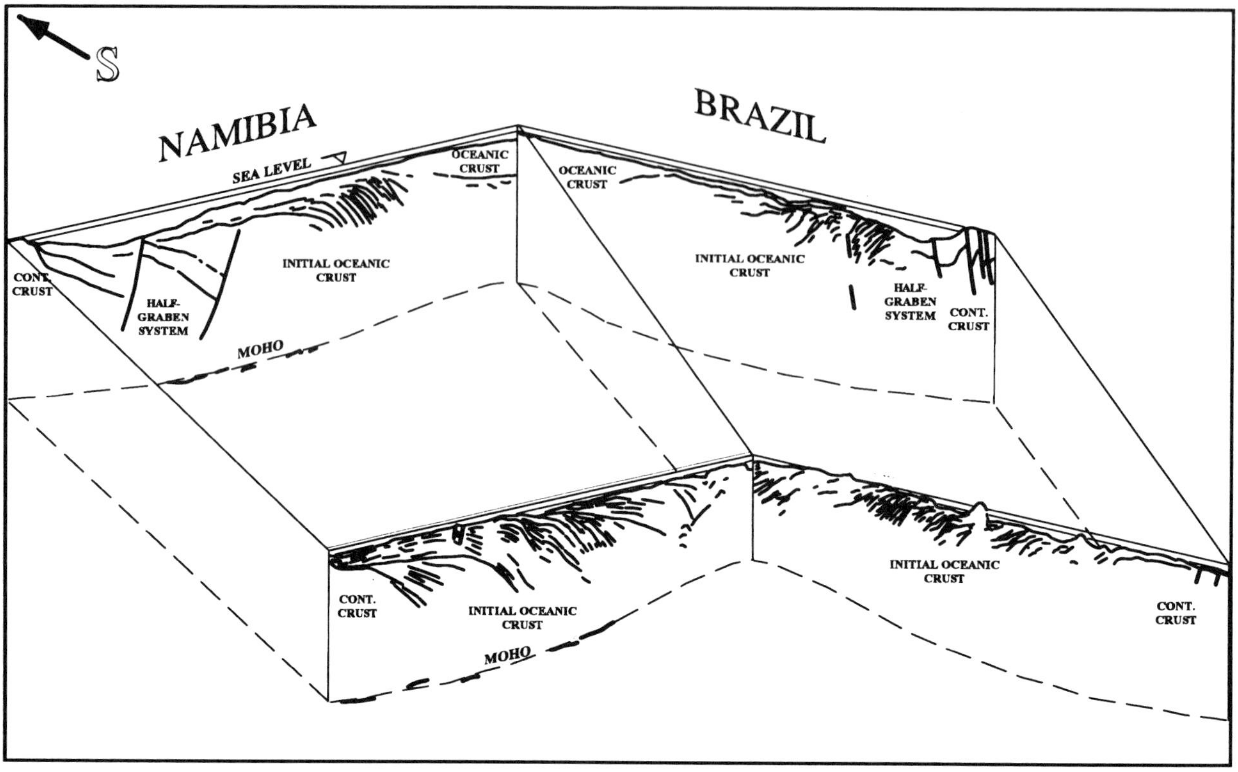

Figure 15. Three-dimensional reconstruction for the conjugate Pelotas and Walvis margins for the Albian (viewed from the north). Subsidence in both margins started at the continental/initial oceanic crust transition. The initial oceanic crust kept close to sea level or in shallow water depths almost until the Turonian, at least in the northern portion of the Pelotas and Walvis basins.

Jorge and Walvis-Orange SDR's provinces, the anomalous large SDR's wedges observed in the northern portion of the Pelotas and Walvis margins could be explained by the presence of the Tristan da Cunha hotspot.

It is proposed that the Santos-San Jorge and Walvis-Orange SDR's provinces were emplaced as an elevated (above sea level) mid-oceanic ridge during an early stage of opening. Magnetic anomaly G coincides roughly with the landward edge of the SDR's (*Hark and Hinz,* 1995; *Hinz et al.,* 1999), indicating a Barremian age for the volcanic emplacement along both margins. Considering the broad distribution of SDR's provinces in passive margins (Figures 9 and 16), a subaerial mid-oceanic ridge with associated volcanics appears to be the first stage of the oceanic basin development.

The Pelotas and Walvis basins SDR's show a dip component parallel to the coast towards the north, with older volcanic wedges in the south being overlain by younger wedges to the north (Figure 12). The geometry of the dipping wedge is different in strike and dip directions. The dipping wedges in Pelotas and Walvis basins extend further along strike than along dip direction and the cause for the strike component of the SDR's may be related to the direction of opening of the South Atlantic.

The conglomerates of the Cassino Formation (*Chang et al.*, 1992) are the only evidence so far obtained for syn-rift sedimentation in the Pelotas Basin. However, Upper Jurassic to Lower Cretaceous syn-rift sedimentation is present along the Argentinean margin (e.g., *Uliana et al.,* 1989; *Soares et al.,* this volume). In offshore Namibia, syn-rift sedimentation is documented mostly in the Luderitz and Orange basins (*Light et al.,* 1992). Although evidence for syn-rift sedimentation prior to the SDR's emplacement is not overwhelming in the Pelotas and Walvis basins, there would be time enough for rift development and syn-rift deposition during the Neocomian before the emplacement of the volcanic wedge. Rift extension and syn-rift deposition along the southeastern Brazilian margin spans in average 15 my. SDR's emplacement is considered to be a short-term event (around 2 Ma) that happened in the Pelotas and Walvis basins during the Barremian/Hauterivian, thus post-dating the interval of terrigenous sedimentation in continental lacustrine environments. Therefore, there would be enough time for development of rift extension

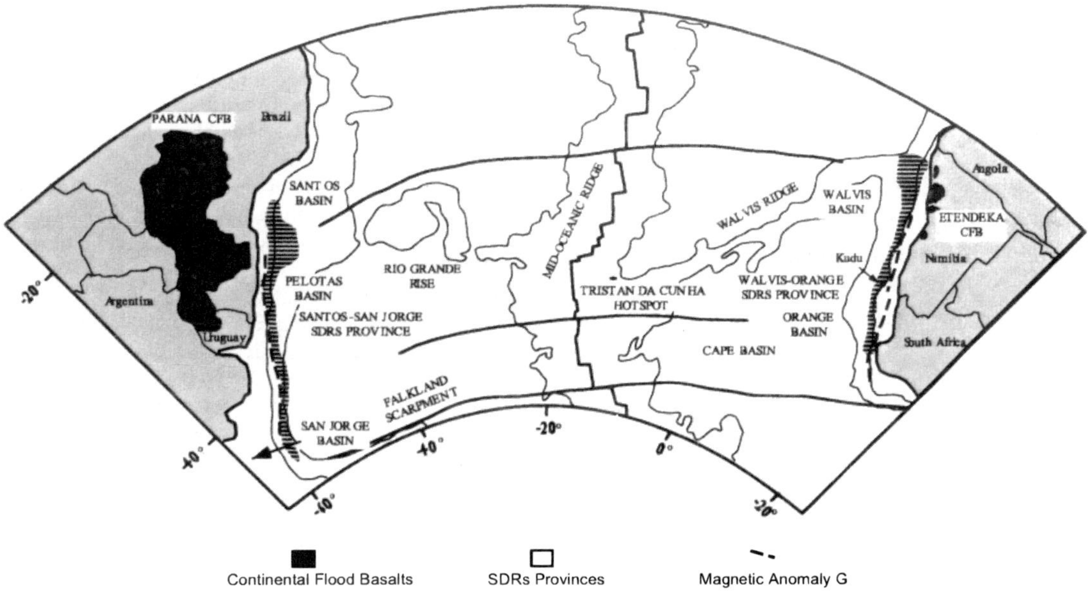

Figure 16: The volcanic systems consisting of SDR's (from San Jorge basin to the Santos basin) and the Paraná basalts on the South American side, and the SDR's (Orange to Walvis basins) and the Etendeka basalts, together with Magnetic anomaly (G), clearly reflect conjugate features at the two margins.

and sedimentary deposition in local troughs during the Neocomian in the Pelotas Basin prior to the SDR's emplacement farther offshore. Moreover, the massive volcanic volumes could partly override the basinward portions of the rift section in the conjugate margins.

DISCUSSION AND CONCLUSIONS

The structure and evolution of continental margins, especially those that contain large thickness of sediments, are difficult to resolve. We have elected to consider a North Atlantic continental margin and two conjugate South Atlantic continental margins in a single study.

Although earlier studies obtained crustal sections across other parts of the US East Coast margins, these were neither as deeply penetrating as the EDGE studies, nor did they lie close together. Thus, along strike continuity of the structure obtained along the EDGE line could not be established in previous studies, and the absence of a comparable study off Morocco hampered the establishment of structural relations with a conjugate margin.

The South Atlantic studies, on the other hand, enable us to observe the along strike variations in the crust on both margins, as well as to compare the structure at conjugate margins. However, no seismic refraction studies were available on the Brazilian side and it was difficult to obtain the detailed deep crustal structure on either side.

The Pelotas and Walvis basins are particularly well imaged examples of volcanic margins. These basins provide a unique opportunity to investigate the evolution of conjugate volcanic rifted margins. The Pelotas and Walvis basins SDR's provinces are symmetrical. Antithetic faults and half-grabens probably mostly infilled by volcanics (Tectonic Hinge Zone) with a narrower SDR's wedge (around 50 Km) are present in the southern portion of these basins. In the North, a broad SDR's wedge is present in both margins, reaching a width of more than 200 Km, which may be the most extensive suite of SDR's known worldwide.

The volcanic complex of the Paraná-Etendeka continental flood basalts and the SDR's of the Pelotas and Walvis basins are the result of the rifting and subsequent breakup of the South American and African plates under an initial influence of the Tristan da Cunha hotspot. The SDR's wedges of Pelotas and Walvis basins were emplaced after continental breakup at least partially over an extended continental crust. Two major SDR's provinces characterize the South Atlantic, the Santos-San Jorge (South America) and the Walvis-Orange (Africa) provinces. These provinces form a broad and symmetrical volcanic complex. It is proposed that these SDR's provinces were emplaced as a subaerial oceanic ridge that would represent the initial stage of formation of the oceanic crust in most passive margins. As the breakup continued with the divergence of the plates, the oceanic ridge would propagate to the North, generating a northward component for the SDR's.

The seaward dipping reflectors have been emplaced during the early stages of subaerial sea floor spreading.

The smooth oceanic basement is also attributed to subaerial sea floor spreading. It is believed that these observations also appear to apply to other margins, notably the Norwegian, Argentine, Brazilian and Namibian margins.

There is no compelling reason to suggest that the events at the time of early opening, or the crustal structures that were developed, were markedly different in the North and in the South Atlantic oceans. Thus, we believe that the two studies taken together give a complementary picture of the formation of passive volcanic margins. In the following we summarize some of the findings.

1. Landward of the SDR's, the presence of continental crust is evident from the highly reflective crust in the North Atlantic and well developed normal faulting in the South Atlantic.
2. Along the EDGE line, the highly reflective continental crust terminates at a steeply dipping interface. We believe that intrusives associated with an Initial Oceanic Crust exist on the basinward side of the interface. This interface is not resolved in the South Atlantic data.
3. Enormous volumes of volcanic material characterize both the North Atlantic and the South Atlantic margins. The SDR's extend along major parts of the Argentine and Brazilian margins as well as the West African and US East Coast margins. However, specific evidence of hot spot activity exists only in the South Atlantic - Walvis Ridge, Rio Grande Rise (and the buried seamounds in the Pelotas Basin). The relationship between hot spot activity and the opening of the Atlantic Ocean is not clear.
4. The SDR's are emplaced symmetrically at the conjugate South Atlantic margins.
5. Along strike variations in volcanic activity are similar along the South Atlantic conjugate margins and reflect the northward progression of volcanic activity related to the northward progression of opening.
6. The early opening of the ocean takes place at the axis of a subaerial ridge. Between this ridge and the continent on either side a basin is formed due to the subsidence under the load of volcanic layers. The subaerial ridge exists until the spreading changes from subaerial to submarine.

Acknowledgments. The authors acknowledge critical reviews by Carlos Cramez, A.W. Bally and Webster Mohriak. Zhong Min Lang carried out the post stack migrations on the EDGE profiles as well as the calculation of synthetic seismograms as part of his Master's dissertation at Rice University. The data set used in this study for the South Atlantic was kindly released from Petrobras (Pelotas Basin) and PGS (Walvis Basin) for a Ph.D. thesis (Abreu, 1998).

REFERENCES

Abreu, V.S., Geologic evolution of conjugate volcanic passive margins: Pelotas Basin (Brazil) and Offshore Namibia (Africa); implication for global sea-level changes, Ph.D. Thesis, Rice University, Houston, 355 p., 1998.

Chalmers, J.A. and K.H. Laursen, Labrador Sea: the extent of continental and oceanic crust and the timing of the onset of seafloor spreading, Marine and Petroleum Geology, v. 12, n. 2, p. 205-217, 1995.

Chang, H. K., Kowsmann, R.O., Figueiredo, A.M.F. and Bender, A., Tectonics and stratigraphy of the East Brazil Rift system: an overview, Tectonophysics, v. 213, p. 97 – 138, 1992.

Fontana, R.L., 1996. SDR (Seaward-dipping reflectors) e a transição crustal na Bacia de Pelotas, Anais do 39 Congresso Brasileiro de Geologia, vol. 5, p. 425-430, 1996.

Gladczenko, T.P., K. Hinz, O. Eldholm, H. Meyer, S. Neben, and J. Skogseid, South Atlantic volcanic margins: *Jour. Geol. Soc., London,* **154**, 465-470, 1997.

Glover III, Lynn, R.E. Sheridan, W. Steven Holbrook, J. Ewing, M. Talwani, R.W. Hawman, and P. Wang, Paleozoic collisions, Mesozoic rifting, and structure of the Middle Atlantic states continental margin: An 'EDGE' Project report, *GSA Special Paper* **314**, 107-135, 1997.

Grow, J.A., and R.E. Sheridan, U.S. Atlantic Continental Margin: A typical Atlantic-type or passive continental margin, in: Sheridan, R.E., and Grow, J.A., eds., The Atlantic Continental Margin: U.S. Boulder Colorado, Geological Society of America, Geology of North America, **I-2**, 1-7, 1988.

Hark, H.U., and K. Hinz, Uruguay, in Kulke, H. (ed.), *Regional Petroleum Geology of the World, Part II: Africa, America, Australia and Antarctica*, Berlin, 549-577, 1995.

Hinz, K., A hypothesis in terrestrial catastrophes: Wedges of very thick oceanward dipping layers beneath passive continental margins, *Geol. J.*, **22**, 3-28, 1981.

Hinz, K., S. Neben, B. Schreckenberger, H.A. Roeser, M. Block, K.G. Souza, and H. Meyer, The Argentine continental margin north of 48 ° S: sedimentary successions, volcanic activity during breakup, *Marine and Petroleum Geology*, **16**, n. 1,1- 25, 1999..

Holbrook, W. Steven, G.M. Purdy, J.A. Collins, R.E. Sheridan, D.L. Musser, L. Glover III, M. Talwani, J.I. Ewing, R. Hawman, and S.B. Smithson, Deep Velocity Structure of Rifted Continental Crust, US Mid-Atlantic Margin, from Wide-Angle Reflection/ Refraction Data, *Geophysical Research Letters,* **19**, **16**, 1699-1702, 1992

Holbrook, W.S., G.M. Purdy, R.E. Sheridan, L. Glover III, M. Talwani, J. Ewing, and D. Hutchinson, Seismic structure of the US Mid-Atlantic continental margin, *JGR,* **99**, **B9**, 17,871-17,891, 1994

Kosloff, D., and D. Kessler, Seismic numerical modeling, in: Desaubies, Y., Tarantola, A., Zinn-Justin, Y. (eds.), Oceanographic and Geophysical Tomography, Elsevier, Amsterdam, 251-312, 1990

LASE Study Group, Deep structure of the US East Coast passive margin from large aperture seismic experiments (LASE), *Mar. Pet. Geol.,* **3**, 234-242, 1986.

Le Pichon, X., and J.C. Sibuet, Passive Margins: A Model of Formation, *JGR, 86, B5,* 3708-3720, 1981.

Light, M.P.R., M.P. Maslanyj, and N.L. Banks, New geophysical evidence for extensional tectonics on the divergent margin offshore Namibia, in Storey, B.C., Alabaster, T, and Pankhurst, R.J. (eds.), Magmatism and the causes of continental breakup: *Geological Society Special Publication 68*, 257-270, 1992.

Lizzaralde, D., and W.S. Holbrook, US mid-Atlantic margin structure and early thermal evolution, *JGR*, **102, B10**, *22,855*-22,875, 1997

Lohmann, H. H., J. Hoffmann-Rothe, and K. Hinz, Argentine, in Kulke, H. (ed.), Regional Petroleum Geology of the World, Part II: Africa, America, Australtia and Antartctica, Berlin, 549-577, 1995.McKenzie, D. P., Some remarks on the development of sedimentary basins, *Earth Planet. Sci. Lett.*, **40**, 25-32, 1978.

Mohriak, W.U., M. Bassetto and I.S. Vieira, Deep seismic constraints on the crustal architecture of sedimentary basins in the Brazilian margin: tectonic and exploratory implications. Boletim de Resumos Expandidos, *V Simpósio Nacional de Estudos Tectônicos - SNET 95*, Gramado - RS, p. 246 – 248, 1995.

Mohriak, W.U., M. Bassetto, and I.S. Vieira, Crustal architecture and tectonic evolution of the Sergipe-Alagoas and Jacuípe basins, offshore northeastern Brazil. *Tectonophysics*, **288**, 199 – 220, 1998.

Montadert, L., O. De Charpal, D.C. Roberts, P. Guennoc, and J.C. Sibuet, Northeast Atlantic passive margins: Rifting and subsidence processes, in *Deep Drilling Results in the Atlantic Ocean: Continental Margins and Paleoenvironment, Maurice Ewing Ser.*, 3, edited by M. Talwani, W.W. Hay, and W.B.F. Ryan, 164-186, AGU, Washington D.C., 1979.

Ojeda, H.A.O., Structural framework, stratigraphy, and evolution of Brazilian marginal basins, *AAPG Bulletin*, **66**, 732-749, 1982.

Rabinowitz, P.D. and J. LaBreque, The Mesozoic South Atlantic Ocean and evolution of its continental margins, *Journal of Geophysical Research*, **84**, n.B11, 5973-6002, 1979.

Sheridan, Robert E., D.L. Musser, L. Glover III, M. Talwani, J. Ewing, W.S. Holbrook, G.M. Purdy, R. Hawman, and Scott Smithson, Deep seismic reflection data of EDGE US mid-Atlantic continental-margin experiment: Implications for Appalachian sutures and Mesozoic rifting and magmatic underplating, *Geology, 21,* 563-567, 1993.

Souza, K. G., R.L. Fontana, J. Mascle, J.M. Macedo, W.U. Mohriak, and K. Hinz,. The southern Brazilian margin: an example of a South Atlantic volcanic margin. Third International Congress of the Brazilian Geophysical Society, Rio de Janeiro, RJ, November 7 - 11th, 1993. vol. 2, p. 1336 – 1341, 1993.

Talwani, M., J. Ewing, R.E. Sheridan, W.S. Holbrook, and L. Glover, III, The EDGE experiment and the US Coast Magnetic anomaly, in NATO/ARW Series book "Rifted Ocean-Continent Boundaries," E. Banda, M. Talwani, and M. Torne (eds.), 155-181, 1995.

Turner, S., M. Regelous, S. Kelley, C. Hawkesworth, and M. Mantovani, Magmatism and continental break-up in the South Atlantic: high precision 40 Ar - 39 Ar geochronology. *Earth and Planetary Science Letters*, **121**, 333 – 348, 1994.

Uliana, M.A., K.T. Biddle, and J. Cerdan, Mesozoic extension and the formation of Argentine sedimentary basins, in Tankard, A.J., and Balkwin, H.R. (eds.), Extensional Tectonics and Stratigraphy of the North Atlantic margins, *AAPG Memoir* 46, 599-614, 1989.

White, R.S., and D. P. McKenzie, Magmatism at rift zones: The generation of volcanic continental margins and flood basalts: *JGR*, **94**, 7685-7729, 1989

Zalán, P.V., S. Wolff, M.A.M. Astolfi, I.S. Vieira, J.C.J Conceição, V.T. Appi, E.V.S Neto, J.R. Cerqueira, and A. Marques, The Paraná Basin, Brazil. In: Leighton, M.W., Kolata, D.R., Oltz, D.S., and Eidel, J.J. (eds.), Interior Cratonic Basins, AAPG Memoir 51, p. 681 – 701, 1990.

Manik Talwani , Rice University and Houston Advanced Research Center

Vitor Abreu, Unocal Corporation

The Oceanic Segment of the Southern Brazilian Margin: Morpho-Structural Domains and Their Tectonic Significance

Marcelo Bassetto [1]; Fernando F.Alkmim [2]; Peter Szatmari [3]; Webster U. Mohriak[1]

A descriptive and evolutionary analysis of the main morpho-structural features of the oceanic domain of the southern portion of the Brazilian Continental Margin is supported by regional seismic profiles and potential field data from the Brazilian governmental LEPLAC (Plano de Levantamento da Plataforma Continental Brasileira) Project. The several morpho-structural elements can be differentiated, as for example: the dominant structural pattern of the acoustic basement, including extensional faulting and long-wavelength folding, crustal thickness changes, fracture zones location, distribution of volcanic centers, and development of wedges of seaward-dipping reflectors. Four broad distinct morpho-structural domains, separated by fracture zones and oceanic lineaments. Domain I is located south of the Porto Alegre Lineament; Domain II corresponds to the area between the Porto Alegre Lineament and the Rio Grande Fracture Zone; Domain III spans the area of the São Paulo Plateau; and Domain IV is located to the east of this plateau, towards the abyssal portions of the oceanic crust. These domains are defined by their distinct regional morphologic and structural characteristics. Sometimes these elements are well imaged in the seismic profiles, corroborated by gravity and magnetic anomalies, and eventually identified as prominent features at the sea bottom physiography. Using a multidisciplinary approach based on bathymetric maps, regional seismic interpretation, magnetic data analysis, and gravity models, this work attempts to characterize these elements in a descriptive and evolutionary view, identifying their role in the tectonic development of this portion of the South Atlantic.

[1] E&P Department, Exploration Management, PETROBRAS, Cinelândia, Rio de Janeiro, Brazil.

[2] Department of Geosciences, Universidade Federal de Ouro Preto, Morro do Cruzeiro, Ouro Preto, Brazil.

[3] CENPES, Research Center, Exploration Division, PETROBRAS, Cidade Universitária, Rio de Janeiro, Brazil.

Atlantic Rifts and Continental Margins
Geophysical Monograph 115

1. INTRODUCTION

In 1988, the Brazilian Navy's commission CIRM (Comissão Interministerial de Recursos Mar), with the participation of PETROBRAS, Mineral Research National Department (DNPM), Universities of the Marine Geology Program (PPMG) and other members of the Brazilian scientific community, began the development of the LEPLAC Project (Plano de Levantamento da Plataforma Continental Brasileira). This project, whose main objective is to estab-

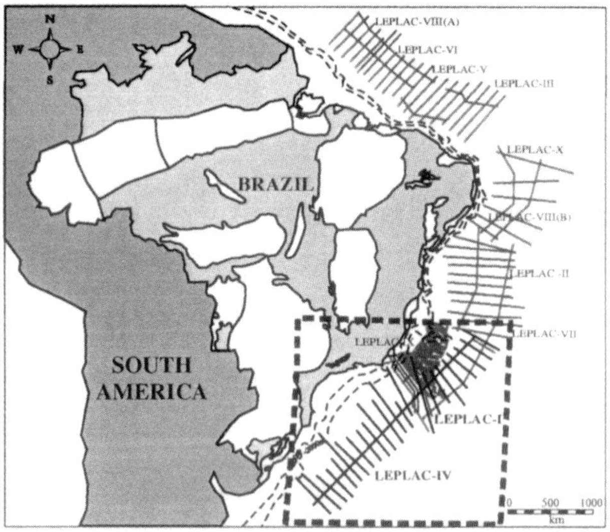

Figure 1. Location map showing the area of this study and the location of the geophysical dataset of the LEPLAC Project.

lish the outer limit of the legal territorial waters along the Brazilian Continental Margin, has been an invaluable source of geophysical data for regional geologic studies, including an analysis of the deep-water, distal portions of the margin, in the oceanic crustal domains basins (figure 1).

The LEPLAC I and IV subprojects cover an area adjacent to the Campos, Santos and Pelotas marginal basins, localized between the parallels 20°30'S and 37°30'S, and meridians 32°W and 52°W (figure 2). This portion of the margin comprises a great amount of oceanic features that registers, in a peculiar way, some of the most important processes involved in the initial stages of the South Atlantic's evolution.

Combining the geophysical data of the LEPLAC Project with regional gravity data obtained by satellite (GEOSAT) and bathymetric data from the PETROBRAS regional dataset, a morpho-structural analysis was conducted in this sector of the South Atlantic. The main results obtained are presented in this work.

Based on the description of the main physiographic elements and crustal structures, four morpho-structural domains were characterized in the oceanic portion of the southern margin of Brazil. This work attempts to demonstrate that the main physiographic and structural differences between these domains may be due to a singular oceanic spreading geometry, lasting from the crustal rupture (Barremian) to the upper Cretaceous.

2. THE GEOPHYSICAL DATA BASE

The geophysical dataset of the LEPLAC Project in this area were acquired in the period from February/ 89 to April/ 89, under coordination of PETROBRAS and DHN (Hydrography and Navigation Directory of the Brazilian Navy). The seismic sections total 7756 km of profiles, distributed along 24 dip (NW-SE) sections spaced of approximately 80 km, and 1 long strike (NE-SW) section, divided in 3 continuous segments transversal to the other profiles [*Bassetto*, 1996].

The general parameters used for the seismic acquisition were: fonts arrangement: 8 airguns of 540 in3 and 4500 PSI; spread of 0-170-2545 m; SP interval: 50 m; groups interval: 25 m; 20 hydrophones/group; 90 groups; 96 record channels; sampling interval: 4 ms; record length: 10 to 12 s.

The seismic processing method basically included the following steps: demultiplexation with record length from 10 to 12 s and sampling interval of 4 ms; sum of two adjacent traces (with trace's distance of 50 m); data arrangement in CDP families (in intervals of 25 m); spherical divergence compensation; instruments' answers and pre-stack deconvolution; NMO and internal MUTE corrections; external muteness; final stacking with sum of 2 adjacent traces; CDP traces distance of 50 m; migration for finites differences; post-stack migration; frequency filter of 5-30 Hz and slope of 36 dB/octave; dynamic equalization (gain) with RMS of 4000, windows of 200/1600 ms; and final section with CDP traces distance of 250 m.

In addition to the seismic acquisition, gravity and magnetic profiles were acquired, but the latter were not corrected for diurnal variation. This study also used the gravity free-air dataset from the GEOSAT Program (U.S.Navy) and the bathymetric and magnetic dataset of PETROBRAS, related to several other seismic surveys.

3. GENERAL ASPECTS OF THE DEVELOPMENT OF THE SOUTH ATLANTIC

The fascinating subject of the Western Gondwanland rupture and the whole tectonic process involved in the Atlantic formation have been treated, in the last decades, by dozens of publications. These studies were based on the general acceptance of the plate tectonics tenets and mechanisms, combined with the technological advances and increment of the geological and geophysical data acquisition.

It is generally accepted that the whole process of the continental rupture is marked by several tectonic pulses, of a predominantly extensional nature, creating distinct margin types along the north, central and southern portion of the South Atlantic Ocean. These areas are characterized by distinctive patterns of tectonic evolution. The beginning of the continental rifting and separation process probably started in the Mid-Jurassic (about 170 Ma), in the southern and central portions of the Atlantic. It is currently interpreted that this process developed from the south to the north in both cases, towards the Equatorial Atlantic and Arctic Ocean, respectively [*Cande et al.*, 1988].

In the Central Atlantic, the continental rupture process involved, in general terms, the present-day continental margins of the eastern North and Central America, northwestern Africa and southwestern Europe. In the North Atlantic, the rifting started in the Upper Cretaceous (70 to 65 Ma), evolving up to a continental rupture along the margins of northeastern North America (Labrador Sea), southwestern and southeastern Greenland, and northeastern Europe (North Sea).

In the South Atlantic, the rift episode affected initially affected the margins of Antarctic, and subsequently, the margins of south-southeastern South America and south-southwestern Africa (about 150 to 130 Ma). In the Equatorial Atlantic, this process started by Early Cretaceous (115 to 110 Ma), resulting in the separation of north-northeastern South America and western Africa by Aptian/Albian times [*Gorini*, 1977].

In the specific case of the Brazilian Continental Margin, two distinct types of margins are recognized, as a result of its tectonic evolution. The Equatorial Margin type is characterized by marginal basins affected by a complex deformation pattern, involving local transpressive and transtensional features due to strike-slip motion between the south-American and African lithospheric plates. The Eastern or Atlantic Margin type is characterized by development of extensional basins due to divergent and semi-orthogonal motion between the lithospheric plates [*Conceição et al*, 1988; *Mohriak et al.*, 1990; *Souza*, 1991; *Chang et al.*, 1992].

Beginning with crustal rupture of the Brazilian Eastern Margin, fourteen main sedimentary basins were developed (figure 2). Three of them are located in continental domains (Recôncavo, Tucano and Jatobá basins), and the others are located along the coast (Pelotas, Santos, Campos, Espírito Santo, Mucuri, Cumuruxatiba, Jequitinhonha, Camamu-Almada, Jacuípe, Sergipe-Alagoas and Pernambuco-Paraíba basins).

Among many regional studies addressing the tectonic development of these basins, we emphasize the ones originated from the REMAC Project, Global Recognition of the Brazilian Continental Margin [*e.g. Asmus & Porto*, 1972; *Asmus & Ponte*, 1973; *Ponte & Asmus*, 1976; *Asmus & Porto* 1980; *Asmus & Guazelli*, 1981; *Guazelli & Carvalho*, 1981; *Ponte & Asmus*, 1982], which resulted in the first regional study of the entire margin. Many papers derived from this project offers a wide view of the shallow and deep structure of the marginal basins, and some discuss themes linked to the origin and evolution of sedimentary basins worldwide.

Other outstanding publications on the Brazilian Continental Margin involve the Deep Sea Drilling Project [e.g. *Kumar et al.*, 1977; *Perch-Nielsen et al.*, 1982; *Barker et al.* 1982; *Roberts et al.*, 1984]. Seminal papers by *Rabinowitz & La Brecque* [1979] and *Austin & Uchupi* [1982] were based on the magnetic anomaly maps of the Atlantic. Ojeda [1982], from a wide study based in regional seismic sections, attempted to establish a model for the whole Brazilian Continental Margin, using an analysis of the structural and stratigraphic framework of the marginal basins. *Asmus & Baisch* [1983], *Conceição et al.* [1988] and *Nürnberg & Müller* [1991] discussed alternative models for the Eastern Brazilian Margin.

Figure 2. Location map showing the Brazilian sedimentary basins and the area of this study.

Chang et al. [1992], in a synthesis on the evolution of the eastern Brazilian rift system, recognized the occurrence of three main phases of rifting before the continental rupture. The syn-rift Phase I, in the Jurassic (150 my), would have caused a great depression in the Gondwana supercontinent, causing the development of intracratonic basins; the syn-rift Phase II, in the Neocomian (140 My), would have corresponded to a period of rapid tectonic subsidence in the areas of the Sergipe-Alagoas and Recôncavo-Tucano basins, besides an intense basaltic volcanism in the south-southeastern marginal basins. The syn-rift Phase III, in the Barremian (120 My), was characterized by fluvio-deltaic and lacustrine deposits on structural lows and carbonate platforms on marginal highs. These rift phases were followed by a transitional evaporitic phase, in the Aptian, with the formation of thick deposits of salt under restrict oceanic conditions. Later, in the Albian, thick carbonate shelf deposits were developed in a marine environment, and were followed by transgressive sequences that characterize a progressive deepening of the marginal basins due to thermal subsidence.

4. MAIN COMPONENTS OF THE PHYSIOGRAPHY IN THE SOUTHEASTERN BRAZILIAN MARGIN

The main physiographic domains presented in the South-Southeastern Brazilian Continental Margin correspond to the continental shelf, the continental slope, the continental rise, the abyssal plain and, farther off the coast, the oceanic basin [*Zembruscki et al.,* 1972].

The continental shelf has a very regular bathymetry, with a low declivity gradient and approximately 300 m of maximum depth. The maximum width is about 220 km near to the latitude 24° S, in the Santos Basin, and the minimum width is about 90 km, at the latitude 31°30' S, in the Pelotas Basin (figure 3).

The continental slope is a relatively narrow and steep, with water-depths ranging from 100 to 300 m (on the shelf), to about 1000 to 3000 m at it's outer boundary, in the continental rise. Locally, the continental slope is interrupted by marginal plateaus and eroded as terraces, both with varied topographic expression, and also by erosional channels and sedimentary wedges of regional extent. Some of the most prominent physiographic features of the area are found in this domain, as for example the São Paulo Plateau, the Rio Grande Terrace and the remarkable sedimentary prograding wedge known as Rio Grande Cone.

Eastwards of the continental rise, the abyssal plain mark the passage to the Argentine Oceanic Basin, at the southern portion of the area, and to the Brazil Oceanic Basin, at the northern portion. In the abyssal plains, the physiography of the area is locally affected by important regional features, as the Jean Charcot seamounts and the São Paulo Ridge. Farther away from the coastal domain, the eminently monotonous plain of the oceanic basins is interrupted by structural highs of igneous nature and submarine channels, such as the Rio Grande Rise and the Vema Channel.

In addition to these anomalous physiographic features, there are a number of remarkable crustal discontinuities with a general E-W trend, which correspond to fracture zones and oceanic lineaments. Among these E-W discontinuities, the Rio de Janeiro and Rio Grande fracture zones and the Porto Alegre and Chuí Lineaments are the most important.

Another prominent regional linear trend in the area is known as Cruzeiro do Sul "Fracture Zone" [*Souza,* 1991; *Souza et al.,* 1993]. It corresponds to a lineament with a NW-SW direction, extending from the southern portion of the Rio Grande Rise, crossing the São Paulo Plateau and advancing towards the continental crust in the Cabo Frio region [*Mohriak et al.,* 1995].

The large amount of evaporites present in the Campos and Santos basins is responsible for a huge number of halotectonic structures in the deep water salt diapir province, giving rise to a unique hilly texture of the seabotton physiography.

5. ANOMALOUS REGIONAL TECTONIC AND SEDIMENTARY FEATURES

5.1. Fracture Zones and Oceanic Lineaments

The South Atlantic seafloor is interrupted by a large number of subparallel crustal discontinuities, identified as oceanic fracture zones, occurring on both sides of the Mid-Atlantic Ridge (figure 4). These discontinuities represent inactive segments of transform faults, genetically associated with the sea-floor spreading axis [*Turcotte,* 1974; *De Long et al.,* 1977; *East Pacific Rise Study Group,* 1985; *Eittreim and Ewing,* 1985; *Lowrie et al.,* 1986; *Severinghaus & MacDonald,* 1988; *Kamesh,* 1993]. They can generally be detected by steps in the bathymetry, by displacements of the magnetic anomalies in the oceanic basement, and by the alignment of gravity anomalies. The area of this study is crossed by the Rio de Janeiro and Rio Grande fracture zones, and by the Porto Alegre and Chuí lineaments.

The Rio de Janeiro Fracture Zone has not much expression on the seismic sections. It is characterized by an anomalous concentration of volcanic cones and punctual highs of the igneous basement, aligned in an E-W direction along a zone with a mean width of about 150 km, between the parallels 23°S and 24°S. In the northern part of this fracture zone, in shallow waters region, the Cabo Frio High represents the limit of the Campos and Santos basins. Towards the deep water province, the limit of these basins is marked by striking differences in stratigraphic evolution and salt tectonics structural styles [*Mohriak et al.,* 1995].

The Rio Grande Fracture Zone, the most important crustal discontinuity in the area, occurs between the parallels 26°20'S and 28°30'S and is marked by a transtensional fault system. The most important morphologic expression of this discontinuity is close to the São Paulo Ridge, at the southern limit of the São Paulo Plateau (figure 6). At this limit, the Rio Grande Fracture Zone marks the southern transitional/oceanic boundary in the area of the plateau, suggesting the originally transformant character of this discontinuity. Northwards of the São Paulo Ridge (around 26°30'S), this fracture zone seems to be associated with the Jean Charcot seamounts, beyond the eastern limit of the São Paulo Plateau. In shallow waters, this fracture zone is aligned with the Florianópolis High, which separates the Pelotas and Santos basins (figure 3).

The Porto Alegre Lineament, located in the area between latitudes 32°S and 31°S (figure 3), is characterized by the association of a pair of normal faults with large offsets. It is characterized by an anomalous concentration of volcanic highs, seamounts and guyots along an E-W trend. This lineament also marks the southern limit of an important re-

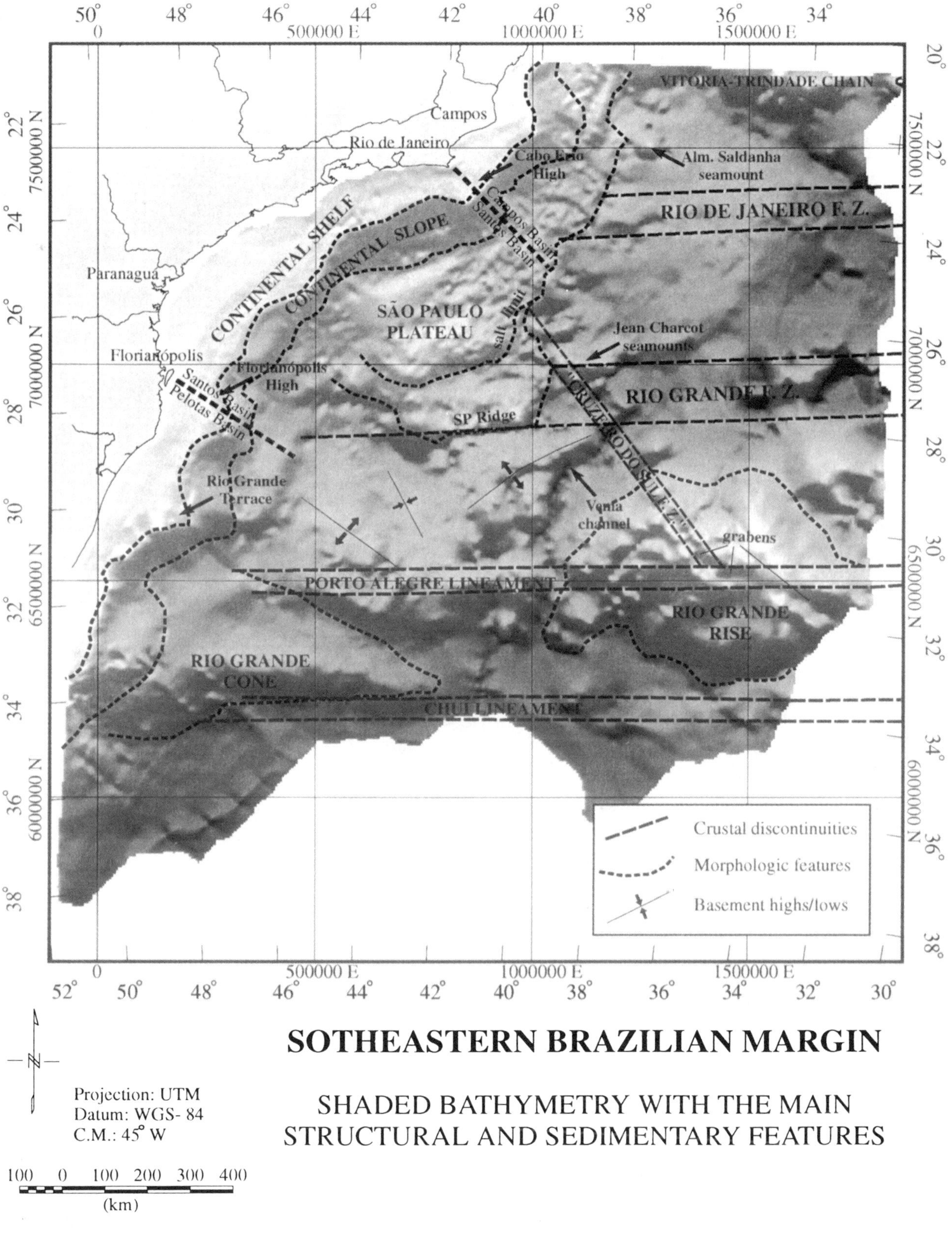

Figure 3. Shaded bathymetric map of the Southeastern Brazilian Continental Margin with the main regional features.

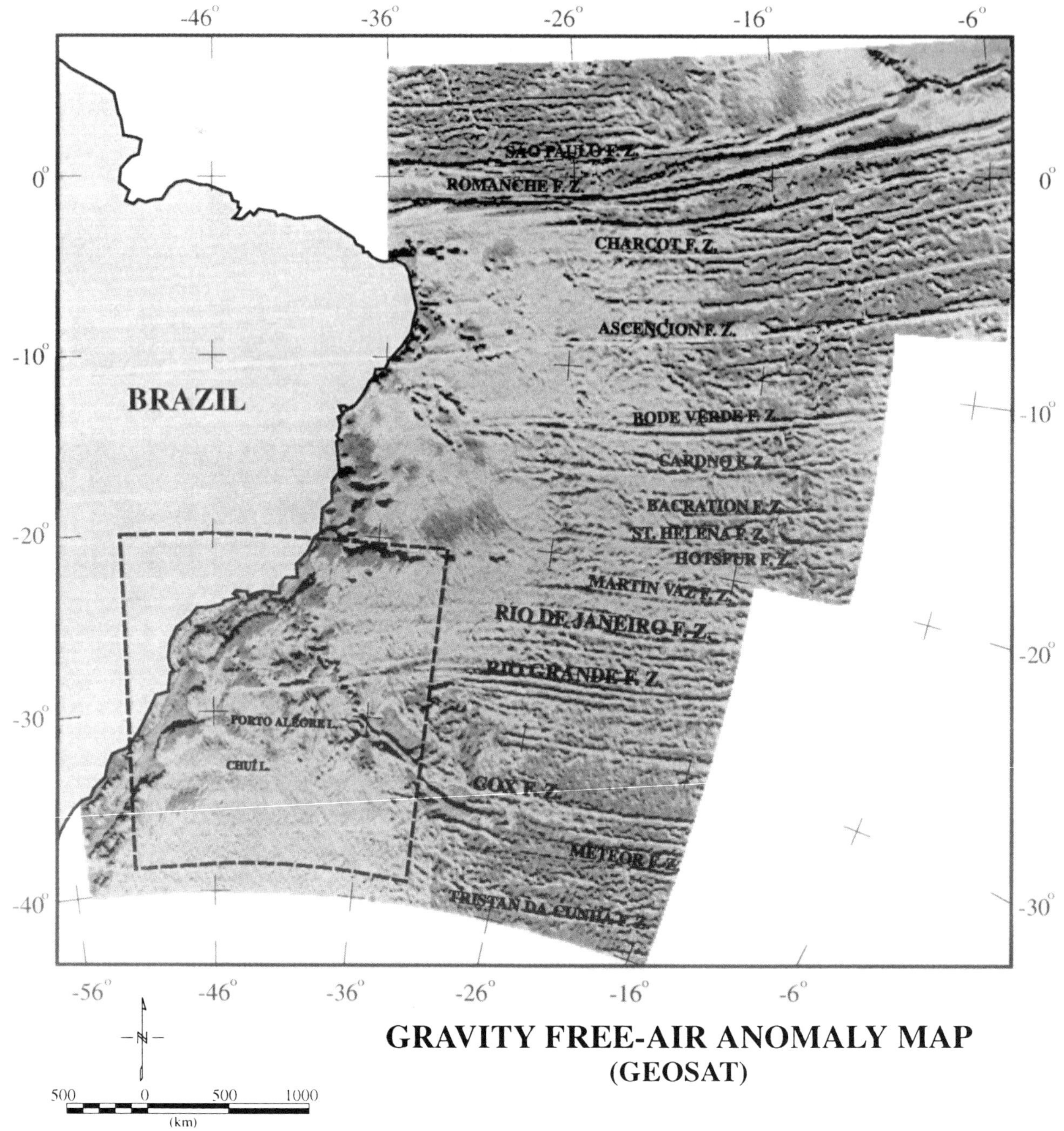

Figure 4. Gravity Free-Air anomaly map showing the location of the main oceanic fracture zones of the South Atlantic.

gional elevation of the oceanic basement in the area, which is onlapped by the northern border of the Rio Grande Cone sedimentary wedge (figures 3 and 5).

The Chuí Lineament, positioned between latitudes 34°S and 35°S, apparently represents the prolongation of the Cox Fracture Zone (figure 4). It is marked by a pair of normal faults dipping to the south, with offsets up to 700 m in the oceanic basement. Such steps on the basement do not have much expression in the bathymetry and may represent the effects of differential thermal subsidence that affected the crustal segments. Although this discontinuity is not associated to any structural high, it locally represents the southern limit of the Rio Grande Cone, besides exerting a strong influence on the contour of the continental fall (figure 3).

SECTION 59

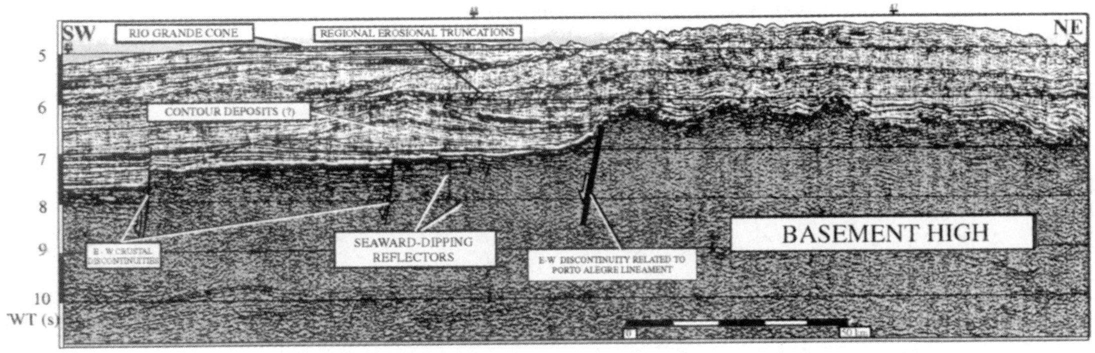

Figure 5. Detail of the seismic section 59 (strike), showing the location of a basement high and its influence on the deposition of the Rio Grande Cone, to the south. It shows the local sedimentary complexity involving contour deposits and a series of erosional truncations in the area.

The Cruzeiro do Sul "Fracture Zone" is characterized by a set of grabens and horsts, aligned to the NW-SE direction (figure 3). These features are observed from the southeastern border of the Rio Grande Rise to the vicinity of the southeastern limit of the São Paulo Plateau, in the area of the Jean Charcot seamounts [*Souza,* 1991]. This discontinuity is well expressed both on the gravity and on the bathymetric maps of the area, specially where it crosses the Rio Grande Rise. At this point, it controls the linear NW-SE trending graben, which presents a very clear displacement to the west, surely controlled by the Porto Alegre Lineament (figure 3). The supposed age for the Cruzeiro do Sul "Fracture Zone" is Eocene, as indicated by magnetic anomalies correlation [*Souza,* 1991]. It may have originated from a NW-SE reactivation of structures that affected both the continental and oceanic crust [*Souza et al.,* 1993; *Szatmari and Mohriak,* 1995].

Cande et al. [1988] suggested that the fossil oceanic spreading center located south of the Rio Grande Rise, in the South American Plate, corresponded to another one at the southern part of the Walvis Ridge, in the African Plate. According these authors, the paleo spreading axis was active from the crustal rupture (Aptian) up to the Late Cretaceous/Early Tertiary, at the age of the C30 oceanic magnetic anomaly. The positioning of the spreading axis, with a NE-SW direction and oblique to the present-day N-S axis, suggests a very particular spreading geometry for this part of the South Atlantic, marked by the change of the seafloor spreading direction from NW-SE, approximately from Aptian to the end of Cretaceous, to E-W, from Paleocene to the present.

In this context, it is possible that the Cruzeiro do Sul "Fracture Zone" originates along an original paleo transform fault, approximately perpendicular to this fossil NE-SW oceanic spreading center, which was responsible for the weakness crustal zone during the inception of the South Atlantic, in spite of the Eocene age attributed to the NW-SE trending graben [*Souza,* 1991].

5.2. Sedimentary Features: Erosion Terraces and the Rio Grande Cone

In this portion of the Continental Margin three important erosion terraces can be identified, all sculptured on the upper portion of the continental slope, eastwards of the continental shelf break. The most pronounced feature of the area is the Rio Grande Terrace, with bathymetry varying from 250 to 500 m and occupying a considerable area of the slope, estimated around 5,500 km^2 (figure 3).

The Rio Grande Cone is another singular sedimentary feature, characterized by a voluminous wedge of sedimentary progradation developed in the southern portion of the Pelotas Basin, approximately between the latitudes 32°S and 35°S (figure 3). The origin of the Rio Grande Cone is attributed to an anomalous local sedimentation after the Upper Eocene. On seismic sections, the youngest sedimentary horizons of the cone present their terminations in "onlap" against a major erosional unconformity, related to an important basement structural high in the area, apparently conditioned by the Porto Alegre Lineament (figure 5).

5.3. The São Paulo Plateau

The largest and most expressive anomalous feature of the physiography, not only in the area of this study, but even in the whole Brazilian Continental Margin, is the São Paulo Plateau. It corresponds to an extensive high located from 22°S and 28°S latitudes and from 39°W to 48°W longitudes, and it is characterized by dominantly low bathymet-

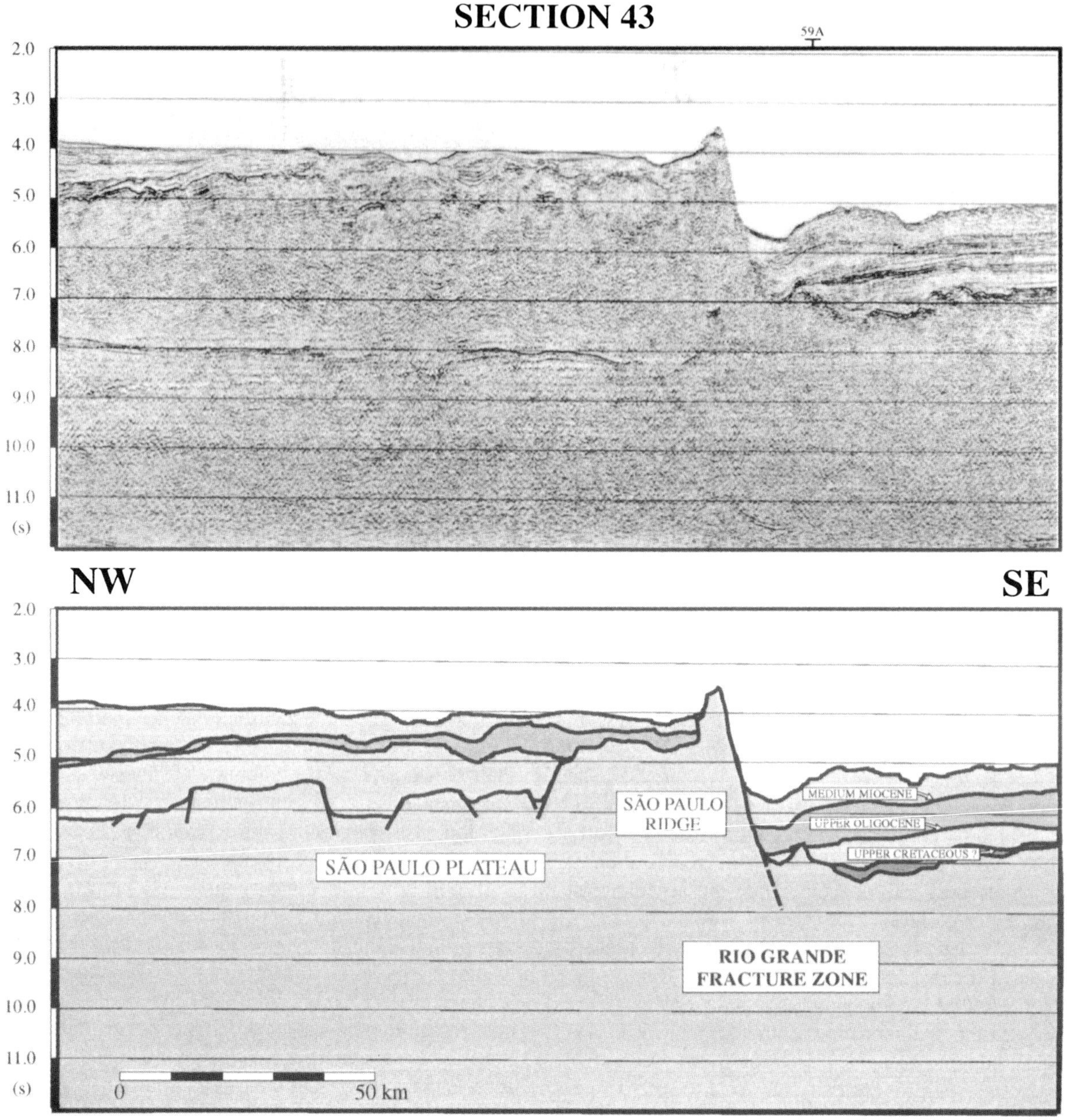

Figure 6. Detail of the seismic section number 43, showing the local aspect of the São Paulo Ridge, at the southern border of the São Paulo Plateau, developed along the Rio Grande Fracture Zone.

ric gradients on the continental slope and parts of the continental rise of the Santos and Campos basins (figure 3).

The São Paulo Plateau presents and is affected by diapiric-halokynetic structures, which strongly affect the post salt sediments and are responsible for the corrugated textural aspect of the bathymetry. The seismic interpretation of the subsalt horizons is hampered by the presence of the evaporites, but the base of the salt is considered relatively flat, in spite of being locally affected by reactivation of rift-phase normal faults, particularly during the Albian. The relief of the basement is also irregular, affected by the rift faults and by the presence of volcanic highs, specially in the deep water province [*Demercian et al.,* 1993].

The southern limit of the plateau is approximately rectilinear, and coincides with an E-W crustal discontinuity, which corresponds to the Rio Grande Fracture Zone. Parallel to this fracture zone, the southern border of the São Paulo Plateau is characterized by development of a very

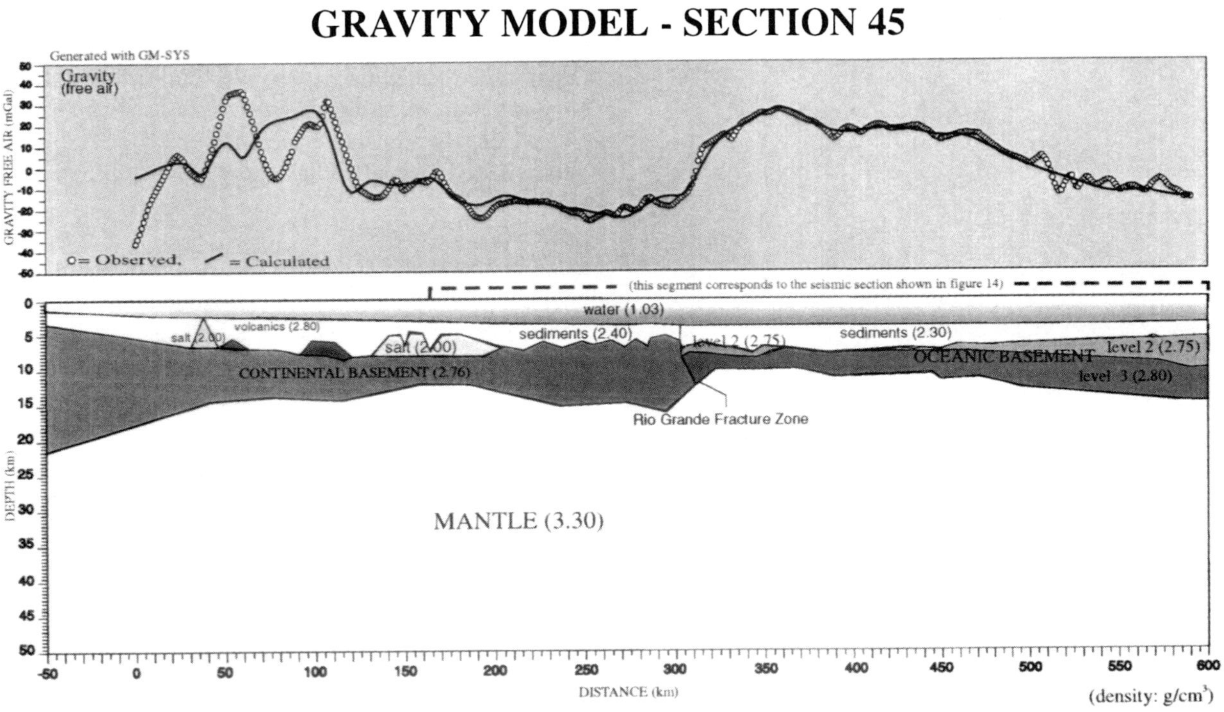

Figure 7. Gravity model based on the seismic section 45 (dip), showing the abrupt limit between the southern border of the São Paulo Plateau, on the left, and the thinned oceanic crust, on the right part of the model. The densities used for each block are indicated in parenthesis.

prominent submarine chain, represented by the São Paulo Ridge (figures 3 and 6).

Besides the São Paulo Ridge, the southern portion of the São Paulo Plateau is characterized by a progressive structural uplift, from north to south, towards the Rio Grande Fracture Zone. This regional elevation is apparently independent of the volcanic chain, and it is also responsible for the important internal unevenness of the plateau. This structural high, which seems to have strongly influenced the sedimentation patterns and even the occurrence of the evaporites in the Santos Basin, is conditioned in part by the accumulation of volcanic material in the area. This magmatism may be responsible for an increased crustal thickness, as indicated by gravity models and maps (figure 8). The structural high may in part correspond to the isostatic rebound caused by the large vertical slip due to normal faults along the main discontinuity of the Rio Grande Fracture Zone (figures 6 and 11).

The eastern limit of the São Paulo Plateau is characterized, in general, by an abrupt termination of the sedimentary layers, which are often deformed by diapirs, walls or allochthonous tongues of evaporites, against a dominantly planar and regular oceanic sedimentary column, marked by planar-parallel superposition of the horizons.

Several factors as the presence of thick evaporite deposits, the intense deformation of the sediments caused by halokinesys, and the extreme depth of the rift sequence make the seismic interpretation difficult. The seismic sections of the LEPLAC Project, partly due to the intrinsic characteristics of the project, unfortunately do not present an adequately resolved deep crustal horizons in areas overlain by salt.

Even today, the origin and crustal nature of the São Paulo Plateau, as well as the outer limit of the rift sequence sedimentary blocks, is a matter of debate. Some authors [e.g. *Leyden et al.*, 1971; and *Kumar et al.*, 1977] mentioned the existence of seismic horizons and magnetic anomalies correlated with a typical oceanic crust within this plateau. However, most researchers [*e.g. Kowsmann et al.*, 1982; *Austin & Uchupi*, 1982; *Conceição et al.*, 1988; *Chang et al.*, 1992; *Souza et al.*, 1993] consider that the São Paulo Plateau was developed on a severely extended continental crust highly intruded by volcanic rocks, with aspects of a transitional crust.

Along this southern boundary of the São Paulo Plateau, the high local gradient of the gravity anomalies, as observed in the gravity model of the section 45 (figure 7), suggests the occurrence of a direct and abrupt contact be-

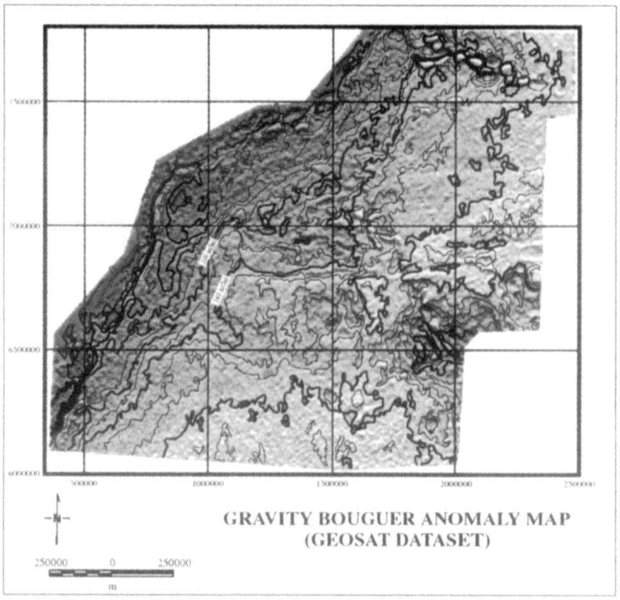

Figure 8. Gravity Bouguer anomaly map (Geosat dataset) of the southeastern region of Brazil.

tween the continental and oceanic crusts, respectively at the north and south of the Rio Grande Fracture Zone. This fact may indicate that, at least in this part of the margin, the actual crustal limit could have been originated from a transform plate limit, representing an exception along the Eastern Brazilian Margin, whose nature is essentially extensional.

The analysis of the Bouguer anomaly map of the area suggests the distinction of three amplitude levels of Bouguer anomalies not only in the São Paulo Plateau, but along almost the whole Eastern Brazilian Margin (figure 8). The proximal portions of the margin, including the continental platform and part of the continental slope, present Bouguer anomalies varying from 0 to 100 mGal, probably corresponding to extended crustal regions, predominantly continental in nature and preferential sites for thick rift sedimentary sequences. The areas of the slope, located farther from the coast, show anomaly levels between 100 and 140 mGal and appear to correspond to crustal portions of intermediate to transitional nature. In spite of being marked by abundant occurrences of volcanic rocks, these areas do not configure a typical oceanic crust. Areas located farther from the coast and characterized by Bouguer anomalies with amplitudes generally exceeding 140 mGal, probably refer to regions of pure oceanic crust.

According to this generic interpretation, the São Paulo Plateau would present, in general terms, the outer limit of the rift sequence bounded by the 100 mGal isogalic curve. This would restrict the supposed continental portion of the São Paulo Plateau to a wide belt approximately 400 km from the coastline. In this case, the continental crust is supposed to be located westwards the 100mGal isogal, whereas the portion between this isogal and the eastward limit of the plateau can be considered a transitional crust, intensively intruded by volcanic bodies and devoid of a significant rift sequence.

5.4. The São Paulo Ridge

Extending for about 220 km at the southern boundary of the São Paulo Plateau, the São Paulo Ridge is an important igneous submarine linear chain, very prominent in the bathymetry and developed along the Rio Grande Fracture Zone. It was surely constructed as a result of the magmatism along this crustal discontinuity (figures 3 and 6) and presents its roots on the igneous basement, being more prominent in the area between the meridians 40° W and 42°30' W. In this area, it is characterized by a 2,000 m scarp near the southern flank of the ridge. To the east and west from this area, the ridge gradually looses expression on the basement relief and on the bathymetry (figures 11, 12 and 13).

During the Aptian, the sedimentation along the Brazilian Continental Margin was affected by a series of restrictions to the free marine circulation. This resulted in the formation of thick evaporite deposits in the Santos, Campos, Espírito Santo, Bahia State and Sergipe-Alagoas basins, covering the older sediments of the rift sequence.

Several authors attributed the origin of such restrictions to the relief escarpments of the São Paulo Ridge, in the South American Tectonic Plate, and the Walvis Ridge, in the African Plate [*Leyden et al.,* 1976; *Gambôa,* 1978]. In this case, these chains would have established a natural fence to the advance of the sea coming from the south at the age of the continental rupture, favoring the formation of thick evaporite deposits in a restricted marine environment. This evolutionary phase of the Continental Margin is known as Transitional Evaporitic Phase [*Chang et al.,* 1992], and preceded the drift or thermal phase of subsidence in these basins.

Near the southern escarpment of the São Paulo Ridge (figure 6), a very wide and deep erosive channel was developed by the action of the seabotton currents towards the bathymetric fence represented by this chain. This channel is 5 km wide and reaches about 500 m in depth in some places.

5.5. The Rio Grande Rise

In the abyssal domains of the Continental Margin, very far from the coast and beyond the area covered by the

LEPLAC, the regional maps show the development of a very large and prominent elevation, known as the Rio Grande Rise, which forms a submarine igneous chain with steep scarps. It covers an area totaling more than 200,000 km2 and is divided into two major portions: the eastern elevation and the western elevation, both located between the longitudes 40°W and 30°W (figures 3).

In this study, the Rio Grande Rise forms the physiographic limit between the oceanic basins of Brazil and Argentina. It provides a sharp change from the monotonously flat nature of the oceanic domain, with a high concentration of seamounts, valleys and scarps along normal faults, with relieves exceeding 4000 m. It is dissected by a major, NW-SE-trending central graben which extends from the southern portion of the elevation towards the area of the Jean Charcot seamounts, eastwards of the São Paulo Plateau. *Souza* [1991] refers to this discontinuity as the "Cruzeiro do Sul Fracture Zone", suggesting that it was formed during Eocene times in oceanic crust, but extended into continental crust through the São Paulo Plateau towards the coastal region off Cabo Frio, located to the north of the Rio de Janeiro city. Onshore, this lineament is marked by a number of alkaline intrusions that date from Late Cretaceous to Early Tertiary [*Mohriak et al.*, 1995].

Both to the south and north, the São Paulo Rise is limited by E-W-trending oceanic crust discontinuities, defined respectively as the Chuí lineament and the Rio Grand Fracture Zone. To the west, near the continental slope, there is a lesser elevation roughly trending N-S, which may be considered as a small western extension of the rise. The Vema channel runs between this feature and the main portion of the Rio Grande Rise, forming the main passage of Antarctic bottom currents from the Argentine to the Brazilian abyssal basin (figures 3 and 13).

Many authors attribute to the Rio Grande Rise an origin common with the Walvis Ridge in the African Plate [*Kumar,* 1979; *Barker et. al.,* 1982, *Cande et al.,* 1988; *O'Connor & Duncan,* 1990; *Wannesson et al.,* 1991; *Nürnberg & Müller,* 1991]. Generally they relate the construction of these chains to intense pulses of volcanism associated with the divergent displacement of the south-American and African plates over the Tristan da Cunha hot spot, stationary in the vicinity of the Mid-Atlantic Ridge.

The estimated age of the Rio Grande Rise is about 85 Ma., based on the drilling of the well number 516 of the Deep Sea Drilling Project [*Barker et al.,* 1982]. It supposedly formed close to the Mid-Atlantic Ridge, in a subaereal environment. Later, the whole area was affected by thermal subsidence like a typical oceanic lithosphere.

Gambôa [1981] and *Gambôa & Rabinovitch* [1981] subdivided in the Rio Grande Rise in two parts. The eastern portion of the rise, positioned outside the area of this study, would have been generated from a paleo-seafloor spreading center, active during part of the Upper Cretaceous (from 100 to 80 Ma). The western portion would have been generated later, in oceanic crust, under influence of a thermal anomaly during the Middle Eocene.

Cande et al. [1988] identified, close to the southeastern limit of the Rio Grande Rise a fossil oceanic spreading center, with a NE-SW direction, which would have been active during the Paleocene and Eocene, approximately from 68 to 45 Ma. In the African Plate, this paleo-axis of sea-floor spreading would be associated to the Walvis Ridge (figure 9).

5.6. Guyots and Seamounts

Prominent seamounts are observed on the continental slope, continental rise and abyssal plain domains, generally concentrated along fracture zones and oceanic lineaments.

The Jean Charcot seamounts are the most expressive seamounts in this part of the margin, positioned near the external limit of the São Paulo Plateau between the parallels 26°S and 27°S, and meridians 38°W and 40°W (figure 3). These seamounts have their roots on the oceanic basement, reaching heights of 2600 m above the oceanic floor and about 5000 m above the volcanic basement (figure 10).

The age of these seamounts seems to be as the same age of the oceanic crust in the area, considered approximately Aptian, once the post-Aptian sedimentary sequences onlap them. They were possibly to be controlled by an excess volcanic activity, associated with the E-W trending Rio Grande Fracture Zone. Their bathymetric expression indicates that they could have been reactivated during later volcanic events, and may be associated with the formation of the NW trending "Cruzeiro do Sul Fracture Zone" [*Souza et al.,* 1993].

In the southernmost portions of the area, between parallels 31°S and 32°S, and meridians 43°W and 45°W, there are other prominent seamounts and guyots. The distribution pattern of these seamounts and other isolated basement highs, covered by sediments, indicates an association with the E-W trending Porto Alegre Lineament.

6. THE OCEANIC CRUST: SEISMIC EXPRESSION AND ASSOCIATED STRUCTURES

6.1. The Internal Structure of the Oceanic Crust

The acquisition and processing parameters applied to the LEPLAC seismic sections were not optimized to visualize deep features of the crust. However, in many cases, it is possible to identify three different units in the oceanic crust,

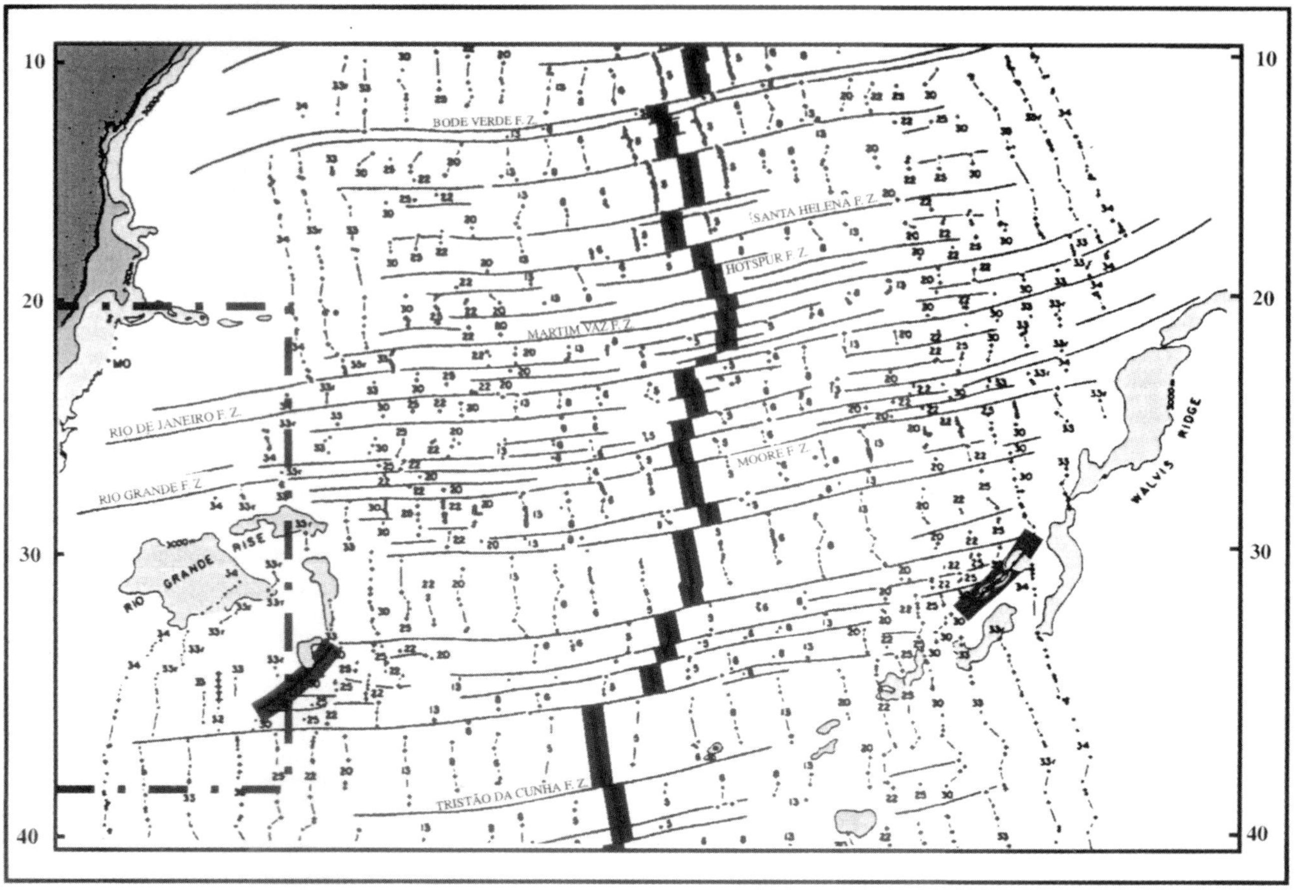

Figure 9. Map of the South-Atlantic showing the position of the fracture zones and the oceanic magnetic anomalies (Cande, 1988).

with distinct seismic characters, which correspond to the units 1, 2 and 3 [*Cann,* 1974].

The Unit 1 corresponds to the hemipelagic and siliciclastic sedimentary column that covers the igneous part of the oceanic crust. Naturally, its thickness is highly affected by the local characteristics of sedimentary supply, by the seabotton currents, responsible by the great amount of erosional truncations, and by structural features of the oceanic basement. The seismic velocities considered for this unit correspond to 1,800 m/s, related to the superficial sediments, and 2,800 m/s, related to the rest of the sedimentary column.

The units 2 and 3 refer to the igneous portion of the oceanic crust, between the base of the sedimentary column (Unit 1) and the Moho discontinuity. In some seismic sections, the Moho is marked by a visible seismic reflection in some intervals.

The Unit 2 corresponds to the portion immediately below the top of the acoustic basement, characterized by the development of subhorizontal reflections generated by the layering of volcanoclastics, extrusive basaltic lava and dikes, preferentially near the oceanic spreading center. Based on refraction analysis [*Leyden,* 1971], the seismic velocity considered for these horizons is around 5.5 km/s.

Unit 3 refers to the crustal level just below the Unit 2 and above the Moho, in the oceanic crust. It is constituted predominantly by intrusive gabros and is characterized by a transparent seismic aspect, with no important internal reflections. For this unit, the seismic velocity is considered to be 6.7 km/s.

In terms of oceanic crustal thickness, the area of this study is considered heterogeneous, influenced by anomalous features such as volcanic highs, concentration of seamounts, positive and negative features of the basement, and crustal discontinuities like fracture zones and oceanic lineaments.

The southernmost portion of the area (latitudes higher than 32°S) is characterized by the predominance of a relatively homogeneous igneous oceanic crust, presenting a total thickness ranging from 5.0 to 6.5 km in this area. In this

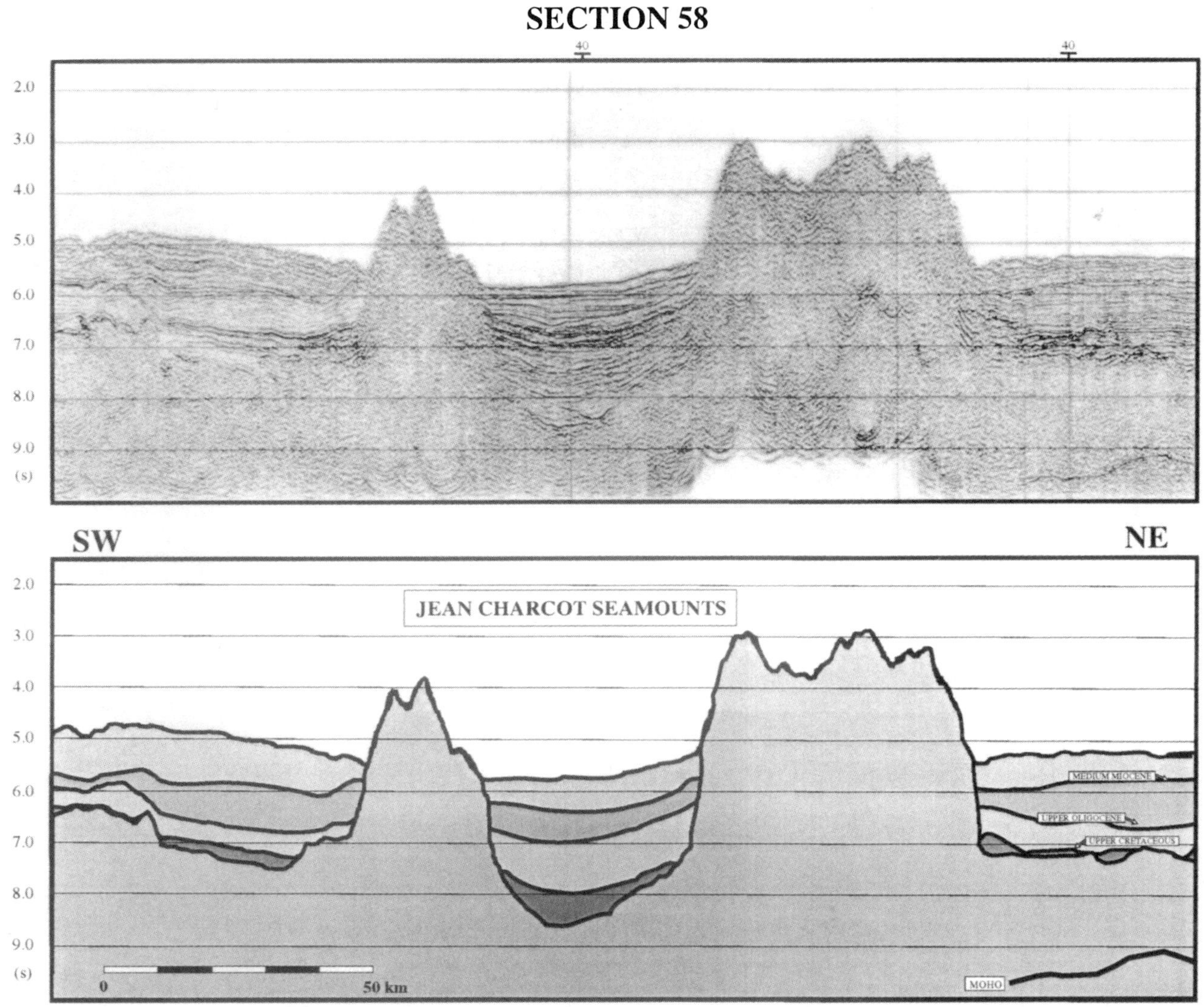

Figure 10. Detail of the seismic section 59 (strike), showing an aspect of the Jean Charcot Seamounts and the isostatic undulation in the basement, caused by their load.

region, due to its poor morphologic expression, the presence of the Chuí Lineament appears to exert not much influence on the crustal thickness, except locally, in areas close to related normal faults with large offsets.

The area between the Porto Alegre Lineament and the Rio Grande Fracture Zone, limited by the latitudes 28°S and 32°, corresponds to the most complex crustal segment in terms of crustal thickness and basement structuration in the area. Between the latitudes 30°S and 32°S, and longitudes 42°W and 46°W, an important basement structural high causes an expressive elevation on the bathymetry, caused by a basement structural high (figure 3). In this area, the thickening of the igneous oceanic crust reaches values up to 10 km (figure 15). Between the latitudes 30°S and 29°S, and the longitudes 42°W and 43°W, the oceanic crust presents is the development of horsts and grabens oriented in the NNW-SSE direction, resulting in a sensible thinning in the oceanic igneous crust, with estimated values of thickness around 4.0 km.

From this zone with horsts and grabens to the southern border of the São Paulo Plateau, the oceanic crust is very thin (about 4.0 km), monotonous and covered by a rather thin sedimentary column. This thinned portion of the crust runs along part of the São Paulo Plateau, between the longitudes 39°W and 43°W. In a more distal position, between the latitudes 28°S and 31°S, and eastwards of the meridian 42°30′W, an important crustal thickening occurs, resulted from an important crustal elevation, related to a basement

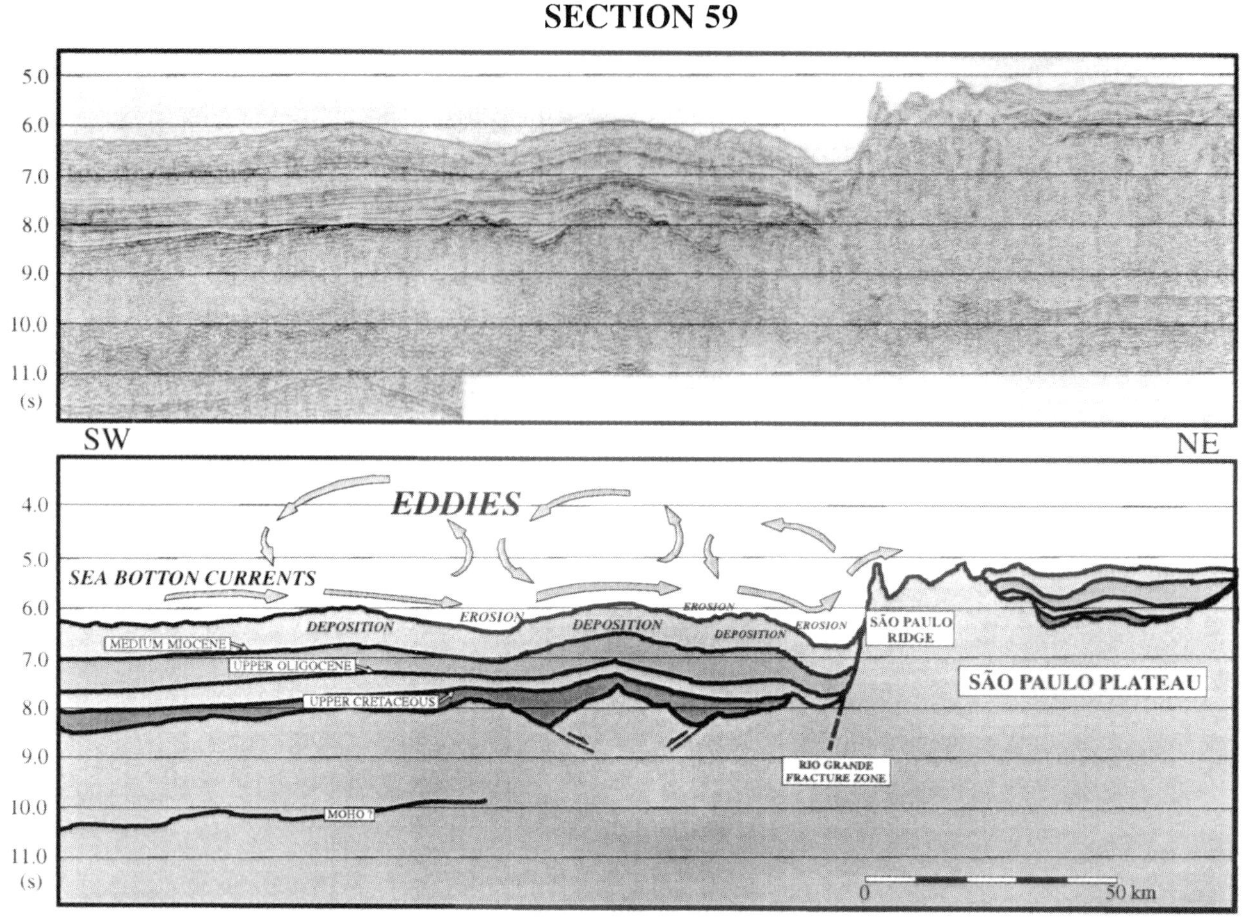

Figure 11. Detail of the seismic section 59 (strike), showing a local aspect of the southern border of the São Paulo Plateau and the effect of the São Paulo Ridge, on the long wavelength undulations observed on the oceanic crust.

arching. This arching or undulation that affects the basement and the whole sedimentary column in this area is possibly related to compressional tectonics in the oceanic crust (figures 3, 12, 13 and 14).

Northwards of the Rio Grande Fracture Zone and eastwards of the São Paulo Plateau, the estimated thickness of the igneous oceanic crust varies from 5.5 to 7.0 km, except in the area affected by the Jean Charcot seamounts, near the latitude 24°S.

6.1) Undulations and Folding Arches on the Seafloor

Remarkable features of the oceanic domain of the Southeastern Brazilian Continental Margin are the well-developed undulations and fold arches that affect the seafloor, part or the entire sedimentary column, and even the igneous basement.

Based on the seismic sections, we separate these undulations and fold arches in two different types, medium and larger undulations, according to the respective wavelength. The medium undulations present wavelengths varying from 5 to 30 km and affect the upper part of the sedimentary cover. The larger undulations present wavelength values higher than 30 km and affect part or the entire sedimentary column and even the basement.

The medium undulations are specially frequent in the area to the south of the São Paulo Plateau, between the latitudes 28°S and 34°S. These features affect specifically the sedimentary levels younger than the Upper Oligocene and their origin is attributed to the action of seabotton currents, through a continuous cut-and-fill process, generating many superimposed channels and resulting in discontinuous reflections on the seismic profiles (figure 11).

The larger undulations correspond to some very prominent and long wavelength shaped features observed on the bathymetry, originated by the action of seabotton currents, by flexural crustal folding or even by compression or transpressive tectonics.

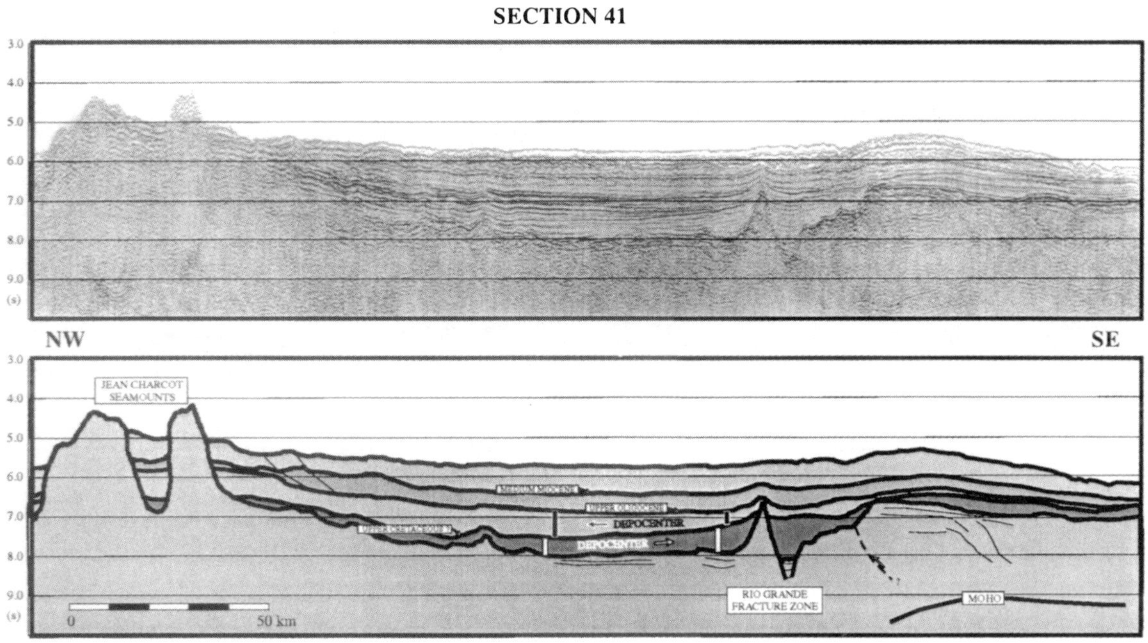

Figure 12. Detail of the seismic section 43 (dip) showing an expressive long wavelength undulation in the oceanic crust, probably caused by a reverse fault zone, interpreted in the rigth side of the section. In the basal sedimentary sequences, an inversion of the depocenter can be observed.

A very expressive example of these large undulations can be observed on the crustal segment south of the southern border of the São Paulo Plateau, between the latitudes 28°S and 29°30'S, and longitudes 39°30'W and 42°30'W. In this area, three successive undulations are observed on the seafloor, affecting part of the sedimentary column and with decreasing wavelength towards the São Paulo Ridge (figure 11).

In this specific case, apparently these undulations do not reflect the basement relief. The most probable origin of these undulations is attributed to the action of the seabotton currents in the area .These undulations are interpreted to have been generated by the fast decrease of the flux velocity of currents that come from the south. These currents were hindered the very prominent rampart represented by the São Paulo Ridge and are interpreted to form a complex system of secondary currents [*G. Karner,* oral communication]. One part of these current flux would cross over the ridge, another part would flow to the east, parallely to the ridge escarpment, and another one would return to the south, forming lateral and vertical vortexes or eddies in the water column. These eddies would be capable to rework the sediments and generating arched features on the sedimentary cover (figure 11).

Near the Jean Charcot seamounts, between the latitudes 25°S and 27°S, and longitudes 39°W and 41°W, a very expressive crustal depression is observed, affecting even the more distal portions of the São Paulo Plateau. The origin of this regional feature is interpreted to be due to the isostatic flexural answer to the overload generated by the Jean Charcot seamounts (figure 10).

One of the most notable example of a basement-involved undulation is observed approximately at the coordinates 28°10´S and 38°40´W, close to the Rio Grande Fracture Zone. This feature has a clear expression on the oceanic domain, where we observe the development of a long-wavelength undulation on the seafloor, affecting the entire sedimentary column (figure 12).

In this region, the sedimentary cover shows a clear thickening related to the basal sequence (Cretaceous) towards the southeast, limited to the top by an expressive erosional truncation. The sedimentary sequence immediately above onlaps this erosional truncation and thickens towards the northwest, in the opposite direction (figure 12), characterizing apparently an inversion of the depocenter. This geometry, if it was not caused simply by the action of seabotton currents, suggests an uplift of the basement in the eastern part of the structure during the Upper Cretaceous, possibly caused by compressional tectonics in the oceanic crust.

This long wavelength undulation affecting the basement and the sedimentary column seems to correspond to an asymmetric basement antiform, developed on the hanging wall of an interpreted reverse fault, with steeper inclination

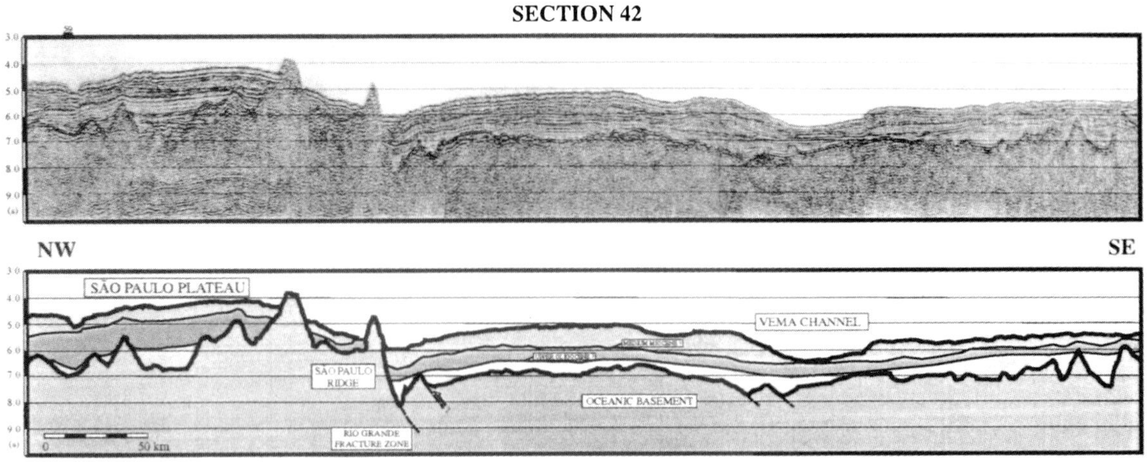

Figure 13. Detail of the seismic section 42, showing another view of the São Paulo Ridge and the Rio Grande Fracture Zone. On the right side of the section, reverse faults are interpreted, and they are probably responsible for the development of long wavelength undulations in oceanic crust.

on the northwestern flank of the structure. If it is confirmed, this compressional structure would have caused a regional compressive fold, limited to the northwest by this reverse fault, with a pronounced dip towards the southeast. In this case, the supposed tectonic efforts would have vergence to the northwest.

The possible development of this reverse fault in oceanic crust is also suggested by the proposed interpretation of the seismic sections 44 and 45 (figures 13 and 14). This reverse fault is caused by a compressive tectonic effort, responsible for the gentler undulation on the basement, as visible on the ocean floor, near the latitude 29°40′S and longitude 42°30′W. Also in this area, the reverse fault separates a thinned crustal segment, to the northwest, from a thicker and elevated crustal segment, to the southeast. The throw of this reverse fault, however, looses amplitude to the south, probably due to the increasing of the distance from the Rio Grande Fracture Zone.

The development of this presumed reverse trend in oceanic crust can be interpreted as the result of the simple perpendicular compression, caused by a tectonic motion towards NW and caused by a paleo oceanic spreading axis, or even by relative transpressive movements, with E-W direction, between two oceanic segments and developed along

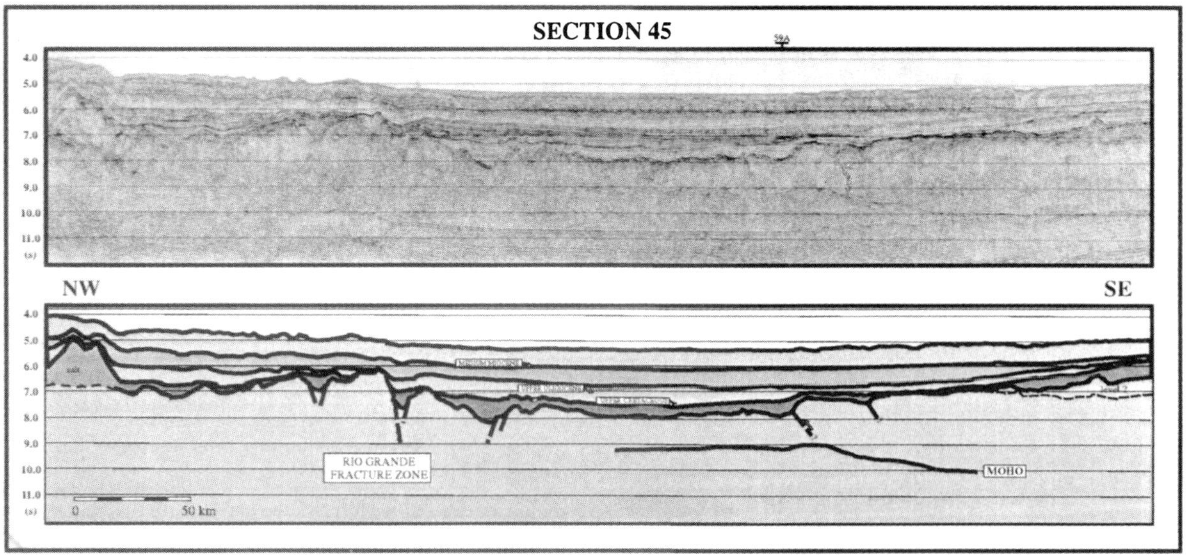

Figure 14. Detail of the seismic section 45 (dip) showing a long wavelength crustal undulation, with low amplitude, possibly caused by a reverse fault interpreted on the right side of the section.

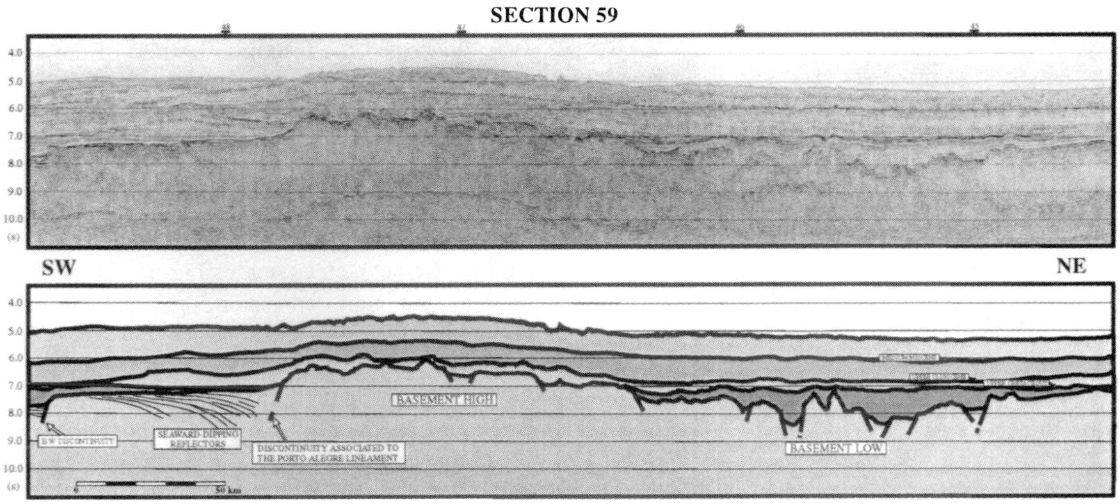

Figure 15. Detail of the seismic section 59 (strike) showing one of the most expressive long wavelength crustal undulation of the area, located northwards of the Porto Alegre Lineament.

the Rio Grande Fracture Zone (figure 22). In both cases, the origin of the tectonic efforts is attributed to the ridge-push mechanism at the oceanic spreading axis.

Another expressive long-wavelength crustal undulation present in the area is observed between the latitudes 29°S and 31°30′S, and meridians 43°W and 45°W (figure 15). According to the seismic interpretation, the positive arching that affects the entire sedimentary column is accompanied by a regional uplift of the basement, responsible for a local crustal thickening. The bathymetric and the gravity Bouguer anomaly maps (figures 3 and 8) also reflect an important elevation of the oceanic floor in the area, elongated to NW and presenting a relative wavelength decreasing to SE. The origin of this crustal thickening is interpreted as a result of the local accumulation of the volcanic material, formed during and subsequent to the crustal break-up process.

Immediately to the north of this area, a long wavelength crustal depression is also observed, caused by the emplacement of a NNW-SSE trend of grabens and horsts, resulting in expressive crustal thinning. This depression is elongated to NNW and is characterized by increasing of the wavelength to SSE (figure 3).

6.2 Seaward-Dipping Reflectors

Several seismic profiles show structures identified as seaward-dipping reflector wedges (SDR) in this portion of the Brazilian Margin. These are constituted by a relatively continuous set of seismic reflections, visibly convex to the top, and generally dipping towards the deep ocean (eastwards). The top of these SDR corresponds also to the top of the oceanic basement, and the base of the wedges may be seated on the continental crust, in the proximal areas, and on the oceanic Unit 3, in the oceanic crust.

Among the several models with respect to the origin and evolution of the SDR, the most widely known were proposed by *Hinz* [1981], *Hinz and Krause* [1982] and *Mutter et al.* [1982]. *Hinz* [1981] considered that the crustal extension and thinning occurred during the process of crustal rupture, and it was followed by the injection of dikes and basalt extrusion through a crustal zone of intense fracturing (figure 16). These volcanic flows might have developed either in subaereal environments, and thus in structurally raised areas, or they might have been accumulated in local structural lows near the crustal fissures, forming thick deposits of volcanic rocks, volcanoclastic and siliciclastic sediments.

At the end of the volcanic activity, perhaps 1 or 2 million years later, the crustal load caused by these deposits, added to the crustal cooling, would cause a local isostatic subsidence and impart to these horizons a crescent dipping towards the source area of the volcanism. As this source area of volcanism is generally associated with the most probable area where the first fragments of oceanic crust were formed after the crustal rupture, the continental/oceanic boundary would be positioned close to the external limit of the SDRs (figure 16).

Due to similarities between the seismic expression of the SDR wedges and fluvio-deltaic deposits associated with rift sequences, the interpretation of these reflections is still contested by some authors in several areas along the Brazilian Margin, including the area of this study. Moreover, interpretations of ultradeep water occurrences of rift se-

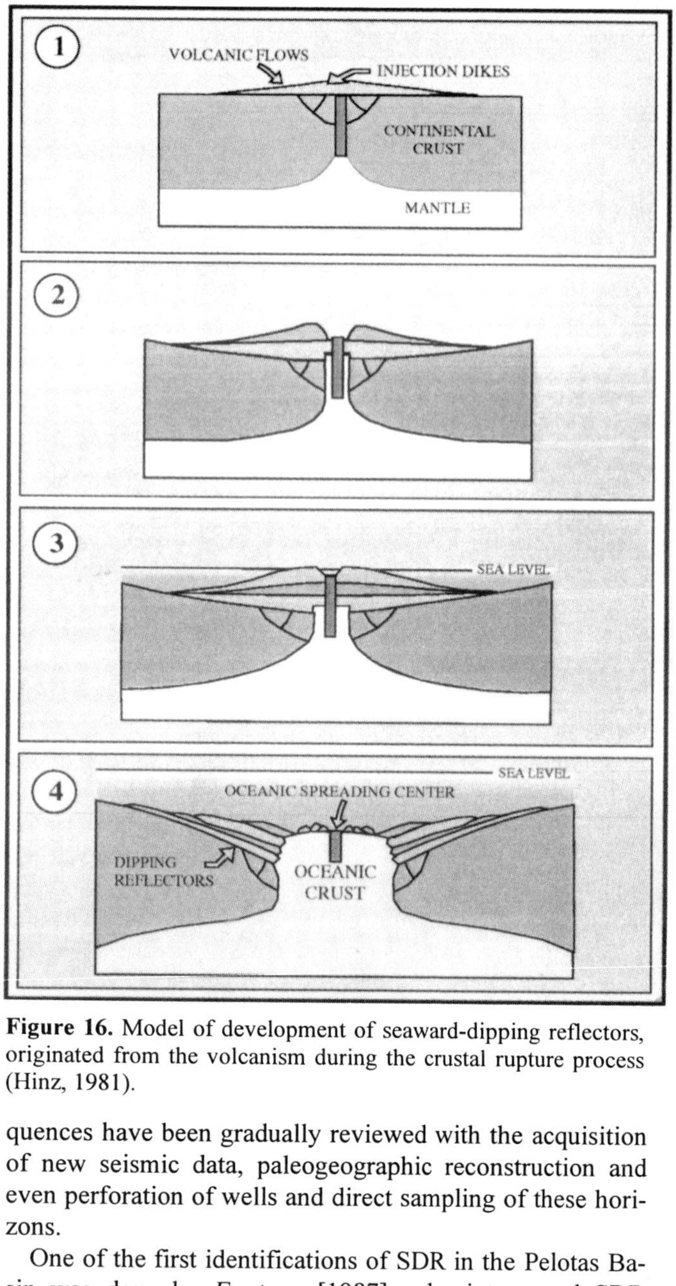

Figure 16. Model of development of seaward-dipping reflectors, originated from the volcanism during the crustal rupture process (Hinz, 1981).

quences have been gradually reviewed with the acquisition of new seismic data, paleogeographic reconstruction and even perforation of wells and direct sampling of these horizons.

One of the first identifications of SDR in the Pelotas Basin was done by *Fontana* [1987], who interpreted SDR wedges along regional seismic sections in deep water regions. The interpretation of the sections of the LEPLAC Project corroborated the expressive development of thick SDR wedges not only in the area of this study, mainly on the vicinity of the continental/oceanic boundary of the Pelotas Basin, but also in Campos Basin and in more distal oceanic regions. The distribution of these sequences (figure 17) shows that their formation is probably controlled by the positioning of fracture zones in the region, considered preferential areas of large scale episodic extrusions of volcanic material.

One of the most notable examples of SDR in the area can be observed in the proximal portion of the seismic section 47 (figure 18), evolving a set of continuous reflectors tracked from shallow to deep water regions, up to the interpreted continental/oceanic boundary. In spite of showing a general tendency of dipping eastwards and a visible convexity to the top along almost the whole sequence, these reflectors show a gentle concavity to the top at the termination of the sequence, suggesting a continuous filling of an ancient basement depression, probably caused by the loading of the superimposed horizons.

The internal structure of the SDR wedge shows an apparent superposition pattern among the dipping reflectors, with onlapping terminations of newer horizons on older ones. Near the continental/oceanic boundary obtained from the gravity model of the section 47 (figure 19), the relations between the reflectors reinforce the aspect of filling of a local crustal depression by the consecutive horizons.

Other notable occurrences of SDR can be observed along the section 48 (figure 20), both in proximal and distal crustal domains. In the proximal portion, the SDR wedge exhibits an estimated thickness of 4.0 km, considering the

Figure 17. Location map of the seaward-dipping reflectors interpreted in the seismic sections.

SECTION 47

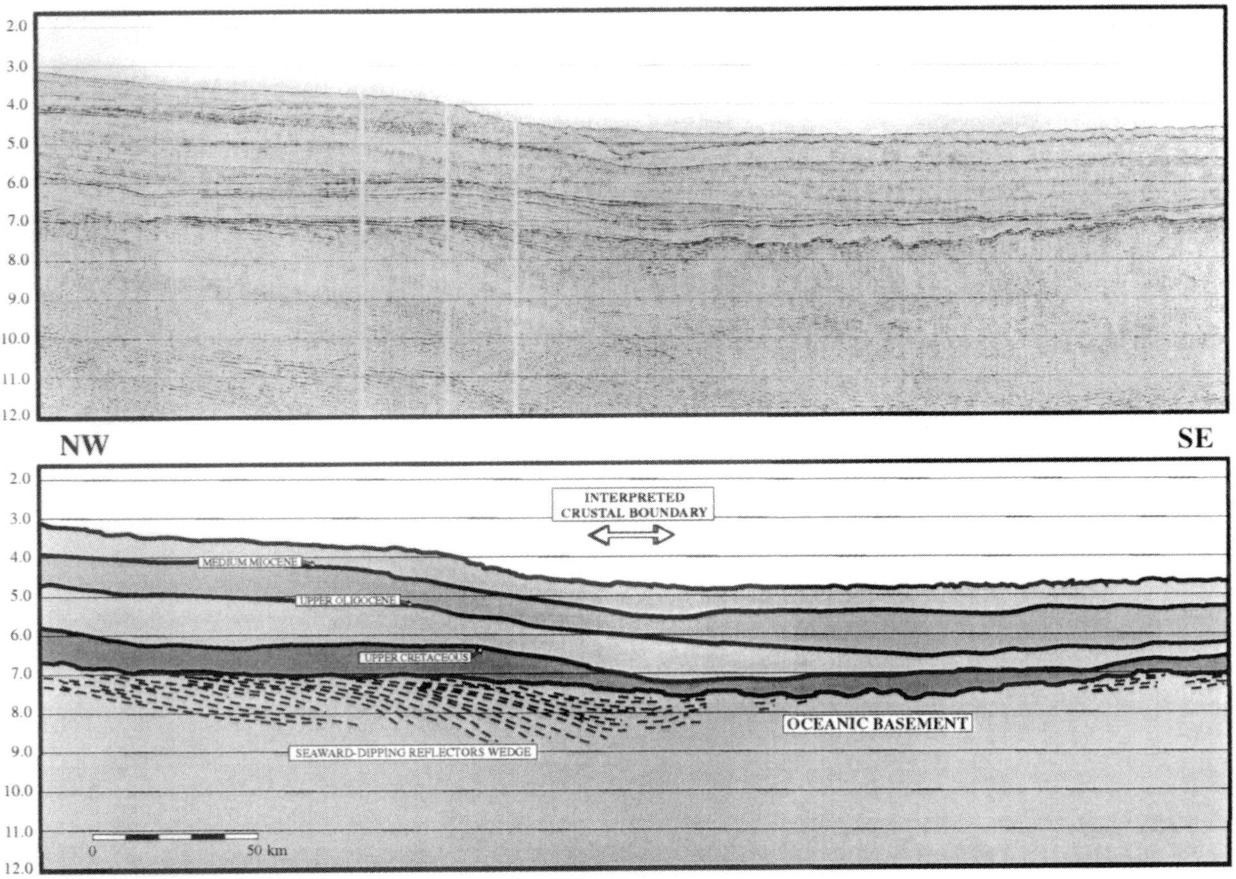

Figure 18. Detail of the seismic section 47 (dip) showing seaward-dipping reflectors occurring close to the continental/ oceanic crustal boundary.

seismic velocity 5.5 km/s, and it is interpreted to occur near the continental/oceanic boundary. In more distal portions, reflections of same characteristics are identified, although they are totally located in an oceanic crustal domain. In this case, these reflectors are supposed to be generated by volcanic pulses after the crustal rupture and probably under influence of the Porto Alegre Lineament.

7. MORPHO-STRUCTURAL DOMAINS AND TECTONIC EVOLUTION

After an analysis of the distribution of the main structural features in the area and the general characteristics of the basement, four structural domains could be distinguished along the area, indicated as domains I, II, III and IV. The domains I and II are located south of the Rio Grande Fracture Zone, the Domain III corresponds to the area of the São Paulo Plateau, and the Domain IV corresponds to the oceanic portion to the east of this plateau (figure 21).

Domain I embodies the entire segment located south of the latitude 32°S, near the Porto Alegre Lineament. This domain spans the continental slope, continental rise and abyssal plain, and presents a relatively thin igneous oceanic crust, with thickness varying from 5.0 to 6.5 km. The bathymetry is affected by anomalous features like the Rio Grande Cone, some dispersed seamounts and steps in the basement, preferentially oriented to NE-SW and E-W. Such steps, also present in the influence zone of the Chuí Lineament, correspond to normal faults that dip to the basin, describing an incipient domino structural pattern in the area. Part of these faults propagates through the sedimentary column, affecting the horizons up to the upper Cretaceous.

Domain II refers to the structurally more complex and diversified area of this southern brazilian margin, evolving the crustal segment between the Porto Alegre Lineament, at latitude 32°S, and the main crustal discontinuity of the Rio Grande Fracture Zone, at the southern limit of the São Paulo Plateau. This domain also corresponds to the area of

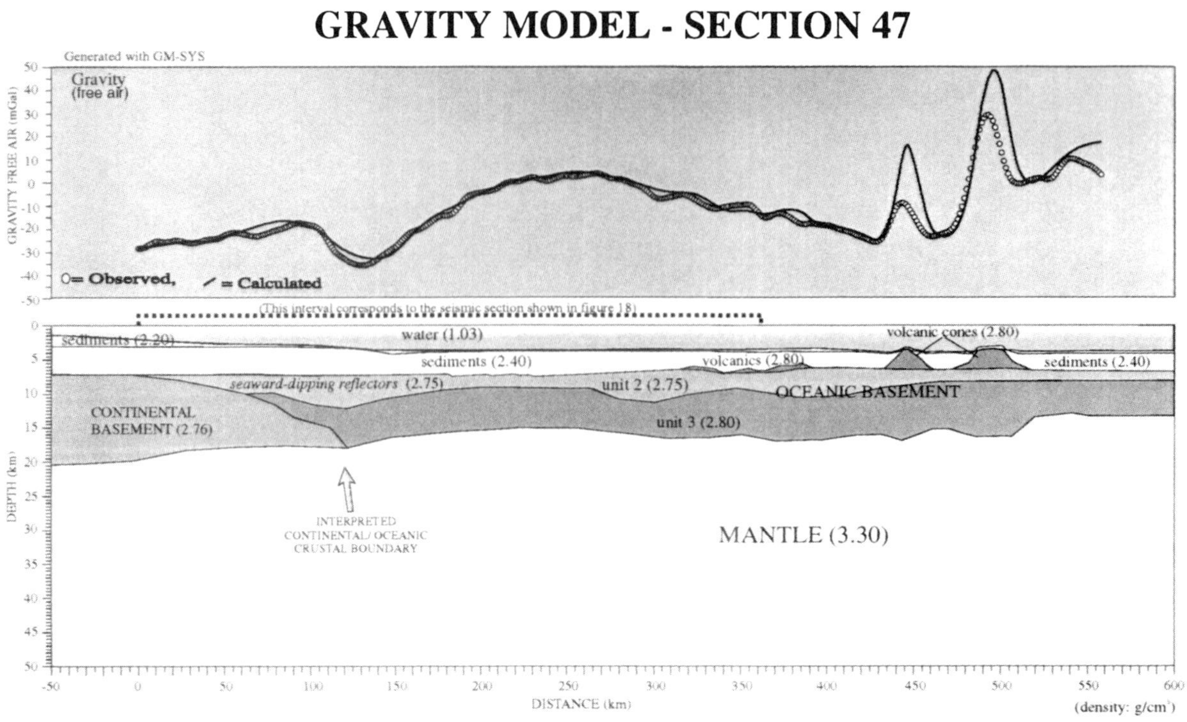

Figure 19. Gravity model obtained from the seismic section 47. The densities related to the blocks are indicated in parenthesis.

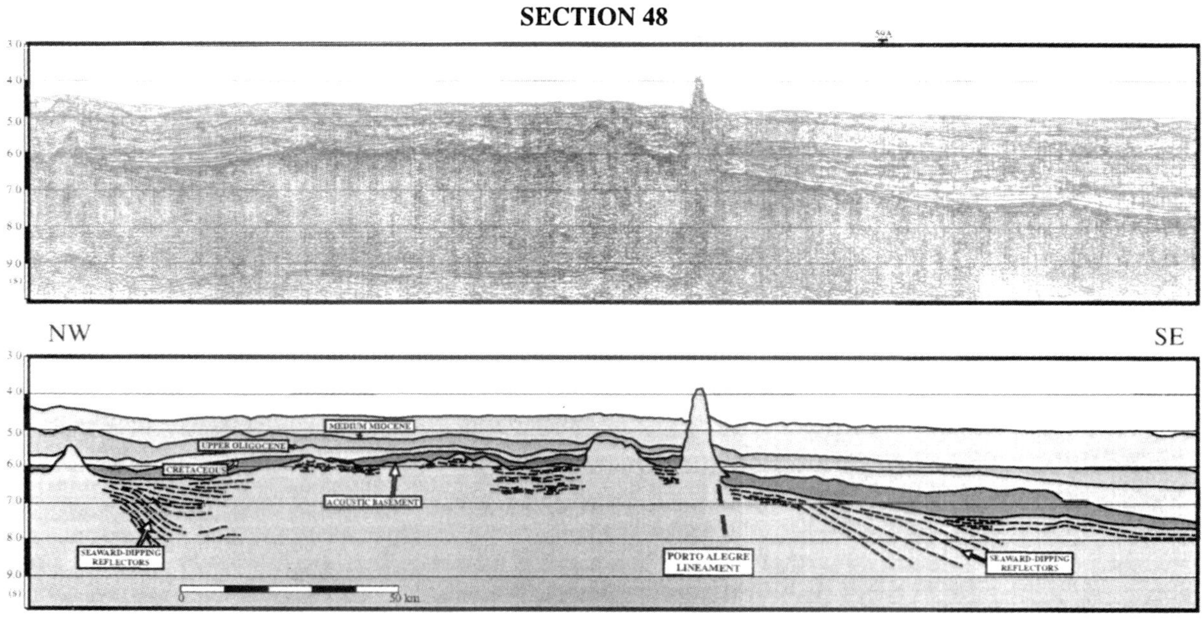

Figure 20. Detail of the seismic section 48 (dip) showing the occurrence of proximal and distal seaward-dipping reflectors wedges.

maximum occurrence of seaward-dipping reflectors in the southern margin (figure 17). The igneous crustal thickness varies from 4.0 to 10.0 km, essentially in the area affected by positive and negative long wavelength undulations of the crust.

Domain III corresponds to the area of the São Paulo Plateau and is constituted, in part, by extended continental crust, and in part by a probably transitional crust, which is characterized by a large number of volcanic constructions, responsible for the very irregular relief in the basement. The sedimentary column of large portions of this plateau is strongly structured by the post-rift salt tectonics.

Domain IV corresponds to the area positioned to the north of the Rio Grande Fracture Zone and to the east of the São Paulo Plateau. It is totally located on oceanic crust, beyond the geographic limits of the continental slope and rise. Between the latitudes 26°20′S and 28°30′S, this crustal segment is crossed by the Rio Grande Fracture Zone and, between 23°S and 24°S by the Rio de Janeiro Fracture Zone. This domain is characterized by a relatively homogeneity in terms of morpho-structural pattern and presents crustal thickness varying from 5.5 to 6.5 km. The area exhibits some local elevations of the basement, volcanic cones and seamounts, particularly the Jean Charcot Seamounts, close to the southeastern border of the São Paulo Plateau.

Integrating the main characteristics of the morpho-structural domains defined along the area, the role of the main anomalous morpho-structural features, and the orientation of the main regional trends, it is possible to formulate two hypotheses with respect to the mechanics and geometry of oceanic spreading in the area (figure 22).

The first hypothesis admits that the development of the area to the south of the Rio Grande Fracture Zone were conditioned, from the continental rupture to the Upper Cretaceous (Maastrichtian), by the ridge-push mechanism from the paleo ocean spreading axis, active at that time. This paleo spreading axis, with a scar identified at the southeastern border of the Rio Grande Rise [*Cande et al.*, 1988], had a NE-SW direction and was responsible by ridge-push tectonic efforts with vergence to NW.

This interpretation also admits that the "Cruzeiro do Sul Fracture Zone" would be developed perpendicular to this paleo axis, as a paleo transform fault with NW-SE direction, probably reactivated somewhat later in the Eocene, according to *Souza et al.* [1995]. After the Upper Cretaceous, this paleo spreading axis was abandoned and another one, with an N-S direction was installed and became active up to the present. The Porto Alegre and Chuí lineaments would correspond, in this case, to discontinuities developed after the Upper Cretaceous. Thus, in general terms, this model considers that, from the rupture to the Upper Cretaceous, the Rio Grande Fracture Zone separated two crustal

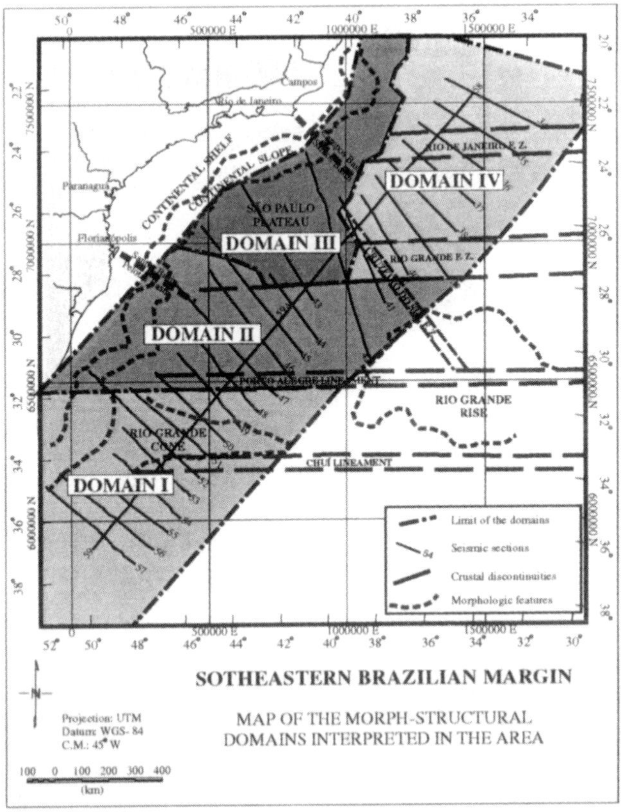

Figure 21. Map showing the limits of the morpho-structural domains interpreted in the area, according to the main regional characteristics of the crustal segments.

segments, conditioned by two distinct spreading axis, with directions NE-SW and N-S, affecting pairs of domains I-II and III-IV, respectively.

A second hypothesis, however, considers that the N-S oceanic spreading axis have been always active in the whole area, including the four domains, with little variations on the spreading directions, but maintaining the sub-parallel relationship between the fracture zones. In this case, anomalous oceanic structures oblique to the fracture zones and lineaments, as the "Cruzeiro do Sul Fracture Zone" and interpreted compressional trends, could be considered the result of the shearing efforts generated from the relative motion between the crustal segments, developed mainly along the Rio Grande Fracture Zone and the Porto Alegre Lineament.

8. CONCLUSIONS

The interpretation of the geophysical dataset of the LEPLAC Project allowed the identification and description of some of the most important morpho-structural features present in the oceanic portion of the Southern Brazilian

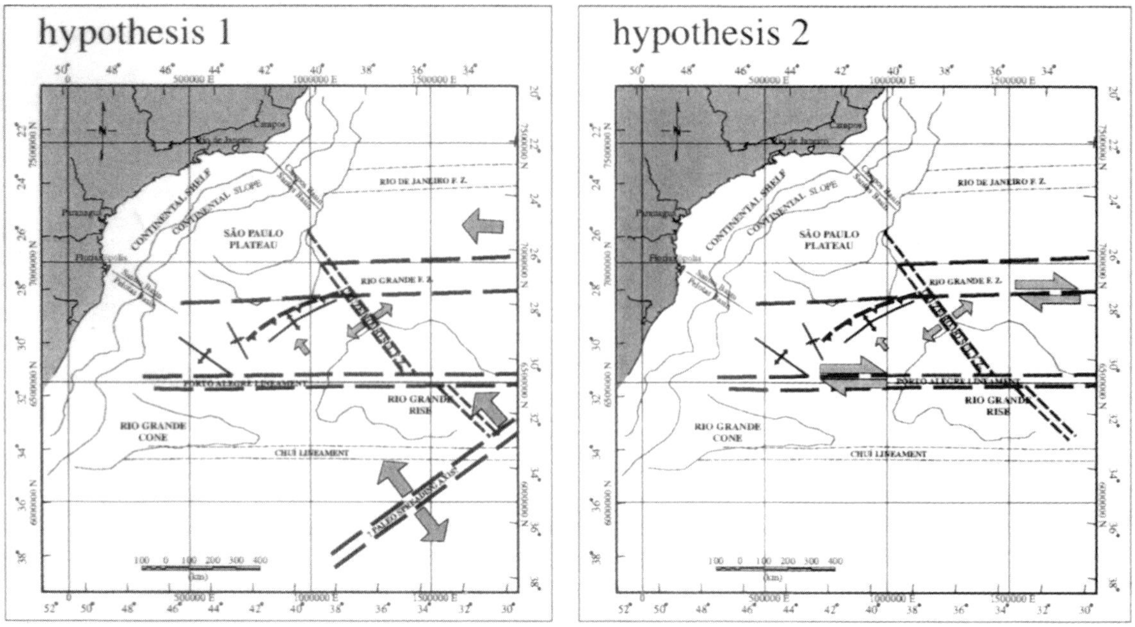

Figure 22. Regional schematic maps showing two alternative models for the evolution of the oceanic portion of the Southern Brazilian Margin. The grey arrows indicate the tectonic transport direction.

Continental Margin, with or without expression on the bathymetry.

Among these features, we can point out: the São Paulo Plateau, the São Paulo Ridge, The Rio Grande and Rio de Janeiro fracture zones, the Porto Alegre and the Chuí lineaments, the Jean Charcot Seamounts, the Rio Grande Cone, the Vema Channel, the "Cruzeiro do Sul Fracture Zone", the Rio Grande Rise, some long wavelength undulations on the seafloor and several other structural features dispersed along the area.

In terms of volcanism, the seismic sections show that, besides the construction of isolated volcanic cones and local highs of the basement, is also noteworthy the occurrence of thick seaward-dipping reflectors (SDR) wedges below the acoustic basement. The origin attributed to these horizons refers to the subaereal volcanic flows, derived from sources preferentially located near the zone of crustal rupture, and also from late pulses of volcanism occurred in clearly oceanic environments. Although the SDR wedges are most commonly located near the continental/oceanic crustal boundary, they are also observed in more distal portions of the margin, apparently controlled by fracture zones.

The main crustal discontinuities present in the area correspond to the Rio Grande and Rio de Janeiro fracture zones, and the Porto Alegre and Chuí lineaments, which trend approximately E-W. The influence of these discontinuities basically refers to steps on the oceanic basement, the construction of volcanic cones, chains and seamounts along trends (e.g. Jean Charcot Seamounts), the control of bathymetric inflections on the continental platform and slope, the limitation of the marginal plateaus (e.g. São Paulo Plateau), basement highs, regional undulations on the oceanic crust and distribution of SDRs in the area.

Another type of notable structural features observed refers to some long wavelength undulations on the oceanic floor, which affect part of the sedimentary column and whose origin is attributed to factors like the action of seabotton currents, flexural folds generated by isostatic overload, or even tectonic crustal folds created by the application of compressive or transpressive efforts in the area.

Considering the main general characteristics of the crust, four structural domains are defined:

- Domain I corresponds to the crustal segment to the south of the Porto Alegre Lineament, characterized by a relatively homogeneous oceanic crust, with thickness varying from 5.0 to 6.5 km and marked by normal faults describing an incipient domino pattern, with direction NE-SW.

- Domain II refers to the most structurally complex and heterogeneous portion of the whole area, positioned between the Porto Alegre Lineament and the Rio Grande Fracture Zone. This domain is marked by anomalous structural features that correspond to seamounts, local highs of the basement, positive and negative long wavelength undulations on the seafloor and wedges of seaward-dipping reflectors. This domain is also by the intense variation of the crustal thickness, from 4.0 to 10.0 km and, in more distal areas, great variations on the bathymetry, specially those located at the western portion of the Rio Grande Rise and submarine channels (e.g. Vema Channel).

- Domain III corresponds to the area of the São Paulo Plateau and it is characterized by a very extended continental crust, in proximal areas, and a transitional crust, in distal areas. In the latter region, it is marked by a very irregular relief of the basement, caused by the intense volcanic activity, and by the post-rift salt tectonics affecting the entire sedimentary column.

- Domain IV is the area eastwards of the São Paulo Plateau and presents an igneous crustal thickness varying from 5.5 to 6.5 km. It is locally affected by steps on the basement, associated to the Rio Grande and Rio de Janeiro fracture zones, and by volcanic cones and seamounts.

Two models are presented to explain the development of the oceanic spreading in the area. The first one admits that this process would be strongly affected, in an initial stage, by the ridge-push efforts related to a NE-SW trending paleo spreading axis, whose scar may be located near the southern limit of the Rio Grande Rise. In this case, it is implied that the area located to the south of the Rio Grande Fracture Zone, correspondent to the domains I and II, would have an oceanic spreading developed towards the northwest, from the epoch of the continental rupture until the abandonment of this paleo axis, in the Late Cretaceous.

The second model to be considered admits that, from the continental rupture to the present, the development of the whole southeastern brazilian margin have been conditioned by the action of a N-S-trended oceanic spreading axis. In this case, some of the relatively anomalous structural features described in the region, as the "Cruzeiro do Sul Fracture Zone" and the interpreted compressional trends have their origin associated with the relative motion between the crustal segments involved, and originated from tectonic reactivation mainly along the Rio Grande Fracture Zone and the Porto Alegre Lineament.

Acknowledgments. We thank the PETROBRAS management for the opportunity and the offering of all the necessary conditions to the development of this work; the Executive Committee of the LEPLAC Project for the cession of the geophysical dataset.

REFERENCES

Asmus, H. E. and R. Porto, Classificação das bacias sedimentares brasileiras segundo a tectônica de placas. *Anais do XXVI Congresso Brasileiro de Geologia*, v.2, pp. 67-90, 1972.

Asmus, H. E. and F. C. Ponte, F. C., The Brazilian marginal basins *in* Nair, A. E. and F. G. Stehli (eds.), The ocean basins and margins: v.1, The South Atlantic, *Plenum Press*, New York, pp. 87-132, 1973.

Asmus, H. E. and R. Porto, Diferenças nos estágios iniciais da evolução da margem continental brasileira: possíveis causas e implicações. *Anais do XXXI Congresso Brasileiro de Geologia*, v.1, pp. 225-239, 1980.

Asmus, H. E. and P. R. Baish, Geological evolution of the Brazilian Continental Margin. *Episodes*, n°4: 3-9, 1983.

Asmus, H. E. and W. Guazelli, Descrição sumária das estruturas da Margem Continental Brasileira e das áreas continentais adjacentes: hipóteses sobre o tectonismo causado e implicações no prognóstico de seu potencial em recursos naturais. *In: PETROBRAS, Série Projeto REMAC IX*, Rio de Janeiro, pp. 187-269, 1981.

Austin, J. A. and Uchupi, E., Continental-oceanic crustal transition off Southwest Africa. *Am. Assoc. Pet. Geol. Bull.*, 66: pp. 1328-1347, 1982.

Barker, P. F., D. A. Johnson, et al.. Tectonic evolution and subsidence history of the Rio Grande Rise. *Initial Reports of Deep Sea Drilling Project, Leg 72*: pp. 953-976, 1982.

Barker, P. F., R. L. Carlson and D. A. Johnson, Regional tectonic setting of the south western Atlantic. *Initial Reports of Deep Sea Drilling Project, Leg 72*: pp. 5-13, 1982.

Cann, J. R.. A model of oceanic crustal structure developed. Geophys. *J. R. Astr. Soc.*, 39: pp. 169-187, 1974.

Conceição, J. C., P. V. Zalán and S. Wollf, Mecanismo, evolução e cronologia do rift sul-atlântico. *Boletim de Geociências da Petrobras*, Rio de Janeiro, v.2 (2/4): pp. 255-265, 1988.

Cande, S. C., J. L. La Brecque and W. F. Haxby, Plate kinematics of the South Atlantic: chron C34 to present. *Journal of Geophysical Research*, v.93 (B11): pp. 13.479-13.492, 1988.

Chang, H. K., R. O. Kowsmann, A. M. F. Figueiredo and A. A. Bender, Tectonics and stratigraphy of the East Brazil Rift system: an overview. *Tectonophysics*, 218: pp. 97- 138, 1992.

De Long, S. E., J. F. Dewey andP. J. Fox, Displacement history of oceanic fracture zones. *Geology*, v. 5: pp. 199-202, 1977.

Demercian, L. S., P. Szatmari and P. R. Cobbold, Style and pattern of salt diapirs due to thin-skinned gravitational gliding, Campos and Santos basins, offshore Brazil. *Tectonophysics*, 228: pp. 393-433, 1993.

East Pacific Rise Study Group, Crustal processes of Mid-Ocean Ridge. *Science*, 2B: pp. 31-40, 1981.

Fontana R. L., Desenvolvimento termomecânico da Bacia de Pelotas e parte sul da plataforma de Florianópolis. *M.S. Thesis*, Univ. Federal de Ouro Preto, Brazil: 128 p., 1987.

Gambôa, L. A. P., Marine geology of the Brazilian Continental Margin and adjacent oceanic basins between the latitudes of 23°S and 37°S. *Ph.D. Thesis*, Columbia University, New York, 1981.

Gambôa, L. A. P. and P. D. Rabinovitch, The Rio Grande Fracture Zone. *In: The western South Atlantic and its tectonic implications. Earth and Planetary Science Letters*, v.52: pp. 410-418, 1981.

Gorini, M. A., The tectonic fabric of the Equatorial Atlantic and adjoining continental margins: Gulf of Guinea to Notheastern Brazil. *Ph.D. Thesis*, Columbia Univ., 365 pp., 1977.

Guazelli, W. and J. C. Carvalho, Estruturas da margem continental leste brasileira e das áreas oceânicas e continentais adjacentes. *Série Projeto Remac*, v.9, pp.117-143, 1981.

Harry, B. L. and D. S. Sawyer, Basaltic volcanism, mantle plumes and the mechanics of rifting: the Paraná flood basalt province of South America. *Geology*, v.20: pp. 207-210, 1992.

Hinz, K., A hypothesis on terrestrial catastrophes. Wedges of very thick oceanward dipping layers beneath passive continental margins. *Geologisches Jahrbuch*, E 22: pp. 3-28, 1981.

Hinz, K. and W. Krause, The continental margin of Queen Maud Land, Antarctica: seismic sequences, structural elements and geological development. *Geologisches Jahrbuch*, E 22: pp. 3-28, 1982.

Holbrook, W. S. and P. B. KELEMAN, Large igneous province on the US Atlantic margin and implications for magmatism during continental breakup. *Nature*, 364: pp. 433-436, 1993.

Kamesh Raj, K. A., Magnetic lineations fracture zones and seamounts in the Central Indian Basin. *Marine Geology*, 109: pp. 195-201, 1993.

Kowsmann, R. O., M. P. Costa, M. P. BOA HORA, H. P. Almeida and P. P. GUIMARÃES, Geologia estrutural do Platô de São Paulo. *In: Anais do XXXII Congresso Brasileiro de Geologia*, Salvador, Brazil, SBG, v.4: pp. 1558-1569, 1982.

Kumar, N. Origin of "paired" aseismic rises: Ceará and Sierra Leone rises in the Equatorial, and the Rio Grande Rise an Walvis Ridge in the South Atlantic. *Marine Geology*, 30: pp. 175-191, 1978.

Kumar, N., L. A. P. Gambôa, B. C. Schreiber and J. Mascle, Geologic history and origin of São Paulo Plateau (southeastern Brasilian margin), comparison with Angolan margin, and the early evolution of the Northern South Atlantic. *In: Initial Reports of Deep Sea Drilling Project, Leg 39*: pp. 927-945, 1977.

Leyden, R., W. J. Ludwig and M. Ewing, Structure of continental margin of Punta Del Este, Uruguay, and Rio de Janeiro, Brazil. *A.A.P.G. Bulletin*, v.55 (12): pp. 2161-2173, 1971.

Lowrie, A., C. Smoot and R. Batiza, Are oceanic fracture zones locked and strong or weak? New evidence for volcanic activity and weakness. *Geology*, v. 14: pp. 242-245, 1986.

Mascle, J., E. Blarez and M. Marinho, The shallow structures of the Guinea and Ivory Coast-Ghana transform margins: their bearing on the Equatorial Atlantic Mesozoic evolution. *Tectonophysics*, 155: pp. 193- 209, 1988.

Mizusaki, A. M. P., A. Thomaz Filho and J. Valença, Volcano-sedimentary sequence of Neocomian age in Campos Basin (Brazil). *Revista Brasileira de Geociências*, 18: pp. 247-251, 1988.

Mohriak, W. U., M. R. Mello, G. D. Karner, J. F. Dewey and J. R. Maxwell, Structural and stratigraphic evolution of the Campos Basin, offshore Brazil. *In: Extensional Tectonics and Stratigraphy of the North Atlantic Margins, AAPG Memoir 46*, Chapter 38, pp. 577-598, 1990.

Mohriak, W. U., J. M. Macedo, R. T. castellani, H. D. Rangel et al., Salt tectonics and structural styles in the deep-water province of the Cabo Frio region, Rio de Janeiro, Brazil. *In: Salt tectonics: a global perspective. AAPG Memoir 65*, pp. 273-304, 1995.

Mohriak, W. U., M. Bassetto and I. S. Vieira,. Deep seismic constraints on the crustal architecture of sedimentary basins in the Brazilian margin: tectonic and exploratory implications. *In: V Simpósio Nacional de Estudos Tectônicos - SNET 95*, Gramado - RS, pp. 19-22, 1995.

Müller, R. D., J. Y. Royen and L. A. Lawyer, Revised plate motions relative to the hot spots from combined Atlantic and Indian Ocean hot spot tracks. *Geology*, v.21: pp. 275-278, 1993.

Mutter, J.; M. Talwani and P. L. Stoffa, Origin of seaward dipping reflectors in oceanic crust of the Norwegian margin by "subaereal seafloor spreading". *Geology*, 10: pp. 353-357, 1982.

Nürnberg, D. and R. D. Müller, The tectonic evolution of South Atlantic from late Jurassic to present. *Tectonophysics*, 191: pp. 27-53, 1991.

O'Connor, J. M. and R. S. Duncan, Evolution of the Walvis Ridge- Rio Grande Rise hot spot system: implications for African and South American plate motions over plumes. *Journal of Geophysical Research*, 95: pp. 17475-17502, 1990.

Ojeda, H. A., Structural framework, stratigraphy and evolution of Brazilian marginal basins. *A.A.P.G. Bulletin*, 66: pp. 732-749, 1982.

Peate, D. W., C. J. Hawkesworth, M. S. M. Mantovani, W. Sukowski, Mantle plumes and flood-basalt stratigraphy in the Paraná, South America. *Geology*, 18: pp. 1223-1226, 1990.

Perch-Nielsen, K., P. R. Supko, S. Boersma, R. L. Carlson et al., Site 356: São Paulo Plateau. *In: Initial Reports of Deep sea Drilling Project, Leg 73*: pp. 141-166, 1982.

Ponte, F. C. and H. E. Asmus, The Brazilian marginal basins: current state of knowledge. *In: Anais da Academia Brasileira de Ciências. Simpósio Internacional sobre as Margens Continentais do Tipo Atlântico*. Rio de Janeiro, RJ, 48 (Supl.): pp. 215-239, 1976.

Rabinovitch, P. D. and J. L. La Brecque, The Mesozoic South Atlantic ocean and evolution of its continental margins. *Journal of Geophysical Research*, 84: pp. 5973-6002, 1979.

Roberts, D. G., A. C. Morton and J. Backman, Late Paleocene-Eocene volcanic events in the northern North Atlantic Ocean. *Initial Reports of Deep Sea Drilling Project, Leg 81*: pp. 913-923, 1984.

Rosendahl, B. R., J. Meyers, H. Groscher and D. Scott, Nature of the transition from continental to oceanic crust and the meaning of reflection Moho. *Geology*, 20: pp. 721-724, 1992.

Severinghaus, J. P. and K. C. Mac Donald, High inside corners at ridge-transform intersections. *Marine Geophysical Researches*, V.9, No.4: pp. 353-367, 1988.

Siedner, G. and J. G. Mitchell, Episodic Mesozoic volcanism in Namibia and Brazil: a K-Ar isochron study on the opening of the South Atlantic. *Earth and Planetary Science Letters*, 30: pp. 292-302, 1976.

Singh, S. C. and D. Mc Kenzie, Layering in the lower crust. Geophysical Journal International, 113: pp. 622-628, 1993.

Souza, K. G., La Marge Continental Brésilienne sud-orientale et les domaines oceaniques adjacents: structure et evolution. *D. Thesis. Université Pierre et Marie Curie*, Paris: 230 p., 1991.

Souza, K. G., R. L. Fontana, J. Mascle, J. Macedo et al., The southern Brazilian margin: an example of a South Atlantic volcanic margin. *In: Anais do 3° Congresso Internacional da Sociedade Brasileira de Geofísica*. SBGf, Rio de Janeiro, v.2: pp. 1336-1341, 1993.

Szatmari, P., J. B. L. Françolin, O. Zanotto and S. Wolff, Evolução tectônica da margem equatorial brasileira. *Revista Brasileira de Geociências*, 17: pp. 180- 188, 1987.

Szatmari, P. and W. U. Mohriak, Plate model of postbreakup tectono-magmatic activity in SE Brazil and the adjacent Atlantic. *In: Anais do V Simpósio Nacional de Estudos Tectônicos - SNET 95*, Gramado, RS, Brazil. pp. 213-214, 1995.

Turcotte, D. L., Are transform faults thermal contraction cracks? *Journal of Geophysical Research*, 79 (17): pp. 2573-2577, 1974.

Wannesson, J., J. C. Ieart and J. Pavat, Structure and evolution of adjoining segments of the west African margin determined from deep seismic profiling. *Geodynamics*, 22: pp. 275-289, 1991.

Warner, M., Basalts, water, or shear zones in the lower continental crust? *Tectonophysics*, 173: pp. 163-174, 1990.

White, R. and D. Mc Kenzie, Magmatism at rift zones: the generation of volcanic continental margins and flood basalts. *Journal of Geophysical Research*, v. 94 (B6): pp. 7685-7729, 1989.

Zalán, P. V., Tectonics and sedimentation of the Piaui- Camocin sub-basin, Ceará Basin, offshore northeastern Brazil. *Ph.D. Thesis. Colorado School of Mines*, 133 pp., 1984.

Zehnder, C. M., J. C. Mutter and P. Buhl, Deep seismic and geochemical constraints on the nature of rift-induced magmatism during breakup of the North Atlantic. *Tectonophysics*, 173: pp. 545-565, 1990.

Zembruscki, S. G., H. T. Barretto, H. T., J. J. C. Palma and M. P. A. Costa, Estudo preliminar das províncias geomorfológicas da Margem Continental Brasileira. *In: Anais do XXVI Congresso Brasileiro de Geologia*, Belém, SBG, v.2: pp. 187-209, 1972.

Bassetto, Marcelo: 65 Chile Avenue, room 1301-E, Rio de Janeiro, Brazil (e-mail: bassetto@ep.petrobras.com.br).

Alkmim, Fernando Flexa: Morro do Cruzeiro, s/n., Depto. Geociências, Ouro Preto, Brazil (e-mail: alkmim@degeo.ufop.br).

Szatmari, Peter: Cidade Universitária, quadra 7, Rio de Janeiro, Brazil (e-mail: szatmari@cenpes.petrobras.com.br).

Mohriak, Webster Ueipass, 65 Chile Avenue, room 1301-E, Rio de Janeiro, Brazil (e-mail: webmohr@ep.petrobras.com.br).

Ocean-Continent Transition and Tectonic Framework of the Oceanic Crust at the Continental Margin off NE Brazil: Results of LEPLAC Project

Paulo Otávio Gomes[1], Benedito S. Gomes[1], Jorge J. C. Palma[2], Koji Jinno[1], and Jairo M. de Souza[1]

In 1992, Brazilian Navy and PETROBRAS carried out a geophysical survey along the continental margin off northeastern Brazil, as part of a governmental plan to delineate the "Legal Continental Shelf" according to the international Law of the Sea. This data set is leading to a better understanding of the crustal transition processes and on the evolution of the oceanic crust over that part of the Brazilian continental margin. On our seismic transects, we show a rifted marginal plateau (Pernambuco Plateau) where crustal extension was controlled by detachment faulting, possibly in a non-volcanic margin setting. Farther north, dealing with the ocean-continent transition nearby a major transform margin, we found a normal passive margin-style transition zone instead of transform-related structures. With the support of multichannel seismic profiles and gravity data derived from GEOSAT altimetry, several well-known oceanic fracture zones and structural lineaments were properly located and correlated. The relationship of these structures with volcanic ridges and extensional, compressive and strike-slip tectonic reactivations suggests that fracture zones at this area behaved either as zones of weakness or as locked transform fault scars. Striking lithospheric flexural deformation is also related to FZs in this region. In the surroundings of the Fernando de Noronha Ridge, lithospheric flexure represents an isostatic response to volcanic loading, while bending across Ascension FZ is likely to have been caused by differential subsidence in crustal segments of contrasting ages. We also correlate some other deformation of the oceanic crust with changes in spreading directions that possibly took place at the Upper Cretaceous.

1. INTRODUCTION

Even though distinguished by outstanding physiographic features and intriguing tectonic structures, the northeastern region off Brazil still remains as the less-studied portion of Brazilian Continental Margin. In this region, the petroleum industry has focused its efforts mainly on the continental shelf and slope of the southern Sergipe-Alagoas basin and the northern Potiguar basin. Hence, the whole area located between these two exploration targets, from 5° S to 10° S , to the east of the 35° W meridian, is characterized by a lack of geophysical and geological data (Fig. 1).

Despite this, important studies performed during the seventies and early eighties by the REMAC Project ("Global Reconnaissance of Brazilian Continental Margin"), produced relevant results [*Chaves et al.*, 1979]. New geophysical advances in this region were achieved in the mid-eighties by the CENTRATLAN programs, jointly

[1] E&P branch, Petróleo Brasileiro S.A.- PETROBRAS –, Rio de Janeiro, Brazil

[2] Departamento de Geologia, Univ. Federal Fluminense, Niterói, Brazil, and Depart. Nacional de Produção Mineral – DNPM

Atlantic Rifts and Continental Margins
Geophysical Monograph 115

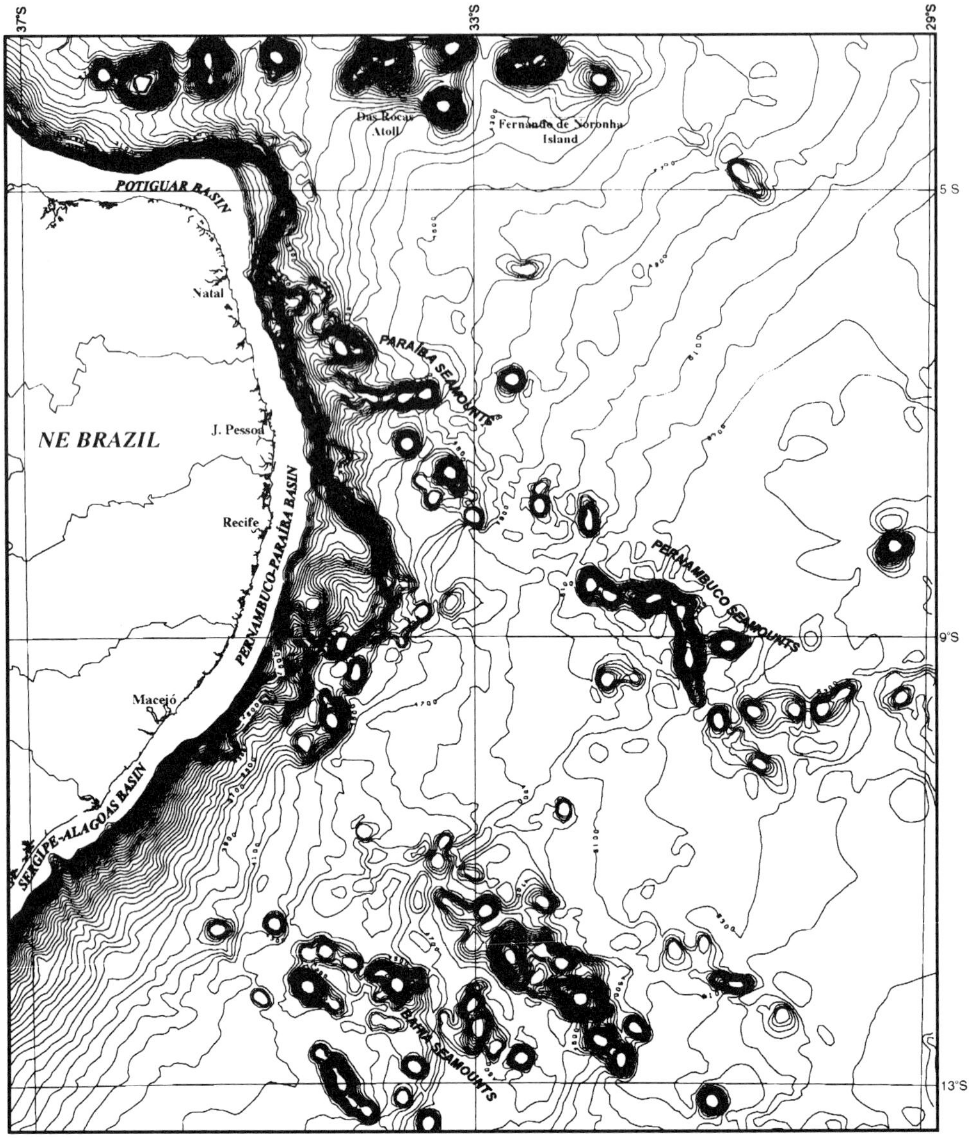

Figure 1. Bathymetric chart of the northeastern continental margin off Brazil.

conducted by U. S. Naval Research Laboratory and Brazilian Navy [*Gorini and Fleming*, 1982]. However, only multichannel seismic reflection, gravity, magnetic and bathymetric data of the LEPLAC Project ("Brazilian Continental Shelf Survey Plan"), collected in the early nineties by a joint-venture between Brazilian Navy and Petróleo Brasileiro S.A. (PETROBRAS), allowed more detailed research. Recent satellite-altimetry data, collected by GEOSAT, were also fundamental to improve the studies of the structural features of northeastern continental margin off Brazil.

The main purpose of this study is to present some interpretations of the new geophysical data in a selected region located between the 4° S and 10° S latitudes, and the 30° W and 35° W longitudes (Fig. 2). The preliminary conclusions of our work were previously presented at the V International Congress of the Brazilian Geophysical Society [*Gomes and Gomes*, 1997; *Gomes*, 1997]. In this revised version, we assemble and develop these conclusions in two sections: first, we present and discuss seismic images that involve crustal extension and ocean-continent transition across the selected area; at the second section, we deal with evolutional features of the oceanic crust, which are mainly related to fracture zones and transverse ridges. The updated position of fracture zones over the northeastern continental margin off Brazil is then

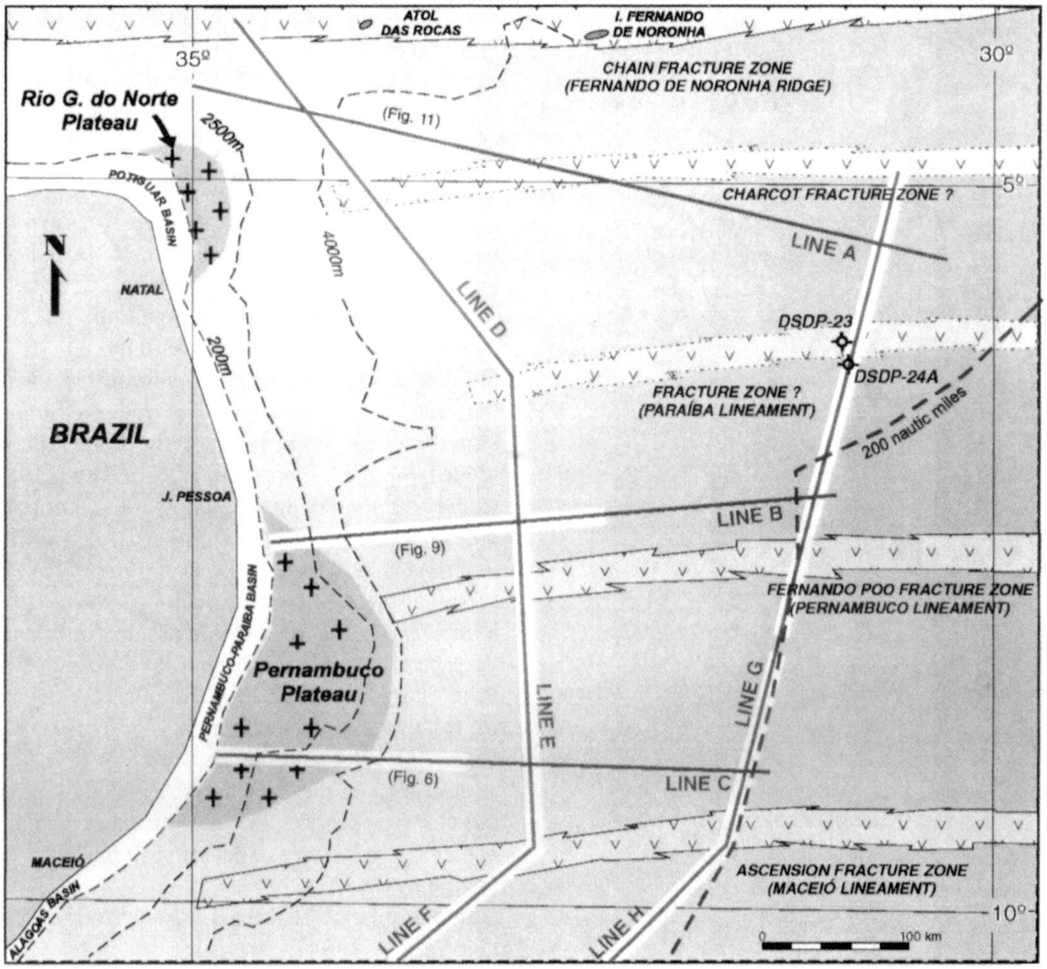

Figure 2. Location map of the geophysical profiles used in this work. The highlighted parts of the dip lines, with their seismic interpretation and gravity modelling, are shown in figures 6, 9 and 11. The strike transects displayed on figures 14 and 15 are also highlighted. The lineament names are those of *Schobbenhaus et al.* [1981], and their suggested connection with fracture zones is mainly based on GEOSAT data.

discussed (Fig. 2), along with magmatic and tectonic features possibly related to these structures. Furthermore, we analyze some flexures of the oceanic lithosphere, which are also related, in different ways, to fracture zones.

2. BRIEF REPORT ON LEPLAC PROJECT

The "United Nations Convention on the Law of the Sea (UNCLOS)" is in force since November 16, 1994. From that date on, countries which ratified the UNCLOS had ten years to establish the outer border of their Legal Continental Shelves (LCS). As one of the signatory countries of the Convention, Brazil started in 1987 an aggressive program of acquisition, processing, and interpretation of geophysical data aiming to define of the outer border of its LCS. Known as LEPLAC Project, this program has acquired, up to this date, over 45,000 line-kilometers of multichannel seismic, gravity, magnetic and bathymetric data along the Brazilian Continental Margin (Fig. 3). With a LCS supported by geophysical data, the country will be able to claim exclusive sovereign rights for the exploration and the exploitation of natural resources in far-away offshore regions, located beyond the 200-nautical mile limit of its Exclusive Economic Zone (EEZ).

3. DATA ACQUISITION AND PROCESSING

At the continental margin off NE Brazil, LEPLAC data were collected during a geophysical survey carried out in 1992 by Brazilian navy's oceanographic vessel "Almirante Câmara". The seismic survey was performed with a shot interval of 50 m, using high-pressure air guns as source.

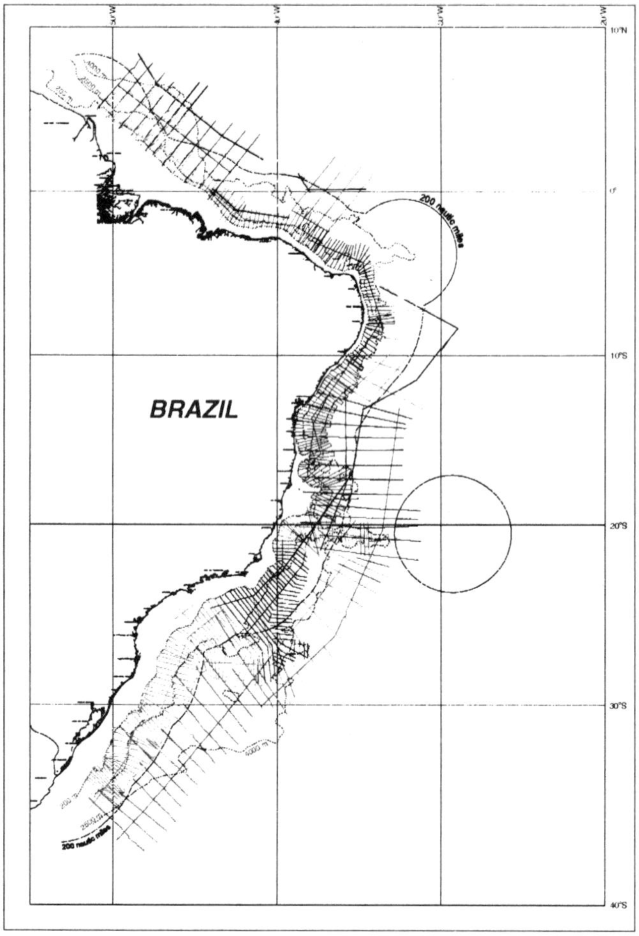

Figure 3. Location map of LEPLAC geophysical profiles. The surveys that cover the continental slope region comprise gravity, magnetic and bathymetric data. The regional lines collected on deeper portions of the continental margin also comprise multichannel seismic reflection data.

Data were recorded with a sampling interval of 4 ms and a record length of 12 seconds. A 2,500 m-long streamer was used, with 96 channels and a group interval of 25 m. Gravity, magnetic and bathymetric data were simultaneously collected along with seismic acquisition.

Seismic data were processed at PETROBRAS, using a standard processing sequence for deep-seismic profiling. After demultiplexing and deterministic source-receiver deconvolution, a predictive deconvolution was applied, with a prediction lag of 16 ms. Correlation-type velocity analysis was performed in two stages: in the former, preliminary velocity functions were determined every 20 km in order to generate a raw stack; in the latter, this brute stack served as a guide to the interpretation of final velocity functions at each 2.5 km. Prior to finite-difference migration with a 10% reduction on the stacking velocities, data were stacked with dip-move-out correction and external mute. Noise attenuation was the next step, with the application of a Radial Predictive Filter, which was followed by frequency filtering and AGC gain.

4. PREVIOUS WORK

The tectonic and sedimentary evolution of the northeastern continental margin off Brazil was studied by the REMAC Project researchers, during the seventies. Tectonic control over sedimentation on the northeastern marginal basins was discussed by *Asmus and Carvalho* [1978]. *França* [1979] dealt with the the geomorphology of the eastern Brazilian oceanic region. Some of the main structures of this portion of the continental margin were identified by *Guazelli and Carvalho* [1981]. The first isopach maps of the region were presented by *Carvalho and Franciscono* [1981].

Still in the seventies, it is important to mention the reconnaisance geophysical study performed by *Summerhayes et al.* [1976] at the continental margin off Sergipe and Alagoas. *Damuth and Hayes* [1977] defined an echo character of the East Brazilian continental margin and its relationship to sedimentary processes. *Kowsmann and Costa* [1976] made the first attempt to establish a seismic stratigraphy in the Pernambuco Plateau region. Genetic and structural aspects of the marginal plateaus on the northeastern region were investigated by *Fainstein and Milliman* [1979].

In a synthesis of the geology up to that date, *Gorini and Carvalho* [1984] presented a map with the main physiographic and structural features of the northeastern continental margin off Brazil (Fig. 4). Those features were incorporated in the "Geological Map of Brazil and Adjoining Oceanic Area including Mineral Deposits" [*Schobbenhaus et al.*, 1981].

UFF (Fluminense Federal University) researchers also studied the ocean-continent transition at Pernambuco and Rio Grande do Norte plateaus [*Costa and Maia*, 1986; *Sperle et al.*, 1989], the evolution of fracture zones and oceanic crust [*Palma et al.*, 1984; *Palma*, 1989; *Sperle et al.*, 1991; *Mello and Dias*, 1996], and the seismic stratigraphy of Pernambuco Plateau and adjoining area [*Alves and Costa*, 1986; *Costa et al.*, 1991; *Alves et al.*, 1993].

5. OCEAN-CONTINENT TRANSITION

The ocean-continent crustal transition changes considerably from the Sergipe-Alagoas basin, located to the south of Ascension Fracture Zone, up to the Rio Grande do Norte Plateau, at the Potiguar Basin (Fig. 2), mainly due to alterations on the thermomechanical behavior and on the rheological characteristics of the rifted crust. While the Sergipe-Alagoas basin transition involved rapid crustal thinning, intense magmatic activity and emplacement of seaward-dipping reflectors wedges (SDRs) [*Mohriak et al.*, 1995; *Mohriak et al.*, this vol.], on the Pernambuco-Paraíba basin a typical marginal plateau evolved in the distal settings of the continental crust, without the volcanism-related outstanding features observed to the south. Farther north, at southern Potiguar basin, the transition occurs in a tectonic setting located very close to a transform continental margin, where shearing stresses are expected to play an important role.

In the following sub-sections, we present and discuss seismic images of LEPLAC Project that cover the ocean-continent transition zone of Pernambuco-Paraíba and Potiguar basins. The images are displayed on interpreted and uninterpreted versions along with gravity modelling and free-air gravity profiles, respectively (figs. 6, 7, 9, 10, 11 and 12). Figure 5 shows the location of geophysical lines over a Bouguer anomaly map, base on LEPLAC, PETROBRAS and GEOSAT-derived data.

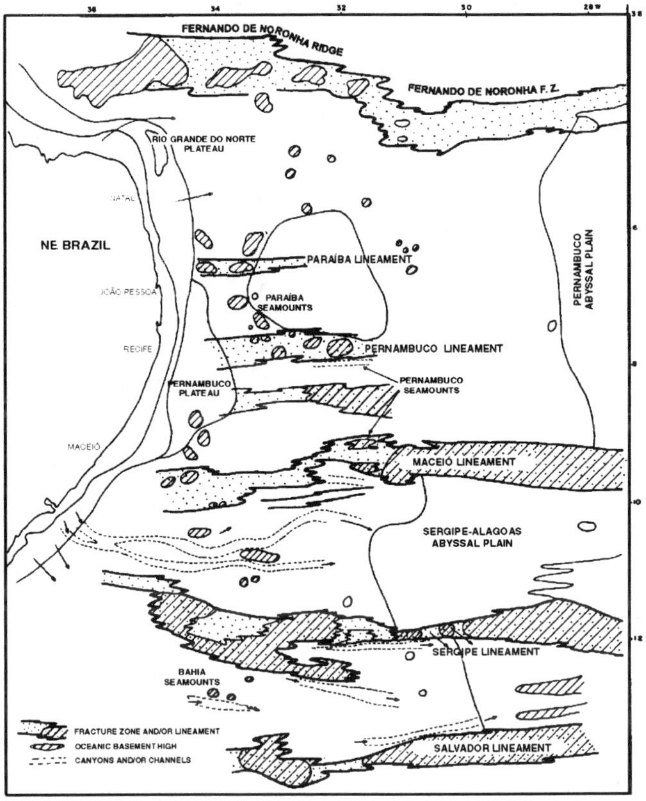

Figure 4. Simplified version of "Geological Map of Brazil and Adjoining Ocean Floor" [*Schobbenhaus et al.*, 1981] showing the main physiographic and structural features of the northeastern continental margin off Brazil [modified from *Gorini and Carvalho*, 1984].

5.1. Pernambuco Plateau

The Pernambuco Plateau is a typical marginal plateau located on the continental slope of the northeastern continental margin off Brazil, between the latitudes of João Pessoa and Maceió (figs. 2 and 4). Considering it as a tectonic subprovince of the Pernambuco-Paraíba basin, *Mabesoone and Alheiros* [1988] inferred a sedimentary thickness up to 4,000 m for the plateau. According to *Fainstein and Milliman* [1979] and *Costa and Maia* [1986], at least part of this marginal feature overlies stretched continental crust.

Two geophysical lines of the LEPLAC Project crossed, respectively, the central-south portion and the northern edge of the Pernambuco Plateau (Fig. 2). Seismic interpretation of line C (figures 6 and 7) indicates a typical rifted continental crust structure, characterized by synthetic tilted blocks. A similar structure was observed by D. L. Bisol [*in*: *Alves and Costa,* 1986] on a PETROBRAS multichannel seismic line located in the northern part of the plateau. Time-depth conversion and sediment isopach calculation resulted on a maximum sedimentary thickness of 4,600 m covering the westernmost tilted block, near the SP 450 of line C.

The interpreted structural geometry resembles those of detachment faulting models [*Lister et al.*, 1986], involving crustal extension along sub-horizontal shear zones. Models of a similar nature were proposed for the evolution of the Sergipe-Alagoas basin, to the south of Pernambuco Plateau, by *Castro Jr.* [1988] and *Guimarães* [1988]. Although based on a reasonable continuous reflection at the central portion of the plateau (Fig. 8), the detachment surface interpreted on line C is inferred, as the model is expected to develop along multiple systems of *décollement*, which complexity prevents from obtaining a good imaging on seismic sections.

The intracrustal horizon of line C (figs. 6 and 8) is somehow similar to the S reflector west of Galicia, Spain [*de Charpal et al.*, 1978], also interpreted as a detachment faulting, active during the rifting processes of the Iberian continental margin [*Hoffmann and Reston*, 1992; *Reston et al.*, 1996]. That region is regarded as one of the best-studied nonvolcanic-type continental margins.

As seaward-dipping reflectors (SDRs) were not identified in the Pernambuco Plateau region, one can infer that the magma supply, during the emplacement of oceanic crust, was limited. These conditions and the relative proximity to a transform margin tectonic environment reduce the probabilities for the occurrence of SDRs at the

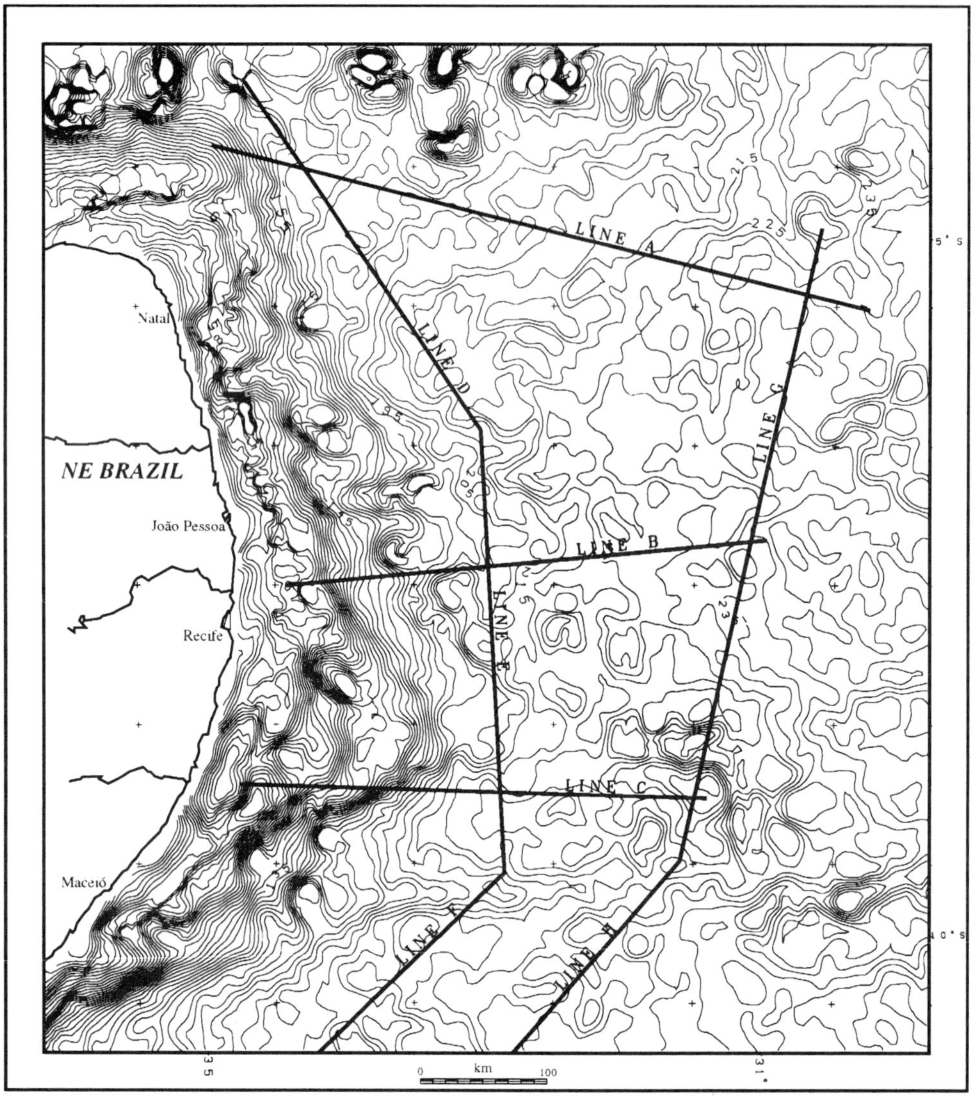

Figure 5. Bouguer anomaly map (3-D) of the northeastern continental margin off Brazil (c. i.=5 mGal; Bouguer density=2.2 g/cm3).

plateau's adjoining area. We do not discard, however, the possibility that new seismic reflection data, collected with a longer streamer and a larger multiplicity, could image this kind of volcanic feature over the region.

So far, the evidence suggests the occurrence of a transitional situation from the south to the north, where a typical volcanic margin, in the Sergipe-Alagoas basin, characterized by a rapid crustal thinning with the emplacement of SDR basaltic wedges [*Mohriak et al.*, 1995; *Mohriak et al.,* this vol.], would be probably succeeded by a nonvolcanic rifted margin, on the Pernambuco Plateau, marked by a more gradual crustal thinning. Depending on additional data, this situation could be compared to that described by *Keen and Potter* [1995] on the east coast off Canada. An important point, in the Brazilian case, is the presence of a structural boundary between the possible volcanic and nonvolcanic provinces, related to the prolongation of the Ascension Fracture Zone towards the continental margin (Fig. 2).

Gravity modelling of line C (Fig. 6) supports the occurrence of stretched continental crust under the central-south segment of the Pernambuco Plateau. According to these models, emplacement of oceanic crust began near the edge of the plateau (marginal escarpment), located near SP-2700. The average crustal thickness on this region is about 19.5 km until SP-1200, and about 15 km from this

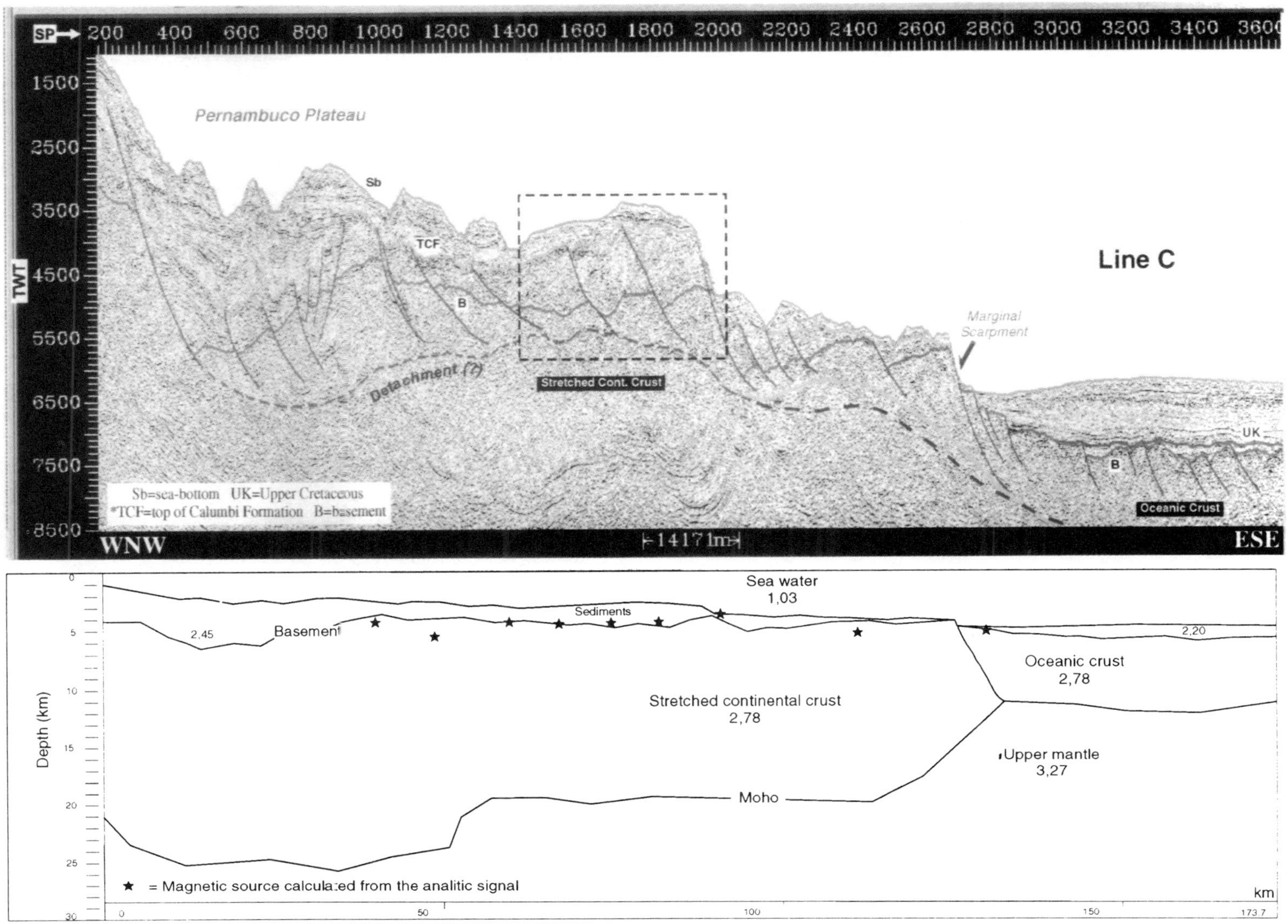

Figure 6. Part of seismic line C with corresponding gravity modelling below, showing the structural architecture and crustal transition along the Pernambuco Plateau (see fig. 2 for location). Due to the poor seismic quality over the plateau region, related to the irregular morphology of sea-bottom and to the presence of multiple reflections, basement interpretation was sometimes inferred (see next figure for an uninterpreted version of this line and fig. 8 for a zoom view of the boxed section). Although based on some strong evidences, the horizon indicated as a detachment is likewise suggestive, since this kind of structure is expected to develop along multiple intracrustal surfaces. The TCF* horizon represents, approximately, a Campanian age. Gravity modelling suggests that oceanic crust spreaded out from the marginal scarpment of the plateau.

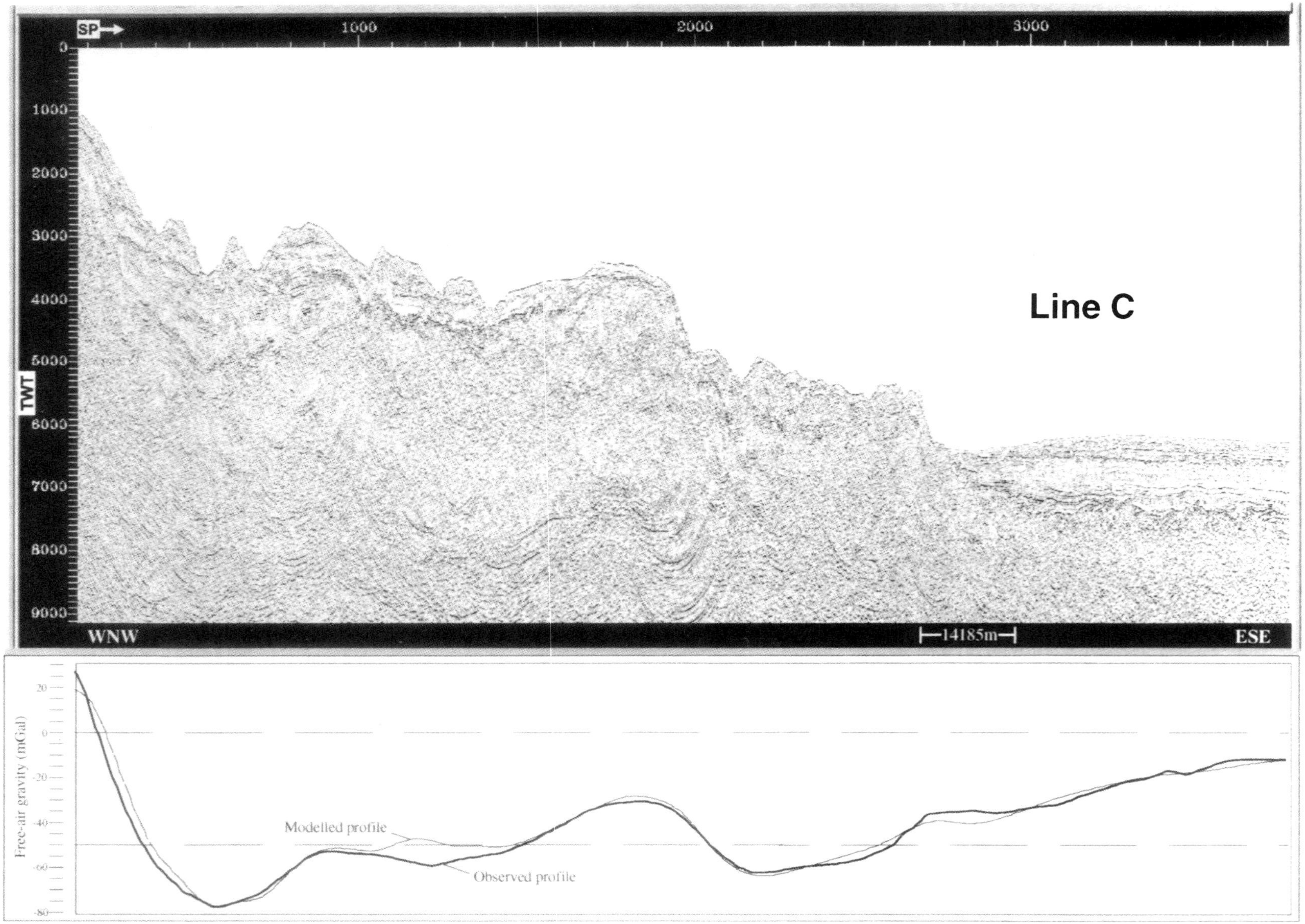

Figure 7. Uninterpreted version of western extension of seismic line C. Due to physiographic complexity (*i.e.*, water depth variations), free-air gravity profiles have no clear distinct patterns for continental and oceanic crustal types. Despite this, an ocean-continent boundary can be pointed out near the SP-2800, as shown in the gravity modelling of the previous figure.

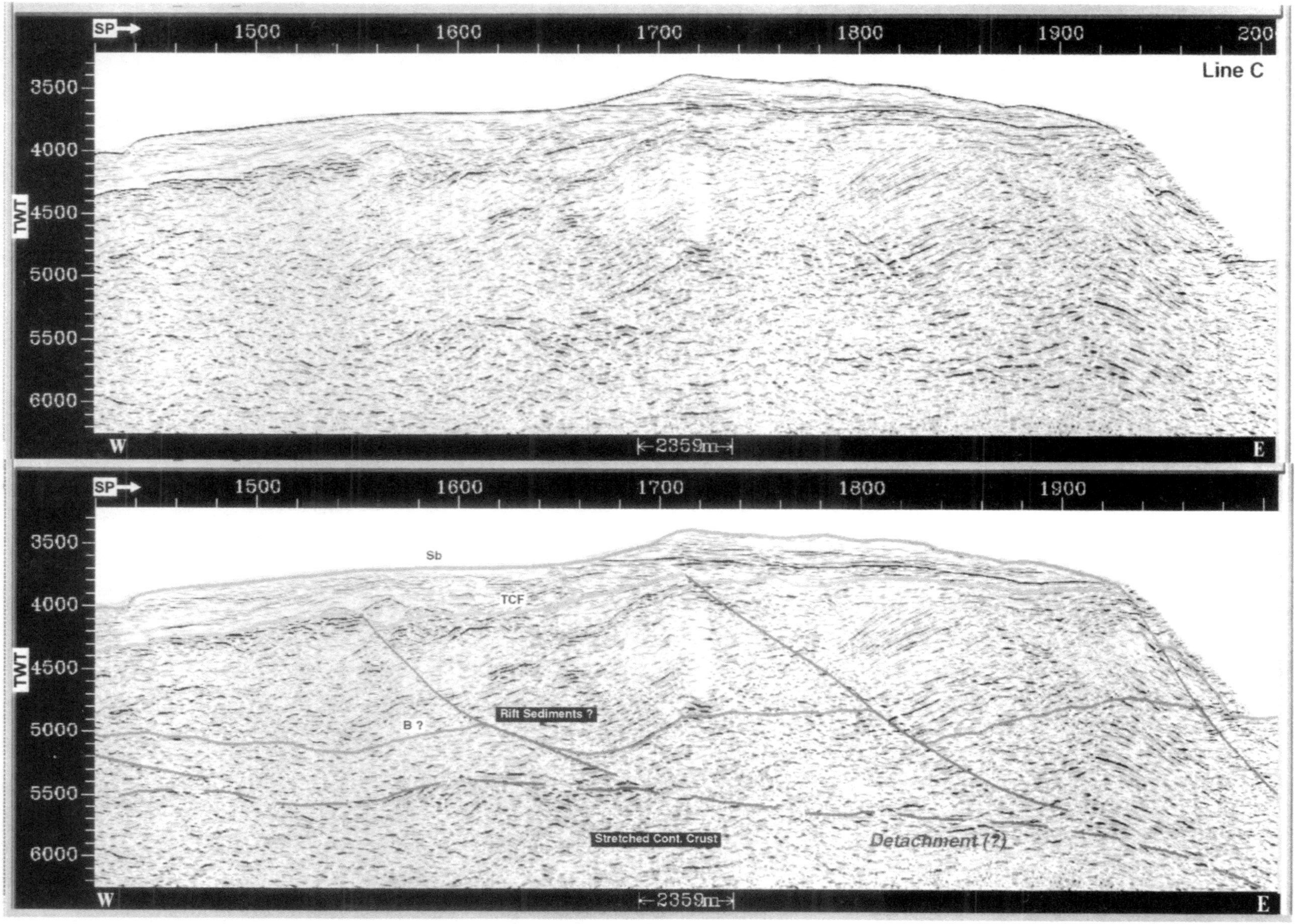

Figure 8. Detailed view of the tilted blocks on the central portion of the Pernambuco Plateau.

point to SP-2450, contrasting with a thickness of 6 km of the oceanic crust, which is nearly constant after the crustal necking that occurs around SP-2850.

Line B (figures 9 and 10), collected about 160 km to the north of line C, covered only the northern edge of the Pernambuco Plateau. The seismic profile itself does not have enough resolution to permit recognition of the crustal nature at the proximal portion of the line. However, the free-air gravity profile (Fig. 9) strongly suggests the presence of an attenuated continental crust beyond the marginal escarpment of the plateau. On the initial part of the line (up to SP-6000), there is a remarkable crustal thinning, with a reduction from 24 km to less than 10 km on the thickness of the crust. In the oceanic crust, the average thickness is about 7 km.

The highly thinned crust on the transitional region could be compensated with the emplacement of a denser material at the crust base (*i. e.*, underplating process), as suggested by many authors for Atlantic margins elsewhere [*Mutter*, 1985; *White et al.*, 1987; *Austin et al.* 1990; *Sheridan et al.*, 1993]. However, to admit the occurrence of such a process in this particular area, it would be necessary the support of additional geophysical data, obtained from wide-angle seismic refraction, for instance.

According to our interpretation, the rifting, of a probable Aptian age at this part of Brazilian Continental Margin [*Feijó*, 1994], propagated to the north of Pernambuco Lineament (Fig. 2), taking domains located beyond the modern continental slope. Because of the lack of data, the continuity of this deep narrow rift to northern regions is still unknown. The strong NW-SE trend, related to the Pernambuco and Paraíba seamounts, could possibly play a structural control role on the rifting process.

5.2. Potiguar Basin

The Potiguar basin, like the other coastal sedimentary basins off equatorial northeastern Brazil, is located near a transform continental margin, where shear stresses are expected to play an important role. It follows that the mechanisms that rule ocean-continent transition on these basins show some peculiarities differing from most of the coastal basins located to the south of 5° S. In a transform setting, one can expect a very rapid ocean-continent transition, without relevant volcanism-related features, such as SDRs. In this environment, fracture zones frequently control the crustal transition zone.

Only the western edge of LEPLAC line A was surveyed over the crustal transition region of Potiguar basin (figs. 11 and 12), to the northeast of Rio Grande do Norte Plateau (Fig. 2), in an area lacking geological and geophysical data. Only three multichannel seismic lines acquired by PETROBRAS cover the same region, and, among them, just two cross LEPLAC line A, reaching the oceanic crust domain.

The WNW-ESE direction of line A, subparallel to the fracture zones, is unfavorable to characterize a transform crustal structure in this region, as done by *Mascle and Blarez* [1987], *Edwards et al.* [1997] and others at the continental margin off Gana, Africa. In order to obtain such a characterization and compare the Potiguar basin with the African transform margin, at least one regional SSW-NNE seismic line would be required in this area.

The ocean-continent transition along line A seems to occur in a normal passive margin style, characterized by rifting and crustal extension. A similar situation across the transform continental margin off Gana, along an ENE-WSW-oriented line, was observed by *Peirce et al.* [1996]. The thickness of the stretched continental crust modelled by these authors, at the western edge of their line, is about 12-14 km. Thickness values of the same order were modelled in the crustal transition region along eastern Potiguar Basin (Fig.11).

The stretched continental crust on the northwestern edge of line A, where the sedimentary pile is affected by listric faulting-related compressive structures (Fig. 12), is located on continental rise domain, beyond the Rio Grande do Norte Plateau (Fig. 2). *Sperle et al.*[1989], based on gravity and magnetic modelling, pointed out the occurrence of transitional crust under the plateau, in an area located about 80 km to the south. The authors did not foresee, however, the continuity of this crustal type beyond the limits of plateau, as suggested by LEPLAC data.

6. OCEANIC CRUST

The most remarkable physiographic-structural features over the oceanic crust domains of the northeastern Brazilian Continental Margin are the alignments of submarine volcanic highs along E-W and NW-SE directions (Fig. 1). Near these volcanic ridges, which are related, in one way or another, to oceanic fracture zones, there are strong signs of tectonic reactivations, involving the action of extensional, compressive and/or strike-slip field stresses. Recurrent tectonics, with related volcanism, seems to have manifested itself in a sporadic way, along with the various evolutional stages of our continental margin.

6.1. Seismic Expression of Fracture Zones

During the generation of oceanic crust, in the various axial segments of a mid-ocean ridge, crustal portions of different ages, with distinct cooling and subsidence evolutional histories are juxtaposed along fracture zones.

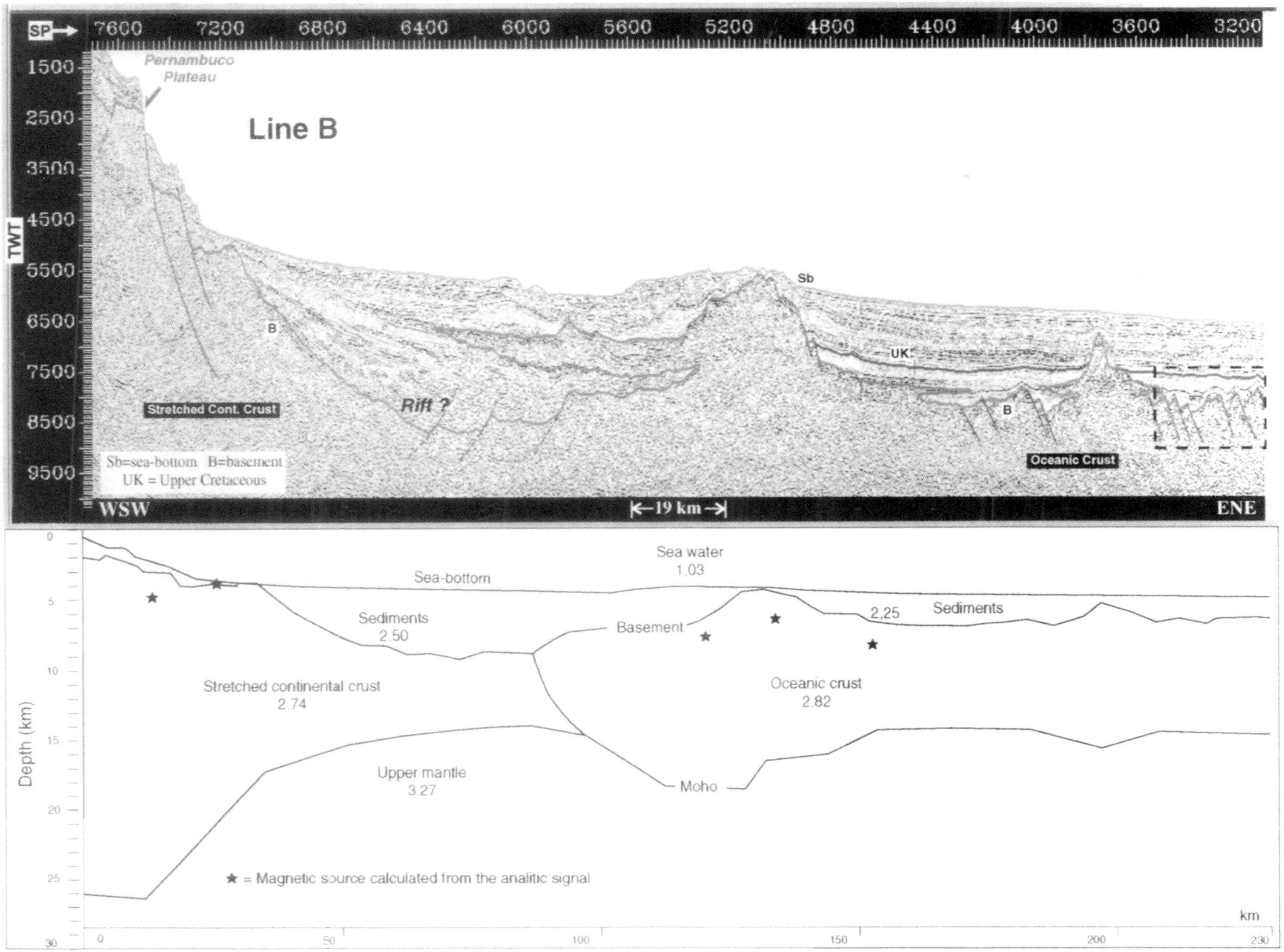

Figure 9. Crustal transition along western extension of seismic line B, northern part of Pernambuco Plateau (see fig. 2 for location). Gravity modelling of this profile, displayed below, indicates that the region situated between SPs 7200 and 6000, where the basement horizon was inferred, has a transitional nature, related to the ocurrence of a stretched continental crust. Two volcanic intrusions with a lacolith shape can be observed near SPs 5200 and 3700. On the bigger one, the southwestern apophysis represents just a simplified interpretation of a region with a different seismofacies, probably composed by a mixture of volcanic rocks, volcanoclastic material and terrigenous/ hemipelagic sediments. Another outstanding feature is the tectonism that affects the oceanic crust on the eastern edge of this section. The region singled-out by a dashed rectangle is zoomed on Figure 21, and an uninterpreted version of this line segment can be analyzed on the next figure.

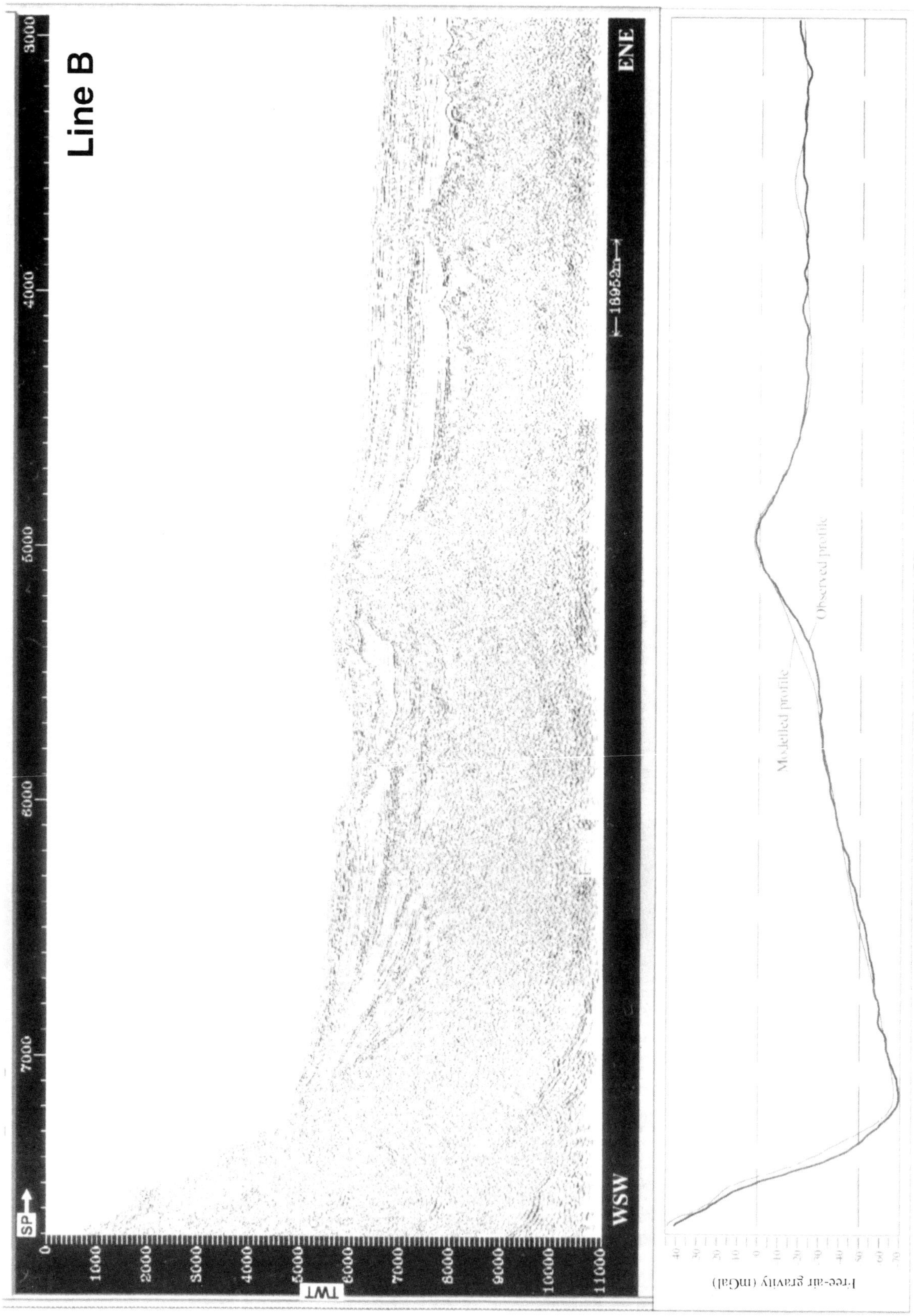

Figure 10. Uninterpreted version of western extension of seismic line B. Compared with those of figure 12, the modelled and observed free-air gravity curves of this profile have a gentler slope in the crustal transition region, located between SPs 5300-7200. This region is supposed to be underlain by highly stretched continental crust.

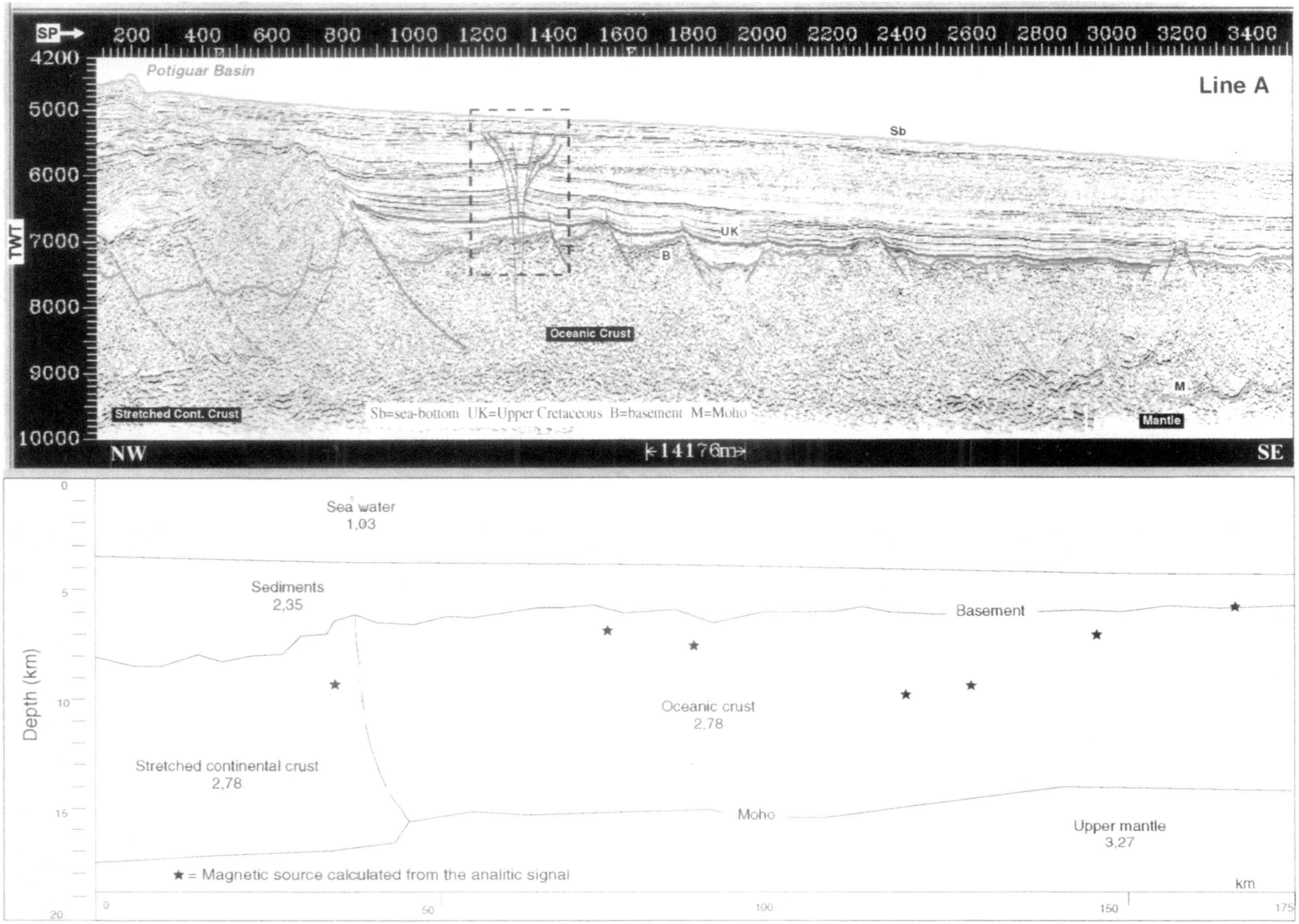

Figure 11. Part of seismic line A with corresponding gravity modelling, showing crustal transition along the eastern Potiguar Basin (see fig. 2 for location). The structure singled-out by a rectangle, probably related to transpression, is discussed at item 6.4. See next figure for the free-air gravity profiles and for an uninterpreted version of the seismic line.

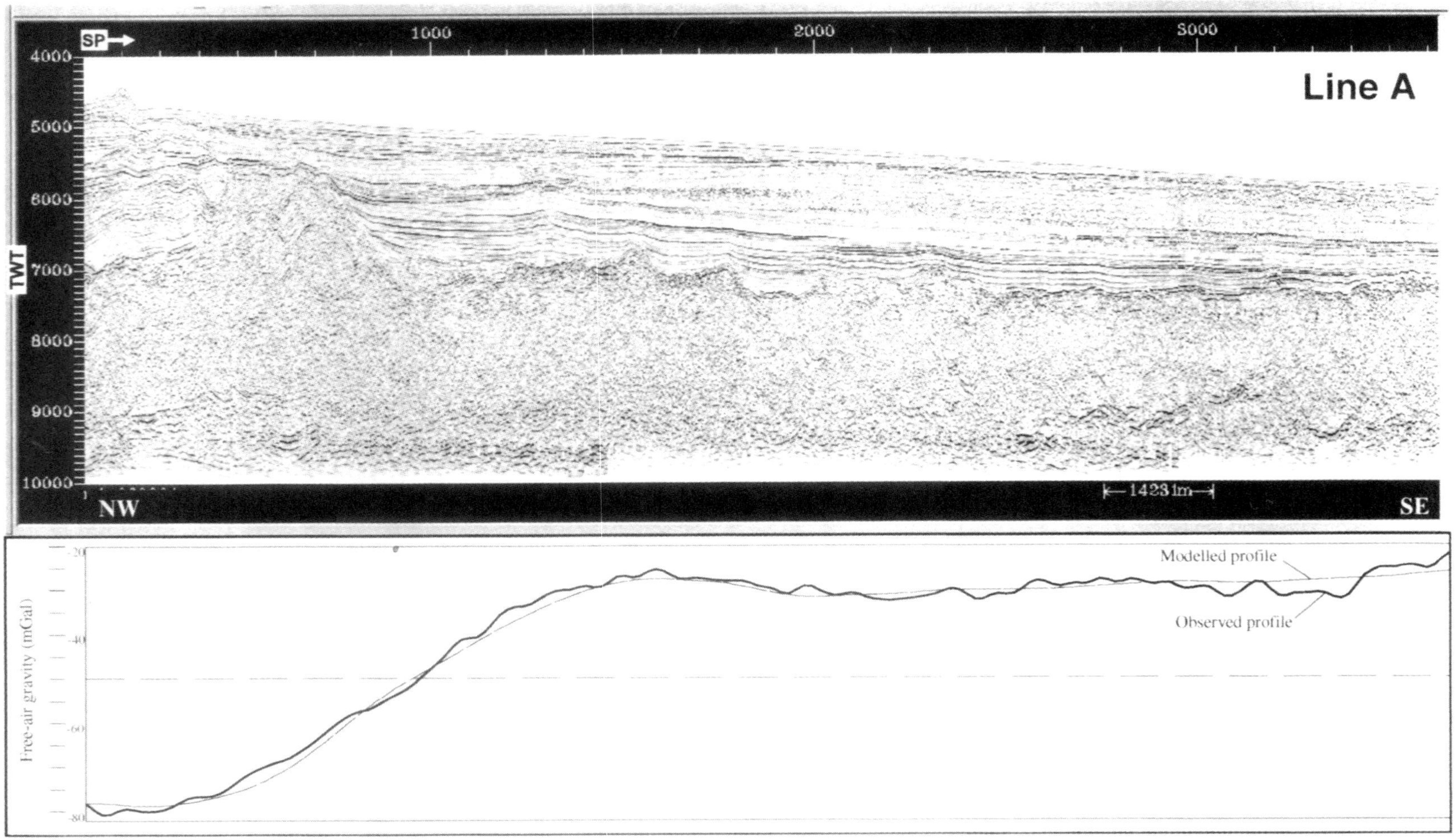

Figure 12. Uninterpreted version of western extension of seismic line A. That mound-like feature with a chaotic internal seismofacies, located around SP-600, between the 5.5 s and 7.0 s TWTs, is related to thrust structures formed by overriding listric-faulted blocks in the continental slope muddy section of Potiguar basin. The interpreted basement reflection (TWT) is probably underestimated at this region. Below the seismic section, the modelled and observed free-air gravity profiles show a remarkable pattern change around SP-1000, related to the emplacement of oceanic crust.

Because of this differential evolution, regional steps on the basement, formed in the vicinities of a MOR, can remain "printed" along the evolving lithospheric plate. These changes on the regional depth of the basement, often associated with submarine volcanic ridges, represent the clearest evidence for the recognition of FZs at seismic sections. The linear ridge-type morphology sometimes works as a barrier to the sedimentation processes over the continental margin/ oceanic basin, leading to the formation of regional steps on the seafloor.

Based on these widespread concepts and assuming the free-air gravity map (Fig. 13) as a reference, we attempted to recognize, over LEPLAC Project seismic data, the lineaments[1]/fracture zones systems of our northeastern continental margin, originally identified in the papers of *Gorini et al.* [1974], *Rezende et al.* [1977], *Asmus and Carvalho* [1978], *Guazelli and Carvalho* [1981] and *Gorini and Carvalho* [1984]: Fernando de Noronha Fracture Zone and Paraíba, Pernambuco and Maceió lineaments. All these structures are displayed on the "Geological Map of Brazil and Adjoining Oceanic Area" [*Schobbenhaus et al.*, 1981], which simplified version, covering just the northeastern continental margin, is shown on Figure 4.

The results of our recognition work over LEPLAC data, between the 3° S and 10° S latitudes, can be checked on Figure 2. Figures 14 and 15 indicate the position of all interpreted structures in mega-regional seismic displays. Figure 14 presents a sequence of three strike seismic lines located on an average distance of 200 km far from the coastline. Connected to each other, these lines form a transect that covers from the south flank of Fernando de Noronha Ridge down to the vicinity of Aracaju's latitude. In a similar way, but at a distal setting, nearby the limit of 200 nautical miles ($\cong$371 km), Fig. 15 allows the identification of the same structures. For the location of these geophysical lines, see figures 2 and 5.

One can observe some discrepancies comparing Figure 2 to the map of Fig. 4, essentially related to the connection between Fernando de Noronha Ridge (FNR) and Chain Fracture Zone. This connection, previously suggested by *Bryan et al.* [1972], was mainly based on the new free-air gravity maps prepared with GEOSAT data (Fig. 13). In the Geological Map of Brazil (Fig. 4), the Fernando de Noronha Fracture Zone, identified by *Gorini* [1974] to the west of 31° W longitude, shows an inflection to northwest on continental margin direction, in order to connect with the FNR. According to *Gorini* [1981], this FZ would represent a South-American extension of Charcot Fracture Zone.

The free-air gravity map (Fig. 13) actually shows that, towards the Brazilian Continental Margin, fracture zones inflect in a curved way to southwest, as a result of global dynamics involved on the opening of the South Atlantic, according to the model of *Sibuet and Mascle* [1978]. Hence, the Charcot FZ, having similar inflection, would be connected with structural features located to the south of FNR, as pointed out on Figure 2. It is important to emphasize, however, that the extension of this fracture zone to the mapped area is only a suggestion, since its expression over the seismic data is not so outstanding as observed in other situations.

Even less expressive, when compared to the other structures, is the seismic expression of the Paraíba Lineament (figs. 14 and 15). Nevertheless, as it is well marked at the free-air gravity map (Fig. 13), we can admit this lineament as a small-offset fracture zone, with gentler regional basement steps.

A typical FZ morphology can be observed at Figure 16, where remarkable basement and seabottom steps mark a boundary between crustal segments. In this example, which shows a detailed view of line E over the Pernambuco Lineament region (Fig. 2), there is a single step related to a drop in excess of 400 m on the seabottom. At the basement, there are multiple steps related to the southern flank of the volcanic edifice. Over one of these steps, located around SP 4200, normal and reverse faulting affect shallow levels of sediment package, implying some kind of tectonic reactivation along this fracture zone. Surprisingly, in spite of the remarkable seismic expression, Pernambuco Lineament is not so obvious on the gravity data (figs. 5 and 13). This lineament could be an extension, towards the Brazilian Continental Margin, of the Fernando Poo FZ, identified by *Emery et al.* [1975] at the African Plate.

The seismic and gravimetric expression of the Maceió Lineament is outstanding enough to characterize it as an occidental extension of the well-known Ascension Fracture Zone. The connection between these two linear structures and its continuity toward the Brazilian Continental Margin was suggested by *Palma* [1989], on a morphological map of the Ascension FZ, and confirmed by *Mello* and *Dias* [1996], based on magnetostratigraphy of the oceanic crust of that region. The GEOSAT data (Fig. 13) emphasize that the extension of this lineament/fracture zone in the direction of Sergipe-Alagoas basin inflects to a southwest direction, as suggested by *Sibuet and Mascle* [1978] in their model for the opening of the South Atlantic, mentioned earlier.

[1] The term "lineament", applied for those transverse structures with no proved connection with the Mid-Atlantic Ridge, was suggested by Asmus and Guazelli [1981] and largely used at the REMAC Project studies.

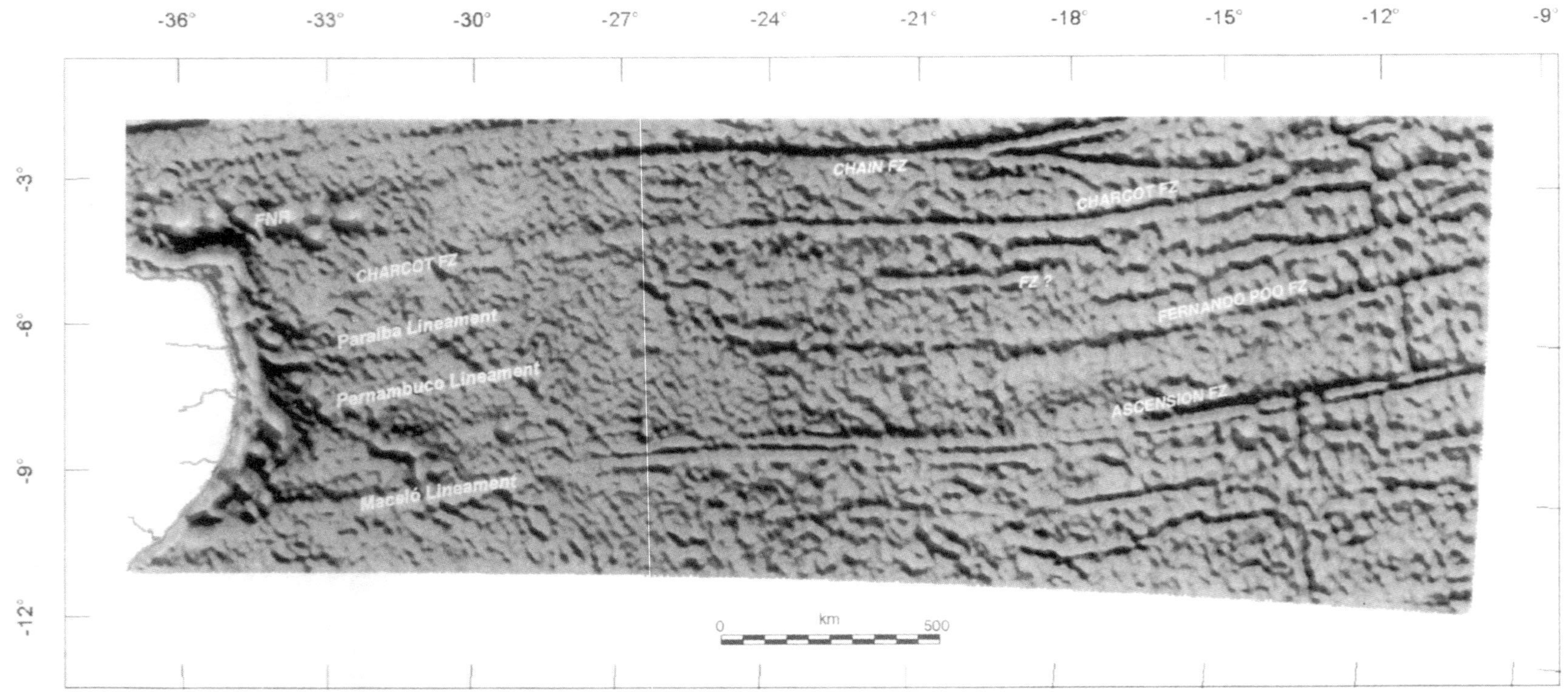

Figure 13. GEOSAT free-air gravity map of the northeastern continental margin off Brazil up to the Mid-Atlantic Ridge. Structural relationship between the lineaments of *Schobbenhaus et al.* [1981] and well-known fracture zones can be easily recognized. The Paraíba Lineament is probably related to a small offset fracture zone, which does not have a widespread designation on the literature (FNR = Fernando de Noronha Ridge).

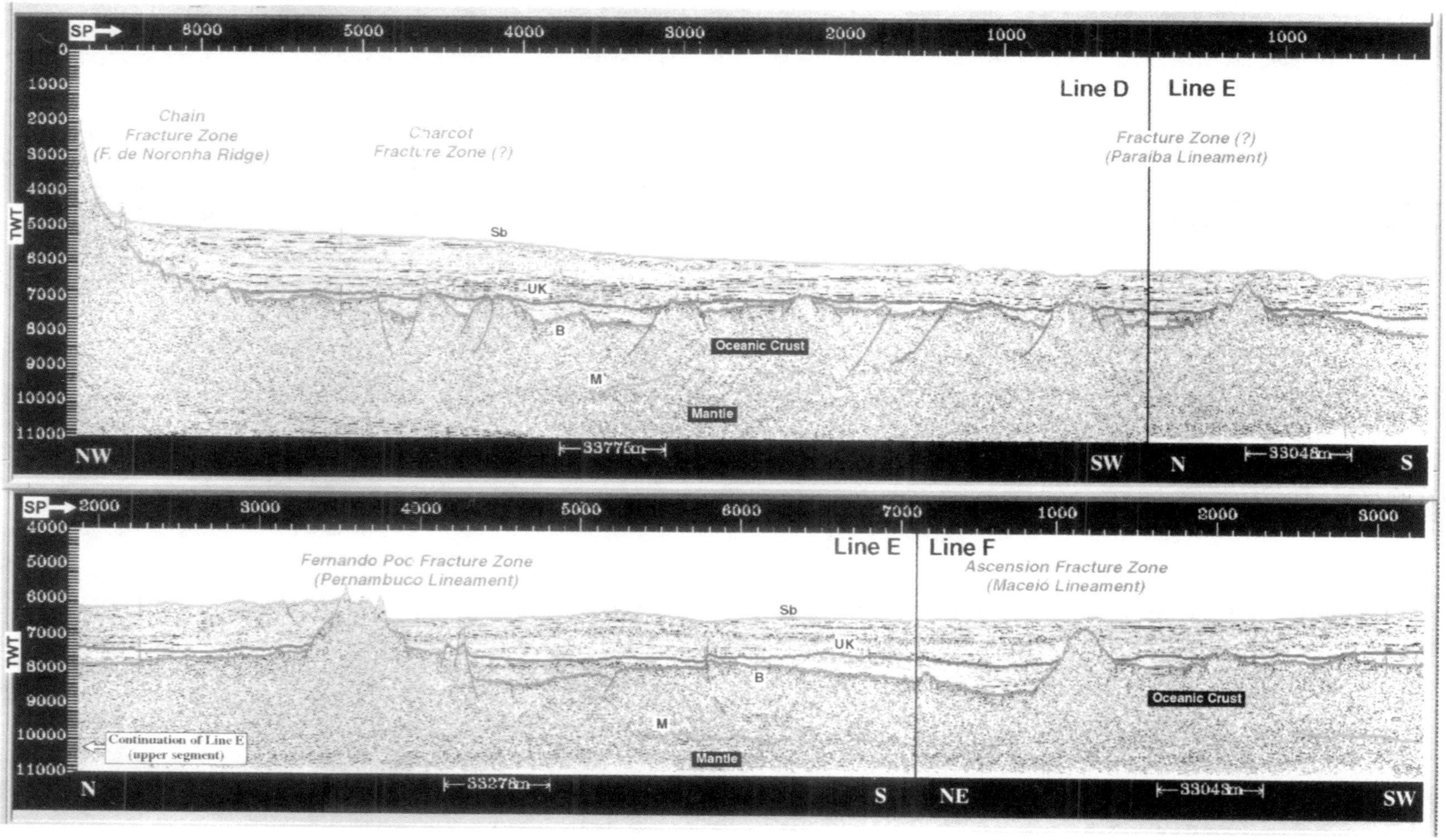

Figure 14. Proximal strike transects of LEPLAC Project (see figure 2 for location). In this regional view, the most remarkable features are the fracture zones with their volcanic highs and related changes on regional depth of basement. These depth changes are not so prominent on small offset fracture zones, such as the two ones located to the south of Fernando de Noronha Ridge, at the upper segment of this figure (Sb=seabottom; UK=Upper Cretaceous; B=basement; M=Moho).

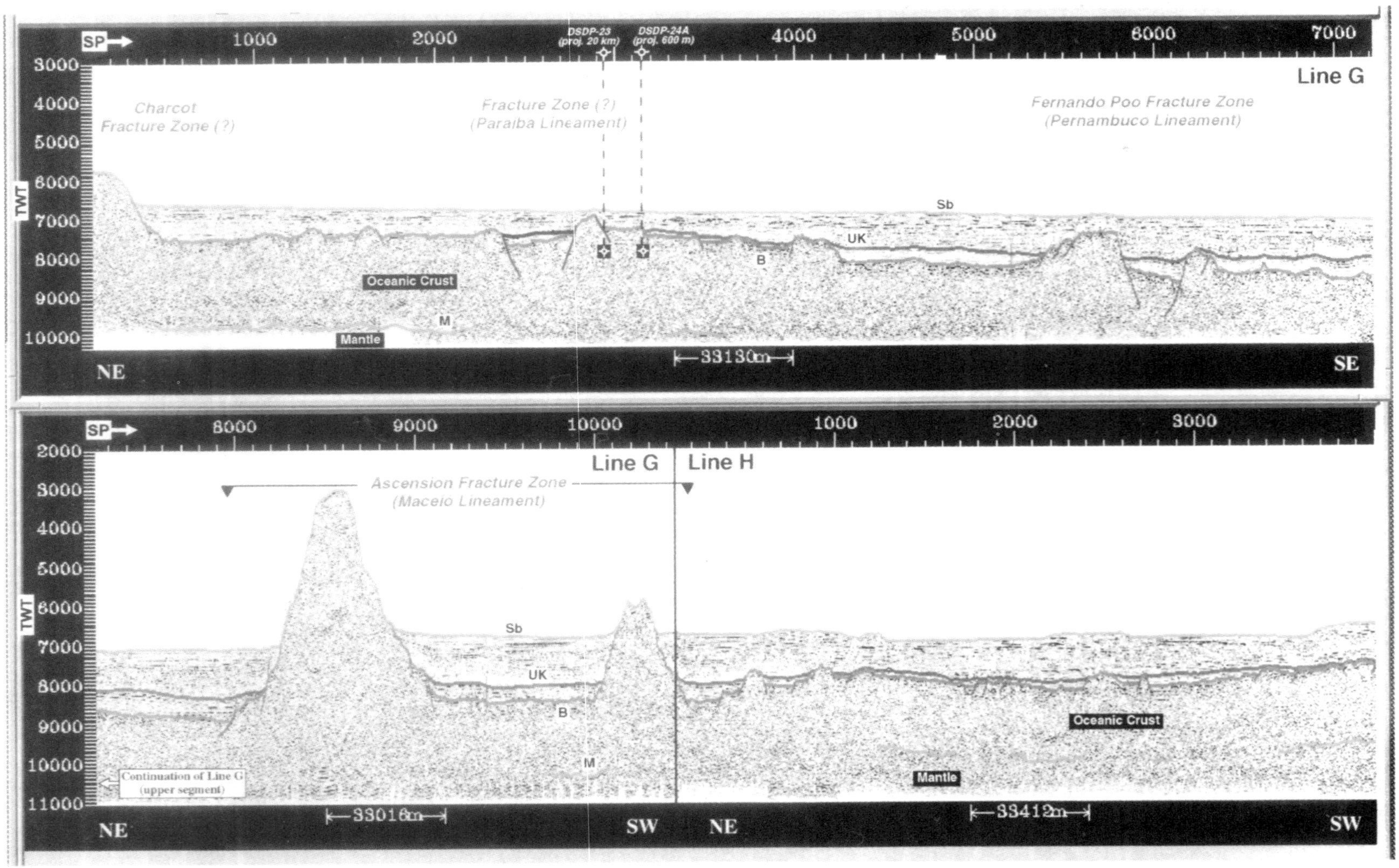

Figure 15. Distal strike transects of LEPLAC Project, with projected DSDP wells (see fig. 2 for location). Again, fault scarps and volcanic highs, with related changes on basement depth, mark the intersection of the seismic lines with fracture zones. The upper portion of sedimentary cover to the south of Ascension FZ (line H) shows erosional features probably related to the action of northward flowing Antarctic Bottom Water (AABW). Actually, contouritic drift-type deposits related to this bottom current appear to be very important in the post-Eocene sedimentary record [*Gomes*, in. prep.] (Sb=seabottom; UK=Upper Cretaceous; B=basement; M=Moho).

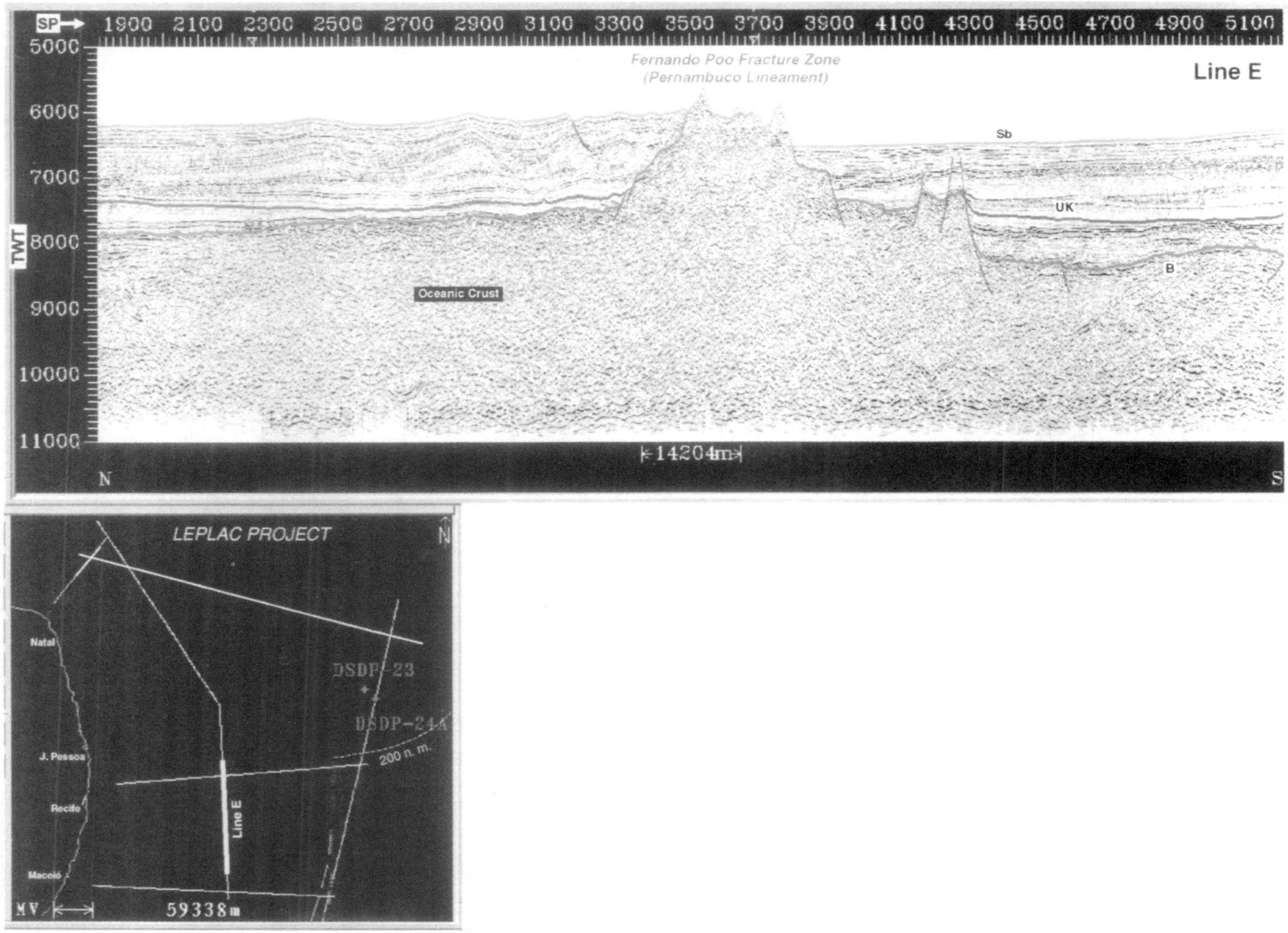

Figure 16. Part of seimic line E that crosses a fracture zone (highlighted on location map beside). A typical morphology can be observed, as seabottom and basement regional depths change remarkably from northern to southern crustal segment. Some faults on the southern segment reach upper layers of the sedimentary cover, suggesting the occurrence of some kind of tectonic reactivation along the fracture zone. On the left half of the figure, those antiform structures located in the intermediate portion of the sedimentary pile were primarily interpreted as a result of a mudflow-like process, where the volcanic high acted as a wall, preventing somehow the sedimentary movement (Sb = sea-bottom; UK = Upper Cretaceous; B = basement).

6.2. Magmatic and Tectonic Reactivation of Fracture Zones

The Fernando de Noronha Ridge (FNR) is a prominent physiographic feature composed by volcanic highs (*e.g.* seamounts, guyots and islands) aligned in an approximate east-west direction. *Gorini and Carvalho* [1984] suggested a structural link between the FNR and the homonymous FZ located to the east, as the occidental extension of the Charcot Fracture Zone. However, as discussed on the previous item, new data collected on northeastern continental margin off Brazil point out to a probable connection with a FZ located to the north, known as Chain Fracture Zone.

The different views regarding the eastern extension of the Fernando de Noronha Ridge reflect the volume and quality of data available for the studies. An important issue to be adressed is the structural relationship of the ridge with some fracture zone. Another relevant question concerns to the origin of the volcanic processes responsible for the building of FNR. We accept here the concepts of *Duncan* [1981] and *O'Connor and Duncan* [1990], according to which the ridge formed by the passage of the South-American plate over a hotspot. Hence, it is possible that a fracture zone, behaving as a zone of weakness, would have controlled the volcanic eruption related to the Fernando de Noronha hotspot. A similar argument was proposed by *Epp* [1984] and *McNutt et al.* [1989] to explain the origin of volcanic ridges at the Pacific Ocean.

The poor seismic coverage of the LEPLAC Project did not allow us to get a precise characterization of the Fernando de Noronha Ridge/Chain FZ. Only the southern flank of a seamount related to the FNR was covered by the NNE edge of line G (Fig. 15). In the same way, just the southern flank of another volcanic edifice component of the ridge was imaged on line D (Fig. 14). Nevertheless, it is possible to discuss the nature of the volcanic and structural features that surround the ridge.

Figure 17 presents a more detailed view of the northern edge of line D, where one can observe the intense volcanic activity near the Fernando de Noronha Ridge, with sills and dikes interspersed in the sedimentary package. These small intrusions are closely related to the volcanic episode that built that submarine edifice and appear to affect several levels of the sedimentary cover on the adjacent area. Besides this, we observed remobilization of old extensional faults and transpression-related features were also identified in the region (Fig. 11). All these seismic evidences suggest that the fracture zone was reactivated in a process that included remarkable magmatic episodes accompanied by strong tectonic effects, some of which with strike-slip components.

Strike-slip tectonics related to fracture zones, expected by *DeLong et al.* [1977] and *Bonatti* [1978], were observed in multichannel seismic reflection profiles at the eastern segment of Romanche FZ [*Lehner and Bakker*, 1983; *Bonatti et al.*, 1994] and at the western portion of Saint Paul FZ [*Silveira*, 1993; *Silveira et al.*, 1994]. Several other structures that suggest the influence of strike-slip tectonics in the proximity of fracture zones can be observed on LEPLAC seismic lines.

Based on LEPLAC data, it would be possible to conclude that the fracture zones located at the Brazilian continental margin behave as intraplate discontinuities. These structures could become, therefore, prefered sites to seismic activity, even at points located far away from the mid-ocean ridge. However, things may not be that simple, because the discussion involving the strength of fracture zones brings back an old controversy of marine geophysics. One group of researchers believe that these structures, behaving as zones of weakness, constitute favourable sites to volcanism and seismicity [*Vogt*, 1974; *Batiza*, 1982]. For another group, FZs are only remnant surface expressions of earlier tectonic activities along transform faults and spreading ridges, without significant weakness relative to the surrounding lithosphere [*Sandwell and Schubert*, 1982; *Bergman and Solomon*, 1992]. *Lowrie et al.* [1986], admitting thermomechanical variations along fracture zones, reconcile both views in a system where seismically active segments would alternate with inactive "strong" segments.

It is important to mention that none of these concepts were supported by multichannel seismic reflection data. The LEPLAC Project data revealed impressive evidence of seismic reactivation in the vicinity of fracture zones, affecting several levels of the sedimentary package (see figs. 16 and 17). Based on this evidence, we support the conclusion of *Silveira* [1993], according to which the recurrent seismicity at fracture zones includes both compressive stresses – with uplifting of the crust - and extensional stresses – with volcanism and normal faulting.

6.3. Lithospheric Flexure at Volcanic Ridges

Accepting magmatic activity along fracture zones, as discussed earlier, one can consider the possibility of crustal thickening related to volcanic stacking. Thus, depending on the effective elastic thickness of the lithosphere at the time of magmatism [*Watts*, 1978], we can expect the occurence of an isostatic instability related to the volcanic edifices that formed along reactivated fracture zones. Observing the line D in a proper scale (Fig. 18), one can notice the presence of a depression (SPs 5700 to 3300) adjacent to a Fernando de Noronha Ridge (FNR) seamount, which is followed by a relative elevation of the basement (SPs 3100

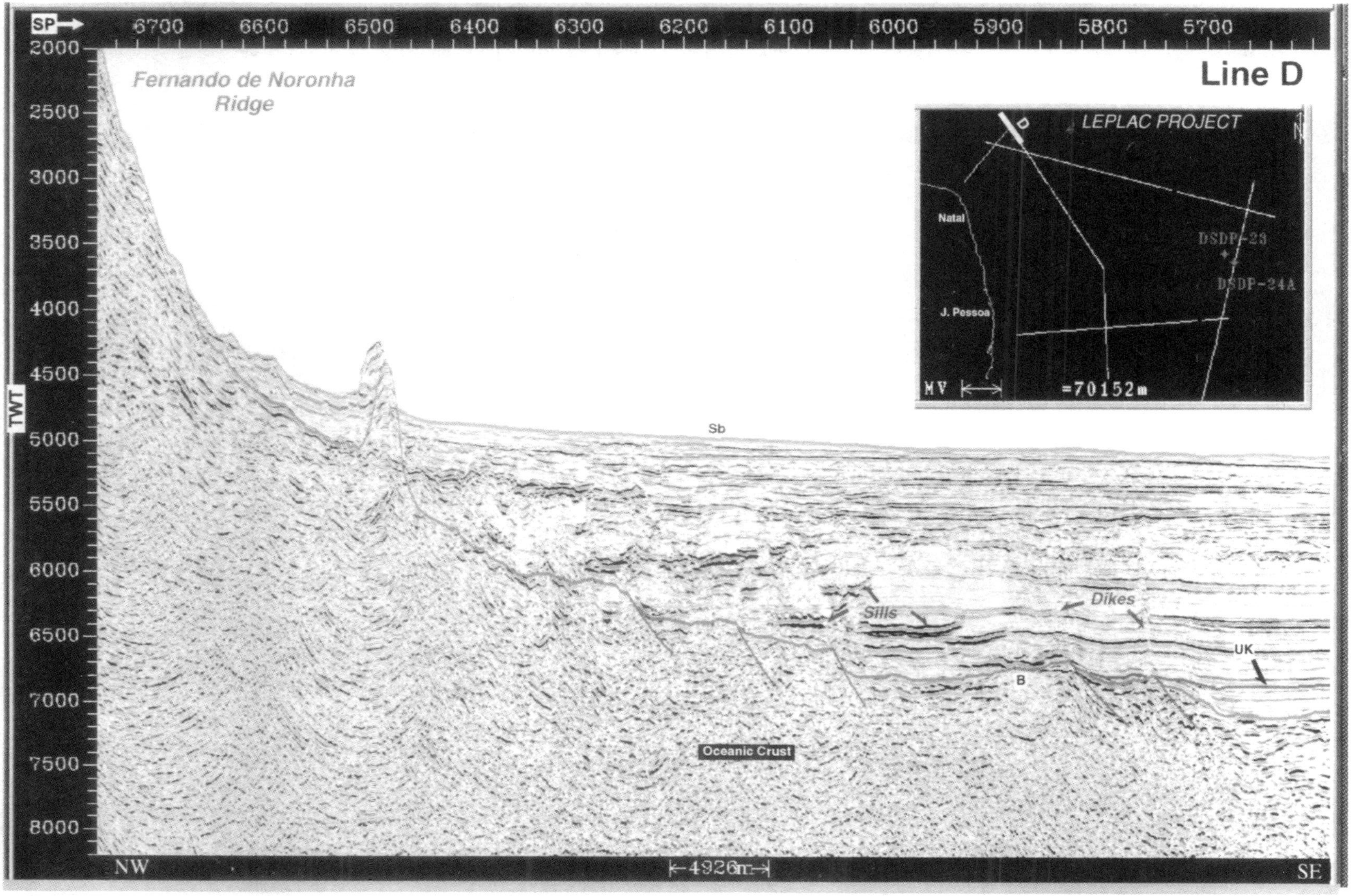

Figure 17. Equatorial edge of seimic line D (highlighted on location map above) showing the southern flank of a volcanic edifice of the Fernando de Noronha Ridge. Outstanding features related to sills and dikes can be observed interspersed within the sediment package, as an evidence of magmatic reactivation along the Chain Fracture Zone (Sb=sea-bottom; UK=Upper Cretaceous; B=basement).

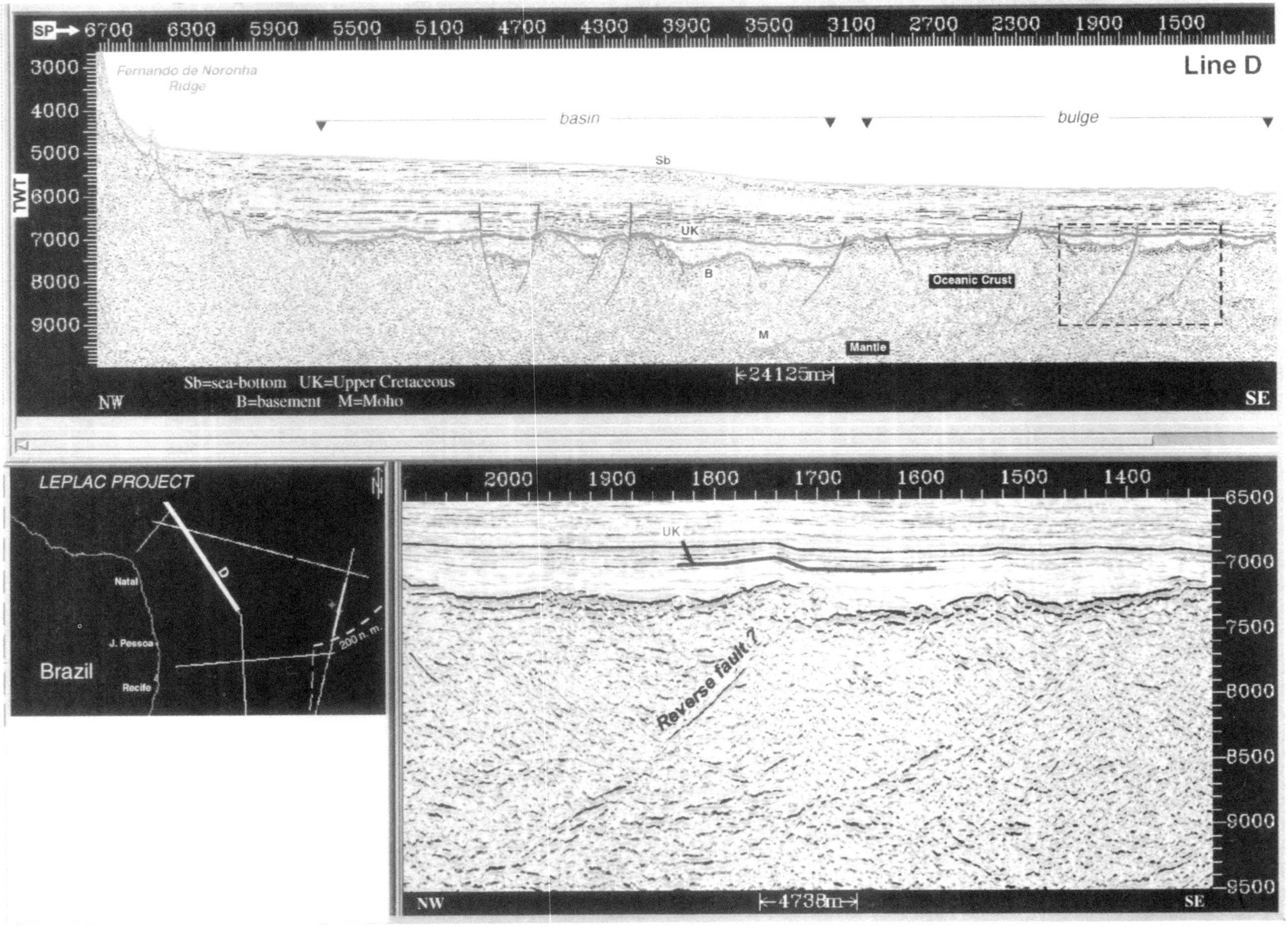

Figure 18. On top, the regional view of seismic line D shows the basin and the bulge adjacent to a Fernando de Noronha Ridge seamount (location is highlighted at the map above). The Moho reflectiondips toward NW (both on time and depht profiles), as an effect of the crustal "sinking" caused by the volcanic pile. To the right, an uninterpreted zoom view of the region singled-out by a rectangle shows the reflection of a fault plane, which throw appears to be tectonically reverted (see text for discussion).

to 1100). That depression was shown on the "Geological Map of Brazil and Adjoining Oceanic Area" [*Schobbenhaus et al.*, 1981] as a NW-SE-elongated basin, bearing sedimentary thicknesses up to 3 km.

Probably, both the basin and the bulge at Figure 18 express flexural-type isostatic compensation to the volcanic overload of the FNR, such as the situation observed across the Hawaiian-Emperor seamount chain (*Watts et al.*, 1985). With this hypothesis, it would be possible to explain an outstanding feature located around SP-1800 of line D, which detailed view is shown at the lower portion of Figure 18. The feature is a strong intracrustal reflection whose configuration, together with the flexure within the sedimentary pile, suggests reverse faulting.

Using a simple lithosphere flexure model caused by a topographic load [*Turcotte and Schubert*, 1982] – in this example a seamount settled over ~ 70 Ma-old oceanic lithosphere with a rigidity of 10^{23} N.m. – , *L.F.S. Braga*[2] (pers. commun.) has estimated that a flexural bulge must occur about 250 km from the center of the load. As the discussed faulting-related feature is located ~ 250 km far from the FNR, it is reasonable to admit its relation with an isostatic compensation process involving lithospheric flexure. One must consider, however, the contribution of alternative factors for the generation of intraplate stresses that are required to arouse the observed tectonic effects. *L.F.S. Braga* suggests, for instance, the existence of another compressive field of stresses superposed to that caused by the overload, which would give plausible conditions to the reactivation of the oceanic crust on the region. This field of stresses would be generated by the presence of differential lithostatic pressures at the ocean-continent transition zone, as proposed by *Turcotte* [1982].

Even so, it is difficult to imagine a compressive stress capable of tearing up the whole crust, as in the case of Figure 18, without the presence of an earlier discontinuity. It is possible, then, to associate the discussed reflection with inversion tectonism, where a fault formed in a oceanic crust extensional setting could be reactivated by compressive stresses, due to the processes mentioned above.

Several faults on line D seem to reactivate shallower post-Eocene levels of the sedimentary column. The shallowest level affected must express the age of the crustal "accommodation", which, in its turn, must get closer to the age of the seamounts located nearby the NW edge of line D.

2 Former senior geophysicist and tectonophysics consultant at E&P/PETROBRAS. Now at *High Sense Geofísica S.A.*.

This discussion lacks support by radiometric age determinations, without which it is impossible to make a reliable estimate about the ages of the seamounts and related tectonism. We just know that the events discussed above are post-Cretaceous and that they must be older than 12.3 Ma, which is the maximum K-Ar age obtained by *Cordani* [1970] for the rock samples from Fernando de Noronha Island, located 250 km to the east of the volcanic edifice imaged by LEPLAC data. One can work with age constraints assuming the hypothesis of plate drift over a hotspot in this region [*O'Connor and Duncan*, 1990], as mentioned in the previous section. Under this theoretical assumption, the discussed seamount would be younger than the 32 Ma Mecejana phonolith [K-Ar age; *Thomaz Filho*, 1983], located onshore about 400 km to the west, in the vicinity of Fortaleza.

Another example of a flexural effect, this time related to the Maceió Lineament/Ascension FZ, can be observed at Figure 19. The lithospheric flexure, in this case, seems to have occurred in a different setting when compared to the Fernando de Noronha Ridge, since the dimension of the volcanic high is much smaller and there are no signs of magmatic reactivation in its proximity. The crustal depression and bulge observed on the figure would represent a flexural response of the lithosphere in the absence of vertical tectonism along fracture zones, according to the "locked fault" model [*Sandwell and Schubert*, 1982], which does not consider FZs as intraplate zones of weakness. In this model, differential subsidence on crustal segments juxtaposed by FZs would be compensated by flexure of the lithosphere, expressed as a regional bending of the segments, like the example shown at Figure 19. To fit the Maceió Lineament/Ascension FZ into this model and still support the "mobility" concept previously proposed, one must accept the composed model of *Lowrie et al.* [1986], which admits the presence of discontinuities along parts of FZs intercalated with "strong" segments (see item 6.2).

Sandwell and Schubert [1982] also point out the occurrence of elastic coupling between fracture zones, i.e., the interference of flexural processes of two FZs relatively close to each other. The crustal segment bounded by Maceió and Pernambuco lineaments (Ascension and Fernando Poo FZs) could have been affected by a similar phenomenon, in which tectonic reactivations are expected to be present, as observed nearby the northern lineament (Fig. 16).

The flexure observed at Figure 19 could also be explained with the model of *Pockalny et al.* [1996], applied to North Atlantic fracture zones. According to this model, FZ-parallel oceanic transverse ridges would form as a flexural response of the lithosphere to normal faulting.

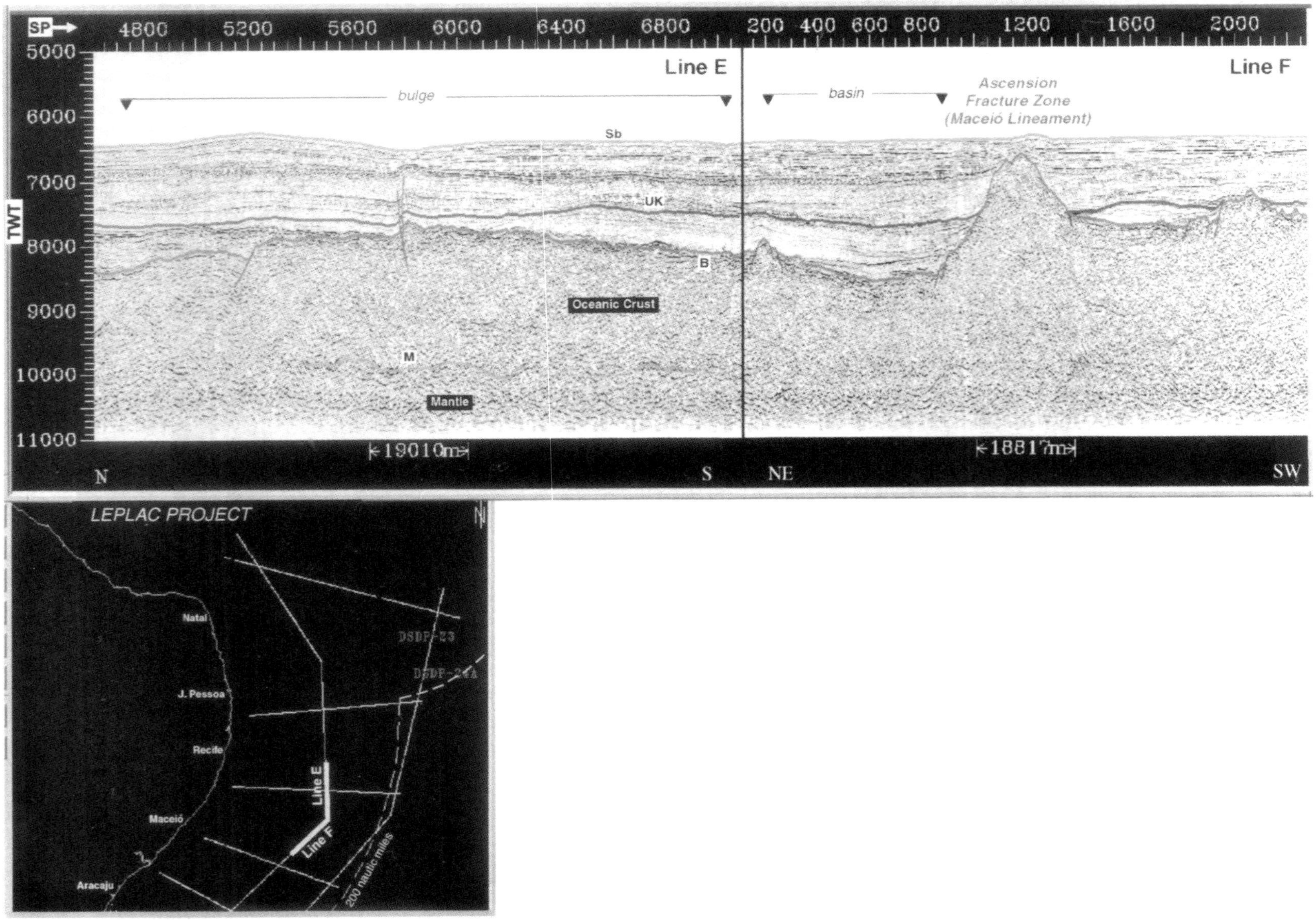

Figure 19. Zig-zag composition of lines E and F (highlighted on the location map beside) showing lithospheric flexure related to a fracture zone (Ascension FZ). According to *Sandwell and Schubert* [1982], the absence of vertical slip along the fracture zone (*i.e.*, tectonic reactivation) leads to a flexural bulge of the crust, due to the differential subsidence of lithospheric segments juxtaposed by the FZ. Observe the interpreted fault with an upthrust geometry near the SP-5800, located about 115 km far away from the fracture zone. This event appears to reach younger sedimentary layers, what means a tectonic reactivation of the oceanic crust. In order to find any kind of relation between this structure and possible intraplate stresses, or even with the flexural effects observed above, further studies are necessary (Sb = sea-bottom; UK = Upper Cretaceous; B = basement; M = Moho).

These, on their turn, would be related to differential responses of FZs due to changes in spreading direction. On Figure's 19 case, the basement high located around SP-1200 seems to constitute, at least morphologically, a linear transverse ridge similar to those studied by the referred authors, which presents a remarkable expression on gravity data (Fig. 13).

6.4. Other Types of Tectonic Reactivation of the Oceanic Crust

As pointed out in section 6.2, several structures related to strike-slip tectonics were identified on the LEPLAC seismic data that cover the oceanic crust domain off NE Brazil. The most expressive reactivations occur to the south of the area discussed in this study, in the intersection region between Ferraz Ridge – a southern component of the Bahia Seamounts [*Cherkis et al.*, 1992] – and the Salvador Lineament/Bode Verde FZ (figs. 1 and 4). In this region transcurrent faults and related folding, with volcanism associated, affect the sedimentary cover up to a level mapped as the base of Eocene [*Gomes*, in. prep.].

Whithin the limits of this study, the most outstanding strike-slip-related seismic feature is found nearby the ocean-continent transition region of line A (Fig. 11). At this place, a positive flower-structure, showed in detail on Figure 20, reaches young levels of the sediment package, which, seismically tied to the DSDP-24A well (figs. 2 and 15), are correlative to the Lower Miocene.

The presence of transpressive structures in the vicinity of a transform continental margin, as exemplified on Fig. 20, is not a surprise, since shear stresses are expected to play an important role in this tectonic setting, as mentioned on item 5.2. A possibility to consider is that the magmatic reactivation process of the Chain FZ, caused by the plate drift over a hotspot, resulting in the Fernando de Noronha Ridge construction (see item 6.2), might has generated the shearing stresses related to the discussed structure.

Remarkable examples of extensional tectonics in oceanic crust, with the presence of tilted blocks, were observed by *Basile and Mascle* [1990] and *Salisbury and Keen* [1993] at the Sierra Leone abyssal plain, Equatorial Atlantic, and in the continental margin off Nova Scotia, Canada, respectively. Similar structures were observed on line B, over continental rise domains off Brazil, nearby the 7° S latitude (figs. 9 and 21).

Unlike the Canadian structures, characterized as listric faults [*Salisbury and Keen, op. cit.*], the tilted blocks of line B (figs. 9 and 21) have a geometric arrangement that resemble a "domino" tectonics [*P. V. Zalán*[3], pers. commun.], where rotational planar faults extinguish themselves at the base of brittle crust, in a possible sub-horizontal shear zone. Although located within asymmetric positions at the South-American and African plates, both LEPLAC and *Basile and Mascle* [*op. cit.*] structures might be contemporaneous, since they affect Upper Cretaceous sediments. The referred authors correlate the tectonism, occurred between Saint Paul and Romanche FZs, with "global opening geometry rearrangements within the South Atlantic".

The tectonic feature shown on Figure 21 is located nearby an important NW-SE alignment which, on the "Geological Map of Brazil and Adjoining Oceanic Area" [*Schobbenhaus et al.*, 1981], is related to the "Pernambuco and Paraíba Seamounts" (figs. 1 and 4). The NW-SE direction, quite impressive on the northeastern continental margin off Brazil, appears as well to the south, between the Ascension and Bode Verde fracture zones, in the Bahia Seamounts region (figs. 1 and 4). All these alignments are well-represented on the gravity maps (figs. 5 and 13) and on the multichannel seismic reflection data of LEPLAC Project.

Mello and Dias [1996] suggest that increases in the oceanic spreading rates and changes in plate motions, which lead to variations in the strike of fracture zones[4], could generate the magmatic reactivation related to the emplacement of those NW-SE aligned seamounts. The authors still point out that these mechanisms involve the propagation of intraplate stresses, which could locally cause tectonic and volcanic activity. Deformations as reported by *Basile and Mascle* [1990], from the Equatorial Atlantic, and the "domino" tectonics observed on LEPLAC data fit perfectly to this setting.

An important point is that all the examples in this section, involving data from Canadian and Brazilian continental margins and from African oceanic basin, express tectonic reactivation of the oceanic crust, since the faults have disturbed a relatively younger sedimentary cover. In the Brazilian example, the perturbation affected sediments of the Upper Cretaceous, an age that may be related to the formation of that NW-SE alignment related to Paraíba and Pernambuco Seamounts. *Sperle et al.* [1991], based on an isostatic analysis, suggested that the

---

[3] Senior geologist and structural geology consultant at E&P/PETROBRAS.

[4] Anomalous FZ structures related to changes in plate motions were also studied by *Kruse et al.* [1996] and *McCarthy et al.* [1996], in the Pacific Ocean. According to the former authors, most of these structures are inherited from the transform fault, and formed during or immediately subsequent to times of changes in spreading directions. Except for locations overprinted by hotspot volcanism, they do not expect any other kind of reactivation along fracture zones.

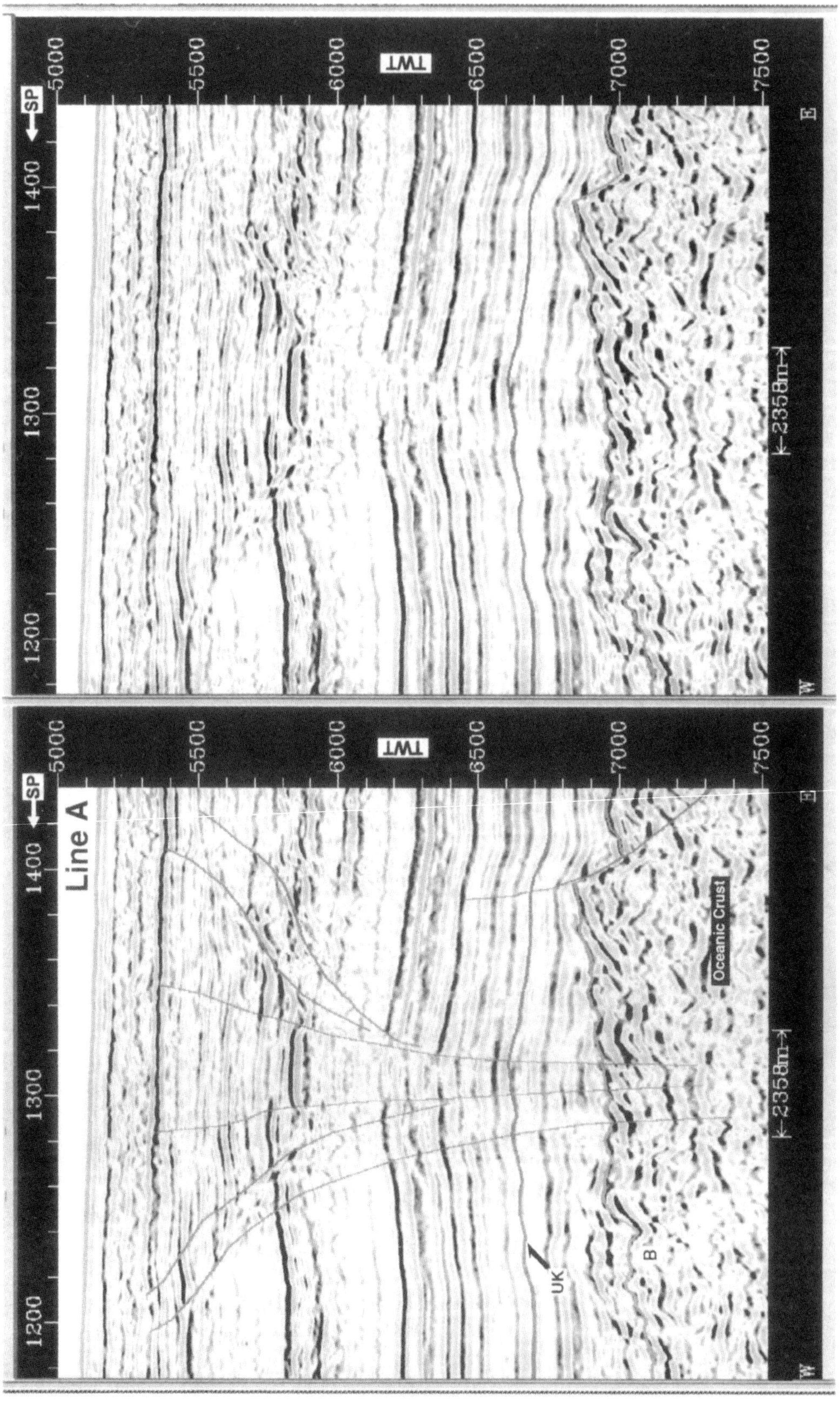

Figure 20. Interpreted (left) and uninterpreted (right) detailed views of the boxed section of Figure 11, showing a probable transpression-related structure (flower-structure).

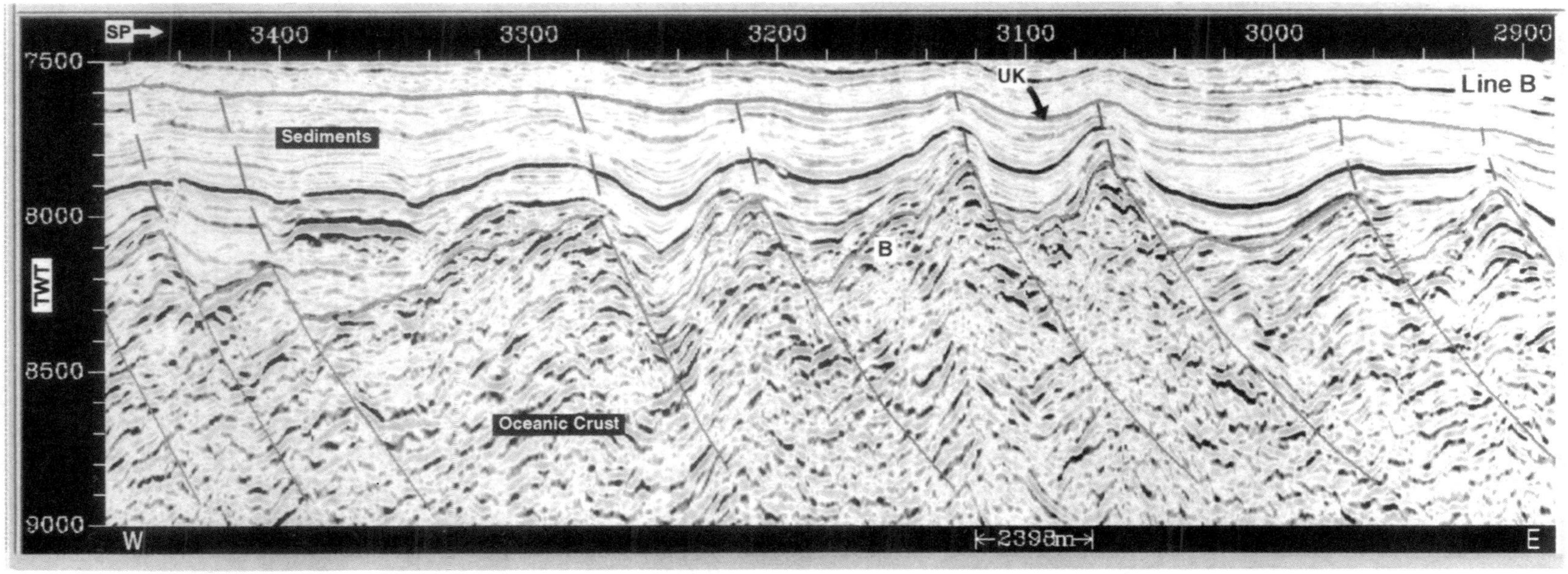

Figure 21. Detailed view of the boxed section of Figure 9, showing a sequence of rotational planar faults in the oceanic crust. This event, which seems to affect Upper Cretaceous sediments, can be interpreted as a tectonic reactivation in an environment not so close to the spreading axis (UK= Upper Cretaceous; B=basement).

Pernambuco Seamounts were formed over a 20-25 Ma old oceanic crust, in a magmatic event that took place around 75-80 Ma, which is compatible with the age proposed for the LEPLAC data tectonic features.

One must still consider the possible relation of some of these NW-SE seamount alignments to the South-American plate drift over hotspots, as suggested by *O'Connor and le Roex* [1992] *and Bryan and Cherkis* [1995]. Actually, for the Brazilian Continental Margin tectonic setting, the activity of mantle plumes can be reconciled with the effects of changes on the seafloor spreading rates and on the relative movement of lithospheric plates. A possible interaction of hotspot thermal anomalies with fracture zones, producing the well-known *en échelon* structures, with alternate NW-SE and E-W alignments, is an issue that deserves further investigation.

7. CONCLUSIONS

The LEPLAC seismic lines collected over the central-south portion and the northern edge of Pernambuco Plateau, off NE Brazil, have a rifted continental crust structure. The southern line suggests that the crustal extension was probably controlled by detachment faulting. Gravity modelling of the same line indicates that the oceanic crust spread out from the marginal scarpment of the plateau. Up to the north, however, this correlation was not so evident, since transitional crust appears to have stretched out beyond the edge of the Pernambuco Plateau, occupying some of the continental rise. Common seismic features found at the volcanic rifted margin to the south (*i.e.*, Sergipe-Alagoas basin), like SDRs wedges, were not observed in this region.

Located in a transform margin tectonic setting, the ocean-continent transition of the Potiguar basin was covered by a single seismic profile, which was collected in a direction that was near-parallel to a fracture zone. In this course, instead of a transform-related structure, the crustal transition seems to occur in a normal passive margin style, with rifting and extension as the driving processes. Again, stretched continental crust lies on deep continental rise environments. Seimic features related to strike-slip tectonics were also observed in this area.

Fracture zones/lineaments systems along the northeastern continental margin off Brazil were very well imaged by the LEPLAC seismic data. With the support of GEOSAT data, we attempted to review the location of these structures and to discuss their termomechanical behaviour. We observed remarkable seismic evidence relating to magmatic and tectonic reactivation in the vicinity of fracture zones. This evidence suggests that the recurrent seimicity at FZs can comprise both compressive stresses – with uplifting of the crust - and extensional stresses – with volcanism and normal faulting.

In the other hand, we observed flexural bending on the oceanic lithosphere adjacent to the Ascension FZ that can be ascribed to a "strong" fracture zone behavior, which seems to fit the "locked fault" model. Hence, we admit that thermomechanical variations along fracture zones could led to an alternation of these inactive "strong" segments with seismically active segments.

Other flexural effects on the lithosphere were observed on the vicinity of Fernando de Noronha Ridge. Over there, volcanic overloading led to a flexural isostatic compensation that appears to have caused regional tectonic reactivations on the oceanic crust. To the south, nearby a NW-SE seamount alignment, we correlate some other deformation of the oceanic crust with changes in spreading directions that possibly took place at the Upper Cretaceous.

Acknowledgments. We thank Alexandre Tagore M. de Albuquerque, coordinator of the Executive Committee for the LEPLAC, and Marcus A. Gorini and Dieter Carl E. H. Muehe, delegates of the Brazilian Scientific Comunity in this committee, for the permission of publishing LEPLAC data. We are specially grateful to our PETROBRAS colleagues Pedro V. Zalán, Luiz A. P. Gamboa, Luiz Fernando S. Braga, Saulo F. Santos, Webster U. Mohriak, Renato M. D. de Matos and Marcelo Basseto, who gave us many helpful suggestions throughout this study. This work was supported by PETROBRAS.

8. REFERENCES

Alves, E.C., and M.P.A. Costa, Interpretação sismo-estratigráfica da porção norte do Platô de Pernambuco e suas possíveis correlações com a Bacia de Pernambuco-Paraíba. In: *XXXIV Cong. Bras. Geol.,* Goiânia, GO, Annals..., SBG, v.1, p. 286-297,1986.

Alves, E.C., M.P.A. Costa, and E.A. Costa, Evolução sedimentar Mesozóica-Cenozóica do Platô de Pernambuco e da área adjacente da Bacia Oceânica do Brasil. In: *III Cong. Int. Soc. Bras. Geof.*, Rio de Janeiro, RJ, Annals..., SBGf, v. 2, p. 1249-1253, 1993.

Asmus, H.E., and J.C. Carvalho, Condicionamento tectônico da sedimentação nas bacias marginais do nordeste do Brasil (Sergipe-Alagoas e Pernambuco-Paraíba). In: *Aspectos Estruturais da Margem Continental Leste e Sudeste do Brasil* (Final Report), Rio de Janeiro: PETROBRÁS/CENPES/ DINTEP (Projeto REMAC Series), p. 7-24, v. 4, 1978.

Asmus, H.E., and W. Guazelli, Descrição sumária das estruturas da Margem Continental Brasileira e das áreas oceânicas e continentais adjacentes – Hipóteses sobre o tectonismo causador, e implicações para os prognósticos do potencial de recursos minerais. In: Asmus, H.E. (ed.). *Estruturas e Tectonismo da Margem Continental Brasileira, e suas Implicações nos Processos Sedimentares e na Avaliação do Potencial de Recursos Minerais* (Final Report). Rio de

Janeiro: PETROBRÁS/CENPES/DINTEP (Projeto REMAC Series), v. 9, p. 187-269, 1981.

Austin, J.A., Jr., P.L. Stoffa, J.D. Phillips, J. Oh, D.S. Sawyer, G.M. Purdy, E. Reiter, and J. Makris, Crustal structure of the southeast Georgia embayment-Carolina trough: Pre-liminary results of a composite seismic image of a continental suture (?) and a volcanic passive margin. *Geology*, v.18, n. 6, p.1023-1027, 1990.

Basile, C., and J. Mascle, Block faulting in oceanic crust: example of intraplate deformation in the Equatorial Atlantic. *Marine Geology*, v. 95, p. 45-50, 1990.

Batiza, R., Abundances, distribution and sizes of volcanoes in the Pacific Ocean and implications for the origin of non-hotspot volcanoes. *Earth and Planet. Sci. Letters*, v. 60, p. 195-206, 1982.

Bergman, E.A., and S.C. Solomon, On the Strengh of Oceanic Fracture Zones and Their Influence on the Intraplate Stress Field. *Journ. Geophys. Res.*, v. 97, n. B11, p. 15,365-15,377, 1992.

Bonatti, E., Vertical tectonism in oceanic fracture zones. *Earth and Planet. Sci. Letters*, v. 37, p. 369-379, 1978.

Bonatti, E., M. Ligi, L. Gasperini, A. Peyve, Y. Raznitsin, and Y.J. Chen, Transform migration and vertical tectonics at the Romanche fracture zone, equatorial Atlantic. *Journ. Geophys. Res.* , v. 99, n. B11, p. 21,779-21,802, 1994.

Bryan, G.M., N. Kumar, and P.J. Castro, The North Brazilian Ridge and the Extension of Equatorial Fracture Zones into the Continent. In: *XXVI Cong. Bras. Geol*, São Paulo, SP, Annals..., SBG, v.2, p. 133-144, 1972.

Bryan, P.C., and N.Z. Cherkis, The Bahia Seamounts: test of a hotspot model and a preliminary South American Late Cretaceous to Tertiary Apparent Polar Wander Path. *Tectonophysics*, v. 241, p. 317-340. 1995.

Carvalho, J.C., and O. Francisconi, Análise de depocentros, e suas associações com a geomorfologia e a estrutura da Margem Cointinental Brasileira. In: Asmus, H.E. (ed.). *Estruturas e Tectonismo da Margem Continental Brasileira, e suas Implicações nos Processos Sedimentares e na Avaliação do Potencial de Recursos Minerais* (Final Report). Rio de Janeiro: PETROBRÁS/CENPES/DINTEP, p.171-186 (Projeto REMAC Series), v. 9, 1981.

Castro Jr., A.C.M., Structural evolution of the Sergipe-Alagoas basin, Brazil. *Rice University*, Houston, Texas, 183 p. (Ph.D. thesis), 1988.

Chaves, H.A.F., S.G. Zembruscki, and A.M.C França, Introdução. In: Chaves, H.A.F. (ed.). *Geomorfologia da Margem Continental Brasileira e das Áreas Oceânicas Adjacentes* (Final Report). Rio de Janeiro: PETROBRÁS/ CENPES/ DINTEP (Projeto REMAC Series), p. 7-24, v. 7, 1979.

de Charpal, O., P. Guennoc, L. Montandert, and D. Roberts, Rifting, crustal attenuation and subsidence in the Bay of Biscay. *Nature*, v. 275, p. 706-711, 1978.

Cherkis, N.Z., D.A. Chayes, and L.C. Costa, The bathymetry and distribution of the Bahia Seamounts, Brazil Basin. *Marine Geology,* Amsterdam, v. 103, p. 335-347, 1992.

Cordani, U.G., Idade do vulcanismo do Oceano Atlântico Sul. *Bol. Inst. Geol. Astron.*, São Paulo, (1): p. 9-76, 1970.

Costa, M.P.A., and M.C.A. Maia, Prováveis contatos de crostas continental/oceânica no Platô de Pernambuco, nordeste do Brasil. In: *XXXIV Cong. Bras. Geol*, Goiânia, GO, Annals..., SBG, v.2, p. 810-823, 1986.

Costa, M.P.A., E.C. Alves, and E.A. Costa, Sismo-estratigrafia da porção da Bacia Oceânica do Brasil adjacente ao Platô de Pernambuco e sua relações com o Platô e a Bacia de Pernambuco-Paraíba. In: *II Cong. Int. Soc. Bras. Geof.*, Salvador, Annals..., SBGf, v. 2, p. 650-655, 1991.

Damuth, J.E., and D.E. Hayes, Echo character of the east Brazilian continental margin and its relationship to sedimentary processes. *Marine Geology*, Amsterdam, v. 24, p. 73-95, 1977.

DeLong, S.E, J.F. Dewey, and P.J. Fox, Displacement hystory of oceanic fracture zones. *Geology*, v. 5, p. 199-202, 1977.

Duncan, R.A., Hotspots in the Southern Oceans - an Absolute Frame of Reference for Motion of the Gondwana Continents. *Tectonophysics*, v. 74, p. 29-42, 1981.

Edwards, R.A., R.B. Whitmarsh, and R.A. Scrutton, The crustal structure across the transform continental margin off Ghana, eastern equatorial Atlantic. *Journ. Geophys. Res.* , v. 102, n. B1, p. 747-772, 1997.

Emery, K. O., E. Uchupi, J. Phillips, C. Bowin, and J. Mascle, Continental Margin Off Western Africa: Angola to Sierra Leone. *AAPG Bull.*, v. 59, n. 12, p. 2209-2262, 1975.

Epp D., Possible Perturbations to Hotspot Traces and Implications for the Origin and Structure of the Line Islands. *Journ. Geophys. Res.*, v. 89, n. B13, p. 11,273 - 11,286, 1984.

Fainstein, R., and J.D. Milliman, Structure and Origin of Three Continental-Margin Plateaus, Northeastern Brazil. *AAPG Bull.*, v. 63, n. 2, p. 218-238, 1979.

Feijó, F.J., Bacia de Pernambuco-Paraíba. *Boletim de Geociências da PETROBRÁS*, Rio de Janeiro, v. 8, n.º 1, p. 143-148, 1994.

França, A.M.C., Geomorfologia da Margem Continental Leste Brasileira e da Bacia Oceânica Adjacente. *In:* Chaves, H.A.F. (ed.). *Geomorfologia da Margem Continental Brasileira e das Áreas Oceânicas Adjacentes* (Final Report). Rio de Janeiro: PETROBRÁS/CENPES/DINTEP, p. 89-128 (Projeto REMAC Series), v. 7, 1979.

Gomes, P.O., O Projeto LEPLAC na Margem Continental Nordeste: Imagens Sísmicas de Flexura Litosférica junto a Zonas de Fratura. In: *V Cong. Int. Soc. Bras. Geof.*, São Paulo, SP, Annals..., SBGf, v. 1, p. 1231-1236, 1997.

Gomes, P.O., Vulcanismo, Tectonismo, Sedimentação e Processos Erosivos no Segmento Nordeste da Margem Continental Brasileira. Rio de Janeiro, M.S. thesis, *Univ. do Estado do Rio de Janeiro (UERJ)*, in prep.

Gomes, P.O., and B.S. Gomes, Transição crustal no Platô de Pernambuco: uma caracterização geofísica a partir de dados do Projeto LEPLAC. In: *V Cong. Int. Soc. Bras. Geof.*, São Paulo, Annals..., SBGf, v. 1, p. 1231-1236, 1997.

Gorini, M.A., The Tectonic Fabric of the Equatorial Atlantic and Adjoining Continental Margins: Gulf of Guinea to Northeastern Brazil. In: Asmus, H.E. (ed.). *Estruturas e Tectonismo da Margem Continental Brasileira, e suas Implicações nos Processos Sedimentares e na Avaliação do*

Potencial de Recursos Minerais (Final Report). Rio de Janeiro: PETROBRÁS/CENPES/DINTEP (Projeto REMAC Series), v. 9, p. 11-116, 1981.

Gorini, M.A., and J.C. Carvalho, Geologia da margem continental inferior brasileira e do fundo oceânico adjacente. In: DNPM (Brasil). *Geologia do Brasil.* Brasília, p. 473-489, 1984.

Gorini, M.A., J.E. Damuth, and G.M. Bryan, The Fernando de Noronha Ridge and its relationship to equatorial Atlantic fracture zones. *Geol. Soc. America, Abstracts with Programs*, Boulder, v. 6, no. 7, p. 762, 1974.

Gorini, M.A., and H.S. Fleming, Projeto CENTRATLAN - Marinha do Brasil e marinha Norte-Americana: Estudos geológicos do Atlântico Sul . In: *XXXII Cong. Bras. Geol*, Salvador, BA, Res. Com., SBG, v.2, p. 96, 1982.

Guazelli, W., and J.C. Carvalho, Estruturas da margem continental leste brasileira e das áreas oceânicas e continentais, adjacentes. In: Asmus, H.E. (ed.). *Estruturas e Tectonismo da Margem Continental Brasileira, e suas Implicações nos Processos Sedimentares e na Avaliação do Potencial de Recursos Minerais* (Final Report). Rio de Janeiro: PETROBRÁS/ CENPES (Projeto REMAC Series), v. 9, p. 117-144, 1981.

Guimarães, P.T.M., Basin analysis and structural development of the Sergipe-Alagoas Basin, Brazil. *The University of Texas at Austin*, 171 p. (Ph.D. thesis), 1988.

Hoffmann, H.J. and T.J. Reston, Nature of The S reflector beneath the Galicia banks rifted margin: Preliminary results from prestack depth migration. *Geology*, v.20, p.1091-1094, 1992.

Keen C.E., and D.P. Potter, The transition from a volcanic to a nonvolcanic rifted margin off eastern Canada. *Tectonics*, v. 14, n. 2, p. 359-371, 1995.

Kowsmann, R.O., and M.P.A. Costa, Estratigrafia sísmica do Platô de Pernambuco. *Rev. Bras. Geoc.*, São Paulo, v. 6, n.º 2, p. 95-101, 1976.

Kruse, S.E., M.C. McCarthy, M.R. Brudzinski, and M.E. Ranieri, Evolution and strengh of Pacific fracture zones. *Journ. Geophys. Res.* , v. 101, n. B6, p. 13,731-13,740, 1996.

Lehner, P., and G. Bakker, Strike-slip tectonics in a Equatorial Fracture Zone (Romanche Fracture). In: Bally,A.W. (ed.). *Seismic expression of structural styles - a picture and work atlas*. AAPG, v.3, p.4.2/25-4.2/29, 1983.

Lister, G.S., M.A. Etheridge, and P.A. Symonds, Detachment faulting and the evolution of passive continental margins. *Geology*, v.14, p. 246-250, 1986.

Lowrie, A., C. Smoot, and R. Batiza, Are oceanic fracture zones locked and strong or weak?: New evidence for volcanic activity and weakness. *Geology*, v.14, p.242-245, 1986.

Mascle, J., and E. Blarez, Evidence for transform margin evolution from the Côte d'Ivoire-Ghana continental margin. *Nature*, v. 326, p. 378-381, 1987.

Mabesoone, J.M., and M.M. Alheiros, Origem da bacia sedimentar costeira Pernambuco-Paraíba. *Rev. Bras. Geoc.*, São Paulo, v. 18, n.º 4, p. 476-482, 1988.

McCarthy, M.C., S.E. Kruse, M.R. Brudzinski, and M.E. Ranieri, Changes in plate motions and the shape of Pacific fracture zones. *Journ. Geophys. Res.*, v. 101, n. B6, p. 13,715-13,730, 1996.

McNutt, M., K. Fischer, S. Kruse, and J. Natland, The Origin of the Marquesas Fracture Zone Ridge and its Implications for the Nature of Hotspots. *Earth and Planet. Sci. Letters*, v. 91, p. 381-393, 1989.

Mello, S.L.M., and M.S. Dias, Magnetoestratigrafia da crosta oceânica entre as zonas de fratura de Ascensão e Bode Verde. *Rev. Bras. Geof.*, Rio de Janeiro, v. 14 (3), p. 237-252, 1996.

Mohriak, W.U., J.H.L. Rabelo, R.D. Matos, and M.C. Barros, Deep seismic reflection profiling of sedimentary basins offshore Brazil: geological objectives and preliminary results in the Sergipe Basin. *J. Geodynamics*, v. 20, No. 4, p. 515-539, 1995.

Mutter, J. C., Seaward-dipping reflectors and the continent-ocean boundary at passive continental margins. *Tectonophysics*, Amsterdam, v. 114, p. 117-131, 1985.

O' Connor, J.M., and R.A. Duncan, Evolution of the Walvis Ridge-Rio Grande Rise Hot Spot System: Implications for African and South American Plate Motions Over Plumes. *Journ. Geophys. Res.* , Washington, v. 95, n. B11, p. 17,475-17.50, 1990.

O' Connor, J.M., and A.P. Le Roex, South Atlantic hot spot-plume systems: 1. Distribution of volcanism in time and space. *Earth and Planet. Sci. Letters*, v. 113, p. 343-364, 1992.

Palma, J.J.C., Mapeamento morfoestrutural do segmento ocidental da Zona de Fratura de Ascensão e porções adjacentes da crista da Cordilheira. Rio de Janeiro, M.S. thesis, *Univ. Fed. do Rio de Janeiro (UFRJ)*, 150 p., 1989.

Palma, J.J.C., J. Brozena, S.L.M. Mello, and J.C. Carvalho, Mapeamento morfoestrutural preliminar da Zona de Fratura de Ascensão desde a Cordilheira Mesoatlântica até a margem continental nordeste brasileira. *In: XXXIII Cong. Bras. Geol.*, Rio de Janeiro, RJ, Annals..., SBG, v. 4, p. 1643-1654, 1984.

Peirce, C., R.A. Whitmarsh, R.A. Scrutton, B. Pontoise, F. Sage, and J. Mascle, Côte d'Ivoire-Ghana margin: seismic imaging of passive rifted crust adjacent to a transform continental margin. Geophys. J. Int., v. 125, p. 781-795, 1996.

Pockalny, R.A., G. Pascal, and R. Buck, Oceanic transverse ridges: A flexural response to fracture-zone-normal extension. *Geology*, v.24, no. 1, p. 71-74, 1996.

Reston, T.J., C.M. Krawczyk, and D. Klaeschen, The S reflector west of Galicia (Spain): Evidence from prestack depth migration for detachment faulting during continental breakup. *Journ. Geophys. Res.* , v. 101, n. B4, p. 8075-8091, 1996.

Rezende, W.M., J.O. Ferradaes, P.J.M. Castro, and R. Porto, The South Atlantic ocean transform fault system and related features at the brazilian continental margins. In: *Offshore Technology Conference*, Texas. Annals..., v.1, p. 85-92, 1977.

Salisbury, M.H., and C.E. Keen, Listric faults imaged in oceanic crust. *Geology*, v.21, p.117-120, 1993.

Sandwell, D., and G. Schubert, Lithospheric Flexure at Fracture Zones. *Journ. Geophys. Res.*, v. 87, n. B6, p. 4675- 4667, 1982.

Schobbenhaus, C., D.A. Campos, G.R. Derze, and H.E. Asmus, Mapa geológico do Brasil e da área oceânica adjacente incluindo depósitos minerais. *DNPM* (Brasil), escala 1:2.500.000, 1981.

Sheridan, R.E., D.L. Musser, L. Glover, M. Talwani, J.I. Ewing, W.S. Holbrock, G.M. Purdy, R. Hawman, and S. Smithson, Deep seismic reflection data of EDGE U.S. mid–Atlantic continental margin experiment: Implications for Appalachian sutures and Mesozoic rifting and magmatic underplating. *Geology*, v.21, n. 6, p. 563-567, 1993.

Sibuet, J.C., and J. Mascle, Plate Kinematics Implications of Atlantic Equatorial Fracture Zone Trends. *Journ. Geophys. Res.*, v. 83, n. B7, p. 3401-3421, 1978.

Silveira, D.P., Interessantes Estruturas Geológicas Interpretadas em Linhas Sísmicas do "LEPLAC Equatorial". In: *III Cong. Int. Soc. Bras. Geof.*, Rio de Janeiro, RJ, Annals..., SBGf, v. 2, p. 1231-1236, 1993.

Silveira, D.P., B.S. Gomes, C.R. Suarez, and P.O. Gomes, Projeto LEPLAC: Interpretação Integrada dos Dados Geofísicos do "LEPLAC EQUATORIAL". In: *XXXVIII Cong. Bras. Geol.,* Baln. Camboriú, Annals..., SBG, v. 2, p. 35-37, 1994.

Sperle, M., S.L.M. Mello, and M.P.A. Costa, Modelagem gravimétrica e magnética do limite de crostas continental/ oceânica no Platô do Rio Grande do Norte. In: *I Cong. Int. Soc. Bras. Geof.*, Rio de Janeiro. Annals..., SBGf, v. 2, p. 863-867, 1989.

Sperle, M., S.L.M. Mello, and L.R. Gomes, Análise isostática na região dos Montes Submarinos de Pernambuco. In: *II Cong. Int. Soc. Bras. Geof.,* Salvador. Annals..., SBGf, v. 2, p. 644-649, 1991.

Summerhayes, C.P., R. Fainstein, and J.P. Ellis, Continental margin off Sergipe and Alagoas, Northeastern Brazil: a reconaissance geophysical study of morphology and structure. *Marine Geology*, Amsterdam, v. 20, n. 4, p. 819-828, 1976.

Thomaz Filho, A., Datação radiométrica das rochas ígneas básicas da Bacia do Ceará. Rio de Janeiro: *PETROBRÁS/CENPES* (Int. Rep.), 45 p., 1983.

Turcotte, D.L., The state of stress at passive continental margins. In: *Dynamics of Passive Margins. Geodyn. Ser.*, vol. 6, AGU, Washington, D. C., p. 141-146, 1982.

Turcotte, D.L., and G. Schubert, *Geodynamics - Applications of Continuum Physics to Geological Problems*. John Wiley & Sons, Inc., U.S.A., 450 p., 1982.

Vogt, P.R., Volcano spacing, fractures, and thickness of the lithosphere. *Earth and Planet. Sci. Letters*, v. 21, p. 235-252, 1974.

Watts, A.B., An Analysis os Isostasy in the World's Oceans – 1. Hawaiian-Emperor Seamount Chain. *Journ. Geophys. Res.* , v. 83, n. B12, p. 5989-6004, 1978.

Watts, A.B., U.S. Ten Brink, P. Buhl, and T.M. Brocher, A multichannel seismic study of lithospheric flexure across the Hawaiian-Emperor seamount chain. *Nature*, v. 315, p. 105-111, 1985.

White, R.S., G.D. Spence, S.R. Fowler, D.P. McKenzie, G.K. Westbrook, and A.N. Bowen, Magmatism at rifted continental margins. *Nature*, v. 330, p. 439-444, 1987.

[1] PETROBRAS/E&P/GEREX/GECON Av. República do Chile, 65, sala 1002-A, Rio de Janeiro, RJ, Brazil, 20035-900 (pauloog@ep.petrobras.com.br)

[2] LAGEMAR-UFF Av. Litorânea S/N – Inst. de Geociências-4º andar, Niterói, RJ, Brazil, 24210-340 (jorge@igeo.uff.br)

Tectonic Evolution of the Rift Basins in the Northeastern Brazilian Region

Webster U. Mohriak; Marcelo Bassetto; Ines S. Vieira

Exploration & Production, PETROBRAS , Rio de Janeiro, Brazil

The transition from onshore failed rifts to offshore sedimentary basins along divergent continental margins is discussed on the basis of a regional, multidisciplinary integration of deep seismic reflection profiling, potential fied methods, geological data, and tectonic analysis. The following themes are addressed: a) the geologic evolution of the onshore and offshore rift systems of the Brazilian northeastern margin; b) the potential field methods response to the deep crustal structures; c) the seismic expression of major structural features in the rifts and within the continental and oceanic crusts; d) a possible geodynamic model for the evolution of the rift system; and e) analogies with a number of failed rifts and passive margin systems in the North Atlantic. The sedimentary basins in northeastern Brazil include a series of asymmetric grabens, such as the onshore Recôncavo-Tucano-Jatobá rift system (RTJ) and the offshore Jacuípe-Sergipe-Alagoas basins (JSA). Pre-rift sediments include Paleozoic to Jurassic/Early Cretaceous sediments deposited above a basement that includes Archean rocks to Late Proterozoic metasediments. The main rift phase (Neocomian to Barremian) terminated in the onshore rifts with fluvial deposits above a major regional unconformity. No further sedimentation is observed in the Recôncavo and Tucano basins, in a marked contrast to the geodynamic evolution of the Sergipe Basin, which is characterized by renewed phases of basement-involved faulting from Aptian to Early Albian, followed by a thermal phase of subsidence. The overall picture of two branches of a rift system, with different geodynamic evolution following the inception of oceanic crust, may be associated with a regional lithospheric extension during the Neocomian, first distributed over a wide region, and subsequently, focussing along a deeper mantle weak zone, local of a later plate rupture.

INTRODUCTION

In the past two decades, new improvements in seismic reflection acquisition and more sophisticated processing techniques have resulted in better imaging of major rift structures in the deeper portions of sedimentary basins. The present-day trend of advancing petroleum exploration towards new frontiers has resulted in the acquisition of seismic profiles along regional lines extending from the platform towards the oceanic crust, both in the Brazilian [*Mohriak and Latgé,* 1991; *Mohriak et al.,* 1993, *Mohriak et al.,* 1995a] and in the African margin [e.g., *Rosendahl et al.*, 1991; *Wannesson et al.,* 1991; *Meyers et al.,* 1996]. The program of deep seismic reflection profiles in northeastern

Atlantic Rifts and Continental Margins
Geophysical Monograph 115

Brazil include also a regional profile across the onshore Central Tucano Basin [*Mohriak et al.,* 1997].

This paper presents an integration of the large potential field dataset, both onshore and offshore the northeastern region, and the analysis of a composite regional transect based on gravity and deep seismic reflection profiles. A reconnaissance gravity profile, acquired in the onshore region by Petrobras in 1995, extends from the São Francisco (or Paramirim) craton towards the southern portion of the Sergipe sub-basin. This profile partly coincides with a regional deep seismic profile that crosses the central Tucano Basin. The transect was extended towards the deep-water portions of the Sergipe-Alagoas Basin along another deep seismic profile acquired by Petrobras in the early 1990's. This resulted in a unique transect across the northeastern margin of Brazil, reaching the oceanic crust in the deep-water regions. Based on the interpretation and modeling of the gravity and seismic data, we propose a new model for the crustal architecture of this segment of the margin, and suggest that the development of the onshore and offshore rifts may be associated with two zones of crustal necking characterized by different amounts of mantle uplift, and that the architecture of the rifts may be related to the depth of crustal necking.

The main objective of this paper is to analyze the deep structure of the rift basins in northeastern Brazil and discuss a possible geodynamic model of evolution. This geodynamic model bears some important analogies with other sedimentary basins in the North Atlantic, particularly the Jeanne d'Arc Basin in the Canadian passive continental margin, and the Newark-Baltimore Canyon Trough, offshore the United States. These geodynamic models bear direct implication on tectonic studies, basin analysis and petroleum exploration [e.g., *Mohriak et al.,* 1995b].

REGIONAL TECTONIC FRAMEWORK

Figure 1 shows the location map of the Brazilian sedimentary basins and major regional arches. The Eastern Brazilian margin sedimentary basins extend from the southernmost territorial waters (Pelotas Basin) towards the northeastern region (Pernambuco-Paraíba Basin), which is also characterized by a system of onshore rifts (Recôncavo, Tucano and Jatobá basins). Figure 2 shows a simplified tectonic map of the NE Brazilian region with the main sedimentary basins, both onshore and offshore. The sedimentary basins include asymmetric grabens that constitute the onshore Recôncavo-Tucano-Jatobá rift system (RTJ) and the offshore Jacuípe-Sergipe-Alagoas basins (JSA). These basins are separated by basement highs and transfer faults associated with opening of a rift system in a NW direction, oblique to the N-S general trend of the master faults [*Milani and Davison,* 1988]. The rift basins are crossed by a regional transect A-A' that will be described and analyzed in this paper.

Figure 3 shows a simplified tectono-stratigraphic chart for the Tucano and the Sergipe- Alagoas basins. During the Precambrian, the northeastern region of Brazil was affected by lithospheric convergence associated with the Brazilian/ Pan-African orogeny, resulting in compressional tectonics [*Inda et al.,* 1984; *Davison and Santos,* 1989, *D'el Rey Silva and McClay,* 1995]. During the Paleozoic, the northeastern region was affected by a relatively small but regional intracratonic subsidence during the Paleozoic, as indicated by sedimentary sequences identified along the borders of the Mesozoic rifts [*Machado Jr. et al.,* 1990]. The Paleozoic sequence is represented by sparse outcrops of several formations associated with diverse environments of deposition, ranging from continental siliciclastics (Lower Paleozoic), marine shales (Devonian), glacial deposits (Carboniferous) and desertic (Permian) [*Inda et al.,* 1984].

Pre-rift sediments in these basins include Paleozoic to Jurassic/Early Cretaceous sediments deposited above a basement that ranges from Archean rocks to Late Proterozoic metasediments [*Magnavita,* 1992]. The main rift phase in northeastern Brazil was implanted during the Early Cretaceous (Neocomian, about 140-135 Ma). Major faults were formed along the eastern border of the Recôncavo and Tucano basins, resulting in deposition of conglomerates of the Salvador Fm., concomitant with deposition of fluvial and lacustrine sediments of the Candeias Fm. [*Bueno,* 1994]. The equivalent sedimentation in the Sergipe Basin corresponds to lacustrine sediments included in the Barra de Itiúba Fm. [*Chagas et al.,* 1993].

Subsidence rates decreased towards the Barremian, when the previously formed lake basin was filled by prograding delta fans (Ilhas Group). The rift phase terminated in the onshore rifts with fluvial deposits of the Massacará Group (figure 3). During the Aptian, an unconformable post-rift phase of fluvial conglomerates (Marizal Fm.) covered most of the basin. Figure 4 shows the breakup unconformity in the Tucano Basin, which is characterized in the outcrop by sub-horizontal coarse siliclastic rocks overlying an angular unconformity that separates the Marizal Fm. from the tilted Neocomian rocks of the Ilhas Group. Subsequently to the deposition of the Marizal Fm., no further sedimentation is observed in the RTJ rift system. On the contrary, the offshore rifts were still actively subsiding during the Aptian, particularly in the Alagoas Basin, which is characterized by very thick syn-rift sedimentary troughs [*Chagas et al.,* 1993]. The Sergipe Basin is also characterized by renewed phases of basement-involved faulting and syn-rift sedimentation in the Aptian (Maceió Fm.) and possibly, even in Early Albian times.

The onshore and offshore rift basins in northeastern Brazil have similar lithostratigraphic relationships, including pre-rift and rift sequences in the Sergipe-Alagoas Basin. However, they differ markedly for the post-Aptian sequences, particularly because of the occurrence of evaporites in the

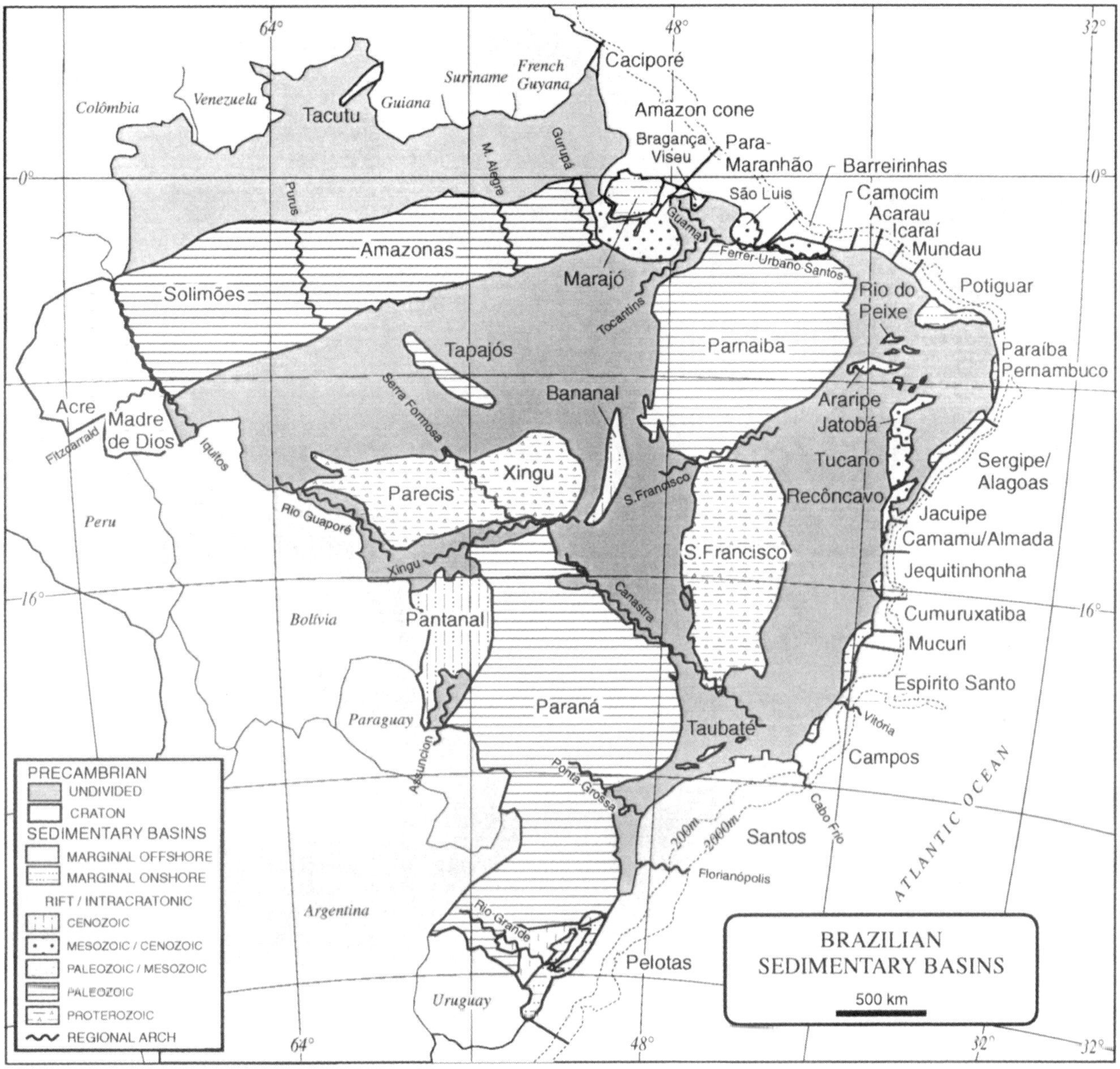

Figure 1: Regional location map of the Brazilian sedimentary basins.

offshore basins (figure 3). The post-Aptian geological history of the onshore rift basins is rather different from the offshore rifts, because the latter evolved as passive continental margins after the inception of oceanic crust.

The breakup of Gondwana in the Mesozoic probably started in the southern portion of the South American plate, and propagated towards the north, begining in the Late Jurassic and probably culminating with a Late Cretaceous final separation of Africa and South America, along transform fault zones in the equatorial *margins [Rabinowitz and LaBreque,* 1979; *Ojeda,* 1982; *Matos,* 1992]. During the Neocomian rift phase, volcanism was voluminous in the southern margin of Brazil (Espírito Santo, Campos, Santos and Pelotas basins), but almost absent in the sedimentary basins along the northeastern margin, particularly in the Recôncavo-Tucano-Jatobá rift system [*Chang et al.,* 1992]. The post-rift phase of subsidence in the offshore basins (figure 3) includes a marine carbonate sequence (Riachuelo Fm.), transgressive sequences of predominantly pelitic rocks (Cotinguiba and Calumbi formations), and prograding se-

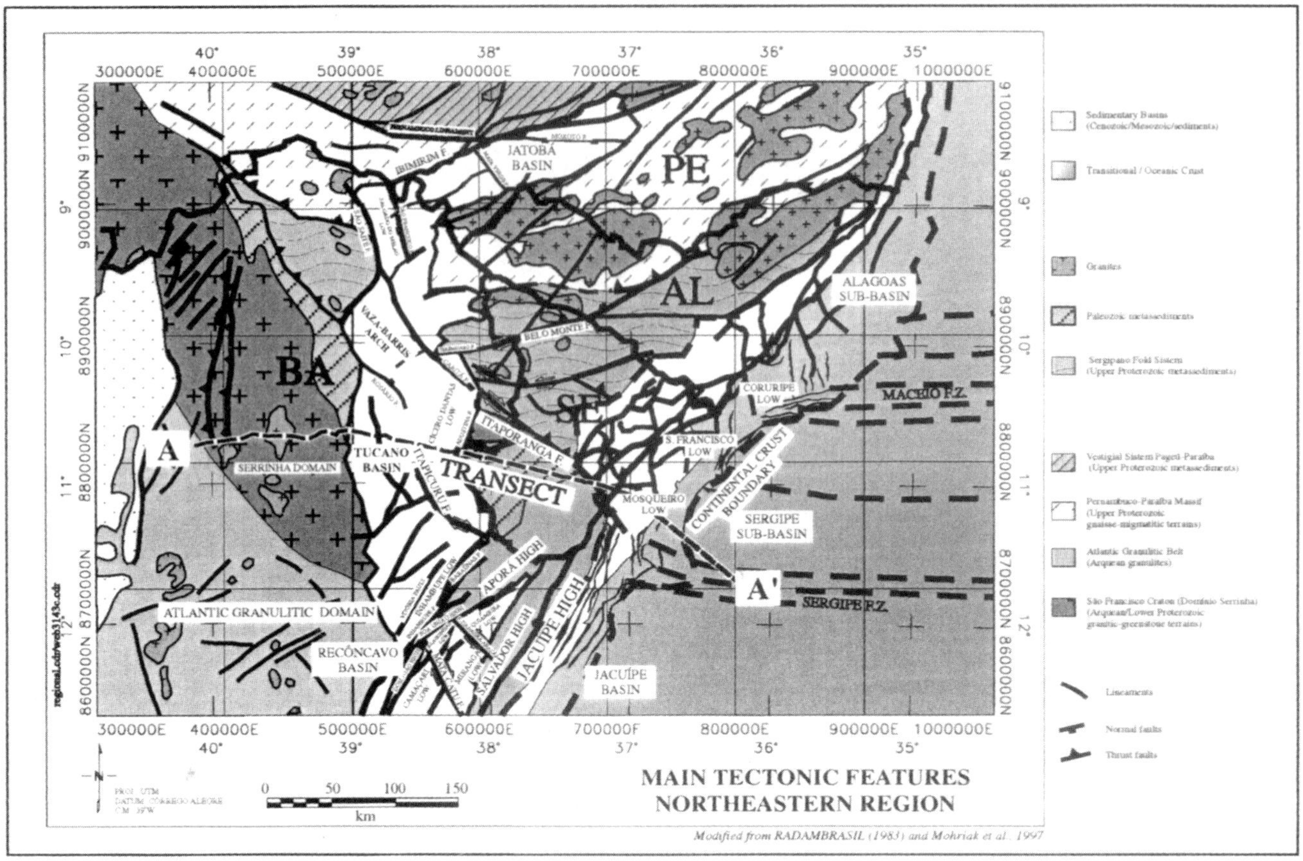

Figure 2-Regional, simplified geological map of the northeastern region of Brazil showing the Recôncavo-Tucano-Jatobá rift system and the main tectonic features offshore the Sergipe-Alagoas and Jacuípe basins. The transect A-A'corresponds to a reconnaisance gravity survey onland, extending from the city of Senhor do Bonfim(Bahia) towards the city of Aracaju (Sergipe). The gravity profile was integrated with gravity maps offshore along a regional deep seismic profile extending from the Mosqueiro Low in the Sergipe sub-basin towards the oceanic basin. Major east-west fracture zones in the oceanic crust are aligned with important structural features in the continental crust. The transect has correspondent seismic profiles both onshore (Tucano Basin) and offshore (Sergipe sub-basin).

quences of siliciclastic rocks associated with the Piaçabuçu Fm. [*Guimarães,* 1988; *Chagas et al.,* 1993; *Cainelli,* 1992].

RIFT ARCHITECTURE INTERPRETATION: TECTONIC PROBLEMS

The architecture of the Tucano Basin has been the subject of several papers based on field geology [e.g., *Radambrasil,* 1983] and seismic interpretation [e.g., *Magnavita and Cupertino,* 1988; *Magnavita,* 1992].

Early models of the rift geometry suggested a graben geometry with major faults controlling the sedimentary sequences, and incorrectly indicated both the western and the eastern border faults dipping to the east [*Radambrasil,* 1983]. Figure 5 shows an schematic geological section based on this interpretation, which would result in a slight tilt towards the west, with location of younger sequences depocenter near the western border fault. This in fact is the case for the North Tucano sub-basin, but is not applicable to the whole rift system. Geological and geophysical data accumulated during the exploratory activity in the rift do not corroborate this interpretation either for the Recôncavo or the Central Tucano sub-basins.

Following widely accepted concepts [e.g., *Rosendahl, 1987*], the rift geometry is presently interpreted to be controlled by master faults that change polarity along transfer zones. In the Tucano Basin, the southern and central segments are controlled by border faults in the eastern margin that dip towards the west, and the opposite margin corresponds to a flexural margin. The Vaza- Barris Arch (figure 2 and figure 6) corresponds to an accomodation zone where the rift master faults change polarity, and the north Tucano Basin is characterized by a depocenter in the western margin, controlled by a border fault that dips towards the east.

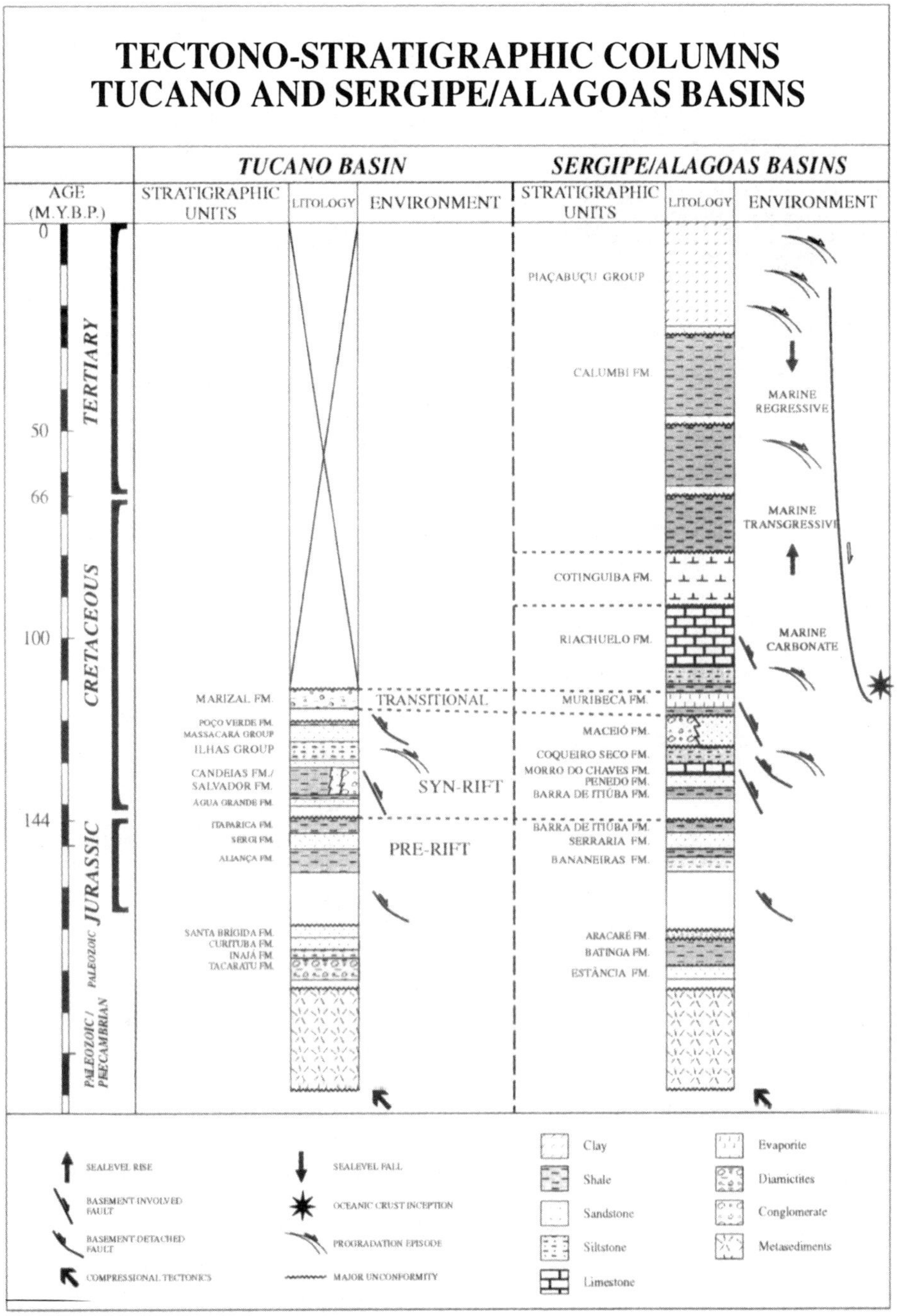

Figure 3: Correlation of the tectono-sedimentary columns of the Tucano and Sergipe-Alagoas basins. The pre-rift sedimentary sequences include Paleozoic Upper Jurassic/ Lower Cretaceous strata. The syn-rift sequences in the South Tucano basin include Candeias Fm., Ilhas Group, and and Massacará Group. The syn-rift sequence in the Sergipe Basin includes part of the Barra de Itiúba Fm., Morro do Chaves, Coqueiro Seco and Maceió formations. The transitional phase of sedimentation in the South Tucano Basin (Aptian) is associated with conglomerates, sandstones and siltstones of the Marizal Fm., corresponding to the Muribeca Fm. in the Sergipe Basin, which includes evaporites.

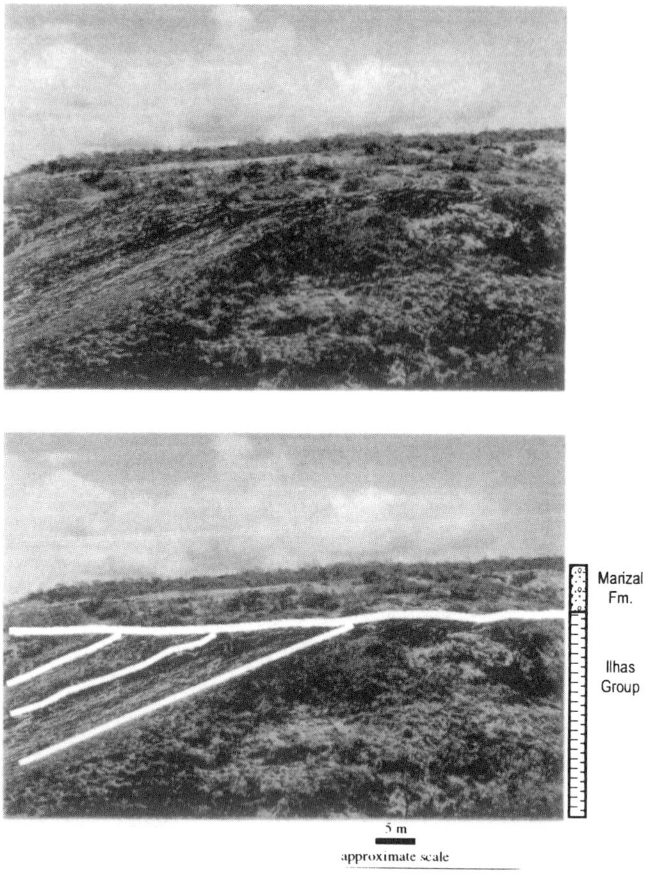

Figure 4: Photograph showing breakup angular unconformity as expressed in an outcrop in the Tucano Basin. Top: uninterpreted outcrop; bottom: with an interpretation.

The Central Tucano Basin deep structure is not imaged well in seismic profiles because the basement may reach depths beyond the resolution of normal industry profiles. Controversial interpretations on total sedimentary thickness have been discussed in the geological literature [e.g., *Aragão and Peraro,* 1994], but no borehole has ever penetrated through the depocenter. Although potential field interpretation methods (quantitative modeling of gravity and magnetic anomalies) are useful tools, they are characterized by non-unique solutions.

One of the few regional seismic profiles that have been acquired in this portion of the basin [figure 7, based on *Magnavita,* 1992] is characterized by a set of westward-dipping faults that progressively advance towards the basin depocenter (Cícero Dantas Low). This interpretation will be compared with the recently acquired regional deep seismic profile along the transect, which is located close to previous section. We discuss this transect by firstly analyzing the geological unities that outcrop on the surface, then we discuss the tectonic framework of the Central Tucano Basin in a continental scale, addressing geodynamic models, and subsequently, we discuss the interpretation of potential field and seismic data used to constrain the rift geometry in particular tectonic compartments.

Metasedimentary rocks of the Palmares Fm. (Upper Precambrian) outcrop eastwards of the Central Tucano Basin, and these are separated from the sedimentary rocks of the Sergipe-Alagoas Basin by the Atlantic granulitic belt of Archean age (figure 2 and figure 8). This belt separates the onshore from the offshore rifts. Figure 9 shows the reconstruction of the South Atlantic rift in the Early Cretaceous, with the location of three geological transects (A – A', B – B', C – C') shown in figure 10. Figure 11 shows the reconstruction of the South Atlantic in the Aptian, with inception of oceanic crust ridges.

The bifurcation of the rift system in two branches near Salvador (figure 2 and figure 9) may be associated with a RRR triple junction. The rift axis in the offshore region abruptly changes to a more northerly direction near this junction. The western branch of the rift formed aborted basins onshore, namely the Recôncavo-Tucano-Jatobá rift system (RTJ). The eastern branch evolved to form the continental passive margin basins along the South Atlantic margins (figure 11), including the Jacuípe-Sergipe-Alagoas rift system (JSA) along the northeastern Brazilian margin [*Asmus and Porto,* 1980, *Chang et al.,* 1992, *Matos,* 1992], and the Gabon-Rio Muni basins in Western Africa [*Teisserenc and Villermin,* 1989; *Wannesson et al.,* 1991; *Meyers et al.,* 1996]. Other models assume that the Sergipe-Alagoas basin was formed on a microplate (Sergipe Microplate) that rotated anticlockise during the breakup of Gondwana [*Milani et al.,* 1988].

Regional gravity anomaly maps indicate that the onshore basins in northeastern Brazil are characterized by large gravity lows, in a marked contrast to the gravity signature of the offshore rifts [*Ussami et al.,* 1986; *Milani,* 1987; *Milani and Davison,* 1988]. This has led to propositions of alternative, radically different geodynamic models to explain these basins (figure 10). *Ussami et al.* [1986] interpreted the RTJ and JSA rifts as resulting from extensional tectonics associated with a single low angle detachment fault connecting the depocenters of the onshore rifts with a mantle uplift underlying the offshore ones, following the simple shear model of *Wernicke* [1981; 1985]. *Castro Jr.* [1987], on the other hand, interpreted a system of double detachments with changes of polarity along transfer zones dividing the sub-basin (figure 10). *Davison* [1988] and *Milani and Davison* [1988] questioned these interpretations and disputed the evidences for lithospheric simple shear as the mechanism of formation of these rifts. They also pointed out that the gravity maps failed to show the positive Bouguer anomalies flanking the Recôn-

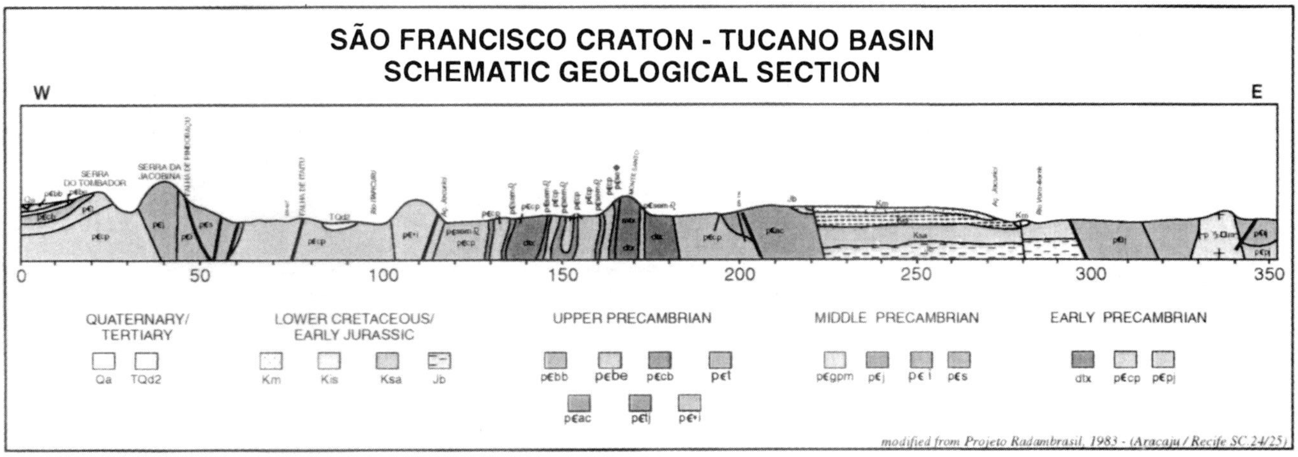

Figure 5: Schematic geological section from the São Francisco Craton, crossing the central Tucano Basin near the transect. It shows an early interpretation of a deep graben controlled by eastward-dipping faults [adapted from *Radambrasil, 1983*].

cavo and Tucano basins because they were oversimplified, and that more detailed gravity surveys indicated positive flanking anomalies up to 50 mGal, indicative of crustal thinning along the eastern margin of the rift.

Matos [1992] emphasized that the field geology and geophysical data showed no clear evidence to support low angle crustal detachments below the Tucano graben, associated with reactivation of precambrian thrusts as postulated by *Ussami et al.* [1986], but admitted that the gravity anomaly in the rifts should be indicative of a flat Moho topography, which should be linked to the mantle uplift offshore. *Chang et al.* [1992] addressed these problems and emphasized that these interpretations lacked deep seismic control and quantitative modelling of the gravity field anomalies. This work will show both the quantitative interpretation of the gravity data that resulted in a gravity model along the transect A-A' (figure 2), and the interpretation of the deep seismic data along portions of the transect, both in the onshore Central Tucano Basin, and in the offshore Sergipe Basin.

INTERPRETATION OF POTENTIAL FIELD DATA

Figure 12 shows a shaded-relief Bouguer anomaly map of the northeastern region of Brazil, and the location of the regional transect (A-A') that extends from the craton towards the oceanic crust in the Northeastern Brazilian margin, crossing the Central Tucano and the Sergipe basins. The regional Bouguer anomaly map was generated by integrating gravity data onshore and the extensive dataset offshore. The Bouguer correction was obtained considering a density of 2.67 g/cc for topographic highs onland and 2.20 g/cc for water, in the offshore region. The onshore portion of the transect is based on a 340 km long recconnaissance gravity profile along 65 stations, which was acquired by Petrobras

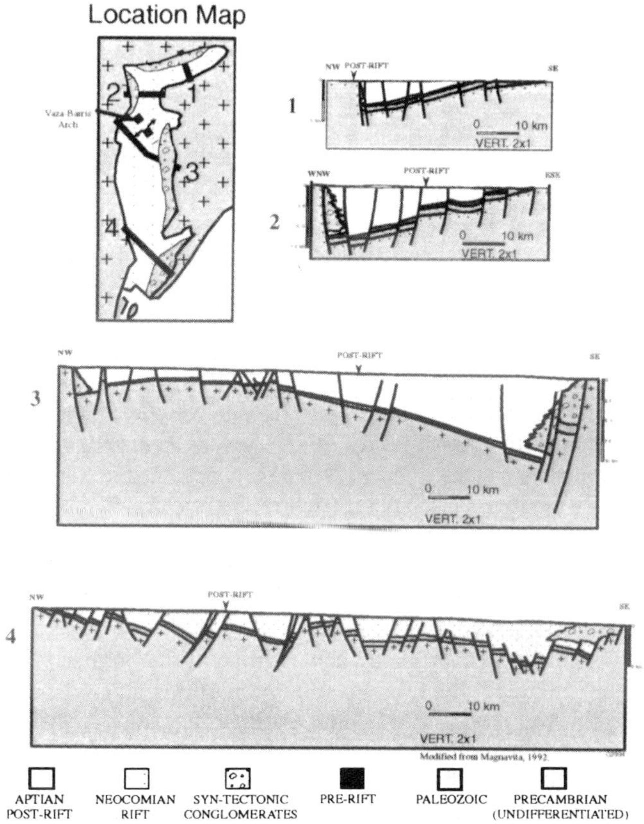

Figure 6: Schematic geological sections across the RTJ rift system, showing inversion of polarity of the rift master faults along the Vaza Barris Arch. Shaded basin infill in the location map indicate local rift depocenters [adapted from *Magnavita, 1992*].

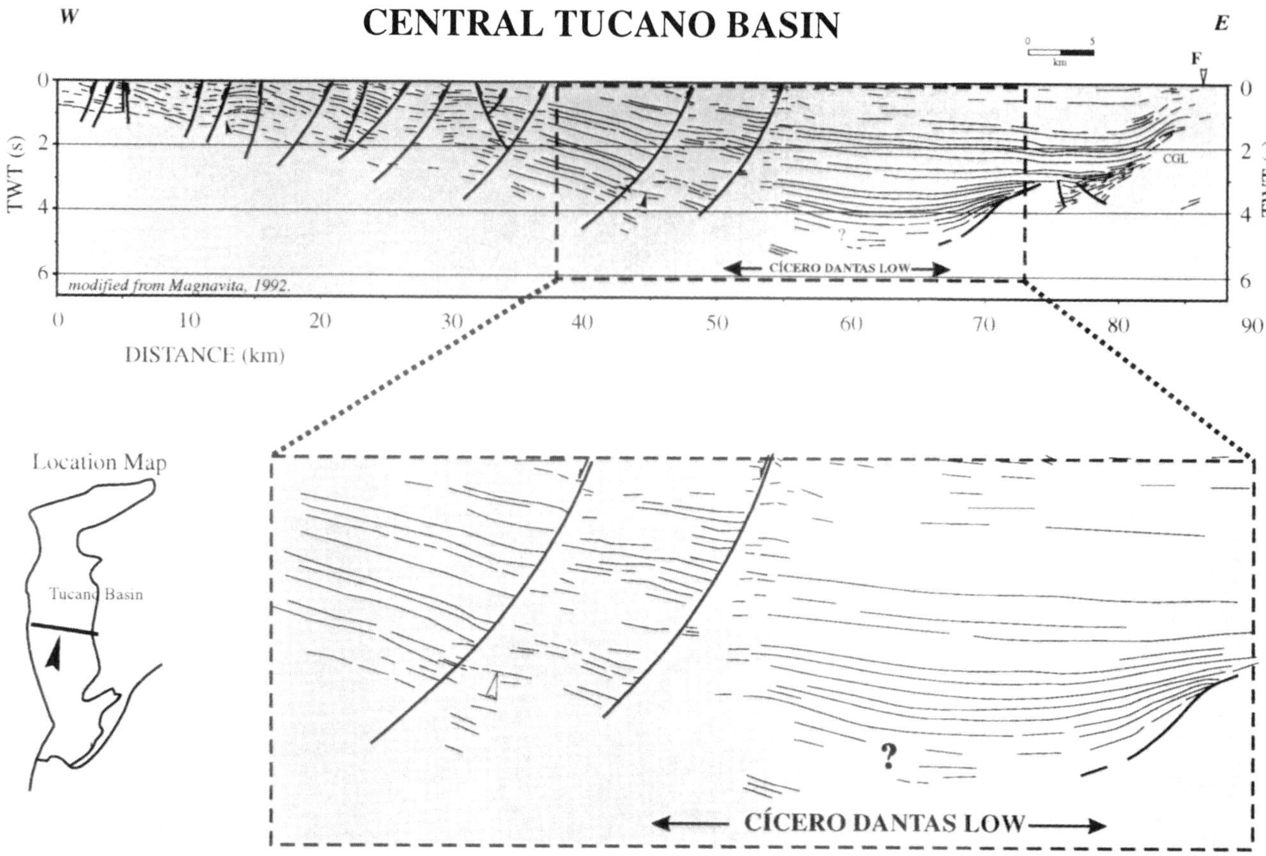

Figure 7: Regional E-W cross section crossing the Central Tucano Basin near the transect. It shows an interpretation based on industry seismic profiles acquired in the 90's, with problematic imaging of deep structures below 5 sec TWT.

[*Gonzaga*, 1995], eastwards of the middle Proterozoic rocks of the São Francisco Basin. It extends from the São Francisco Craton, running almost E-W from the city of Filadelfia, Bahia, towards Aracaju city. It crosses the central Tucano Basin near the city of Cícero Dantas and reaches the Sergipe Basin near the coastal plain (figure 2). The gravity profile was extended in the offshore region from the Mosqueiro Low (in the platform) towards the deep water region of the Sergipe Basin, resulting in a unique transect across different tectonic compartments in the northeastern region of Brazil.

The total magnetic field anomaly map of the northeastern region of Brazil (figure 13) is marked by strong anomalies in the offshore region. The strong positive magnetic anomaly that forms en-echelon segments from the Jacuípe towards the Pernambuco-Paraíba basins may be interpreted as an equivalent to the ECMA in the North Atlantic, marking the transition from continental to oceanic crust [*Mohriak et al.*, 1995a; *Mohriak et al.*, 1998].

Figure 14 shows the analytical signal map, which is a direct measure of the gradient of the magnetic field, and is based on derivatives of the total magnetic anomaly map. It shows the main depocenters and the regional basement elevations that limit the syn-rift troughs, both onshore and offshore along the northeastern margin.

The interpretation of the potential field data in the onshore region indicates that the RTJ rifts are separated from the JSA rifts by a major gravity anomaly that trends roughly N-S from Salvador to the Vaza-Barris Arch (figure 2 and figure 12). This anomaly corresponds to shallow Precambrian blocks as indicated by magnetic and derivative maps (figure 13 and figure 14). The Estância platform, south of the transect, is also marked by positive anomalies both onshore and offshore. The rift blocks eastwards of the Jacuípe High (south of the Sergipe Fracture Zone, figure 2) are restricted to the slope region.

The gravity signature of the Sergipe-Alagoas Basin is characterized by NE-trending gravity lows in Alagoas and in northern Sergipe, but roughly N-S south of the Aracaju city (figure 2 and figure 12). Towards the shelf-edge and slope of the Sergipe and Jacuípe basins, there is a rapid increase in the Bouguer anomaly values, which also show contours with a predominantly NE trend. In distal regions of oceanic crust,

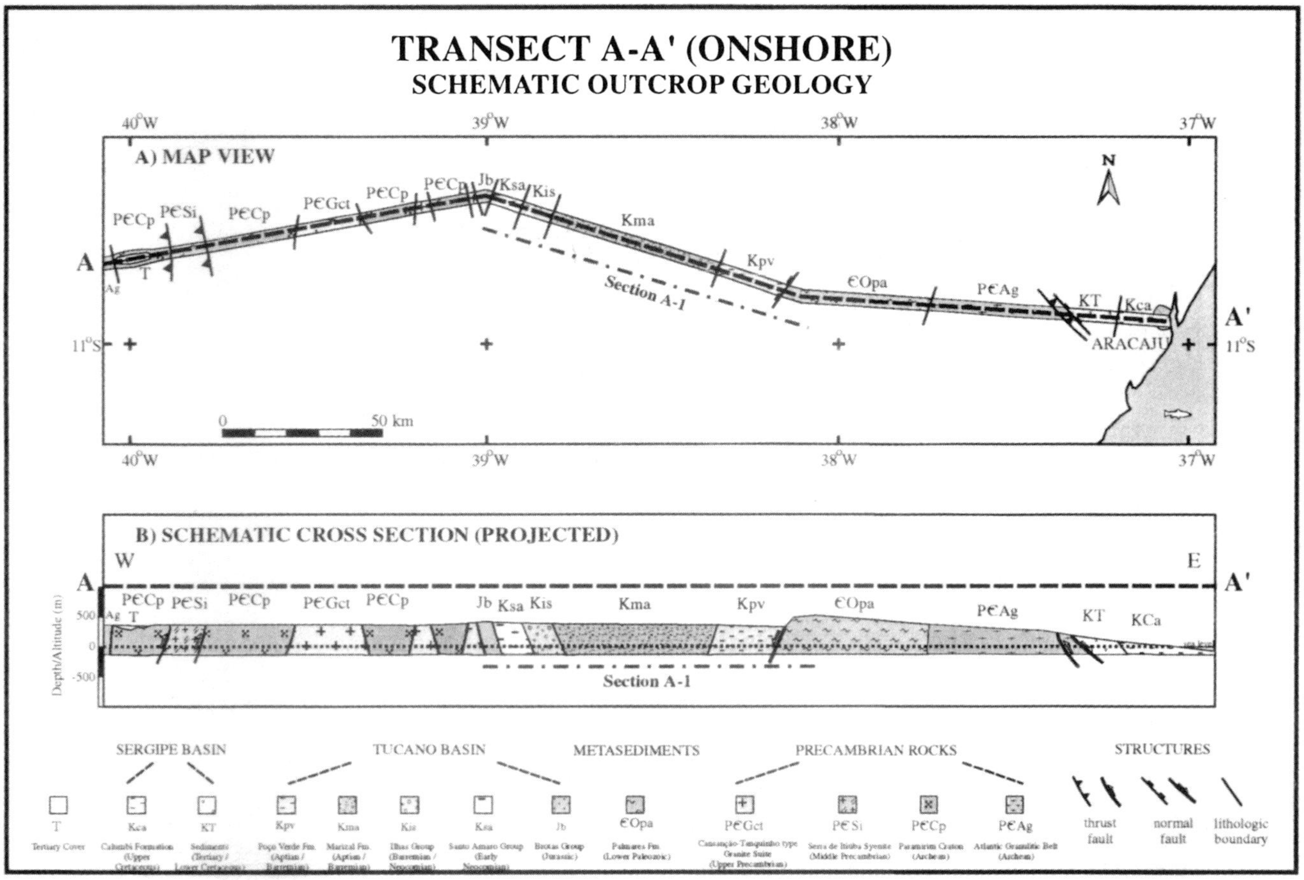

Figure 8: Onshore portion of the transect, from the São Francisco Craton, across the Central Tucano Basin, and reaching the Sergipe-Alagoas Basin near the coastline. Based on *Radambrasil* [*1983*]. The seismic profile in the Tucano Basin is indicated along the map view (from 38 to 39 °W) and along the cross-section.

some E-W-and NW-SE-trending anomalies are indicative of fracture zones and oceanic lineaments (figure 2 and figure 12). The deep water region of northeastern Brazil is characterized by arcuate segments of the positive axis of the total magnetic field anomalies, resembling the East Coast Magnetic Anomaly [*Mohriak et al.,* 1995b; *Mohriak et al.,* 1998].

Figure 15 (top) shows the gravity profile (Bouguer anomaly) as measured in the transect (0 to 340 km from the origin), and the seaward extension of the gravity anomaly towards the deep-water region of the Sergipe Basin, as measured along a regional deep seismic profile from the Mosqueiro low in the platform towards the oceanic crust domain. Figure 15 (bottom) shows the topography along the profile, which is rather flat in the Tucano Basin (altitudes ranging from 400 to 500 m above sea-level), and slightly higher adjacent to the eastern border of the basin (up to 700 m above sea-level). From this point eastwards, the topography declines steadily towards the coastline (approximately 340 km from the origin of the transect, figure 15). The shelf in the Sergipe Basin is rather narrow, and the slope is extremely steep, with bathymetry varying abruptly from the shelf-edge (370 km from the origin) towards the deep basin (390 km from the origin). The profile reaches water depths exceeding 3,000 m near the easternmost portion of the profile (450 km from the origin).

The Bouguer gravity anomalies plotted along the profile (figure 16) show the following characteristics: 1) the first 125 km of the profile have smooth Bouguer anomalies (ranging from-40 to-60 mGal); 2) from 125 to 200 km, the gravity anomalies are highly influenced by the Tucano Basin trough, filled with comparatively low density sedimentary rocks, resulting in readings as low as-150 mGal; 3) there is a rapid increase in the gravity anomaly readings from about km 200 to km 225, near the eastern boundary of the graben, reaching almost +10 mGal; 4) the gravity profile from km 225 to km 310 has a convex upwards shape, with a maximum of about + 25 mGal; 5) from 300 to 350 km, the anomaly decreases to-40 mGal, as the coastal plain of the Sergipe

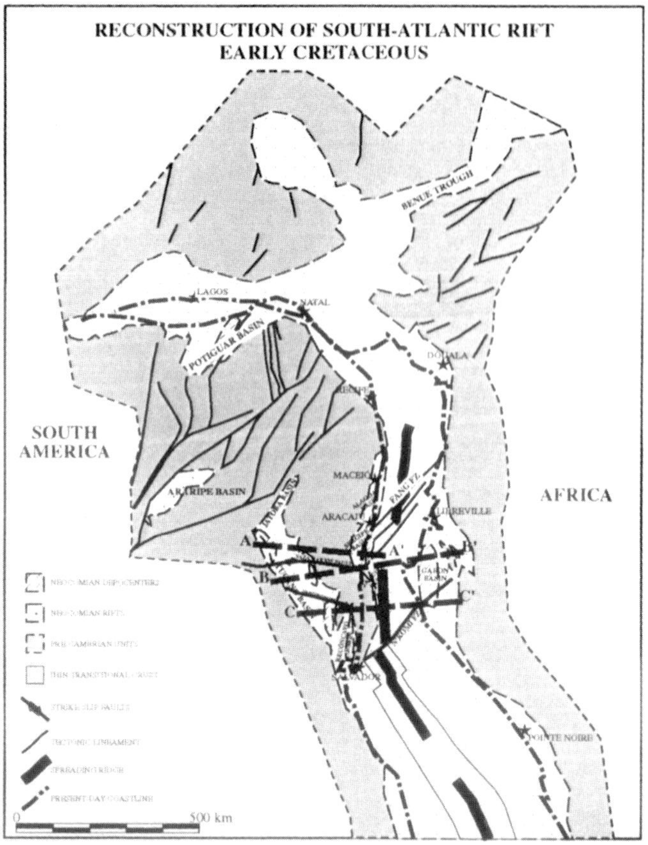

Figure 9: Reconstruction of South Atlantic rift system at the Early Cretaceous. A – A' , B – B', and C – C' correspond to schematic geological sections with alternative hypothesis of detachments for basin formation.

sub-basin is approached; 6) a remarkable increase in the Bouguer anomaly is observed in the platform, reaching more than +100 mGal near the shelf break; and 7) the Bouguer anomaly flattens in the deep water region, towards the end of the profile, with values ranging from 100 to 150 mGal.

These gravity characteristics allow to divide the transect into the following tectonic compartments: (figure 16): compartments I-IV, onshore, and compartments V-VI, in the offshore regions. Compartment I corresponds to a mean gravity anomaly around-50 mgal; comparment II corresponds to a negative anomaly coinciding with the Central Tucano Basin; compartment III corresponds to an increase in thc anomaly to values above 0 mGal; and compartment IV corresponds to another decrease in the Bouguer anomaly near the coastline. Compartment V, in the platform and slope, corresponds to a rapid increase in the Bouguer anomaly, and Compartment VI corresponds to a very high Bouguer anomaly in the deep water region. The interpretation of these compartments will be discussed later.

The gravity anomalies as measured along the profile were quantitatively modeled assuming the following parameters for the crustal architecture:

BODY	VELOCITY (m/s)	DENSITY (g/cm^3)
Water (Bouguer correction)	1500	2.20
Mesozoic/Cenozoic sediments	2400	2.30-2.40
Proterozoic metasediments	2550	2.65
Continental basement	5500	2.78
Oceanic basement	5500	2.80
Underplating	7500	2.95
Mantle	8000	3.25

The modeling of the regional transect was obtained by using the GM-SYS program, developed by Northwest Geophysical Associates Inc. (NGA). This program is considered appropriate for calculating the gravity response of bi-dimensional bodies in geological models.

The gravity model (figure 16) is useful to constrain the Moho topography from the osnhore region to the offshore region, assuming that the depth to basement and sedimentary sequences are better evaluated by the seismic interpretation.

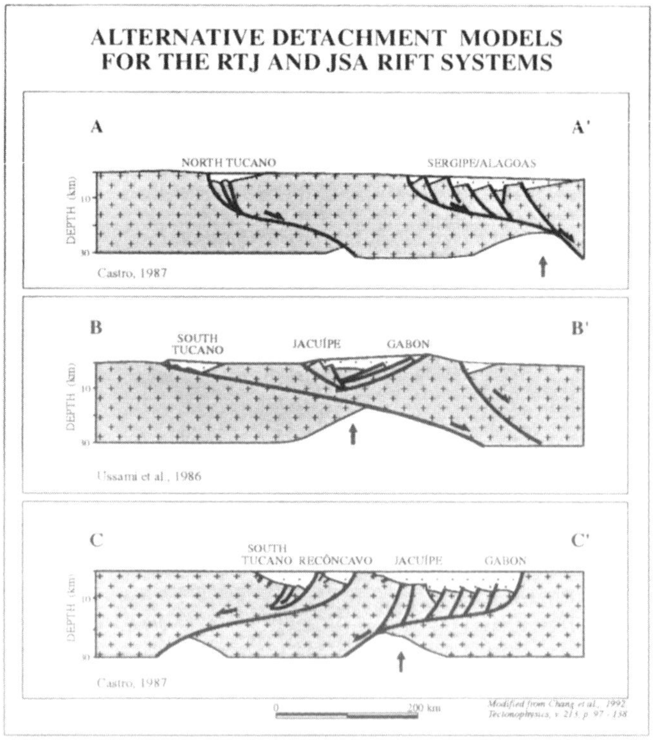

Figure 10: Schematic sections showing alternative models involving crustal detachments in the RTJ and JSA rift systems. [top and bottom, from *Castro Jr., 1987*; middle, from *Ussami et al., 1986*].

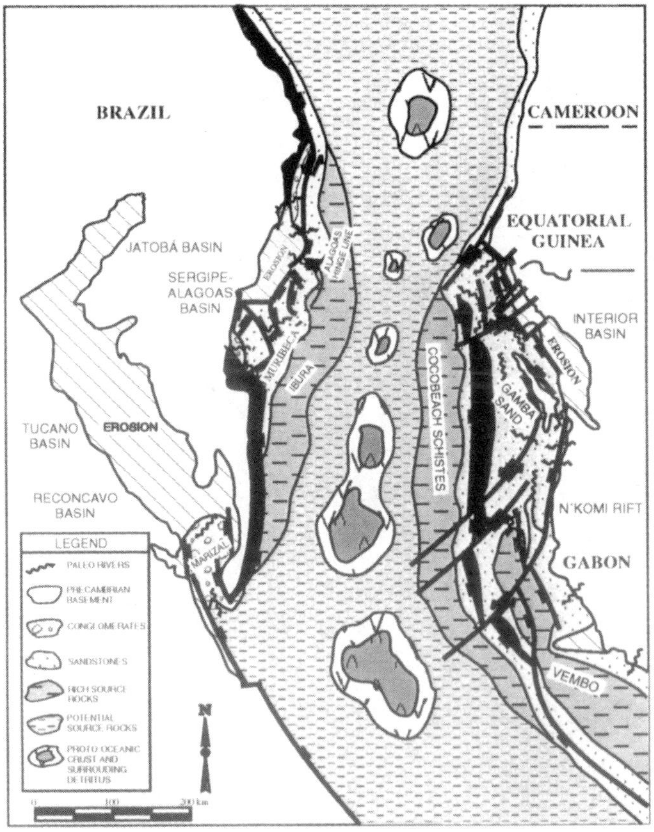

Figure 11: Palinspastic reconstruction of the rift system between northeastern Brazil and western Africa at the time of oceanic crust inception (late Aptian), showing bifurcation of the spreading axis in two branches, one towards the RTJ aborted rift system, and the other which evolved as the JSA/ GRM passive continental margin basins [modified from *Bradley and Fernandez, 1991*).

The model indicates that the Moho is rather flat (with values ranging around 35-30 km) for most of the onshore portion of the profile, with a rather small, local uplift near the eastern flank of the Central Tucano Basin. The minimum negative Bouguer anomaly (about-150 mGal, observed near the depocenter) indicates that the huge gravity low is not compensated by a coaxial mantle uplift. The resultant Moho topography is characterized by a steady rise from the western portion of the Tucano Basin towards the region immediately east of the rift depocenter [*Mohriak et al.,* 1997]. Apparently, it forms a local apex at the footwall of the rift boundary fault. This gravity high coincides with precambrian horsts that extend northwards from Salvador, corresponding to the Jacuípe and Aporá highs.

The gravity model also indicates that the Sergipe sub-basin is characterized by unthinned crust near the western boundary of the rift, but there is an extremely rapid thinning near the shelf-break. The deep water province of the Sergipe Basin is characterized by an extremely rapid crustal thinning eastwards of the depocenter in the Mosqueiro Low (about 340 km from the origin). The model suggests that from the platform to the shelf-edge, the Moho rises from 30 km to slightly less than 20 km deep, and that an extremely thinned crust occurs beyond the shelf edge (figure 16, 350 to 400 km from the origin). From this point basinwards, the Moho rises slowly, smoothly approaching the transition to a predominantly volcanic (oceanic) crust, which occurs rather close to the coastline (420 km from the origin). This model results in a crust/mantle interface at depths of about 15-20 km in regions interpreted as pure oceanic crust, which is perfectly compatible with results recently obtained in the West African sedimentary basins [*Meyers et al.,* 1996].

INTERPRETATION OF DEEP SEISMIC REFLECTION PROFILES

The crustal structure of the rifts along the transect are analyzed by comparing the gravity model (figure 16) with actual seismic observations from deep seismic profiles in the Tucano Basin (125 to 230 km from the origin) and in the Sergipe sub-basin (350 to 450 km from the origin).

A deep-seismic line onshore (profile A-1, see figure 2 for location) was acquired in the central Tucano Basin using powerful explosive sources to image the deeper portions of the basin. The seismic section extends for about 108.5 km with a shotpoint interval of 35.0 m. It was processed at Petrobras to a time length of 16 s TWT using standard migration techniques.

The regional profile offshore (profile A-2, see figures 2 for location) extends for about 111.2 km in the deep water region of the Sergipe-Alagoas Basin. It is included in the deep seismic grid of the Brazilian northeastern margin, which used special acquisition equipment, particularly a 6,000 m long cable, and an airgun source with a volume of 8,832 cu.in. (144.7 l), operating at 2,000 psi. The processing included migration to a time length of 18 s TWT using standard techniques [*Mohriak et al., 1995a*].

A. Tucano Basin Rift Architecture

Figure 17 shows the regional seismic profile in the Tucano Basin (crossed at its central portion, south of the Vaza-Barris Arch). The interpretation of the profile (figure 15) indicates that the basin floor dips consistently towards the master fault along the eastern border of the rift, with a major depocenter located between 70 and 90 km from the origin of the profile (shotpoints 2000-2500). According to regional maps [*e.g., Milani, 1987; Magnavita and Cupertino, 1988; Milani and Davison, 1988, Bueno et al., 1994*] the master faults trend N25E, dipping NW with a high angle (about 60°).The deeper portions of northwestern extremity of the seismic profile, indicating that the west-

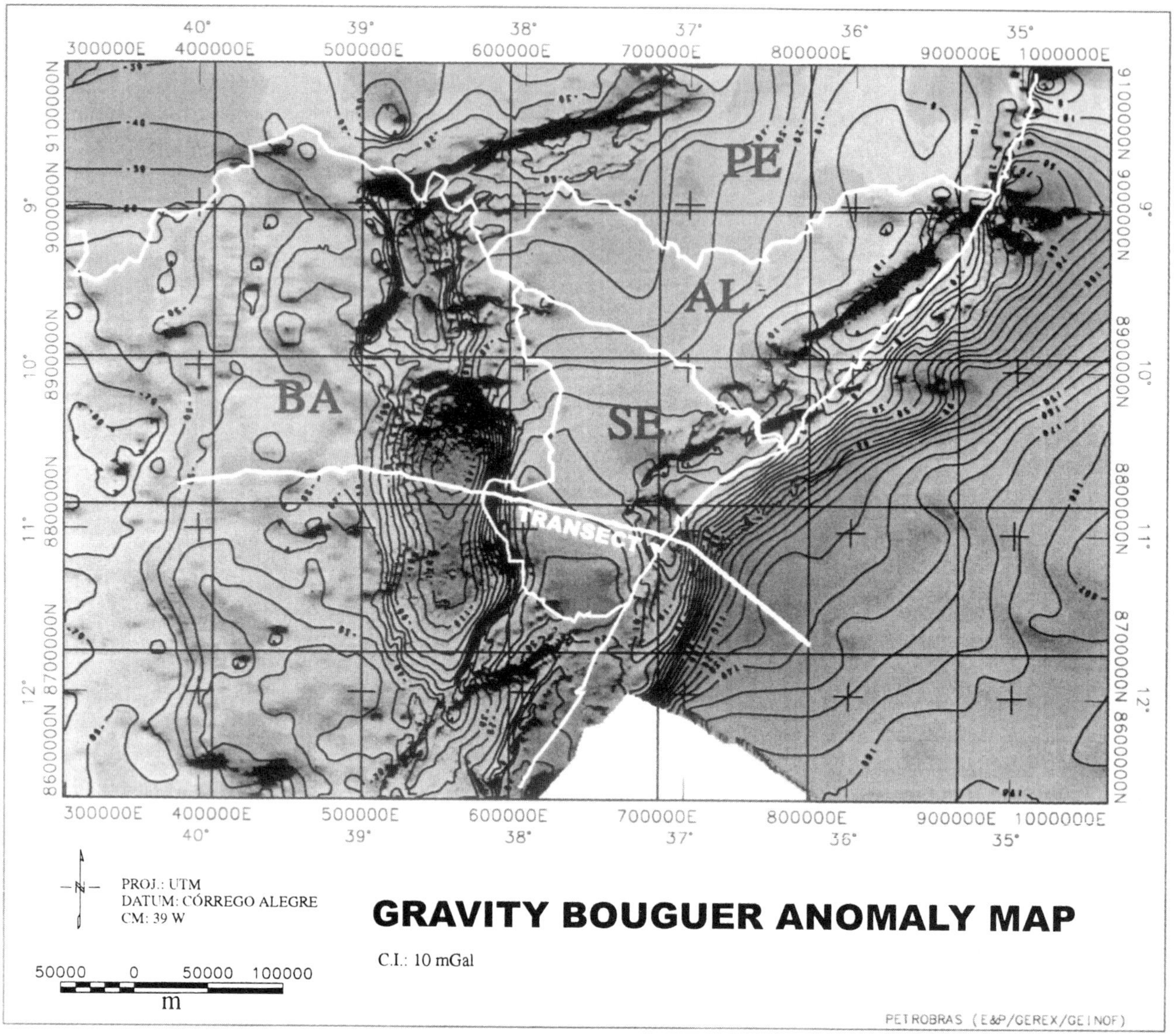

Figure 12: Shaded-relief Bouguer anomaly map of the northeastern region of Brazil. A regional transect A-A', based on a recconnaissance gravity survey onshore, and integration of gravity data offshore, extends from the São Francisco Craton, crosses the central Tucano Basin, and reaches the oceanic crust of the Sergipe Basin at its easternmost extremity.

dipping reflectors are apparently offset by a continuous band of reflectors that dip east and extends to mid crust depths. Similar seismic fabrics have been also observed in other regions affected by major precambrian orogenies, and they are usually interpreted as extensional shear zones [*Serpa and Voogd, 1987*], basic intrusions formed during lithospheric extension [*Singh and McKenzie, 1993*], granitic intrusions formed during lithospheric convergence [*Scheirer and Hobbs, 1990*], or suture zones in the lower crust [*McBride et al., 1996*].

The structural fabric of the western-dipping array of reflectors in the seismic section is clearly traversed by a subhorizontal band of reflectors, extending from km 10 to km 30 of the profile, at time depths of about 13 s TWT (figure 17). This image is the first characterization of deep seismic reflectors in the transition between the lower continental crust and the upper mantle available in the São Francisco craton, possibly corresponding to the seismic Moho [*Mohriak et al., 1997*]. It shows remarkable similarities with equivalent horizons interpreted as Moho in other cratonic regions elsewhere

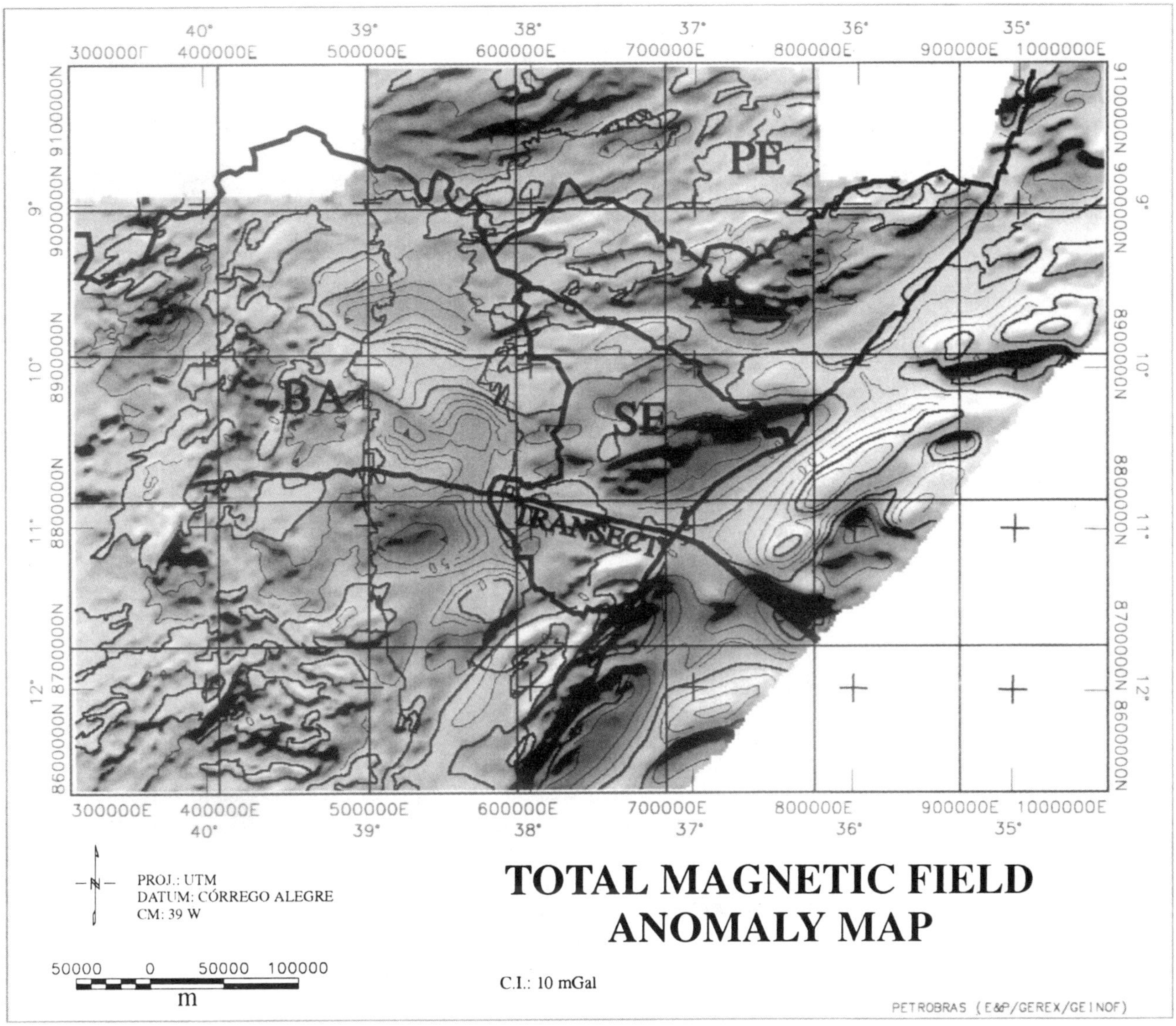

Figure 13: Shaded magnetic (total field) anomaly map of the northeastern region of Brazil. The regional transect A-A' is also shown.

in the world. In these regions, combined seismic refraction and reflection indicate that the Moho is characterized by an abrupt change of seismic velocity from less than 7.0 km/s to values around 8.0 km/s, corresponding to a zone of flat-lying, high-impedance reflectors around 10-13 s TWT, at the base of sub-horizontal to slightly dipping arrays of reflectors [*e.g., Klemperer et al., 1986*]. Depth conversion of this reflector, assuming a basement velocity of 5,000 km/s for the upper crust (0 to 8 s TWT) and 6,500 m/s for the lower crust (8 to 13 s TWT) results in a mantle transition at a depth of about 36 km, which is consistent with worldwide observations of Moho in cratonic regions [*Christensen and Mooney, 1995*].

The general aspect of eastward dipping reflectors in the seismic section changes completely between km 55 and 70 of the profile, as the seismic signature of the sedimentary layers looses coherency due to the upwards propagation of major westward and eastward-dipping basement-involved faults that affect the whole stratigraphic section (figure 17 and figure 18). This results in a large structure that might correspond to an uplifted block due to transcurrent movement along NW-trending transfer faults. We hypothesize that

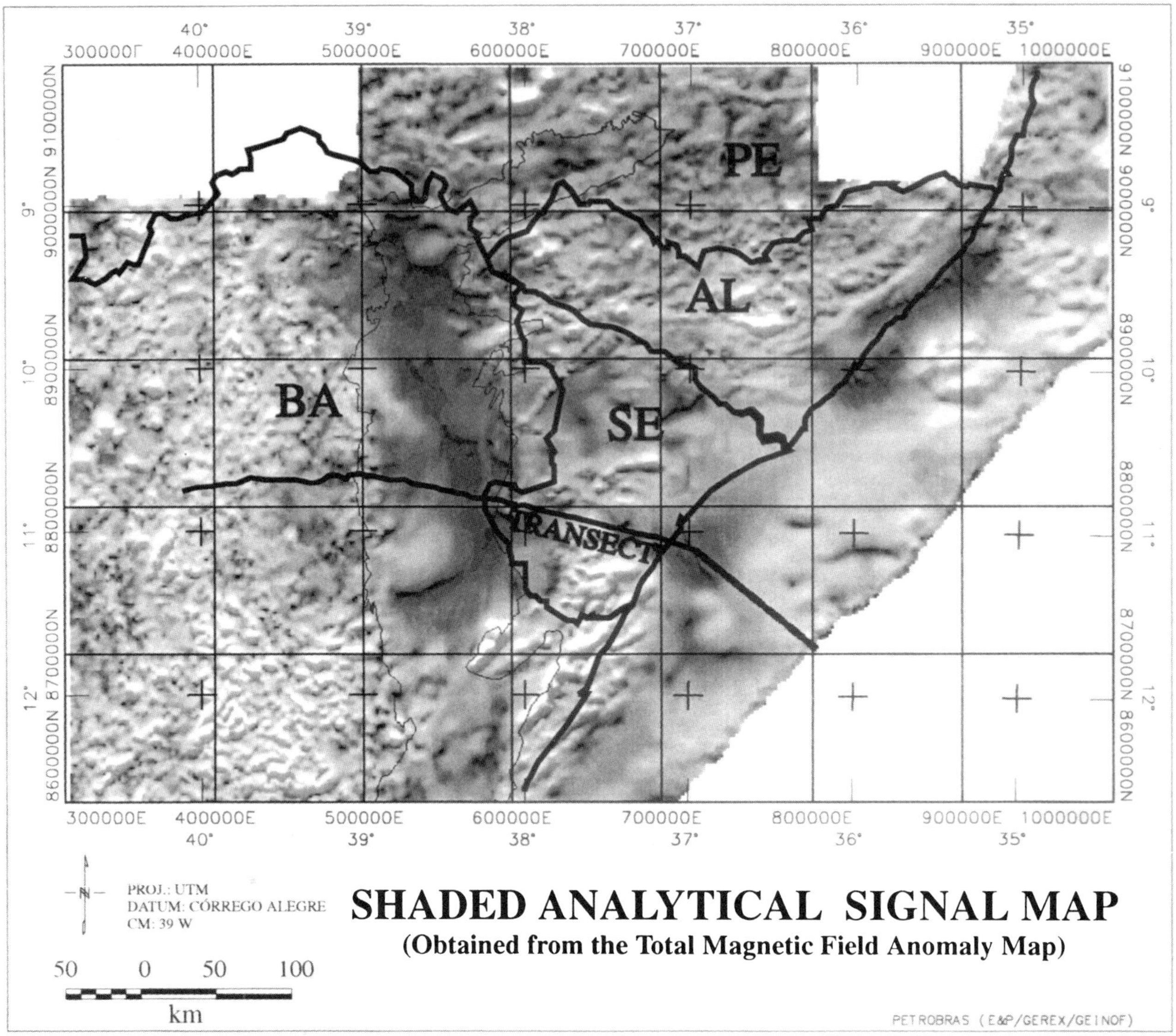

Figure 14: Shaded Analytical signal map (derived from total magnetic field) of the northeastern region of Brazil. Profile A-A' is also shown.

it might be associated with the northwest prolongation of the Itapicuru transcurrent fault. This fault crosses most of the southern portion of the central Tucano Basin, and is responsible for a major offset of the depocenter north of the Aporá High [*Milani, 1987; Milani and Davison, 1988, Magnavita, 1992*]. A downfaulted block is observed near km 70 of the profile, resulting from collapse of sedimentary layers between the major eastward dipping fault that limits the horst, and a compensatory normal fault that dips westward.

The depocenter of the Tucano Basin (shotpoints 2000-2500, 70 to 90 km from the origin of the profile) is comparatively much less faulted in its upper portions of the sedimentary succession than in the segment of the profile westwards of shotpoint 1,500. The lower sequences of the sedimentary layers in the depocenter are distinguished by sigmoidal reflectors forming sag basins that pinch-out and disappear towards a high-angle, west-dipping inner border fault (figure 18). Younger sequences onlap this step towards the easternmost boundary of the Tucano Basin, which is controlled by a major border fault that aparently comes close to the surface. Late syn-rift sediments dip consistently towards the easternmost boundary fault (near km 100 of the profile). The Marizal Fm., which is identified as a veneer of post-rift sediments of Aptian age, is not affected by any significant

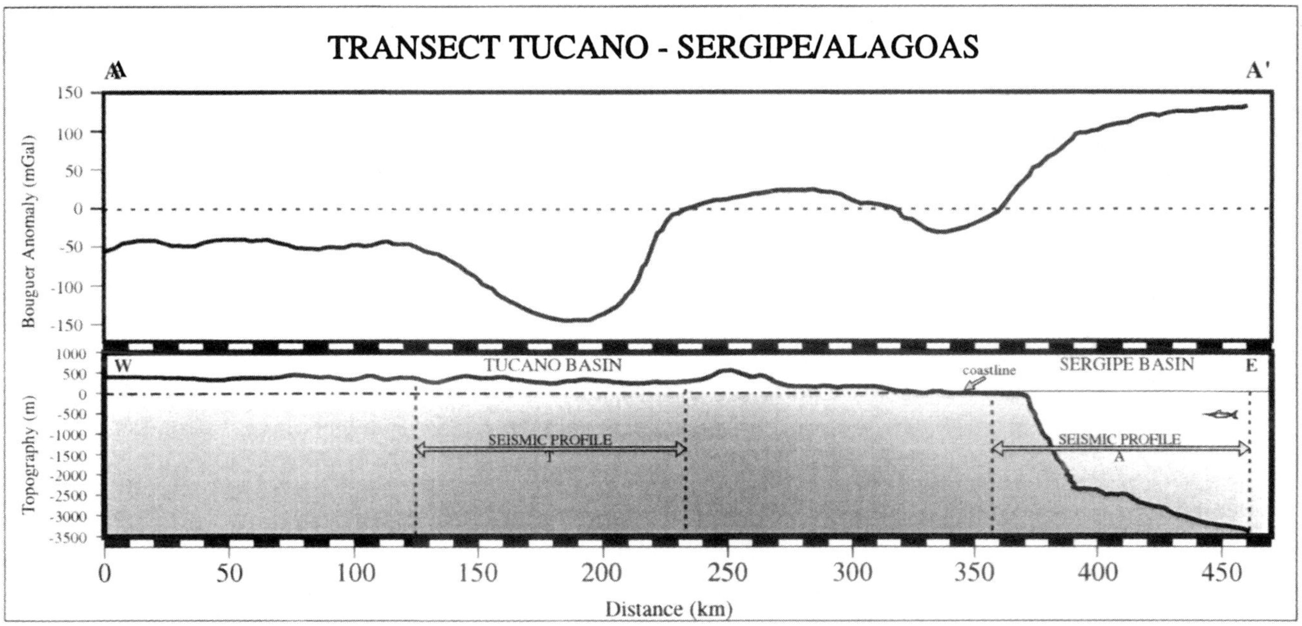

Figure 15: Topographic profile and gravity (Bouguer anomaly) profile along the regional transect.

fault along this portion of the profile. These sediments form a small, shallow depocenter towards the center of the syn-rift trough, westward of the eastern boundary fault (center at shotpoint 2000, 70 to 90 km from the origin, 0.0 to 0.5 s TWT).

The easternmost extremity of the profile runs along the footwall block of the boundary fault. Its is expressed as a high-angle fault formed later than the inner normal fault that controlled the first subsidence phases. In the region near the boundary faults, the seismic profile shows a rather transparent upper crust overlying a series of arcuate reflectors (from 4 to 6 s TWT), similar to the reflectors observed in the western portion. The lower crust oblique reflectors observed between 7 and 11 s TWT (near the eastern portion of the seismic profile, shotpoints 2500-3000, see figure 17) have consistent dips to the west, and climb towards the footwal block. We interpret the Moho, which is not clearly identified in the seismic section (probably because of seismic energy absorption in the sedimentary layers), as climbing steadily towards the east. An approximate depth-conversion of these reflectors resulted in a crust-mantle transition around 30 km.

Regional geological maps [*e.g., Radambrasil, 1983*] show that the eastern portion of the profile is characterized by north-dipping thrust and reverse faults affecting the Upper Proterozoic metasedimentary rocks, indicating a predominantly southern vergence in this segment of the Brasiliano Orogenic Belt [*Neves et al., 1977*]. The footwall block, which has a cover of Upper Proterozoic/Lower Paleozoic metasedimentary rocks (figure 2 and figure 3), was apparently uplifted during the Mesozoic extensional episode, as a result of a flexural rebound due to the major normal fault to the west [*Magnavita et al., 1994*]. This uplift is still preserved in the present-day topography (figure15-bottom), reaching altitudes of more than 500 m eastwards of the Tucano seismic profile.

B. *Sergipe sub-basin Rift Architecture*

Although there is a gap of about 100 km between the Tucano Basin profile and the coastline (figure 15), the shallow structure of the Sergipe sub-basin is well constrained by exploratory boreholes and geophysical data. Industry seismic profiles in the onshore and continental platform of the southern Sergipe Basin indicate a rather thin sedimentary cover of Upper Cretaceous/ Tertiary rocks onlapping a major regional unconformity at the top of the rift blocks. These rift blocks are controlled by small antithetic and synthetic faults, and basement is rather shallow. The sedimentary rocks thicken rapidly in the hanging wall block of the Vaza-Barris fault in the Mosqueiro Low, and the syn-rift sediments dip towards the west, whereas the post-rift sediments prograde from the coastline towards the shelf-edge with a small basinward dip.

Figure 19 shows the seismic section A-2 with an schematic interpretation. Towards the platform and the slope of the Sergipe sub-basin, the seismic profile shows major crustal faults that control syn-rift sediments in the Mosqueiro low, as indicated by westward dipping arrays of reflectors around 4-5 s TWT near the western portion of the seismic section (figure 19, shotpoints 1200-1400).

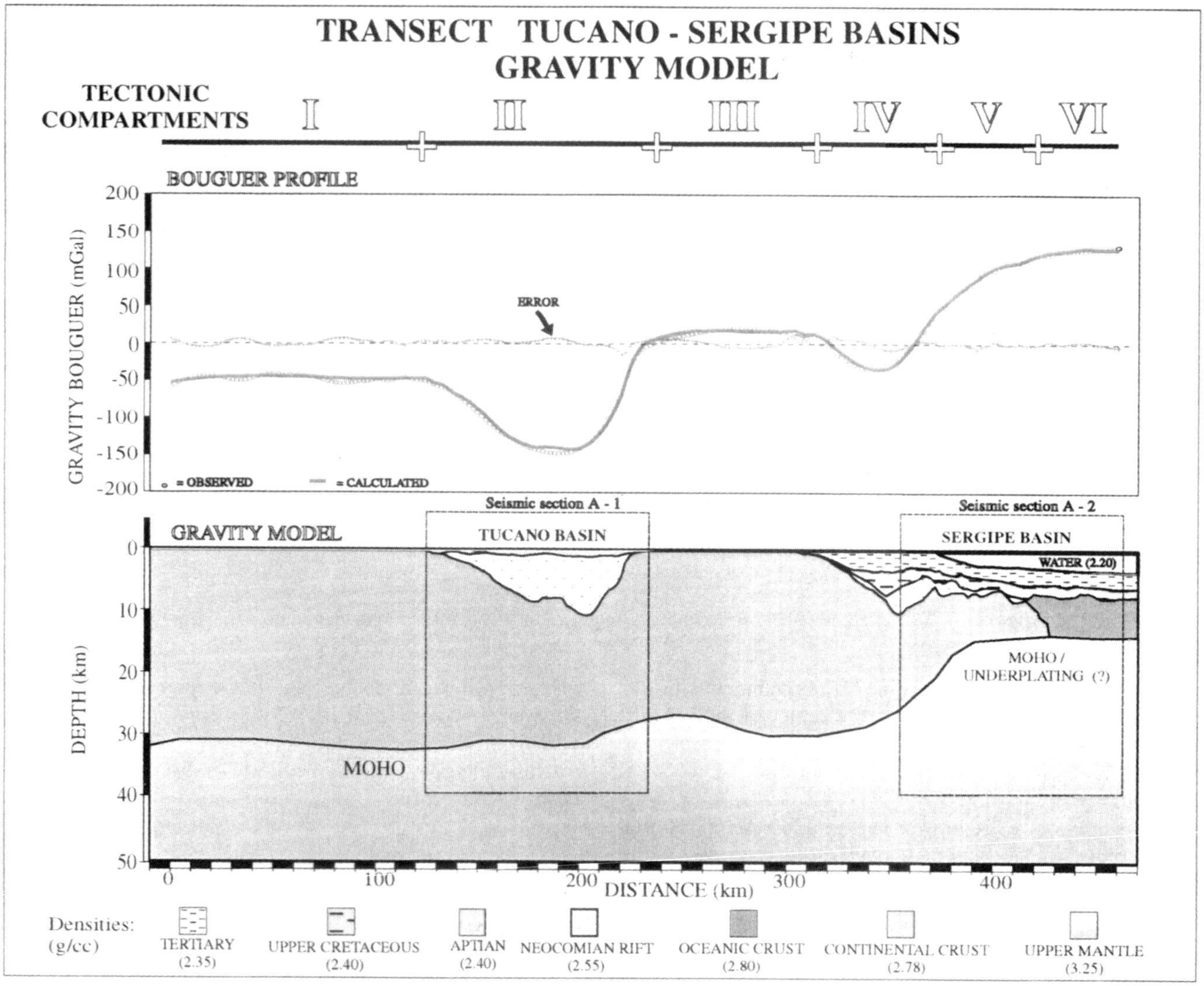

Figure 16: Bouguer anomaly plot and gravity model of the regional transect. Compartments I to VI are identified along the transect.

The maximum sedimentary thickness in the Sergipe-AlagoasBasin is also highly controversial, and some workers have suggested basement depths exceeding 15 km in the Mosqueiro low and in the deep-water region [*e.g., Destro, 1994*]. However, normal industry seismic profiles most often do not provide reliable images of the rift structure in the deep-water region, and the JSA rift system has depocenters that are beyond the reach of the exploratory drill. More recently, deep seismic reflection profiles have been used to elucidate this problem. Previous interpretations [*e.g., Pontes et al., 1991*] assumed that arrays of coherent, strong reflectors observed in deep seismic sections corresponded to the sediment/basement interface, resulting in extremely deep troughs in the distal portions of the Sergipe-Alagoas Basin. However, integrated studies based on regional seismic and potential field data, as well as comparison with other regional deep seismic profiles worldwide, indicate that the reflectors observed around 7 to 9 s TWT in the deep-water province actually correspond to lower crustal horizons near the seismic Moho [*Mohriak et al., 1993; Mohriak et al., 1995a; Mohriak et al., 1998*].

The interpretation of the deep seismic profile in the Sergipe sub-basin (figure 20) indicates that the rift sediments rapidly disappear towards the deep-water region. Near the eastern extremity of the profile, we suggest that the sigmoidal reflectors, observed between 6 and 7 s TWT, correspond to seaward-dipping volcanic reflectors rather than sedimentary layers associated with clastic progradation. The strong

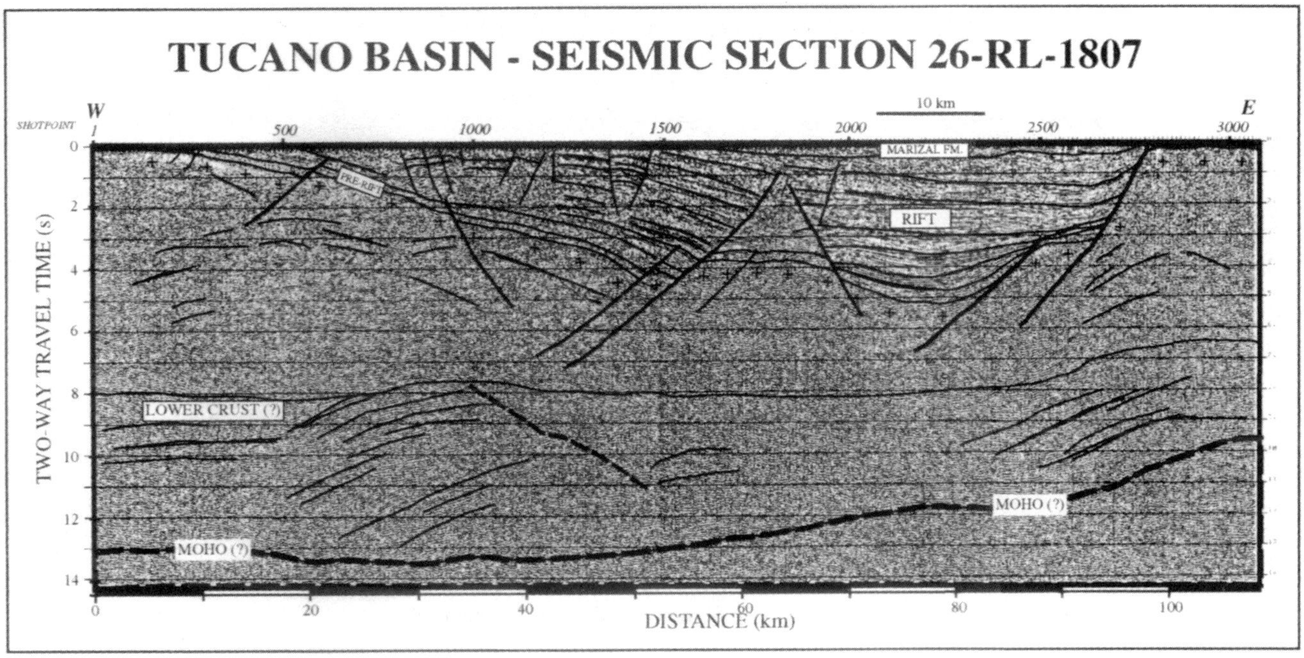

Figure 17: Seismic profile A-1 in the central part of the Tucano Basin.

deep seismic reflector observed in the deep-water region around 9 s TWT (figure 19), which was previously interpreted as top of basement in the Sergipe Basin [*Pontes et al., 1991*], has been re-interpreted as a detachment surface in the lowermost crust, slightly above the seismic Moho [*Mohriak et al., 1995a*]. The crustal faults beyond the shelfbreak apparently detach on this strong reflector, which possibly corresponds to the crust-mantle transition in the deep-water region. Alternatively, the array of reflectors at around 7 to 9 s TWT might correspond to the top of an underplated layer and the petrologic Moho might be positioned slightly deeper [*Mohriak et al., 1995a; Mohriak et al., 1998*].

The seismic profile also shows arrays of strong reflectors with a sigmoidal geometry that dip seaward (as for example, near shotpoint 2600, 6.5 to 7.5 s TWT), and others that dip landward (as for example, near shotpoint 2600, 7.5 to 8.5 s TWT). Although there are some currents of opinion that interpret these features as possible rift structures in the deep-water Brazilian margin [e.g. Pontes et al., 1991; S. F. Santos, personal communication, 1994], there is evidence that they are associated with igneous structures [*Mohriak et al., 1995a*]. The eastward dipping reflectors possibly correspond to SDR (seaward-dipping reflectors) zones. These features are associated with magmatic activity near the boundary between the continental and oceanic crust, and are predominantly formed by volcanic rocks originated by extrusion during early phases of spreading at oceanic ridges [*e.g., Hinz 1981; Mutter et al 1982; Holbrook et al., 1994*].

DISCUSSION OF RESULTS

This paper analyzed the deep structure of the sedimentary basins in NE Brazil by integrating potential field data and gravity modelling with regional deep seismic reflection profiles both onland (Tucano Basin) and offshore (Sergipe Basin). A key point is that the seismic expression of the lower crust and the Moho show significant similarities with worldwide regional deep seismic profiles, particularly with the African conjugate margin. We also stress that the transition to oceanic crust in the northeastern margin of Brazil is marked by thick wedges of seaward-dipping reflectors, and the proximity to a pure oceanic crust in distances less than 50 km from the coastline may have important exploratory implications [*Mohriak et al., 1995b*].

Figure 16 shows a depth-converted geoseismic transect based on the gravity model combined with the interpretation of the deep seismic profiles in the Tucano and in the Sergipe basins. The seismic images of some of these compartments are shown in figures 17 and 19.

Based on integration of different geophysical data (mainly gravity and seismic), the transect may be divided in the following tectonic compartments (figure 16): a) Compartment I: unextended cratonic region westwards of the central Tucano Basin.; b) Compartment II: central Tucano Basin, sub-divided into a monoclinal ramp (westwards of the rift trough), the main depocenter, and the eastern platform near the main boundary fault; c) Compartment III:

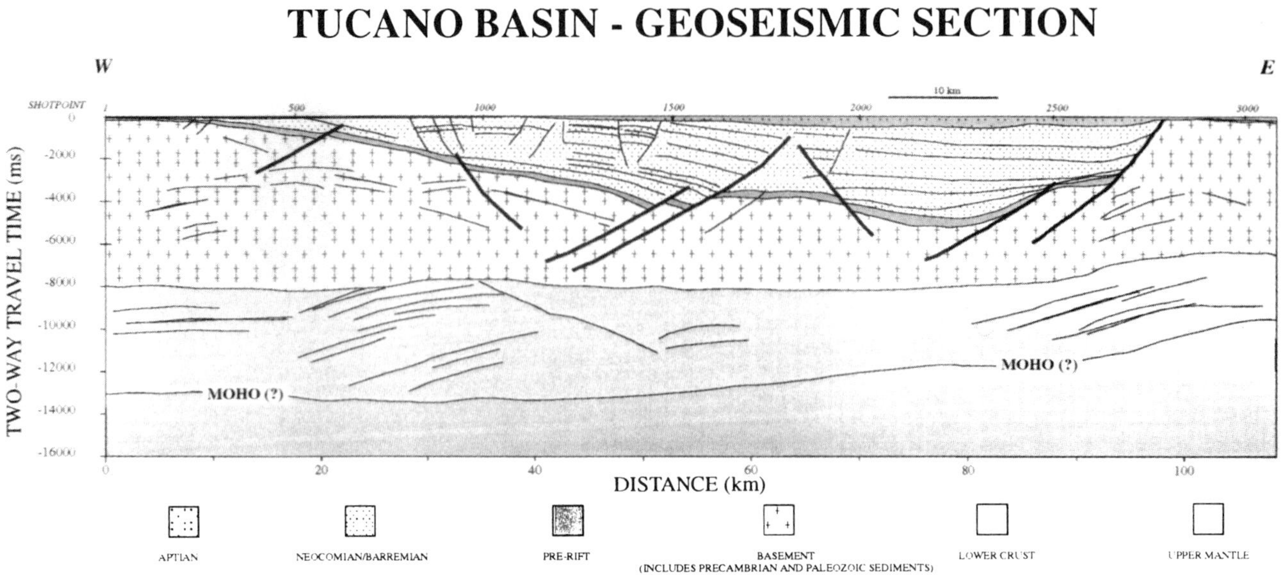

Figure 18: Line interpretation of Central Tucano profile

uplifted Precambrian block between the central Tucano Basin and the western boundary of the Sergipe Basin; d) Compartment IV: south Sergipe Basin, subdivided into a shallow basement province (onland and in the platform), and the deep basement trough corresponding to the Mosqueiro Low depocenter; e) Compartment V: slope basin and deep-water province of the Sergipe Basin; and f) Compartment VI: volcanic wedge associated with seaward-dipping reflectors and the transition to a pure oceanic crust.

The geological transect shows that the lower crust underlying the São Francisco (or Paramirim) craton onshore (figure 2) apparently conserved its original thickness almost unchanged during the Mesozoic extensional processes. Probably influenced by the lower crustal fabric inherited from previous tectonic events, the major syn-rift faults in the central Tucano Basin affect the upper crust and show a predominant dip to the west.

The western portion of the Central Tucano Basin seismic profile (figure 17 and figure 18) is characterized by a very thin sedimentary cover overlying an eastward-dipping basement. The depocenter of the basin, which may reach a total sediment thickness exceeding 16 km [*Aragão and Peraro, 1994*], is crossed near the easternmost extremity of the profile, adjacent to the Adustina high-angle master fault, in the Cícero Dantas Low (figure 2). However, along the profile we estimated maximum basement depth around 10 km, based on the seismic data and gravity model. In any case, the minimum estimate of 10,000 m of syn-rift sediments in the depocenter ranks the Central Tucano Basin as one of the deepest troughs formed during the break-up of Gondwana.

The Bouguer anomaly for the Central Tucano Basin depocenter is strongly negative, suggesting that the low-density sediments are responsible for most of the anomaly. In general, the depocenters of the Recôncavo-Tucano-Jatobá grabens are associated with major gravity lows, indicating a negligible mantle uplift underlying these rifts. This conclusion is at variance with the interpretation of *Milani and Davison* [*1988*], but reasonably agrees with the gravimetric interpretation of *Ussami et al.* [*1986*]. However, the gravity model based on the deep seismic transects differs both from Ussami et al.'s and Milani and Davison's models in several aspects.

The compartment III block, which has a cover of Late Proterozoic/ Early Paleozoic metasedimentary rocks, was apparently uplifted during the final phases of lithospheric convergence in the Late Precambrian/ Early Paleozoic, or most probably during the Mesozoic extensional episode, as a result of a flexural rebound due to the major normal fault to the west. The block between the Tucano and the Sergipe basins is underlain by a slightly uplifted Moho, and the gravity model indicated a local apex located slightly eastwards of the central Tucano Basin depocenter.

The Sergipe-Alagoas Basin is characterized by gravity anomaly lows onshore and in the platform, indicating that the proximal rift blocks are underlain by almost unthinned crust. An extremely rapid crustal thinning occurs from the shelf-break toward the slope. The Sergipe sub-basin deep seismic profile indicates that the rift layers are tilted landwards, and that the rift blocks are bounded by major normal faults that cut most of the crust, detaching at Moho depths. The rift blocks beyond the shelf-break have been affected by a major post-rift episode of erosion.

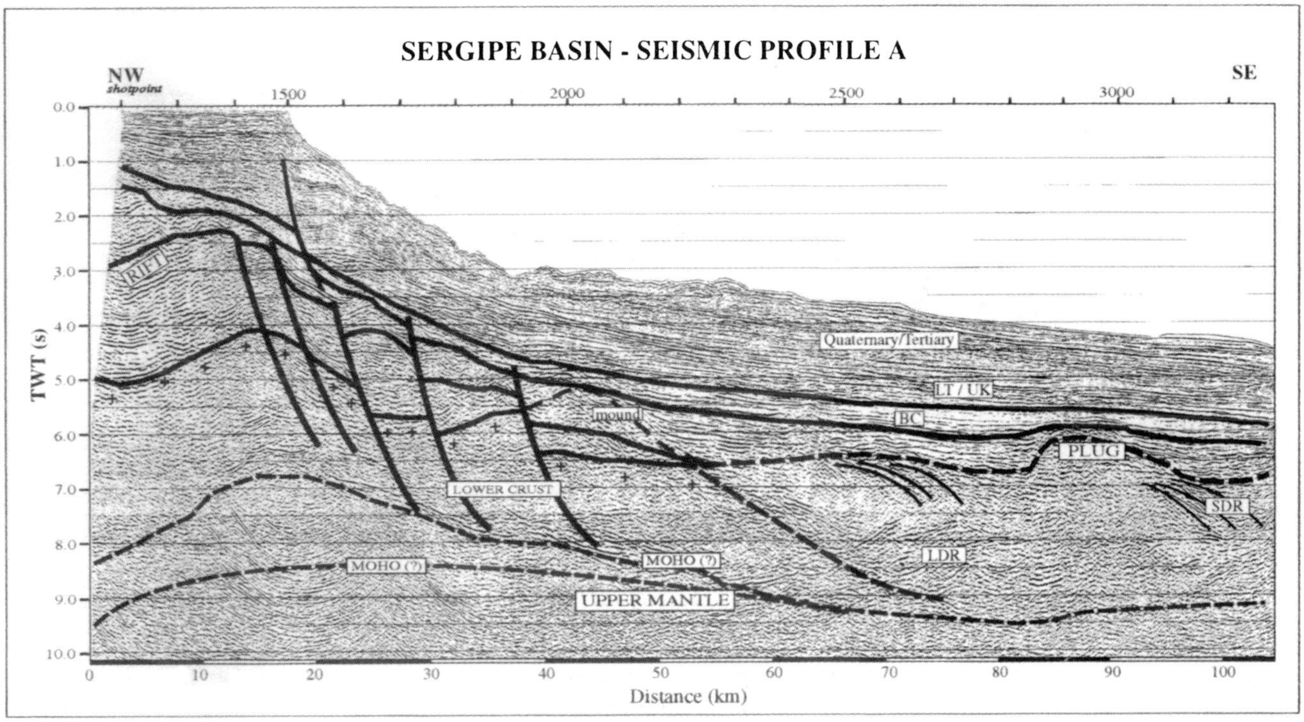

Figure 19: Seismic profile A-2 with a line interpretation.

The Moho topography shows a minimum near the eastern extremity of the Sergipe sub-basin, but it rises from the Mosqueiro low towards the shelf edge. Towards the east, as the depocenter of the Sergipe sub-basin is approached, the lower crust is radically thinned. The Moho climbs from about 30 km near the coastline to a depth of about 20-15 km near the shelf edge. This results in an almost complete attenuation of the lower crust, which was probably highly extended in a ductile manner. The upper crust, characterized by brittle behavior, is also thinned by the extensional processes. Major normal faults dip basinward, detaching at the lower crust/ upper mantle interface, which in the deep-water region probably coincides with the seismic Moho at depths of about 15 km. An alternative intepretation, also indicated on figure 16, would place the Moho climbing less rapidly from about 30 km in the platform to about 20 km near the eastern extremity of the profile, and the strong deep seismic reflectors would correspond to underplated layers or a detachment zone slightly above the petrologic Moho.

Figure 19 show typical structures of compartments IV to VI in the Sergipe Basin, particularly the thinning of the syn-rift sediments towards the deep-water region. Figure 20 shows an schematic interpretaton of this portion of the transect. The boundary between continental and oceanic crust is marked by volcanic intrusions and wedges of seaward-dipping reflectors interpreted as volcanic rocks extruded during oceanic crust inception and in the early phases of spreading. These sigmoidal reflectors, previously interpreted as rift-phase sediments, mark the transition to a pure oceanic crust, and are probably associated with a predominantly igneous rock sequence overlying the mantle rocks [*Mohriak et al., 1995a; Mohriak et al., 1998*]. This results in a very narrow margin architecture, characterized by a rapid transition from eroded rift blocks, to predominantly volcanic layers, and finally to pure oceanic crust not far from the shelf-break.

The integration of gravity and seismic data suggests two zones of crustal necking along the transect, one with a small amplitude, located eastwards of the Central Tucano depocenter, and coinciding with an uplifted block in the footwall of the rift border fault. The other is located eastwards of the Mosqueiro low in the Sergipe sub-basin, near the present day shelf edge. The latter is also associated with magmatic features, such as seaward-dipping reflectors and igneous plugs near the boundary with oceanic crust. The possibility of a Moho uplift eastward of the rift depocenter in the Tucano Basin poses interesting question about the dislocation of the extensional axis from the onshore region during the Neocomian, towards the offshore region during the Barremian and Aptian, resulting in final rupturing of the Gondwana by inception of oceanic crust in post-Aptian

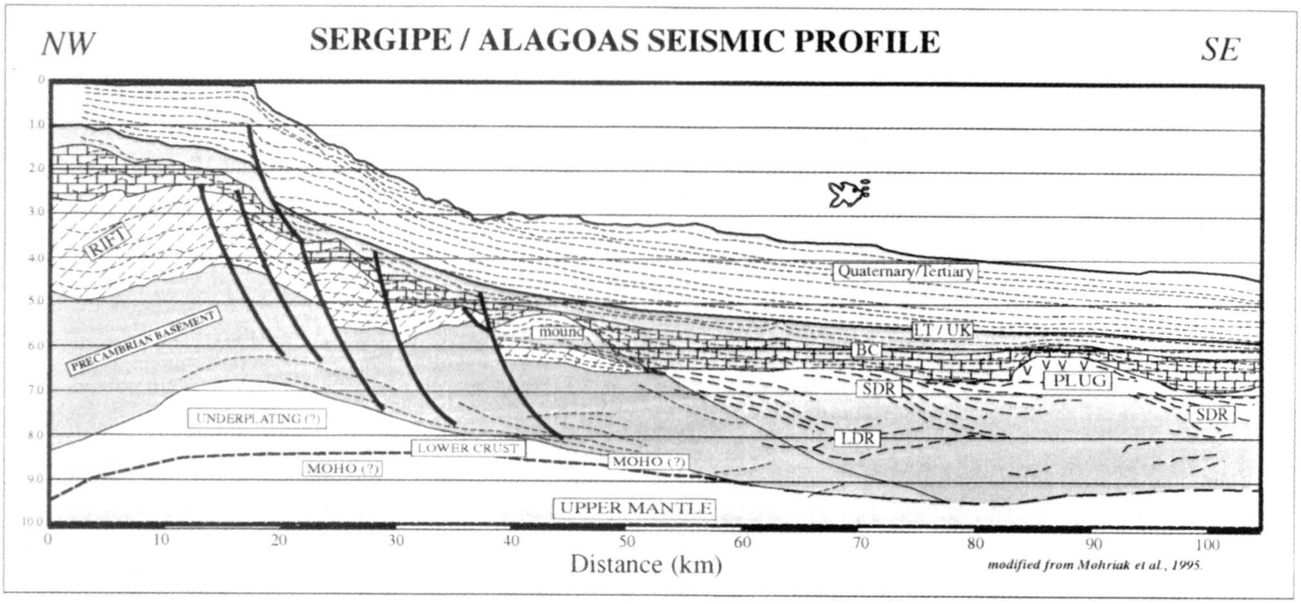

Figure 20: Line interpretation of the Sergipe profile. Near the western limit of the section, there are several sets of layered, sub-parallel reflectors with a landward dip (low frequency events from 5 to 3 s TWT, between shotpoints 1100 and 1400), which are interpreted as pre-rift and Neocomian/Aptian rift sediments that fill the Mosqueiro Low. Major crustal faults controlling rift depocenters are associated with synthetic master faults that detach at lower crust horizons near the seismic Moho. Both landward-dipping and seaward-dipping reflectors (LDR and SDR) are observed in the transition zone between the rifted continental crust and the oceanic crust. A large volcanic plug, probably associated with tectonic activity along the Sergipe fracture zone, occurs at the southeastern extension of the profile. Pure oceanic crust, characterized by hyperbolic reflectors at the top of the volcanic basement, occurs slightly farther basinwards, beyond the last shotpoint of the profile.

times, with a spreading ridge separating the JSA (Brazil) and the Gabon rifts (Africa).

The overall picture of two branches of a rift system with different geodynamic evolution and different crustal architectures (figure 16) has tectonic analogues in many other sedimentary basins worldwide. We suggest that the Tucano and Sergipe basins evolved as a result of regional lithospheric extension during the Neocomian, first distributed over a wide region, perhaps forming rifts along pre-existing zones of weakness in the crust, and subsequently, focussing along a deeper mantle weak zone, local of a posterior plate rupture. This mechanism has been proposed for the Newark Basin (onshore Eastern United States) and the offshore Baltimore Canyon Trough, with a reasonable fit between geological observation and modelling based on numerical methods [*e.g., Harry and Sawyer, 1991*]. Figure 21 shows a schematic geodynamic model based on analogies with the geological model proposed for the Newark Basin-Baltimore Canyon Trough rift system.

There are several similarities between the Newark – Baltimore Canyon model and the development of the Tucano-Sergipe rift system, but also important differences. Some of the problems still remaining in the model include: a) The Tucano rift is much deeper than equivalent rifts along the Eastern Brazilian margin, and is controlled by a westward-dipping fault, whereas most other rifts in the margin are controlled by eastward-dipping faults. b) There is indication of major post-rift uplift and erosion affecting rift-phase stratigraphic sequences; c) The geodynamic models proposed for the Newark Basin-Baltimore Canyon Trough and other rift systems indicate that zones of crustal weaknesses are preferential sites for magmatic intrusions and extrusions, whereas the Tucano rift is almost devoid of any volcanic rocks. The Sergipe Basin is also characterized by absence of volcanism during the Neocomian rift phase, but there is evidence that basinwards, the transition from continental to oceanic crust is associated with thick wedges of seaward-dipping reflectors.

Another geodynamic mechanism that has been recently been advocated to explain the different crustal architectures of rift systems along passive margins is the depth of necking concept, as suggested by *Keen* [*1997*] to explain the deep rift trough of the Jeanne D'Arc Basin. The basement in the depocenter of this basin is extremely deep in comparison with the shallower rift sequences of the Carson Basin, along a regional transect extending from the Newfoundland margins towards the oceanic crust in Eastern Canada. The depth of necking model assumes that the

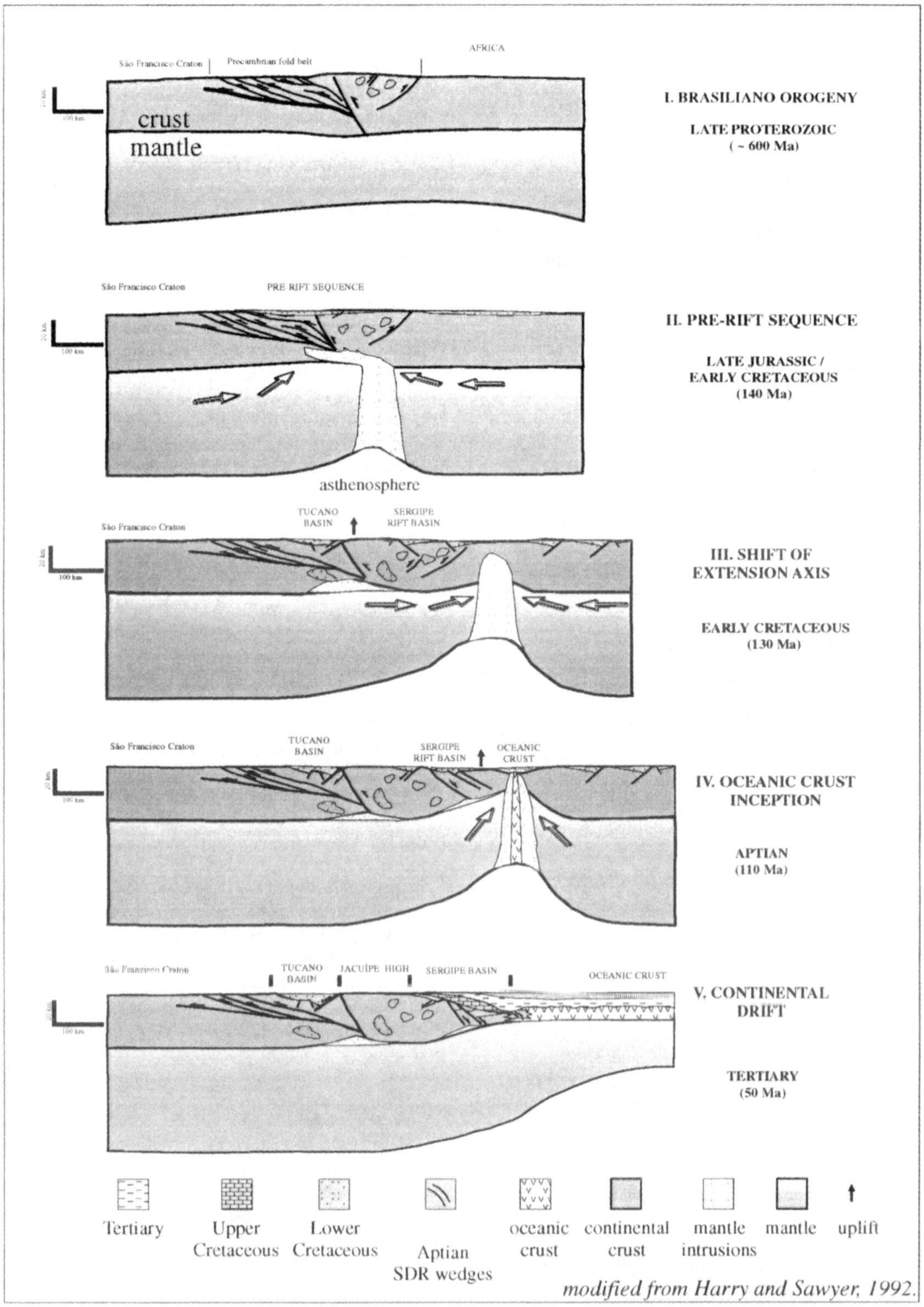

Figure 21: Schematic geodynamic model for the tectonic evolution of the Tucano-Sergipe rift basins.

Moho topography controls the depocenters during the formation of rifts, as observed along the depth-converted profile from the Tucano to the Sergipe basin (figure 16). The crustal architecture associated with the deep level of necking is associated with sedimentary basins characterized by deep syn-rift basins and negligible post-rift or thermal phase of subsidence, and they are associated with a relatively flat Moho topography. The shallow level of necking is characterized by a substantial phase of post-rift or thermal phase of subsidence, and the basins are associated with a marked rise in the Moho topography due to crustal thinning. We suggest that the combination of the two possible mechanisms discussed above (depth of necking and migration of spreading axis) may form deep syn-rift troughs in the landward portion of the adjacent rifted continental margins, and the seaward basins may be characterized by relatively small syn-rift sequences and a substantial thickness of post-rift sediments.

Acknowledgments. We thank M. A. P. Brito for processing the deep seismic profile in the Tucano Basin. We are grateful to W. S. Wanderley and W. L. Machado for their expertise and cooperation in depth-converting the regional deep seismic profiles in the Tucano and Sergipe-Alagoas basins and improving the quality of reproduction of the seismic sections. We also thank F. R. Barros for his contribution in acquisition and interpretation of potential methods in the Tucano and in the Sergipe-Alagoas Basin. Dr. R. M. Darros de Matos and Dr. L. P. Magnavita critically read a preliminary version of this manuscript and contributed with several suggestions that greatly improved the paper. We thank Dr. S. R. Michelucci for his constant cooperation and incentive to prepare this work, and Dr. A. Spadini for constructive observations. We also thank Petrobras management, particularly Dr. L. R. Guardado and Dr. C. F. Lucchesi, for authorization to publish. And last but not least, we express our thanks to Dr. A. Bally for his wholehearted cooperation in reviewing the manuscript and for many insightful comments.

REFERENCES

Aragão, M. A. N. F. and A. A. Peraro, Elementos estruturais do rifte Tucano-Jatobá. *Boletim do Terceiro Simpósio sobre o Cretáceo do Brasil*, Unesp-Rio Claro, pp. 161-165, 1994.

Asmus, H. E. and R. Porto, Diferenças nos estágios iniciais da evolução da margem continental brasileira: possíveis causas e implicações. Anais do XXXI Congresso Brasileiro de Geologia, v.1, pp.225-239, 1980.

Bueno, G. V., Projeto Tucano-Análise de bacias sedimentares. Petrobras-Dexba/ Direx e Cenpes-Divex, Salvador, Ba, 139 p., 1994.

Cainelli, C., Sequence stratigraphy, canyons, and gravity mass flow deposits in the Piaçabuçu Formation, Sergipe-Alagoas Basin, Brasil. *PhD thesis, University of Texas at Austin*, 233 p., 1992.

Castro Jr., A. C. M., The northeastern Brazil and Gabon basins: a double rifting system associated with multiple crustal detachment surfaces. *Tectonics*, v. 6, pp. 727-738, 1987.

Chagas, L. S., D. S. N. Chiossi, R. M. Cerqueira, G. P. Hamsi Jr., G. Marques and G. Lisboa, Evolução tectono-sedimentar do rift da Bacia de Sergipe-Alagoas e novas perspectivas exploratórias. *Relatório Interno Petrobras*, Depex/ Denest/ Serint, 76 p., 1993.

Chang, H. K., R. O. Kowsmann, A. M. F. Figueiredo and A. A. Bender, Tectonics and stratigraphy of the East Brazil Rift system: an overview. *Tectonophysics*, v. 213, pp. 97-138, 1992.

Christensen, N.I. and W. D. Mooney, Seismic velocity structure and composition of the continental crust: a global view. *Journal of Geophysical Research*, v. 100, n. B7, pp. 9761-9788, 1995.

Davison, I., Comment on "The northeastern Brazil and Gabon basins: a double rifting system associated with multiple crustal detachment surfaces", by A.C.M. Castro Jr., and Reply. *Tectonics*, v. 7, n. 6, pp. 1385-1394, 1988.

Davison, I. and R. A. Santos, Tectonic evolution of the Sergipano Fold Belt, NE Brazil, during the Brasiliano Orogeny. *Precambrian Research*, vol. 45, p. 319-342, 1989.

D'el-Rey Silva, L. J. H. and K. R. McClay, Stratigraphy of the southern part of the Sergipano Belt, NE Brazil: tectonic implications. *Revista Brasileira de Geociências*, vol. 25, n. 3, pp. 185-202, 1995.

Destro, N., Tectonism, stratigraphy and sedimentation in the Sergipe and Alagoas Basins, NE Brazil: an overview. *14th International Sedimentological Congress*, Recife-Pe, pp. G.25-G.26, 1994.

Gonzaga, P. M., Levantamento Gravimétrico-Equipe Gravimétrica 32: Bacia Paraíba-Pernambuco e Perfil Tucano/ Sergipe-Alagoas. *Relatório Interno Petrobras*, Depex-Digef-Segeot, Rio de Janeiro, 11 p., 1995.

Guimarães, P. T. M., Basin analysis and structural development of the Sergipe-Alagoas Basin, Brazil. *PhD thesis*, The University of Texas at Austin, 171 p., 1988.

Harry, D. L., and D. S. Sawyer, A dynamic model of extension in the Baltimore Canyon Trough region. *Tectonics*, vol. 11, n. 2, pp. 420-436, 1991.

Hinz, K., A hypothesis on terrestrial catastrophes: wedges of very thick oceanward dipping layers beneath passive continental margins. *Geologisches Jahrbuch*, E-22, pp. 3-28, 1981.

Holbrook, W. S., G. M. Purdy, R. E. Sheridan, L. Glover III, M. Talwani, J. Ewing and D. Hutchinson, Seismic structure of the U.S. Mid-Atlantic continental margin. *Journal of Geophysical Research*, v. 99, n. B9, pp. 17,871-17,891, 1994.

Inda, H. A. V., H. D. Schorscher, M. A. Dardenne, C. Schobbenhaus, N. L. E. Haralyi, P. C. A. Branco, R. Ramalho, O craton do São Francisco e a faixa de dobramentos Araçuaí. In: C. Schobbenhaus, D. A. Campos, G. R. Derze, e H. E. Asmus (coord.), *Geologia do Brasil*, DNPM, Brasilia, 1984, pp. 193-248, 1984.

Keen, C. E., Crustal Architecture, extension and tectono-stratigraphic evolution of rifted margins in the North Atlantic. *5th International Congress of the Brazilian Geophysical Society*, Expanded Abstracts, vol. 1, pp. 1-2, 1997.

Klemperer, S. L., T. A. Hauge, E. C. Hauser, J. E. Oliver and C. J. Potter, The Moho in the northern Basin and Range province, Nevada, along COCORP 40°N seismic reflection transect. *Geol. Soc. of America Bull.*, v. 97, pp. 603-618, 1986.

Machado Jr., D. L., L. K. Dehira, C. D. R. Carneiro and F. F. M. Almeida, Reconstruções paleoambientais do Juro-Cretáceo no nordeste oriental brasileiro. *Revista Brasileira de Geociências*, vol. 19, n. 4, p. 470-485, 1990.

Magnavita, L.P., Geometry and kinematics of the Recôncavo-Tucano-Jatobá rift, NE Brazil, *Phd Thesis*, University of Oxford, England, 492 p., 1992.

Magnavita, L. P. and J. D. Cupertino, A new approach to the geologic configuration of the Lower Cretaceous Tucano and Jatobá basins, northeastern Brazil. *Revista Brasileira de Geociências*, v. 18 (2), pp. 222-230, 1988.

Magnavita, L. P., I. Davison and N. J. Kusznir, Rifting, erosion, and uplift history of the Recôncavo-Tucano-Jatobá rift, northeast Brazil.*Tectonics*, vol. 13, n. 2, pp. 367-388, 1994.

Matos, R. M. D., The northeastern Brazilian rift system. *Tectonics*, v. 11, n. 4, pp. 766-791, 1992.

McBride, J. H., D. B. Snyder, R. W. England and R. W. Hobbs, Dipping reflectors beneath old orogens: a perspective from the British Caledonides. *GSA Today*, vol. 6, n. 4, pp. 1-6, 1996.

Meyers, J. B., B. R. Rosendahl and J. A. Austin Jr., Deep-penetrating MCS images of the South Gabon Basin: implications for rift tectonics and post-breakup salt remobilization. *Basin Research*, v. 8, pp. 65-84, 1996.

Milani, E. J., Aspectos da evolução tectônica das bacias do Recôncavo e Tucano Sul, Bahia, Brasil. *Ciência-Técnica-Petróleo, Seção Exploração de Petróleo*, Petrobras/ CENPES, v. 18, 61 p., 1987.

Milani, E. J. and I. Davison, Basement control and transfer tectonics in the Recôncavo-Tucano-Jatobá rift, northeast Brazil. *Tectonophysics*, v. 154, pp. 47-70, 1988.

Milani, E. J., M. C. Lana and P. Szatmari, Mesozoic rift basins around the northeast Brazilian microplate (Reconcavo-Tucano-Jatoba, Sergipe-Alagoas). *In: W. Manspeizer (ed.), Triassic-Jurassic rifting*, Part B, Elsevier, New York, pp. 833-858, 1988.

Mohriak, W. U. and M. A. L. Latgé, Deep Seismic Survey of Brazilian Passive Margin Basins: The Southeastern Region. *In: Second International Geophysical Congress of the Brazilian Geophysical Society*, Salvador-BA, Vol. II, pp. 621-626, 1991.

Mohriak, W. U., M. C. Barros, H. L. Rabelo, R. D. Matos, Deep seismic survey of Brazilian Passive Basins: the northern and northeastern regions. *In: Third International Congress of the Brazilian Geophysical Society*, Rio de Janeiro, RJ, November 7-11th, 1993, Expanded Abstracts, vol. 2, pp. 1134-1139, 1993.

Mohriak, W. U., J. H. L. Rabelo, R. D. Matos and M. C. Barros, Deep seismic reflection profiling of sedimentary basins offshore Brazil: geological objectives and preliminary results in the Sergipe Basin. *Journal of Geodynamics*, vol. 20, n. 4, pp. 515-539. 1995a.

Mohriak, W. U., M. Bassetto and I. S. Vieira,. Deep seismic constraints on the crustal architecture of sedimentary basins in the Brazilian margin: tectonic and exploratory implications. *V Simpósio Nacional de Estudos Tectônicos-SNET 95*, Boletim de Resumos Expandidos, Gramado-RS, 19-22/11/1995, pp. 246-248, 1995b.

Mohriak, W. U., M. Bassetto and I. S. Vieira, Tectonic evolution of the rift basins in the Northeastern Brazilian margin. *5th International Congress of the Brazilian Geophysical Society*, Expanded Abstracts, vol. 1, pp. 4-7, 1997.

Mohriak, W. U., M. Bassetto and I. S. Vieira, Crustal architecture and tectonic evolution of the Sergipe-Alagoas and Jacuípe basins, offshore northeastern Brazil. *Tectonophysics*, 288, pp. 199-220, 1998.

Mutter, J. C., Seaward dipping reflectors and the continent-ocean boundary at passive continental margins. *Tectonophysics*, v.114, pp.117-131, 1985.

Neves, B. B. B., A. N. Sial and J. P. T. Albuquerque, Vergência centrífuga residual no sistema de dobramentos sergipanos. *Revista Brasileira de Geociências*, v. 7, pp. 102-114, 1977.

Ojeda, H. A. O., Structural framework, stratigraphy, and evolution of Brazilian marginal basins. *AAPG Bulletin*, v. 66, pp. 732-749, 1982.

Pontes, C. E. S., F. C. C.Castro, J. J. G. Rodrigues, R. R. P. Alves, R. T. Castellani, S. F. Santos e M. B. Munis, Reconhecimento tectônico e estratigráfico da Bacia Sergipe-Alagoas em águas profundas. *In: Congresso Brasileiro de Geofísica*, 2, Salvador, BA, SBGF, Salvador, BA, Boletim de Resumos Expandidos, vol. II, pp. 638-643, 1991.

Rabinowitz, P.D. and J. LaBreque, The Mesozoic South Atlantic Ocean and evolution of its continental margins. *Journal of Geophysical Research*, v.84, n.B11, pp.5973-6002, 1979.

RADAMBRASIL, Projeto Radambrasil-Folhas SC-24/25, Aracajú/ Recife. MME/ SG, Rio de Janeiro, v. 30, 855 p., 1983.

Rosendahl, B. R., Architecture of continental rifts with special reference to East Africa. *Ann. Review Earth Planetary Sciences*, 15: pp. 445-503, 1987.

Rosendahl, B. R., H. Groschel-Becker, J. Meyers and K. Kaczmarick, Deep seismic reflection study of a passive margin, southeastern Gulf of Guinea. *Geology*, v. 19, pp. 291-295, 1991.

Scheirer, D. S. and R. W. Hobbs, Seismic attenuation in the continental crust SW of England. *Geophysical Journal International*, v. 103, pp. 533-540, 1990.

Serpa, L. and B. Voogd, Deep seismic reflection evidence for the role of extension in the evolution of continental crust. *Geophysical Journal fo the Royal Astronomical Society*, v.89, pp.55-60, 1987.

Singh, S. C. and D. McKenzie, Layering in the lower crust. *Geophysical Journal International*, v. 113, pp. 622-628, 1993.

Teisserenc, P. and J. Villermin, Sedimentary Basin of Gabon: geology and oil systems. *In: Edwards J. D. and P. A. Santogrossi (editors), Divergent/Passive Margin Basins. Am. Assoc. Pet. Geol. Mem.*, 48: 117-199, 1989.

Ussami, N., G. D. Karner and M. H. P. Bott, Crustal detachment during South Atlantic rifting and formation of Tucano-Gabon basin system. *Nature*, v. 322, pp. 629-632, 1986.

Wanderley Filho, J.R., and J. C. S. V. Graddi, Projeto Jacuípe. *Relatório Interno Petrobras*, Salvador, Ba, 28 p., 1993.

Wannesson, J., J. C. Icart, and J. Ravart, Structure and evolution of adjoining segments of the west African margin determined from deep seismic profiling. *In: R. Meissner and others (eds.), Continental Lithosphere: Deep Seismic Reflections, American Geophysical Union*, Washington, D.C., pp. 275-289, 1991.

Mohriak, Webster Ueipass, Avda. Chile, 65 - room 1301-E, Rio de Janeiro, RJ, Brazil. E-mail: webmohr@ep.petrobras.com.br

Bassetto, Marcelo, Avda. Chile, 65 - room 1301-E, Rio de Janeiro, RJ, Brazil. E-mail: bassetto@ep.petrobras.com.br

Vieira, Ines Santos, Avda. Chile, 65 - room 1301-E, Rio de Janeiro, RJ, Brazil. E-mail: isvieira@ep.petrobras.com.br

Towards Dynamic Restoration of Geologic Profiles: Some Lessons from Analogue Modelling

Hemin A. Koyi

Hans Ramberg Tectonic Laboratory, Department of Earth Sciences, Uppsala University, Sweden

Restoration of geologic profiles aims to achieve geometric consistency and kinematic compatibility between deformed and restored sections. During restoration, major folds and faults can be fairly easily detected and accounted for. However, mass transfer and volume changes cannot be measured as easily and therefore cannot be accurately accounted for during the restoration. Accurate restoration requires not only correct geometric and kinematic restoration, but also consideration of strain balancing (e.g. penetrative strain) and material flow into and out of the restored profile. Two examples from analogue models are described here to illustrate; first, the significance of area change in restoration, and second, how to account for it. Since their initial dimensions and material properties are well documented, analogue modelling can be a useful tool for testing restoration of profiles in extensional and contractional provinces. Models simulating both extensional and contractional areas show that mass transfer is a significant component of deformation. Modelling of a ductile material simulating rock salt in extensional areas reveals that the three-dimensional flow plays a significant role in the deformation. Similarly, models of contractional areas illustrate that tectonic compaction can account for 15% of the shortening. Therefore, change in area due to three-dimensional flow of ductile substrates and tectonic compaction should be accounted for during restoration of geologic profiles. As a first step towards dynamic restoration of geologic profiles, a preliminary approach to restore the 3-D flow of rocksalt by calculation of 3-D differential loading is presented here. Since passive margins are areas where non-evaporitic sediments can undergo both extension (at the trailing updip areas) and compression (in the leading downdip deformation front) during their gravitational gliding above a layer of rock salt or lubricant shale, the results of the current study can give insights to restoring deformation in passive margins.

INTRODUCTION

On many passive margins, non-evaporitic sediments suffer extreme extension as they gravitationally glide basinward above a layer of ductile salt [*Duval et al.*, 1992; *Liro and Coen,* 1995; *Mohriak et al.,* 1995]. During their basinward gliding, these sediments undergo shortening at the leading edge of the deformational front. In order to restore deformation in these areas, both extensional and contractional structures need to be accounted for. Section balancing and restoration techniques are widely used in the analysis of deformation in extensional and compressional terrain [*Suppe,* 1985; *Ramsey and Huber,* 1987; *Marshak and Woodward,* 1988; *Rowan and Kligfield,* 1989; *Woodward et al.,* 1985, 1986, 1989; *Moretti et al.,* 1990; *Schultz-Ela,* 1992; *Rowan,* 1993; *Mitra,* 1994; *Diegel et al.,* 1995; *Hooper and More,* 1995; *Buddin et al.,* 1997;

Atlantic Rifts and Continental Margins
Geophysical Monograph 115

William et al., 1997; *McBride*, 1998; *McBride et al.*, 1998a and b]. Balanced sections and palinspatic reconstruction have been used to study salt tectonics in the Gulf of Mexico [*Worrall and Snelson* 1989; *Hossack and McGuinness* 1990; *Wu et al.* 1990; *Rowan*, 1993; *Hossack*, 1995; *Rowan et al.*, 1995; *McBride et al.*, 1998].

Our knowledge of salt and thrust tectonics has greatly improved in recent years by studying analogue models and reflection seismic data [*Mulugeta and Koyi*, 1987, 1992; *Mulugeta*, 1988; *Dahlen, et al.*, 1983; *Colletta et al.*, 1991; 1992, *Hale et al.*, 1992; *Koyi et al.*, 1992a; *Liu et al.*, 1992; *Vendeville and Jackson*, 1992 a and b; *Jackson et al.*, 1995; *Mohriak et al.*, 1995; *Rowan*, 1993; 1995; and many others]. In analogue models, comparison of the deformed stages with well-documented initial stages of the models, allows accurate restoration and enables quantification of area loss in shortened models through the measurement of amount of layer-parallel shortening (layer-parallel thickening and area/volume loss) and monitoring of the three-dimensional flow of ductile material in extensional models. In this short paper, only one of the limitations of restoration (change in area/volume), which has a significant impact on the outcome of the restoration, is discussed. Examples from model analogues, with known initial dimensions, are presented to emphasise the significance of penetrative strain (in fold/thrust belts) and 3-D flow of material (e.g. rock salt or overpressure shale) in extensional areas during restoration.

Some restoration softwares have been applied on physical analogues to study the evolution of diapiric structures [e.g., *Lin* 1992; *Schultz–Ela* 1992]. Several assumptions and limitations are inherent in any cross section restoration, and these should be fully understood in order to evaluate the results. Some of the limitations of section restoration (i.e. two dimensional) in extensional areas with salt structures were addressed by Koyi [1994], Williams et al. [1997] and recently by McBride [1998]. The increasing availability of 3-D seismic data, should be a driving element for 3-D restoration of extensional areas with salt tectonics.

Map view restoration of folded and faulted areas have been carried out by using strain trajectories *[Cobbold*, 1979; *Gratier et al.*, 1989], interactive techniques [*Gratier et al.*, 1991] where defined rigid elements of a folded surface were flattened to form a horizontal initial surface, vertical shear restoration in three dimensions [*Verrall*, 1981]; and rotation of rigid triangles whose nodal points define deformed surfaces in 3-D space [*Williams et al.*, 1997]. As an attempt towards 3-D restoration of normal faults, Buddin et al. [1997] presented a new software which extends the 2-D inclined and vertical shear geometric techniques into the third dimension to enable the manipulation of 3-D surfaces and volumes rather than lines and areas on a cross section.

CONTRACTION

Bed length and area balancing have been used extensively to interpret the deformation history of and the amount of shortening in contractional areas from cross sections [*Dahlstrom*, 1969; *Hossack*, 1979; *Baker et al.*, 1988; *Dixon*, 1982; *Cooper et al.*, 1983; *Woodward et al.*, 1985; *Mitra*, 1994; *Jawal et.al.*, 1997; *Mukul and Mitra*, 1998 and many others]. Since most fold/thrust belts are deeply eroded and records of syntectonic sedimentation is limited, restoration of profiles of these areas is mainly confined to a single reconstruction of the undeformed geometry to a flattened datum [*Boyer and Elliot*, 1982; *Woodward et al.*, 1985]. In these methods, either the length or the area of the shortened layers is kept constant during the construction of the balanced cross sections. However, deformation in fold and thrust belts is taken up by three main components: layer-parallel shortening, folding and thrusting. Section balancing only takes account of the deformation by folding and thrusting, whereas, shortening by layer-parallel shortening is not considered in these approaches.

Fold belts in deep water regions are addressed by many workers [e.g., *Weimer and Buffler*, 1992]. These areas suffer significant amount of shortening as a consequence of updip extension, which is compensated by downdip contraction. Figure 1a shows a regional seismic profile illustrating a deep-water fold belt in the Atlantic continental margin. Figure 1b shows a simplified line interpretation of the seismic data, and Figure 1c shows a schematic line-interpretation [modified from *Guimarães et al.*, 1989] of a regional profile in a deep-water fold belt in the South Atlantic equatorial margin. Some of these structures displayed in the seismic image might be interpreted as related to fault bend folding and to general models of thrust belts (Fig. 2).

Several papers have addressed penetrative strain during restoration of geologic profiles by accounting for small scale strain to estimate the amount of layer-parallel strain in shortened areas [e.g. *Hossack*, 1979; *Woodward et al.*, 1985; *Mitra*, 1978, 1994]. Investigating the deformation in the Cambrian quartzites in the Appalachian fold belt, Mitra [1978] quantified the extent of deformation by pressure solution and concluded that pressure solution was an important deformation mechanism in fold and thrust belts. Marshak and Engelder [1985] emphasized the role of cleavage formation and pressure solution in fold and thrust belts to account for volume loss strain. However, since their initial dimensions are well documented, analogue models enable a more precise estimate of the spatial and temporal distribution of penetrative strain in fold and thrust belts. This paper will discuss in detail a sandbox model to illustrate the significance of penetrative strain in restoring profiles in compressional areas.

Model

Sandbox models have been used to study different aspects of accretionary wedges [*Davis et al.* 1983; *Dahlen et al.* 1984; *Karig* 1986; *Zhao et al.* 1986; *Mulugeta,* 1988; *Mulugeta and Koyi* 1987, 1992; *Colletta et al.* 1991; *Liu et al.,* 1992; *Koyi,* 1995; *Gutscher et al.,* 1996]. Loose sand, which has a low friction and shows a Coulomb behaviour, is an appropriate analogue material to simulate deformation in the upper brittle crust. The rheology of sand is approximated by the Coulomb equation:

$$\tau = \tau_o + \sigma \tan \phi \qquad (1)$$

Where τ is shear strength, τ_o is cohesive strength, σ is the normal stress and ϕ is the angle of internal friction. Figure 3 shows the Coulomb strength of the sand used in the experiment.

Here, a sandbox model is used to illustrate the significance of layer-parallel shortening component of strain and distribution of penetrative strain in contractional areas. The model, consisting of passively layered loose sand, was shortened by 47% from one end (Fig. 2). The model contained thin layers of passively coloured homogeneous sand making a constant thickness of 5 mm [*Mulugeta and Koyi,* 1992; *Koyi,* 1995]. Sequential sections were prepared by means of a vacuum cleaner throughout the deformation at every 1.5% shortening for photographing. These sections were used to quantify area loss and penetrative strain in the model.

In the experimental approach, the additional complicating factors of erosion, deposition, material anisotropy due to facies changes, time variations in pore fluid pressure ratios across the wedge and changes in friction coupling along the decollement surface were deliberately left out. The model described here was deformed on a horizontal and rigid planar substrate, and therefore does not account for natural decollement slope or irregularities. For details on deformation and sectioning of this model, see [*Mulugeta and Koyi,* 1992].

Length change

In a profile of the final stage of the deformed model, bed-length balancing was conducted for three layers located at different stratigraphic horizons (Fig. 4). For each layer, deformation was partitioned to quantify each of the three strain components; layer-parallel shortening, kink folding and thrusting. This was done by measuring the length of segments of each deformed layer and adding them together. Consequently, the effects of imbrication and folding were taken out by joining the displaced segments of the layers together and unfolding them (Fig. 4). By comparing this "restored" length with the deformed length of the layers, the amount of shortening by thrusting and folding was calculated (Fig. 4). However, the "restored" layers showed to be shorter than the initial length of the layers. Therefore, the "restored" length was compared to the initial (undeformed) length of the model to calculate the amount of shortening taken up by layer-parallel shortening.

The above procedure enabled quantifying the change of mode of deformation with depth. Earlier, Mulugeta and Koyi [1992] and Koyi [1995] documented that layer-parallel shortening dominates at deeper levels of model accretionary wedges whereas imbrication prevails at shallower levels. The current analysis demonstrate that shallow layers could be restored back to 83% of their initial length (Fig. 4). The remaining 17% of the initial length, which was not restorable, was consumed by layer-parallel shortening during deformation of the model. The deep layer, on the other hand, could only be restored to 59% of its initial length. The entire unrestored 41% of the layer length was consumed by layer-parallel shortening. This comparison does not only illustrate that the mode of deformation in contractional areas changes with depth, but it also points out that bed length balancing of different layers in natural profiles of accretionary wedges may give different percentages of shortening. As deeper layers could be restored to shorter initial lengths than shallower layers, they show more percentage bulk shortening. Layer-parallel shortening is recoverable in the model because its initial dimensions are well documented. However, in nature, where initial stages are poorly known, it would be difficult to restore the amount of layer-parallel shortening which increases with depth. Therefore, to determine the bulk shortening of natural examples, bed length balancing should be exercized on layers located at different stratigraphic horizons. The above analysis shows that for the same lithology, shallow-located layers better represent the amount of shortening since they undergo less layer-parallel shortening than deeper layers. However, an initially highly porous rock lo-

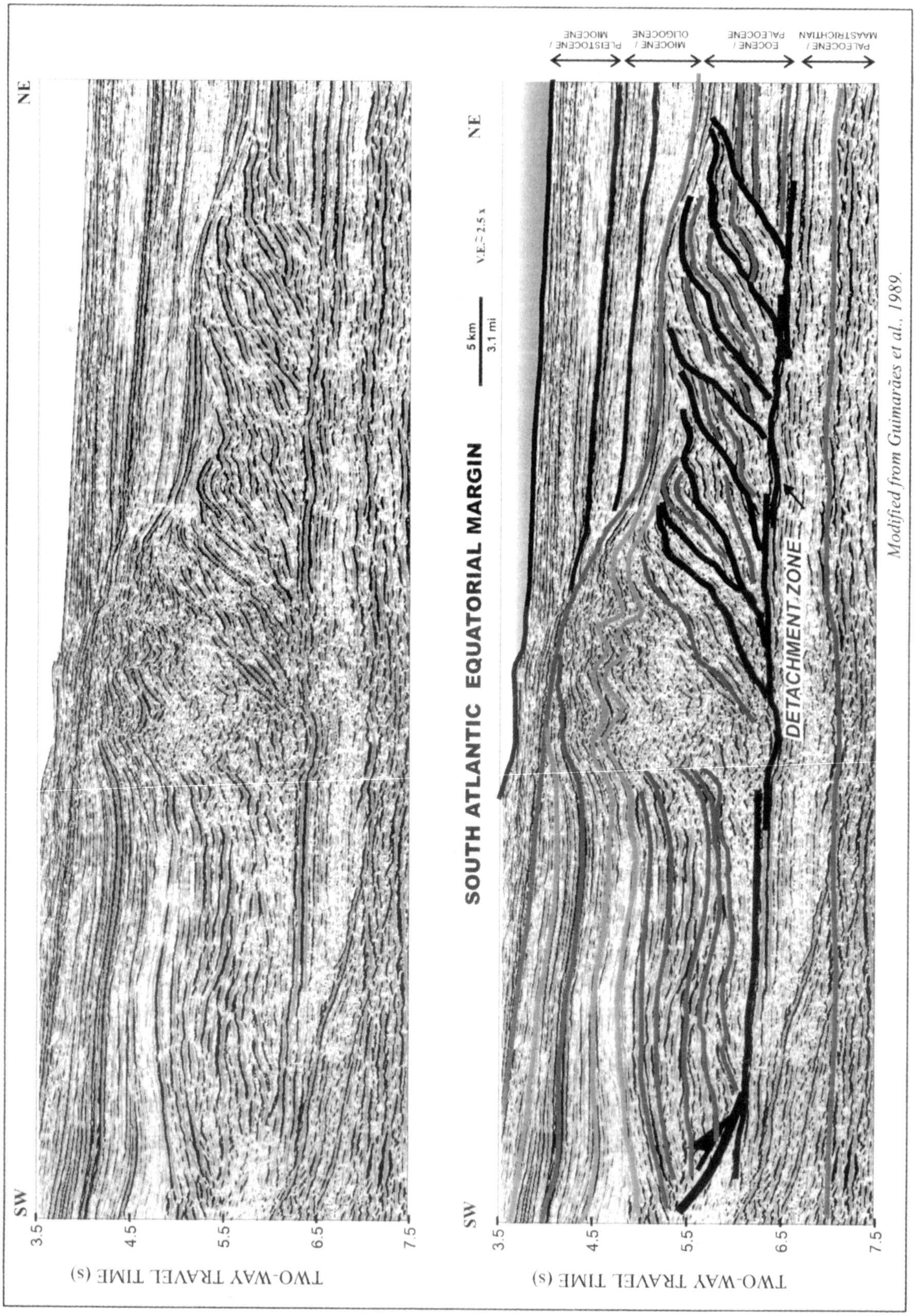

Figure 1-A: Regional seismic profile in the South Atlantic continental margin, showing a deep water fold belt;
Figure 1-B: Simplified interpretation of the regional seismic profile in the South Atlantic continental margin.

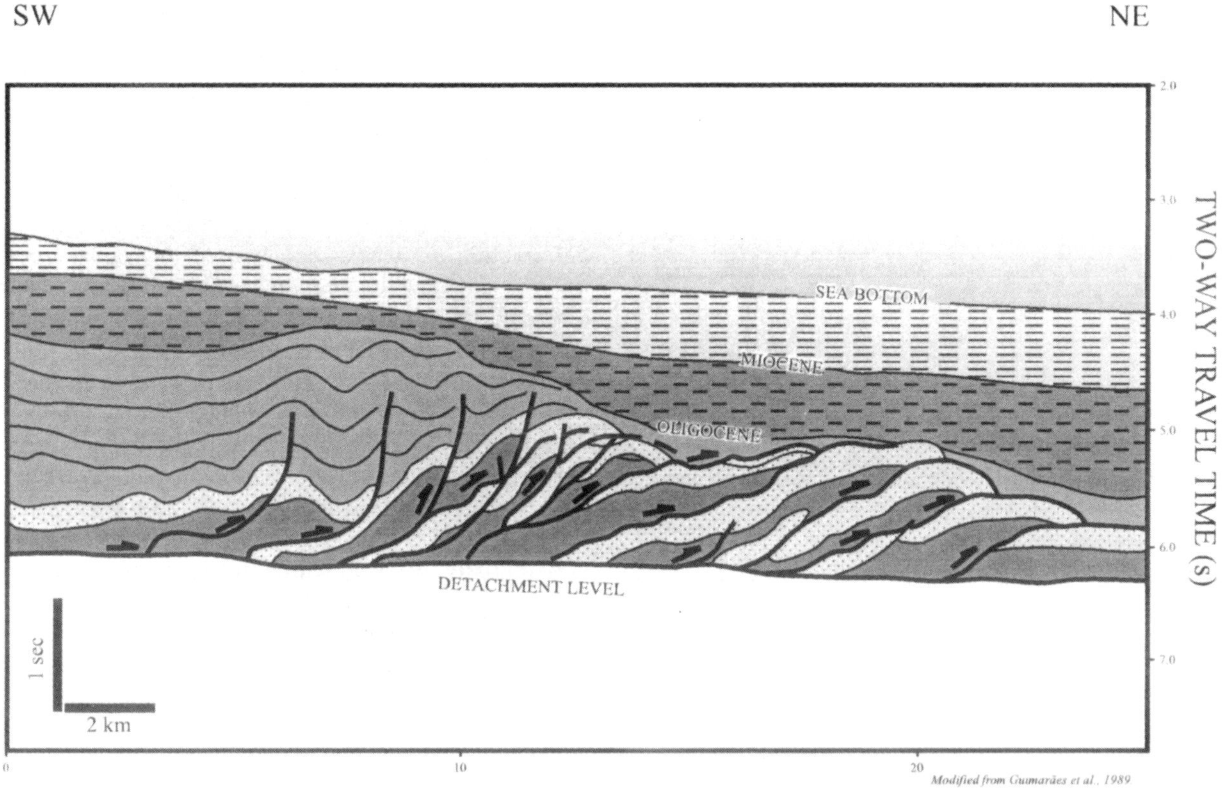

Figure 1-C: Schematic line drawing interpretation of a deep water fold belt in the equatorial margin of the South Atlantic (modified from Guimarães et al, 1989).

Figure 2: A photograph of a sandbox model after 40% shortening, showing seven imbricate sheets (numbered). The numbers indicate the sequence at which the imbricates have formed. Note the thickening of the deeper layers. Arrow shows direction of shortening (from left to right).

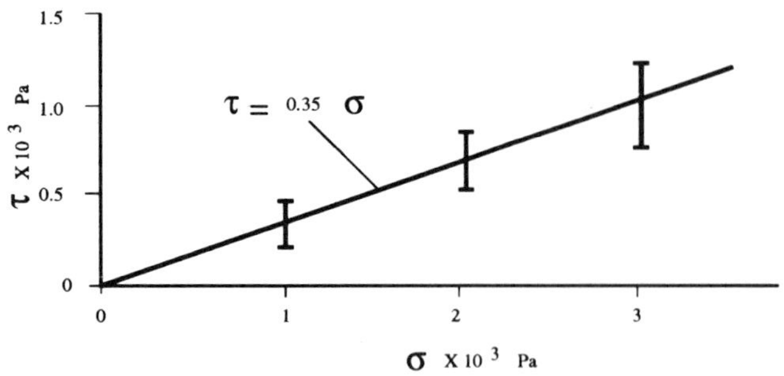

Figure 3. A plot of the mechanical property, shown by shear strength (τ) versus normal stress (σ), for the loose sand used in the sandbox model. The angle of internal friction (φ) of the sand is 18° (modified after Mulugeta, 1988).

cated at a shallow level may undergo a considerable amount of layer-parallel shortening during compression.

Area change

In order to quantify the amount of area (volume) loss and its lateral variation in the model during deformation, the area of the model wedge was measured between two stages of imbricate formation. In other words, the area of the wedge was measured at the onset of formation of an imbricate (at 29.5% bulk shortening) and throughout the deformation of a new imbricate (in this case imbricate number 4, Fig. 2), (at 38% bulk shortening). During the same period (between 29.5% and 38% bulk shortening), the area of that specific imbricate (number 4) was also measured to compare area loss within the wedge and the imbricate during the same period of deformation.

These measurements show that the entire wedge looses 3.9% of its area of which 3% was taken up by the youngest of the imbricates (in this case, imbricate number 4). This indicates that 76% of the total area loss within the wedge was accommodated within the youngest of the imbricates at the toe where as only 23% of the area loss was taken up by the rest of the wedge (Fig. 5). The newly accreted materials at the toe of the wedge undergo more tectonic compaction, whereas the rear part of the wedge, which contains older and already compacted stack of imbricates, accommodates only a small amount of tectonic compaction with deformation. As was also suggested for Nankai accretionary prisms [*Karig and Lundberg,* 1990], where shortening of the soft sediments in the prism toe is pervasive and is not only accommodated by displacement along imbricate thrusts and shear bands, the active toe area in the model wedge stands for most of the penetrative strain which the wedge experiences. In general, cumulative compaction in the model is highest at the rear of the wedge, whereas incremental compaction is highest at the toe area. In other words, as the wedge grows, incremental compaction is preferentially localized within the newly accreted materials at the toe of the model. The rear portion of the wedge is transported as a relatively rigid block without much strain as the wedge accretes material at the toe. Erosion or change in the basal slope or friction may relocalise the deformation and thereby alter this scenario.

EXTENSION

Different hypothesis for the complexities of structures associated with salt tectonics in the South Atlantic have been discussed recently by several researchers. Figure 6a shows a regional seismic profile in the Santos Basin, where reflectors dipping seawards are characterized in the post-salt successions, suggesting pseudo-downlaps [*Mohriak et al.,* 1995]. Figure 6b shows a line-drawing interpretation of this profile, which is characterized by thickening of post-Aptian to Early Tertiary strata from the platform toward deep water, and younging of the stratigraphic successions in the same direction. Several different interpretations have

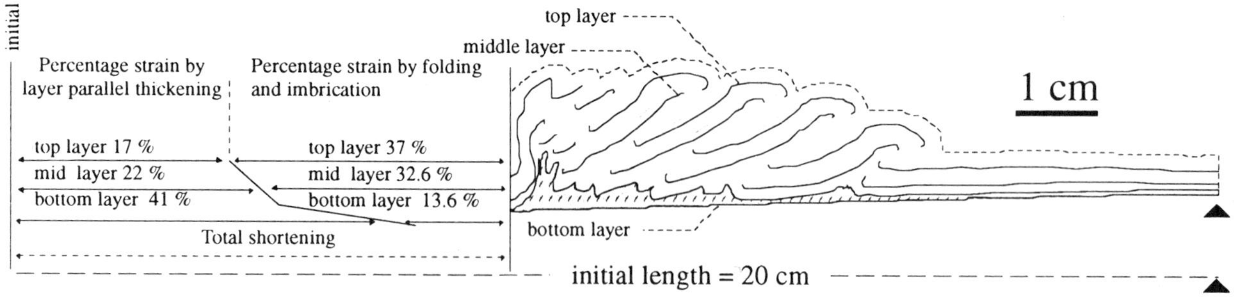

Figure 4. A line drawing of model profile at 47% bulk shortening. Three layers (top, middle and bottom) are length-balanced in order to quantify the three components of strain (layer-parallel shortening, folding and thrusting). Note that the top layer deforms mainly by folding and thrusting in contrast to the bottom layer which deforms mainly by layer-parallel shortening.

been suggested for the Santos profile. Current interpretation assumes that extension was associated with the early halokinesis and faulting of the Albian overburden rocks during episodes of massive clastic progradation [*Mohriak et al.,* 1995], whereas others assume almost no extension [*Ge et al.* 1997].

Ge et al. [1997] restored the Santos profile without incorporating regional extension of the overburden, and based on physical models, interpreted the pseudo-downlaps as caused by collapse of overburden on a salt basin that was totally evacuated. According to their models, the extension, as proposed by others [e.g. *Demercian et al.,* 1993; *Mohriak et al.,* 1995; *Szatamari et al.,* 1996] would either violate basin mechanics (for example, requiring up-slope gravity gliding) or generate unrealistic strains, and balancing of the Albian gap (Fig. 6b) would require at least 25 km of downdip shortening. Although some contractional structures have been reported in deep water region of the Santos Basin [*Cobbold et al.,* 1995], their magnitude seems to be far too low to compensate for the suggested extension. Get et al.'s [1997] model also suggested that an extensional origin for the Albian gap would require an initial salt layer to be an absurd 36 km thick, assuming conserved area. However, other models suggested that regional extension of the overburden during episodes of progradation might account for these structures. Szatmari et al. [1996], based on physical modelling of the Santos Basin structure, suggested that extension of the overburden plays an important role in creating the antithetic faults. Further, the effects of massive clastic progradation on salt structures characterized by antithetic faults have also been observed in both the African margin [*Liro and Coen,* 1995] and the Gulf of Mexico [*Diegel et al.,*1995].

The above discussion illustrates that overburden extension associated with massive clastic progradation constitutes an open problem in passive margins studies. One of the major reasons for such different results in 2-D restorations is the uncertainty about the amount of salt loss/gain versus the amount of regional extension in the profile. There are two extreme end-members: 1, no extension, but maximum salt loss/gain in the restored profile, and 2, no salt loss/gain in the profile, but maximum extension. These two end members are applied to the Quenguela graben, Kwanza Basin [*Masson,* 1972; *Duval et. al.* 1992]. Assuming no regional extension in this basin, Masson [1972] needed to reduce the salt area by 77% to restore the profile. Duval et. al. [1992], restored the same profile by assuming maximum regional extension with minimum loss of salt. Any solution between these end-members may satisfy a more appropriate two-dimensional restoration. By outlining the volume of salt in the restored area, three-dimensional restoration narrows the uncertainty associated with salt loss considerably and improves the likelihood of a more realistic approach.

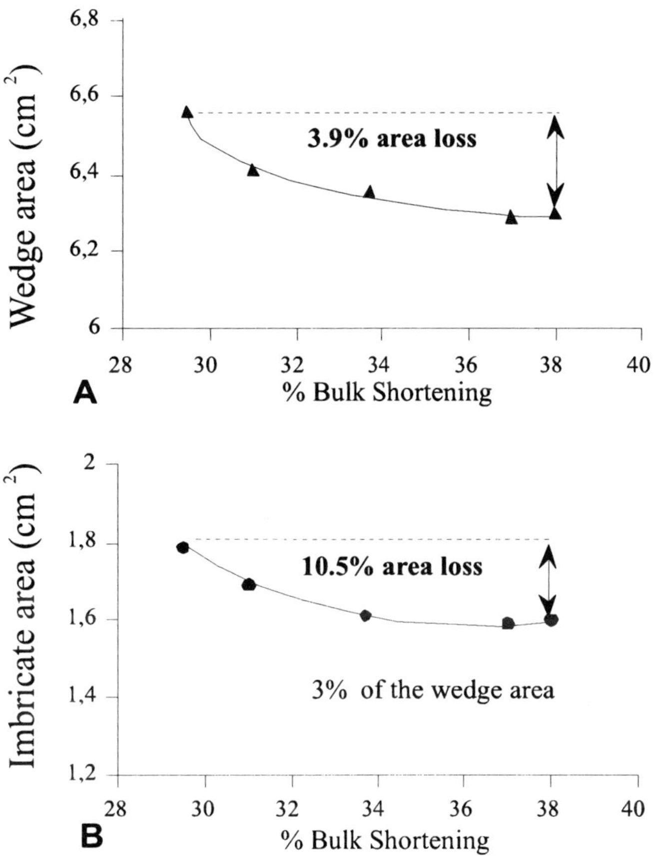

Figure 5. Plots of (A) wedge area and (B) imbricate area with bulk shortening. imbricates. At 38% shortening, the wedge looses ca. 4% of its initial area. During the same period, the youngest imbricate looses 10.5% of its initial length which is equal to ca. 3% of the total area of the wedge. This suggests that much of the area loss within the wedge is taken up within the youngest of the imbricates at the toe.

Restoration of profiles in extensional areas, where a ductile layer (e.g. rocksalt or overpressure shale) is involved, faces a similar problem when area change is concerned. However, area change here is mainly due to flow of the ductile material in and out of the restored section. Worrall and Snelsen [1989] used structural restorations to study the evolution of allochthonous salt in the Gulf of Mexico. In their restoration, Worrall and Snelsen [1989] correctly allowed the areas of salt between the deformed and restored profiles to change according to the overburden loading. Below, an analogue model, simulating the influence of basement faults on salt diapirs in the Danish Basin [*Koyi and Petersen,* 1993], is used to illustrate the out-of-section (along strike) flow of a ductile material during the evolution of the model.

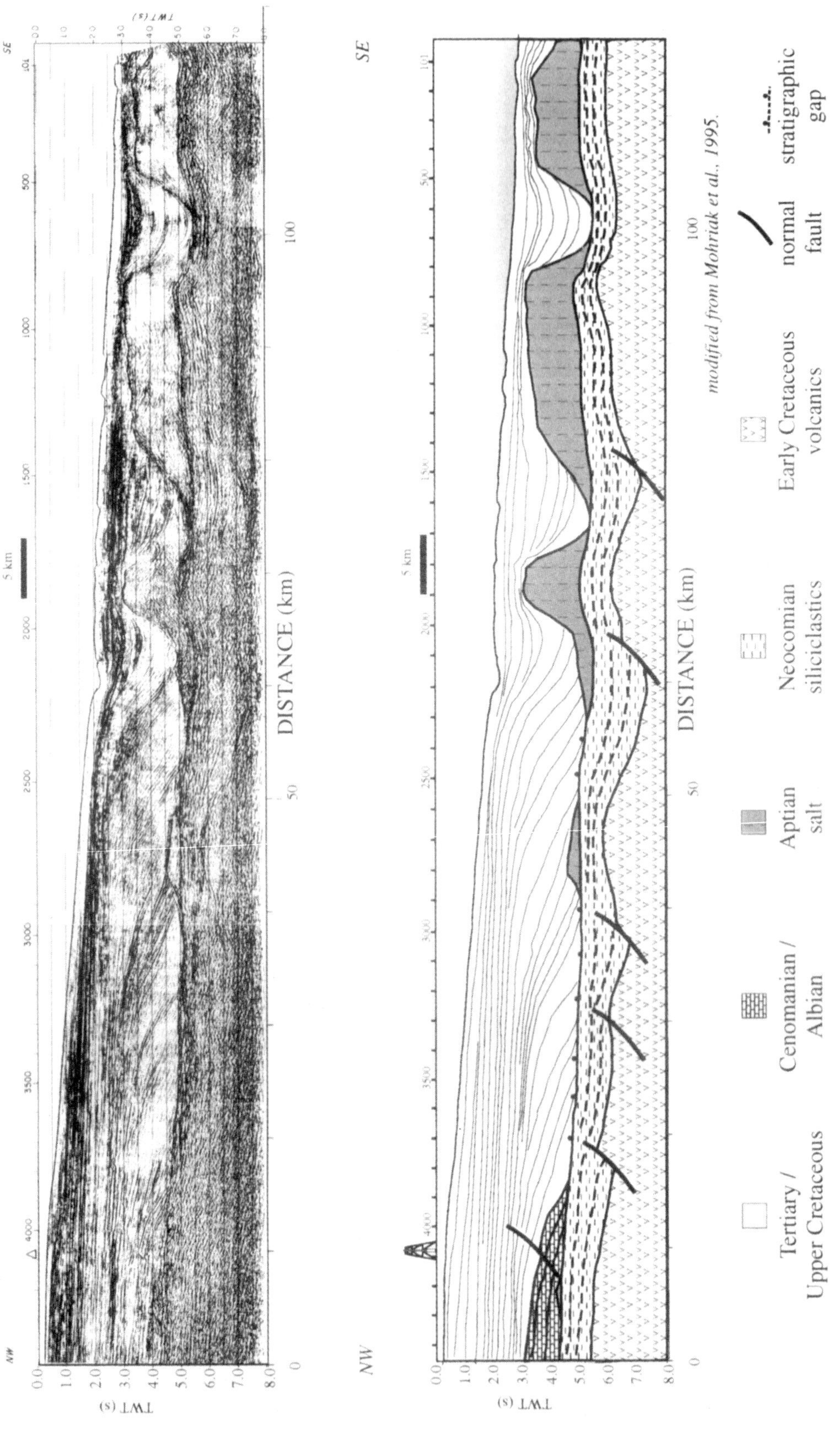

Figure 6-A: Regional seismic profile in the Santos Basin, showing the Cabo Frio fault zone.
Figure 6-B: Line interpretation of the regional profile in the Santos Basin (modified from Mohriak et al., 1995)

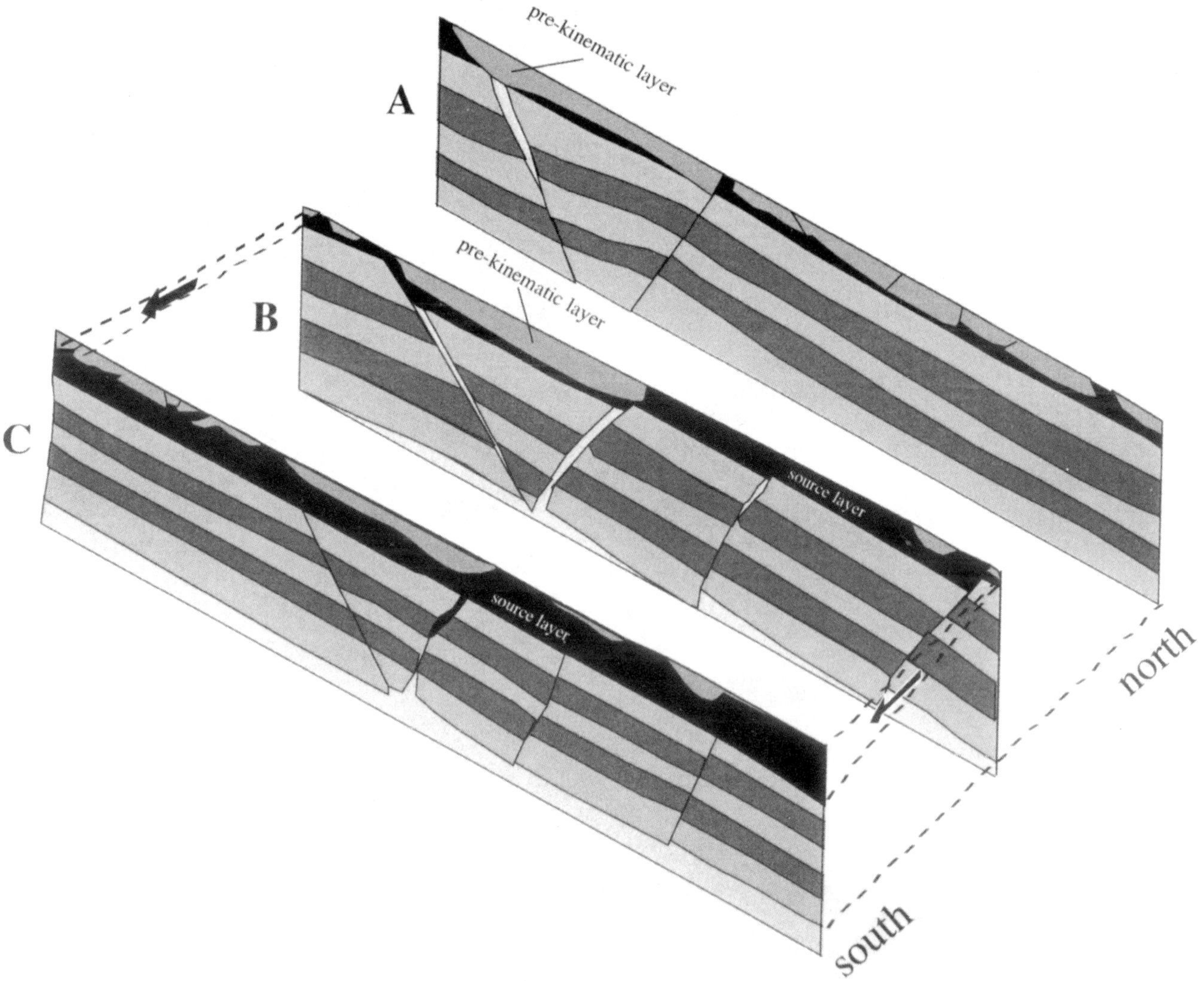

Figure 7. Three restored profiles (A, B and C) of a scaled analogue model that started with an initially uniformly thick viscous source layer of Rhodorsil gomme (black) overlain by a uniformly thick prekinamtic layer of cohesive sand. These profiles were restored separately with the restoration software LOCACE (Courtesy of Dr. Jake Hossack). Note that due to flow of the source material in and out of the profiles (arrows), the source layer restored to a variable thickness and the prekinematic layer (grey layer above the source layer) erroneously appears to have started discontinuous.

Three parallel sections of this model (Fig. 7) were restored using LOCACE [*Moretti and Larrere,* 1989]. In one of the restored sections (Figure 7C), the area of the ductile salt analogue was 220% of the initial area. This two-fold increase in the area of the ductile layer is due to flow of the ductile material into the section during the evolution of the diapirs. On the contrary, in another section (Figure 7A), the restored area of the source layer was only 40% of the initial area. In this section, 60% of the initial source layer flowed out of the section due to higher load to areas of lower load. The software used to restore the sections could not account for and restore this three-dimensional flow of the ductile material into and out of the restored sections.

Applied to scaled analouges with known sedimentation and deformation histories, restoration can be successful [*Lin,* 1992; *Schultz-Ela,* 1992]. If the initial thickness of the ductile substratum is known (as in models), the rigid basement can be shifted vertically until the initial thickness of the ductile substratum is restored to its true value after restoration of the overburden. In nature, where the initial parameters are not (or poorly) known, restoration of the thickness of the ductile substratum and its three-dimensional flow is more difficult. Restoring the salt thickness incorrectly leads to errors in the evolution history

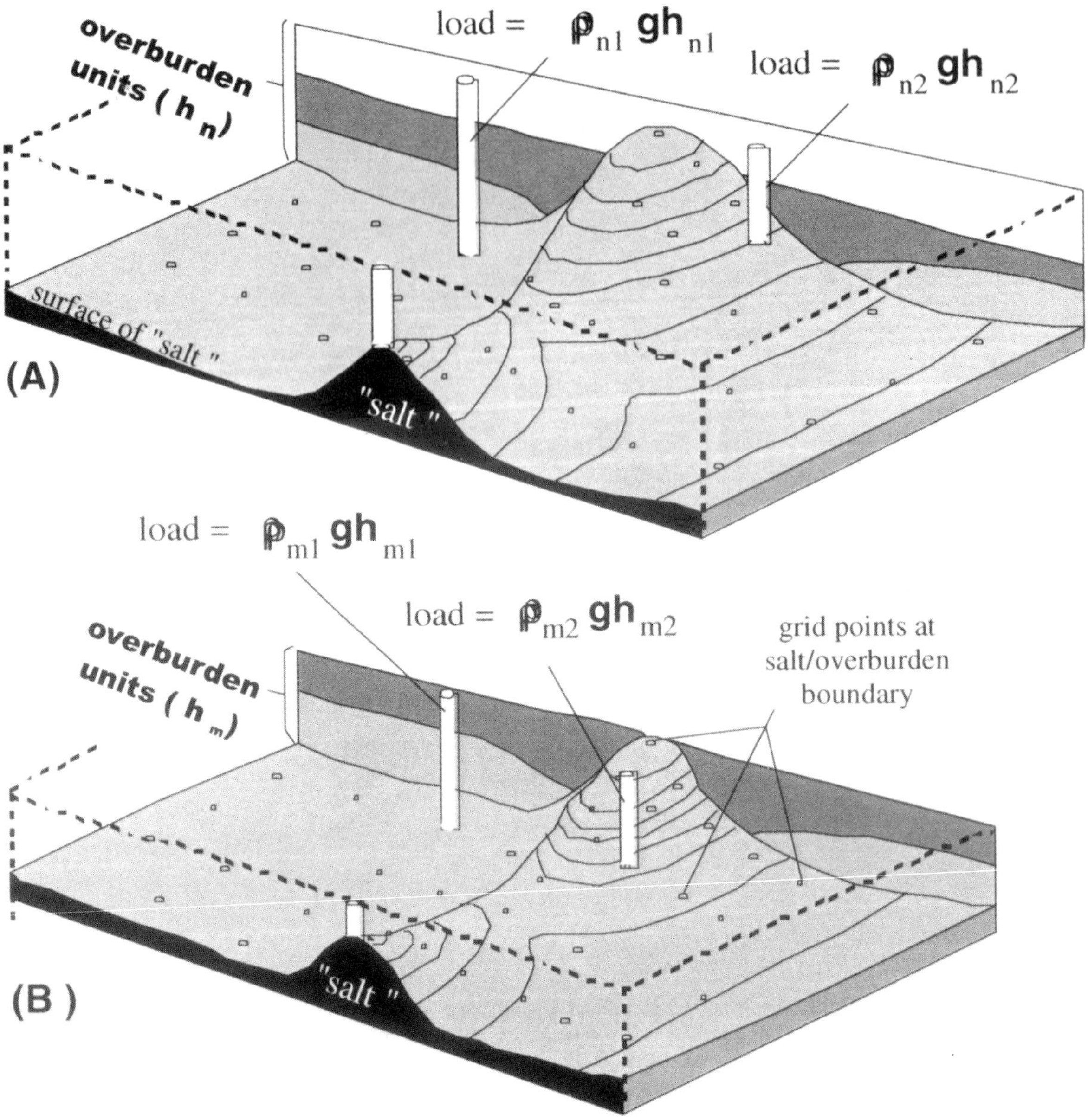

Figure 8. Two schematic block diagrams showing a late (A) and an earlier stage (B) of the evolution of two salt structures. In the starting diagram (A), a grid of points is chosen on the overburden/salt boundary where the load in each point is calculated. In (B), the top layer of overburden is removed and the salt/overburden boundary is adjusted according to the new differential loading on this surface.

of the studied area because flow of the ductile material in and out of the section plane influences the sedimentation and deformation of the overburden units.

Restoring 3-D flow

Rowan [1993] used a systematic approach for restoration of salt structures in profiles by accounting for the loading of the removed overburden units and incorporating isostatic subsidence. Here, this approach is taken a step further by accounting for the three dimensional flow of salt during restoration as a result of three-dimensional overburden loading. It may be possible to restore the flow of the ductile material in and out of restored profiles to a certain extent if parallel profiles are restored interactively. To achieve that, interactive restoration of serial profiles could be incorporated in the restoration software. Parallel profiles give an idea of the relative thickness change of overburden units and the resulting differential loading across the profiles. By including the effect of differential loading across the parallel profiles during the restoration procedure, at least part of the flow of the ductile material could be restored. With the increasing availability of 3-D seismic data, it should not be difficult to structurally manipulate 3-D surfaces and maintain a detailed kinematics in order to exercise a reliable 3-D restoration of an area with salt structures. Below, a preliminary approach for 3-D restoration of salt flow by using 3-D seismic image, is presented.

Since the driving force for the movement of rocksalt is the differential loading between areas of lower and those of higher overburden load, a 3-D restoration of salt flow is directly controlled by the differential loading on the overburden/salt interface. Calculating this differential loading is achieved by defining a grid of points on the interface between the rocksalt and its overburden, where the load that the overburden exerts on the underlying rocksalt is measured (Figure 8A). In the next step, a layer of overburden is removed, the underlying layers are decompacted, unfolded and rotated until the surface of the top layer is horizontal (Figure 8B), which is a normal procedure in many 2-D restorations of diapiric structures. Any extension that has occurred during the deposition of the removed layer is roughly restored by joining the top surfaces. After that, new measurements of the load at the same grid points are taken. Then, the 3-D flow of the rocksalt is approached by allowing the rocksalt to adjust to the new pressure field (Figure 8B). The difference in differential loading between two stages of restoration is the key factor which allows the flow of salt "back" to its initial location before the deposition of the removed layer. This procedure can be repeated until the entire overburden is removed. It should be emphasized here that the above approach does not account for any deformation (e.g. faulting) in the sub-salt (basement) layers.

CONCLUSIONS

Analysis of sequential sections of a sandbox model suggests that layer-parallel shortening, which is a significant component of strain in fold-and-thrust belts, vary both laterally and with depth. Model results also show that area loss is dominant at the active deformation front of a fold-and-thrust belt.

Three-dimensional restoration of salt flow can be restored on 3-D seismics by calculating the differential loading on the overburden salt interface during the sequential removal of each overburden unit.

Testing restorations using analogue models where the initial and final stages are known and checking them with the geological history of the restored sections would clarify the limitation of the applied methods and may help to minimize errors during restoration of 3-D salt flow.

Acknowledgements. Thanks are due to Prof. C. J. Talbot, Dr. G. Mulugeta and M. Friberg for reading and commenting on the manuscript. Review comments by Drs. A.W. Balley, Mark Rowan and Webster Mohriak are greatly appreciated. Special thanks are due to Dr. Webster Mohriak for kindly providing final copies of figures 1 and 6. Funded by the Swedish Natural Research Council (NFR). Dr. Jake Hossack provided the restored version of the model profiles in figure 7.

REFERENCES

Baker, D.M., J. R. Lillie, S. R. Yeats, G. D. Johnson, M. Yousuf, M., and Zamin, A.S.H., Development of the Himalayan frontal thrust zone: Salt Range, Pakistan, *Geology*, 16, 3-7, 1988.

Boyer, S.E. and D. Elliott, Thrust systems. *AAPG Bull.* 66, 1196-1230, 1982.

Buddin, T. S., S. J. Kane, G. D. Williams, and S. S. Egan, A sensitivity analysis of 3-dimensional restoration techniques using vertical and inclined shear constructions, *Tectonophysics*, 269, 33-50, 1997.

Colletta, B., J. Letouzey, R. Pinedo, J. F. Ballard, P. and Balé, Computerized X-ray tomography analysis of sandbox models: Examples of thin-skinned thrust systems. *Geology*, 19, 1063-1067, 1991.

Cobbold, P. R., Removal of finite deformation using strain trajectories, *Journal of Structural Geology*, 1, 67-72, 1979.

Cobbold, P.R., P. Szatmari, L.S. Demercian, D. Coelho and E.A. Rossello, Seismic and experimental evidence for thin-skinned horizontal shortening by convergent radial gliding on evaporites, deep-water Santos Basin, Brazil, in Salt Tectonics: a global perspective, edited by M. P. A. Jackson, D. G. Roberts, and S. Snelson, *AAPG Memoir*, 65, 305-321, 1995.

Cooper, M.A., Garton, M.R. and Hossack, J.R., The origin of the Basse Normandie duplex, Boulonnais, France, *Journal of Structural Geology*, 5, 139-152, 1983.

Dahlen, F. A., J. Suppe, and D. M. Davis, Mechanics of fold and-thrust belts and accretionary wedges (continued): Cohesive Coulomb Theory, *J. Geophy. Res.*, 89, 10087-10101, 1984.

Dahlen, F. A., Non-cohesive critical Coulomb wedges: and exact solution, *J. Geophy. Res.*, 89 (10), 125-33, 1984.

Dahlstrom, C. D. A., Balanced cross sections, *Can. J. Earth. Sci.*, 6: 743-747, 1969.

Davis, D., J. Suppe, and F. A. Dahlen, Mechanics of fold-and-thrust belts and accretionary wedges, *J.Geophy. Res.*, 88, 1153-1172, 1983.

Demercian, L.S., P. Szatmari and P.R. Cobbold, Style and pattern of salt diapirs due to thin-skinned gravitational gliding, Campos and Santos basins, offshore Brazil. *Tectonophysics*, 228, 393 – 433, 1993.

Diegel, F.A., J.F. Karlo, D.C. Schuster, R.C. Schoup, and P.R. Tauvers, Cenozoic structural evolution and tectono-stratigraphic framework of the northern Gulf Coast continental margin, *in Salt Tectonics: a global perspective*, edited by M. P. A. Jackson, D. G. Roberts, and S. Snelson, *AAPG Memoir*, 65, 109-151, 1995.

Duval, B., C. Cramez and M.P.A. Jackson, Raft tectonics in the Kwanza Basin, Angola, *Marine and Petroleum Geology*, 9, 389-404, 1992.

Ge, H., M.P.A. Jackson, B.C. Vendeville, Kinematics and dynamics of salt tectonics driven by progradation. *AAPG Bulletin*, 81, 393-423, 1997.

Gratier, J. P., G. Menard, and A. Arpin, Strain displacement compatibility and restoration of the Chaines Subalpines of the western Alps, in Alpine Tectonics, edited by M. P. Coward, D. Dietrich and R. G. Park, *Geological Society of London Special Publication*, 45, 65-81, 1989.

Gutscher, M.A., N. Kukowiski, J. Malavieille, and S. Lallemand, Cyclical behaviour of thrust wedges: Insight from basal friction sandbox experiments, *Geology*, 24, 135-138, 1996.

Hale, D., N. H. Ross, and J. Stefani, Imaging salt with turning seismic waves, *Geophysics*, 57 (11), 1453–1462, 1992.

Hooper, R. J., and C. More, Evaluation of some salt-related overburden structures in the U.K. Southern North Sea, in Salt Tectonics: a global perspective, edited by M. P. A. Jackson, D. G. Roberts, and S. Snelson, *AAPG Memoir* 65, 251-259, 1995

Hossack, J., The use of balanced cross section in the calculation of orogenic contraction, *Journal of Geological Society of London*, 136, 705-711, 1979.

Hossack, J., Geometrical rules of section balancing of salt structures, in Salt Tectonics: a global perspective, edited by M. P. A. Jackson, D. G. Roberts, and S. Snelson, *AAPG Memoir*, 65, 29-40, 1995.

Jackson, M. P. A., D. G. Roberts, and S. Snelson, Salt Tectonics: a global perspective, *AAPG Memoir* 65, 454 pp., 1995

Jaswal, T. J., R. J. Lillie, and R. D. Lawrence, Structure and evolution of the northern Potwar deformed zone, pakistan, *AAPG Bull.*, 81, 308-328, 1997.

Karig, D.E., Physical properties and mechanical state of accreted sediments in the Nankai Trough, Southwest Japan Arc, *Geol. Soc. Am. Memoir* 166, 117-133, 1986.

Karig, D.E. and N. Lundberg, Deformation bands from the toe of the Nankai Accretionary prism, *J. Geophy. Res.*, 95 (B6), 9099-9109, 1990.

Koyi, H., Estimation of salt thickness and restoration of cross-sections with diapiric structures: a few critical comments on two powerful methods, *Journal of Structural Geology*, 16, 1121-1128, 1994.

Koyi, H., Mode of internal deformation in sand wedges, *Journal of Structural Geology*, 17, 293-300, 1995.

Koyi, H. and K. Petersen, The influence of basement faults on the development of salt structures in the Danish Basin, *Marine and Petroleum Geology*, 10, 82-94, 1993.

Lin, H. T., Experimental study of syndepositional and postdepositional gravity spreading of a brittle overburden and viscous substratum, *MSc–Thesis, Department of Geological Sciences*, The University of Texas at Austin, Austin, Texas, 196pp, 1992.

Liro, L.M. and R. Coen, Salt deformation history and postsalt structural trends, offshore southern Gabon, West Africa, in Salt Tectonics: a global perspective, edited by M. P. A. Jackson, D. G. Roberts, and S. Snelson, *AAPG Memoir*, 65, 323-331, 1995.

Liu, H., K. R. McClay, and D. Powell, Physical models of thrust wedges, in *Thrust Tectonics*, edited by K. R. McClay, 71-81, 1992.

Marshak, S. and T. Engelder, Development of cleavage in a fold-thrust belt in eastern New York, *Journal of Structural Geology*, 7, 345-359, 1985.

Marshak, S., and N. B. Woodward, Introduction ot cross section balancing, in Basic Methods in *Structural Geology,* edited by S. Marshak, and G. Mitra, Engelwood Cliffs, NJ, Prentice Hall, 303-332, 1988.

Masson, P., L'exploration du pétrole en Angola, *Rev. Assoc. Franç. Tech. Pétrole*, 212, 21-34, 1972.

McBride, B. C., The evolution of allochthonous salt along a mega regional profile across the northern Gulf of Mexico Basin, *AAPG Bull.*, 82 (5B), 1037-1054, 1998.

McBride, B. C., M. G. Rowan, and P. Weimer, The evolution of allochthonous salt system, northern Green Canyon and Ewing Bank (offshore Louisiana), northern Gulf of Mexico, *AAPG Bull.,* 82 (5B), 1013-1036, 1998a.

McBride, B. C., P. Weimer, and M. G. Rowan, The effect of allochthonous salt on the petroleum systems of northern Green Canyon and Ewing Bank (offshore Louisiana), northern Gulf of Mexico, *AAPG Bull.*, 82 (5B), 1083-1112, 1998b.

McGuiness, D. B. and J.R. Hossack, The development of allochthonous salt sheets as controlled by the rates of extension, sedimentation, and salt supply, *Fourteenth Annual Research Conference*, Gulf Coast Section, SEPM Foundation, Houston, Texas, 127-149, 1993.

Mitra, S., Microscopic deformation mechanisms and flow laws in quartzites within the South Mountain anticline, Journal of Geology, 86, 129-152, 1978.

Mitra, G., Strain variation in thrust sheets across the sevier fold-and-thrust belt (Idaho-Uta-Wyoming): implications for section restoration and wedge taper evolution, *Journal of Structural Geology*, 6, 51-61, 1994.

Mohriak, W. U., J. M. Macedo, R. T. Castellani, H. D. Rangel, A. Z. N. Barros, M. A. L. Latgé, J. A. Ricci, A. M. P. Mizusaki, P. Szatmari, L. S. Demercian, J. G. Rizzo, and J. R. Aires, Salt tectonics and structural styles in the deep-water province of the Cabo Frio region, Rio de janeiro, Brazil, in Salt Tectonics: a global perspective, edited by M. P. A. Jackson, D. G. Roberts, and S. Snelson, *AAPG Memoir*, 65, 273-304, 1995.

Moretti, I., M. Larrere, LOCACE, computer-aided construction of balanced geological cross sections, *Geobyte*, 4, 16-24, 1989.

Moretti, I., S. Wu, and A. W. Bally, Computerized balanced cross section LOCACE to reconstruct an allochthonous salt sheet, offshore Louisiana, *Marine and Petroleum Geology*, 371-377, 1990.

Mukul, M. and G. Mitra, Finite strain and strain variation analysis in the Sheeprock thrust sheet; an internal thrust sheet in the

Provo salient of the Sevier fold-and-thrust belt, central Utah, *Journal of Structural Geology*, 20, 385-405, 1998.

Mulugeta, G., Modelling the geometry of Coulomb thrust wedges, *Journal of Structural Geology*, 10, 847-859, 1988.

Mulugeta, G., and H. Koyi, Three-dimensional geometry andkinematics of experimental piggyback thrusting, *Geology*, 15, 1052-1056, 1987.

Mulugeta, G., and H. Koyi, Episodic accretion and strain partitioning in a model sand wedge, *Tectonophysics*, 202, 319-333, 1992.

Ramsey, J. G., and M. I. Huber, The techniques of modern structural geology, volume 2: Folds and Fractures, *Academic Press*, Oxford, 700 pp., 1987.

Ratcliff, W. D., Post stack depth migration improves salt body imaging, *Oil and Gas Journal*, 90 (43), 51–57, 1992.

Rowan M. J., A systematic technique for sequential restoration of salt structures, *Tectonophysics*, 228, 331-348, 1993.

Rowan M. J., Structural styles and evolution of allochthonous salt, central Louisiana outer shelf and upper slope, in Salt Tectonics: a global perspective, edited by M. P. A. Jackson, D. G. Roberts, and S. Snelson, *AAPG Memoir*, 65, 199-228, 1995.

Rowan, M. J., and R. Kligfield, Cross section restoration and balancing as an aid to seismic interpretation in extensional terranes, *AAPG Bull.*, 73, 955-966, 1989.

Rowan, G. M., T. Villami, P. B. Flemmings, and P. Weimer, Use of cross section restoration to determine paleobathymetry and sea-floor paleotopography in the Gulf of Mexico, *Geology*, 24, 299-302, 1996.

Rowan, M. J., and P. Weimer, Salt-sediment interaction, northern Green canyon and Ewing bank (offshore Louisiana), northern Gulf of Mexico, *AAPG Bull.*, 82 (5B), 1055-1082, 1998.

Schults-Ela, D. D., Restoration of cross sections to constrain deformation processes of extensional terranes, *Marine and Petroleum Geology*, 9, 372-388, 1992.

Suppe, J., Principles of structural geology, *Prentice Hall*, N. Jersey, 537 pp., 1985.

Szatmari, P., M.C. Guerra, and M.A. Pequeno, Genesis of large counter-regional normal fault by flow of Cretaceous salt in the South Atlantic, Santos Basin, Brazil, in Salt Tectonics, edited by G.I.Alsop, D.J. Blundell, and I. Davison, *Geological Society of London, Special Publication 100*, 259-264, 1996.

Vendeville, B. C., and M. P. A. Jackson, The rise of diapirs during thin–skinned extension. *Marine and Petroleum Geology*, 9, 331–353, 1992a.

Vendeville, B. C., and M. P. A. Jackson, The fall of diapirs during thin–skinned extension. *Marine and Petroleum Geology*, 9, 354–371, 1992b.

Verall, P., Structural interpretations with application to North Sea problems, course Notes, *J. A. P. E. S.*, 3, 156 pp, 1981.

Williams, G. D., S. J. Kane, T. S. Buddin, and J. A. Richards, Rostoration and balance of complex folded and faulted rock volumes: flextural flattening jigsaw, fitting and decompaction in three dimensions, *Tectonophysics*, 273, 203-218, 1997.

Woodward, N. B., S. E. Boyer, and J. Suppe, An outline of balanced cross sections, *University of Tennessee Department of Geological Sciences Studies in Geology*, 11 (second edition), 170 pp., 1985.

Woodward, N. B., D.R. Grey, and D.B. Spears, Including strain data in balanced cross-sections, *Journal of Structural Geology*, 8, 313-324, 1986.

Worrall, D. M., and S. Snelson, Evolution of the northern Gulf of Mexico, with emphasis on Cenozoic growth faulting and the role of salt, in The Geology of North America; An overview, edited by A. W. Balley and A. R. Palmer, *Geological Society of America Decade of North American Geology*, A, 97-138, 1989.

Wu, S., A. W. Bally, and C. Cramez, Allochthonous salt, structure and stratigraphy of the north-eastern Gulf of Mexico, Part II: Structure, *Marine and Petroleum Geology*, 7, 334-370, 1990.

Zhao, W. L., D. M. Davis, F. A. Dahlen, and J. Suppe, Origin of convex accretionary wedges: evidence from Barbados, *Journal of Geophysical Research*, 91, 10246-10258, 1986.

Hemin A. Koyi, Hans Ramberg Tectonic Laboratory, Department of Earth Sciences, Uppsala University, Sweden.

Tectonic Evolution of the Equatorial South Atlantic

Renato M. Darros de Matos

Petrobras – E&P-RNCE, Natal - RN - Brazil 59074-600

The Brazilian and West African Equatorial margins comprise an unique example of a transform margin, characterized by a series of on- and offshore Mesozoic-Cenozoic basins, distributed over 2,200 km along the South Atlantic. Although a multi-stage stretching process evolved throughout the Aptian-Cenomanian, an almost instantaneous Aptian stage was responsible for a widespread fracturing of the Equatorial Atlantic. Conventional extensional processes can not explain the kinematics and rift geometry of the Equatorial South Atlantic basins. Accepted pure-shear or simple-shear rift mechanisms, typical of divergent margins, cannot be promptly used in basins generated as a response to major transform motions along a continental-scale plate boundary. The commonly accepted causal processes for rifting, such as passive/active or diffuse/discrete rifting, can not accommodate the South Atlantic Equatorial data set. Even though shearing signatures and pull-apart features are easily recognized throughout the margin, their magnitude and basin architecture varies significantly as a function of the distance from the main transform faults. These factors resulted in significant differences in thermal evolution, tectonic subsidence, facies distribution and uplift history. The tectonic evolution of the sedimentary basins along the Equatorial Atlantic is better understood by considering three stages: pre, syn and post-transform movements. These are related to kinematic and dynamic controls provided by the emplacement of fractured swells as proto mid ocean ridges, followed by the creation of oceanic crust and the onset of transform shearing between Africa and Brazil.

INTRODUCTION

Active transform margins are characterized by sharp topographic differences between ocean and continent, which are defined by active strike-slip faulting, a poorly developed shelf, and very deep sedimentary basins (*Mascle,* 1976; *Twiss* and *Moores*, 1992; *Basile et al.*, 1993). The Equatorial Atlantic provide may provide a very good example of an ancient transform margin with a long history of transform movements, leading to the development of very complex sedimentary basins on each side of the ocean (Figures 1 and 2). These basins are characterized by multiple subsidence phases and associated structural styles.

The stepwise opening of the South Atlantic resulted in considerable diachronic deformation in the Equatorial and eastern Brazilian margins, recorded in the northern and northeastern portion of Brazil as a series of magmatic and sedimentary events. While the eastern Brazilian margin and the central portion of the Western African margin had a longer rift stage (Neocomian-Barremian), the stretching stage in the Equatorial Margin was shorter. During the Aptian stage, an almost instantaneous period of oblique

Atlantic Rifts and Continental Margins
Geophysical Monograph 115

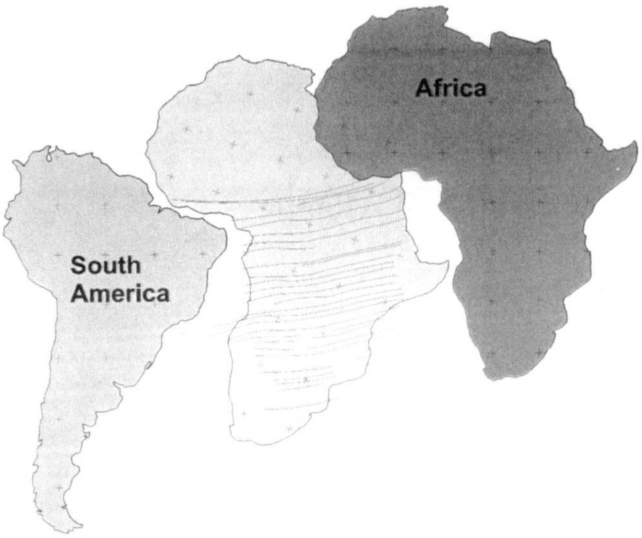

Figure 1: South America and Africa at present and about 90 Mybp. Continuous lines between Africa and South America represent gravity / topographic anomalies, interpreted as fracture zones based on Geosat gravity map of Sandwell & Smith (1995).

extension was responsible for a widespread fracturing of the Equatorial Atlantic, even though a multi-stage stretching process evolved throughout the Aptian-Cenomanian interval.

The well-known divergent margin rift architecture, commonly defined by tilted half-grabens, with depocenters controlled by planar/listric faults, and filled with a divergent wedge of siliciclastics sediments, normally with fanglomerates adjacent to the main border fault, are not promptly recognized in the Equatorial margin domain. In many cases it is difficult to recognize a divergent seismic pattern in the rift sequence, as illustrated in the offshore Potiguar and Ceará basins in Brazil, where conglomerates and even coarse to medium-grained alluvial sediments may be missing adjacent to main border faults. Another major difference between the transform and the divergent margin segments of the South Atlantic includes the lack of a significant pre-rift section, as well as a transitional evaporitic section between the rift and drift stages. Local tectonics and magmatism had a key role in the post-break up, thermal subsidence phase. Abrupt variations in the temporal and spatial location of magmatism as well as the spatial distribution of releasing and restraining bends through the region led to a quite different drift stage from basin to basin. Another key factor was deformation rate.

The purpose of this paper is to provide an overview of the tectono-stratigraphic evolution of the Equatorial Brazilian Margin, and its African counterpart, as well as to discuss a kinematic model to describe the Atlantic opening (Figure 2). Emphasis is placed on the interpretation of three-dimensional geometries of the basins, and particularly, in the description of their structural styles and tectono-stratigraphic sequences.

The Equatorial portion of the Brazilian Margin encompass a series of offshore basins, from east to west, the Potiguar, Ceará, Barreirinhas, Pará-Maranhão and Foz do Amazonas (Figures 2, 3, and 4), some of which extend to the onshore (e.g. the Potiguar and Barreirinhas basins). Among them, there are three aborted rifts: (1) the onshore Potiguar graben, (2) the Marajó basin (Mexiana, Limoeiro and Gurupá grabens) and (3) Gurupi Graben System (São Luís, Bragança Viseu and Ilha Nova grabens; Azevedo, 1991) as illustrated in Figures 4, 5 and 6.

The Equatorial African margin consists of a series of open marine basins that make up the northern part of the Gulf of Guinea (Figure 2), with a wide variety of mostly offshore basins, named, from west to east: The Sierra Leone-Liberia, Côte-D'Ivoire-Ghana (Tano-Salt Pond-Accra-Keta basins, Figures 2, 4 and 5), the Dohomey Embayment (Togo-Benin-western Nigeria basins, Figures 2, and 6) and Benue Trough/Niger Delta (Figure 6). The entire region was developed under the kinematic influence of the St. Paul, Romanche and Chain Fracture Zones, with a special influence of the Romanche Fracture zone, which kept a much longer history of transform movements between the Côte d'Ivoire-Ghana margin and the abyssal plain of the Gulf of Ghinea, as illustrated in Figure 7, a Santonian reconstruction of the Equatorial South Atlantic.

Main Physiographic Features of the Equatorial Brazilian Continental Margin

The physiography of the Brazilian Equatorial margin is characterized by a relatively narrow shelf on the eastern side, between 30 and 50 km wide, broadening towards west, where the shelf-width ranges between 50 and 300 km. A key feature is the Amazon Cone, a 900 by 400 km sedimentary feature, cut by canyons, fed by the Amazon river (Figure 3). The shelf break is characterized by a steeper slope, in general at depths between 50 and 2000 meters. Several submarine canyons cut deeply into the continental shelf down to the abyssal plain. Another important feature of the Brazilian Equatorial Atlantic is the presence of many seamounts or guyots, related to fracture zones and / or hot-spots tracks.

TECTONO- STRATIGRAPHIC EVOLUTION:

Deformation associated with the South Atlantic opening began during Jurassic-Triassic time (225-160 Ma), as re-

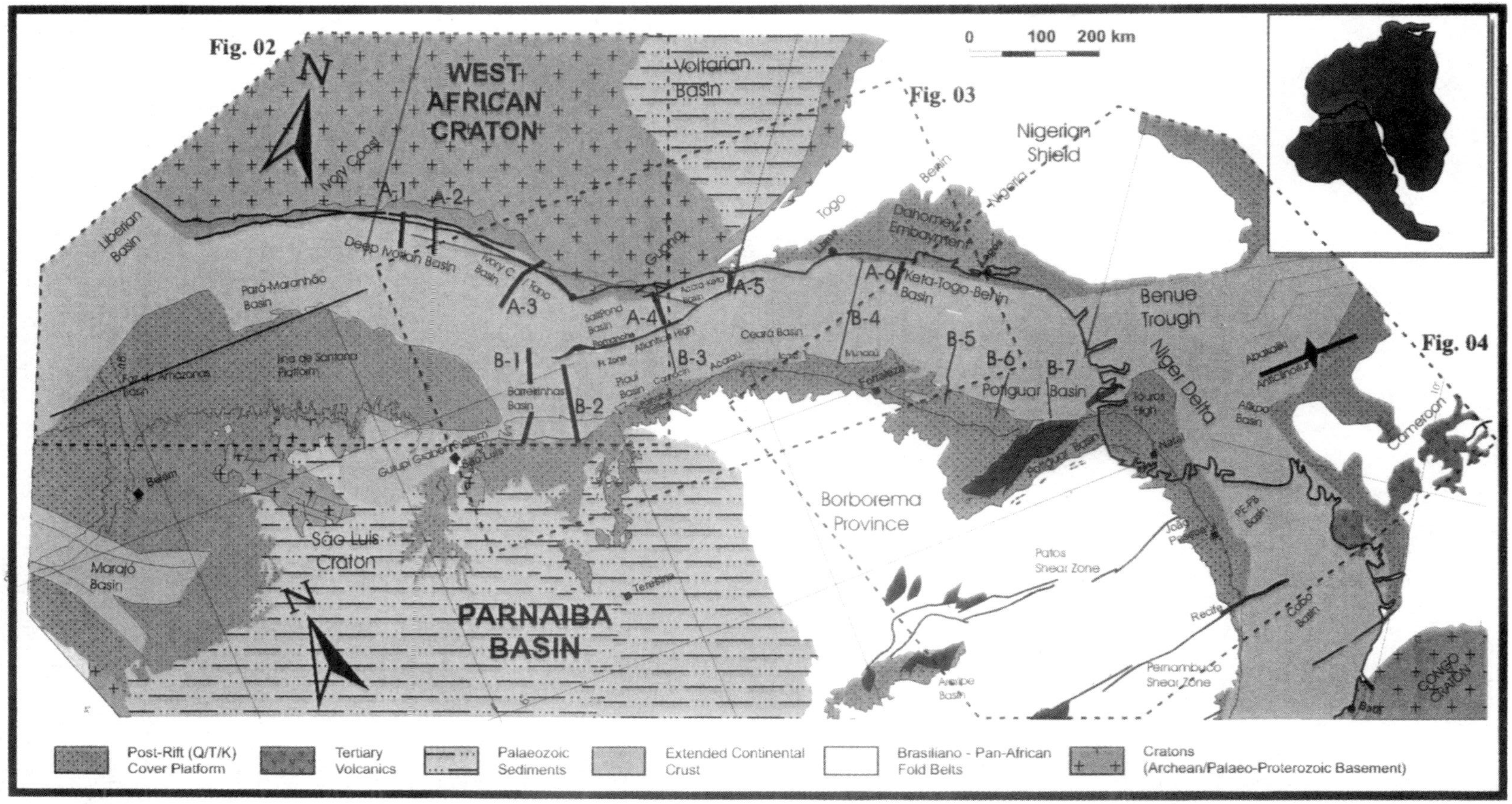

Figure 2: Pre-drift reconstruction of north-northeast Brazil and West Africa (Aptian Time) and the distribution of sedimentary basins. The main geotectonic features around the Equatorial path are the cratons (West African and São Luis), the Palaeozoic basins (Voltarian and Parnaíba) and the Brasiliano-Pan African terranes (Nigerian Shield and Borborema Province). Dashed polygons represent the position of zoomed maps illustrated on Figs. 4, 5 and 6. A-1 to A-6, and B-1 to B-7 represent the position of geologic cross-sections or seismic line drawing of the African (A) and Brazilian (B) side of the Equatorial Atlantic.

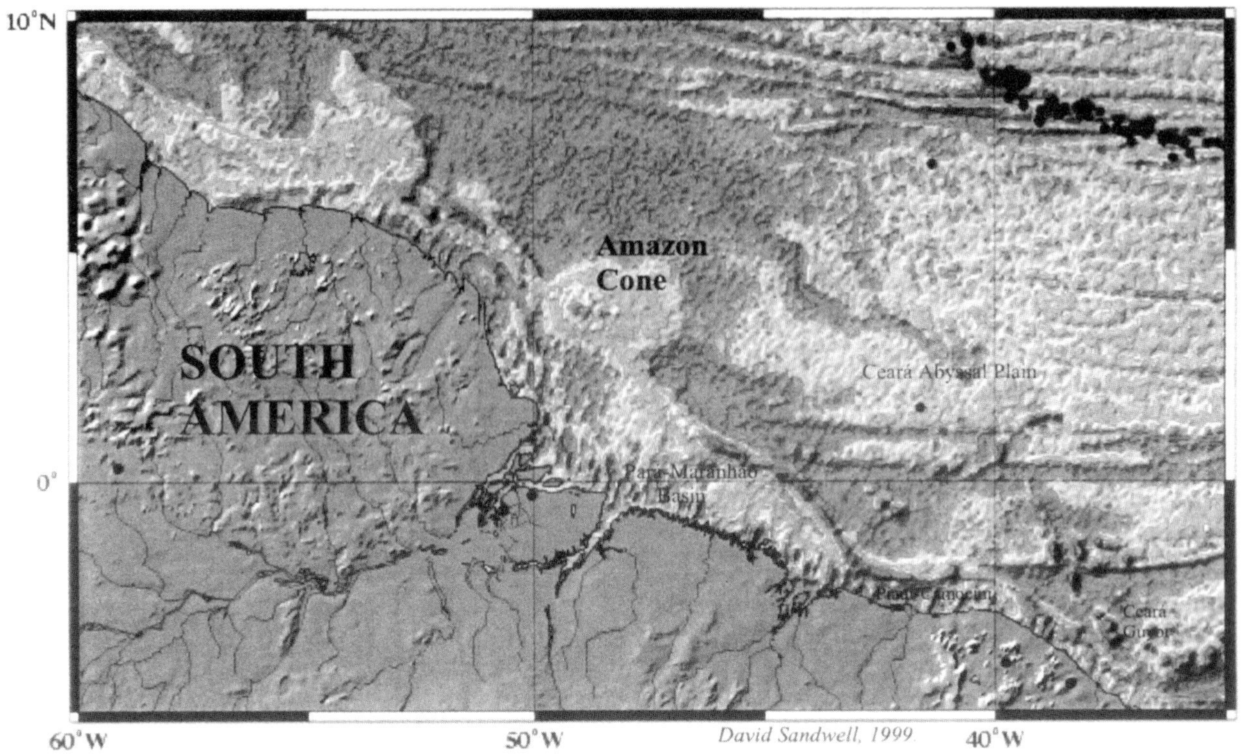

Figure 3. Main features of the equatorial Brazilian continental margin, as illustrated by enhanced gravity anomalies from Geosat data (provided by D. Sandwell). Topography is shown onshore. Lineaments are due to anomalies weaker or stronger than the normal gravity field.

corded by magmatic events and sedimentation in southern and northern Brazil. However, the major rifting event occurred considerably later, during the Cretaceous (160-115 Ma). Rifting has been dated as Early Cretaceous (Neocomian-Barremian time) in the southern South Atlantic, and slightly younger in the Equatorial Atlantic (Aptian-Albian-Cenomanian), when transtensional conditions prevailed throughout the Equatorial domain. Seawaters from the Equatorial and southern Atlantic Oceans only mixed during mid to late Albian (*Koutsoukos,* 1992), dating the final separation between the South America and Africa.

Figure 8 is a composite image of stratigraphic columns of the Brazilian side of the Equatorial Atlantic, illustrating the lack of known pre-Aptian (syn-stretching) sediments in the Equatorial Atlantic, with the exception of the eastern and westernmost basins of the Brazilian Equatorial Atlantic margin, which recorded diachronous deformations associated with the Central Atlantic and the Southern Atlantic, respectively. The Potiguar basin is a key segment of the Equatorial margin because it is part of a series of initially intracontinental rift basins associated with the intersection of the Southern and Equatorial branches of the Atlantic rift system, developed during the Neocomian (*Matos*, 1992, Figure 6). On the other hand, the Marajó basin is an older rift system, recording the southernmost effect of the Triassic-Jurassic extension associated with the opening of the Central Atlantic Ocean, which led to the emplacement of the Cassiporé dyke swarm observed on the Guyana Shield. Late Jurassic extensional-related magmatism in the proto-Equatorial region occurred in the Borborema Province (around the Potiguar basin). It is represented by a dyke swarm (the alkaline Rio Ceará Mirim magmatism; Gomes et al., 1981), which yielded K/Ar ages ranging between 150 and 120 Ma (late Jurassic - Neocomian), with two peaks of activity at 145 and 130 Ma (*Oliveira,* 1992). Magmatism in the Benue Trough (Africa) is represented by the Burashika volcanic complex (147 + 7 Ma, K/Ar, Popoff , 1988) and earlier granitic intrusions of the Jos Plateau (Nigeria), with Triassic-Jurassic ages (213 to 141 Ma Rb-Sr, Vail, 1989).

The Paleozoic intracratonic Parnaíba basin recorded effusive and intrusive magmatic events, composed of up to 500 meters of tholeitic basalts of the Sardinha Formation within the sedimentary pile, and by intrusive dyke swarms

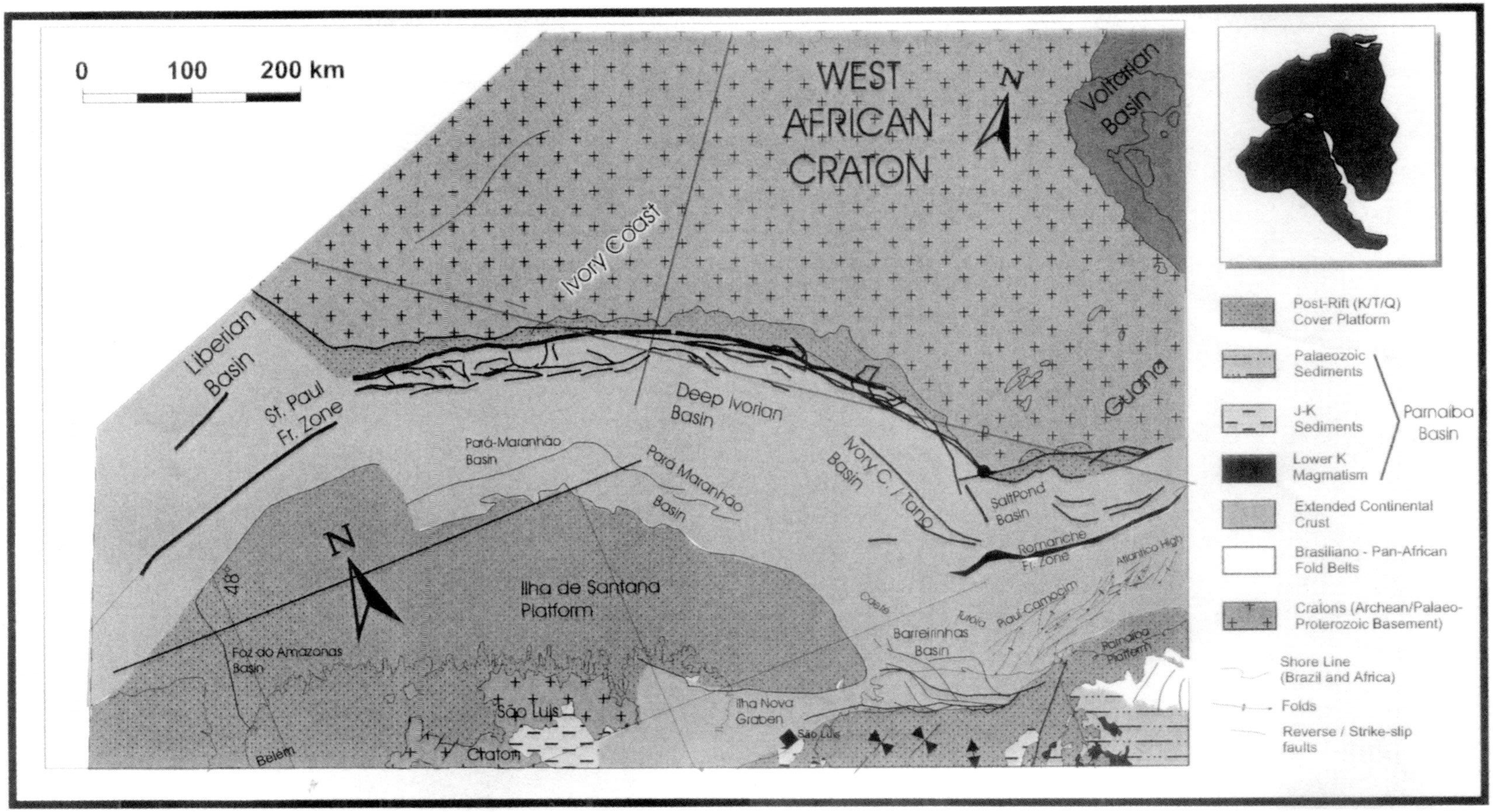

Figure 4: Pre-drift reconstruction and tectonic framework of the sedimentary basins between North Brazil and West Africa during the Aptian Time. Location as in Fig. 2.

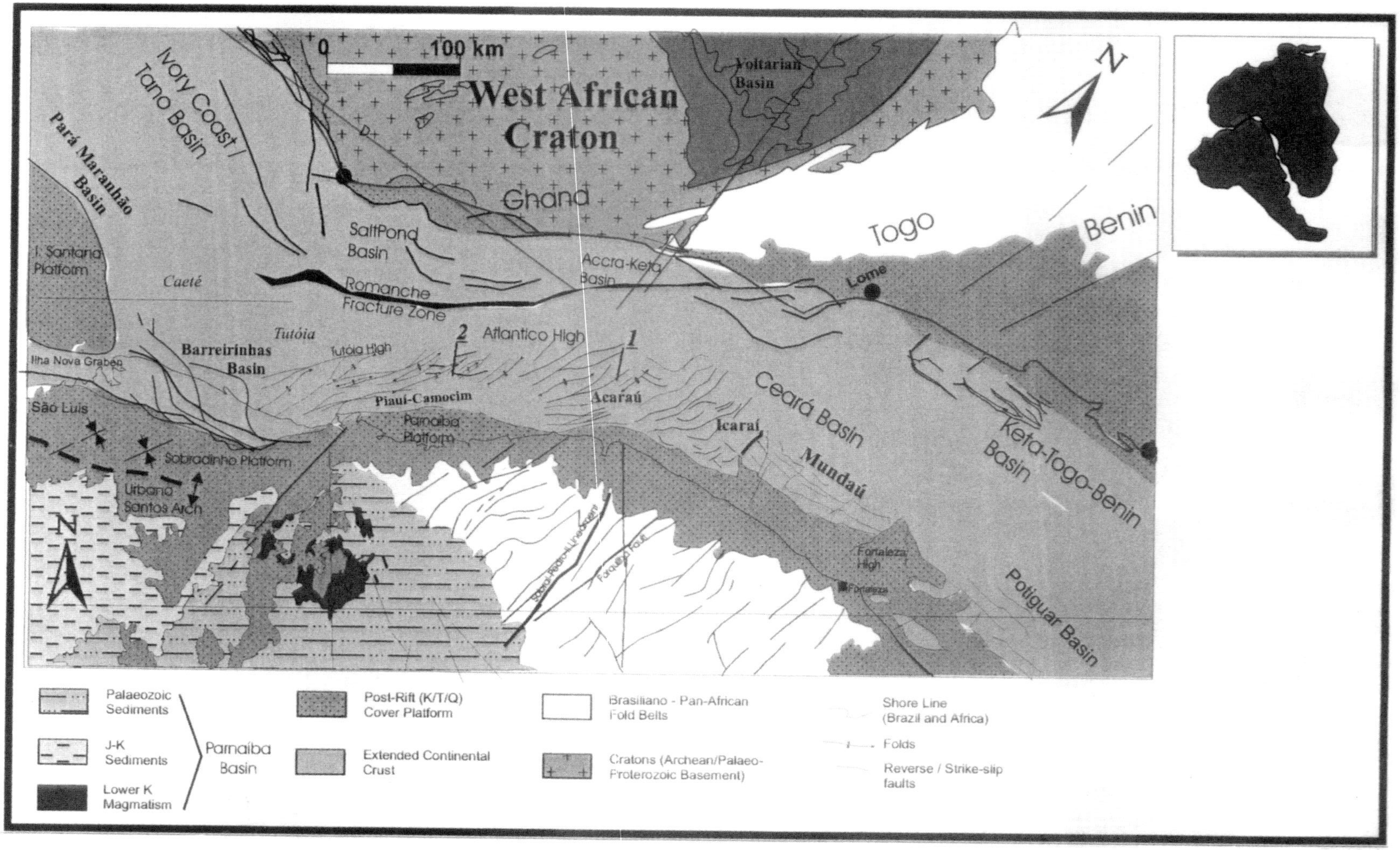

Figure 5: Pre-drift reconstruction and tectonic framework of the sedimentary basins between N-NE Brazil and West Africa during the Aptian Time. Location as in Figure 2. Note the structural framework of the Piauí-Camocim, Acaraú and Icaraí basins in Brazil, defined by an complex array of normal faults, overprinted by NE trending folds and reverse/strike slip faults, as illustrated in Figs. 18 and 19 (location as 1 and 2 respectively).

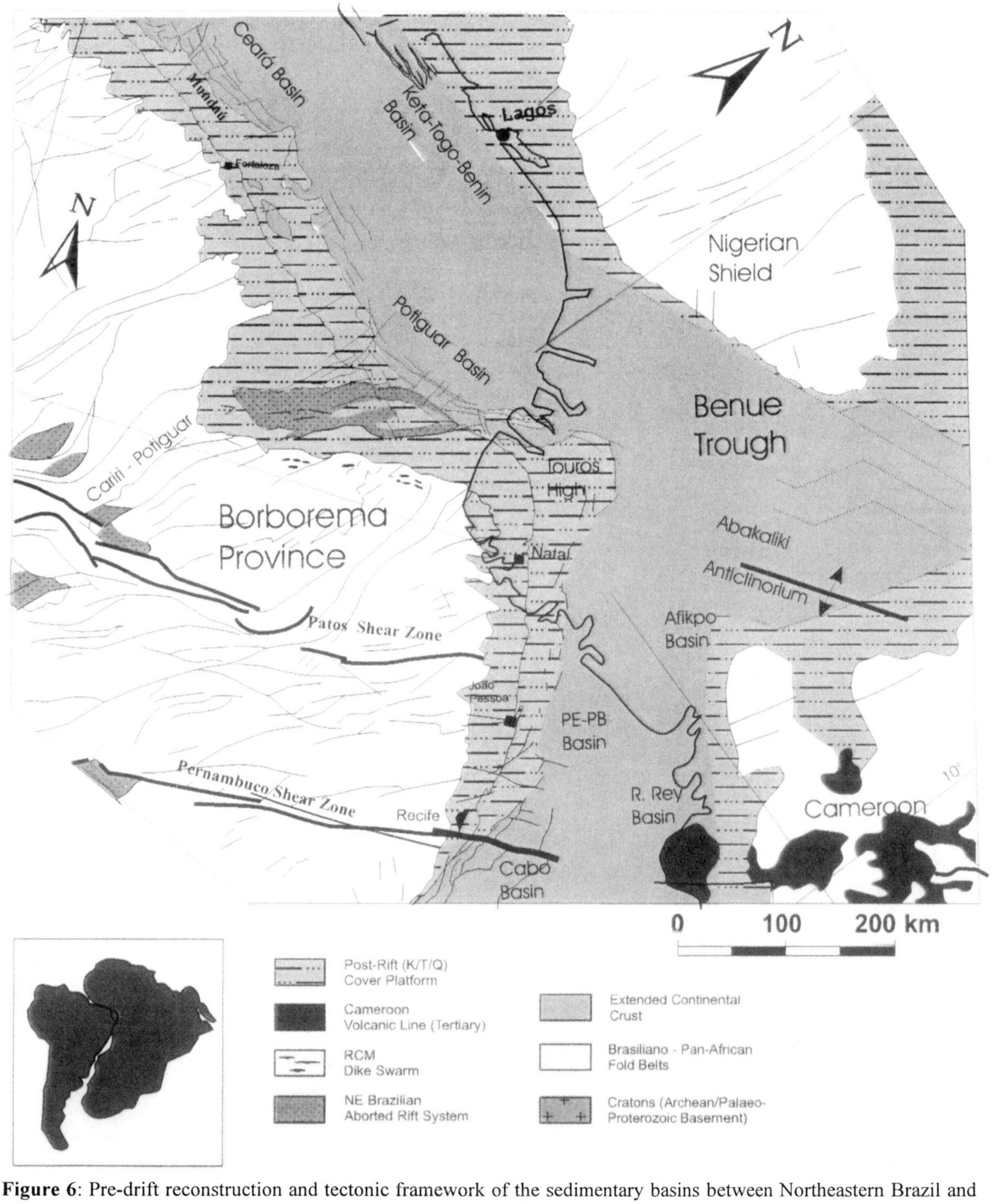

Figure 6: Pre-drift reconstruction and tectonic framework of the sedimentary basins between Northeastern Brazil and West Africa during the Aptian Time. The Potiguar basin - Touros High marks the junction between the Southern and the Equatorial branches of the South Atlantic rift system. The Potiguar basin recorded deformations associated with the Southern branch (Cariri-Potiguar rift valley of Matos, 1992), and the Equatorial branch. Location as in Fig. 2.

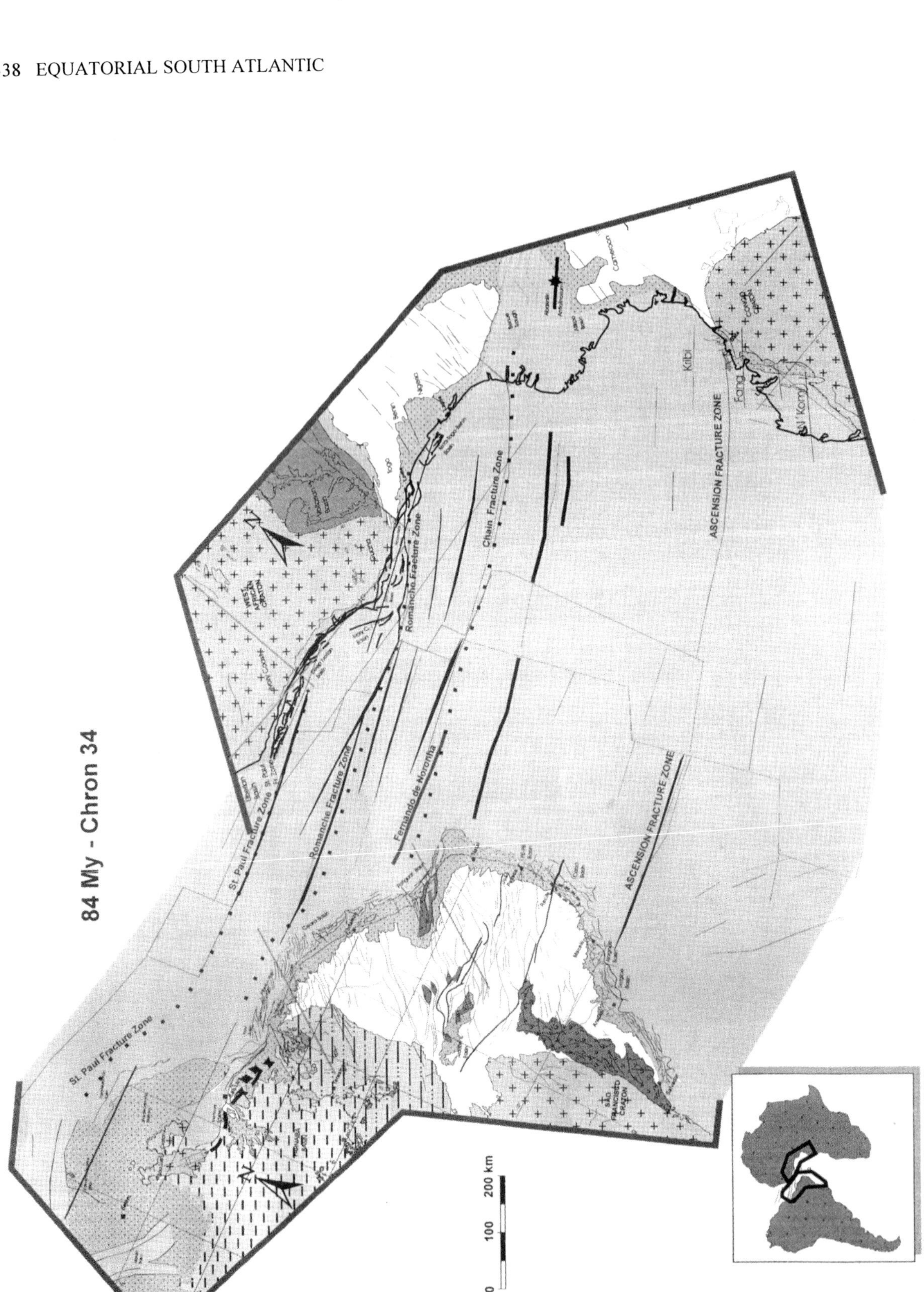

Figure 7: Pre-drift reconstruction of and tectonic framework of the sedimentary basins between North- Northeast Brazil and West Africa at 84 Mybp (Santonian/Campanian transition), and the location of transform faults and spreading centers (Adapted and modified from Scotese et al., 1988 and Sandwell and Smith, 1995).

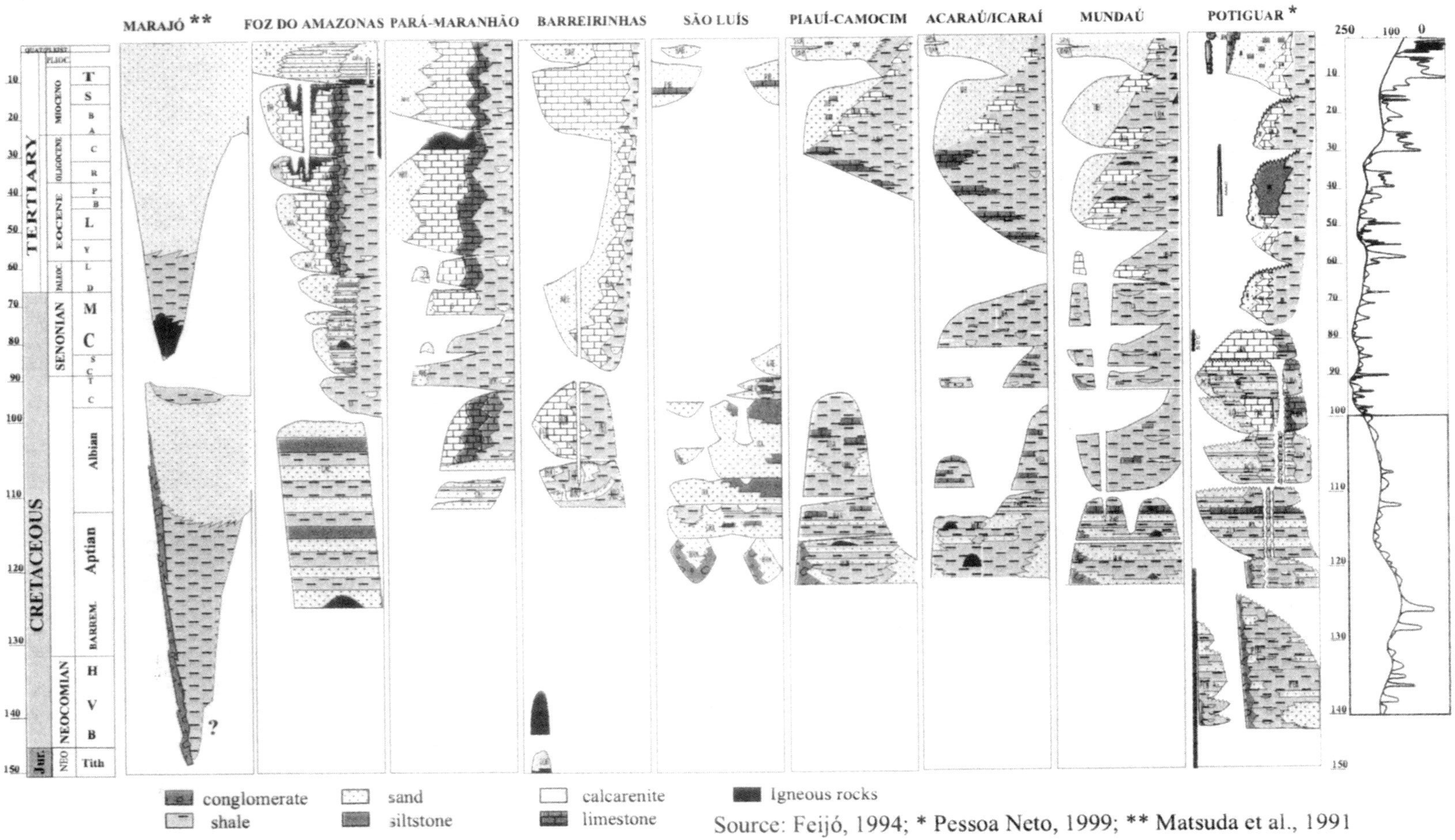

Figure 8: Chart demonstrating the spatial and temporal distribution of sediments in the Equatorial Brazilian Atlantic since Late Jurassic times. Note the distribution of an almost instantaneous stretching phase during the Aptian, responsible for widespread fracturing in the Equatorial Atlantic, and diachronous post-transtension / syn Transform deformation, recorded as a series of diachronous unconformities throughout the margin, not necessarily associated with sea level variations.

in the eastern portion of the basin. The oldest magmatic event is Late Triassic - Jurassic (*Almeida et al,* 1988), and the second episode correspond to an Early Cretaceous event (*Caldasso and Hama,* 1978), synchronous with the Rio Ceará Mirim magmatism.

GEODYNAMIC MECHANISMS

A controversial issue regarding the tectonic evolution of the African and Brazilian Equatorial margins is based on the relative importance given to mantle dynamics versus plate dynamics (or passive versus active rifting). The Potiguar basin (Brazil) and Benue Trough (Africa) have been used as an example of a triple junction during the break-up of the South Atlantic, which implies an active type model of basin development. The Brazilian Equatorial margin dataset provides a good chronological control of deformation and magmatism in transform margins. Rifting is diachronous, and despite some pre and syn-extension dyke intrusions in the Potiguar basin, voluminous magmatism clearly post-dates the opening of the Equatorial Atlantic. There is a significant time-lag for the onset of rifting between the Potiguar basin and the Benue Trough (Matos, 1992). By the time extensional deformation started in Benue, the onshore Potiguar basin and a series of intracontinental rift basins of Brazil (Cariri-Potiguar rift valley) were already aborted. The issue of a pre-rift uplift is also controversial because of the long term epirogenetic response of the Borborema Province / Nigerian Shield, which correspond to a heritage of the Pan-African / Brasiliano orogeny (*Azevedo,* 1991).

The idea of a hot spot (St. Helena) triggering the onset of rifting in the Equatorial Atlantic (*O'Connor and Duncan,* 1990) is controversial because true evidence for an active hot spot in the region comes only from the Tertiary record. In order to add corroborative evidence for an active plume system, some authors have considered magmatic data from distant basins, such as the Cabo basin (Wilson and Guiraud, 1992). However, this basin is located about 500 km away from the Potiguar-Benue junction (Figure 6). These data would suggest a very broad zone of diffuse volcanism, and might provide evidence for a quite broad "mushroom head" or extensive underplating. The issue is not whether we have enough evidence for an ancient cold weak plume, but whether it was strong enough to trigger rifting or it just weakened the lithosphere, favoring the concentration of stresses in this region, allowing plate tectonic forces to play a key role in the process.

The intracontinental and offshore basins of the easternmost region of the Equatorial Atlantic can be grouped in two ages: Neocomian-Barremian and Aptian. The site and architecture of the first group of basins were strongly controlled by upper crustal heterogeneities (*Matos,* 1992), while the second group (Aptian basins in the Equatorial Atlantic) were not significantly influenced by the Proterozoic heritage, indicating different processes of lithospheric thinning between the Equatorial and the Southern branches, which could indicate deep-seated processes on the Equatorial margin. Equatorial fragmentation was a response to stresses induced by plate tectonic forces. These were controlled by large-scale lateral movements of the crust, triggering transform movements throughout the region and climaxing with the complete opening of the South Atlantic Ocean. This progressive, kinematically driven deformation led to a stack of complex structures on a regional scale, very different from those found in typical extensional terrains.

Conventional mechanisms of basin formation cannot be promptly used when modeling sedimentary basins generated as a response to major transform motions of a continental-scale plate boundary. The commonly adopted causal processes for passive margins, such as passive/active or diffuse/discrete rifting, or volcanic/non-volcanic margins can not be easily applied in the Equatorial basins. One of the biggest challenges in the understanding of transform margins is the quantification of stretching prior to break-up. Another key factor is the deformation rate during the syn-transform stage. Divergent margins show relatively longer rifting stages (between 10 and 25 Ma in the eastern Brazilian Margin) prior to break up, allowing lithospheric stretching at reasonable rates, resulting in mantle decompression and in some cases the outpouring of continental flood basalts. Although this multi-stage rifting process in the Brazilian/African Equatorial margin evolved throughout the Aptian-Cenomanian interval, an almost instantaneous Aptian rift stage was responsible for the widespread fracturing of the Equatorial Atlantic (Figure 8). Stretching prior to sea-floor spreading was very fast (Aptian time), with no evidence of volcanic margins in the sense defined in orthogonal extensional regimes (*White and McKenzie,* 1989; *White,* 1992). These are characterized by syn-rift igneous activity and also by thick wedges of seaward-dipping reflectors marking the transition from continental to oceanic crust (Hinz, 1981).

The tectonic evolution of the Equatorial Atlantic is better understood if considering three stages: pre, syn and post-transform movements (Figure 9). A multi-stage basin development has been recognized in both sides of the Equatorial Atlantic: a rift stage (Aptian - early Albian) and a shear-dominated stage (early Albian-Cenomanian), as discussed by *Zalán* and *Warme* (1985); *Azevedo* (1991), *Szatmari et al.* (1987), *Cremonini et al.* (1996), *and Gui-*

raud et al. (1997). The main unconformities in the sedimentary succession are clearly associated with events that pre-date, are synchronous or post-date the activation of transform faults. Mainly on the basis of changes in structural style, facies association, inferred deformation rate, strain analysis, and plate kinematics studies, the tectonic evolution of the Equatorial Atlantic is described considering the kinematic and dynamic controls on the activation of transform faults, the emplacement of oceanic crust and the onset of drifting of Africa and South America, as summarized in Figure 9. The three stages associated with the development of the Equatorial Atlantic margin are described as follows.

Pre-Transform

The Pre-Transform stage in the Equatorial branch is recorded by two sedimentary cycles:

(1) *Pre-Transtension*: Jurassic and early Cretaceous (pre-Barremian) sediments that pre-date the main stretching phase in the Equatorial Atlantic. This phase is represented by the rift sediments of the Marajó and onshore Potiguar basins, and are related to the opening of the Central and Southern Atlantic, respectively;

(2) *Syn-Transtension*: The onset of intracontinental shear and fracturing under transtensional conditions of the Equatorial Atlantic has been dated as late Barremian, coeval with the occurrence of the oldest sediments of the Benue Trough in Africa (Aptian shales of the Bima Formation, *Popoff,* 1988), and simultaneous with late syn-rift stages of the eastern margin (*Matos,* 1992). Stress in the Equatorial domain began to dissipate as the entire region started to break apart. Transtensional conditions created a series of NW-SE trending en-echelon depocenters throughout the Equatorial domain. This latter stage has been called syn-rift. However, these sediments show major deviations from a classical rift signature, which is characterized by tilted half-grabens, with depocenters controlled by planar/listric faults, and normally filled with a divergent wedge of siliciclastics sediments, with a predominance of fanglomerates adjacent to the main border fault (*Chang et al.,* 1992). These features are not recognized in the Equatorial margin domain. In many cases it is difficult to see a divergent seismic pattern, as illustrated in the Potiguar and Ceará basins in Brazil and many others equatorial basins (Figures 10 to 14), where conglomerates and even coarse to medium grained alluvial sediments may be missing adjacent to main border faults. In many basins of the Equatorial Atlantic, geologic restorations indicate relatively wide and shallow "syn-rift" precursory basin, which very often resembles a sag-type basin, and not a half-graben. Stratal geometries adjacent to the border fault systems characterize either synclines or sets of parallel strata dipping basinward, truncated by a diffuse fault pattern (Figures 10 to 12). Figure 15 illustrates schematically typical equatorial Afro-Brazilian basin geometries compared with a classical rift-basin architecture.

While some basins were generated under pull-apart conditions, some segments of the margin located on releasing bends experienced almost orthogonal extension, which led many authors to characterize a rift succession throughout the transform margin segment. However, the lack of classical rift structures and associated sedimentary facies indicate that this may not be the case for the Brazilian Equatorial margin. Figure 15 illustrates a migrated seismic section of the Ceará basin (Mundaú sub-basin), developed under transtensional conditions, which displays an unique border fault geometry.

Another major difference between transform and divergent margins is the lack of an expressive pre-rift section, as well as an evaporitic transitional section between the rift and drift stages, which are common features in the eastern Brazilian margin. Detailed pre-drift reconstructions demonstrated diffuse deformation distributed through a transtensional path in the proto-Equatorial region. Basins 200-250 km wide (Afro-American basins) developed before the breakup of Western Gondwana (Figures 4, 5 and 6). Obliquity of the stress field during the early stages of transform motion was a key controlling factor on the resultant basin architecture.

The rigidity of the West-African/São Luiz Craton in the west, and the Transversal Zone (*Matos,* 1992; *Matos* and *Waick,* 1998) in the east, prevented the onset of transtensional deformation in the Equatorial Atlantic until the early Aptian, when the entire equatorial domain began to develop a series of en-echelon depocenters. The Transversal Zone is the northernmost and youngest rifted segment of the Southern Branch (eastern Brazilian margin), a roughly orthogonal structure between the Pernambuco shear zone and the Benue Trough (Touros High in Brazil, Figures 2 and 6), which record a very fast and focused stretching phase (5 My, *Matos,* 1998), responsible for overcoming the last link between Africa and Brazil. This transversal zone acted as a giant accommodation zone, balancing extension between the Southern and the Equatorial branches (*Matos,* 1992).

GEODYNAMIC EVOLUTION OF THE EQUATORIAL ATLANTIC

STAGE	KINEMATIC / DYNAMIC	EVENTS	AGE	PHASE
PRE - TRANSFORM	PRE-TRANSTENSION	Rifts in the Central and in the Southern Atlantic (Marajó and Potiguar grabens, respectively)	PRE-BARREMIAN	PRE - STRETCHING
	SYN-TRANSTENSION	Transtensional conditions created a series of NW-SE trending enechelon depo-centers throughout the Equatorial domain	BARREMIAN TO APTIAN	STRETCHING
SYN - TRANSFORM	PURE-SHEAR DOMINATED TRANSTENSION	Deformation characterized by broad regions dominated by extension, limited by discrete shear zones	ALBIAN TO CENOMANIAN	WRENCH
	WRENCH DOMINATED TRANSTENSION	The divergent movement was accommodated by relatively narrow zones, accounting for most of the the slip between Africa and Brazil		
	WRENCH DOMINATED TRANSPRESSION	A large transpressive belt (Piauí-Acaraú and Accra basins) as the result of an overall shortening and up-lift around a restraining bend of the Equatorial Atl.		
	PASSIVE TRANSFORM MARGIN	Oceanic/continental crust contact through an active Transform Fault	CENOMANIAN TO RECENT	DRIFT
POST - TRANSFORM	PASSIVE MARGIIN	Almost continuous sedimen-tation as the result of cooling-thermal contraction of the lithosphere		

Table 1. Geodynamic evolution of the Equatorial Atlantic considering events that predate, are synchronous or posdate the activation of transform faults.

Figure 9: Geodynamic evolution of the Equatorial Atlantic considering events that pre-date, are synchronous or post-date the activation of transform faults.

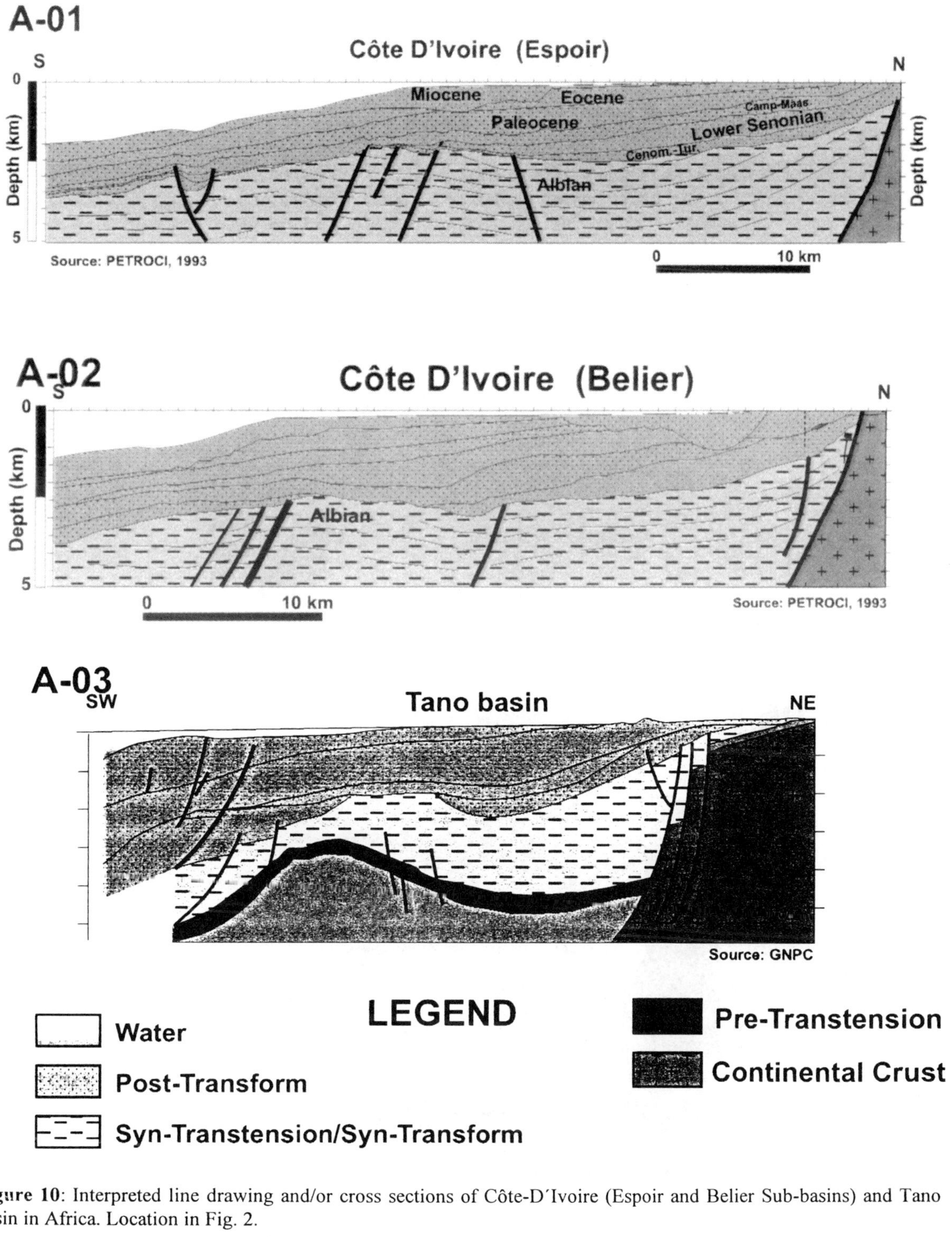

Figure 10: Interpreted line drawing and/or cross sections of Côte-D'Ivoire (Espoir and Belier Sub-basins) and Tano basin in Africa. Location in Fig. 2.

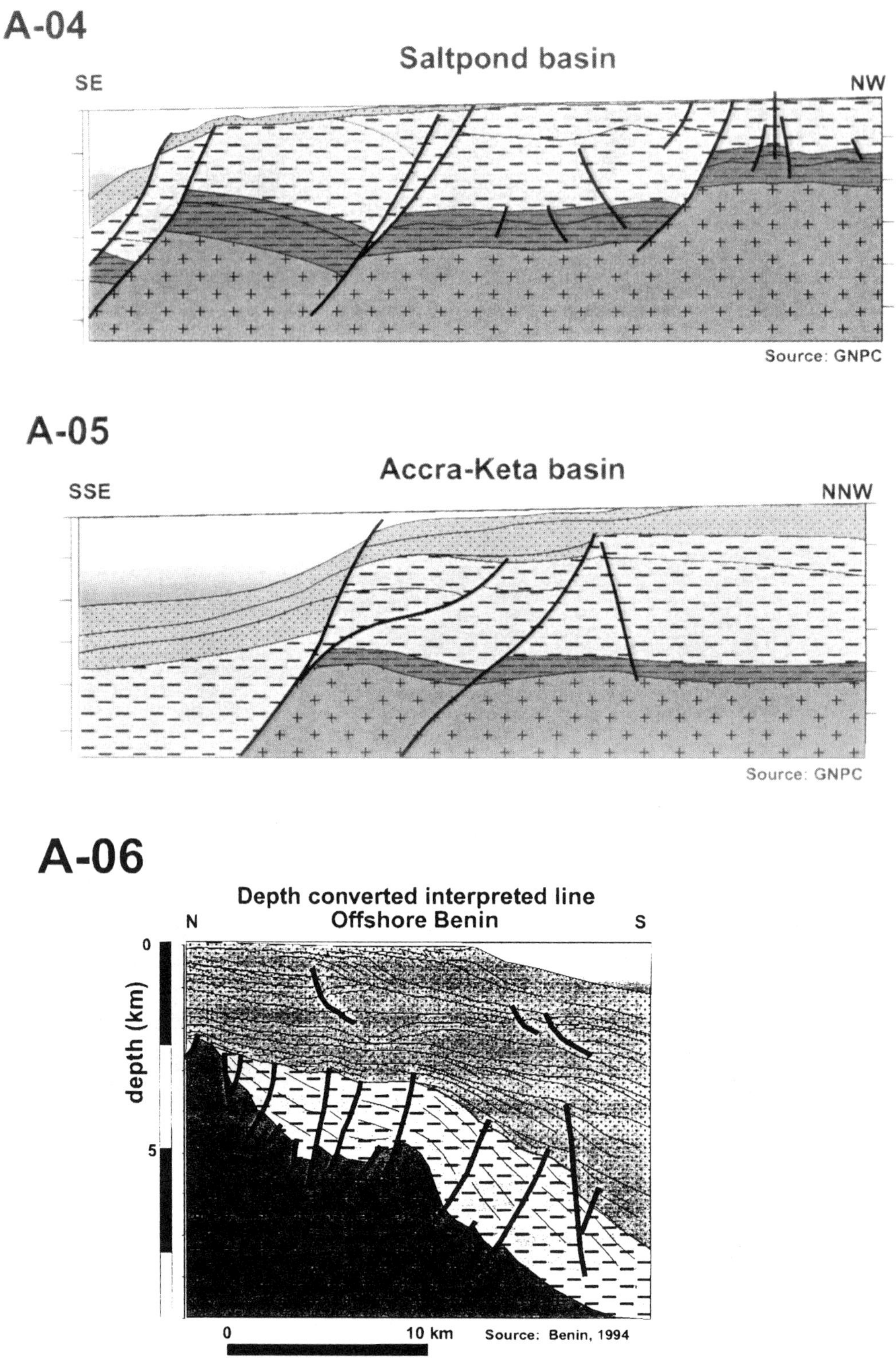

Figure 11: Three interpreted line drawing and/or cross sections from offshore Ghana, Togo and Benin, represented by the Saltpond, Accra-Keta and Benin basins. Location in Fig. 2 and legend in Fig. 10.

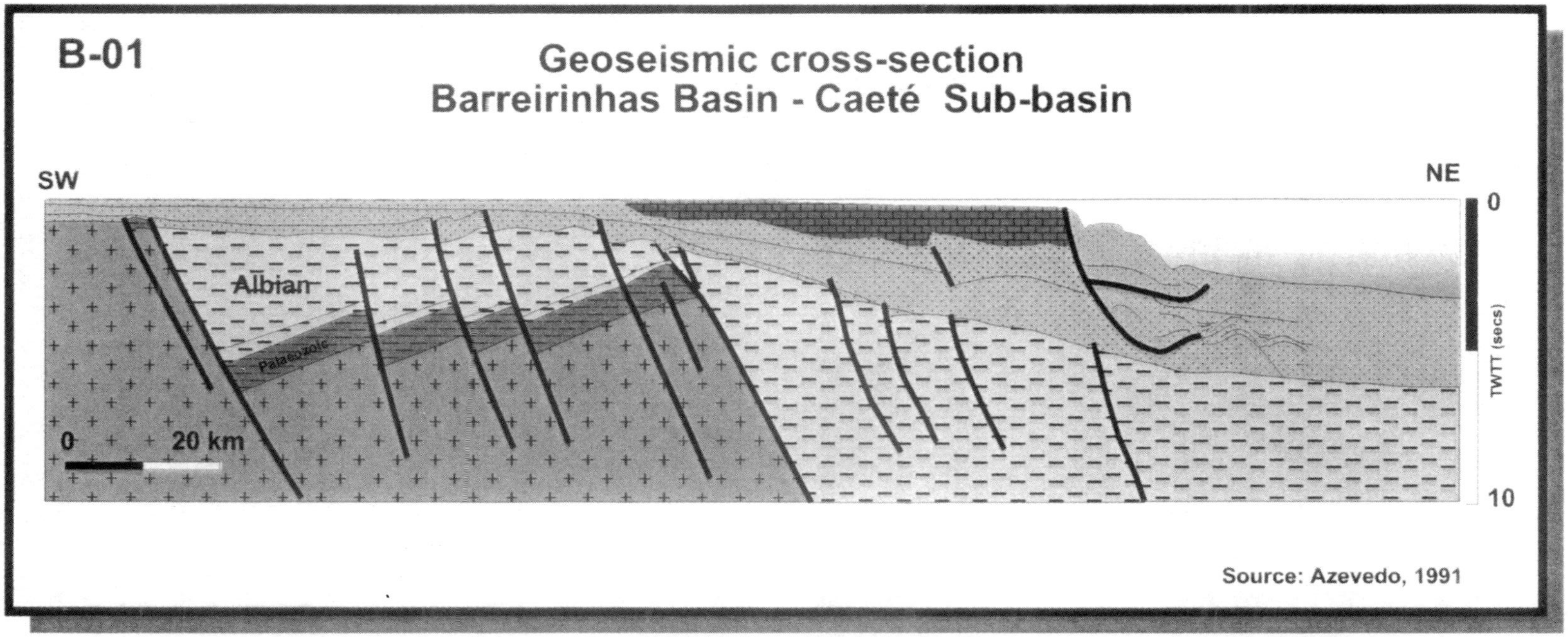

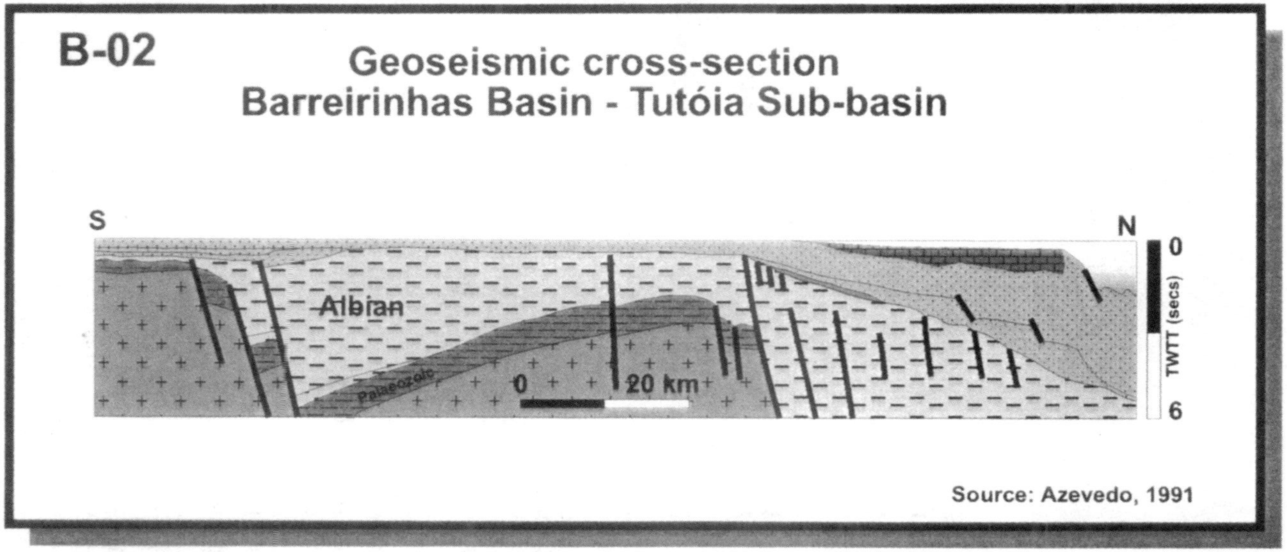

Figure 12: Cross-sections from offshore Barreirinhas basin in Brazil. Location in Fig. 2 and legend in Fig. 10. The Tutóia and Caeté sub-basins display rift-type structures, defined by tilted half-grabens, as expected under transtensional conditions.

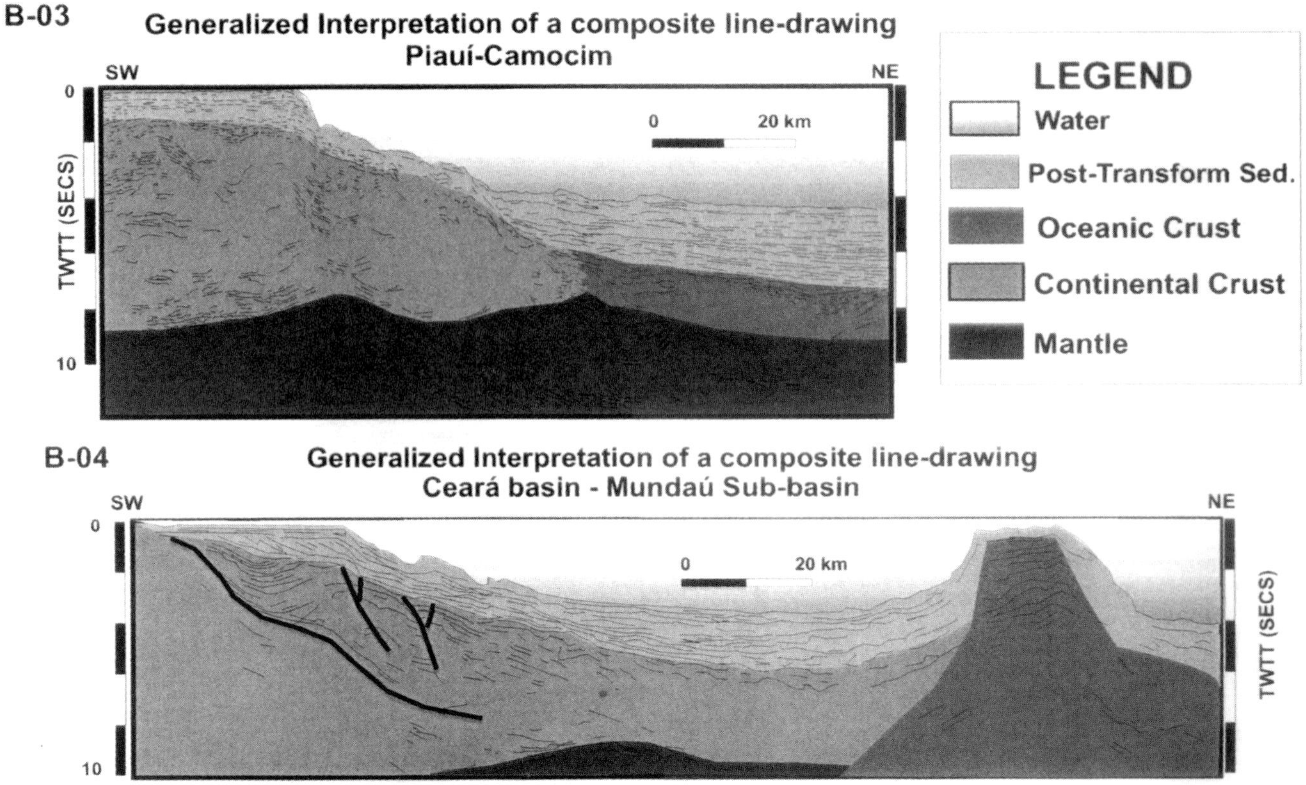

Figure 13: Line drawings from the Piauí and Ceará basins in Brazil, represented by the Camocim and Mundaú sub-basins. Location in Fig. 2. The Piauí-Camocim basin was heavily deformed and faulted during the opening of the Equatorial Atlantic, while the Mundaú sub-basin was less deformed due to its location in a releasing bend of the Equatorial shear corridor.

The Albian Transition

The Albian stage encompasses a typical transtensional sequence throughout the Equatorial branch/Benue trough (*Szatmari et al.*, 1987; *Benkhelil,* 1989), while a shallow carbonate platform megasequence was developed in the eastern Brazilian Margin (*Chang et al.*, 1988). The Lower Albian record display mixed continental / marine environments in many equatorial basins, such as Potiguar, Piauí-Camocim, and Barreirinhas basins, suggesting a progressive marine influence into the region. Deep water planktonic morphotypes of middle Albian age confirm the presence of deep oceanic circulation between the Equatorial and the eastern margins (*Koutsoukos,* 1992), providing evidence for the onset of major transform movements in the Equatorial Margin.

Syn-Transform

There is a clear boundary between the transtensional stage and the syn-transform stage. As stretching increased and the equatorial transtensional path widened, the lithosphere became increasingly thinned, particularly at some releasing bends, where the emplacement of fractured swells acted as proto mid-ocean-ridges, opening windows of oceanic crust in the continental crust. This led to the onset of transform faulting linking the proto-spreading centers throughout the region, defining a lithospheric boundary between Africa and South America. This new boundary forced deformation to become focused at greater depths, but probably still with a diffuse signature at shallower depths, producing distinct styles of deformation at different structural levels. Upper crustal heritage was not responsible for focusing the deformation. Except for some local exceptions, the structural fabric of the São Luis / West African cratons and the Brasiliano / Pan-African belts of Africa and Brazil did not control the overall geometry of the Equatorial margin, which reinforces a lithospheric / kinematic control on the resultant equatorial transform margin geometry.

The term wrench is used here to kinematically characterize an overall tectonic regime (as used by *Milnes,* 1994),

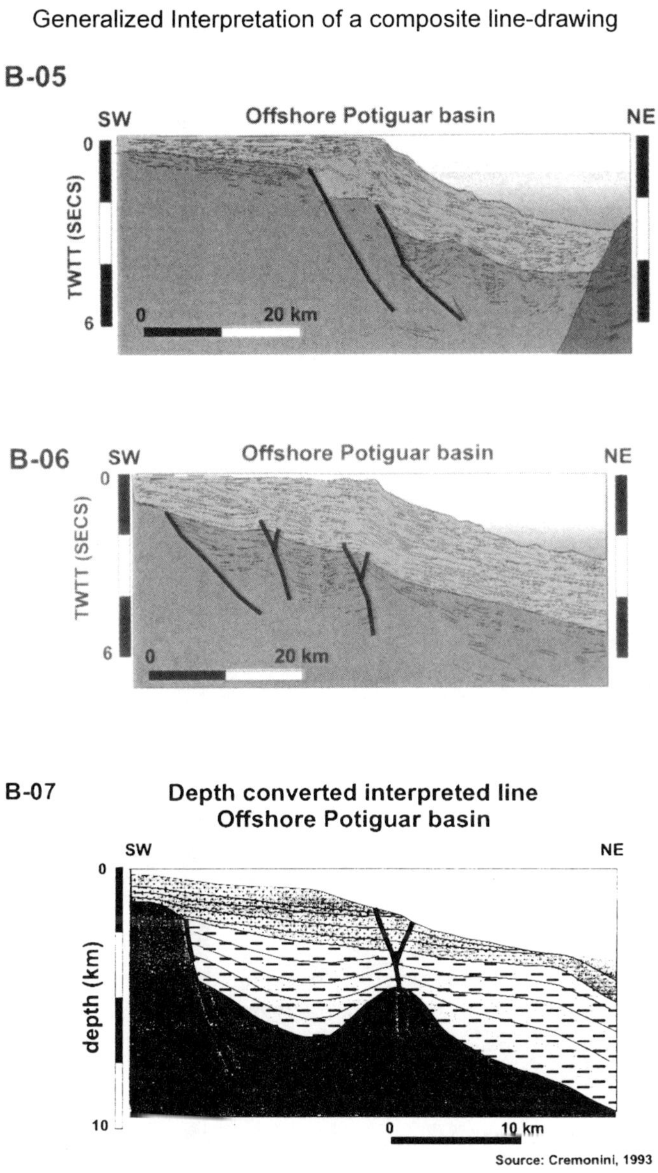

Figure 14: Two line drawings and a geologic cross-section from the Offshore Potiguar basin in Brazil. Location in Fig. 2. Legend of B-05 and B-06 in Fig. 13, and B-07 in Fig. 10.

instead of a dynamic meaning as developed by Anderson (1942). As expected in wrench tectonics, the structures are not evenly distributed along the margin. A regional rotational strain regime, kinematically controlled, produced complex structural patterns, characterized by conjugate strike-slip fault systems, with associated normal and reverse faults. Oblique slip faults and folded structures of many different types, whose orientation and scale are the result of simultaneous extension and contraction in the horizontal plane, are shown in map-view in Figures 4 and 5.

The Albian reconstruction (105 Mybp, Figure 17) provides an early view of the final plate boundary, where an alternance between straight and curved segments is quite obvious, generating major releasing and restraining transform segments. Deformation partitioning as a function of the coaxiality of the progressive deformation (*Fossen et al,* 1994) in the Equatorial Atlantic led to basin segments dominated by transtension or transpression. The Barreirinhas (Tano in Africa) and Mundaú-Potiguar (Keta-Togo-Benin in Africa) segments of the Equatorial margin evolved under pure shear-dominated transtension, characterized by broad regions dominated by extension, limited by discrete shear zones, recognized on conventional seismic data as discrete oblique faults, and flower-structures. In the Pará-Maranhão (Côte-d'Ivoire) and East Potiguar (Benue) basins deformation evolved through wrench-dominated transtension, where the divergent movement was accommodated by relatively narrow zones, accounting for most of the the slip between Africa and Brazil. A large transpressive belt was developed in the Piauí-Camocim-Acaraú (Accra) basins as the result of an overall shortening and uplift around a left overstep of this dextral transform system. This segment evolved under wrench dominated transpression, generating major oblique/strike slip faults, en echelon folds and shale ridges (*Zalán* and *Warme,* 1985). Figures 18 and 19 show some examples of these structures.

It is important to note that the direction of tectonic transport along the entire margin was kinematically controlled by the oblique divergent movement of Africa and Brazil, as suggested by the spreading directions inferred from Albian reconstructions in the Central and South Atlantic (*Pindell et al.,* 1988; *Scotese et al.,* 1988; *Unternehr et al.,* 1988). The lithospheric boundary was defined by an array of transform faults that provided a network of right oversteps (jogs), mechanically responsible for accommodating this rotational deformation. Some basins located at the heart of major transtensional segments may display a local tectonic style close to a pure extensional basin, such as tilted half-grabens, and normal faults, as shown in Figure 12. However, the entire margin was clearly marked by the overall obliquity of the movement between Africa and Brazil.

Even though shearing signatures and pull-apart features are easily recognized throughout the Brazilian Equatorial margin, their magnitude and architecture vary significantly according to the distance from the main transform faults. These faults are diachronous by nature. As the African and South American continents progressively moved apart, the

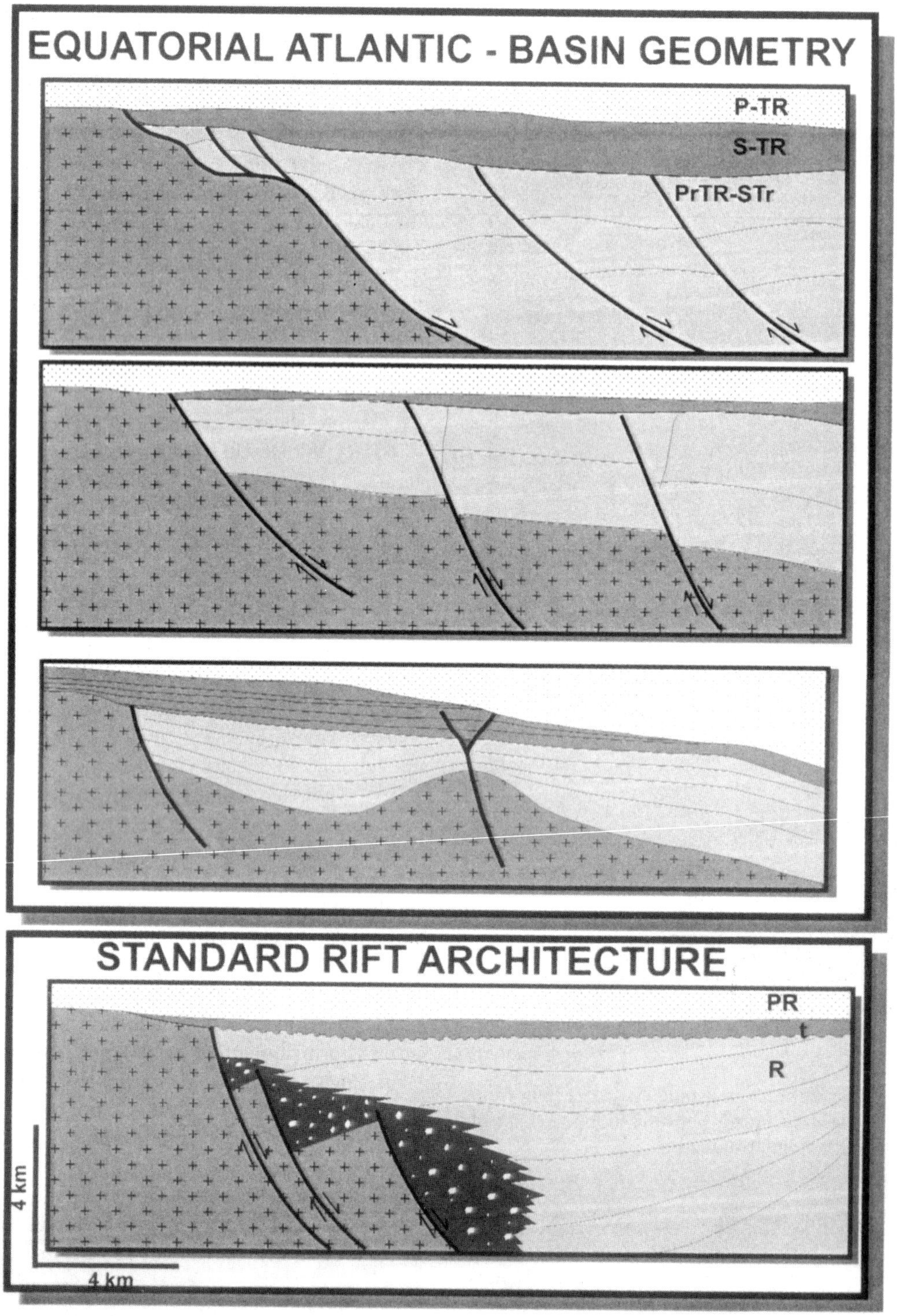

Figure 15: Schematic structural sections across the Equatorial domain compared with a typical rift basin architecture observed in the Potiguar or Recôncavo basins. Note the lack of coarse to medium conglomerates associated with the fault border system in the Equatorial domain, as well as the lack of expressive wedges - syn-rift type - basin filling. Legend: P-TR = Post-Transform section; S-TR = Syn-Trasnform section, and PrTR / S-Tr = Pre-Transform / Syn-Transtension section, PR = Pre-rift, t = transitional and R = Rift.

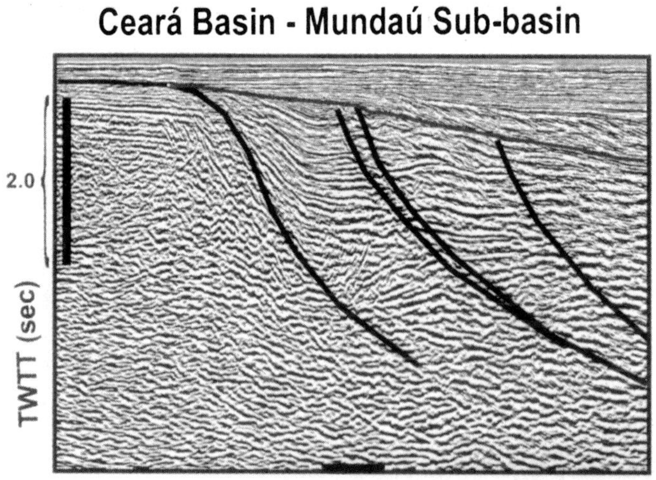

Figure 16. Migrated seismic section of the Ceará Basin (Mundaú sub-basin- Brazil), illustrating a typical basin architecture developed under transtensional conditions at a releasing bend of the Equatorial Atlantic.

deformation zone became wide, and newly formed transform faults were generated, leading to a migration of the active transform faults towards newly formed active spreading centers. Therefore, as this process evolved, some newly formed faulted blocks were developed towards the active transform fault, while some older basin compartments became inactive and were sealed by younger sediments. This resulted in significant differences between basins in terms of their subsidence histories, geometries, and facies distributions, according to the distance from an active transform fault and / or any newly active spreading center. Moreover, the age for juxtaposition of continental crust against oceanic crust across a transform fault will differ from basin to basin, according with the lateral distance from a spreading center. Shearing of continental crust against continental crust on transform faults was kinematically controlled. Shearing of continental crust against newly formed oceanic crust was strongly influenced by the proximity of spreading centers.

The dynamic juxtaposition of spreading centers against the continental margin on the opposite side of a transform fault causes heating and thermal uplift (*Pierce et al,* 1996) The thermal uplift was responsible for anomalous stress fields, leading to deformation, recorded as important unconformities, which may or may not be amplified by syncronous eustatic sea level variations. An example of this type of uplift has been proposed for the marginal ridge along the Côte d'Ivoire-Ghana margin, as described by *Mascle et al.* (1997).

Therefore, local tectonics and magmatism exhibit important roles in the "post-break-up" subsidence histories. This kinematically controlled, shear-dominated fragmentation process led to diachronous pseudo-break-up unconformites. This is markedly different from the divergent segments of the Southern Atlantic, where the break-up unconformity marks the change from a mechanically driven subsidence (rift stage) to a thermally driven subsidence (passive margin stage). In the Equatorial Atlantic margin, major unconformities may not be correlated with the emplacement of the oceanic crust (juxtaposed across a transform fault) but related to the thermal effect of dynamic features, such as a spreading center, or the development of a new transform fault, leading to tectonic uplift and heating increase around the fault path.

Another key issue that deserves attention is the use of small tilted-blocks as indicators of syn-rift sedimentation. This paradigm is not recommended in the Equatorial margin, because releasing bends associated with active transform faults provided enough opportunities to develop normal faults and pull-apart sub-basins, even long after stretching ceased. During the process of strain concentration, new sets of roughly E-W trending strike-slip fault systems were developed by the basinward migration of active transform faults, overprinting the previously deposited sediments, above either continental or oceanic crust, identified on seismic lines as flower-structures, folds and /or uplifted tilted blocks.

The Passive Transform Margin - Syn-Transform

In typical divergent passive margins, the drift stage is characterized by almost continuous sedimentation as the result of cooling and thermal contraction of the lithosphere where eustasy played an important role on the final facies distribution. Local tectonics and magmatism had a key hole in the post-break-up subsidence in the Equatorial margin. In the sedimentary basins along the Brazilian Equatorial margin, the proximity from a transform fault defines when and where a transform margin became a passive margin. As the continental shelf of some basins developed quite away from any transform fault, these basins display basin architecture and facies distribution of typical passive margins. However, basins that were developed close to and coeval with a transform fault and/or fracture zone display a signature of a passive transform margin, or a passive margin developed with a strong wrench component. The Côte d'Ivoire-Ghana margin provides a good example of the dynamic evolution of passive transform margins, developing diachronous marginal ridges between the Deep Ivorian Basin and the Gulf of Guinea (*Basile et al,* 1993). This process indicates an ocean-ward migration of active transform faults through time, leading to local tectonic uplifts, heating increases and often local unconfor-

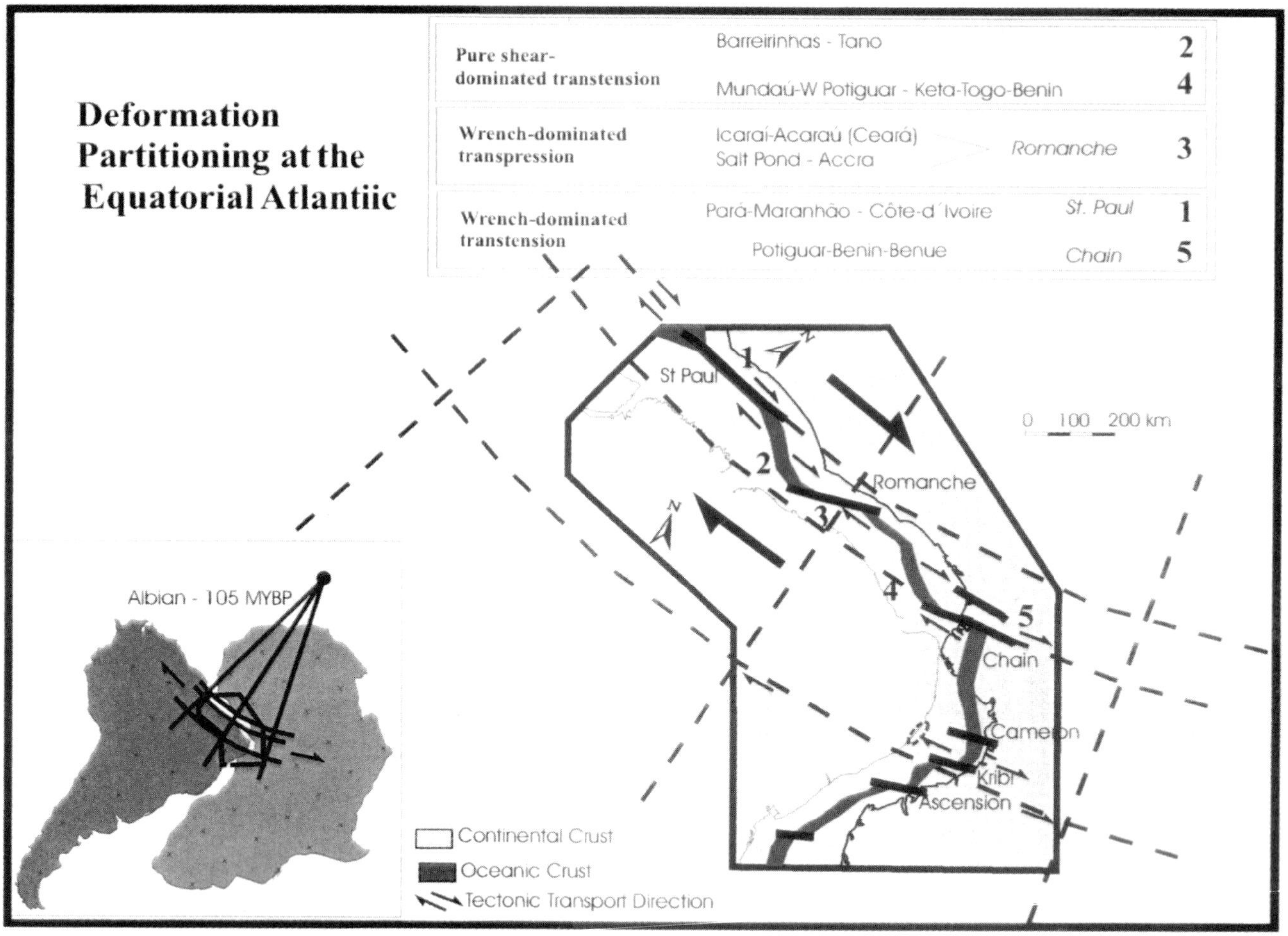

Figure 17: Palaeogeography and early sea-floor spreading immediately after the breakup of South Atlantic (Albian) and how the deformation was partitioned at the Equatorial Atlantic as a function of the progressive deformation, controlled by the site of an early set of transform faults, defining major releasing and restraining bends in the Equatorial shear corridor. Deformation was kinematically controlled, defining anomalous local stress fields, changing through space and time. Pink arrows represent plate kinematics between Brazil and Africa, while the dashed lines indicate the tectonic transport direction regarding a pole of rotation located northward.

mities. Thermal exchanges between the continental crust of the Ivorian basin and the oceanic crust of the Gulf of Guinea resulted in a marginal ridge uplift about 130 km long (*Guiraud et al,* 1997). On both sides of the Equatorial Atlantic, the progressive ocean-ward migration of active transform faults left behind a series of ridges, bouded by oblique to strike-slip faults, very often interpreted as normal faults, covered by a wedge of younger, siliciclastic to carbonatic, platformal sediments. The 84 My reconstruction of Figure 7 indicates an assymetric distribution of spreading center between Africa and Brazil, clearly close to Africa, which explains the continuous influence of the Romanche Fracture Zone over the Côte d'Ivoire-Ghana segment of the African margin (*Mascle and Basile,* 1997).

Post- Transform - Passive Margin

A typical passive margin environment is only reached after the thermal effect of any nearby spreading center is finished; and the tectonic influence of a fracture zone / transform fault becomes a minor control on the available space generated by the thermal cooling of the lithosphere and sea level variations. The subsidence curves of transform margin segments indicate a flattening of the tectonic component of the total subsidence during the post-transform stage. However, some segments of the Equatorial margin are affected by large sedimentary input associated with deltas that prograded from the platform toward deep water region (e.g., Amazon cone and the Niger delta).

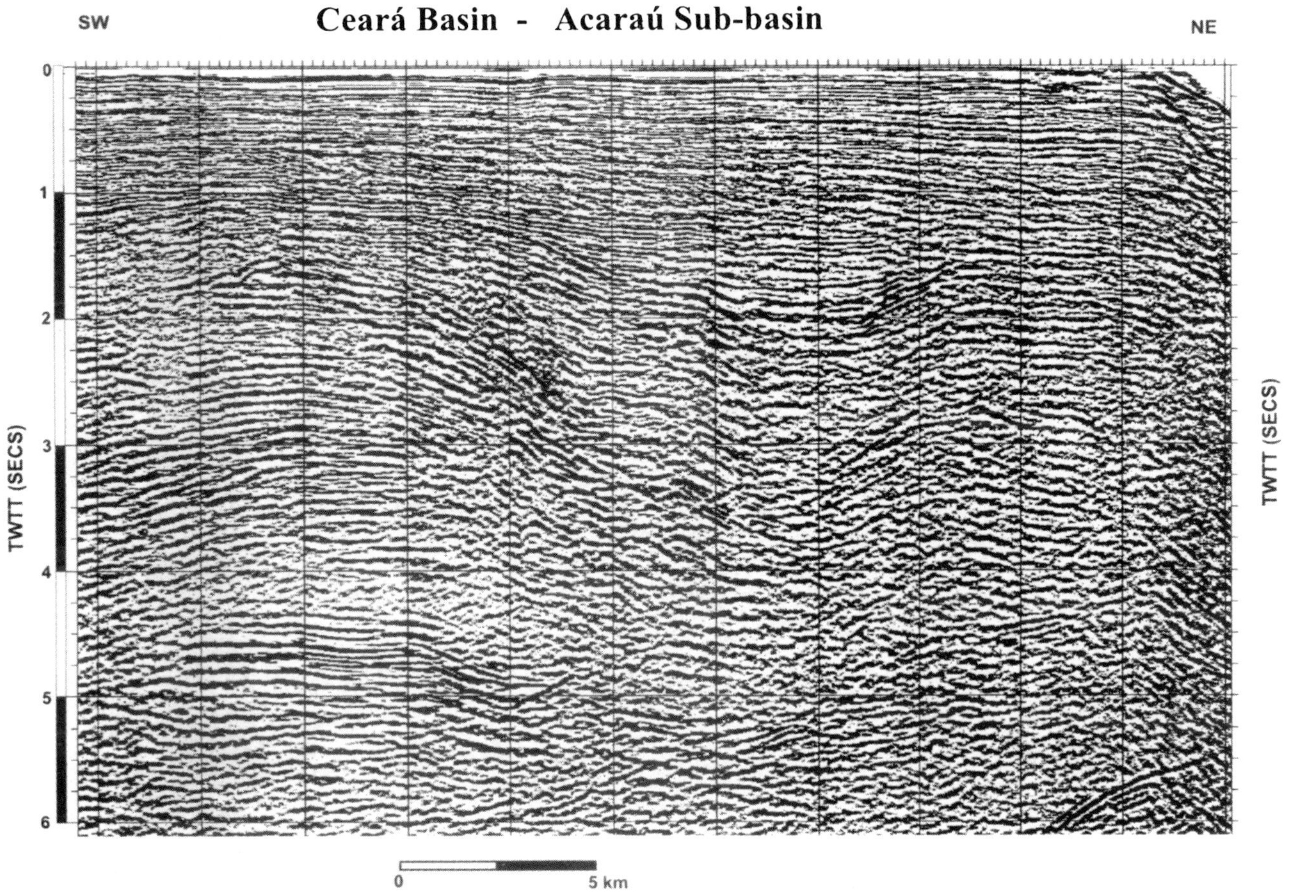

Figure 18: Migrated seismic section from Ceará basin - Acaraú Sub-basin in Brazil. Location in Fig. 5. Folded Albian-Aptian sequences as the result of a wrench-dominated transpression associated with the Romanche Fracture Zone.

The upper portions of these deltas are characterized by extensional, gravity-driven tectonics, whereas the slope toe of the progradational wedge is characterized by several thrusts associated with downdip contraction of the wedge, generating toe-thrust structures (*Silva et al.,* 1999).

CONCLUSIONS

The integrated analysis of the South American and African equatorial basins may provide important clues regarding the kinematic evolution of transform margins. The opening of the equatorial Atlantic was dynamically controlled by lithospheric stresses, resulting in complex displacements throughout the margin. This dynamic element of lithospheric deformation produced an unique segment among the Circum-Atlantic basins, which can not be adequately explained by conventional extensional processes, such as passive or active rifting, or by mechanisms such as pure-shear or simple-shear, typical of divergent margins. For example, the Brazilian and African data set do not corroborate the idea of a plume-generated triple junction over the easternmost portion of the Equatorial Atlantic. Rifting was diachronous between the Benue and Potiguar basins, and voluminous magmatism clearly post-dated the opening of the Equatorial Atlantic.

The fragmentation process started as diffuse magmatic and sedimentary events, which began in Late Barremian time, with the initiation of lithospheric stretching in the equatorial Atlantic, triggered by the onset of transtensional deformation. This event climaxed during the Aptian, when almost instantaneous extension was responsible for the widespread fracturing of the Equatorial Atlantic, generating relatively wide and shallow precursory basins, with no evidence of volcanic margins or widespread development of typical syn-rift basins, as expected in orthogonal extensional regimes. The equatorial splitting evolved during the Albian-Cenomanian interval, through a multi-stage basin development, which is better understood if the kinematic

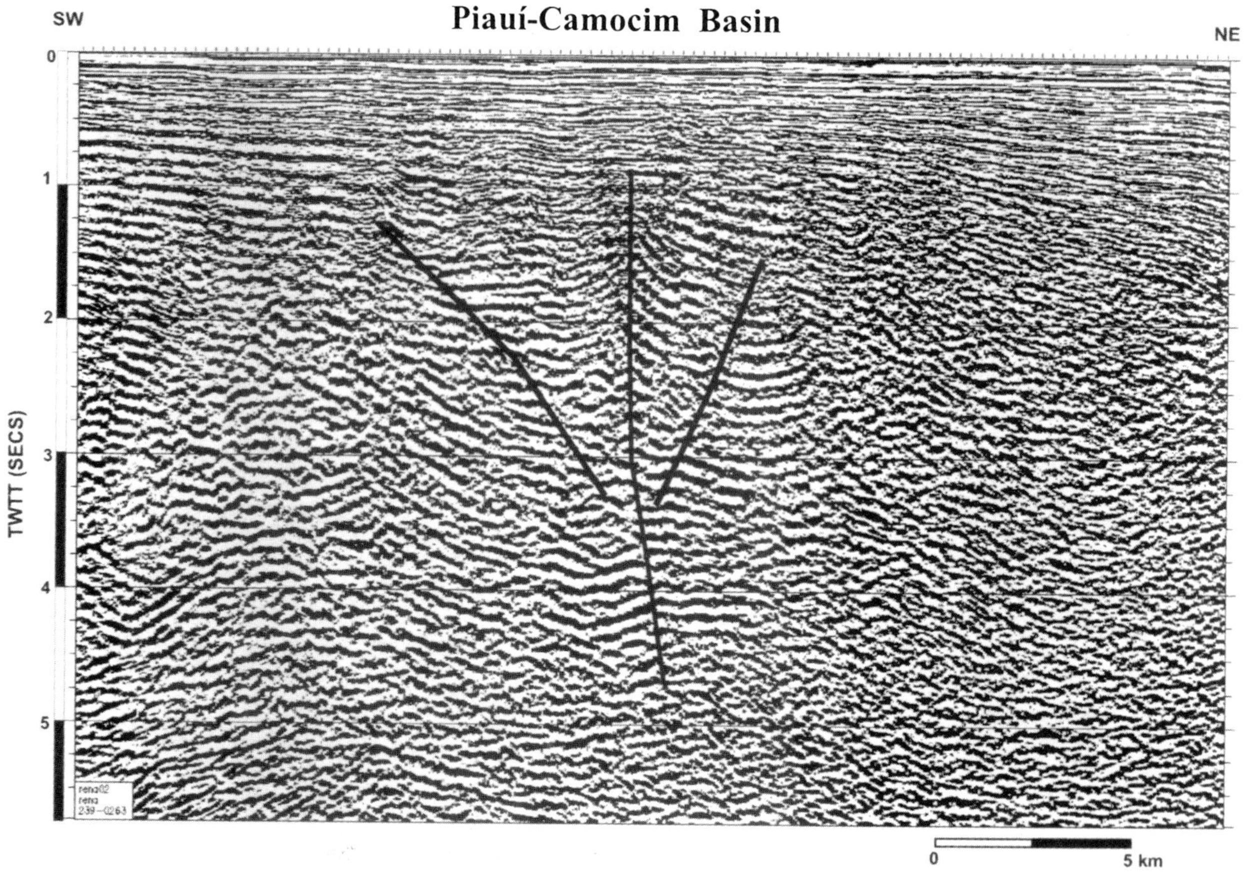

Figure 19: Migrated seismic section from the Piauí-Camocim Basin in Brazil. Location in Fig. 5. Seismic signature of strike-slip faults associated with the Atlântico High and the Romanche Fracture Zone.

and dynamic controls are considered along with the chronologic activation of transform faults, the emplacement of oceanic crust and the onset of drifting between Africa and South America. The geodynamic evolution on the Equatorial Atlantic is addressed by recognizing tectonic stages that pre-date, are synchronous or post-date the activation of transform faults in the Equatorial Atlantic.

Prior to the activation of transform faulting, transtensional conditions created a series of en-echelon basins throughout the Equatorial domain, referred to here as a syn-stretching phase. The term syn-rift is avoided due to major deviations from a classical rift signature. The basin architectures observed in most of the Equatorial margin are usually characterized by the lack of expressive divergent wedges of sediments next to the main border fault systems, even though some basins generated under pull-apart conditions. The Albian Stage recorded a diachronous transition from transtensional (shallow) deformation to deep-seated deformational processes, represented by the activation of transform faults, and the emplacement of oceanic crust, defining the lithospheric boundary between Africa and South America. The boundary geometry was controlled by an array of transform faults that provided a network of right oversteps, which evolved to major releasing and restraining transform segments, kinematically controlled by the oblique divergent movement of Africa and Brazil. A regional rotational strain regime, kinematically controlled, produced complex structural patterns, unevenly distributed in the Equatorial branch, a characteristic of the wrench tectonics structural style. The resultant deformation was partitioned into domains dominated by pure-shear or wrench dominated transtension and/or transpression.

Local tectonics and magmatism played a key role in the "post-break-up" subsidence. The resultant basin architecture, facies distribution and subsidence history for any particular basin vary significantly according to its paleogeographic position and to its proximity to active transform faults. The dynamic juxtaposition of continental crust against oceanic crust or spreading centers on the opposite side of a transform fault caused diachronous deformations, recorded on the sedimentary record as important unconformities, amplified or not by synchronous eustatic sea level variations.

Acknowledgments. The author is grateful to Petrobras for permission to publish this paper. Webster Mohriak greatly contributed with several suggestions to improve early versions of the manuscript. Dr. Barry Katz is thanked for many helpful suggestions to improve the final text. David Sandwell kindly provided the Geosat image of the Equatorial portion of the Brazilian margin.

REFERENCES

Almeida, F. F. M., C. D. R Carneiro, D. L Machado Jr, and L. K. Dehira, Magmatismo pós-paleozóico do Nordeste Oriental do Brasil, *Rev. Bras. Geoc.*, 18,4, 451-462, 1988.

Anderson, E. M., *The Dynamics of Faulting and Dyke Formation with Application to Britain*. Oliver & Boyd, Edinburgh, 1942.

Azevedo, R. P., *Tectonic Evolution of Brazilian Equatorial Continental Margin Basins*, Ph.D. thesis, Imperial College, London , 455p., 1991.

Basile, C., J. Mascle, M. Popoff, J. P. Bouillin, and G. Mascle, The Ivory Coast - Ghana transform margin: a marginal ridge structure deduced from seismic data, *Tectonophysics*, 222, 1-19, 1993.

Benin. Ministry of Energy, Mines and Hydraulics. *Hydrocarbon prospects for exploration and production*. Cotonou, 88p., 1994.

Benkhelil, J., The origin and evolution of the Cretaceous Benue Trough (Nigeria). *Journal of African Earth Sciences*, 8, 2-4, 251-282, 1989.

Caldasso, A. L. S. and M. Hama, Posicionamento estratigráfico das rochas básicas da Bacia do Parnaíba. *Congresso Brasileiro de Geologia*, 30, Recife, SBG, 2, 567-581, 1978.

Chang, H.K., R. O. Koswmann, and A. M. F. Figueiredo, New concepts on the development of East Brazilian Marginal Basins. *Episodes*, 11, 3, 194-202, 1988.

Chang, H. K., R. O. Kowsmann, A. M. F. Figueiredo, and A. Bender, Tectonics and stratigraphy of the East Brazil Rift system: an overview. *Tectonophysics*, 213, 97 - 138, 1992.

Cremonini, O. A., J. P. de M. Goulart, and U. M. Soares, O Rifte Potiguar: Novos dados e implicações tectonicas. *Boletim do 4o. Simp. sobre o Cretáceo do Brasil*, Unesp, 89-93, 1996.

Feijó, F. J., *Boletim de Geociências da Petrobrás*, 1, 8, 249p., 1994.

Fossen, H., B. Tikoff, and C. Teyssier, C., Strain modeling of transpressional and transtensional deformation, *Norsk Geologisk Tidsskrift*, 74, 134-145, 1994.

Ghana National Petroleum Corporation. *Petroleum exploration opportunities*. [S. l.] [199-]: 6p.

Gomes, J.R.C., and C. M. P.P. Gatto, et als., *Projeto Radambrasil, Folhas SB 24-25, Jaguaribe e Natal, Geologia, geomorfologia, pedologia, vegetação e uso potencial da terra*. Rio de Janeiro, MME/Sec. Geral, 27,300p., 1981.

Guiraud, M, Benkhelil, J. Mascle, C. Basile, G. Mascle, J. P. Bouillin, and M. Cousin, Synrift to syntransform deformation along the Côte d'Ivoire-Ghana transform margin: evidence from deep-sea dives, *Geo-Marine Letters*, 17, 70-78, 1977.

Hinz, K., A hypothesis on terrestrial catastrophes: wedges of very thick oceanward dipping layers beneath passive continental margins, *Geologisches Jahrbuch*, E-22, 3-28, 1981.

Koutsoukos, E. A. M., Late Aptian to Maastrichtian foraminiferal biogeography and palaeoceanography of the Sergipe basin, Brazil. *Palaeogeography, Palaeoclimatology, Palaeocology*, 92, 295-324, 1992.

Mascle, J., Atlantic-type continental margins: distinction of two basic structural types. *Anais da Academia Brasileira de Cências*, 48, 191-197, 1976.

Mascle, J. and C. Basile, The Côte d 'Ivoire – Ghana continental margin (Eastern Equatorial Atlantic): a reference model for transform margins. *5th International Congress of the Brazilian Geophysical Society*, Expanded Abstracts, vol. 1, p. 22, 1977.

Mascle, J., P. Lohmann, P., and P. Clift, Development of a passive transform margin: Côte d'Ivoire-Ghana transform margin - ODP Leg 159 preliminary results. *Geo-Marine Letters*, 17, 4-11, 1977.

Matos, R. M. D., The Northeast Brazilian Rift System. *Tectonics*, 11,4, 766-791, 1992.

Matos, R.M.D., The Transversal Zone: a key feature between NE Brazil and W Africa. In: M. R. Mello and P.O. Yilmaz (eds.), Extended Abstract Volume, *1998 AAPG International Conference and Exhibition*, Rio Janeiro, Brazil, 426-427, 1998.

Matos, R.M.D. and R. N. Waick., A unique transform margin: the Equatorial Atlantic. In: M. R. Mello and P.O. Yilmaz (eds.), Extended Abstract Volume, *1998 AAPG International Conference and Exhibition*, Rio Janeiro, Brazil, 798-799, 1988.

Matsuda, N. S., C. R. N. da Rosa, and H. P. Lima. Bacia do Marajó, Estágio Atual da Exploração Petrolífera, Petrobrás Internal Report., 1991.

Milnes, A. G., Aspects of `strike-slip´ or wrench tectonics – an introductory discussion,. *Norsk Geologisk Tidsskrift*, 74, 129-133, 1994.

O'Connor, J. M., and R. A. Duncan, Evolution of the Walvis Ridge-Rio Grande Rise hot spot system: Implications for African and South American plate motions over plumes. *J. Geophys. Res.*, 95, 17,475-15,502, 1990.

Oliveira, D. C., *O papel do enxame de diques Rio Ceará Mirim na evolução tectônica do Nordeste Oriental (Brasil): Implicações na formação do rifte Potiguar*. M.Sc. Dissertation, 172 p., UFOP, Ouro Preto, 1992.

Pessoa Neto, O., *Análise estratigráfica integrada da plataforma mista (siliciclástica-carbonática) do Neogeno da Bacia Potiguar, Nordeste do Brasil*, M.Sc. Dissertation, 220 p., URGS, Porto Alegre, 1999.

Pierce, C., R. B. Whitmarsh, R. A. Scrutton, B. Pontoise, R. Sage, and J. Mascle, Côte d'Ivoire-Ghana margin: seismic imaging of passive rifted crust adjacent to a transform continental margin. *Geophys. J. Int.* , 125, 781-795, 1996.

Pindell, J. L. et al., A plate-kinematic framework for models of Caribbean evolution. *Tectonophysics,* 155, 121-138, 1988.

Popoff, M., Du Gondwana à l'Atlantique sud : les connexions du fossé de la Bénoué avec les bassins du Nord-Est brésilien jusqu'a l'ouverture du golfe de Guinée au Crétacé inféreur, *In* : The West African Connection. In: Sough J. and Rodgers J. (eds) *J. Afric. Earth Sci., Special Publication*, Oxford, G.B. 7, 2, 409-431, 1988.

Scotese, C. R., L. M. Gahagan, and R. L. Larson, Plate tectonic reconstructions of the Cretaceous and Cenozoic basins, *Tectonophysics*, 155, 27-48, 1988.

Sandwell D. T. and W. H. F. Smith, *Map of marine gravity anomaly from satellite altimetry*. Scripps Institute of Oceanography, La Jolla, California, 1995.

Silva, S.R.P., R. R. Maciel, and M. C. G. Severino, Cenozoic tectonics of Amazon Mouth Basin. *Geo-Marine Letters*, 18, 256-262, 1999.

Societe Nationale D'Operations Petrolieres de La Côte D'Ivoire. *Côte D'Ivoire: 1993 petroleum evaluation concessions*. Abidjon, 15 p., 1993.

Szatmari, P., J. B. L. Françolin, O. Zanotto, and S. Wolff, 1987. Evolução Tectônica da margem equatorial Brasileira. *Revista Brasileira de Geociências, SBG*, 17, 2, 180-88, 1987.

Twiss, R. J., and E. M. Moores, *Structural Geology*. W. H. Freeman and Company, 532p, 1992.

Unternehr, P., D. Curie, J. L. Olivet, J. Goslin, and P.Beuzart, South Atlantic fits and intraplate boundaries in Africa and South America, *Tectonophysics*, 155, 169-179, 1988.

Vail, J. R., Ring complexes and related rocks in Africa, *Journal of African Earth Science*, 8, 19-40, 1989.

White, R.S., Crustal structure and magmatism of North Atlantic continental margins, *Journal of the Geological Society*, London, 149, 841-854, 1992.

White, R.S. and D. P. McKenzie, Magmatism at rift zones: the generation of volcanic continental margins and flood basalts, *Journal of Geophysical Research*, 94, 7685 - 7730, 1989.

Wilson, M. and R. Guiraud, Magmatism and rifting in Western and Central Africa, from Late Jurassic to recent times, *Tectonophysics*, 213, 203-225, 1992.

Zalán, P.V., and J. E. Warme, Tectonics and sedimentation of the Piauí-Camocim sub-basin, offshore Northeastern Brazil. *Petrobrás-Cenpes, Ciência Técnica Petróleo, seção exploração de petróleo*, 17, 71p., 1985.

Renato M. Darros de Matos, Petrobras – E&P – RNCE, Natal – RN – Brazil 59074-600
e-mail: renatom@ep-rnce.petrobras.com.br